Sarcopenia – Age-Related Muscle Wasting and Weakness

Gordon S. Lynch
Editor

Sarcopenia – Age-Related Muscle Wasting and Weakness

Mechanisms and Treatments

Springer

Editor
Gordon S. Lynch
Department of Physiology
Basic and Clinical Myology Laboratory
The University of Melbourne, Victoria
Australia
gsl@unimelb.edu.au

ISBN 978-90-481-9712-5 e-ISBN 978-90-481-9713-2
DOI 10.1007/978-90-481-9713-2
Springer Dordrecht Heidelberg London New York

Printed on acid-free paper

Springer is part of Springer Science+Business Media (www.springer.com)

Dedicated to my mentor
Professor John A. Faulkner,
with great respect and affection

Contents

Contributors

Stephen E. Alway
Department of Exercise Physiology, and Center for Cardiovascular and Respiratory Sciences, West Virginia University School of Medicine, Morgantown, WV 26506, USA
salway@hsc.wvu.edu

Josep M. Argilés
Departament de Bioquímica i Biologia Molecular, Universitat de Barcelona, Barcelona
jargiles@ub.edu

Susan V. Brooks
Departments of Biomedical Engineering and Molecular and Integrative Physiology, University of Michigan, Ann Arbor, MI 48109-2200, USA

Sílvia Busquets
Departament de Bioquímica i Biologia Molecular, Facultat de Biologia, Universitat de Barcelona, Barcelona

Kathleen O'Connell
Department of Biology, National University of Ireland, Maynooth, Co. Kildare, Ireland

Alexander Cristea
Department of Neuroscience, Clinical Neurophysiology, Uppsala University, Sweden

Carol S. Davis
Departments of Molecular and Integrative Physiology, University of Michigan, Ann Arbor, MI 48109-2200, USA

Osvaldo Delbono
Departments of Internal Medicine, Section on Gerontology and Geriatric Medicine, Department of Physiology and Pharmacology, Molecular Medicine and Neuroscience Programs, Wake Forest University School of Medicine, Winston-Salem, NC 27157, USA
odelbono@wfubmc.edu

Pamela Donoghue
Conway Institute, University College Dublin, Belfield, Ireland

Philip Doran
Department of Biological Chemistry, University of California, Los Angeles, CA, USA

John A. Faulkner
Departments of Biomedical Engineering and Molecular and Integrative Physiology, University of Michigan, Ann Arbor, MI 48109-2200, USA
jafaulk@umich.edu

Roger A. Fielding
Nutrition Exercise Physiology and Sarcopenia Laboratory, Jean Mayer USDA Human Nutrition Research Center on Aging, Tufts University, 711 Washington Street, Boston, MA 02111, USA
roger.fielding@tufts.edu

Joan Gannon
Department of Biology, National University of Ireland, Maynooth, Co. Kildare, Ireland

Luc E. Gosselin
Department of Exercise and Nutrition Sciences, University at Buffalo, 211 Kimball Tower, Buffalo, NY 14214-8028, USA
gosselin@buffalo.edu

Miranda D. Grounds
School of Anatomy & Human Biology, the University of Western Australia, Nedlands Western Australia, 6009, Australia
mgrounds@anhb.uwa.edu.au

Russell T. Hepple
Faculty of Kinesiology and Faculty of Medicine, University of Calgary, Calgary, Canada
hepple@ucalgary.ca

Malcolm J. Jackson
School of Clinical Sciences, University of Liverpool, UK
m.j.jackson@liverpool.ac.uk

Ravi Kambadur
School of Biological Sciences, Nanyang Technological University,
60 Nanyang Drive, Singapore
Kravi@ntu.edu.sg

René Koopman
Basic and Clinical Myology Laboratory, Department of Physiology,
The University of Melbourne, Australia
rkoopman@unimelb.edu.au

Lars Larsson
Department of Clinical Neurophysiology, Uppsala University Hospital,
Entrance 85, 3rd Floor, 751 85 Uppsala, Sweden
and Department of Biobehavioral Health, the Pennsylvania State University,
PA, USA
lars.larsson@neurofys.uu.se

Bertrand Lèger
Basic and Clinical Myology Laboratory, Department of Physiology,
The University of Melbourne, Parkville, 3010, Australia
Bertrand.Leger@crr-suva.ch

Francisco J. López-Soriano
Departament de Bioquímica i Biologia Molecular, Facultat de Biologia,
Universitat de Barcelona, Barcelona

Gordon S. Lynch
Department of Physiology, Basic and Clinical Myology Laboratory,
The University of Melbourne, Victoria, Australia
gsl@unimelb.edu.au

Carlos B. Mantilla
Departments of Physiology & Biomedical Engineering
and Anesthesiology, College of Medicine, Mayo Clinic,
Joseph 4W-184, St. Marys Hospital, 200 First Street SW,
Rochester, MN 55905, USA
mantilla.carlos@mayo.edu

Anne McArdle
School of Clinical Sciences, University of Liverpool, UK
mdcr02@liverpool.ac.uk

Craig McFarlane
Singapore Institute for Clinical Sciences, Singapore

Chris D. McMahon
AgResearch Limited, Ruakura Research Centre, Hamilton, New Zealand
chris.mcmahon@agresearch.co.nz

Christopher L. Mendias
Departments of Orthopaedic Surgery and School of Kinesiology,
University of Michigan, Ann Arbor, MI 48109-2200, USA

Kay Ohlendieck
Department of Biology, National University of Ireland, Maynooth,
Co. Kildare, Ireland
kay.ohlendieck@nuim.ie

Marcel Orpi
Departament de Bioquímica i Biologia Molecular, Facultat de Biologia,
Universitat de Barcelona, Barcelona

Donato A. Rivas
Nutrition Exercise Physiology and Sarcopenia Laboratory Jean Mayer
USDA Human Nutrition Research Center on Aging, Tufts University,
711 Washington Street, Boston, MA 02111, USA

Stephen M. Roth
Department of Kinesiology, School of Public Health, University of Maryland,
College Park, MD 20742, USA
sroth1@umd.edu

Aaron P. Russell
Centre for Physical Activity and Nutrition, School of Exercise
and Nutrition Sciences, Deakin University, Burwood 3125, Australia
aaron.russell@deakin.edu.au

James G. Ryall
The Laboratory of Muscle Stem Cells and Gene Regulation,
National Institute of Arthritis, Musculoskeletal and Skin Diseases,
National Institutes of Health (NIH), Bethesda, MD, USA
ryallj@mail.nih.gov

Roberto Serpe
Departament de Bioquímica i Biologia Molecular, Facultat de Biologia,
Universitat de Barcelona, Barcelona

Mridula Sharma
Department of Biochemistry, National University of Singapore

Thea Shavlakadze
School of Anatomy & Human Biology, the University of Western Australia,
Nedlands Western Australia, 6009, Australia
tshavlakadze@anhb.uwa.edu.au

Gary C. Sieck
Departments of Physiology & Biomedical Engineering and Anesthesiology,
College of Medicine, Mayo Clinic, Joseph 4W-184, St. Marys Hospital,
200 First Street SW, Rochester, MN 55905, USA
sieck.gary@mayo.edu

Parco M. Siu
Department of Health Technology and Informatics,
The Hong Kong Polytechnic University, Hung Hom, Kowloon, Hong Kong, China
htpsiu@inet.polyu.edu.hk

Ladora V. Thompson
University of Minnesota, Medical School Program in Physical Therapy,
Department of Physical Medicine and Rehabilitation,
420 Delaware St, SE, Minneapolis, MN 55455, USA
thomp067@umn.edu

David E. Vaillancourt
Department of Kinesiology and Nutrition and Departments of Bioengineering
and Neurology, University of Illinois at Chicago, Chicago, IL, USA

Luc J.C. van Loon
Department of Human Movement Sciences, Maastricht University Medical
Centre, 6200 MD, Maastricht, The Netherlands
L.vanLoon@HB.unimaas.nl

Lex B. Verdijk
Department of Human Movement Sciences, Nutrition and Toxicology Research
Institute Maastricht (NUTRIM), Maastricht University Medical Centre,
Maastricht, The Netherlands

Overview of Sarcopenia

Gordon S. Lynch

Abstract Some of the most serious consequences of ageing are its effects on skeletal muscle. 'Sarcopenia' involves a progressive age-related loss of muscle mass and associated muscle weakness that renders frail elders susceptible to serious injury from sudden falls and fractures and losing their functional independence. Not surprisingly, sarcopenia is a significant global public health problem, especially in the developed world. There is an urgent need to better understand the mechanisms underlying age-related muscle wasting and to develop therapeutic strategies that can attenuate, prevent, or ultimately reverse skeletal muscle wasting and weakness. Research and development in academic and research institutions and in large and small pharma is being directed to sarcopenia and related issues to develop and evaluate novel therapies. This book provides the latest information on sarcopenia from leading international researchers studying the cellular and molecular mechanisms underlying age-related changes in skeletal muscle and identifying strategies to combat sarcopenia and related muscle wasting conditions and neuromuscular disorders. The range of interventions for sarcopenia is extensive and not all can be covered in this first volume. While not covering every possible theme, the selected topics provide important insights into the some of the mechanisms underlying sarcopenia and serve as the basis for subsequent complementary volumes that will eventually provide a definitive resource for understanding age-related muscle wasting and weakness and therapeutic approaches to combat sarcopenia.

Keywords Ageing • Aging, cancer cachexia • Cytokine • Geriatrics • Gerontology • Growth factors • Hormones • Inflammation • Muscle injury and repair • Muscle wasting • Muscle weakness • Neuromuscular • Sarcopenia • Senescence • Skeletal muscle

---

G.S. Lynch (✉)
Department of Physiology, Basic and Clinical Myology Laboratory, The University of Melbourne, Victoria, Australia
e-mail: gsl@unimelb.edu.au

G.S. Lynch (ed.), *Sarcopenia – Age-Related Muscle Wasting and Weakness*,
DOI 10.1007/978-90-481-9713-2_1,

1 Defining Sarcopenia

Some of the most serious consequences of ageing are its effects on skeletal muscle particularly the progressive loss of mass and function which impacts on quality of life, and ultimately on survival. Although the term 'sarcopenia' was originally coined to describe the progressive loss of muscle mass with advancing age (Rosenberg 1989; Evans and Campbell 1993; Evans 1995), only recently have consensus definitions of 'sarcopenia' been established. Updated definitions of sarcopenia were published in 2010 by the European Working Group on Sarcopenia in Older People (Cruz-Jentoft et al. 2010), by the Special Interest Group on cachexia-anorexia in chronic wasting diseases within The European Society for Clinical Nutrition and Metabolism (ESPEN, Muscaritoli et al. 2010), and by Evans (2010) who all proposed that the accompanying deterioration of muscle function or muscle weakness should be included in the definition of sarcopenia. A slightly different view was proposed by Narici and Maffulli (2010) who suggested that although muscle weakness was an inevitable consequence of sarcopenia, the two terms should not be used interchangeably because of the implication that they were proportional. Instead, they proposed that sarcopenia should be used uniquely to describe age-related loss of muscle mass and that its relation to the loss of muscle strength be discussed separately (Narici and Maffulli 2010).

Regardless of these slight variations in definition, most groups agree that there are several criteria for the clinical diagnosis of sarcopenia, such as the presence of low muscle mass accompanied by low muscle strength and/or low physical performance (Janssen et al. 2002; Cruz-Jentoft et al. 2010). The definition of sarcopenia provided by Evans (2010) describes these structural and functional criteria comprehensively; i.e.

> Sarcopenia is the age-associated loss of skeletal muscle mass and function. The causes of sarcopenia are multifactorial and can include disuse, changing endocrine function, chronic diseases, inflammation, insulin resistance, and nutritional deficiencies. Whereas cachexia may be a component of sarcopenia, the two conditions are not the same. The diagnosis of sarcopenia should be considered in all older patients who present with observed declines in physical function, strength, or overall health. Sarcopenia should specifically be considered in patients who are bedridden, cannot independently rise from a chair, or who have a measured gait speed <1.0 m/s. Patients who meet these initial criteria should further undergo body composition assessment using dual-energy X-ray absorptiometry with sarcopenia being defined as an appendicular lean/fat mass 2 SD less than that of young adult. A diagnosis of sarcopenia is consistent with a gait speed of <1 m/s and an appendicular lean/fat ratio <2 SD of the average of a young adult (Evans 2010).

This definition serves as an appropriate starting point for understanding the underlying mechanisms of sarcopenia and for developing safe and effective interventions.

2 The International Health Problem of Sarcopenia

Sarcopenia is a highly significant public health problem affecting the developed world. The true magnitude of the health problems associated with age-related musculoskeletal disability is being realized worldwide as the number and proportion of older persons in the population continues to escalate. Sarcopenia imposes a significant but modifiable economic burden on healthcare services in most industrialized nations (Lynch 2004a). In 2000 it was estimated that healthcare costs in the United States associated with sarcopenia were $18.5 US billion; or ~1.5% of total healthcare expenditure (Janssen et al. 2004). The Centers for Disease Control and Prevention (CDC) later predicted that there were ~34 million people in the United States aged 65 years and older, or ~13% of the total population, and that this would increase to 70 million people by 2030, or ~20% of the total population (Thompson 2007). Furthermore, 1.5 million people in the United States aged 65 years and older were institutionalized and 33% of these people had been admitted to long-term healthcare facilities because of their inability to perform activities of daily living (Thompson 2007).

Sarcopenia affects all elderly and does not discriminate based on ethnicity, gender, or wealth. Frail elders who have lost significant muscle mass and strength often require assistance for accomplishing even the most basic tasks of independent living, and they are also at increased risk of serious injury from sudden falls and subsequent fractures. The loss of functional independence is painful not only for the individual but also for family members and carers. Sarcopenia has a dramatic impact on the lives of the elderly and places increasing demands on public health care systems worldwide. Not surprisingly, there is an acute awareness among researchers and clinicians in academic and research institutions and in the pharmaceutical industry about the importance of sarcopenia and the urgent need to develop novel therapies that can attenuate, prevent, and potentially reverse age-related muscle wasting and weakness.

3 Overview of Our Current Understanding of the Cellular and Molecular Mechanisms Underlying Sarcopenia

Several reviews have summarized the cellular and molecular mechanisms underlying age-related muscle wasting and weakness (Ryall et al. 2008) and this textbook provides in-depth discussions on some of these different contributing factors. The loss of muscle mass and strength is thought to be attributed to the progressive atrophy and loss of individual muscle fibres associated with some loss of motor units, and a reduction in muscle 'quality' due to the infiltration of fat and other non-contractile tissue. Thus, the age-related changes in skeletal muscle are neuromuscular in origin and associated with a complex interaction of factors affecting neuromuscular transmission, protein synthesis and degradation, muscle architecture, fibre composition, increased generation of reactive oxygen species, myonuclear apoptosis, altered excitation-contraction coupling, and metabolism (Lynch et al. 2007; Ryall et al. 2008; Arnold et al. 2010; Wenz et al. 2009). Sarcopenia is mechanistically different from the acute muscle atrophies as a consequence of disuse, cachexia, denervation and other conditions (Edström et al. 2006; Combaret et al. 2009).

Age-related changes in skeletal muscle can be exacerbated by the normally decreasing levels of physical activity with advancing age and also by metabolic changes and oxidative stresses that can result in the accumulation of intracellular damage from free radicals (Meng and Yu 2010). Although physical activity (especially strength training) and good nutrition can help slow the rate of these neuromuscular impairments (Aagaard et al. 2010), even very active Masters athletes and otherwise healthy older adults also exhibit a progressive loss of muscle mass, strength and (especially) power output (Runge et al. 2004; Yamauchi et al. 2009) that can affect their performance of everyday tasks (Korhonen et al. 2003, 2006; Cristea et al. 2008). Age-related changes in circulating muscle anabolic hormones and growth factors, also contribute to the emergence of the sarcopenic phenotype and the subsequent loss of functional independence and quality of life (Orr and Fiatarone Singh 2004; Bain 2010; Kovacheva et al. 2010; Perrini et al. 2010; Scicchitano et al. 2009). Other conditions can accelerate the progression of muscle atrophy in older adults, including co-morbid diseases such as cancer, kidney disease, diabetes, and peripheral artery disease (Buford et al. 2010). Although age-related changes in skeletal muscle structure and function are inevitable, pharmacological approaches to attenuate, halt or reverse the deleterious effects of advancing age on skeletal muscle are realistic possibilities (Borst 2004; Lynch 2004, 2008; Gullett et al. 2010). Since sarcopenia is considered a neuromuscular syndrome (Tseng et al. 1995; Koopman et al. 2009) drugs for sarcopenia could induce neural and/or muscle-specific effects and I have described these approaches in detail elsewhere (Lynch 2002, 2004b, 2008). The list of different interventions for sarcopenia is extensive and not all can be covered in this first volume. Instead, this text will cover some of the main signalling pathways thought responsible for age-related muscle wasting and weakness and just some of the interventions proposed to counteract these effects.

This text serves to introduce the reader to some of the significant age-related changes in skeletal muscle and to identify the different factors affecting neuromuscular transmission, muscle structure and fibre composition, excitation-contraction coupling, and skeletal muscle metabolism. Contributions have been sought from leading researchers in the field to describe these different factors and mechanisms responsible for the deleterious changes to skeletal muscle as a consequence of advancing age. While there sometimes may be conflicting views among researchers about the relative importance of these different contributing mechanisms, each chapter provides a concise and timely update about the age-associated changes in the structural, functional and biochemical properties of skeletal muscle and taken together they provide a basis for identifying novel approaches to tackle sarcopenia.

The chapters cover diverse topics ranging from insights into the mechanisms of the neuromuscular deficit, including age-related changes in the neuromuscular junction and neurotransmission, alterations in motor unit properties, actomyosin structure and interaction, and excitation-contraction coupling; alterations in metabolic properties including mitochondrial function and some of the factors regulating fibrosis and nuclear apoptosis. The book discusses the mechanisms regulating the balance between protein synthesis and protein degradation and how these processes are affected during aging as well as understanding genetic variation and proteomic profiling of skeletal muscles during aging. Other topics describe the role of exercise in counteracting some of the effects of aging on skeletal muscle, how contraction-mediated injury contributes to age-related muscle wasting and weakness, and the role of different signaling pathways in regulating skeletal muscle mass and how these pathways can be modified during aging. While not covering every possible theme, the selected topics provide important insights into the some of the mechanisms underlying sarcopenia and generous reference lists for pursuing topics further. It is expected that the introductory themes provided in this text will serve as the basis for subsequent volumes that will eventually provide the definitive resource for understanding all of the signalling pathways implicated in age-related muscle wasting and weakness and describing the comprehensive list of drugs and approaches to combat sarcopenia.

References

Aagaard, P., Suetta, C., Caserotti, P., Magnusson, S. P., Kjær, M. (2010). Role of the nervous system in sarcopenia and muscle atrophy with aging: strength training as a countermeasure. *Scandinavian Journal of Medicine & Science in Sports, 20*, 49–64.

Arnold, A. S., Egger, A., Handschin, C. (2010). PGC-1alpha and myokines in the aging muscle – a mini-review. *Gerontology* (in press) DOI: 10.1159/000281883.

Bain, J. (2010). Testosterone and the aging male: to treat or not to treat? *Maturitas, 66*, 16–22.

Borst, S. E. (2004). Interventions for sarcopenia and muscle weakness in older people. *Age and Ageing, 33*, 548–555.

Buford, T. W., Anton, S. D., Judge, A. R., Marzetti, E., Wohlgemuth, S. E., Carter, C. S., Leeuwenburgh, C., Pahor, M., Manini, T. M. (2010). Models of accelerated sarcopenia: critical pieces for solving the puzzle of age-related muscle atrophy. *Ageing Research Reviews, 9*, 369–383.

Combaret, L., Dardevet, D., Béchet, D., Taillandier, D., Mosoni, L., Attaix, D. (2009). Skeletal muscle proteolysis in aging. *Current Opinion in Clinical Nutrition and Metabolic Care, 12*, 37–41.

Cristea, A., Korhonen, M. T., Häkkinen, K., Mero, A., Alén, M., Sipilä, S., Viitasalo, J. T., Koljonen, M. J., Suominen, H., Larsson, L. (2008). Effects of combined strength and sprint training on regulation of muscle contraction at the whole-muscle and single-fibre levels in elite master sprinters. *Acta Physiologica, 193*, 275–289.

Cruz-Jentoft, A. J., Baeyens, J. P., Bauer, J. M., Boirie, Y., Cederholm, T., Landi, F., Martin, F. C., Michel, J. P., Rolland, Y., Schneider, S. M., Topinková, E., Vandewoude, M., Zamboni, M. (2010). Sarcopenia: European consensus on definition and diagnosis: Report of the European Working Group on Sarcopenia in Older People. *Age Ageing*, April 13, 1–12.

Edström, E., Altun, M., Hägglund, M., Ulfhake, B. (2006). Atrogin-1/MAFbx and MuRF1 are downregulated in aging-related loss of skeletal muscle. *The Journals of Gerontology. Series A: Biological Sciences and Medical Sciences, 61*, 663–674.

Evans, W. J. (1995). What is sarcopenia? *The Journals of Gerontology. Series A: Biological Sciences and Medical Sciences, 50A*, 5–8.

Evans, W. J. (2010). Skeletal muscle loss: cachexia, sarcopenia, and inactivity. *The American Journal of Clinical Nutrition, 91*, 1123S–1127S.

Evans, W. J. & Campbell, W. W. (1993). Sarcopenia and age-related changes in body composition and functional capacity. *Journal of Nutrition 123*(2 Suppl), 465–468.

Gullett, N. P., Hebbarm G., Ziegler, T. R. (2010). Update on clinical trials of growth factors and anabolic steroids in cachexia and wasting. *The American Journal of Clinical Nutrition, 91*, 1143S–1147S.

Janssen, I., Heymsfield, S. B., Ross, R. (2002). Low relative skeletal muscle mass (sarcopenia) in older persons is associated with functional impairment and physical disability. *Journal of the American Geriatrics Society, 50*, 889–896.

Janssen, I., Shepard, D. S., Katzmarzyk, P. T., Roubenoff, R. (2004). The healthcare costs of sarcopenia in the United States. *Journal of the American Geriatrics Society, 52*, 80–85.

Koopman, R., Ryall, J. G., Church, J. E., Lynch, G. S. (2009). The role of β-adrenoceptor signaling in skeletal muscle: therapeutic implications for muscle wasting disorders. *Current Opinion in Clinical Nutrition and Metabolic Care, 12*, 601–606.

Korhonen, M. T., Mero, A., Suominen, H. (2003). Age-related differences in 100-m sprint performance in male and female master runners. *Medicine and Science in Sports and Exercise, 35*, 1419–1428.

Korhonen, M. T., Cristea, A., Alén, M., Häkkinen, K., Sipilä, S., Mero, A., Viitasalo, J. T., Larsson, L., Suominen, H. (2006). Aging, muscle fiber type, and contractile function in sprint-trained athletes. *Journal of Applied Physiology, 101*, 906–917.

Kovacheva, E. L., Sinha-Hikim, A. P., Shen, R., Sinha, I., Sinha-Hikim, I. (2010). Testosterone supplementation reverses sarcopenia in aging through regulation of myostatin, c-Jun NH2-terminal kinase, Notch and Akt signalling pathways. *Endocrinology, 151*, 628–638.

Lynch, G. S. (2002). Novel therapies for sarcopenia: ameliorating age-related changes in skeletal muscle. *Expert Opinin on Therapeutic Patents, 12*, 11–27.

Lynch, G. S. (2004a). Tackling Australia's future health problems: developing strategies to combat sarcopenia–age-related muscle wasting and weakness. *Internal Medicine Journal, 34*, 294–296.

Lynch, G. S. (2004b). Emerging drugs for sarcopenia: age-related muscle wasting. *Expert Opinion on Emerging Drugs, 9*, 345–361.

Lynch, G. S. (2008). Update on emerging drugs for sarcopenia – age-related muscle wasting. *Expert Opinion on Emerging Drugs, 13*, 655–673.

Lynch, G. S, Schertzer, J. D, Ryall, J. G. (2007). Therapeutic approaches for muscle wasting disorders. *Pharmacology & Therapeutics, 113*, 461–487.

Meng, S. J. & Yum L. J. (2010). Oxidative stress, molecular inflammation and sarcopenia. *International Journal of Molecular Sciences, 11*, 1509–1526.

Muscaritoli, M., Anker, S. D., Argilés, J., Aversa, Z., Bauer, J. M., Biolo, G., Boirie, Y., Bosaeus, I., Cederholm, T., Costelli, P., Fearon, K. C., Laviano, A., Maggio, M., Fanelli, F. R., Schneider, S. M., Schols, A., Sieber, C. C. (2010). Consensus definition of sarcopenia, cachexia and pre-cachexia: joint document elaborated by Special Interest Groups (SIG) "cachexia-anorexia in chronic wasting diseases" and "nutrition in geriatrics". *Clinical Nutrition, 29*, 154–159.

Narici, M. V. & Maffulli, N. (2010). Sarcopenia: characteristics, mechanisms and functional significance. *British Medical Bulletin, 95*, 139–159.

Orr, R. & Fiatarone Singh, M. (2004). The anabolic androgenic steroid oxandrolone in the treatment of wasting and catabolic disorders: review of efficacy and safety. *Drugs, 64*, 725–750.

Perrini, S., Laviola, L., Carreira, M. C., Cignarelli, A., Natalicchio, A., Giorgino, F. (2010). The GH/IGF1 axis and signaling pathways in the muscle and bone: mechanisms underlying age-related skeletal muscle wasting and osteoporosis. *The Journal of Endocrinology, 205*, 201–210.

Rosenberg, I. (1989). Summary comments: epidemiological and methodological problems in determining nutritional status of older persons. *American Journal of Clinical Nutrition 50*, 1231–1233.

Runge, M., Rittweger, J., Russo, C. R., Schiessl, H., Felsenberg, D. (2004). Is muscle power output a key factor in the age-related decline in physical performance? A comparison of muscle cross section, chair-rising test and jumping power. *Clinical Physiology and Functional Imaging 24*, 335–340.

Ryall, J. G., Schertzer, J. D., Lynch, G. S. (2008). Cellular and molecular mechanisms underlying age-related skeletal muscle wasting and weakness. *Biogerontology, 9*, 213–228.

Scicchitano, B. M., Rizzuto, E., Musaro, A. (2009). Counteracting muscle wasting in aging and neuromuscular diseases: the critical role of IGF-1. *Aging, 1*, 451–457.

Thompson, D. D. (2007). Aging and sarcopenia. *Journal of Musculoskeletal & Neuronal Interactions, 7*, 344–345.

Tseng, B. S., Marsh, D. R., Hamilton, M. T., Booth, F. W. (1995). Strength and aerobic training attenuate muscle wasting and improve resistance to the development of disability with aging. *The Journals of Gerontology. Series A: Biological Sciences and Medical Sciences, 50A*, 113–119.

Wenz, T., Rossi, S. G., Rotundo, R. L., Spiegelman, B. M., Moraes, C. T. (2009). Increased muscle PGC-1alpha expression protects from sarcopenia and metabolic disease during aging. *Proceedings of the National Academy of Sciences of the United States of America, 106*, 20405–20410.

Yamauchi, J., Mishima, C., Nakayama, S., Ishii, N. (2009). Force-velocity, force-power relationships of bilateral and unilateral leg multi-joint movements in young and elderly women. *Journal of Biomechanics, 42*, 2151–2157.

Muscle Wasting in Cancer and Ageing: Cachexia Versus Sarcopenia

Josep M. Argilés, Sílvia Busquets, Marcel Orpi, Roberto Serpe, and Francisco J. López-Soriano

Abstract The aim of this chapter is to summarize and evaluate the different mechanisms and catabolic mediators involved in cancer cachexia and ageing sarcopenia since they may represent targets for future promising clinical investigations. Cancer cachexia is a syndrome characterized by a marked weight loss, anorexia, asthenia and anemia. In fact, many patients who die with advanced cancer suffer from cachexia. The degree of cachexia is inversely correlated with the survival time of the patient and it always implies a poor prognosis. Unfortunately, at the clinical level, cachexia is not treated until the patient suffers from a considerable weight loss and wasting. At this point, the cachectic syndrome is almost irreversible. The cachectic state is often associated with the presence and growth of the tumour and leads to a malnutrition status due to the induction of anorexia. In recent years, age-related diseases and disabilities have become of major health interest and importance. This holds particularly for muscle wasting, also known as sarcopenia, that decreases the quality of life of the geriatric population, increasing morbidity and decreasing life expectancy. The cachectic factors (associated with both depletion of fat stores and muscular tissue) can be divided into two categories: of tumour origin and humoural factors. In conclusion, more research should be devoted to the understanding of muscle wasting mediators, both in cancer and ageing, in particular the identification of common mediators may prove as a good therapeutic strategies for both prevention and treatment of wasting both in disease and during healthy ageing.

Keywords Cancer cachexia • Mediators • Muscle wasting • Metabolic changes • Cytokines • Ageing • Sarcopenia

---

J.M. Argilés (✉), S. Busquets, M. Orpi, R. Serpe, and F.J. López-Soriano
Departament de Bioquímica i Biologia Molecular, Universitat de Barcelona, Barcelona
e-mail: jargiles@ub.edu

G.S. Lynch (ed.), *Sarcopenia – Age-Related Muscle Wasting and Weakness*,
DOI 10.1007/978-90-481-9713-2_2, © Springer Science+Business Media B.V. 2011

1 Introduction

Perhaps the most common manifestation of advanced malignant disease is the development of cancer cachexia. Indeed, cachexia occurs in the majority of cancer patients before death, and it is responsible for the deaths of 22% of cancer patients (Warren 1932). The abnormalities associated with cancer cachexia include anorexia, weight loss, muscle loss and atrophy, anemia and alterations in carbohydrate, lipid and protein metabolism (Argiles et al. 1997). The degree of cachexia is inversely correlated with the survival time of the patient and it always implies a poor prognosis (Harvey et al. 1979; Nixon et al. 1980; DeWys 1985). Perhaps one of the most relevant characteristics of cachexia is that of asthenia (or lack of muscular strength), which reflects the great muscle waste that takes place in the cachectic cancer patient (Argiles et al. 1992). Asthenia is also characterized by a general weakness as well as physical and mental fatigue (Adams and Victor 1981). In addition, lean body mass depletion is one of the main trends of cachexia, and it involves not only skeletal muscle but it also affects cardiac proteins, resulting in important alterations in heart performance.

At the biochemical level, different explanations can be found to account for cancer-induced cachexia (Fig. 1). First, the presence and growth of the tumour is invariably associated with a malnutrition status due to the induction of anorexia (decreased food intake). In addition, the presence of the tumour promotes important metabolic disturbances, which include a considerable nitrogen flow from the skeletal muscle to the liver. Amino acids are used there for both acute-phase protein (APP) synthesis and gluconeogenesis. Both tumoural and humoural (mainly

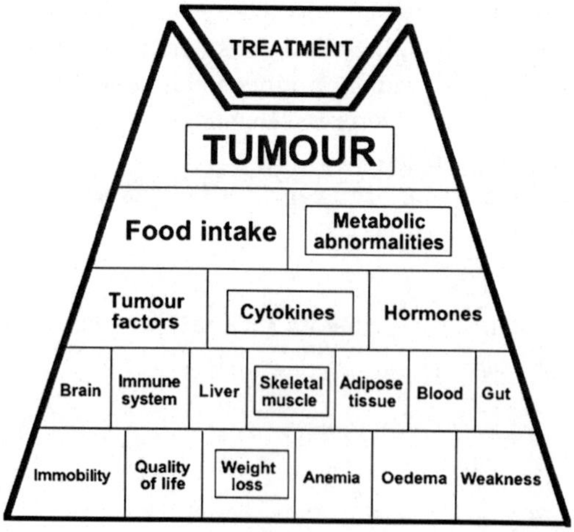

Fig. 1 Cancer cachexia: the pyramid. Cancer cachexia is a complex pathological condition characterized by many metabolic changes involving numerous organs. These changes are triggered by alterations in the hormonal milieu, release of different tumour factors and a systemic inflammatory reaction characterized by cytokine production and release

cytokines) factors are associated with depletion of fat stores and muscular tissues. Indeed cells of the immune system release cytokines that act on multiple target cells such as bone marrow cells, myocytes, hepatocytes, adipocytes, endothelial cells and neurons, where they produce a complex cascade of biological responses leading to the wasting associated with cancer cachexia. Among the cytokines that have been involved in this cachectic response are tumour necrosis factor-α (TNF), interleukin-1 (IL-1), interleukin-6 (IL-6) and interferon-γ (IFN-γ). Interestingly, these cytokines share the same metabolic effects and their activities are closely interrelated, showing in many cases synergistic effects.

The aim of the present chapter is to summarize and evaluate the different mechanisms and catabolic mediators (both humoural and tumoural) involved in cancer cachexia and ageing sarcopenia since they may represent targets for future promising clinical investigations.

2 Cancer: An Inflammatory Disorder

The presence of the tumour clearly elicits a systemic inflammatory response that triggers anorexia and hypermetabolism and neuroendocrine alterations. This systemic inflammatory response is triggered by different mediators either generated by the tumour or by non-tumoural cells of the patient. Mainly, two basic hypotheses can explain this phenomenon. First, the so-called endotoxic hypothesis, by which the tumour burden results in an enhanced translocation of intestinal bacteria into the peritoneum and consequently a release of endotoxin which finally triggers the cytokine cascade. Second, the tumour hypothesis involves either specific tumour-derived compounds or cytokines produced by the tumour which trigger the inflammatory response.

All together, the systemic inflammatory response generates many alterations that affect the patient's metabolism activating among others muscle protein breakdown, and consequently, wasting.

2.1 Hypermetabolism

As anorexia is not the only factor involved in cancer cachexia, it becomes clear that metabolic abnormalities leading to a hypermetabolic state must have a very important role. Interestingly, during cachectic states there is an increase in brown adipose tissue (BAT) thermogenesis in both humans and experimental animals. Until recently, the uncoupling protein-1 (UCP1) protein (present only in BAT) was considered to be the only mitochondrial protein carrier that stimulated heat production by dissipating the proton gradient generated during respiration across the inner mitochondrial membrane and therefore uncoupling respiration from adenosine-5′-triphosphate (ATP) synthesis. Interestingly, two additional proteins sharing the same function, UCP2 and UCP3, have been described. While UCP2 is expressed ubiquitously, UCP3 is expressed abundantly and specifically in skeletal muscle in humans and also in BAT of rodents. Our research group has demonstrated that both UCP2 and UCP3 mRNAs are elevated in skeletal muscle during tumour growth and that tumour necrosis factor-α (TNF-α) is able to mimic the increase in gene expression (Busquets et al. 1998). Indeed, injection of low doses of TNF-α either peripherally or into the brain of laboratory animals, elicits rapid increases in metabolic rate which are not associated with increased metabolic activity but rather with an increase in blood flow and thermogenic activity of BAT, associated with UCP1. In addition, TNF-α is able to induce uncoupling of mitochondrial respiration as shown in isolated mitochondria (Busquets et al. 2003).

2.2 *Muscle Wasting*

The loss of muscle mass is a hallmark of cancer cachexia and it is essentially caused by an increase of myofibrillar protein (especially myosin heavy-chain (Acharyya et al. 2004) degradation (Llovera et al. 1994, 1995; Busquets et al. 2004), sometimes accompanied by a decrease in protein synthesis (Smith and Tisdale 1993; Eley and Tisdale 2007). The enhanced protein degradation is caused by an activation of the ubiquitin-dependent proteolytic system (Temparis et al. 1994; Baracos et al. 1995; Costelli et al. 1995). This enhanced proteolysis may be caused by tumour factors such as proteolysis-inducing factor (Lorite et al. 1998; Belizario et al. 1991) or by cytokines (Mahony et al. 1988; Tracey et al. 1990). Thus, administration of TNF-α to rats results in an increased skeletal muscle proteolysis associated with an increase in both gene expression and higher levels of free and conjugated ubiquitin, both in experimental animals (Bossola et al. 2001) and humans (Baracos 2000). Other cytokines such as interleukin-1 or interferon-γ are also able to activate ubiquitin gene expression. Therefore, TNF-α, alone or in combination with other cytokines (Alvarez et al. 2002), seems to mediate most of the changes concerning nitrogen metabolism associated with cachectic states (Pajak et al. 2008). In addition to the massive muscle protein loss, and similar to that observed in skeletal muscle of chronic heart failure patients suffering from cardiac cachexia (Sharma and Anker 2002), muscle DNA is also decreased during cancer cachexia, leading to DNA fragmentation and, thus, apoptosis (van Royen et al. 2000; Belizario et al. 2001). Interestingly, TNF-α can mimic the apoptotic response in the muscle of healthy animals (Carbo et al. 2002).

The therapy against wasting during cachexia has concentrated on either increasing food intake or normalizing the persistent metabolic alterations that take place in the patient. It is difficult to apply a therapeutic approach based on the neutralization of the potential mediators involved in muscle wasting (i.e. TNF-α, IL-6, IFN-γ, proteolysis-inducing factor) because many of them are simultaneously involved in promoting the metabolic alterations and the anorexia present in the cancer patients (Argiles et al. 2007). Bearing this in mind, it is obvious that a good understanding of the molecular mechanisms involved in the signalling of these mediators may be very positive in the design of the therapeutic strategy. This is especially relevant because different mediators may be sharing the same signalling pathways. There are currently few studies describing the role of cytokines and tumour factors in the signalling associated with muscle wasting. Penner et al. (2001) reported an increase in both NF-κB and AP-1 transcription factors during sepsis in experimental animals. The increase in NF-κB observed in skeletal muscle during sepsis can be mimicked by TNF-α. Indeed, TNF-α addition to C2C12 muscle cultures results in a short-term increase in NF-κB (Fernandez-Celemin et al. 2002; Li et al. 1998). Whether or not this increase in NF-κB promoted by TNF-α is associated with increased proteolysis and/or increased apoptosis in skeletal muscle remains to be established. In relation to AP-1 activation, TNF-α has been shown to increase c-jun expression in C2C12 cells (Brenner et al. 1989). Interestingly, overexpression of c-jun mimics the observed effect of TNF-α upon differentiation; indeed, it results in decreased myoblast differentiation (Thinakaran et al. 1993). Tumour mediators, proteolysis-inducing factor (PIF) in particular, also seem to be able to increase NF-κB expression in cultured muscle cells, this possibly being linked with increased proteolysis (Wyke and Tisdale 2005). Other reports, using experimental cancer models, have also suggested that NF-κB is involved in the signalling of muscle wasting (Wyke et al. 2004; Cai et al. 2004). In our laboratory, we have recently demonstrated increased activation of AP-1 in the skeletal muscle of tumour-bearing rats, therefore suggesting that this factor is involved in the muscle events that take place during cancer cachexia (Costelli et al. 2005a). Indeed, the intramuscular administration of adenoviruses carrying TAM 67 (a negative-dominant of c-jun [AP-1]) resulted in an improvement of the muscle weight during tumour growth (Moore-Carrasco et al. 2006). Other transcriptional factors that have been reported to be involved in muscle changes associated with catabolic conditions include c/EBPβ and δ (which are increased in skeletal muscle during sepsis (Penner et al. 2002), PW-1 and PGC-1. TNF-α decreases MyoD content in cultured myoblasts (Guttridge et al. 2000) and blocks differentiation by a mechanism which seems to be independent of NF-κB and which involves PW-1, a transcriptional factor related to p53-induced apoptosis (Coletti et al. 2002). The action of the cytokines on muscle cells therefore seems to rely most likely on satellite cells blocking muscle differentiation or, in other words, regeneration. Finally the transcription factor PGC-1 has been associated with the activation of both UCP-2 and UCP-3 and increased oxygen consumption by cytokines in cultured myotubes (Puigserver et al. 2001). This transcription factor is involved as an activator of peroxisomal proliferator-activated receptor (PPAR)-γ in the expression of uncoupling proteins. Very recent investigations have revealed a role for PPAR-γ and PPAR-δ in experimental muscle wasting (Fuster et al. 2007).

Muscle wasting is invariably associated with DNA fragmentation in many catabolic states. One of the first reports showing apoptosis in skeletal muscle was in experimental cancer cachexia (van Royen et al. 2000; Sumi et al. 1999). Recently, the same phenomenon has been observed in cancer patients (Busquets et al. 2007). Our laboratory has also described the activation of muscle apoptosis during sepsis

(Almendro et al. 2003). In diabetes (Lee et al. 2004), chronic heart failure (Vescovo and Dalla Libera 2006) and chronic obstructive pulmonary disease (Agusti et al. 2002), apoptosis is also activated in muscle tissue. Recent work on the molecular mediators involved in the intracellular activation of the proteasome has clearly shown that caspase-3 is essential for the activation of proteolysis (Lee et al. 2004; Agusti et al. 2002). Indeed, caspase-3 cleaves actomyosin to actin, which can be degradated by the ubiquitin-proteasome-dependent system (Du et al. 2004). In this cleavage, caspase-3 generates a characteristic 14-kDa actin fragment, which is a marker for muscle proteolysis (Workeneh et al. 2006). In this way, the activation of caspase-3 seems to be associated with myofibril degradation, a process that precedes active protein degradation by the proteasome. Interestingly, caspase-3 is an enzyme involved in apoptosis which is activated by caspase-8 as a result of an apoptotic stimulus such as TNF-α (Benn and Woolf 2004; Adams et al. 2001). In this activation process, the apoptosome (cytochrome c, APAF-1 and caspase-9) is also involved, along with caspase-12 (Benn and Woolf 2004). Interestingly, Fernando et al. (2002) have shown that caspase-3 activity is required for skeletal muscle differentiation. Indeed, during differentiation, reorganization of myofibrillar proteins is essential and possibly linked with the activity of caspase-3. Another interesting observation is that during wasting there is an enhanced myoblast/satellite cell proliferation (Ferreira et al. 2006). All these observations are of utmost importance as inhibitors of caspase-3 in skeletal muscle during wasting could be a potential way of blocking proteolysis (Argiles et al. 2008).

In skeletal muscle, anabolic signals influence protein synthesis and accumulation by activation of phosphatidylinositol-3-kinase (PI3K) which is involved in the phosphorylation of the Akt-mTOR signalling pathway leading to protein anabolism (Latres et al. 2005). Interestingly, the PI3K activation is also associated with the phosphorylation – and therefore inactivation – of the FOXO transcription factor (Sandri et al. 2004). FOXO is known to participate in the transcription of Atrogin-1 and Murf-1, specific ubiquitin ligases involved in muscle proteolysis (Sandri et al. 2004). Therefore, the PI3K signalling pathway is linked with both synthesis and degradation of muscle proteins. For instance, both insulin-like growth factor-1 (IGF-1) and insulin act by activating PI3K (Latres et al. 2005; Kirwan and del Aguila 2003). In catabolic conditions, muscle insulin sensitivity is often hampered (type II diabetes) (Wang et al. 2006) or muscle IGF-1 expression is reduced (cancer) (Costelli et al. 2006). Interestingly, PI3K is linked with caspase-3; indeed, activation of caspase-3 is associated with a suppressed activity of the kinase (Lee et al. 2004). Thus, when PI3K activity is low, both apoptotic and ubiquitin-proteaseome proteolysis pathways are activated, suggesting that PI3K participates in the inhibition of caspase-3. Apparently normal protein turnover in skeletal muscle under healthy conditions does not seem to be linked with a protein breakdown activated by caspase-3 (Du et al. 2004). Indeed, inhibition of caspase-3 with the specific compound Ac-DEVD-CHO in isolated epitrochlearis muscle from rats, does not lead to an inhibition of basal proteolysis (Du et al. 2004). The excessive protein breakdown of myofibrillar proteins in catabolic conditions can, however, be blocked with the mentioned inhibitor. This idea is supported by experiments carried out in muscles from acutely-induced diabetes (Du et al. 2004). Bearing all this in mind, it seems clear that excessive proteolysis (the fraction of protein breakdown which is activated during catabolic conditions) is linked with activation of the apoptotic enzyme caspase-3 and, as mentioned above, inhibition of this enzyme could be a potential therapeutic target for the treatment of muscle wasting associated with chronic diseases.

In addition to the abovementioned PI3K signalling pathway, other factors are related to the activation/inhibition of caspase-3. Indeed, the intracellular levels of calcium have a role in proteolysis not only by activating the calpain-dependent system (specific calcium-dependent proteases) (Costelli et al. 2005b) but also in the activation of caspase-3 (Benn and Woolf 2004; Choi et al. 2006). From this point of view, some studies have shown that calcium can either directly activate caspase-3 or indirectly by favouring a release of mitochondrial cytochrome c, which, in term, activates the apoptosome, which then acts on caspase-3 (Benn and Woolf 2004). From this point of view, an increased entry of calcium into the mitochondria, either by the calcium release from the endoplasmic reticulum or by the entry of extracellular calcium, results in an activation of caspase-3, apoptosis and finally skeletal muscle proteolysis (Benn and Woolf 2004; Hajnoczky et al. 2006). Interestingly, there is another way that calcium can activate caspase-3; indeed, calcium is essential for calpain activation and calpains are able to activate caspase-12, which acts on caspase-3 (Benn and Woolf 2004; Bajaj and Sharma 2006). From the point of view of proteolysis, calpains have been shown to also act before the ubiquitin-proteasome-dependent proteolytic pathway, in a similar manner to that described for caspase-3 (Costelli et al. 2005b; Williams et al. 1999). In fact, calpains have been proposed to act on myofibrils to promote their breakage to myosin, which is then degraded by the proteasome (Costelli et al. 2005b). In a way, therefore, both calpain and caspase-3 activation seem to be essential for ATP-dependent degradation of myofibrillar proteins.

Recent studies have shown that alterations in the muscular dystrophy-associated dystrophin glycoprotein complex may have an important role in muscle wasting during cancer (Acharyya et al. 2005; Glass 2005). Finally, necdin, a protein which has a key role in fetal and postnatal physiological myogenesis is selectively expressed in muscles of cachectic mice and this seems to be linked to a protective response of the tissue against tumour-induced wasting, inhibition of myogenic differentiation and in muscle regeneration (Sciorati et al. 2009).

Moreover, myostatin, a transforming growth factor-β super-family member well characterized as a negative regulator of muscle growth and development, has been implicated in several forms of muscle wasting including the severe cachexia observed as a result of conditions such as AIDS and liver cirrhosis. McFarlane et al. (2006) have demonstrated that myostatin induces cachexia through a NF-κB independent mechanism, by antagonizing hypertrophy signalling through regulation of the AKT-FoxO1 pathway. Antimyostatin strategies are therefore promising and should be considered in future clinical trials involving cachectic patients (Patel and Amthor 2005; Bonetto et al. 2009).

2.3 Adipose Tissue Dissolution and Hypertriglyceridaemia

Lipid metabolism in cancer has been extensively studied, the main trends being an important reduction in body fat content (particularly white adipose tissue) together with a clear hyperlipaemia. The dissolution of the fat mass is the result of three different altered processes. First, there is an increase in lipolytic activity (Thompson et al. 1981), which results in an important release of both glycerol and fatty acids. Recent studies have shown that the mechanism of increased lipolysis is associated with activation of hormone-sensitive lipase in adipose tissue. In addition, in human cancer cachexia there is a decreased antilipolytic effect of insulin on adipocytes together with an increased responsiveness to catecholamines and atrial natriuretic peptide (Agustsson et al. 2007). Second, an important decrease in the activity of lipoprotein lipase (LPL), the enzyme responsible for the cleavage of both endogenous and exogenous triacylglycerols (present in lipoproteins) into glycerol and fatty acids, occurs in white adipose tissue (Thompson et al. 1981; Lanza-Jacoby et al. 1984; Noguchi et al. 1991) and, consequently, lipid uptake is severely hampered. Finally, adipose tissue de-novo lipogenesis is also reduced in tumour-bearing states (Thompson et al. 1981), resulting in a decreased esterification and, consequently, a decreased lipid deposition.

Hyperlipaemia in cancer-bearing states seems to be the result of an elevation in both triacylglycerols and cholesterol. Hypertriglyceridaemia is the consequence of the decreased LPL activity, which results in a decrease in the plasma clearance of both endogenous (transported as very low-density lipoproteins) and exogenous (transported as chilomicra) triacylglycerols. Muscaritoli et al. (1990) have clearly demonstrated that both the fractional removal rate and the maximum clearing capacity (calculated at high infusion rates when LPL activity is saturated) are significantly decreased after the administration of an exogenous triacylglycerol load to cancer patients. In tumour-bearing animals with a high degree of cachexia, there is also an important association between decreased LPL activity and hypertriglyceridaemia (Lopez-Soriano et al. 1996; Evans and Williamson 1988). Another factor that could contribute to the elevation in circulating triacylglycerols is an increase in liver lipogenesis (Mulligan and Tisdale 1991).

Hypercholesterolaemia is often seen in both tumour-bearing animals and humans with cancer (Dessi et al. 1991, 1992, 1995). Interestingly, most cancer cells show an altered regulation in cholesterol biosynthesis showing a lack of feedback control on 3-hydroxy-3-methylglutaryl CoA reductase, the key enzyme in the regulation of cholesterol biosynthesis. Cholesterol perturbations during cancer include changes in lipoprotein profiles, in particular an important decrease in the amount of cholesterol transported in the high-density lipoproteins (HDL) fraction. This finding has been observed in both experimental animals and human subjects (Dessi et al. 1991, 1992, 1995). HDL plays an important role in the transport of excess cholesterol from extrahepatic tissues to the liver for reutilization or excretion into bile (reverse cholesterol transport). It is thus conceivable that the observed low levels of HDL-cholesterol may be related, at least in part, to a decreased cholesterol efflux to HDL as a consequence of increased utilization and/or storage in proliferating tissues, such as neoplasms. As precursor particles of HDL are thought to derive from lipolysis of triacylglycerol-rich lipoproteins such as very low-density lipoproteins and chylomicra (Eisenberg 1984), and as a significant positive correlation between plasma HDL-cholesterol and LPL activity in adipose tissue has also been reported (Eisenberg 1984), one must also consider the possibility that low HDL-cholesterol concentrations observed during tumour growth may be secondary to the decreased triacylglycerol clearance from plasma, as a result of LPL inhibition. Consequently, elevation of circulating lipid seems to be a hallmark of cancer-bearing states to the extent that some authors have suggested that plasma levels may be used to screen patients for cancer (Rossi Fanelli et al. 1995).

Finally, both cytokines – TNF-α in particular (Zhang et al. 2002; Ryden et al. 2002, 2004) – and tumour factors – lipid-mobilising factor (LMF) (Russell and Tisdale 2005; Russell et al. 2004) and toxohormone L – have been related to all the commented alterations in lipid metabolism during cancer cachexia.

2.4 Liver Inflammatory Response

The result of the enhanced muscle proteolysis is a large release of amino acids from skeletal muscle which takes place specially as alanine and glutamine (Fig. 2). The release of amino acids is also potentiated by an inhibition of amino acid transport into skeletal muscle. While glutamine is basically taken up by the tumour to sustain both its energy and nitrogen demands, alanine is mainly channelled to the liver for both gluconeogenesis and protein synthesis. Increased hepatic production of APP has been suggested to be partly responsible for the catabolism of skeletal muscle protein, the essential amino acids being indeed required for APP synthesis. Despite the increased synthesis of APP, hypoalbuminemia is common in cancer patients, although this does not appear to be due to a decreased in albumin synthesis (Fearon et al. 1998).

The acute-phase response is a systemic reaction to tissue injury, typically observed during infection, inflammation or trauma, characterized by the increased production of a series of hepatocyte-derived plasma proteins known as acute-phase reactants (including C-reactive protein (CRP), serum amyloid A (SAA), α1-antitrypsin, fibrinogen, and complement factors B and C3) and by decreased circulating concentrations of albumin and transferrin. An APP response is observed in a significant proportion of patients with the type of cancer frequently associated with weight loss (i.e. pancreas, lung, esophagus). The proportion of pancreatic patients exhibiting an acute-phase response increases with disease progression (Falconer et al. 1994; Stephens et al. 2008). For many years investigators have been searching for mediators involved in the regulation of APP synthesis. Interestingly the cytokines IL-6, IL-1 and TNF are now regarded as the major mediators of APP induction

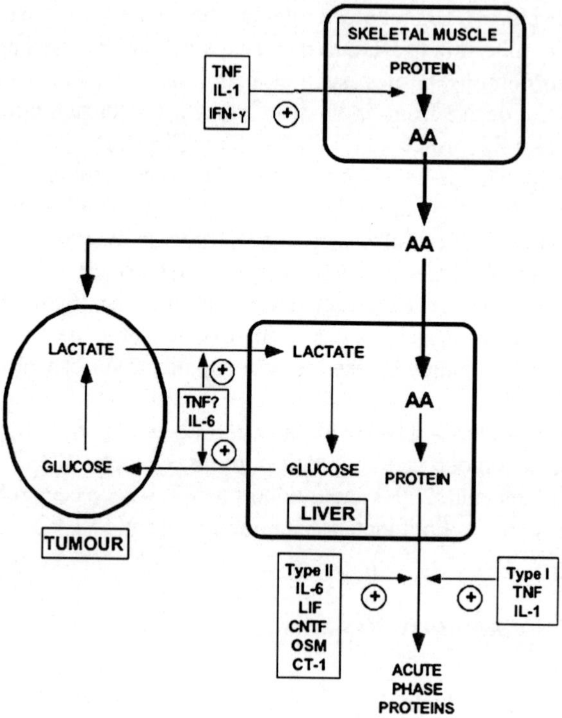

Fig. 2 Cytokines can mimic most metabolic alterations. Most of the metabolic alterations present during cancer cachexia can be mimicked by pro-inflammatory cytokines

in the liver (Moshage 1997; Moses et al. 2009). In fact, APP can be divided into two groups: type I and type II. Type I proteins include SAA, CRP, C3, haptoglobin (rat) and α1-acid glycoprotein, and are induced by IL-1 and TNF. Type II proteins include fibrinogen, haptoglobin (human), α1-antichymotrypsin and α2-macroglobulin (rat), and are induced by IL-6, LIF, OSM (oncostatin M), CNTF and CT-1 (cardiotrophin-1). Unfortunately, the role of APP during cancer growth is still far from understood.

3 Ageing, Inflammation and Sarcopenia

3.1 The Problem

Ageing is an extremely complex biological phenomenon of immense importance. Currently, we have a poor, incomplete understanding of the fundamental molecular mechanisms involved. Discussions on ageing invariably begin by establishing a satisfactory definition for the term ageing and the related word senescence.

Although the term ageing is commonly used to refer to postmaturational processes that are deteriorative and lead to an increased vulnerability, the more correct term for this is senescence. Ageing could refer to any time-dependent process. In this proposal, the terms ageing and senescence are used interchangeably. All aging changes have a cellular basis, and ageing is perhaps best studied, fundamentally at the cellular level under defined and controlled environmental conditions.

In recent years, age-related diseases and disabilities have become of major health interest and importance. This holds particularly for the Western community, where the dramatic improvement of medical health, standard of living and hygiene have reduced the main causes of death prevalent in previous eras, most notably infectious diseases. Thanks to the discovery and development of antibiotics, vaccines and improved hygiene, the average life span has dramatically increased and has resulted in a conversion of the age-pyramid structure from a population numerically dominated by the younger generations to one in which the elderly have become of significant importance. Simple prediction of human life span from the average decline in kidney function results in a maximum life span of 120–140 years. Although the age statistics are inaccurate and records of previous centuries are missing, anecdotal evidence does not indicate a change in maximum life span.

Weight loss is a major problem that increases mortality in the geriatric population. Feelings of well-being and the pleasure derived from eating affect the quality of older individuals' lives positively. The connection between eating and good heath has been understood for hundreds of years and trascends all cultures. Furthermore, it is understood that when elderly people stop eating their death is imminent. Treating malnutrition and weight loss can help to ameliorate many medical conditions. Rehabilitation time after hip fractures has been shown to be shortened with nutritional support (Bastow et al. 1983). In hospitalized geriatric patients, low serum albumin concentrations with weight loss predict those patients at highest risk of death (McMurtry and Rosenthal 1995).

Weight loss in geriatric patients is not unusual (Fig. 3). Of nursing home residents, 30–50% have substandard body weight and midarm muscle circumferences, and low albumin concentrations (Abbasi and Rudman 1994). Morley and Kraenzle (1994) found that 15–21% of 1,156 nursing home residents had lost more than 5 lb over a period of 3–6 months. According to Schneider et al. (2002) weight loss in the elderly leads to cachexia with a preferential loss of lean versus adipose tissue. The same authors report that the elderly show an increased resting energy expenditure that may be one of the underlying causes of the weight loss. Wasting and cachexia are associated with severe physiological, psychological, and immunological consequences, regardless of the underlying causes (Chandra 1983). Cachexia has been associated with an increased number of infections, decubitus ulcers, and even deaths (Pinchcofsky-Devin and Kaminski 1986). Wallace et al. (1995) reported that involuntary weight loss exceeded 13% in a group of 247 community-residing male veterans of 65 years of age or older. They also found involuntary weight loss of more than 4% of body weight to be an important independent predictor of increased mortality (Wallace et al. 1995). Goodwin et al. (1983),

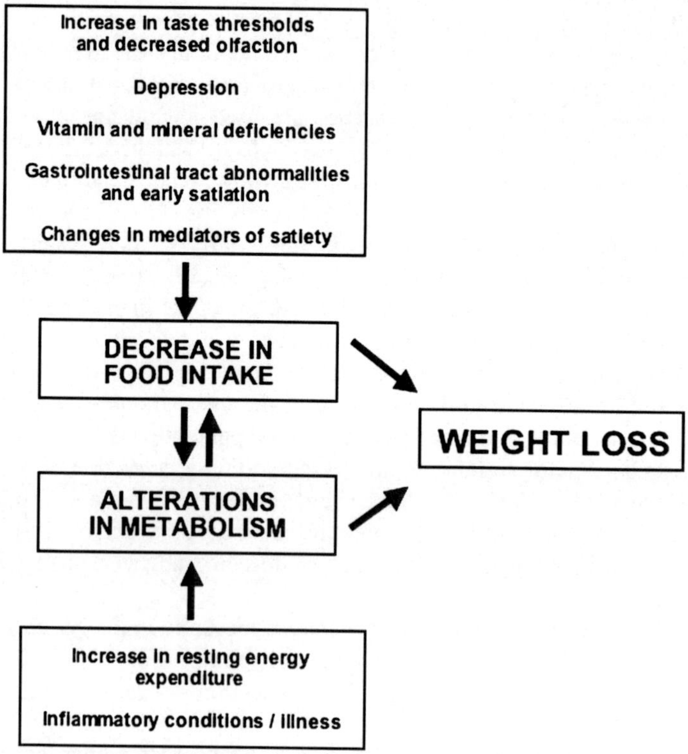

Fig. 3 Factors involved in ageing malnutrition. The main factors that contribute to the malnutrition commonly observed in geriatric patients

Braun et al. (1988) and Morley and Silver (1988) found that malnutrition may also cause or exacerbate cognitive and mood disorders. Others have found that weight loss and cachexia are also predictive of morbidity and mortality (Marton et al. 1981; Rabinovitz et al. 1986). In the elderly, medical, cognitive and psychiatric disorders may diminish self-reliance in activities of daily living, thus reducing quality of life and increasing the frequency of secondary procedures, hospitalizations, and the need for skilled nursing care (Aubertin-Leheudre et al. 2008). Therefore, adequate weight and nutrition are necessary for a good quality of life and for optimal health in nursing home settings.

3.2 *Cachexia and Sarcopenia are Driven by Different Factors*

As can be seen in Fig. 4, the factors involved in the etiology of cachexia are different from those involved in sarcopenia. While proinflammatory cytokines, hypermetabolism and malnutrition play an important role in cachexia, hormonal changes and physical inactivity are the main triggering factors in sarcopenia.

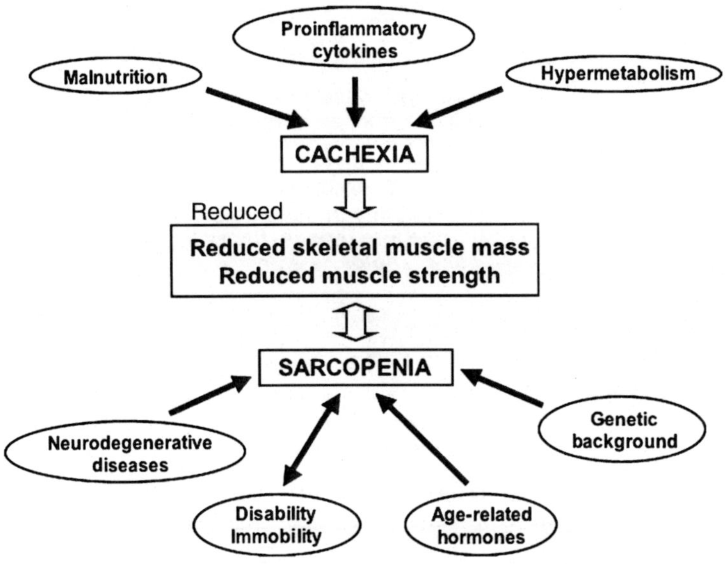

Fig. 4 Diferential factors involved in sarcopenia and cachexia. The factors involved in cancer cachexia are very different from those behind sarcopenia. Thus, in cancer, proinflammatory cytokines play a very important role together with the hypermetabolic state and anorexia, while in sarcopenia endocrine changes and neurodegenerative alterations are very important

3.3 Age-Related Muscle Wasting: Mechanisms

Despite numerous theories and intensive research, the principal molecular mechanisms underlying the process of ageing are still unknown. Most, if not all, attempts to prevent or stop the onset of typical degenerative diseases associated with ageing have so far been futile. Solutions to the major problems of dealing with age-related diseases can only come from a systematic and thorough molecular analysis of the ageing process and a detailed understanding of its causes. Thus, effective measures to prevent the onset of age-related disease and disabilities depend on solid fundamental scientific knowledge and a detailed mechanistic insight.

Some of the mechanisms and determinants involved in muscle wasting (Fig. 5) during ageing involve hormonal changes. Glucocorticoids seem to be involved in the emergence of muscle atrophy with advancing age (Dardevet et al. 1995, 1998; Savary et al. 1998). These hormones seem to interfere with other anabolic ones such as insulin or IGF-I (Dardevet et al. 1998, 1996; Vary et al. 1997, 1999, 1998; Sinaud et al. 1999). Some studies have suggested that exercise can delay the onset of muscle wasting in aged experimental animals (Mosoni et al. 1995; Slentz and Holloszy 1993; Lambert et al. 2002). Other investigations have shown that treatment with β2-agonists can delay the onset of wasting associated with ageing (Carter and Lynch 1994). Bearing in mind the fact that the regenerative potential of skeletal muscle, and overall muscle mass, decline with age, this may be influenced

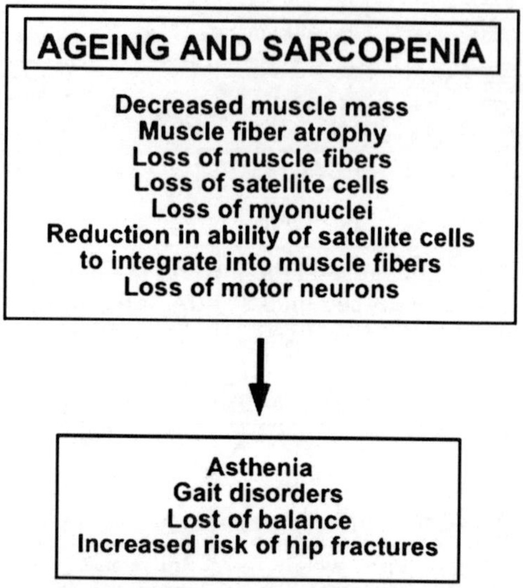

Fig. 5 Main events that take place in skeletal muscle leading to sarcopenia. The reduction in muscle mass is accompanied by a clear atrophy involving changes that affect not only muscle fibers but also satellite cells, all of it leading to a considerable degree of asthenia

by autocrine growth factors intrinsic to the muscle itself. Extrinsic host factors that may influence muscle regeneration include hormones, growth factors secreted in a paracrine manner by accesory cells, innervation, and antioxidant mechanisms (Cannon 1995) (Fig. 6). An inflammatory response ensues in which distinctive populations of macrophages infiltrate the affected tissue: some of these macrophages are involved in phagocytosis of damaged fibers; other macrophages arriving at later times may deliver growth factors or cytokines that promote regeneration. These include fibroblast growth factor and IGF-I, which are important regulators of muscle precursors cell growth and differentiation, as well as nerve growth factor (NGF), which is essential for maintenance or restablishment of neuronal contact. Other cytokines, including IL-1, TNF, IL-15 and CNTF, have a strong influence on the balance between muscle protein synthesis and breakdown. Beyond the severe reduction in life quality for a large fraction of the ageing population suffering from muscle wasting, the age-related loss of muscle mass leaves the affected individuals more prone to risk factors that adversely affect their health including social isolation, stress, depression and accidents.

Among the factors that could be involved in modulating protein turnover in skeletal muscle during ageing, hormonal status may play a very important role. From this point of view, alterations in the somatotropic (GH/IGF-1) axis with a decrease in both mediators during ageing could be either be a symptom of declining neuroendocrine function, a cause of age-related alterations in body composition and functionality or protective mechanism against age-associated disease (bartke 372 22). Thus, insulin resistance phenomena may alter the rates of protein synthesis

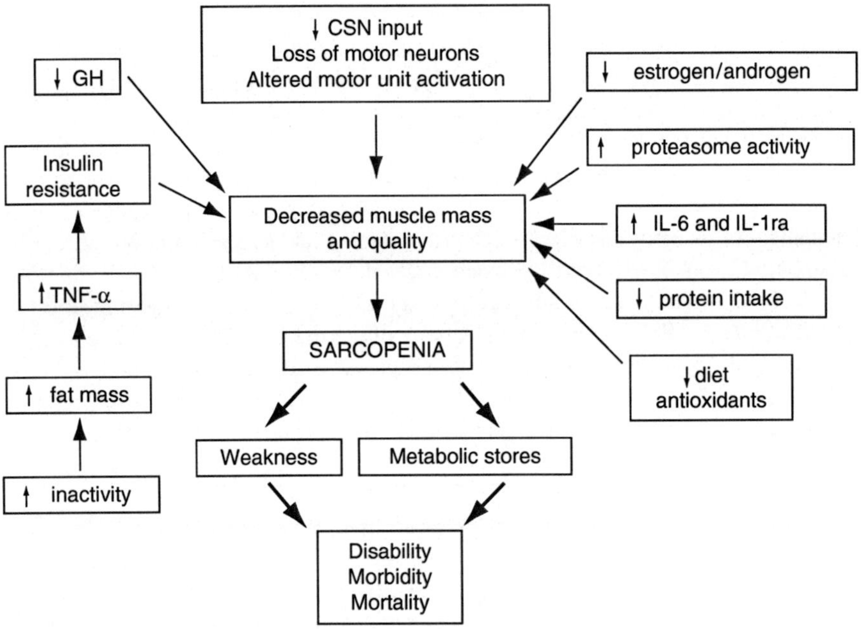

Fig. 6 Etiology of sarcopenia. The etiology of sarcopenia involves many different factors, including hormonal changes, cytokine alterations and alterations in food intake, that result in protein and vitamin deficiencies

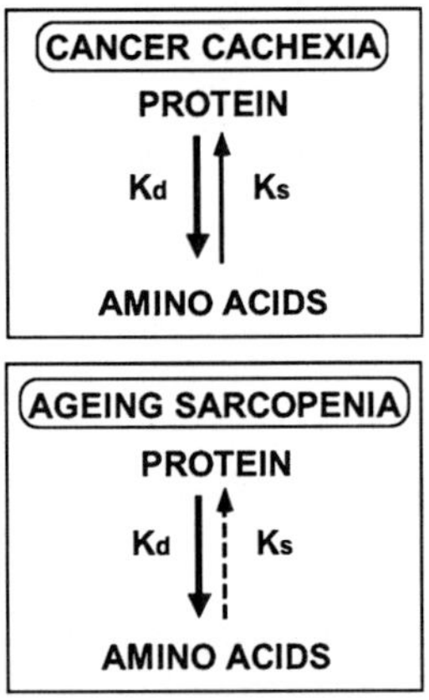

Fig. 7 Differences in protein turnover in cancer cachexia and muscle sarcopenia. Interestingly, while in cancer cachexia protein degradation is the main factor involved in ageing, sarcopenia includes a dramatic decrease in the rate of myofibrillar protein synthesis

in skeletal muscle. It has been reported that glucocorticoids that induce the ubiquitin-dependent muscle proteolysis in fasted or acidotic young rats, do not induce such proteolysis in aged rats (Dardevet et al. 1995) (Fig. 7). Similarly, a reduced sensitivity to a variety of hormones and growth factors in aged tissues has been reported (Carlin et al. 1983; Harley et al. 1981; Plisko and Gilchrest 1983). It may then be suggested that a defect in signal transduction could be related to the ubiquitin system in aged cells.

Several other mechanisms have been postulated to explain the skeletal muscle weakness associated with ageing and it appears that sarcopenia is only partially explained by the loss in muscle mass. Thus, apoptosis has been implicated as a mechanism of loss of muscle cells in normal ageing and plays an important role in sarcopenia (Dirks Naylor and Leeuwenburgh 2008). In the apoptotic events, both caspase-2 ad oxidative stress seem to play an important role in triggering physiological cell death (Braga et al. 2008). A body of evidence suggest that ion channels and their ability to respond to growth factors such as IGF-I could be a key factor underlying skeletal muscle impairment with ageing (Delbono 2000, 2002; Renganathan et al. 1998). In this context, the reduction in L-type Ca^{2+} channels expression in ageing mice reduced peak cytosolic Ca^{2+} with subsequent decrease in skeletal muscle force (Delbono 2002). On the other hand, K^{+} channels are essential to both induce myogenesis and proliferation of muscle cells (Fischer-Lougheed et al. 2001; Grande et al. 2003). K^{+} channels are modulated by IGF-I and the overexpression of human IGF-I exclusively in skeletal muscle increases the number and prevents age-related decline in the sarcoplasmic reticulum dihydropyridine-sensitive voltage-gated L-type Ca^{2+} channel (Delbono 2002; Gamper et al. 2002). Taking all of this into consideration, it is clear that ion channels are involved in the age-related decline in muscle force. Concerning neuronal activity important changes in ion channel expression occurs during ageing. It is not clear what is the relationship between the observed changes and the decreased of synaptic contacts, ion balances or neuronal loss. However, several hypothesis have been evaluated such as the Ca^{2+} theory and the effects of reactive oxygen/nitrogen species in ion channel activity in the aged brain (Foster and Kumar 2002; Dirksen 2002; Annunziato et al. 2002). However, it seems quite clear that changes in nerve ion channel expression may modify behavioral, feeding, learning and cognitive conducts during ageing those affecting muscle wasting in sarcopenia. Di Giulio et al. (2009) have recently found an altered mitochondrial status in skeletal muscles during ageing with a tight correlation between muscle total mitochondrial volume and sarcopenia. Therefore, hypoxia could well be involved in the muscle wasting process associated with ageing. In addition, ageing seems to be related to increased frequency of mutations in mitochondrial DNA. These mutations originate mitochondrial dysfunction and seem to be intimately related with the apoptotic process (Fig. 8). Additionally, the mentioned mutations lead to a decreased rate of electronic transport which results in increased ROS production, therefore increasing even more the mitochondrial damage (Fig. 8) (Thompson 2009; Hiona and Leeuwenburgh 2008).

Cytokines seem to play a key role in muscle wasting, at least during pathological conditions thus, cytokines are best known as mediators of host defense to invasive stimuli (Fig. 9). However, some of them (TNF, IL-1 and IL-6 in particular) may modulate clearance and repair processes in skeletal muscle following injury and may also be involved with the sustained viability of muscle cells. Muscle repair also

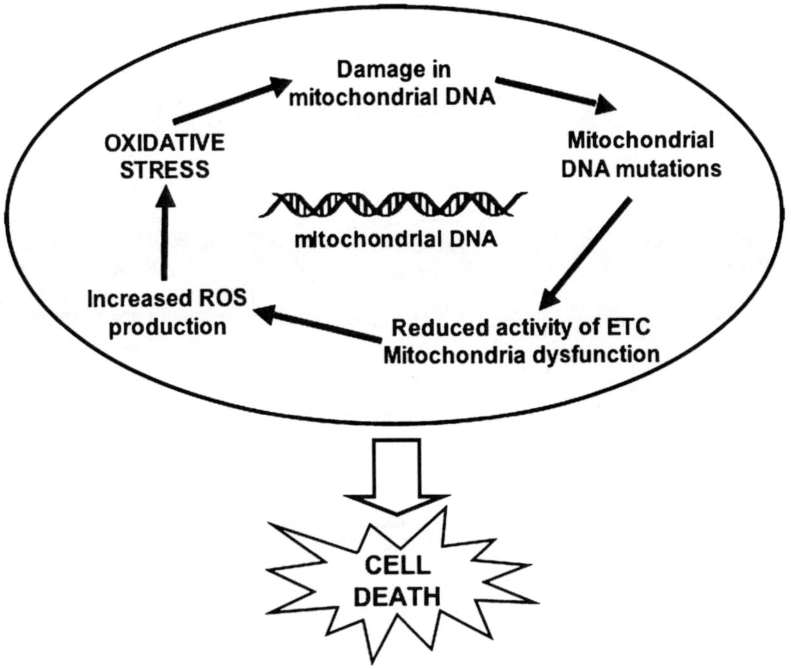

Fig. 8 Mitochondrial mutations and oxidative stress. Mitochondrial DNA mutations may play a key role in triggering sarcopenia. These mutations would generate mitochondrial dysfunction and activation of mitochondrial apoptosis. The problem is under a positive feedback since mitochondrial dysfunction generates an increase in reactive oxygen species (ROS) due to a deficient electron transfer machanism, and this generates more ROS and, therefore, increased mitochondrial dysfunction

requires neuronal contact influenced by other cytokines (such as NGF and CNTFr) as well as angiogenesis and connective tissue matrix formation. Successful muscle ageing will depend, in part, on how well a muscle repairs itself after damage. Age-related loss of muscle mass or function may be the cumulative result of repeated episodes of incomplete repair. Abnormal production or sensitivity to cytokines by aged cells may contribute to these changes in muscle mass and function. Grounds (Grounds 2002) has recently suggested that inflammatory cytokines could be involved in sarcopenia by interfering with IGF-I signaling in skeletal muscle. Cytokines – interleukins in particular – appears to stimulate both corticotropin-releasing factor (CRF) and prostaglandin $E_{1\alpha}$ production which behave as powerful anorectic agents, thus contributing to the decrease in food intake associated with aging (Morley 2001). In addition, cytokines inhibit the release of orexigenic peptides such as neuropeptide Y. It becomes thus clear that cytokines alter the balance between orexigenic and anorexigenic signals in brain and therefore contribute significantly to the alterations observed in appetite in the elderly (Morley 2001). Interestingly, many cytokines also cause an elevation in availability of leptin which, in turn, further contributes to the decline in food intake (Morley 2001; Lee et al. 2007).

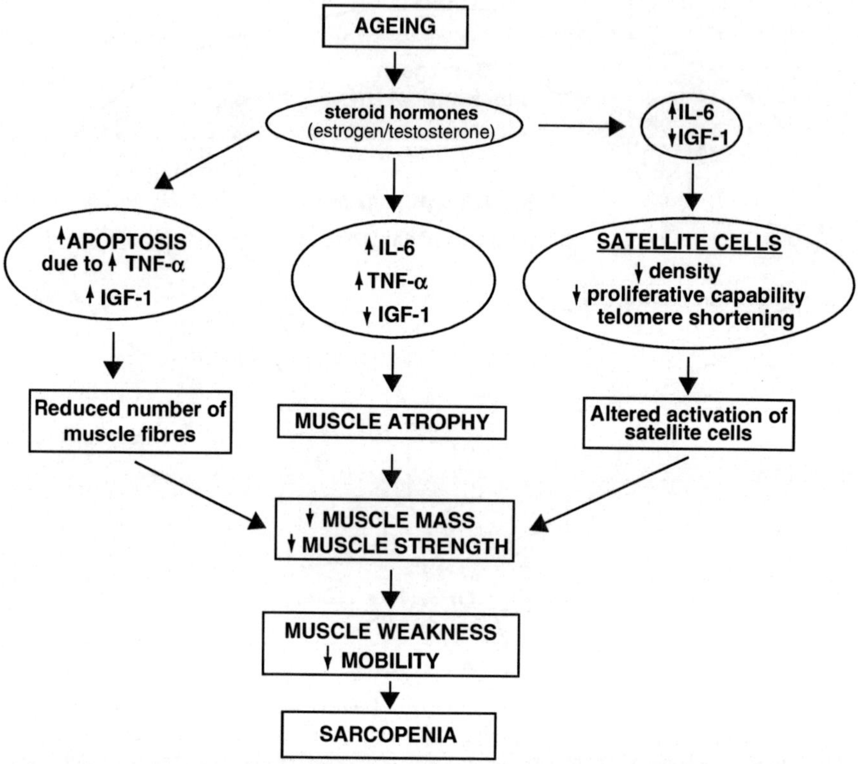

Fig. 9 Role of cytokines in myofiber alterations associated with sarcopenia. Some cytokines may influence muscle repair mechanisms following injury, and may, therefore, be involved in the maintenance of muscle integrity

4 Conclusions

Cancer cachexia is a complex pathological condition characterized by many metabolic changes involving numerous organs. These changes are triggered by alterations in the hormonal milieu, release of different tumour factors and a systemic inflammatory reaction characterized by cytokine production and release. In fact, the macrophage-derived proinflammatory cytokines (IL-1, IL-6, TNF-α) have key roles in inducing metabolic changes associated with many pathophysiological conditions, not only immune and inflammatory reactions but also in the development of cachexia. In fact, the balance between these and the anti-inflammatory cytokines such as IL-1ra, IL-10 and TGF is pivotal for the fine tuning of many biochemical processes. For instance, in chronic myelogenous leukemia, high cellular (leukocyte) levels of IL-1β and low levels of IL-ra are seen in advanced disease and correlate with reduced survival (Harley et al. 1981).

A complex interaction of pro-cachectic and anti-cachectic cytokines or cytokine-neutralizing molecules probably determines the critical presentation and course of cachexia. Intervening in this sequence of events to modify the host responses may prove to be a beneficial treatment strategy for cachexia. Currently tested anti-proinflammatory cytokines have produced interesting results.

Bearing in mind all the information presented here, it can indeed be concluded that no definite mediator of cancer cachexia has yet been identified. However, among all the possible mediators considered here, TNF-α is one of the most relevant candidates. Indeed, TNF-α can mimic most of the abnormalities found during cancer cachexia: weight loss, anorexia, increased thermogenesis, alterations in lipid metabolism and adipose tissue dissolution, insulin resistance and muscle waste including activation of protein breakdown. However, TNF-α alone cannot explain all the cachectic metabolic alterations present in different types of human cancers and experimental tumours. Another important drawback is the fact that TNF-α circulating concentrations are not always elevated in cancer-bearing states and, although it may be argued that in those cases local tissue production of the cytokine may be high, cachexia does not seem to be a local tumour effect. Consequently, both tumour-produced and humoural factors must collaborate in the full induction of the cachectic state. In the particular case of ageing sarcopenia, investigations are needed to elucidate not only mechanisms involved in the wasting process but also to clarify the role of the different factors involved in the complex etiology of sarcopenia.

In conclusion, and because metabolic alterations often appear early after the onset of tumour growth, the scope of appropriate treatment, although not aimed at achieving immediate eradication of the tumour mass, could influence the course of the patient's clinical state or, at least, prevent the steady erosion of dignity that the patient may feel in association with the syndrome. This would no doubt contribute to improving the patient's quality of life and, possibly, prolong survival. Although exploration of the role that cytokines play in the host response to invasive stimuli is an endeavour that has been underway for many years, considerable controversy still exists over the mechanisms of lean tissue and body fat dissolution that occur in the patient with either cancer or inflammation and whether humoural factors regulate this process. A better understanding of the role of cytokines interfering with the molecular mechanisms accounting for protein wasting in skeletal muscle is essential for the design of future effective therapeutic strategies. In any case, understanding the humoural response to inflammation and modifying cytokine actions pharmacologically may prove very effective, and no doubt future research will concentrate on this interesting field.

References

Abbasi, A. A. & Rudman, D. (1994). Undernutrition in the nursing home: prevalence, consequences, causes and prevention. *Nutrition Reviews, 52*, 113–122.

Acharyya, S., Ladner, K. J., Nelsen, L. L., Damrauer, J., Reiser, P. J., Swoap, S., Guttridge, D. C. (2004). Cancer cachexia is regulated by selective targeting of skeletal muscle gene products. *The Journal of Clinical Investigation, 114*, 370–378.

Acharyya, S., Butchbach, M. E., Sahenk, Z., Wang, H., Saji, M., Carathers, M., Ringel, M. D., Skipworth, R. J., Fearon, K. C., Hollingsworth, M. A., Muscarella, P., Burghes, A. H., Rafael-Fortney, J. A., Guttridge, D. C. (2005). Dystrophin glycoprotein complex dysfunction: a regulatory link between muscular dystrophy and cancer cachexia. *Cancer Cell, 8*, 421–432.

Adams, M. & Victor, M. (1981). Asthenia. In Adams R, Victor M (Eds.), *Principles of Neurology* (pp. 341–345). New York: McGraw-Hill.

Adams, V., Gielen, S., Hambrecht, R., Schuler, G. (2001). Apoptosis in skeletal muscle. *Frontiers in Bioscience, 6*, D1–D11.

Agusti, A. G., Sauleda, J., Miralles, C., Gomez, C., Togores, B., Sala, E., Batle, S., Busquets, X. (2002). Skeletal muscle apoptosis and weight loss in chronic obstructive pulmonary disease. *American Journal of Respiratory and Critical Care Medicine, 166*, 485–489.

Agustsson, T., Ryden, M., Hoffstedt, J., Van Harmelen, V., Dicker, A., Laurencikiene, J., Isaksson, B., Permert, J., Arner, P. (2007). Mechanism of increased lipolysis in cancer cachexia. *Cancer Research, 67*, 5531–5537.

Almendro, V., Carbo, N., Busquets, S., Figueras, M., Tessitore, L., Lopez-Soriano, F. J., Argiles, J. M. (2003). Sepsis induces DNA fragmentation in rat skeletal muscle. *European Cytokine Network, 14*, 256–259.

Alvarez, B., Quinn, L. S., Busquets, S., Quiles, M. T., Lopez-Soriano, F. J., Argiles, J. M. (2002). Tumor necrosis factor-alpha exerts interleukin-6-dependent and -independent effects on cultured skeletal muscle cells. *Biochimica et Biophysica Acta, 1542*, 66–72.

Annunziato, L., Pannaccione, A., Cataldi, M., Secondo, A., Castaldo, P., DI Renzo, G., Taglialatela, M. (2002). Modulation of ion channels by reactive oxygen and nitrogen species: a pathophysiological role in brain aging? *Neurobiology of Aging, 23*, 819–834.

Argiles, J. M., Garcia-Martinez, C., Llovera, M., Lopez-Soriano, F. J. (1992). The role of cytokines in muscle wasting: its relation with cancer cachexia. *Medicinal Research Reviews, 12*, 637–652.

Argiles, J. M., Alvarez, B., Lopez-Soriano, F. J. (1997). The metabolic basis of cancer cachexia. *Medicinal Research Reviews, 17*, 477–498.

Argiles, J. M., Busquets, S., Moore-Carrasco, R., Figueras, M., Almendro, V., Lopez-Soriano, F. J. (2007). Targets in clinical oncology: the metabolic environment of the patient. *Frontiers in Bioscience, 12*, 3024–3051.

Argiles, J. M., Lopez-Soriano, F. J., Busquets, S. (2008). Apoptosis signalling is essential and precedes protein degradation in wasting skeletal muscle during catabolic conditions. *The International Journal of Biochemistry & Cell Biology, 40*, 1674–1678.

Aubertin-Leheudre, M., Lord, C., Labonte, M., Khalil, A., Dionne, I. J. (2008). Relationship between sarcopenia and fracture risks in obese postmenopausal women. *Journal of Women and Aging, 20*, 297–308.

Bajaj, G. & Sharma, R. K. (2006). TNF-alpha-mediated cardiomyocyte apoptosis involves caspase-12 and calpain. *Biochemical and Biophysical Research Communications, 345*, 1558–1564.

Baracos, V. E. (2000). Regulation of skeletal-muscle-protein turnover in cancer-associated cachexia. *Nutrition, 16*, 1015–1018.

Baracos, V. E., Devivo, C., Hoyle, D. H., Goldberg, A. L. (1995). Activation of the ATP-ubiquitin-proteasome pathway in skeletal muscle of cachectic rats bearing a hepatoma. *The American Journal of Physiology, 268*, E996–E1006.

Bastow, M. D., Rawlings, J., Allison, S. P. (1983). Benefits of supplementary tube feeding after fractured neck of femur: a randomised controlled trial. *British Medical Journal (Clinical Research Ed.), 287*, 1589–1592.

Belizario, J. E., Katz, M., Chenker, E., Raw, I. (1991). Bioactivity of skeletal muscle proteolysis-inducing factors in the plasma proteins from cancer patients with weight loss. *British Journal of Cancer, 63*, 705–710.

Belizario, J. E., Lorite, M. J., Tisdale, M. J. (2001). Cleavage of caspases–1, –3, –6, –8 and –9 substrates by proteases in skeletal muscles from mice undergoing cancer cachexia. *British Journal of Cancer, 84*, 1135–1140.

Benn, S. C. & Woolf, C. J. (2004). Adult neuron survival strategies–slamming on the brakes. *Nature Reviews. Neuroscience, 5*, 686–700.

Bonetto, A., Penna, F., Minero, V. G., Reffo, P., Bonelli, G., Baccino, F. M., Costelli, P. (2009). Deacetylase inhibitors modulate the myostatin/follistatin axis without improving cachexia in tumor-bearing mice. *Current Cancer Drug Targets, 9*(5), 608–616.

Bossola, M., Muscaritoli, M., Costelli, P., Bellantone, R., Pacelli, F., Busquets, S., Argiles, J., Lopez-Soriano, F. J., Civello, I. M., Baccino, F. M., Rossi Fanelli, F., Doglietto, G. B. (2001). Increased muscle ubiquitin mRNA levels in gastric cancer patients. *American Journal of Physiology: Regulatory, Integrative and Comparative Physiology, 280*, R1518–R1523.

Braga, M., Sinha Hikim, A. P., Datta, S., Ferrini, M. G., Brown, D., Kovacheva, E. L., Gonzalez-Cadavid, N. F., Sinha-Hikim, I. (2008). Involvement of oxidative stress and caspase 2-mediated intrinsic pathway signaling in age-related increase in muscle cell apoptosis in mice. *Apoptosis, 13*, 822–832.

Braun, J. V., Wykle, M. H., Cowling, W. R. 3rd (1988). Failure to thrive in older persons: a concept derived. *The Gerontologist, 28*, 809–812.

Brenner, D. A., O'HARA, M., Angel, P., Chojkier, M., Karin, M. (1989). Prolonged activation of jun and collagenase genes by tumour necrosis factor-alpha. *Nature, 337*, 661–663.

Busquets, S., Sanchis, D., Alvarez, B., Ricquier, D., Lopez-Soriano, F. J., Argiles, J. M. (1998). In the rat, tumor necrosis factor alpha administration results in an increase in both UCP2 and UCP3 mRNAs in skeletal muscle: a possible mechanism for cytokine-induced thermogenesis? *FEBS Letters, 440*, 348–350.

Busquets, S., Aranda, X., Ribas-Carbo, M., Azcon-Bieto, J., Lopez-Soriano, F. J., Argiles, J. M. (2003). Tumour necrosis factor-alpha uncouples respiration in isolated rat mitochondria. *Cytokine, 22*, 1–4.

Busquets, S., Figueras, M. T., Fuster, G., Almendro, V., Moore-Carrasco, R., Ametller, E., Argiles, J. M., Lopez-Soriano, F. J. (2004). Anticachectic effects of formoterol: a drug for potential treatment of muscle wasting. *Cancer Research, 64*, 6725–6731.

Busquets, S., Deans, C., Figueras, M., Moore-Carrasco, R., Lopez-Soriano, F. J., Fearon, K. C., Argiles, J. M. (2007). Apoptosis is present in skeletal muscle of cachectic gastro-intestinal cancer patients. *Clinical Nutrition, 26*, 614–618.

Cai, D., Frantz, J. D., Tawa, N. E., Melendez, P. A., Oh, B. C., Lidov, H. G., Hasselgren, P. O., Frontera, W. R., Lee, J., Glass, D. J., Shoelson, S. E. (2004). IKKbeta/NF-kappaB activation causes severe muscle wasting in mice. *Cell, 119*, 285–298.

Cannon, J. G. (1995) Cytokines in aging and muscle homeostasis. *The Journals of Gerontology. Series A: Biological Sciences and Medical Sciences*, 50 Spec No, 120-3.

Carbo, N., Busquets, S., van Royen, M., Alvarez, B., Lopez-Soriano, F. J., Argiles, J. M. (2002). TNF-alpha is involved in activating DNA fragmentation in skeletal muscle. *British Journal of Cancer, 86*, 1012–1016.

Carlin, C. R., Phillips, P. D., Knowles, B. B., Cristofalo,V. J. (1983). Diminished in vitro tyrosine kinase activity of the EGF receptor of senescent human fibroblasts. *Nature, 306*, 617–620.

Carter, W. J. & Lynch, M. E. (1994). Comparison of the effects of salbutamol and clenbuterol on skeletal muscle mass and carcass composition in senescent rats. *Metabolism, 43*, 1119–1125.

Coletti, D., Yang, E., Marazzi, G., Sassoon, D. (2002). TNFalpha inhibits skeletal myogenesis through a PW1-dependent pathway by recruitment of caspase pathways. *The EMBO Journal, 21*, 631–642.

Costelli, P., Garcia-Martinez, C., Llovera, M., Carbo, N., Lopez-Soriano, F. J., Agell, N., Tessitore, L., Baccino, F. M., Argiles, J. M. (1995). Muscle protein waste in tumor-bearing rats is effectively antagonized by a beta 2-adrenergic agonist (clenbuterol). Role of the ATP-ubiquitin-dependent proteolytic pathway. *The Journal of Clinical Investigation, 95*, 2367–2372.

Costelli, P., Muscaritoli, M., Bossola, M., Moore-Carrasco, R., Crepaldi, S., Grieco, G., Autelli, R., Bonelli, G., Pacelli, F., Lopez-Soriano, F. J., Argiles, J. M., Doglietto, G. B., Baccino, F. M., Rossi Fanelli, F. (2005a). Skeletal muscle wasting in tumor-bearing rats is associated with MyoD down-regulation. *International Journal of Oncology, 26*, 1663–1668.

Costelli, P., Reffo, P., Penna, F., Autelli, R., Bonelli, G., Baccino, F. M. (2005b). Ca(2+)-dependent proteolysis in muscle wasting. *The International Journal of Biochemistry and Cell Biology, 37*, 2134–2146.

Costelli, P., Muscaritoli, M., Bossola, M., Penna, F., Reffo, P., Bonetto, A., Busquets, S., Bonelli, G., Lopez-Soriano, F. J., Doglietto, G. B., Argiles, J. M., Baccino, F. M., Rossi Fanelli, F. (2006). IGF-1 is downregulated in experimental cancer cachexia. *American Journal of Physiology: Regulatory, Integrative and Comparative Physiology, 291*, R674–R683.

Chandra, R. K. (1983). Nutrition, immunity, and infection: present knowledge and future directions. *Lancet, 1*, 688–691.

Choi, S. E., Min, S. H., Shin, H. C., Kim, H. E., Jung, M. W., Kang, Y. (2006). Involvement of calcium-mediated apoptotic signals in H2O2-induced MIN6N8a cell death. *European Journal of Pharmacology, 547*, 1–9.

Dardevet, D., Sornet, C., Taillandier, D., Savary, I., Attaix, D., Grizard, J. (1995). Sensitivity and protein turnover response to glucocorticoids are different in skeletal muscle from adult and old rats. Lack of regulation of the ubiquitin-proteasome proteolytic pathway in aging. *The Journal of Clinical Investigation, 96*, 2113–2119.

Dardevet, D., Sornet, C., Vary, T., Grizard, J. (1996). Phosphatidylinositol 3-kinase and p70 s6 kinase participate in the regulation of protein turnover in skeletal muscle by insulin and insulin-like growth factor I. *Endocrinology, 137*, 4087–4094.

Dardevet, D., Sornet, C., Savary, I., Debras, E., Patureau-Mirand, P., Grizard, J. (1998). Glucocorticoid effects on insulin- and IGF-I-regulated muscle protein metabolism during aging. *The Journal of Endocrinology, 156*, 83–89.

Delbono, O. (2000). Regulation of excitation contraction coupling by insulin-like growth factor-1 in aging skeletal muscle. *The Journal of Nutrition, Health & Aging, 4*, 162–164.

Delbono, O. (2002). Molecular mechanisms and therapeutics of the deficit in specific force in ageing skeletal muscle. *Biogerontology, 3*, 265–270.

Dessi, S., Batetta, B., Pulisci, D., Accogli, P., Pani, P., Broccia, G. (1991). Total and HDL cholesterol in human hematologic neoplasms. International Journal of Hematology, 54, 483–486.

Dessi, S., Batetta, B., Anchisi, C., Pani, P., Costelli, P., Tessitore, L., Baccino, F. M. (1992). Cholesterol metabolism during the growth of a rat ascites hepatoma (Yoshida AH-130). *British Journal of Cancer, 66*, 787–793.

Dessi, S., Batetta, B., Spano, O., Bagby, G. J., Tessitore, L., Costelli, P., Baccino, F. M., Pani, P., Argiles, J. M. (1995). Perturbations of triglycerides but not of cholesterol metabolism are prevented by anti-tumour necrosis factor treatment in rats bearing an ascites hepatoma (Yoshida AH-130). *British Journal of Cancer, 72*, 1138–1143.

Dewys, W. (1985). Management of cancer cachexia. Seminars in Oncology, 12, 452–460.

DI Giulio, C., Petruccelli, G., Bianchi, G., Cacchio, M., Verratti, V. (2009). Does hypoxia cause sarcopenia? Prevention of hypoxia could reduce sarcopenia. *Journal of Biological Regulators and Homeostatic Agents, 23*, 55–58.

Dirks Naylor, A. J. & Leeuwenburgh, C. (2008). Sarcopenia: the role of apoptosis and modulation by caloric restriction. *Exercise and Sport Sciences Reviews, 36*, 19–24.

Dirksen, R. T. (2002). Reactive oxygen/nitrogen species and the aged brain: radical impact of ion channel function. *Neurobiology of Aging, 23*, 837–839. discussion 841–2.

Du, J., Wang, X., Miereles, C., Bailey, J. L., Debigare, R., Zheng, B., Price, S. R., Mitch, W. E. (2004). Activation of caspase-3 is an initial step triggering accelerated muscle proteolysis in catabolic conditions. *The Journal of Clinical Investigation, 113*, 115–123.

Eisenberg, S. (1984). High density lipoprotein metabolism. *Journal of Lipid Research, 25*, 1017–1058.

Eley, H. L. & Tisdale, M. J. (2007). Skeletal muscle atrophy, a link between depression of protein synthesis and increase in degradation. *The Journal of Biological Chemistry, 282*, 7087–7097.

Evans, R. D. & Williamson, D. H. (1988). Tissue-specific effects of rapid tumour growth on lipid metabolism in the rat during lactation and on litter removal. *The Biochemical Journal, 252*, 65–72.

Falconer, J. S., Fearon, K. C., Plester, C. E., Ross, J. A., Carter, D. C. (1994). Cytokines, the acute-phase response, and resting energy expenditure in cachectic patients with pancreatic cancer. *Annals of Surgery, 219*, 325–331.

Fearon, K. C., Falconer, J. S., Slater, C., Mcmillan, D. C., Ross, J. A., Preston, T. (1998). Albumin synthesis rates are not decreased in hypoalbuminemic cachectic cancer patients with an ongoing acute-phase protein response. *Annals of Surgery, 227*, 249–254.

Fernandez-Celemin, L., Pasko, N., Blomart, V., Thissen, J. P. (2002). Inhibition of muscle insulin-like growth factor I expression by tumor necrosis factor-alpha. *American Journal of Physiology. Endocrinology and Metabolism, 283*, E1279–E1290.

Fernando, P., Kelly, J. F., Balazsi, K., Slack, R. S., Megeney, L. A. (2002). Caspase 3 activity is required for skeletal muscle differentiation. *Proceedings of the National Academy of Sciences of the United States of America, 99*, 11025–11030.

Ferreira, R., Neuparth, M. J., Ascensao, A., Magalhaes, J., Vitorino, R., Duarte, J. A., Amado, F. (2006). Skeletal muscle atrophy increases cell proliferation in mice gastrocnemius during the first week of hindlimb suspension. *European Journal of Applied Physiology, 97*, 340–346.

Fischer-Lougheed, J., Liu, J. H., Espinos, E., Mordasini, D., Bader, C. R., Belin, D., Bernheim, L. (2001). Human myoblast fusion requires expression of functional inward rectifier Kir2.1 channels. *The Journal of Cell Biology, 153*, 677–686.

Foster. T. C. & Kumar, A. (2002). Calcium dysregulation in the aging brain. *The Neuroscientist, 8*, 297–301.

Fuster, G., Busquets, S., Ametller, E., Olivan, M., Almendro, V., DE Oliveira, C. C., Figueras, M., Lopez-Soriano, F. J., Argiles, J. M. (2007). Are peroxisome proliferator-activated receptors involved in skeletal muscle wasting during experimental cancer cachexia? Role of beta2-adrenergic agonists. *Cancer Research, 67*, 6512–6519.

Gamper, N., Fillon, S., Huber, S. M., Feng, Y., Kobayashi, T., Cohen, P., Lang, F. (2002). IGF-1 up-regulates K+ channels via PI3-kinase, PDK1 and SGK1. *Pflugers Archiv, 443*, 625–634.

Glass, D. J. (2005). A signaling role for dystrophin: inhibiting skeletal muscle atrophy pathways. *Cancer Cell, 8*, 351–352.

Goodwin, J. S., Goodwin, J. M., Garry, P. J. (1983). Association between nutritional status and cognitive functioning in a healthy elderly population. *JAMA, 249*, 2917–2921.

Grande, M., Suarez, E., Vicente, R., Canto, C., Coma, M., Tamkun, M. M., Zorzano, A., Guma, A., Felipe, A. (2003). Voltage-dependent K+ channel beta subunits in muscle: differential regulation during postnatal development and myogenesis. *Journal of Cellular Physiology, 195*, 187–193.

Grounds, M. D. (2002). Reasons for the degeneration of ageing skeletal muscle: a central role for IGF-1 signalling. *Biogerontology, 3*, 19–24.

Guttridge, D. C., Mayo, M. W., Madrid, L. V., Wang, C. Y., Baldwin, A. S., Jr. (2000). NF-kappaB-induced loss of MyoD messenger RNA: possible role in muscle decay and cachexia. *Science, 289*, 2363–2366.

Hajnoczky, G., Csordas, G., Das, S., Garcia-Perez, C., Saotome, M., Sinha Roy, S., Yi, M. (2006). Mitochondrial calcium signalling and cell death: approaches for assessing the role of mitochondrial Ca2+ uptake in apoptosis. *Cell Calcium, 40*, 553–560.

Harley, C. B., Goldstein, S., Posner, B. I., Guyda, H. (1981). Decreased sensitivity of old and progeric human fibroblasts to a preparation of factors with insulinlike activity. *The Journal of Clinical Investigation, 68*, 988–994.

Harvey, K. B., Bothe, A., Jr., Blackburn, G. L. (1979). Nutritional assessment and patient outcome during oncological therapy. *Cancer, 43*, 2065–2069.

Hiona, A. & Leeuwenburgh, C. (2008). The role of mitochondrial DNA mutations in aging and sarcopenia: implications for the mitochondrial vicious cycle theory of aging. *Experimental Gerontology, 43*, 24–33.

Kirwan, J. P. & Del Aguila, L. F. (2003). Insulin signalling, exercise and cellular integrity. *Biochemical Society Transactions, 31*, 1281–1285.

Lambert, C. P., Sullivan, D. H., Freeling, S. A., Lindquist, D. M., Evans, W. J. (2002). Effects of testosterone replacement and/or resistance exercise on the composition of megestrol acetate stimulated weight gain in elderly men: a randomized controlled trial. *The Journal of Clinical Endocrinology and Metabolism, 87*, 2100–2106.

Lanza-Jacoby, S., Lansey, S. C., Miller, E. E., Cleary, M. P. (1984). Sequential changes in the activities of lipoprotein lipase and lipogenic enzymes during tumor growth in rats. *Cancer Research, 44*, 5062–5067.

Latres, E., Amini, A. R., Amini, A. A., Griffiths, J., Martin, F. J., Wei, Y., Lin, H. C., Yancopoulos, G. D., Glass, D. J. (2005). Insulin-like growth factor-1 (IGF-1) inversely regulates atrophy-induced genes via the phosphatidylinositol 3-kinase/Akt/mammalian target of rapamycin (PI3K/Akt/mTOR) pathway. *The Journal of Biological Chemistry, 280*, 2737–2744.

Lee, C. E., Mcardle, A., Griffiths, R. D. (2007). The role of hormones, cytokines and heat shock proteins during age-related muscle loss. *Clinical Nutrition, 26*, 524–534.

Lee, S. W., Dai, G., Hu, Z., Wang, X., Du, J., Mitch, W. E. (2004). Regulation of muscle protein degradation: coordinated control of apoptotic and ubiquitin-proteasome systems by phosphatidylinositol 3 kinase. *Journal of the American Society of Nephrology, 15*, 1537–1545.

Li, Y. P., Schwartz, R. J., Waddell, I. D., Holloway, B. R., Reid, M. B. (1998). Skeletal muscle myocytes undergo protein loss and reactive oxygen-mediated NF-kappaB activation in response to tumor necrosis factor alpha. *The FASEB Journal, 12*, 871–880.

Lopez-Soriano, J., Argiles, J. M., Lopez-Soriano, F. J. (1996). Lipid metabolism in rats bearing the Yoshida AH-130 ascites hepatoma. *Molecular and Cellular Biochemistry, 165*, 17–23.

Lorite, M. J., Thompson, M. G., Drake, J. L., Carling, G., Tisdale, M. J. (1998). Mechanism of muscle protein degradation induced by a cancer cachectic factor. *British Journal of Cancer, 78*, 850–856.

Llovera, M., Garcia-Martinez, C., Agell, N., Marzabal, M., Lopez-Soriano, F. J., Argiles, J. M. (1994). Ubiquitin gene expression is increased in skeletal muscle of tumour-bearing rats. *FEBS Letters, 338*, 311–318.

Llovera, M., Garcia-Martinez, C., Agell, N., Lopez-Soriano, F. J., Argiles, J. M. (1995). Muscle wasting associated with cancer cachexia is linked to an important activation of the ATP-dependent ubiquitin-mediated proteolysis. *International Journal of Cancer, 61*, 138–141.

Mahony, S. M., Beck, S. A., Tisdale, M. J. (1988). Comparison of weight loss induced by recombinant tumour necrosis factor with that produced by a cachexia-inducing tumour. *British Journal of Cancer, 57*, 385–389.

Marton, K. I., Sox, H. C., Jr., Krupp, J. R. (1981). Involuntary weight loss: diagnostic and prognostic significance. *Annals of Internal Medicine, 95*, 568–574.

McFarlane, C., Plummer, E., Thomas, M., Hennebry, A., Ashby, M., Ling, N., Smith, H., Sharma, M., Kambadur, R. (2006). Myostatin induces cachexia by activating the ubiquitin proteolytic system through an NF-kappaB-independent, FoxO1-dependent mechanism. *Journal of Cellular Physiology, 209*, 501–514.

McMurtry, C. T. & Rosenthal, A. (1995). Predictors of 2-year mortality among older male veterans on a geriatric rehabilitation unit. *Journal of the American Geriatrics Society, 43*, 1123–1126.

Moore-Carrasco, R., Garcia-Martinez, C., Busquets, S., Ametller, E., Barreiro, E., Lopez-Soriano, F. J., Argiles J. M. (2006). The AP-1/CJUN signaling cascade is involved in muscle differentiation: implications in muscle wasting during cancer cachexia. *FEBS Letters, 580*, 691–696.

Morley, J. E. (2001). Anorexia, sarcopenia, and aging. *Nutrition, 17*, 660–663.

Morley, J. E. & Kraenzle, D. (1994). Causes of weight loss in a community nursing home. *Journal of the American Geriatrics Society, 42*, 583–585.

Morley, J. E. & Silver, A. J. (1988). Anorexia in the elderly. *Neurobiology of Aging, 9*, 9–16.

Moses, A. G., Maingay, J., Sangster, K., Fearon, K. C., Ross, J. A. (2009). Pro-inflammatory cytokine release by peripheral blood mononuclear cells from patients with advanced pancreatic cancer: relationship to acute phase response and survival. *Oncology Reports, 21*, 1091–1095.

Moshage, H. (1997). Cytokines and the hepatic acute phase response. *The Journal of Pathology, 181*, 257–266.

Mosoni, L., Valluy, M. C., Serrurier, B., Prugnaud, J., Obled, C., Guezennec, C. Y., Mirand, P. P. (1995). Altered response of protein synthesis to nutritional state and endurance training in old rats. *The American Journal of Physiology, 268*, E328–E335.

Mulligan, H. D. & Tisdale, M. J. (1991). Lipogenesis in tumour and host tissues in mice bearing colonic adenocarcinomas. *British Journal of Cancer, 63*, 719–722.
Muscaritoli, M., Cangiano, C., Cascino, A., Ceci, F., Giacomelli, L., Cardelli-Cangiano, P., Mulieri, M., Rossi-Fanelli, F. (1990). Plasma clearance of exogenous lipids in patients with malignant disease. *Nutrition, 6*, 147–151.
Nixon, D. W., Heymsfield, S. B., Cohen, A. E., Kutner, M. H., Ansley, J., Lawson, D. H., Rudman, D. (1980). Protein-calorie undernutrition in hospitalized cancer patients. *The American Journal of Medicine, 68*, 683–690.
Noguchi, Y., Vydelingum, N. A., Younes, R. N., Fried, S. K., Brennan, M. F. (1991). Tumor-induced alterations in tissue lipoprotein lipase activity and mRNA levels. *Cancer Research, 51*, 863–869.
Pajak, B., Orzechowska, S., Pijet, B., Pijet, M., Pogorzelska, A., Gajkowska, B., Orzechowski, A. (2008). Crossroads of cytokine signaling–the chase to stop muscle cachexia. *Journal of Physiology and Pharmacology, 59*(Suppl 9), 251–264.
Patel, K. & Amthor, H. (2005). The function of Myostatin and strategies of Myostatin blockade-new hope for therapies aimed at promoting growth of skeletal muscle. *Neuromuscular Disorders, 15*, 117–126.
Penner, C. G., Gang, G., Wray, C., Fischer, J. E., Hasselgren, P. O. (2001). The transcription factors NF-kappab and AP-1 are differentially regulated in skeletal muscle during sepsis. *Biochemical and Biophysical Research Communications, 281*, 1331–1336.
Penner, G., Gang, G., Sun, X., Wray, C., Hasselgren, P. O. (2002). C/EBP DNA-binding activity is upregulated by a glucocorticoid-dependent mechanism in septic muscle. *American Journal of Physiology: Regulatory, Integrative and Comparative Physiology, 282*, R439–R444.
Pinchcofsky-Devin, G. D. & Kaminski, M. V., Jr. (1986). Correlation of pressure sores and nutritional status. *Journal of the American Geriatrics Society, 34*, 435–440.
Plisko, A. & Gilchrest, B. A. (1983). Growth factor responsiveness of cultured human fibroblasts declines with age. *Journal of Gerontology, 38*, 513–518.
Puigserver, P., Rhee, J., Lin, J., Wu, Z., Yoon, J. C., Zhang, C. Y., Krauss, S., Mootha, V. K., Lowell, B. B., Spiegelman, B. M. (2001). Cytokine stimulation of energy expenditure through p38 MAP kinase activation of PPARgamma coactivator-1. *Molecular Cell, 8*, 971–982.
Rabinovitz, M., Pitlik, S. D., Leifer, M., Garty, M., Rosenfeld, J. B. (1986). Unintentional weight loss. A retrospective analysis of 154 cases. *Archives of Internal Medicine, 146*, 186–187.
Renganathan, M., Messi, M. L., Delbono, O. (1998). Overexpression of IGF-1 exclusively in skeletal muscle prevents age-related decline in the number of dihydropyridine receptors. *The Journal of Biological Chemistry, 273*, 28845–28851.
Rossi Fanelli, F., Cangiano, C., Muscaritoli, M., Conversano, L., Torelli, G. F., Cascino, A. (1995). Tumor-induced changes in host metabolism: a possible marker of neoplastic disease. *Nutrition, 11*, 595–600.
Russell, S. T. & Tisdale, M. J. (2005). The role of glucocorticoids in the induction of zinc-alpha2-glycoprotein expression in adipose tissue in cancer cachexia. *British Journal of Cancer, 92*, 876–881.
Russell, S. T., Zimmerman, T. P., Domin, B. A., Tisdale, M. J. (2004). Induction of lipolysis in vitro and loss of body fat in vivo by zinc-alpha2-glycoprotein. *Biochimica et Biophysica Acta, 1636*, 59–68.
Ryden, M., Dicker, A., Van Harmelen, V., Hauner, H., Brunnberg, M., Perbeck, L., Lonnqvist, F., Arner, P. (2002). Mapping of early signaling events in tumor necrosis factor-alpha -mediated lipolysis in human fat cells. *The Journal of Biological Chemistry, 277*, 1085–1091.
Ryden, M., Arvidsson, E., Blomqvist, L., Perbeck, L., Dicker, A., Arner, P. (2004). Targets for TNF-alpha-induced lipolysis in human adipocytes. *Biochemical and Biophysical Research Communications, 318*, 168–175.
Sandri, M., Sandri, C., Gilbert, A., Skurk, C., Calabria, E., Picard, A., Walsh, K., Schiaffino, S., Lecker, S. H., Goldberg, A. L. (2004). Foxo transcription factors induce the atrophy-related ubiquitin ligase atrogin-1 and cause skeletal muscle atrophy. *Cell, 117*, 399–412.
Savary, I., Debras, E., Dardevet, D., Sornet, C., Capitan, P., Prugnaud, J., Mirand, P. P., Grizard, J. (1998). Effect of glucocorticoid excess on skeletal muscle and heart protein synthesis in adult and old rats. *The British Journal of Nutrition, 79*, 297–304.

Sciorati, C., Touvier, T., Buono, R., Pessina, P., Francois, S., Perrotta, C., Meneveri, R., Clementi, E., Brunelli, S. (2009). Necdin is expressed in cachectic skeletal muscle to protect fibers from tumor-induced wasting. *Journal of Cell Science, 122*, 1119–1125.

Schneider, S. M., Al-Jaouni, R., Pivot, X., Braulio, V. B., Rampal, P., Hebuterne, X. (2002). Lack of adaptation to severe malnutrition in elderly patients. *Clinical Nutrition, 21*, 499–504.

Sharma, R. & Anker, S. D. (2002). Cytokines, apoptosis and cachexia: the potential for TNF antagonism. *International Journal of Cardiology, 85*, 161–171.

Sinaud, S., Balage, M., Bayle, G., Dardevet, D., Vary, T. C., Kimball, S. R., Jefferson, L. S., Grizard, J. (1999). Diazoxide-induced insulin deficiency greatly reduced muscle protein synthesis in rats: involvement of eIF4E. *The American Journal of Physiology, 276*, E50–E61.

Slentz, C. A. & Holloszy, J. O. (1993). Body composition of physically inactive and active 25-month-old female rats. *Mechanisms of Ageing and Development, 69*, 161–166.

Smith, K. L. & Tisdale, M. J. (1993). Increased protein degradation and decreased protein synthesis in skeletal muscle during cancer cachexia. *British Journal of Cancer, 67*, 680–685.

Stephens, N. A., Skipworth, R. J., Fearon, K. C. (2008). Cachexia, survival and the acute phase response. *Current Opinion in Supportive and Palliative Care, 2*, 267–274.

Sumi, T., Ishiko, O., Honda, K., Hirai, K., Yasui, T., Ogita, S. (1999). Muscle cell apoptosis is responsible for the body weight loss in tumor-bearing rabbits. *Osaka City Medical Journal, 45*, 25–35.

Temparis, S., Asensi, M., Taillandier, D., Aurousseau, E., Larbaud, D., Obled, A., Bechet, D., Ferrara, M., Estrela, J. M., Attaix, D. (1994). Increased ATP-ubiquitin-dependent proteolysis in skeletal muscles of tumor-bearing rats. *Cancer Research, 54*, 5568–5573.

Thinakaran, G., Ojala, J., Bag, J. (1993). Expression of c-jun/AP-1 during myogenic differentiation in mouse C2C12 myoblasts. *FEBS Letters, 319*, 271–276.

Thompson, L. V. (2009). Age-related muscle dysfunction. *Experimental Gerontology, 44*, 106–111.

Thompson, M. P., Koons, J. E., Tan, E. T., Grigor, M. R. (1981). Modified lipoprotein lipase activities, rates of lipogenesis, and lipolysis as factors leading to lipid depletion in C57BL mice bearing the preputial gland tumor, ESR-586. *Cancer Research, 41*, 3228–3232.

Tracey, K. J., Morgello, S., Koplin, B., Fahey, T. J., 3RD., Fox, J., Aledo, A., Manogue, K. R., Cerami, A. (1990). Metabolic effects of cachectin/tumor necrosis factor are modified by site of production. Cachectin/tumor necrosis factor-secreting tumor in skeletal muscle induces chronic cachexia, while implantation in brain induces predominantly acute anorexia. *The Journal of Clinical Investigation, 86*, 2014–2024.

Van Royen, M., Carbo, N., Busquets, S., Alvarez, B., Quinn, L. S., Lopez-Soriano, F. J., Argiles, J. M. (2000). DNA fragmentation occurs in skeletal muscle during tumor growth: A link with cancer cachexia? *Biochemical and Biophysical Research Communications, 270*, 533–537.

Vary, T., Dardevet, D., Grizard, J., Voisin, L., Buffiere, C., Denis, P., Breuille, D., Obled C (1999). Pentoxifylline improves insulin action limiting skeletal muscle catabolism after infection. *The Journal of Endocrinology, 163*, 15–24.

Vary, T. C., Dardevet, D., Obled, C., Pouyet, C., Breuille, D., Grizard, J. (1997). Modulation of skeletal muscle lactate metabolism following bacteremia by insulin or insulin-like growth factor-I: effects of pentoxifylline. *Shock, 7*, 432–438.

Vary, T. C., Dardevet, D., Grizard, J., Voisin, L., Buffiere, C., Denis, P,, Breuille, D., Obled, C. (1998). Differential regulation of skeletal muscle protein turnover by insulin and IGF-I after bacteremia. *The American Journal of Physiology, 275*, E584–E593.

Vescovo, G. & Dalla Libera, L. (2006). Skeletal muscle apoptosis in experimental heart failure: the only link between inflammation and skeletal muscle wastage? *Current Opinion in Clinical Nutrition and Metabolic Care, 9*, 416–422.

Wallace, J. I., Schwartz, R. S., Lacroix, A. Z., Uhlmann, R. F., Pearlman, R. A. (1995). Involuntary weight loss in older outpatients: incidence and clinical significance. *Journal of the American Geriatrics Society, 43*, 329–337.

Wang, X., Hu, Z., Hu, J., Du, J., Mitch, W. E. (2006). Insulin resistance accelerates muscle protein degradation: Activation of the ubiquitin-proteasome pathway by defects in muscle cell signaling. *Endocrinology, 147*, 4160–4168.

Warren, S. (1932). The inmediate cause of death in cancer. *The American Journal of the Medical Sciences, 184*, 610–613.

Williams, A. B., Decourten-Myers, G. M., Fischer, J. E., Luo, G., Sun, X., Hasselgren, P. O. (1999). Sepsis stimulates release of myofilaments in skeletal muscle by a calcium-dependent mechanism. *The FASEB Journal, 13*, 1435–1443.

Workeneh, B. T., Rondon-Berrios, H., Zhang, L., Hu, Z., Ayehu, G., Ferrando, A., Kopple, J. D., Wang, H., Storer, T., Fournier, M., Lee, S. W., Du, J., Mitch, W. E. (2006). Development of a diagnostic method for detecting increased muscle protein degradation in patients with catabolic conditions. *Journal of the American Society of Nephrology, 17*, 3233–3239.

Wyke, S. M., Russell, S. T., Tisdale, M. J. (2004). Induction of proteasome expression in skeletal muscle is attenuated by inhibitors of NF-kappaB activation. *British Journal of Cancer, 91*, 1742–1750.

Wyke, S. M. & Tisdale, M. J. (2005). NF-kappaB mediates proteolysis-inducing factor induced protein degradation and expression of the ubiquitin-proteasome system in skeletal muscle. *British Journal of Cancer, 92*, 711–721.

Zhang, H. H., Halbleib, M., Ahmad, F., Manganiello, V. C., Greenberg, A. S. (2002). Tumor necrosis factor-alpha stimulates lipolysis in differentiated human adipocytes through activation of extracellular signal-related kinase and elevation of intracellular cAMP. *Diabetes, 51*, 2929–2935.

Age-Related Remodeling of Neuromuscular Junctions

Carlos B. Mantilla and Gary C. Sieck

Abstract Remodeling of neuromuscular junctions (NMJs) and ensuing structural and functional plasticity occurs with aging. Age-related changes result from reductions in physical activity, loss of motor neurons, and decreased muscle fiber size (sarcopenia). The properties of motor neurons and muscle fibers are precisely matched. In addition, motor unit recruitment in a selective manner is a primary mechanism by which the nervous system controls muscle contraction. Thus, it is essential to consider motor unit (and muscle fiber) type in any age-related plasticity. The following chapter examines changes in motor unit properties associated with aging and how these affect structural and functional remodeling at NMJs.

Keywords Aging • Morphological adaptations • Motor units • Muscle fiber type • Plasticity • Recruitment • Skeletal muscle

1 Introduction

The neuromuscular junction provides the sole link between a motor neuron and muscle fibers. Within a motor unit (Fig. 1), the mechanical and biochemical properties of muscle fibers are relatively uniform, and it is clear that the motor neuron plays an important role in influencing these properties through the neuromuscular junction. This influence is imparted either through activity levels or nerve-derived trophic factors (Mantilla and Sieck 2008; Delbono 2003). As a result, the mechanical and metabolic properties of muscle fibers and motor neurons are precisely matched (Burke et al. 1971; Sieck et al. 1989) – an essential feature of neuromotor control and functional performance of a skeletal muscle across a range of physiological behaviors.

C.B. Mantilla and G.C. Sieck (✉)
Departments of Physiology and Biomedical Engineering and Anesthesiology, College of Medicine, Mayo Clinic, St. Marys Hospital, Joseph 4W-184, 200 First Street SW, Rochester, MN 55905, USA
e-mail: mantilla.carlos@mayo.edu; sieck.gary@mayo.edu

G.S. Lynch (ed.), *Sarcopenia – Age-Related Muscle Wasting and Weakness*,
DOI 10.1007/978-90-481-9713-2_3, © Springer Science+Business Media B.V. 2011

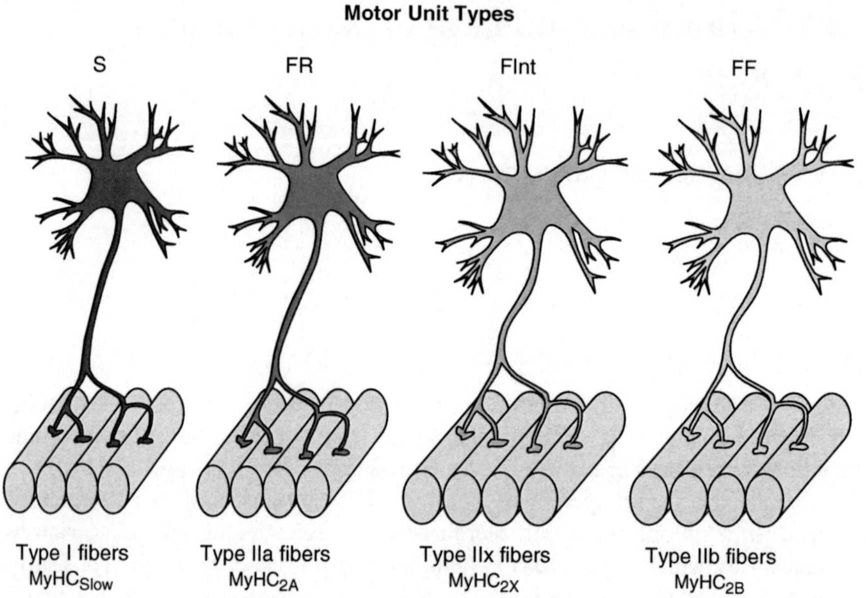

Fig. 1 Motor units (i.e., a motor neuron and the muscle fibers it innervates) are classified based on the mechanical and fatigue properties of muscle fibers. Four types are commonly described: (1) slow-twitch, fatigue resistant (type S), (2) fast-twitch, fatigue resistant (type FR), (3) fast-twitch, fatigue-intermediate (type FInt), and (4) fast-twitch, fatigable (type FF), which generally correspond to the expression of specific myosin heavy chain (MyHC) isoforms in the muscle fibers (type I fibers - $MyHC_{Slow}$, type IIa fibers - $MyHC_{2A}$, type IIx fibers - $MyHC_{2X}$ and type IIb fibers - $MyHC_{2B}$). Motor unit recruitment order is generally matched to their mechanical and fatigue properties; thus, type S and FR motor units are recruited first and more often than type FInt and FF units

In most skeletal muscles, motor units exhibit considerable functional diversity in terms of size, mechanical and fatigue properties (Burke et al. 1971; Sieck et al. 1989). Accordingly, recruitment of specific motor unit types is a major mechanism in neural control of muscle force generation and fatigue resistance (Clamann 1993).

1.1 Synaptic Plasticity

More than 60 years ago, Donald Hebb introduced a conceptual framework (Hebbian Theory) to describe the basic mechanisms for changes in synaptic efficacy (synaptic plasticity). Central to his theory was the observation that synaptic efficacy improves when the fidelity between pre- and post-synaptic activity increases. Conversely, when fidelity between pre- and postsynaptic activity is disrupted, synaptic transmission worsens. Synaptic plasticity has both structural and functional correlates. For examples, structurally, there may be axonal terminal sprouting or retraction, changes in the size and distribution of synaptic vesicle pools, and/or changes in the extent of pre- and postsynaptic apposition and overlap. Functionally, synaptic plasticity is reflected by enhanced evoked postsynaptic potentials, persistent changes in presynaptic neurotransmitter release or postsynaptic excitability (long-term facilitation or depression), and changes in safety factor for neurotransmission resulting in either improved neurotransmission fidelity or neurotransmission failure.

1.2 Aging and Synaptic Plasticity

With aging and senescence, there is a decrease in muscle activity often accompanied by unloading of limb muscle fibers. However, inactivity alone may not drive synaptic plasticity at the neuromuscular junction if fidelity of neuromuscular transmission (i.e., extent of correlation between pre- and postsynaptic activity) is maintained. Other age-related changes may drive synaptic plasticity. For example, an age-related loss of motor neurons amounts to denervation of some muscle fibers, consequently there may be axonal sprouting of spared motor neurons and re-innervation of muscle fibers and an increase in motor unit innervation ratio (Gordon et al. 2004; Balice-Gordon 1997). Age-related muscle fiber atrophy (i.e., sarcopenia) is also associated with concomitant changes in neuromuscular junction morphology, which may relate to removal of shared trophic influences (Vandervoort 2002; Delbono 2003). The effects of age-related inactivity, motor neuron loss and sarcopenia all depend on motor unit and/or muscle fiber type (Macaluso and De Vito 2004). Thus, it is likely that synaptic plasticity is a part of the normal aging process necessary to maintain muscle performance.

2 Motor Unit Properties and Recruitment

The concept of the motor unit was introduced by Charles Sherrington in 1925 and forms the cornerstone of neuromotor control. A motor unit comprises a motor neuron and the group of muscle fibers it innervates (Fig. 1). In adult mammals, each muscle fiber is innervated by only a single motor neuron, while each motor neuron can innervate multiple muscle fibers. The number of muscle fibers innervated by a motor neuron (innervation ratio) varies widely from very small innervation ratios in hand and eye muscles (<10 fibers per motor neuron) to very large innervation ratios in trunk and proximal limb muscles (>500 fibers per motor neuron). Innervation ratio is inversely related to the fine control of force gradation with motor unit recruitment. Together with average muscle fiber cross-sectional area, innervation ratio determines the size of a motor unit and maximal force contributed by the motor unit. The level of force contributed by a motor unit is also dependent on the frequency of motor neuron discharge rate (frequency coding of force). Force-frequency properties of muscle fibers comprising motor units vary depending on contractile protein composition, which forms the basis of muscle fiber type classification (Fig. 1; see below).

2.1 Motor Unit and Muscle Fiber Type Classification

Motor unit and muscle fiber type classification are concordant since they both relate to the mechanical and fatigue properties of muscle fibers. Different muscle fiber type classification schemes have been proposed, but the most commonly accepted scheme is based on the expression of different myosin heavy chain (MyHC) isoforms. Accordingly, in adult mammals, four muscle fiber types are classified: (1) type I (fibers expressing $MyHC_{Slow}$), (2) type IIa (fibers expressing $MyHC_{2A}$), (3) type IIx (fibers expressing $MyHC_{2X}$) and (4) type IIb (fibers expressing $MyHC_{2B}$). In single fiber studies, MyHC isoform expression has been shown to correlate with maximum isometric force, Ca^{2+} sensitivity (related to force at submaximal activation underlying the force-frequency relationship), maximum velocity of shortening, cross-bridge cycling rate, ATP consumption rate, mitochondrial volume density, and fatigue resistance (Geiger et al. 1999, 2000; Han et al. 2001, 2003; Sieck et al. 2003).

Since motor units comprise a relatively homogenous group of muscle fibers, classification of four motor unit types is based on the mechanical and fatigue properties of their constituent muscle fibers: (1) slow-twitch, fatigue resistant (type S; comprising type I fibers), (2) fast-twitch, fatigue resistant (type FR; comprising type IIa fibers), (3) fast-twitch, fatigue-intermediate (type FInt; comprising type IIx fibers), and (4) fast-twitch, fatigable (type FF; comprising type IIb fibers) (Fig. 1). As mentioned above, innervation ratio varies across muscles, but within a muscle, innervation ratio is generally greater for type FInt and FF motor units compared to type S and FR units. Muscle fiber size also varies across muscles, but within a muscle type IIx and IIb fibers are generally larger than type I and IIa fibers. Thus, there are differences in motor unit size across muscles and within a muscle, but generally type FInt and FF motor units are larger than type S and FR motor units. There are also differences in specific force (i.e., force per unit cross-sectional area) of different muscle fiber and motor unit types. Generally, type IIx and IIb fibers (type FInt and FF motor units) have greater specific force than type I and IIa fibers (type S and FR motor units). Consequently, because of their greater innervation ratio, larger fiber size and greater specific force, type FInt and FF motor units contribute greater forces than type S and FR units.

2.2 Motor Unit Recruitment

In muscles of heterogeneous muscle fiber type composition, motor unit recruitment order is generally matched to their mechanical and fatigue properties; thus, type S and FR motor units are recruited first followed by type FInt and FF units. In models where this recruitment order was assumed and where the force contributed by each motor unit type was known, it was predicted that the forces required during most sustained motor behaviors (e.g., standing in the medial gastrocnemius (Walmsley et al. 1978) or quiet breathing in the diaphragm muscle (Sieck and Fournier 1989)) could be accomplished by recruitment of only type S and FR motor units (Fig. 2). In these models, recruitment of type FInt and FF motor units was required only during high force, short duration motor behaviors (e.g., jumping in the medial gastrocnemius and coughing/sneezing in the diaphragm).

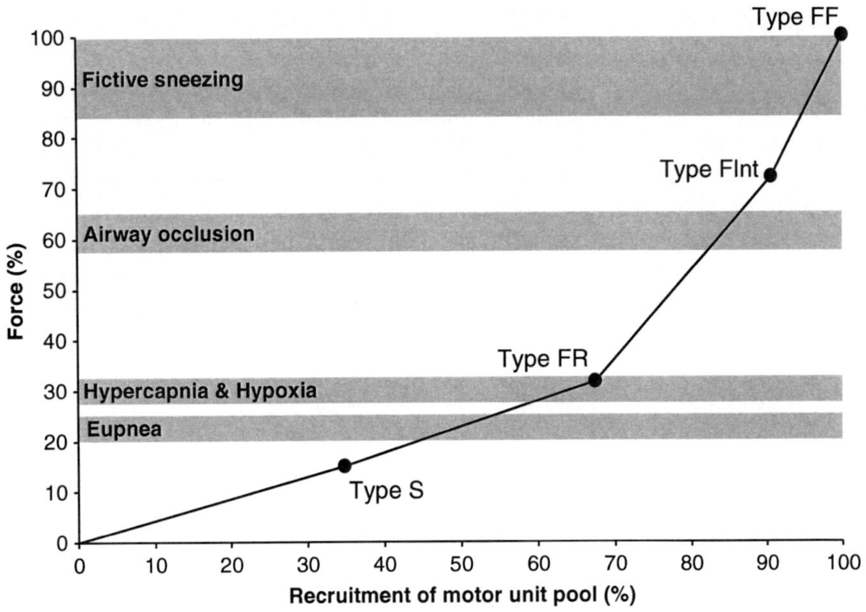

Fig. 2 Model of motor unit recruitment for the rat diaphragm muscle. Motor units were assumed to be recruited in order: type S → type FR → type FInt → type FF with complete activation of one motor unit type before the next type is recruited. Data is derived from previous studies reporting diaphragm muscle fiber type composition, force generated by type-identified fibers, and innervation ratio in adult male rats (Miyata et al. 1995; Zhan et al. 1997; Geiger et al. 2000; Sieck 1994). The relative force developed during different ventilatory (e.g., eupnea and hypercapnia & hypoxia) and non-ventilatory tasks (e.g., airway occlusion and fictive sneezing). Based on the model, the inspiratory effort necessary to accomplish ventilatory demands imposed during eupnea requires recruitment of all of the type S motor units and some of the type FR motor units, while chemical airway irritation (i.e., fictive sneezing) would result in recruitment of most diaphragm motor units

2.3 Aging Effects on Motor Unit Properties

Clearly, age affects the mechanical properties of muscle fibers and consequently motor units. Generally muscles become weaker with age and this effect may reflect changes in muscle fiber cross-sectional area, MyHC content per half-sarcomere, and/or specific force. The cross-sectional area of type IIx and/or IIb fibers decreases with age (Maxwell et al. 1973). This may be the result of motor neuron loss and consequent denervation-induced atrophy (Xie et al. 2003). It may also reflect decreased neuromuscular activity, mechanical unloading or altered trophic influences (Delbono 2003). MyHC content per half-sarcomere varies across muscle fiber types (Geiger et al. 2003, 2000), but does not appear to be affected by aging (Lowe et al. 2004b). However, with aging there is an increase in the proportion of fibers co-expressing MyHC isoforms, something that is relatively rare in young adults (Andersen et al. 1999). Specific force decreases with age, and this effect is especially pronounced at type IIx and IIb muscle fibers (i.e., type FInt and FF motor units) (Gosselin et al. 1994). Thus, muscle fiber weakness appears to reflect the combined influence of decreased fiber cross-sectional area and specific force. With respect to other mechanical properties of muscle fibers, converging evidence indicates that maximum velocity of shortening, cross-bridge cycling rate and ATP consumption rate are unaffected by aging across fiber types, but there may be differences across muscles (Lowe et al. 2004a). Importantly, there appears to be no age-related change in fatigability across muscle fiber types (Gonzalez and Delbono 2001), although maximum oxidative capacity is reduced in type II fibers of aged individuals (Proctor et al. 1995).

2.4 Aging Effects on Motor Unit Recruitment

Based on converging indirect evidence it appears that with aging, there is a decrease in the number of type FInt and FF motor units due to the specific loss of these motor neurons (Hashizume et al. 1988; Caccia et al. 1979; Ishihara et al. 1987; Hashizume and Kanda 1995). This conclusion is based on the observation of a reduction in the number of retrogradely labeled motor neurons which appears to be most pronounced in fast-twitch hind limb muscles (Ishihara et al. 1987; Hashizume and Kanda 1995). In the same studies, it was observed that there were fewer type II fibers (no distinction was made between type IIa, IIx or IIb fibers) in hind limb muscles showing fewer motor neurons. In separate studies that did not estimate the number of motor neurons, selective reduction in the proportion of type IIx and IIb fibers was observed (Caccia et al. 1979). Selective loss of type FF and FInt motor units is also indirectly supported by the observation of an age-related increase in the proportion of type S and FR motor units in the rat plantaris muscle (Pettigrew and Gardiner 1987; Pettigrew and Noble 1991). The underlying basis for a selective loss of motor neurons is not yet resolved, but such an effect would definitely impact the ability to accomplish motor behaviors that require generation of greater forces (Fig. 2). As a result of motor neuron loss, some type IIx and IIb fibers would be denervated, and with subsequent reinnervation by remaining motor axons (mostly those of type S and FR motor units), there may be fiber type conversion as reflected by an increase in the proportion of fibers co-expressing different MyHC isoforms (Larsson et al. 1991). Sprouting and reinnervation of adjacent muscle fibers should lead to an increase in motor unit innervation ratios. Indirect evidence for such an increase in innervation ratios stems from analysis of changes in EMG during incremental force steps relative to the maximum evoked EMG response (M-wave) (Galea 1996). Age-related changes in the specific force of type IIx and IIb fibers together with the decrease in the overall proportion of these motor unit types would tend to decrease the diversity of motor unit properties within a muscle. An increase in the innervation ratio of type S and FR motor units would result in increased force production by these units, but it is unclear whether this increased force is required for the normal recruitment of these motor unit types (e.g., standing or quiet breathing). It is possible that an age-related increase in force generation by type S and FR motor units partially offsets any age-related effects on type FInt and FF motor units, but it is unlikely that recruitment of type S and FR motor units can completely compensate for the forces required during high-force generating behaviors (e.g., jumping or coughing/sneezing). With aging, there appears to be a selective preservation of mechanical properties of motor units required for low force, sustained motor behaviors. In some cases, the advantage of such preservation is quite obvious, e.g., recruitment of type S and FR motor units in the diaphragm muscle to sustain ventilation or a similar recruitment of motor units in anti-gravity muscles to sustain posture.

3 Structural Properties of Neuromuscular Junctions

The structural properties of neuromuscular junctions are matched to the functional demands of muscle fibers such that within a motor unit type the structure of neuromuscular junctions is relatively uniform but there is considerable variability across different muscle fiber types (Fig. 3). The matching of pre- and post-synaptic specializations at the neuromuscular junction also depends on muscle fiber type. For example, presynaptically, there are differences in the distribution and size of synaptic vesicle pools and terminal surface area. Postsynaptically, there are differences in the number and depth of junctional folds and apposition of subcellular organelles such as mitochondria. Finally, the overlap of pre- and post-synaptic structures varies across fiber types.

3.1 Fiber Type Differences in Neuromuscular Junction Structure

Within a muscle, neuromuscular junctions at type I and IIa fibers are smaller with less complex branching patterns than those at type IIx and/or IIb fibers (Prakash and Sieck 1998; Mantilla et al. 2004; Prakash et al. 1995, 1996b; Sieck and Prakash 1997).

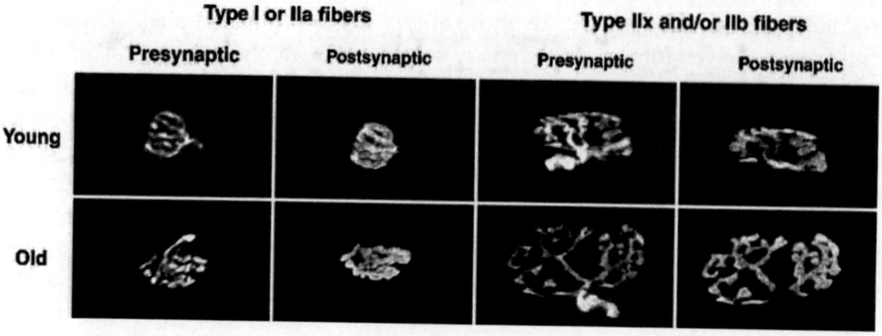

Fig. 3 Structural characteristics of a neuromuscular junction (NMJ) vary across muscle fiber types. Pre-synaptic terminals and motor end-plates at the diaphragm muscle of young (6 months) and old rats (24 months) were labeled with the neuronal ubiquitin decarboxylase PGP9.5 and α-bungarotoxin, respectively (Prakash and Sieck 1998). Note the differences in size and complexity (number and length of branches) across fiber types, with NMJs present at type I or IIa fibers being smaller and less complex than those at type IIx and/or IIb fibers. With aging there is considerable fragmentation and expansion of both pre- and post-synaptic elements

However, it is difficult to extrapolate across muscles since neuromuscular junctions at type I fibers in the soleus muscle are larger and more complex than neuromuscular junctions at type IIx and/or IIb fibers in the extensor digitorum longus muscle (Reid et al. 2003). Within a muscle, fiber size is an important determinant of neuromuscular junction area and complexity. For example, in the rat diaphragm muscle, the area of neuromuscular junctions among type I fibers varies directly with fiber cross-sectional area (Prakash and Sieck 1998; Sieck and Prakash 1997).

Fiber type dependent differences in gross structural properties of neuromuscular junctions are also reflected at pre- and post-synaptic elements. For example, both axon terminal and motor end-plate surface areas are ~75–90% greater at type IIx and/or IIb fibers than at type I and IIa fibers in the rat diaphragm (Sieck and Prakash 1997; Prakash et al. 1996b; Rowley et al. 2007; Mantilla et al. 2004). At all muscle fibers, the surface area of axon terminals is smaller than their corresponding motor end-plate and the extent of this difference varies across muscle fiber types (Prakash et al. 1996b). For example, at type I diaphragm fibers, the surface area of the presynaptic terminal more closely approximates that of the motor end-plate, with nearly 95% overlap. By comparison, at type IIb fibers, the presynaptic terminal only overlaps ~70% of the motor end-plate. These differences in the extent of overlap may reflect phenotypic differences in the ability of nerve terminal branches to invade motor end-plate gutters during development (Prakash et al. 1995) or remodeling (Prakash et al. 1996a, 1999). It is also possible that the increased fragmentation of neuromuscular junctions at muscle fibers of greater size results in greater branch termination limiting invagination of the axon terminal into motor end-plate gutters. In either case, these differences in extent of overlap may have significant physiological implications, impacting neuromuscular transmission.

Fiber type differences in neuromuscular junction remodeling vary depending on a number of factors including hormonal environment and activity. For example, the areas of both pre- and postsynaptic elements of neuromuscular junctions at type I diaphragm fibers decreased after 3 weeks of hypothyroidism induced by propylthiouracil (Prakash et al. 1996a). In contrast, after 2 weeks of diaphragm inactivity induced by either tetrodotoxin phrenic nerve blockade or spinal cord hemisection at C_2 the areas of both pre- and postsynaptic elements of neuromuscular junctions at type IIx and/or IIb diaphragm fibers increased while those at type I fibers decreased (Prakash et al. 1999). At type IIx and/or IIb fibers, the extent of overlap between pre- and postsynaptic elements of the neuromuscular junction increased to ~90% after 2 weeks of diaphragm inactivity induced by tetrodotoxin phrenic nerve blockade or spinal cord hemisection at C_2. Surprisingly, the similar structural changes induced by tetrodotoxin phrenic nerve blockade and spinal cord hemisection at C_2 yielded markedly different effects on neuromuscular transmission. Following inactivity induced by spinal cord hemisection at C_2 neuromuscular transmission with repetitive activation was markedly improved, whereas there was substantially greater neuromuscular transmission failure following tetrodotoxin phrenic nerve blockade. These functional differences are closely related to ultrastructural differences at the neuromuscular junction that form the basis of neuromuscular transmission (see below).

3.2 Ultrastructural Properties of Presynaptic Terminals

The total number of synaptic vesicles undergoing repeated cycles of endo- and exocytosis (i.e., cycling) is greater at type IIx and/or IIb fibers compared to type I and IIa fibers (Mantilla et al. 2004, 2007; Rowley et al. 2007). Ultrastructurally, synaptic vesicles at presynaptic terminals segregate into a pool of vesicles docked at specialized sites for neurotransmitter release – active zones – i.e., readily releasable, a pool immediately adjacent to active zones (within 200 nm) and a more distant, reserve pool (Sudhof 2004). Consistent with greater overall size of the cycling synaptic vesicle pool size, the densities of synaptic vesicles in both the immediately adjacent pool and the reserve pool are greater at presynaptic terminals of type I and IIa fibers compared to type IIx and/or IIb fibers. The size (length) and distribution of individual active zones does not vary across presynaptic terminals at the different fibers types (Fig. 4). Similarly, the number of synaptic vesicles docked at each active zone (i.e., readily releasable) is consistent across fiber types (Rowley et al. 2007). However, fiber type differences in presynaptic terminal surface area yield greater total number of active zones per presynaptic terminal at type IIx and/or IIb fibers than at type I and IIa fibers, and thus, a greater total number of synaptic vesicles in the readily releasable pool at type IIx and/or IIb fibers compared to type I and IIa fibers (Mantilla et al. 2004; Rowley et al. 2007). Consistent with these ultrastructural properties, quantal release at type IIx

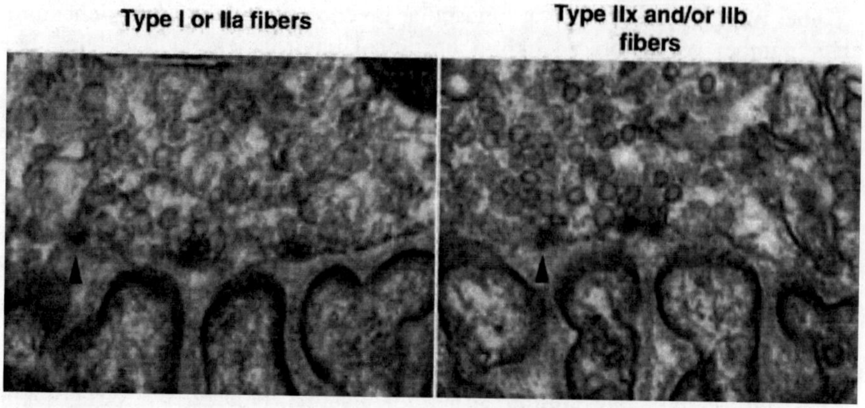

Fig. 4 Ultrastructural elements of type-identified NMJs. Presynaptic terminals at type-identified rat diaphragm muscle fibers (Mantilla et al. 2004, 2007; Rowley et al. 2007) are full of synaptic vesicles (*top*), which cluster around active zones (*arrowheads*) opposing postsynaptic folds (*bottom*). The number and distribution of synaptic vesicles docked at each active zone is consistent across fiber types

and/or IIb fibers is greater than at type I and IIa fibers (see below). Mitochondrial volume density is also greater at presynaptic terminals innervating type I and IIa fibers compared to those innervating type IIx and/or IIb fibers, possibly reflecting the metabolic requirements of increased activation of these presynaptic terminals. Taken together with the greater number of cycling vesicles at type I and IIa fibers, these ultrastructural differences may contribute to a greater ability of presynaptic terminals at type I or IIa fibers to sustain neuromuscular transmission with repeated activation (Johnson and Sieck 1993; Ermilov et al. 2007; Rowley et al. 2007).

3.3 *Ultrastructural Properties of Postsynaptic Motor End-Plates*

At type I and IIa fibers, motor end-plates display less complex postsynaptic folding but increased gutter depth compared to type IIx and/or IIb fibers (Fahim and Robbins 1982; Fahim et al. 1983; Rowley et al. 2007). In addition, there is frequent interposition of cellular organelles (e.g., mitochondria, endoplasmic reticulum or myonuclei) between the motor end-plate and underlying myofibrils at type I and IIa fibers, but not at type IIx and/or IIb fibers. Indeed, these features of motor end-plate ultrastructure can be used to grossly distinguish among muscle fiber types. The density and location of cholinergic receptors at the crest of postsynaptic folds does not appear to differ across motor end-plates at different muscle fiber types (Oda 1984; Fertuck and Salpeter 1974). However, the given the larger surface area of motor end-plates at type IIx and/or IIb fibers, the actual number of cholinergic receptors is greater at these fibers.

3.4 Aging Effects on Neuromuscular Junction Structure

With aging, there is increased fragmentation of nerve terminals at all muscle fibers resulting in increased number of nerve terminal branches and greater complexity (Prakash and Sieck 1998; Courtney and Steinbach 1981; Andonian and Fahim 1989; Fahim and Robbins 1982; Fahim et al. 1983). The increased number of terminal branches may reflect sprouting as motor neurons degenerate with aging. In the rat diaphragm muscle, the number of nerve terminal branches increased with aging at all muscle fiber types, but this increase was more pronounced at type IIx and/or IIb fibers (Prakash and Sieck 1998). Whereas individual branch length remained relatively unchanged, total branch length increased as a result of the greater number of branches at type IIx and/or IIb fibers. The increased branching complexity associated with aging resulted in a significant increase in the planar surface area of presynaptic terminals at type IIx and/or IIb fibers (Prakash and Sieck 1998). This occurs despite an actual age-related reduction in the cross-sectional area of type IIx and/or IIb diaphragm muscle fibers. Similar fiber type-dependent changes in nerve terminal size and complexity also occur in muscles of different fiber type composition (e.g., soleus muscle - composed of primarily slow-twitch fibers, and extensor digitorum longus muscle – composed of predominantly fast-twitch fibers) in both rats and mice (Andonian and Fahim 1989; Fahim and Robbins 1982; Fahim et al. 1983). In mice hind limb muscles, there are age-related reductions in the number and density of mitochondria and synaptic vesicles at presynaptic terminals (Fahim and Robbins 1982). Indeed, aging nerve terminals appear to be increasingly occupied by smooth endoplasmic reticulum, cisternae, microtubules, neurofilaments and coated vesicles. Accordingly, with age, acetylcholine content decreases at presynaptic terminals in the rat diaphragm muscle, most likely reflecting increased acetylcholine leakage despite increased synthesis rate and choline availability (Smith and Weiler 1987). Increased acetylcholine leakage may result from the overall increase in presynaptic terminal area, but is not associated with increased frequency of miniature end-plate potentials (indicative of spontaneous synaptic vesicle release) or a change in Ca^{2+} sensitivity for synaptic vesicle release (Smith 1988).

On the postsynaptic side, aging is also associated with increased branching and complexity of junctional folds (Wokke et al. 1990; Rosenheimer and Smith 1985; Prakash and Sieck 1998). In the rat diaphragm muscle, there is a corresponding age-related increase in motor end-plate surface area, predominantly at type IIx and/or IIb fibers (Prakash and Sieck 1998; Arizono et al. 1984). In addition, there is an age-related increase in subsarcolemmal vesicles and appearance of lipofuscin deposits (Fahim and Robbins 1982). With aging, there is a gradual decrease in the number of cholinergic receptors at the motor end-plate and appearance of extra-junctional receptors (Courtney and Steinbach 1981; Smith et al. 1990). These changes may be the result of motor neuron loss and consequent denervation of some muscle fibers.

The incidence of nerve terminals projecting beyond the motor end-plate markedly increases with aging (Prakash and Sieck 1998). This may reflect some general stimulus for terminal sprouting, consistent with age-related motor neuron loss and denervation and/or the reduced activity levels associated with aging. In agreement with some impact of age-related inactivity, similar morphological changes occur at an earlier age in the extensor digitorum longus muscle compared to the soleus and diaphragm muscles of rats (Kelly and Robbins 1983; Rosenheimer and Smith 1985). In older animals, the extent of overlap between nerve terminals and motor end-plates (expressed as a percent of end-plate area) remains greatest at type I and IIa fibers. At type IIx and/or IIb fibers, the extent of overlap between nerve terminals and motor end-plates increases with age but remains below that of type I and IIa fibers (Prakash and Sieck 1998).

4 Functional Properties of Neuromuscular Junctions

In mammals, functional properties of a neuromuscular junction depend on both presynaptic release of acetylcholine and cholinergic receptor-induced postsynaptic responses. These functional properties of neuromuscular junctions are matched to the demands of muscle fibers particularly as they relate to activation level and susceptibility to neuromuscular transmission failure. For example, functional properties of neuromuscular junctions at type I or IIa fibers must meet the functional demands of higher activity levels and subsequent metabolic demands. These motor units are often recruited to accomplish motor behaviors where failure cannot be tolerated and therefore fidelity of the postsynaptic contractile response must be maintained. Functional properties of neuromuscular junctions can be assessed using a variety of techniques, including assessment of electromyographic recordings of evoked compound muscle action potentials, assessment of force loss during nerve vs. direct muscle stimulation, and microelectrode measurements of synaptic potentials and/or currents. Important in all these measures are dependencies on fiber type, muscle fiber size (cross-sectional area) and frequency of activation.

4.1 Fiber Type Differences in Neuromuscular Transmission

As mentioned above, there are significant structural differences in presynaptic terminals across fiber types that will affect the release of acetylcholine. For example, the total number of active zones at type IIx and/or IIb fibers is substantially greater than at type I or IIa fibers. Thus, while quantal size as reflected by average miniature end-plate potential (mEPP) amplitude normalized for membrane input resistance (or capacitance) does not vary across fiber types, quantal content (defined as the ratio of EPP to mEPP) is significantly greater at type IIx and/or IIb fibers in diaphragm muscle (Rowley et al. 2007; Ermilov et al. 2007). Comparing across muscle predominantly composed of type I or IIa (rat soleus muscle) vs. type IIx and/or IIb fibers (rat extensor digitorum muscle), several investigators have also confirmed higher quantal content at type IIx and/or IIb fibers (Reid et al. 1999; Wood and Slater 1997).

The safety factor for neuromuscular transmission is defined as the ratio of EPP amplitude to action potential activation threshold in muscle fibers. The action potential activation threshold is highly dependent on the density of Na^+ channels near the motor end-plate. Comparing across muscles predominantly comprising a single fiber type, several studies (Ruff 1996; Harrison et al. 1997) have reported that Na^+ channel density is much higher at type IIx and/or IIb fibers (e.g., extensor digitorum longus muscle) than at type I fibers (e.g. soleus muscle). The impact of higher Na^+ channel density and Na^+ input current would be mitigated by the larger membrane surface area of type IIx and/or IIb muscle fibers that results in increased membrane capacitance. For example, type IIx and/or IIb fibers in the rat diaphragm are ~2-fold larger than type I or IIa muscle fibers, and accordingly, their membrane capacitance would be ~4-fold higher. To maintain the same threshold for action potential generation, Na^+ channel density and Na^+ input currents must be proportionally higher in type IIx and/or IIb fibers. Indeed, in the rat diaphragm, we found that the action potential activation threshold was lower at type IIx and/or IIb fibers compared to type I and IIa fibers (Ermilov et al. 2007). Thus, if anything, it appears that higher Na^+ channel density at type IIx and/or IIb fibers in the rat diaphragm muscle more than compensates for differences in fiber size.

Due to both differences in EPP amplitude and action potential activation threshold, the safety factor for neuromuscular transmission is higher at type IIx and/or IIb fibers compared to type I or IIa fibers (Ermilov et al. 2007). However, during repetitive stimulation, EPP amplitude progressively declines across all muscle fiber types, but this decline is much greater at type IIx and/or IIb fibers (Rowley et al. 2007). The decline in quantal content varies across stimulation frequencies at type IIx and/or IIb fibers but not at type I or IIa fibers. With continuous stimulation, the decline in quantal content is dynamic, being very steep during the initial few pulses followed by a much slower decrement until a plateau is reached where there is no difference between fiber types (Rowley et al. 2007). It appears that the initial rapid decline in quantal content reflects both a depletion of the readily releasable pool of synaptic vesicles and a decrease in the probability of synaptic vesicle release. The slower decline in quantal content during repetitive stimulation likely reflects a balance between synaptic vesicle depletion and repletion at the readily releasable pool. Synaptic vesicles at the readily releasable pool are replenished either by recruitment from the immediately adjacent pool of vesicles (reserve pool) or by recycling of released vesicles. Since the density of vesicles in the immediately available pool is greater at type I and IIa fibers compared to type IIx and/or IIb fibers this could contribute to the maintenance of quantal content in these fibers compared to type IIx and/or IIb fibers. Based on the presynaptic uptake of styryl dyes (e.g., FM4-64 or FM1-43), the extent of synaptic vesicle recycling is greater at type I and IIa fibers compared to type IIx and/or IIb fibers (Mantilla et al. 2004; Rowley et al. 2007). Both mechanisms likely contribute to the replenishment of the readily releasable pool at type I and IIa fibers during repetitive stimulation reducing susceptibility to neuromuscular transmission failure.

It is unclear whether the threshold for action potential generation changes with repetitive activation, although this seems unlikely given the duration of the action potential refractory period in muscle fibers. Thus, with repetitive stimulation, it is clear that the safety factor declines and that type IIx and/or IIb muscle fibers become more susceptible to neuromuscular transmission failure. This has been confirmed in the diaphragm muscle using a glycogen-depletion technique where neuromuscular transmission failure is reflected by the failure to activate muscle fibers and thus deplete their glycogen stores (Johnson and Sieck 1993).

4.2 *Effects of Aging on Neuromuscular Transmission*

No study to date has directly examined age-related changes in safety factor across different fiber types. Miniature end-plate potential amplitude appears to be unaffected by aging (Banker et al. 1983; Kelly and Robbins 1983). In the mouse extensor digitorum longus and soleus muscles, EPP amplitude was reported to increase with age (Kelly and Robbins 1983). The age-related change in EPP amplitude is not related to an increase in cholinergic receptor density at motor end-plates (Courtney and Steinbach 1981; Smith et al. 1990). Thus, it is likely that quantal content increases with age in these limb muscles. These investigators found that EPP amplitude in soleus muscle fibers was greater than in extensor digitorum longus muscle fibers. In the mouse, the extensor digitorum longus muscle comprises predominantly type IIx and/or IIb fibers and the soleus comprises predominantly type I and IIa fibers. Thus, these observations were in contrast to other reports where in a mixed muscle EPP amplitude (and quantal content) of type I and IIa fibers was lower than that of type IIx and/or IIb fibers (Ermilov et al. 2007; Rowley et al. 2007). Minimal age-related changes in EPP amplitude and quantal content were reported for the mouse diaphragm muscle (Banker et al. 1983; Kelly and Robbins 1987). In this sense, it is possible that the age-related increase in EPP amplitude and quantal content are limited to limb muscles and do not reflect a general response across all muscles.

An age-related decrease in cross-sectional area, in particular at type IIx and/or IIb fibers, would be associated with increased input resistance and, thus, a lower action potential activation threshold. Furthermore, with aging, there is either no change or an increase in the density of Na^+ channels at skeletal muscle fibers (Desaphy et al. 1998), which together with the change in fiber cross-sectional area would tend to lower action potential activation threshold. However, further work needs to be performed to confirm that there are age-related changes in action potential threshold at these fibers. If there is and age-related reduction in the threshold for action potential generation, it is unclear whether this would be sufficient to mitigate reductions in EPP amplitude. There is little age-related change in type I and IIa fiber cross-sectional areas, but there may be changes in Na^+ conductance in these fibers; thus, it is difficult to predict whether the action potential activation threshold is affected. Without a change in action potential threshold, any change in EPP amplitude would result in a decrease in safety factor at these fibers. Comparing across muscles of a predominant fiber type composition, it was reported that age-related changes at the neuromuscular junction are well compensated across fiber types with minimal impact on safety factor for neuromuscular transmission (Banker et al. 1983; Kelly and Robbins 1987). However, it is very important to compare across muscle fibers in a single muscle since type I muscle fibers in the soleus muscle are generally much larger than type I fibers found in mixed muscles. The safety factor for neuromuscular transmission decreases in the rat diaphragm muscle with increasing age (Kelly 1978), but fiber type differences were not examined.

5 Conclusions

At present, there is surprisingly little direct information about the effects of aging on the long-term plasticity of NMJs at different fiber types, especially in muscle of mixed fiber type composition. Yet aging is clearly associated with changes that could affect remodeling of the NMJ rendering it less resilient to perturbations induced by disease or injury. Indeed, reductions in physical activity and the resulting unloading of limb muscles affect NMJ structure and function to a greater extent in older animals. Aging-related loss of motor neurons results in functional denervation of muscle fibers that are then re-innervated by axons sprouting from remaining neighboring motor neurons. The combined reduction in motor neuron number and enlargement of motor unit size leads to loss of motor unit diversity. Importantly, aging results in a disproportionate loss of those motor units able to generate greater forces, which are also those that display the greatest reduction in muscle fiber size. With aging, there appear to be effective compensatory mechanisms that provide preservation of motor units required for low force, sustained motor behaviors, which may be advantageous for example in the maintenance of adequate ventilation or posture. However, aging-related changes in neuromuscular plasticity may be at the expense of maintaining structural and functional diversity in motor unit properties.

References

Andersen, J. L., Terzis, G., Kryger, A. (1999). Increase in the degree of coexpression of myosin heavy chain isoforms in skeletal muscle fibers of the very old. *Muscle and Nerve, 22*, 449–454.

Andonian, M. H. & Fahim, M. A. (1989). Nerve terminal morphology in C57BL/6NNia mice at different ages. *Journal of Gerontology, 44*, B43–B51.

Arizono, N., Koreto, O., Iwai, Y., Hidaka, T., Takeoka, O. (1984). Morphometric analysis of human neuromuscular junction in different ages. *Acta Pathologica Japonica, 34*, 1243–1249.

Balice-Gordon, R. J. (1997). Age-related changes in neuromuscular innervation. *Muscle and Nerve. Supplement, 5*, S83–S87.

Banker, B. Q., Kelly, S. S., Robbins, N. (1983). Neuromuscular transmission and correlative morphology in young and old mice. *Journal de Physiologie, 339*, 355–377.

Burke, R. E., Levine, D. N., Zajac, F. E., 3rd (1971). Mammalian motor units: physiological-histochemical correlation in three types in cat gastrocnemius. *Science, 174*, 709–712.

Caccia, M. R., Harris, J. B., Johnson, M. A. (1979). Morphology and physiology of skeletal muscle in aging rodents. *Muscle and Nerve, 2*, 202–212.
Clamann, H. P. (1993). Motor unit recruitment and the gradation of muscle force. *Physical Therapy, 73*, 830–843.
Courtney, J. & Steinbach, J. H. (1981). Age changes in neuromucular junction morphology and acetylcholine receptor distribution on rat skeletal muscle fibres. *Journal de Physiologie, 320*, 435–447.
Delbono, O. (2003). Neural control of aging skeletal muscle. *Aging Cell, 2*, 21–29.
Desaphy, J. F., De Luca, A., Imbrici, P., Conte Camerino, D. (1998). Modification by ageing of the tetrodotoxin-sensitive sodium channels in rat skeletal muscle fibres. *Biochimica et Biophysica Acta, 1373*, 37–46.
Ermilov, L. G., Mantilla, C. B., Rowley, K. L., Sieck, G. C. (2007). Safety factor for neuromuscular transmission at type-identified diaphragm fibers. *Muscle and Nerve, 35*, 800–803.
Fahim, M. A. & Robbins, N. (1982). Ultrastructural studies of young and old mouse neuromuscular junctions. *Journal of Neurocytology, 11*, 641–656.
Fahim, M. A., Holley, J. A., Robbins, N. (1983). Scanning and light microscopic study of age changes at a neuromuscular junction in the mouse. *Journal of Neurocytology, 12*, 13–25.
Fertuck, H. C. & Salpeter, M. M. (1974). Localization of acetylcholine receptor by 125I-labeled alpha-bungarotoxin binding at mouse motor endplates. *Proceedings of the National Academy of Sciences of the United States of America, 71*, 1376–1378.
Galea, V. (1996). Changes in motor unit estimates with aging. *Journal of Clinical Neurophysiology, 13*, 253–260.
Geiger, P. C., Cody, M. J., Sieck, G. C. (1999). Force-calcium relationship depends on myosin heavy chain and troponin isoforms in rat diaphragm muscle fibers. *Journal of Applied Physiology, 87*, 1894–1900.
Geiger, P. C., Cody, M. J., Macken, R. L., Sieck, G. C. (2000). Maximum specific force depends on myosin heavy chain content in rat diaphragm muscle fibers. *Journal of Applied Physiology, 89*, 695–703.
Geiger, P. C., Bailey, J. P., Zhan, W. Z., Mantilla, C. B., Sieck, G. C. (2003). Denervation-induced changes in myosin heavy chain expression in the rat diaphragm muscle. *Journal of Applied Physiology, 95*, 611–619.
Gonzalez, E. & Delbono, O. (2001). Recovery from fatigue in fast and slow single intact skeletal muscle fibers from aging mouse. *Muscle and Nerve, 24*, 1219–1224.
Gordon, T., Thomas, C. K., Munson, J. B., Stein, R. B. (2004). The resilience of the size principle in the organization of motor unit properties in normal and reinnervated adult skeletal muscles. *Canadian Journal of Physiology and Pharmacology, 82*, 645–661.
Gosselin, L. E., Johnson, B. D., Sieck, G. C. (1994). Age-related changes in diaphragm muscle contractile properties and myosin heavy chain isoforms [published erratum appears in Am J Respir Crit Care Med 1994 Sep;150(3):879]. *American Journal of Respiratory and Critical Care Medicine, 150*, 174–178.
Han, Y. S., Proctor, D. N., Geiger, P. C., Sieck, G. C. (2001). Reserve capacity of ATP consumption during isometric contraction in human skeletal muscle fibers. *Journal of Applied Physiology, 90*, 657–664.
Han, Y. S., Geiger, P. C., Cody, M. J., Macken, R. L., Sieck, G. C. (2003). ATP consumption rate per cross bridge depends on myosin heavy chain isoform. *Journal of Applied Physiology, 94*, 2188–2196.
Harrison, A. P., Nielsen, O. B., Clausen, T. (1997). Role of Na(+)-K+ pump and Na+ channel concentrations in the contractility of rat soleus muscle. *The American Journal of Physiology, 272*, R1402–R1408.
Hashizume, K. & Kanda, K. (1995). Differential effects of aging on motoneurons and peripheral nerves innervating the hindlimb and forelimb muscles of rats. *Neuroscience Research, 22*, 189–196.
Hashizume, K., Kanda, K., Burke, R. E. (1988). Medial gastrocnemius motor nucleus in the rat: age-related changes in the number and size of motoneurons. *The Journal of Comparative Neurology, 269*, 425–430.

Ishihara, A., Naitoh, H., Katsuta, S. (1987). Effects of ageing on the total number of muscle fibers and motoneurons of the tibialis anterior and soleus muscles in the rat. *Brain Research, 435*, 355–358.

Johnson, B. D. & Sieck, G. C. (1993). Differential susceptibility of diaphragm muscle fibers to neuromuscular transmission failure. *Journal of Applied Physiology, 75*, 341–348.

Kelly, S. S. (1978). The effect of age on neuromuscular transmission. *Journal de Physiologie, 274*, 51–62.

Kelly, S. S. & Robbins, N. (1983). Progression of age changes in synaptic transmission at mouse neuromuscular junctions. *Journal de Physiologie, 343*, 375–383.

Kelly, S. S. & Robbins, N. (1987). Statistics of neuromuscular transmitter release in young and old mouse muscle. *Journal de Physiologie, 385*, 507–516.

Larsson, L., Ansved, T., Edstrom, L., Gorza, L., Schiaffino, S. (1991). Effects of age on physiological, immunohistochemical and biochemical properties of fast-twitch single motor units in the rat. *Journal de Physiologie, 443*, 257–275.

Lowe, D. A., Husom, A. D., Ferrington, D. A., Thompson, L. V. (2004a). Myofibrillar myosin ATPase activity in hindlimb muscles from young and aged rats. *Mechanisms of Ageing and Development, 125*, 619–627.

Lowe, D. A., Warren, G. L., Snow, L. M., Thompson, L. V., Thomas, D. D. (2004b). Muscle activity and aging affect myosin structural distribution and force generation in rat fibers. *Journal of Applied Physiology, 96*, 498–506.

Macaluso, A. & DE Vito, G. (2004). Muscle strength, power and adaptations to resistance training in older people. *European Journal of Applied Physiology, 91*, 450–472.

Mantilla, C. B. & Sieck, G. C. (2008). Trophic factor expression in phrenic motor neurons. *Respiratory Physiology and Neurobiology, 164*, 252–262.

Mantilla, C. B., Rowley, K. L., Fahim, M. A., Zhan, W. Z., Sieck, G. C. (2004). Synaptic vesicle cycling at type-identified diaphragm neuromuscular junctions. *Muscle and Nerve, 30*, 774–783.

Mantilla, C. B., Rowley, K. L., Zhan, W. Z., Fahim, M. A., Sieck, G. C. (2007). Synaptic vesicle pools at diaphragm neuromuscular junctions vary with motoneuron soma, not axon terminal, inactivity. *Neuroscience, 146*, 178–189.

Maxwell, L. C., Faulkner, J. A., Lieberman, D. A. (1973). Histochemical manifestations of age and endurance training in skeletal muscle fibers. *American Journal of Physics, 224*, 356–361.

Miyata, H., Zhan, W. Z., Prakash, Y. S., Sieck, G. C. (1995). Myoneural interactions affect diaphragm muscle adaptations to inactivity. *Journal of Applied Physiology, 79*, 1640–1649.

Oda, K. (1984). Age changes of motor innervation and acetylcholine receptor distribution on human skeletal muscle fibres. *Journal of the Neurological Sciences, 66*, 327–338.

Pettigrew, F. P. & Gardiner, P. F. (1987). Changes in rat plantaris motor unit profiles with advanced age. *Mechanisms of Ageing and Development, 40*, 243–259.

Pettigrew, F. P. & Noble, E. G. (1991). Shifts in rat plantaris motor unit characteristics with aging and compensatory overload. *Journal of Applied Physiology, 71*, 2363–2368.

Prakash, Y. S. & Sieck, G. C. (1998). Age-related remodeling of neuromuscular junctions on type-identified diaphragm fibers. *Muscle and Nerve, 21*, 887–895.

Prakash, Y. S., Smithson, K. G., Sieck, G. C. (1995). Growth-related alterations in motor endplates of type-identified diaphragm muscle fibres. *Journal of Neurocytology, 24*, 225–235.

Prakash, Y. S., Gosselin, L. E., Zhan, W. Z., Sieck, G. C. (1996a). Alterations of diaphragm neuromuscular junctions with hypothyroidism. *Journal of Applied Physiology, 81*, 1240–1248.

Prakash, Y. S., Miller, S. M., Huang, M., Sieck, G. C. (1996b). Morphology of diaphragm neuromuscular junctions on different fibre types. *Journal of Neurocytology, 25*, 88–100.

Prakash, Y. S., Miyata, H., Zhan, W. Z., Sieck, G. C. (1999). Inactivity-induced remodeling of neuromuscular junctions in rat diaphragmatic muscle. *Muscle and Nerve, 22*, 307–319.

Proctor, D. N., Sinning, W. E., Walro, J. M., Sieck, G. C., Lemon, P. W. R. (1995). Oxidative capacity of human muscle fiber types: Effects of age and training status. *Journal of Applied Physiology, 78*, 2033–2038.

Reid, B., Slater, C. R., Bewick, G. S. (1999). Synaptic vesicle dynamics in rat fast and slow motor nerve terminals. *The Journal of Neuroscience, 19*, 2511–2521.

Reid, B., Martinov, V. N., Nja, A., Lomo, T., Bewick, G. S. (2003). Activity-dependent plasticity of transmitter release from nerve terminals in rat fast and slow muscles. *The Journal of Neuroscience, 23*, 9340–9348.

Rosenheimer, J. L. & Smith, D. O. (1985). Differential changes in the endplate architecture of functionally diverse muscles during aging. *Journal of Neurophysiology, 53*, 1567–1581.

Rowley, K. L., Mantilla, C. B., Ermilov, L. G., Sieck, G. C. (2007). Synaptic vesicle distribution and release at rat diaphragm neuromuscular junctions. *Journal of Neurophysiology, 98*, 478–487.

Ruff, R. L. (1996). Sodium channel slow inactivation and the distribution of sodium channels on skeletal muscle fibres enable the performance properties of different skeletal muscle fibre types. *Acta Physiologica Scandinavica, 156*, 159–168.

Sieck, G. C. (1994). Physiological effects of diaphragm muscle denervation and disuse. *Clinics in Chest Medicine, 15*, 641–659.

Sieck, G. C. & Fournier, M. (1989). Diaphragm motor unit recruitment during ventilatory and nonventilatory behaviors. *Journal of Applied Physiology, 66*, 2539–2545.

Sieck, G. C. & Prakash, Y. S. (1997). Morphological adaptations of neuromuscular junctions depend on fiber type. *Canadian Journal of Applied Physiology, 22*, 197–230.

Sieck, G. C., Fournier, M., Enad, J. G. (1989). Fiber type composition of muscle units in the cat diaphragm. *Neuroscience Letters, 97*, 29–34.

Sieck, G. C., Prakash, Y. S., Han, Y. S., Fang, Y. H., Geiger, P. C., Zhan, W. Z. (2003). Changes in actomyosin ATP consumption rate in rat diaphragm muscle fibers during postnatal development. *Journal of Applied Physiology, 94*, 1896–1902.

Smith, D. O. (1988). Muscle-specific decrease in presynaptic calcium dependence and clearance during neuromuscular transmission in aged rats. *Journal of Neurophysiology, 59*, 1069–1082.

Smith, D. O. & Weiler, M. H. (1987). Acetylcholine metabolism and choline availability at the neuromuscular junction of mature adult and aged rats. *Journal de Physiologie, 383*, 693–709.

Smith, D. O., Williams, K. D., Emmerling, M. (1990). Changes in acetylcholine receptor distribution and binding properties at the neuromuscular junction during aging. *International Journal of Developmental Neuroscience, 8*, 629–642.

Sudhof, T. C. (2004). The synaptic vesicle cycle. *Annual Review of Neuroscience, 27*, 509–547.

Vandervoort AA (2002). Aging of the human neuromuscular system. *Muscle and Nerve, 25*, 17–25.

Walmsley, B., Hodgson, J. A., Burke, R. E. (1978). Forces produced by medial gastrocnemius and soleus muscles during locomotion in freely moving cats. *Journal of Neurophysiology, 41*, 1203–1216.

Wokke, J. H., Jennekens, F. G., Vanden Oord, C. J., Veldman, H., Smit, L. M., Leppink, G. J. (1990). Morphological changes in the human end plate with age. *Journal of the Neurological Sciences, 95*, 291–310.

Wood, S. J. & Slater, C. R. (1997). The contribution of postsynaptic folds to the safety factor for neuromuscular transmission in rat fast- and slow-twitch muscles. *Journal de Physiologie, 500*, 165–176.

Xie, Y., Yao, Z., Chai, H., Wong, W. M., WU, W. (2003). Expression and role of low-affinity nerve growth factor receptor (p75) in spinal motor neurons of aged rats following axonal injury. *Developmental Neuroscience, 25*, 65–71.

Zhan, W. Z., Miyata, H., Prakash, Y. S., Sieck, G. C. (1997). Metabolic and phenotypic adaptations of diaphragm muscle fibers with inactivation. *Journal of Applied Physiology, 82*, 1145–1153.

Aging-Related Changes Motor Unit Structure and Function

Alexander Cristea, David E. Vaillancourt, and Lars Larsson

Abstract Aging has profound effects on skeletal muscle structure and function with significant consequences for both the individual and society. In this short review aging-related changes in the structure and function of the final functional unit in the motor system, i.e., the motor unit, are discussed. This review does not aim to give an overview of all aspects associated with aging-related changes in the motor unit, but will focus on specific changes in motor unit structure and function, such as the spatial organization of the muscle fibers innervated by a single alpha motoneuron, i.e., motor unit fiber, as well as aging-related motor unit transitions, and changes in motor unit physiological and firing properties.

Keywords Myosin • Motoneuron • Glycogen-depletion • Firing pattern

1 Introduction

Aging has profound effects on the motor system resulting in impaired coordination, balance, speed and force with significant negative consequences for morbidity and mortality in elderly citizens. That is, falls are a major cause of morbidity and mortality

L. Larsson (✉)
Department of Clinical Neurophysiology, Uppsala University Hospital, Entrance 85, 3rd Floor, 751 85, Uppsala, Sweden
e-mail: lars.larsson@neurofys.uu.se
and
Department of Neuroscience, Clinical Neurophysiology, Uppsala University, Sweden
and
Department of Biobehavioral Health, the Pennsylvania State University, PA, USA

A. Cristea
Department of Neuroscience, Clinical Neurophysiology, Uppsala University, Sweden

D.E. Vaillancourt
Department of Kinesiology and Nutrition, Departments of Bioengineering and Neurology, University of Illinois at Chicago, Chicago, IL, USA

G.S. Lynch (ed.), *Sarcopenia – Age-Related Muscle Wasting and Weakness*,
DOI 10.1007/978-90-481-9713-2_4,

in the growing population of elderly citizens and an improved understanding of the mechanisms underlying the impaired motor function is needed for several reasons (Mahoney et al. 1999). First, the growing population of elderly persons and the consequent enormous social and economic impact of aging-related problems in general, along with motor handicap and dependency in particular, have led to a need for focused research efforts to increase the quality of life in the aged. Second, normal aging processes may have a negative influence on the progression of various neuromuscular disorders. Aging may also play an important etiological role in some neuromuscular diseases, such as amyotrophic lateral sclerosis, muscular dystrophies, mitochondrial myopathies and the post-polio syndrome. Third, a significant proportion of frail elderly people have to use all their muscle power even to rise from a chair, and an additional small impairment in muscle function may dramatically change their life from an independent to a dependent one (Bean et al. 2007; Frontera et al. 2008). The cause of falls and fall-related injuries in old age are complex and involve multiple risk factors (Nevitt et al. 1991). It is important, however, to separate the different components of the fall, i.e., what initiates the fall and the ability to restore standing balance when it is disturbed; or to safely arrest a fall that may occur when standing balance cannot be recovered. It is widely accepted that somatosensory function has a strong impact on posture and locomotion in humans. The aging-related impairments in somatosensation, vision and vestibular function will, accordingly, have a negative effect on maintaining postural balance. The difficulty in recovering from an impending fall which is impaired in old age, especially in old women, is not related to an impairment of the sensory process or to the motor planning that leads to the initiation of muscle contraction (Schultz et al. 1997). The sources of these aging and gender differences have been reported to lie primarily in events after depolarization of skeletal muscle, i.e., in force-generation and contractile speed (Schultz et al. 1997). In support of this, the decline in mobility and lower extremity disability have been reported most influential in predicting falls in the elderly (Welle et al. 1993; Robbins et al. 1989; Mahoney et al. 1994, 1999), and the predictive value of the muscle weakness is increased further when muscle force is measured at a speed of movement resembling more functional limb velocities.

It is becoming increasingly evident that the aging-related muscle wasting represent a wasting condition that is different from other types of muscle atrophy, i.e., in response to denervation, microgravity, bed rest, immobilization in plaster or the cachexia associated with cancer, renal failure or chronic obstructive lung disease. This review will focus on aging-related changes at the motor unit level. The motor unit concept was first introduced by Lidell and Sherrington (1925) and in a report 5 years later Eccles and Sherrington (1930) emphasized the importance of the motor unit as being the "final functional unit" that forms the basis of all graded muscle contractions. However, the spatial organization of the muscle fibers belonging to a single motor unit, motor unit fibers, was not resolved until the introduction of the single muscle fiber EMG (Stålberg and Ekstedt 1973a) and glycogen-depletion techniques (Edstrom and Kugelberg 1968; Burke et al. 1971; Brandstater and Lambert 1969, 1973; Doyle 1969) almost half a century later. Visual evaluation of spatial organization and enzyme-histochemical properties of motor unit fibers indicated that motor unit fibers were randomly distributed within a defined territory, motor unit territory, and that motor unit fibers expressed metabolic homogeneity (Burke and Tsairis 1973; Edstrom and Kugelberg 1968). However, with the introduction of quantitative enzyme-histochemical methods and computer assisted detailed analyses of spatial arrangement of motor unit fibers, it was shown that there is inhomogeneity within motor units, but these differences are significantly smaller than the differences observed between different normal motor units from young individuals (Monti et al. 2001; Ansved et al. 1991; Bodine-Fowler et al. 1990; Martin et al. 1988). Further, a systematic difference in the metabolic properties of motor unit fibers along the superficial deep axis within the motor unit territory in the fast-twitch tibialis anterior muscle suggests a biological etiology of this inhomogeneity (Larsson 1992). This also challenges the concept that motoneuron properties, firing frequency and/or trophic factors, are the sole factors controlling muscle fiber properties within single motor units (Larsson 1992). One factor of importance for the inhomogeneity observed along the superficial-deep axis of the tibialis anterior muscle is the impact of baseline cell tension, i.e., *tensegrity*, on cell structure and function (Ingber 1991, 1993, 1997, 2002a, b) in a large muscle with a complex muscle fiber arrangement such as the rat tibialis anterior muscle. In addition, aging-related changes in tensegrity may have a significant impact on motor unit structure and function.

A number of different classifications of motor units into different types have been used, based on biochemical, enzyme-histochemical, contractile and fatigue properties (Burke 1981). However, it is important to emphasize that all motor unit types display considerable variation of e.g. physiological and biochemical properties, representing a continuum rather than discrete units (Burke 1981; Kugelberg and Lindegren 1979). However, there is a need to group units into specific categories to systemize and communicate experimental observations (Henneman et al. 1965; McPhederan et al. 1965; Wuerker et al. 1965; Burke 1981). In this short review, motor units will be classified based on contractile properties into the fast- or slow-twitch types or as types I, IIa, IIx or IIb based on motor unit fiber myosin heavy chain (MyHC) isoform expression.

2 Effects of Aging on Motor Unit Organization

In human skeletal muscle, motor unit fiber organization was first resolved using the single fibre EMG technique introduced by Stålberg and Ekstedt (1973b), i.e., electrophysiological data supporting previous experimental animal studies using the glycogen depletion technique. The glycogen depletion technique was originally proposed by Krnjevic' and Miledi (1958) and successfully used first by Edström and Kugelberg (1968), Burke et al. (1971), Brandstater and Lambert (1969, 1973), and Doyle and Mayer (1969). After isolating single motor units by ventral root teasing or intracellular stimulation of the motoneuron soma in the ventral horn, single motor unit contractile and physiological properties were characterized and followed by the depletion of the motor unit fibers of their glycogen by prolonged repetitive stimulation. Serial cross-sections of the muscle were subsequently stained for glycogen, and different enzyme- or immuno-histochemical methods, allowing the identification of the motor unit fibers and characterizing them according to metabolic properties (mitochondrial enzyme activities), myofibrillar ATPase type, MyHC isoform expression etc. By using the glycogen-depletion method, it was for the first time demonstrated that motor unit fibers were scattered in the muscle, resulting in considerable overlap in territories of different motor units and different motor units being highly intermingled in non-pathological muscle. In response to denervation reinnervation via collateral sprouting motor unit fibers are reorganized, resulting in groups or whole fascicles innervated by a single alpha motoneuron and the fibre type grouping observed in enzyme-histochemically stained muscle cross-sections from patients with chronic denervation and reinnervation (Kugelberg et al. 1970).

During aging, several studies have documented a loss of myelinated neurones in both peripheral nerves and ventral roots in humans as well as in different experimental animal models, although some studies in animals have reported and unaltered number of myelinated neurones during aging (for refs. see Larsson and Ansved 1995). Conflicting results may at least in part be related to methodological differences, but more importantly different muscles and rodent strains are affected differently by aging (Lionikas et al. 2003, 2005a, 2006). The aging-related loss of α-motoneurones is a slow process and the ultimate loss of the neurone is preceded by a gradual loss in motoneuron structure and function, referred to as the "sick" motoneuron (McComas et al. 1971). The decreased protein synthesis in the neuronal cell body and the decreased rate of axonal transport in old age may accordingly be a reflection of this "sickness" (Knox et al. 1989; Komiya 1980; McMartin and O'Connor 1979). The mechanisms underlying the aging-related loss of alpha motoneurons are not known. However, it is interesting to note that non-enzymatic glycosylation (glycation), a post-translational modification regarded as one of the biochemical bases underlying the pathophysiology of aging (Brownlee 1995), has been forwarded as a potential mechanism underlying the progressive loss of motoneurons in patients with amyotrophic lateral sclerosis (Shinpo et al. 2000). In addition, glycation has recently been shown to induce impairment of myosin function, i.e., resembling aging-related changes in myosin function (Ramamurthy et al. 2001). It may therefore be speculated that an aging-related decrease in the prevention of advanced glycation endproduct formation by antioxidants, such as glutathione, may provide an aging-related factor of importance in the loss of large motoneurons and sarcopenia, as well as for regulation of muscle contraction at the motor protein level (Larsson 2003).

The aging-related changes in the number, structure and function of alpha motoneurones have significant effects on the spatial organization of both fast- and slow-twitch motor units. An increased innervation ratio, i.e., muscle fibers per motor unit, and expansion of the motor unit territory were observed in the old fast- and slow-twitch motor units over and above the borders of the motor units in the young animals (Larsson et al. 1991a). Kanda and Hashizume (1991) reported that the motor unit function in the rat medial gastrocnemius was restored after nerve crush injury even in old age, suggesting that aged motoneurons have a preserved capacity for axonal regeneration and reinnervation. In fact, aged reinnervated motor units of both fast- and slow-twitch types produced very large tensions, indicating that the ability to reinnervate and hold extra fibers was well-maintained. Further, slow-twitch motor units may have an advantage compared with fast-twitch units in innervating previously denervated muscle fibers even in old age and that the largest, most rapidly conducting motoneurons with the lowest oxidative enzyme activity, i.e., innervating fast-twitch muscle fibers, are preferentially lost during aging (see Einsiedel et al. 1992; Einsiedel and Luff 1992; Hashizume et al. 1988; Ishihara et al. 1987; Kanda and Hashizume 1989; Pettigrew and Gardiner 1987). By calculating interfiber and nearest neighbor distances between all motor unit fibers within specific motor units, by using a computer-assisted algorithm where all motor unit fibers are given their x,y-coordinates in the muscle cross-section, significant rearrangements of the spatial distribution was observed, although mechanisms may vary according to muscle type (Ansved et al. 1991; Edstrom and Larsson 1987). For instance, an increased proportion of short distances, as judged from the nearest-neighbor distances, was observed in fast-twitch motor units from the rat tibialis anterior muscle, whereas the interfiber distances did not differ between young and old animals. A trend towards an increased proportion of short distances was also observed in slow-twitch motor units from the rat soleus, according to nearest-neighbor distance analyses, and interfiber distance distributions revealed a greater proportion of short and long distances in the old slow-twitch motor units. These signs of motor unit fiber rearrangements, together with the increased innervation ratio, the increased size of the motor unit territories and a decrease in the total number of muscle fibers (motor units) suggest an aging-related denervation-reinnervation process. The different patterns of fast- and slow-twitch motor unit fiber rearrangements in old rat skeletal muscle indicate differences in the type of reinnervation (Ansved et al. 1991). Less grouping of fibers is expected after nodal sprouting than after terminal sprouting, and conceivably more readily detected by interfiber distances. Terminal sprouting, on the other hand, would be expected to cause a change in the distribution of nearest neighbor distances (Kugelberg et al. 1970).

In humans, there are electrophysiological evidence of an ongoing denervation-reinnervation process during the aging process reflected by needle-EMG examinations showing motor unit potentials with longer duration, greater amplitude and a larger number of phases and satellite potentials in old than is found in young individuals (e.g. Buchtahl and Rosenfalck 1955; Campbell et al. 1973, Carlson et al. 1964). This is supported by more modern electromyographic techniques such as the macro-, scanning- and single fiber-EMG recordings confirming an increased motor unit size and redistribution of motor unit fibers in aging human skeletal muscle (see Cavanagh et al. 1993; Gilchrist et al. 1992; Stalberg and Antoni 1980; Stalberg and Fawcett 1982; Stalberg and Thiele 1975; Thiele and Stalberg 1975).

3 Effects of Aging on Motor Unit Types

An aging-related increase in the proportion of slow-twitch (type I) fibers, according to enzyme-histochemical mATPase staining, was originally reported in the lateral portion of the human quadriceps muscle (Larsson et al. 1978, 1979). This observation was originally considered controversial and caused significant scientific debate (Larsson 1983). Three decades later, the fast to slow fiber type transitions appear less controversial and fiber type transitions have been reported in human tibialis anterior, quadriceps and biceps brachi muscles (Scelsi et al. 1980; Caccia et al. 1979; Jakobsson et al. 1988; Larsson et al. 1978; Monemi et al. 1998, 1999; Tomonaga 1977). These observations at the protein level have been confirmed at the gene level, demonstrating a significant aging-related decreases in fast, but not slow, MyHC mRNA levels (Balagopal et al. 2001; Welle et al. 2000). However, aging-related fiber type transitions are muscle specific and a slow-to-fast MyHC isoform transition has been observed in human cranial nerve innervated masticatory muscles (Monemi et al. 1999). It is important to emphasize that the magnitude of these aging-related fiber type transitions is small and the physiological importance may be questioned. In the rat slow-twitch soleus muscle, a fast-to-slow fiber type transition is observed during development and maturation and adult rats have almost 100% slow fibers prior to the aging-related loss of muscle fibers (see Larsson and Ansved 1995). In fast-twitch muscles, aging-related fiber type transitions have frequently gone undetected when using conventional enzyme-histochemical myofibrillar ATPase stainings (Larsson et al. 1991a, 1993). However, the identification of a third fast myosin isoform in rat skeletal muscle according to immunocytochemistry (Schiaffino et al. 1989) and electrophoretic separation (Termin et al. 1989) named type IIx and IId, respectively, is of specific interest in this context. By using the set of monoclonal antibodies developed by Schiaffino and co-workers (Schiaffino et al. 1989), i.e., type II, IIa and IIb MyHC antibodies, it was possible to identify the IIx MyHC isoform in glycogen-depleted motor unit fibers. It was shown that the type IIx MyHC motor unit is a specific motor unit with spatial arrangement, physiological, biochemical and morphological properties separating it from the type IIa and IIb MyHC motor units (Larsson et al. 1991b).

It appears as if the IIx MyHC motor unit plays an integral role in the aging-related motor unit transition in fast-twitch muscles (Larsson et al. 1991a). First, it is the dominating motor unit type in the tibialis anterior muscle in old rats, whereas in young animals it only constitutes a small proportion (Larsson et al. 1991a, b). Second, the reorganization of motor units in old age with the appearance of type IIx motor unit fibers in regions, which in the young animals are restricted to type IIb motor units (Figs. 1 and 2). Third, a large number of the IIx MyHC motor units in the old animals co-expressed IIa and IIx or IIx and IIb MyHCs. However, it is important to the emphasize that the motor unit fibers in the glycogen depleted motor unit fibers were characterized by the monoclonal bodies reactive with (a) all type II MyHCs, (b) the type IIa MyHC, and (c) the type IIb MyHC (Larsson et al. 1991a). This means that the type IIx motor unit fibers are identified as those

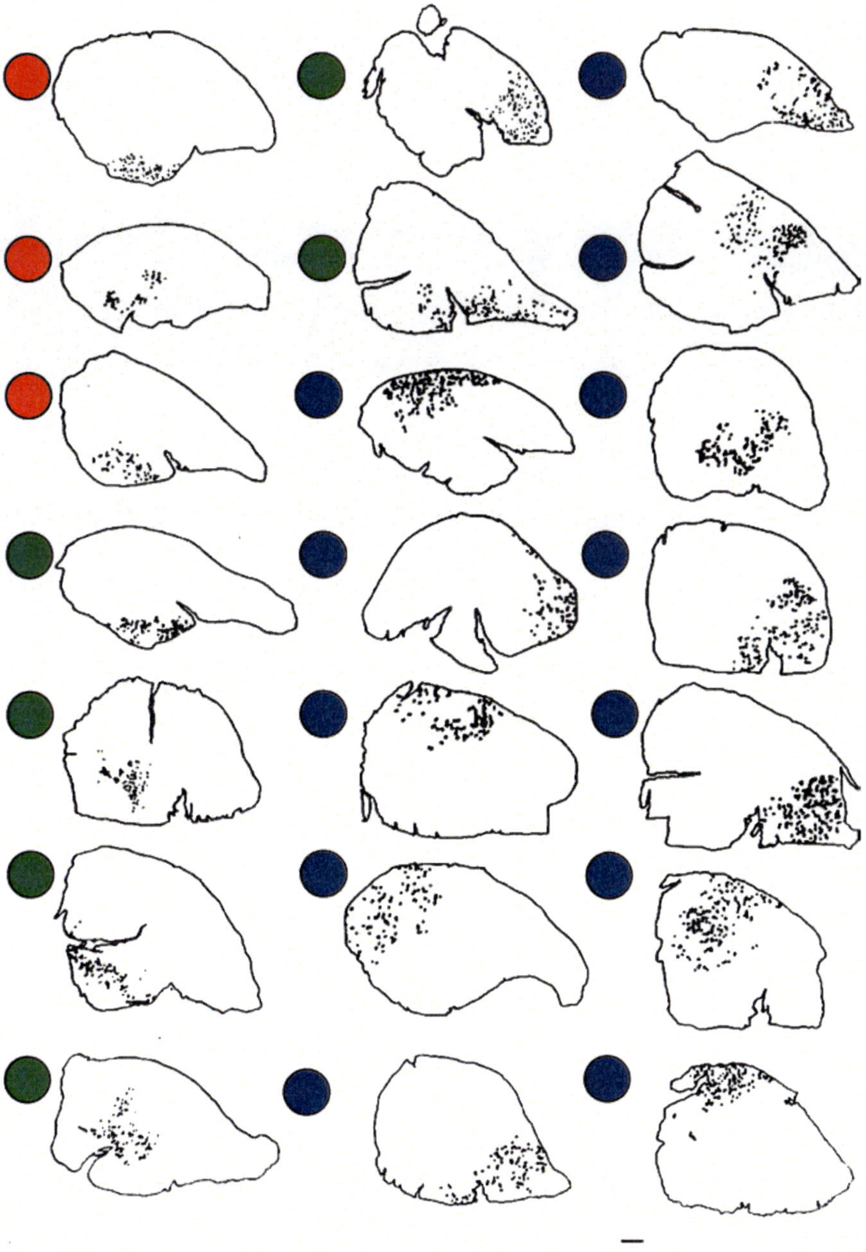

Fig. 1 Camera lucida tracings of glycogen-depleted cross-sections of fast-twitch motor unit fibres from young animals, classified according to their MyHC composition, in 21 cross-sections of tibialis anterior muscle. The superficial part of the muscle is facing the top of the figure. The type IIa, type IIx and IIb MyHC units are identified by red, green and blue filled circles, respectively. The horizontal bar represents 1 mm (The graph is modified from Larsson et al. 1991a, b)

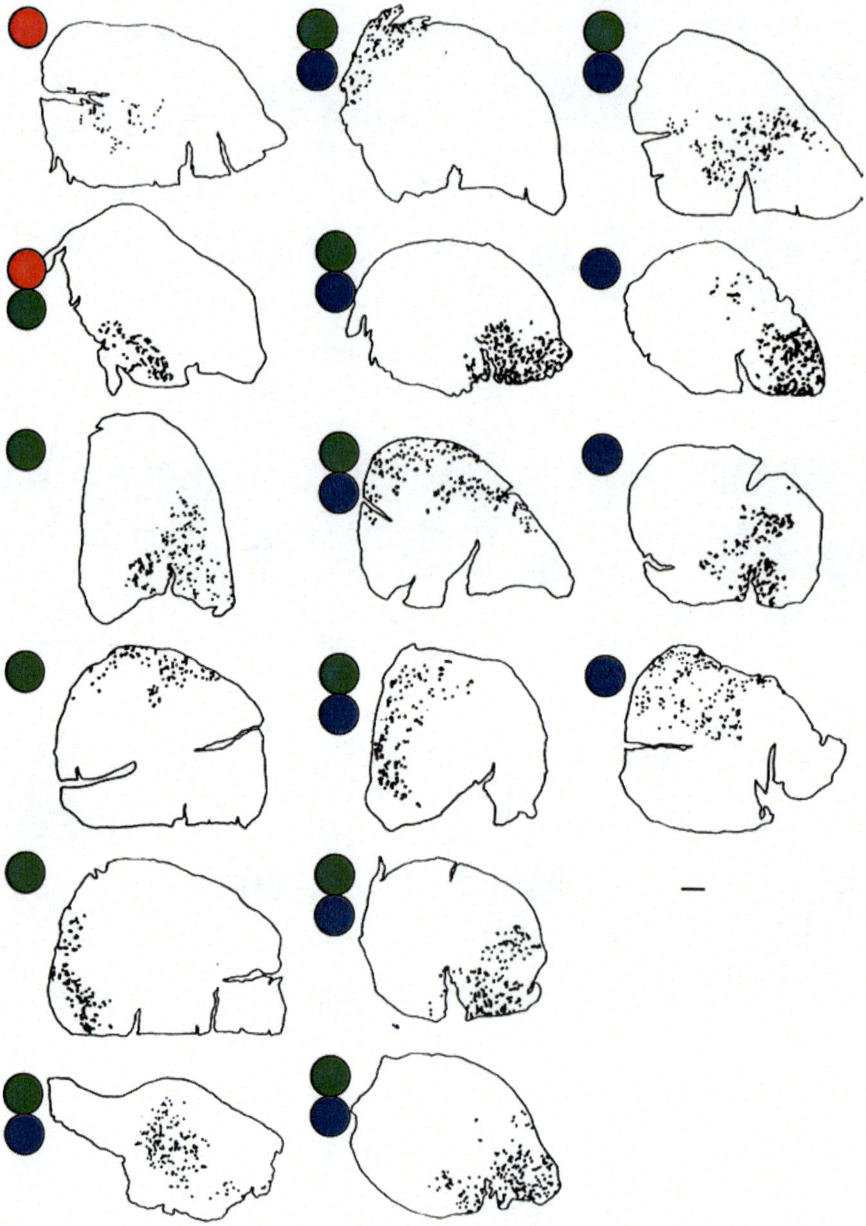

Fig. 2 Camera lucida tracings of glycogen-depleted cross-sections of fast-twitch motor unit fibres from young animals, classified according to their MyHC composition, in 16 cross-sections of tibialis anterior muscle. The superficial part of the muscle is facing the top of the figure. Motor units including type IIa, type IIx and IIb MyHC muscle fibers are identified by red, green and blue filled circles, respectively. Motor units containing more than one type of muscle fiber are identified by two filled circles. The horizontal bar represents 1 mm (The graph is modified from Larsson et al. 1991a, b)

reactive with the general fast MyHC antibody and unreactive with the specific type IIa or IIb MyHC antibodies. Thus, the "hybrid" motor units co-expressing type IIx with another MyHCs, included both muscle fibers expressing the type IIx MyHC and fibers expressing one of the other fast MyHC isoforms and not type IIx fibers co-expressing another fast MyHC isoform. Thus, the MyHC expression in these units are controlled by other mechanisms than the motoneuron properties, although an inhomogeneity of motoneuron properties within different branches of the neuron as a consequence of an aging-related impairment of motoneuron function cannot be completely ruled out. This type of non-uniform MyHC isoform expression among motor unit fibers was never observed in the young animals (Larsson et al. 1991a) (Fig. 1). A further fast-to-slow transformation process resulting in an increased number of type IIa and type I fibers has been reported in old age, but this transformation process appears to be confined to a very old age (Bass et al. 1975; Boreham et al. 1988; Caccia et al. 1979; Ishihara et al. 1987; Kanda and Hashizume 1989; Kovanen and Suominen 1987; Pettigrew and Gardiner 1987). It is therefore suggested that the increased number of the IIx MyHC units in fast-twitch muscles in old age reflects an aging-related motor unit transition from type IIb- to IIx, possibly preceding a transformation to types IIa and I, following the sequence IIb →IIxb →IIx →IIxa →IIa →I (Gorza 1990; Larsson et al. 1991a).

4 Physiological Properties of Motor Units

The muscle wasting associated with old age and the associated decline of maximum contractile force is the most common type of muscle atrophy and impaired muscle function observed in humans, but the rate varies between different individuals, as well as between different muscles in the same individual (Larsson 1982). A more sedentary lifestyle in old age contributes to the impaired muscle function, but cannot account for all the differences in the aging-related loss of muscle force and mass. The underlying mechanisms are complex and involve genetic factors (Lionikas et al. 2005a, b, 2006). Muscle force is proportional to the cross-sectional area and the force generating capacity (maximum force normalized to cross-sectional area, specific force) of the activated muscle tissue, and there is reason to believe that both area and specific force decrease in old age. The loss of alpha motoneurons, the incomplete reinnervation of previously denervated muscle fibers, the following muscle fiber loss and fiber atrophy play an important role in the aging-related decline in muscle force documented in rodents (e.g. Ansved and Larsson 1989; Arabadjis et al. 1990; Edstrom and Larsson 1987; Larsson and Edstrom 1986). An aging-related total muscle fiber loss has also been reported in humans, based on autopsy sections and measurements of fiber numbers in selected regions of the muscle, followed by extrapolation to the whole-muscle cross-sections (Lexell et al. 1983). The aging-related decrease in cross-sectional fiber area appears to affect muscle fibers of the fast-twitch type preferentially in humans (Larsson et al. 1978, 1979; Scelsi et al. 1980; Tomonaga 1977). In addition to the loss of muscle fiber number and size, an altered spatial organization of motor unit fibers may affect the lateral transmission of forces and overall muscle force (see Monti et al. 2001). Divergent results have been reported with regard to aging-related changes in specific force based on measurements at whole muscle or motor unit levels (see Larsson 2003; Larsson and Ansved 1995). This is, at least in part, due to a number of different factors that obscure measurements of aging-related differences in specific force at the muscle and motor unit levels, such as differences in intramuscular fiber orientation, differences in the mechanical leverage provided by the bony anatomy of the joint, the elasticity of the muscle and the muscle tendons, patterns of motor unit recruitment, and activation of antagonist muscles. Single fiber preparations, on the other hand, allow investigation of the function of myofilament proteins in a cell with an intact filament lattice, but without the confounding effects related to intercellular connective tissue or protein heterogeneity between cells of multicellular preparations.

Studies on aging-related changes in specific tension at the single fiber level, using the skinned fiber preparation, have reported an aging-related decline in specific tension varying between 9% and 47% (Frontera et al. 2000; Larsson et al. 1997; Li 1996; Lowe et al. 2001; Thompson and Brown 1999). A decrease in specific force in skinned fibers, where excitation-contraction coupling has been by passed and contractile proteins are activated directly with calcium, could be due to a decrease in the number of cross-bridges in the driving stroke per muscle fiber volume, or to a decrease in the force developed by each cross-bridge, or to a combination of both mechanisms. In very old rodents, there is ultrastructural evidence of myofibrillar loss and also of an increase in intermyofibrillar spaces, which is expected to influence the specific tension (Ansved and Edstrom 1991), a preferential loss of myosin has been reported in rodents (Thompson et al. 2006) and similar observations have also been observed in human skeletal muscle (Cristea and Larsson, unpublished observations, Fig. 3), indicating that an altered myofibrillar protein stochiometry may contribute to the decreased specific force in old age in rodents as well as in humans. Lowe and co-workers (Lowe et al. 2001), using electron paramagnetic resonance spectroscopy analyses, have shown aging-related changes in the function of the myosin head in old animals, i.e., a decreased number of myosin heads in the strong-binding structural state. We have recently observed aging-related changes in 3D myonuclear organization with regional accumulation of myonuclei resulting in an increased variablility of myonuclei domain size, but with no or only minimal changes in average myonuclei domain size (Cristea et al. 2010) (Fig. 4). It is hypothesized that the increased variability in myonuclei domain size may have a negative effect on contractile protein synthesis and transport. This may also result in increased post-translational modifications of contractile proteins. It has been known for many years that skeletal muscle generates free radicals and muscle derived reactive oxygen species and nitric oxide derivatives influence regulation of muscle contraction and induce posttranslational modifications, and that aging exaggerates these effects (Reid and Durham 2002). Myofibrillar proteins, such as myosin, actin and tropomyosin, are highly sensitive to free radical-mediated oxidative stress (Barreiro and Hussain 2010). An aging-related increase in carbonylated

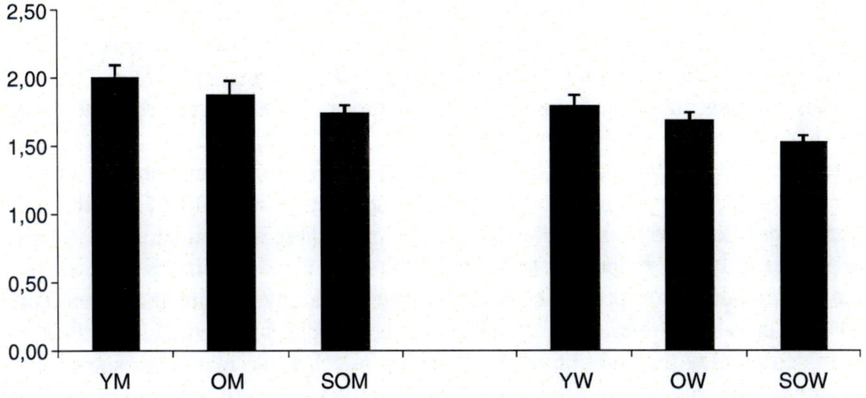

Fig. 3 Myosin:actin ratio determined on 12% SDS-PAGE gels in different age and gender groups. Subjects are divided into young (Y, 24–35 years, n=12), old (O, 65–83, years, n=12) and very old (SO, 89–96 years, n= 9) men (M) and women (W). According to two-way ANOVA there were significant age (p<0.01) and gender (p<0.001) effects, i.e., lower myosin:actin ratios among women and lower ratios in the old and very old individuals. Values are means ± SE

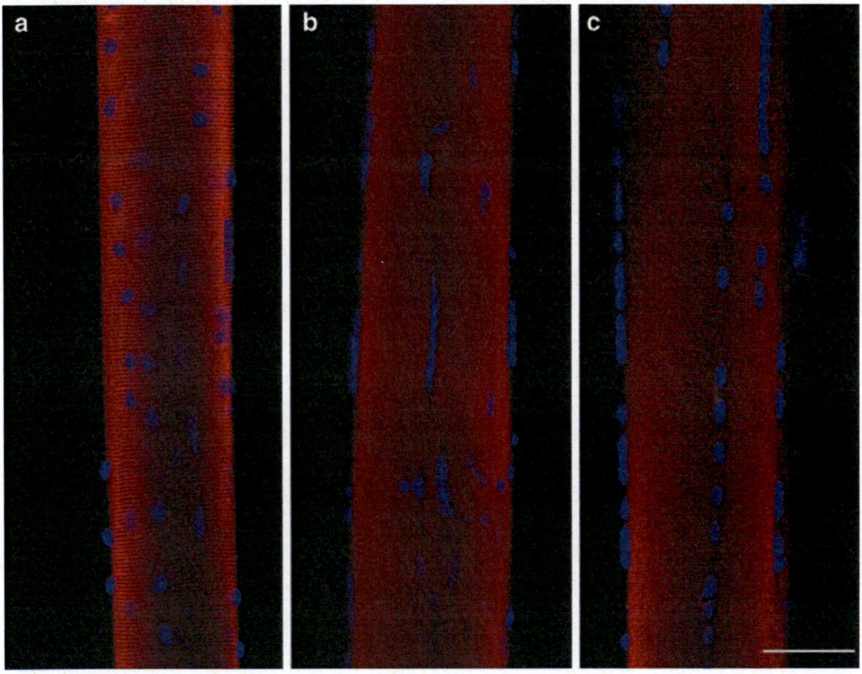

Fig. 4 Confocal microscopy images of myonuclei in muscle fibers expressing the type I MyHC isoform. (**a**) Young women, age 35 years. (**b**) Old women, age 78 years. (**c**) Very old women, age 96 years. The horizontal bar denotes 100 µm

and glycosylated myofibrillar proteins have been reported in old age (Syrovy and Hodny 1992; Barreiro and Hussain 2010; Chen et al. 2010) with negative consequences for muscle function (Ramamurthy et al. 2001; Ramamurthy et al. 2003).

During aging, the decrease in force is accompanied by a decreased contractile speed. Speed of contraction typically refers to (a) the duration of the isometric twitch, i.e., the contraction and half-relaxation times, or (b) the maximum velocity of unloaded shortening determined by isotonic measurements and fitted with the Hill equation or calculated from slack-test data. Typically, there is a close match between the isometric twitch properties and maximum shortening velocity, but the major determinants of twitch properties and shortening velocity are not the same. The capacity for calcium release and recapture by the sarcoplasmic reticulum is the key factor determining the duration of the isometric twitch while maximum shortening velocity is primarily determined by the isoform and ATPase activity of the molecular motor protein myosin. In adult mammalian muscle there is a close co-ordination between SR and contractile proteins (e.g. Brody 1976; Dulhunty and Valois 1983; Kugelberg and Thornell 1983; Salviati et al. 1984), but this close co-ordination may be lost under certain experimental conditions (Brody 1976; Fitts et al. 1980) as well as during the aging process (Salviati et al. 1984). Reliable shortening velocity measurements are very difficult, impossible, to conduct at the motor unit level due to the confounding influence of surrounding motor units and connective tissue. A number of studies have confirmed an aging-related slowing of the isometric twitch both at the whole muscle and motor unit levels in various mammals, including man (for refs. see Larsson and Ansved 1995). An aging-related slowing of the isometric twitch is seen before the senile muscle wasting becomes manifest (Ansved and Larsson 1989; Larsson and Edstrom 1986) and it may be speculated to be due to an aging-related loss of fast-twitch muscle fibers (Campbell et al. 1973; Newton and Yemm 1986). However, the parallel aging-related slowing of the isometric twitch in motor units of both fast- and slow-twitch type indicate that the slowing at the whole muscle level cannot be explained solely by altered motor unit proportions (Edstrom and Larsson 1987; Larsson et al. 1991a; Ansved and Larsson 1994; Larsson and Salviati 1989). Various factors in the contractile apparatus may play a role in the regulation of contractile speed, but an aging-related change in the properties of the SR appears to be a probable explanation for the longer twitch duration in old age (see above). This is supported by aging-related changes in the structure, function, and biochemical properties SR, as well as an altered modulation of the SR calcium release channel (De Coster et al. 1981; Larsson and Salviati 1989; Viner et al. 1996).

5 Motor Unit Firing Properties

Analyses of motor unit firing patterns in young and old individuals indicate that the aging-related motor unit loss may begin as early as age 30 (Kamen and De Luca 1989). Motor unit firing properties include changes in the mechanical force of motor units, mean firing rate, variability of firing rate, doublet firing, and motor unit synchronization. The most consistent models used in humans have been the first dorsal interosseous during abduction force and tibialis anterior during dorsiflexion, mainly because the muscles are easy to access which minimizes experimental complications. The spike-triggered average force technique has been used to estimate the amount of force exerted by subjects during an isometric contraction. In the first dorsal interosseous, Galgangski and colleagues (1993) reported that the force exerted by the motor units discharging at a low rate was greater in older adults compared with young adults. This finding represents the functional consequence of motor unit reorganization that preferentially affects fast-twitch muscle fibers. These findings have been replicated in the first dorsal interosseous muscle and also extended to the tibialis anterior (Fling et al. 2009). Despite these changes, older adults retain the ability to recruit motor units based on size, since larger motor units are recruited at higher levels of force (Fling et al. 2009; Galganski et al. 1993).

As shown by Monster and Chan (1977), the discharge rate of motor units increases when the amount of isometric force produced by muscle is increased. Indeed, the firing rates of aged motor units have been found to be lower than the firing rate of motor units in younger individuals (Howard et al. 1988; Newton et al. 1988; Erim et al. 1999). Kamen and colleagues found the decrease in mean firing rate only occurred at high level contractions (Kamen et al. 1995), which may explain why studies investigating low force contractions have not always found this result (Galganski et al. 1993; Vaillancourt et al. 2003). It could also be that the difference in the control of isometric force versus firing rate influences the ability to detect a change in firing rate with age. Also, it is important to incorporate the recruitment threshold in the firing rate calculation, since motor units with different thresholds could weigh differently in the average (Erim et al. 1999). Firing rate variability has been shown to increase in older adults compared with young adults (Laidlaw et al. 2000; Tracy et al. 2005). The variability of motor unit discharge has been hypothesized to underlie increased force fluctuations in older adults (Enoka et al. 2003). Increased force fluctuations affect the regulation of fine motor tasks such as buttoning a shirt and drinking and eating. In some instances, the motor unit will fire twice with a short interval between pulses (Kamen 2005). It has been suggested that the doublet discharge occurs more frequently during a muscle contraction of high force or high velocity (Garland and Griffin 1999), and this is because higher forces are achieved by doublet firing than the summation of two motor unit twich forces (Clamann and Schelhorn 1988). Christie and Kamen (2006) intestigated doublet discharges in the tibialis anterior and first dorsal interosseous muscle, and found that older adults have a reduced percentage of doublet discharge at 10%, 30%, and 50% of the maximum voluntary contraction.

Another property of motor unit firing that has been investigated is motor unit synchronization. Motor unit synchronization occurs when two different motor units firing simultaneously or near-simultaneously, and the result is increased muscle force. Motor unit synchronization has been measured using different techniques that include cumulative probability distributions (Datta and Stephens 1990; Nordstrom et al. 1992), coherence between motor unit pairs (Semmler 2002), and coherence between the EMG electrode and acceleration of the effector (Halliday et al. 1999). Using the cumulative probability distributions, it has been shown by two different laboratories that motor unit synchronization is not affected by the aging process (Semmler et al. 2000; Kamen and Roy 2000). This technique measures short-term synchronization from common presynaptic input by assessing the number of synchronized discharges per unit time. However, motor unit coherence between motor unit pairs has been shown to be greater in older adults compared with young adults in the 5–9 Hz bandwidth (Semmler et al. 2003). Motor unit to motor unit coherence is a technique performed in the frequency domain, and is thought to represent the common oscillatory input to motor neurons (Conway et al. 1995). Also, when using coherence between surface EMG and acceleration, motor unit synchronization was also found to be increased in the 5–9 Hz bandwidth and these authors related these measures of synchronization to physiological postural tremor in older adults (Sturman et al. 2005).

6 Conclusion and Perspectives

The motor unit and its constituent components, the motoneuron, motor end plate and muscle cells are all affected by aging with significant consequences for motor function, dependence, quality of life, morbidity and mortality in the growing population of elderly citizens. This has resulted in an increasing scientific interest in aging-related changes within the motor unit ranging from the alpha motoneuron soma, axonal transport, motor end plate, muscle fiber extracellular matrix, membrane properties, intermediate filaments, excitation-contraction coupling, muscle metabolism to contractile elements in the sarcomere. There is a significant "safety factor" within several of these components making assumptions regarding the impact of aging-related changes within specific components of the motor unit on overall muscle function very difficult. In addition, there is a highly coordinated regulation of all these systems at the motor unit and muscle fiber level, but this close coordination is not invariable (Wilson and Woledge 1985), and it may, at least in part, be lost during the aging process (Salviati et al. 1984; Larsson et al. 1991a). The mechanism underlying the loss of the coordinated expression of contractile, mitochondrial and sarcoplasmic reticulum proteins in skeletal muscle fibers in old age remains unknown. However, it is becoming increasingly evident that myogenic microRNAs (miRNAs), nestled within introns of myosin genes, modulate a variety of muscle functions by fine-tuning gene expression patterns through repression of mRNA targets (van Rooij et al. 2008, 2009; Small et al. 2010). Misexpression of muscle-specific miRNAs in old age may according play an important role in this complex process, resulting in a less coordinated control within motor units and muscle cells. This altered coordinated control mechanism is forwarded as an important factor underlying the impaired muscle function in old age, representing a significant challenge for future mechanistic studies of the aging-related motor handicap at the motor unit level.

Acknowledgements The scientific work from our group referenced and discussed in this review were supported by grants from the Swedish Medical research Council (08651), the European Commission MyoAge (EC Fp7 CT-223756), King Gustaf V and Queen Victoria's Foundation, NIH (AR45627, AR47318, AG014731), AFM, Thuréus Foundation and the Swedish Sports Research Council.

References

Ansved, T. and Edstrom, L. (1991) Effects of age on fibre structure, ultrastructure and expression of desmin and spectrin in fast- and slow-twitch rat muscles. *J Anat, 174*, 61–79.

Ansved, T. and Larsson, L. (1989) Effects of ageing on enzyme-histochemical, morphometrical and contractile properties of the soleus muscle in the rat. *J Neurol Sci, 93*, 105–24.

Ansved, T., Wallner, P. and Larsson, L. (1991) Spatial distribution of motor unit fibres in fast- and slow-twitch rat muscles with special reference to age. *Acta Physiol Scand, 143*, 345–54.

Arabadjis, P. G., Heffner, R. R., Jr. and Pendergast, D. R. (1990) Morphologic and functional alterations in aging rat muscle. *J Neuropathol Exp Neurol, 49*, 600–9.

Balagopal, P., Schimke, J. C., Ades, P., Adey, D. and Nair, K. S. (2001) Age effect on transcript levels and synthesis rate of muscle MHC and response to resistance exercise. *Am J Physiol Endocrinol Metab, 280*, E203–8.

Barreiro, E. and Hussain, S. N. (2010) Protein carbonylation in skeletal muscles: impact on function. *Antioxid Redox Signal, 12*, 417–29.

Bass, A., Gutmann, E. and Hanzlikova, V. (1975) Biochemical and histochemical changes in energy supply enzyme pattern of muscles of the rat during old age. *Gerontologia, 21*, 31–45.

Bean JF, Kiely DK, LaRose S, Alian J, Frontera WR (2007). Is stair climb power a clinically relevant measure of leg power impairments in at-risk older adults? Archives of Physical Medicine and Rehabilitation *88*, 604–609.

Bodine-Fowler, S., Garfinkel, A., Roy, R. R. and Edgerton, V. R. (1990) Spatial distribution of muscle fibers within the territory of a motor unit. *Muscle Nerve, 13*, 1133–45.

Boreham, C. A., Watt, P. W., Williams, P. E., Merry, B. J., Goldspink, G. and Goldspink, D. F. (1988) Effects of ageing and chronic dietary restriction on the morphology of fast and slow muscles of the rat. *J Anat, 157*, 111–25.

Brandstater, M. E., Lambert, E.H. (1969) A histochemical study of the spatial arrangement of muscle fibers in single motor units within rat tibialis anterior motor units. Bull. *Am. Ass. EMG Electrodiag., 82*, 15–16.

Brandstater, M. E., Lambert, E.H. (1973) Motor unit anatomy. Type and spatial arrangement of muscle fibers. In: *New Developments in Electromyography and Clinical Neurophysiology*, Vol. 1 Karger, Basel, pp. 14–22.

Brody, I. A. (1976) Regulation of isometric contraction in skeletal muscle. *Exp Neurol, 50*, 673–83.

Brownlee, M. (1995) Advanced protein glycosylation in diabetes and aging. *Annu Rev Med, 46*, 223–34.

Buchtahl, F., Rosenfalck, P. (1955) Action potential parameters in different human muscles. Acta Psychiatri. *Neurol. Scand., 30*, 125–131.

Burke, R. E. (1981) Motor units: Anatomy, physiology and functional organization. In Handbook of Physiology. The Nervous System, Motor Control, Vol. II, Section 1, Part 1 American Physiological Society, pp. 345–422.

Burke, R. E., Levine, D. N. and Zajac, F. E., 3rd (1971) Mammalian motor units; physiological histochemical correlation in three types in cat gastrocnemius. Science, *174*, 709–12.

Burke, R. E. and Tsairis, P. (1973) Anatomy and innervation ratios in motor units of cat gastrocnemius. *J Physiol, 234*, 749–65.

Caccia, M. R., Harris, J. B. and Johnson, M. A. (1979) Morphology and physiology of skeletal muscle in aging rodents. *Muscle Nerve, 2*, 202–12.

Campbell, M. J., McComas, A. J. and Petito, F. (1973) Physiological changes in ageing muscles. *J Neurol Neurosurg Psychiatry, 36*, 174–82.

Carlson, K. E., Alston, W., Feldman, D.J. (1964) Electromyographic study of ageing in skeletal muscle. *Arch Phys Med Rehabil, 58*, 423–426.

Cavanagh, P. R., Simoneau, G. G. and Ulbrecht, J. S. (1993) Ulceration, unsteadiness, and uncertainty: the biomechanical consequences of diabetes mellitus. *J Biomech, 26*, 23–40.

Chen, C. N., Brown-Borg, H. M., Rakoczy, S. G., Ferrington, D. A. and Thompson, L. V. (2010) Aging impairs the expression of the catalytic subunit of glutamate cysteine ligase in soleus muscle under stress. *J Gerontol A Biol Sci Med Sci, 65*, 129–37.

Christie, A. and Kamen, G. (2006) Doublet discharges in motoneurons of young and older adults. *J Neurophysiol, 95*, 2787–95.
Clamann, H. P. and Schelhorn, T. B. (1988) Nonlinear force addition of newly recruited motor units in the cat hindlimb. *Muscle Nerve, 11*, 1079–89.
Conway, B. A., Halliday, D. M., Farmer, S. F., Shahani, U., Maas, P., Weir, A. I. and Rosenberg, J. R. (1995) Synchronization between motor cortex and spinal motoneuronal pool during the performance of a maintained motor task in man. *Journal of Physiology, 489*, 917–924.
Cristea, A., Karlsson Edlund, P., Lindblad, J., Qaisar, R., Bengtsson, E., Larsson, L. (2010) Effects of aging and gender on the spatial organization of nuclei in single human skeletal muscle cells. *Aging Cell, 9*, 685–697.
Datta, A. K. and Stephens, J. A. (1990) Synchronization of motor unit activity during voluntary contraction in man. *Journal of Physiology, 422*, 397–419.
De Coster, W., De Reuck, J., Sieben, G. and Vander Eecken, H. (1981) Early ultrastructural changes in aging rat gastrocnemius muscle: A stereological study. *Muscle Nerve, 4*, 111–6.
Doyle, A. M., Mayer, R.F. (1969) Studies of the motor unit in the cat. *Bull. Sch. Med. Univ. Maryland, 54*, 11–17.
Dulhunty, A. and Valois, A. (1983) Indentations in the terminal cisternae of amphibian and mammalian skeletal muscle fibres. *J Ultrastruct Res, 84*, 34–49.
Eccles, J. C., Sherrington, C.S. (1930) Numbers and contraction values of individual raotor units examined in some muscles of the limb. *Proc R Soc Lond B Biol Sci, 106*, 326–357.
Edstrom, L. and Kugelberg, E. (1968) Histochemical composition, distribution and fatiguability of single motor units. Anterior tibial muscle of the rat. *J Neurol Neurosurg Psychiatry, 31*, 424–33.
Edstrom, L. and Larsson, L. (1987) Effects of age on muscle fibre characteristics of fast- and slow-twitch single motor units in the rat. *J Physiol (Lond), 392*, 129–45.
Einsiedel, L., Luff, A. R. and Proske, U. (1992) Sprouting of fusimotor neurones after partial denervation of the cat soleus muscle. *Exp Brain Res, 90*, 369–74.
Einsiedel, L. J. and Luff, A. R. (1992) Effect of partial denervation on motor units in the ageing rat medial gastrocnemius. *J Neurol Sci, 112*, 178–84.
Enoka, R. M., Christou, E. A., Hunter, S. K., Kornatz, K. W., Semmler, J. G., Taylor, A. M. and Tracy, B. L. (2003) Mechanisms that contribute to differences in motor performance between young and old adults. *J Electromyogr Kinesiol, 13*, 1–12.
Erim, Z., Beg, M. F., Burke, D. T. and Luca, C. J. D. (1999) Effects of aging on motor-unit control properties. *Journal of Neurophysiology, 82*, 2081–2019.
Fitts, R. H., Winder, W. W., Brooke, M. H., Kaiser, K. K. and Holloszy, J. O. (1980) Contractile, biochemical, and histochemical properties of thyrotoxic rat soleus muscle. *Am J Physiol, 238*, C14–20.
Fling, B. W., Knight, C. A. and Kamen, G. (2009) Relationships between motor unit size and recruitment threshold in older adults: implications for size principle. *Exp Brain Res, 197*, 125–33.
Frontera, W. R., Suh, D., Krivickas, L. S., Hughes, V. A., Goldstein, R. and Roubenoff, R. (2000) Skeletal muscle fiber quality in older men and women. *Am J Physiol Cell Physiol, 279*, C611–8.
Frontera WR, Reid KF, Phillips EM, Krivickas LS, Hughes VA, Roubenoff R, Fielding RA (2008). Muscle fiber size and function in elderly humans: a longitudinal study. *Journal of Applied Physiology 105*, 637–642.
Galganski, M. E., Fuglevand, A. J. and Enoka, R. M. (1993) Reduced control of motor output in a human hand muscle of elderly subjects during submaximal contractions. *Journal of Neurophysiology, 69*, 2108–2114.
Garland, S. J. and Griffin, L. (1999) Motor unit double discharges: statistical anomaly or functional entity? *Can J Appl Physiol, 24*, 113–30.
Gilchrist, J., Barkhaus, P.E., Bril, V., Daube, J.R., DeMeirsman, J., Howard, J., Jablecki, C., Sanders, D.B., Stålberg, E., Trontelj, J.V., Pezzulo, J. (1992) Single fiber EMG reference values: a collaborative effort. Ad Hoc Committee of the AAEM Special Interest Group on Single Fiber EMG. *Muscle & Nerve, 15*, 151–161.
Gorza, L. (1990) Identification of a novel type 2 fiber population in mammalian skeletal muscle by combined use of histochemical myosin ATPase and anti-myosin monoclonal antibodies. *J Histochem Cytochem, 38*, 257–65.

Halliday, D. M., Conway, B. A., Farmer, S. F. and Rosenberg, J. R. (1999) Load-independent contributions from motor-unit synchronization to human physiological tremor. *Journal of Neurophysiology, 82*, 664–675.

Hashizume, K., Kanda, K. and Burke, R. E. (1988) Medial gastrocnemius motor nucleus in the rat: Age-related changes in the number and size of motoneurons. *J Comp Neurol, 269*, 425–30.

Henneman, E., Somjen, G. and Carpenter, D. O. (1965) Excitability and inhibitability of motoneurons of different sizes. *J Neurophysiol, 28*, 599–620.

Howard, J. E., McGill, K. C. and Dorfman, L. J. (1988) Age effects on properties of motor unit action potentials: ADEMG analysis. *Ann Neurol, 24*, 207–13.

Ingber, D (2002a) Mechanical signaling. *Ann N Y Acad Sci, 961*, 162–3.

Ingber, D. E. (1991) Control of capillary growth and differentiation by extracellular matrix. Use of a tensegrity (tensional integrity) mechanism for signal processing. *Chest, 99*, 34S–40S.

Ingber, D. E. (1993) Cellular tensegrity: defining new rules of biological design that govern the cytoskeleton. *J Cell Sci, 104* (Pt 3), 613–27.

Ingber, D. E. (1997) Tensegrity: the architectural basis of cellular mechanotransduction. *Annu Rev Physiol, 59*, 575–99.

Ingber, D. E. (2002b) Mechanical signaling and the cellular response to extracellular matrix in angiogenesis and cardiovascular physiology. *Circ Res, 91*, 877–87.

Ishihara, A., Naitoh, H. and Katsuta, S. (1987) Effects of ageing on the total number of muscle fibres and motoneurons of the tibialis anterior and soleus muscles in the rat. *Brain Res, 435*, 355–8.

Jakobsson, F., Borg, K., Edstrom, L. and Grimby, L. (1988) Use of motor units in relation to muscle fiber type and size in man. *Muscle Nerve, 11*, 1211–8.

Kamen, G. (2005) Aging, resistance training, and motor unit discharge behavior. *Can J Appl Physiol, 30*, 341–51.

Kamen, G. and De Luca, C. J. (1989) Unusual motor unit firing behavior in older adults. *Brain Res, 482*, 136–40.

Kamen, G. and Roy, A. (2000) Motor unit synchronization in young and elderly adults. *Eur J Appl Physiol, 81*, 403–10.

Kamen, G., Sison, S. V., Duke, C. C. and Patten, C. (1995) Motor unit discharge behavior in older adults during maximal-effort contractions. *Journal of Applied Physiology, 79*, 1908–1913.

Kanda, K. and Hashizume, K. (1989) Changes in properties of the medial gastrocnemius motor units in aging rats. *J Neurophysiol, 61*, 737–46.

Kanda, K. and Hashizume, K. (1991) Recovery of motor-unit function after peripheral nerve injury in aged rats. *Neurobiol Aging, 12*, 271–6.

Knox, C. A., Kokmen, E. and Dyck, P. J. (1989) Morphometric alteration of rat myelinated fibers with aging. *J Neuropathol Exp Neurol, 48*, 119–39.

Komiya, Y. (1980) Slowing with age of the rate of slow axonal flow in bifurcating axons of rat dorsal root ganglion cells. *Brain Res, 183*, 477–80.

Kovanen, V. and Suominen, H. (1987) Effects of age and life-time physical training on fibre composition of slow and fast skeletal muscle in rats. *Pflugers Arch, 408*, 543–51.

Krnjevic, K., Miledi, R. (1958) Motor units in the rat diaphragm. *J Physiol (Lond), 140*, 427–439.

Kugelberg, E., Edstrom, L. and Abbruzzese, M. (1970) Mapping of motor units in experimentally reinnervated rat muscle. *J Neurol Neurosurg Psychiatry, 33*, 319–29.

Kugelberg, E. and Lindegren, B. (1979) Transmission and contraction fatigue of rat motor units in relation to succinate dehydrogenase activity of motor unit fibres. *J Physiol (Lond), 288*, 285–300.

Kugelberg, E. and Thornell, L. E. (1983) Contraction time, histochemical type, and terminal cisternae volume of rat motor units. *Muscle Nerve, 6*, 149–53.

Laidlaw, D. H., Bilodeau, M. and Enoka, R. M. (2000) Steadiness is reduced and motor unit discharge is more variable in old adults. *Muscle Nerve, 23*, 600–12.

Larsson, L. (1982) Ageing in mammalian skeletal muscles. Praeger, New York.

Larsson, L. (1983) Histochemical characteristics of human skeletal muscle during aging. *Acta Physiol Scand, 117*, 469–71.

Larsson, L. (1992) Is the motor unit uniform? *Acta Physiol Scand, 144*, 143–54.

Larsson, L. (2003) In Clinical Neurophysiology of Disorders of Muscle and Neuromuscular Junction, Including Fatigue, Vol. 2 (Ed, Stålberg, E.) Elsevier, Amsterdam, pp. 119–144.

Larsson, L. and Ansved, T. (1995) Effects of ageing on the motor unit. *Prog Neurobiol, 45*, 397–458.

Larsson, L., Ansved, T., Edstrom, L., Gorza, L. and Schiaffino, S. (1991a) Effects of age on physiological, immunohistochemical and biochemical properties of fast-twitch single motor units in the rat. *J Physiol (Lond), 443*, 257–75.

Larsson, L., Biral, D., Campione, M. and Schiaffino, S. (1993) An age-related type IIB to IIX myosin heavy chain switching in rat skeletal muscle. *Acta Physiol Scand, 147*, 227–34.

Larsson, L. and Edstrom, L. (1986) Effects of age on enzyme-histochemical fibre spectra and contractile properties of fast- and slow-twitch skeletal muscles in the rat. *J Neurol Sci, 76*, 69–89.

Larsson, L., Edstrom, L., Lindegren, B., Gorza, L. and Schiaffino, S. (1991b) MHC composition and enzyme-histochemical and physiological properties of a novel fast-twitch motor unit type. *Am J Physiol, 261*, C93–101.

Larsson, L., Grimby, G. and Karlsson, J. (1979) Muscle strength and speed of movement in relation to age and muscle morphology. *J Appl Physiol, 46*, 451–6.

Larsson, L., Li, X. and Frontera, W. R. (1997) Effects of aging on shortening velocity and myosin isoform composition in single human skeletal muscle cells. *Am J Physiol, 272*, C638–49.

Larsson, L. and Salviati, G. (1989) Effects of age on calcium transport activity of sarcoplasmic reticulum in fast- and slow-twitch rat muscle fibres. *J Physiol (Lond), 419*, 253–64.

Larsson, L., Sjodin, B. and Karlsson, J. (1978) Histochemical and biochemical changes in human skeletal muscle with age in sedentary males, age 22–65 years. *Acta Physiol Scand, 103*, 31–9.

Lexell, J., Henriksson-Larsen, K., Winblad, B. and Sjostrom, M. (1983) Distribution of different fiber types in human skeletal muscles: effects of aging studied in whole muscle cross sections. *Muscle Nerve, 6*, 588–95.

Li, X. (1996) Effects of ageing on contractility and myosin composition of rat and man skeletal muscles and muscle cells. Doktorsavhandling vid Karolinska Institutet (Doctoral thesis, Karolinska Institute), Stockholm.

Lidell, E. G. T., Sherrington, C.S. (1925) Recruitment and some factors of reflex inhibition. *Proc R Soc Lond B Biol Sci, 97*, 488–518.

Lionikas, A., Blizard, D. A., Gerhard, G. S., Vandenbergh, D. J., Stout, J. T., Vogler, G. P., McClearn, G. E. and Larsson, L. (2005a) Genetic determinants of weight of fast- and slow-twitch skeletal muscle in 500-day-old mice of the C57BL/6J and DBA/2J lineage. *Physiol Genomics, 21*, 184–92.

Lionikas, A., Blizard, D. A., Vandenbergh, D. J., Glover, M. G., Stout, J. T., Vogler, G. P., McClearn, G. E. and Larsson, L. (2003) Genetic architecture of fast- and slow-twitch skeletal muscle weight in 200-day-old mice of the C57BL/6J and DBA/2J lineage. *Physiol Genomics, 16*, 141–52.

Lionikas, A., Blizard, D. A., Vandenbergh, D. J., Stout, J. T., Vogler, G. P., McClearn, G. E. and Larsson, L. (2006) Genetic determinants of weight of fast- and slow-twitch skeletal muscles in old mice. *Mamm Genome, 17*, 615–28.

Lowe, D. A., Surek, J. T., Thomas, D. D. and Thompson, L. V. (2001) Electron paramagnetic resonance reveals age-related myosin structural changes in rat skeletal muscle fibers. *Am J Physiol Cell Physiol, 280*, C540–7.

Mahoney, J., Sager, M., Dunham, N. C. and Johnson, J. (1994) Risk of falls after hospital discharge. *J Am Geriatr Soc, 42*, 269–74.

Mahoney, J. E., Sager, M. A. and Jalaluddin, M. (1999) Use of an ambulation assistive device predicts functional decline associated with hospitalization. *J Gerontol A Biol Sci Med Sci, 54*, M83–8.

Martin, T. P., Bodine-Fowler, S., Roy, R. R., Eldred, E. and Edgerton, V. R. (1988) Metabolic and fiber size properties of cat tibialis anterior motor units. *Am J Physiol, 255*, C43–50.

McComas, A. J., Sica, R. E. and Campbell, M. J. (1971) "Sick" motoneurones. A unifying concept of muscle disease. Lancet, 1, 321–6.

McMartin, D. N. and O'Connor, J. A., Jr. (1979) Effect of age on axoplasmic transport of cholinesterase in rat sciatic nerves. *Mech Ageing Dev, 10, 241–8.*

McPhederan, A. M., Wuerker, R.B., Henneman, E. (1965) Properties of motor units in a homogeneous red muscle (soleus) of the cat. *J Neurophysiol, 28*, 71–84.

Monemi, M., Eriksson, P. O., Eriksson, A. and Thornell, L. E. (1998) Adverse changes in fibre type composition of the human masseter versus biceps brachii muscle during aging. *J Neurol Sci, 154*, 35–48.

Monemi, M., Kadi, F., Liu, J. X., Thornell, L. E. and Eriksson, P. O. (1999) Adverse changes in fibre type and myosin heavy chain compositions of human jaw muscle vs. limb muscle during ageing. *Acta Physiol Scand, 167*, 339–45.

Monster, A. W. and Chan, H. (1977) Isometric force production by motor units of extensor digitorum communis muscle in man. *J Neurophysiol, 40*, 1432–43.

Monti, R. J., Roy, R. R. and Edgerton, V. R. (2001) Role of motor unit structure in defining function. *Muscle Nerve, 24*, 848–66.

Nevitt, M. C., Cummings, S. R. and Hudes, E. S. (1991) Risk factors for injurious falls: a prospective study. *J Gerontol, 46*, M164–70.

Newton, J. P. and Yemm, R. (1986) Changes in the contractile properties of the human first dorsal interosseus muscle with age. *Gerontology, 32*, 98–104.

Newton, J. P., Yemm, R. and McDonagh, M. J. (1988) Study of age changes in the motor units of the first dorsal interosseous muscle in man. *Gerontology, 34*, 115–9.

Nordstrom, M. A., Fuglevand, A. J. and Enoka, R. M. (1992) Estimating the strength of common input to human motoneurons from the cross-correlogram. *J Physiol, 453*, 547–74.

Pettigrew, F. P. and Gardiner, P. F. (1987) Changes in rat plantaris motor unit profiles with advanced age. *Mech Ageing Dev, 40*, 243–59.

Ramamurthy, B., Hook, P., Jones, A. D. and Larsson, L. (2001) Changes in myosin structure and function in response to glycation. *FASEB Journal, 15*, 2415–22.

Ramamurthy, B., Jones, A. D. and Larsson, L. (2003) Glutathione reverses early effects of glycation on myosin function. *Am J Physiol Cell Physiol, 285*, C419–24.

Reid, M. B. and Durham, W. J. (2002) Generation of reactive oxygen and nitrogen species in contracting skeletal muscle: potential impact on aging. *Ann N Y Acad Sci, 959*, 108–16.

Robbins, A. S., Rubenstein, L. Z., Josephson, K. R., Schulman, B. L., Osterweil, D. and Fine, G. (1989) Predictors of falls among elderly people. Results of two population-based studies. *Arch Intern Med, 149*, 1628–33.

Salviati, G., Betto, R., Danieli Betto, D. and Zeviani, M. (1984) Myofibrillar-protein isoforms and sarcoplasmic-reticulum Ca 2+-transport activity of single human fibres. *Biochem J, 224*, 215–25.

Scelsi, R., Marchetti, C. and Poggi, P. (1980) Histochemical and ultrastructural aspects of m. vastus lateralis in sedentary old people (age 65–89 years). *Acta Neuropathol, 51*, 99–105.

Schiaffino, S., Gorza, L., Sartore, S., Saggin, L., Ausoni, S., Vianello, M., Gundersen, K. and Lomo, T. (1989) Three myosin heavy chain isoforms in type 2 skeletal muscle fibres. *J Muscle Res Cell Motil, 10*, 197–205.

Schultz, A. B., Ashton-Miller, J. A. and Alexander, N. B. (1997) What leads to age and gender differences in balance maintenance and recovery? *Muscle Nerve Suppl, 5*, S60–4.

Semmler, J. G. (2002) Motor unit synchronization and neuromuscular performance. *Exerc Sport Sci Rev, 30*, 8–14.

Semmler, J. G., Kornatz, K. W. and Enoka, R. M. (2003) Motor-unit coherence during isometric contractions is greater in a hand muscle of older adults. *J Neurophysiol, 90*, 1346–9.

Semmler, J. G., Steege, J. W., Kornatz, K. W. and Enoka, R. M. (2000) Motor-unit synchronization is not responsible for larger motor-unit forces in old adults. *Journal of Neurophysiology, 84*, 358–366.

Shinpo, K., Kikuchi, S., Sasaki, H., Ogata, A., Moriwaka, F. and Tashiro, K. (2000) Selective vulnerability of spinal motor neurons to reactive dicarbonyl compounds, intermediate products of glycation, in vitro: implication of inefficient glutathione system in spinal motor neurons. *Brain Res, 861*, 151–9.

Small, E. M., O'Rourke, J. R., Moresi, V., Sutherland, L. B., McAnally, J., Gerard, R. D., Richardson, J. A. and Olson, E. N. (2010) Regulation of PI3-kinase/Akt signaling by muscle-enriched microRNA-486. *Proc Natl Acad Sci U S A, 107*, 4218–4223.

Stalberg, E. and Antoni, L. (1980) Electrophysiological cross section of the motor unit. *J Neurol Neurosurg Psychiatry, 43*, 469–74.

Stalberg, E. and Fawcett, P. R. (1982) Macro EMG in healthy subjects of different ages. *J Neurol Neurosurg Psychiatry, 45*, 870–8.

Stalberg, E. and Thiele, B. (1975) Motor unit fibre density in the extensor digitorum communis muscle. Single fibre electromyographic study in normal subjects at different ages. *J Neurol Neurosurg Psychiatry, 38*, 874–80.

Sturman, M. M., Vaillancourt, D. E. and Corcos, D. M. (2005) Effects of aging on the regularity of physiological tremor. *J Neurophysiol, 93*, 3064–74.

Stålberg, E., Ekstedt, J (1973) Single fibre EMG and microphysiology of the motor unit in normal and diseased human muscle. In New Developments in EMG and Clinical Neurophysiology(Ed, Desmedt, J.) S. Karger, Basel, pp. 113–129.

Syrovy, I. and Hodny, Z. (1992) Non-enzymatic glycosylation of myosin: effects of diabetes and ageing. *Gen Physiol Biophys, 11*, 301–7.

Termin, A., Staron, R. S. and Pette, D. (1989) Myosin heavy chain isoforms in histochemically defined fiber types of rat muscle. *Histochemistry, 92*, 453–7.

Thiele, B. and Stalberg, E. (1975) Single fibre EMG findings in polyneuropathies of different aetiology. *J Neurol Neurosurg Psychiatry, 38*, 881–7.

Thompson, L. V. and Brown, M. (1999) Age-related changes in contractile properties of single skeletal fibers from the soleus muscle. *J Appl Physiol, 86*, 881–6.

Thompson, L. V., Durand, D., Fugere, N. A. and Ferrington, D. A. (2006) Myosin and actin expression and oxidation in aging muscle. *J Appl Physiol, 101*, 1581–7.

Tomonaga, M. (1977) Histochemical and ultrastructural changes in senile human skeletal muscle. *J Am Geriatr Soc, 25*, 125–31.

Tracy, B. L., Maluf, K. S., Stephenson, J. L., Hunter, S. K. and Enoka, R. M. (2005) Variability of motor unit discharge and force fluctuations across a range of muscle forces in older adults. *Muscle Nerve, 32*, 533–40.

Vaillancourt, D. E., Larsson, L. and Newell, K. M. (2003) Effects of aging on force variability, single motor unit discharge patterns, and the structure of 10, 20, and 40 Hz EMG activity. *Neurobiology of Aging, 24*, 25–35.

van Rooij, E., Liu, N. and Olson, E. N. (2008) MicroRNAs flex their muscles. *Trends Genet, 24*, 159–66.

van Rooij, E., Quiat, D., Johnson, B. A., Sutherland, L. B., Qi, X., Richardson, J. A., Kelm, R. J., Jr. and Olson, E. N. (2009) A family of microRNAs encoded by myosin genes governs myosin expression and muscle performance. *Dev Cell, 17*, 662–73.

Welle, S., Bhatt, K. and Thornton, C. A. (2000) High-abundance mRNAs in human muscle: comparison between young and old. *J Appl Physiol, 89*, 297–304.

Welle, S., Thornton, C., Jozefowicz, R. and Statt, M. (1993) Myofibrillar protein synthesis in young and old men. *Am J Physiol, 264*, E693–8.

Wilson, M. G. A., Woledge, R.C. (1985) Lack of correlation between twitch contraction time and velocity of unloaded shortening in fibres of frog anterior tibialis muscle. *J Physiol (Lond), 358*, 81P.

Viner, R. I., Ferrington, D. A., Huhmer, A. F., Bigelow, D. J. and Schoneich, C. (1996) Accumulation of nitrotyrosine on the SERCA2a isoform of SR Ca-ATPase of rat skeletal muscle during aging: a peroxynitrite-mediated process? *FEBS Lett, 379*, 286–90.

Wuerker, R. B., McPhedran, A.M., Henneman, E. (1965) Properties of motor units in a heterogeneous pale muscle (m. gastrocnemius) of the cat. *J Neurophysiol, 28*, 85–99.

Age-Related Decline in Actomyosin Structure and Function

LaDora V. Thompson

Abstract Aging is associated with a progressive decline of muscle mass, strength, and quality, a condition described as sarcopenia of aging. Despite the significance of skeletal muscle atrophy, the mechanisms responsible for the deterioration of muscle performance are only partially understood. The purpose of this chapter is to highlight cellular, molecular, and biochemical changes that contribute to age-related muscle dysfunction, particularly the molecular basis of contraction, changes in muscle protein structure assessed by electron paramagnetic resonance spectroscopy, oxidative damage from reactive oxygen species, and post-translational modifications in key contractile proteins. Age-related changes in the interaction between the contractile proteins, actin and myosin, provide insights into potential molecular mechanisms responsible for changes in muscle contractility with advancing age.

Keywords Actin • Cabonylation • Glycation • In vitro motility assay • Myosin • Nitrotyrosine • Oxidative stress • Post-translation modifications • Proteome • Reactive oxygen species • Single permeabilized muscle fibers

1 Introduction

Skeletal muscle *function* is a key component in performing activities of daily. For the older adult, skeletal muscle *dysfunction* represents a risk factor for frailty, loss of independence, and physical disability. Loss of mobility resulting from muscle weakness predicts major physical disability and all-cause mortality, and is associated with poor quality of life, social needs, and health care needs. The economic impact of sarcopenia and its detrimental correlates are immense.

L.V. Thompson (✉)
University of Minnesota, Medical School Program in Physical Therapy,
Department of Physical Medicine and Rehabilitation,
420 Delaware St, SE, Minneapolis, MN 55455, USA
e-mail: thomp067@umn.edu

G.S. Lynch (ed.), *Sarcopenia – Age-Related Muscle Wasting and Weakness*,
DOI 10.1007/978-90-481-9713-2_5,

The extent of age-related physiological and molecular changes is dependent on many factors. Epidemiological studies suggest multiple contributing factors, including neuronal and hormonal changes, inadequate nutrition, low-grade chronic inflammation, and physical inactivity. Physiological studies indicate that protein structure and function in addition to translational and transcriptional mechanisms are involved. Therefore, understanding the aging process requires systematic, multidisciplinary studies on physiological, biochemical, structural, and chemical changes in specific muscles. The purpose of this chapter is to highlight cellular, molecular, and biochemical changes that contribute to age-related muscle dysfunction, specifically age-related changes in the interaction between the contractile proteins myosin and actin.

2 Muscle Structure, Contraction, Plasticity Review

2.1 Muscle Structure

Skeletal muscles are composed of individual, multinucleated cells, containing specialized structures for excitation–contraction coupling. Each individual skeletal muscle cell, termed fiber, is more or less cylindrical, with diameters between 10 and 100 μm and up to a few centimeters in length. The fiber is composed of a bundle of myofibrils, each being a linear array or a string of sarcomeres (1 μm in diameter) from one end of the fiber to the other. The sarcomere, the primary contractile unit, is composed of interdigitating thick and thin myofilaments. Figure 1 is a schematic of a muscle, an individual fiber and one sarcomere.

The major components of the thick filament are myosin dimers, composed of two myosin heavy chains (Fig. 2). Each of the myosin heavy chains (~220 kD) has a globular "head" domain at the N-terminal and an alpha-helix at the C-terminal "tail" domain. The tail region is periodically interspersed with hydrophobic residues to give a "coiled-coil" type rod. The amino acids in the C-terminal are non-helical, which provide filament backbone stabilization. The tail regions aggregate into bipolar filaments to form the thick filament.

The N-terminal of the myosin heavy chain contains specific regions essential to myosin's contractile and enzymatic activity. The catalytic (ATP hydrolysis) and force-generating (interaction with actin) functions of myosin are located in its "head" region, specific to the subfragment-1 or S1 region (95 kD). In the catalytic domain there is a SH1–SH2 helix consisting of two critical cysteines, SH1 (Cys707) and SH2 (Cys697), both fast-reacting sulfhydryls. The myosin head region also contains the light chain (LC) domain, which contains the essential (ELC) and regulatory (RLC) light chains. The light chains of skeletal muscle myosin have regulatory functions such as calcium regulation, shortening velocity and the extent of actin-activation of myosin ATPase. Figure 2 describes the myosin filament, myosin dimer and the components of the S1 region.

The thin filaments are composed of three different types of protein: actin, tropomyosin, and troponin (Fig. 3). Actin is a globular protein (~40 kD), which contains specific sites of interaction with myosin. The globular actin or G-actin

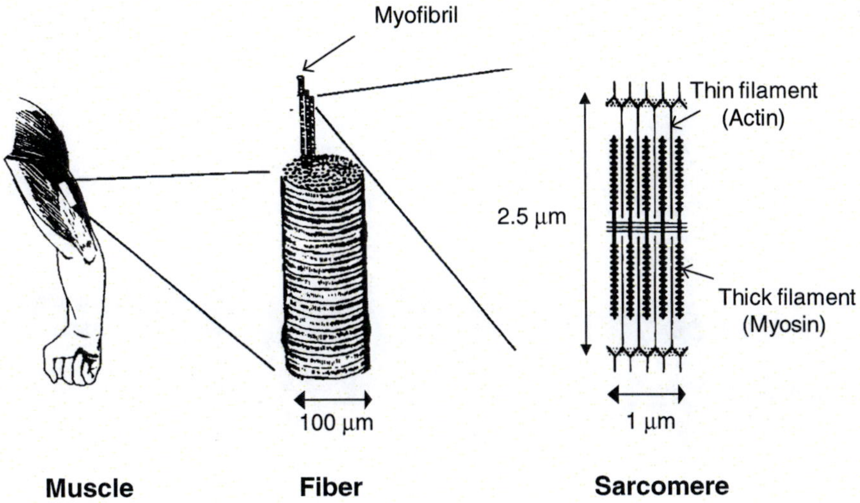

Fig. 1 Skeletal muscle structure. Skeletal muscles are composed of individual muscle fibers arranged in parallel. In this example, the individual fiber is ~100 μm in diameter. An individual muscle fiber is composed of many myofibrils arranged in parallel. The myofibril is made up of the major contractile unit, the sarcomere. The sarcomere is ~1μm in diameter and 2.5 μm in length. Individual sarcomeres are arranged in series to form a single myofibril. The sarcomere contains interdigitating thick and thin filaments. The main components of the thick filaments are myosin molecules. The major components of the thin filaments are actin monomers. With aging, the individual fibers decrease in cross-sectional area by decreasing the number of myofibrils

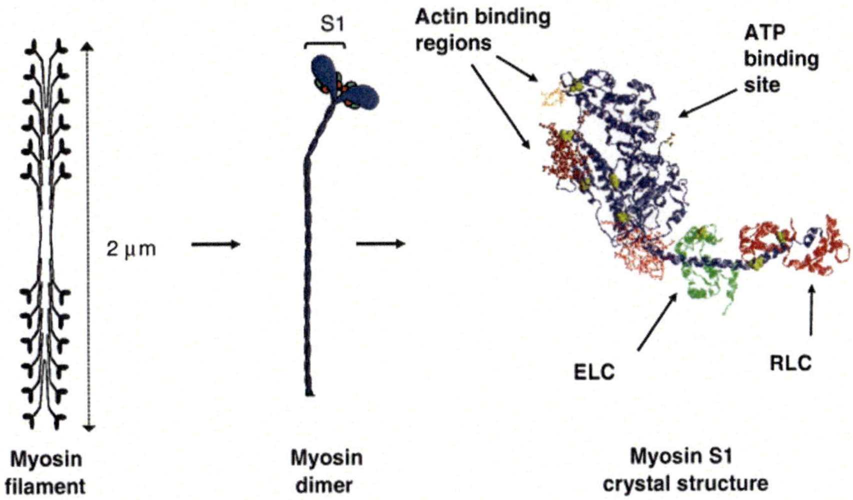

Fig. 2 Structure of skeletal muscle myosin (myosin class II). The myosin filament is composed of myosin molecules aggregated via their "tail" regions into a bipolar filament, ~2 μm in length. Skeletal muscle myosin forms a dimer composed of two heavy chains (MHC) containing regions involved in enzymatic and actin-binding functions of myosin (S1), two pairs of regulatory light chains (*RLC, red*) and two pairs of essential light chains (ELC, green). The crystal structure of myosin head shows that the catalytic and force-generating function of myosin is located in its "head" region, which contains the catalytic domain (with sites for ATP hydrolysis and interaction with actin), and the light chain domain, which contains the ECL and RLC light chains. In the catalytic domain there is a SH1–SH2 helix consisting of two critical cysteines, SH1 (Cys707) and SH2 (Cys697), both fast-reacting sulfhydryls

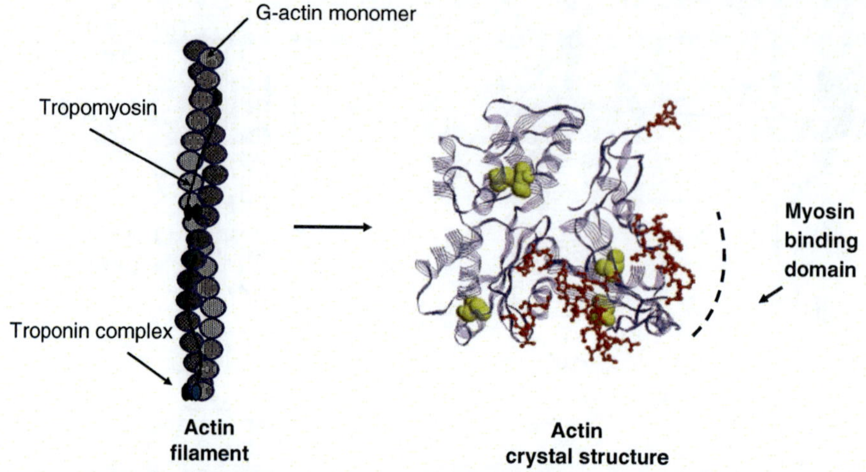

Fig. 3 Structure of skeletal muscle actin. The main component of the actin thin filaments is a helical polymer of globular actin monomers. Interdigitated between the actin strands are rod-shaped proteins termed tropomyosin. Attached to the tropomyosin at regular intervals is the troponin complex. The troponin complex is made up of three subunits: troponin-T (TN-T), which attaches to the tropomyosin; troponin-C (TN-C), which serves as a binding site for Ca^{2+} during excitation–contraction coupling (four Ca^{2+} can bind per TN-C); and troponin-I (TN-I), which inhibits the myosin binding site on the actin. The crystal structure of the actin monomer shows specific sites of interaction with myosin

subunits assembles into long filamentous polymers called F-actin forming two strands of an alpha helix. Interdigitated between the actin strands are rod-shaped proteins termed tropomyosin. There are 6–7 actin molecules per tropomyosin. Attached to the tropomyosin at regular intervals is the troponin complex (Fig. 3), which is made up of three subunits: troponin-T (TN-T), which attaches to the tropomyosin; troponin-C (TN-C), which serves as a binding site for Ca^{2+} during excitation–contraction coupling (four Ca^{2+} can bind per TN-C); and troponin-I (TN-I), which inhibits the myosin binding site on the actin. When Ca^{2+} binds to TN-C, there is a conformational change in the troponin complex such that TN-I moves away from the myosin binding site on the actin, thereby making it assessable to the myosin head. When Ca^{2+} is removed from the TN-C, the troponin complex resumes its inactivated position, thereby inhibiting myosin-actin binding.

2.2 *Molecular Basis of Muscle Contraction*

2.2.1 Force Generation, Structural States of Myosin, and Myosin ATPase

The basic process of muscle contraction is well understood and it is a result of cyclic interactions between myosin and actin, driven by the chemical free energy released from ATP hydrolysis. The interactions of the myosin head with actin, in the presence of ATP during the actomyosin ATPase cycle, results in sliding of thin filaments past thick filaments toward the center of the sarcomere (muscle contraction).

The biochemical steps of ATP hydrolysis during the cyclic interaction of actin with myosin are accompanied by a sequence of *structural transitions* in both proteins. Figure 4 is a schematic of structural and biochemical steps of ATP hydrolysis by myosin in the presence of actin (actomyosin ATPase cycle). Force is generated during transition of the myosin head from the states of weak binding (pre-powerstroke) to the states of strong binding (post-powerstroke) with actin. Particularly, in the absence of ATP and/or the presence of ADP, the myosin head forms a strong and well-ordered complex with actin. Binding of ATP to myosin produces a weaker complex where the catalytic domain and light chain domain of myosin are disordered. The release of phosphate (P_i) from myosin causes a structural transition in the catalytic and light chain binding domains, generates force, and initiates a new cycle (Prochniewicz et al. 2004; Thomas et al. 2009).

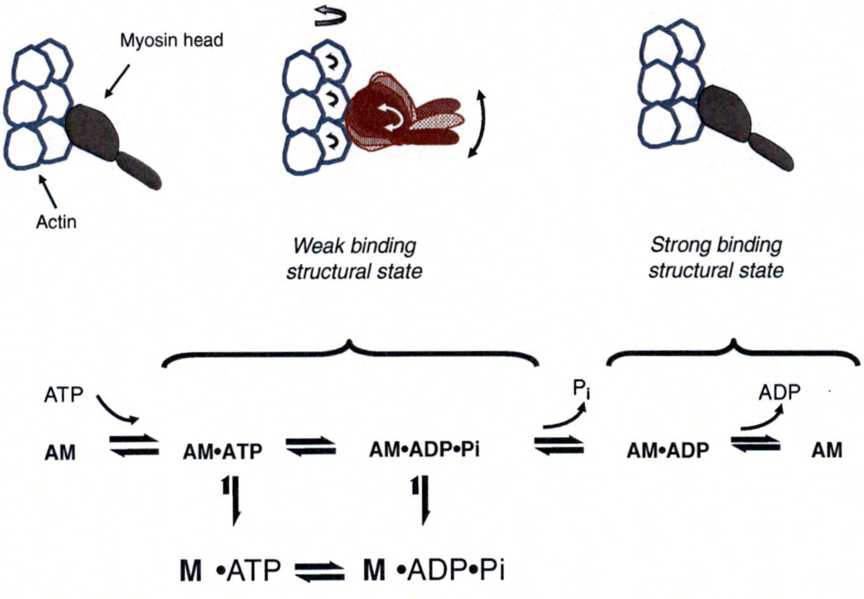

Fig. 4 Schematic of structural changes of the myosin head during muscle contraction, coupled to the ATPase cycle. The primary molecular event responsible for force generation in muscle is a global rotation of myosin heads (cross-bridges) coupled to myosin-catalyzed ATP hydrolysis. Interaction of the myosin head with actin in the presence of ATP results in sliding of thin filaments past thick filaments toward the center of the sarcomere, contracting the sarcomere. The biochemical steps of ATP hydrolysis during this cyclic interaction of actin with myosin are accompanied by a sequence of structural transition in both proteins. The biochemical step associated with the power stroke is the release of the hydrolysis product phosphate (P_i), while the release of ADP follows the execution of the power stroke. In the absence of ATP and/or the presence of ADP, the myosin head forms a strong and well-ordered complex with actin. Binding of ATP to myosin produces a weaker complex where the catalytic domain and light chain domain of myosin are disordered and identified as a weak binding structural state (red, pre-powerstroke). The release of phosphate (P_i) from myosin causes a structural transition (black, strong binding structural state) in the catalytic and light chain binding domains, generates force (post-powerstroke), and initiates a new cycle

The physiologically relevant step of force generation is a transition of the actomyosin complex from the states of weak interaction (AM•ATP and AM•ADP•P_i) to the states of strong interaction (AM•ADP and AM). In other words, force generation occurs when myosin and actin are in the strong-binding structural state (Fig. 4). Relevant to this chapter, force can decline if the system spends too much time in the weak-binding states, or the strong-binding states are weakened. Velocity can decline if the system spends too much time in the strong-binding states.

It is important to note that the hydrolysis of ATP requires myosin. Myosin is an enzyme which catalyzes the hydrolysis of ATP in the presence of actin, providing the source of free energy that drives muscle contraction. In the absence of actin, myosin ATPase activity is low and requires Ca^{2+}. Myosin ATPase activity is positively correlated with the myosin heavy chain isoform, with myosin heavy chain type I (MHC type I) hydrolyzing ATP at a slower rate than myosin heavy chain type II (MHC type II). Thus, any structural or chemical changes in myosin and actin that affect actomyosin ATPase activity by affecting the weak-to-strong actomyosin transition are likely to alter muscle function.

2.3 *Two Parameters of Contractility: Unloaded Shortening Velocity and Specific Force*

The maximal unloaded shortening velocity (V_o) is a mechanical parameter associated with the rate of cross-bridge cycling. V_o is directly related to, and dependent on, the activity of myosin ATPase. ATPase is predominantly determined by the MHC isoform because, as noted above, the MHC contains the catalytic site for ATPase activity. For individual skeletal muscle fibers, V_o varies with the MHC isoform with the hierarchy for V_o is type II > type I, with type IIb > IIx > IIa > I.

Muscle strength or the force-generating capacity varies directly with muscle (or fiber) cross-sectional area, and thus the ratio of force to muscle (or fiber) size is defined as *specific force*. Analysis of *specific force* allows for a comparison of the intrinsic capacity of the contractile unit between samples or after an intervention, even though muscle cross-sectional area may change.

2.4 *Muscle Plasticity*

Skeletal muscles and fibers are considered dynamic because they are capable of changing contractile properties in response to altered functional demands or changes in the pattern of recruitment. This plasticity is reflected by pronounced changes in muscle and single fiber strength, endurance and contractility as a result of an alteration in demand. Generally, increased contractile activities, e.g., chronic stimulation, the removal of a synergist muscle, or progressive resistance exercise training, are responsible for an increase in muscle strength. In contrast, decreased contractile activities, e.g., limb immobilization leads to a decrease in muscle strength. The peripheral adaptation most often observed with a change in muscle strength is an alteration in muscle mass.

3 Age-Related Changes in Contractility

Aging is associated with a progressive decline of muscle mass, strength, and quality, a condition often described as sarcopenia. The prevalence of sarcopenia in older adults is about 25% under the age of 70 years, and increases to 40% in adults 80 years or older (Baumgartner et al. 1998). Sarcopenia is a risk factor for frailty, loss of independence, and physical disability (Roubenoff 2000). The mechanisms responsible for the age-dependent contractile dysfunction are multi-factorial, resulting in altered cellular homeostasis (Prochniewicz et al. 2007; Thompson 2009). Having an understanding of the mechanisms contributing to altered cellular homeostasis leading to muscle dysfunction (e.g., weakness) provides a foundation for the testing of potential interventions (e.g., exercise).

In view of the many investigations elucidating the mechanisms responsible for muscle dysfunction, several points require attention. (1) Age-related deterioration of contractility is progressive, with the extent of changes being variable, depending on the muscle and age of the subjects. (2) The multitude of experimental approaches (e.g., permeabilized fiber preparation, intact fiber preparation, isolated proteins) enables specific hypotheses to be tested about the underlying mechanisms. (3) The experimental results from rodent studies, over the past 30 years, parallel the findings from human biopsy studies.

3.1 Single Fiber Contractility

Single skeletal muscle fiber contractility includes an array of contractile parameters (i.e., force-generating capacity, contraction velocity (V_o), power output) that are sensitive to the protein composition. It is possible to investigate single muscle fiber contractility using the permeabilized or skinned muscle fiber preparation. The permeabilized fiber preparation is a fiber that does not have intact membranes, so force generation and contraction velocity reflect directly the interactions of myosin and actin, exclusive of other factors in excitation–contraction coupling (e.g., sarcoplasmic reticulum Ca^{2+} release).

Two principal contractile parameters, specific force (P_o) and maximal unloaded shortening velocity (V_o), decrease progressively with age in studies using the permeabilized fiber preparation. The extent of deterioration depends on many factors, such as the fiber type composition of the single fiber (i.e., myosin heavy chain isoform type I or type II), the selected muscle (i.e., postural or phasic function), and the age group (i.e., young adult, aged).

Overall with progressive aging, skeletal fibers show a deficit in specific force (P_o) and in maximal unloaded shortening velocity (V_o), independent of the myosin heavy chain isoform (reviewed in Prochniewicz et al. 2007). It is important to note that single skeletal fibers with myosin heavy chain type II show faster age-related declines (or earlier) compared to single skeletal fibers with myosin heavy chain type I, revealing an *aging phenotype* between the fiber types (Fig. 5). The reduction in specific force (force generation normalized for cross-sectional area) with aging suggests *qualitative* as well as *quantitative* deficiencies in myosin and/or actin (defects in contractile protein). The decline in maximal unloaded shortening velocity with aging suggests alterations in the ability of the myosin ATPase to hydrolyze ATP.

The next questions to be answered, why do these occur? One possibility is that the decline in specific force is due to an alteration in the number of cross-bridges producing force, or in other words, it is possible that there is a reduction in the fraction of myosin heads in the strong-binding structural state during contraction. Thus, age-related structural or chemical changes in actin and myosin that affect the weak-to-strong actomyosin transition are potential candidates contributing to

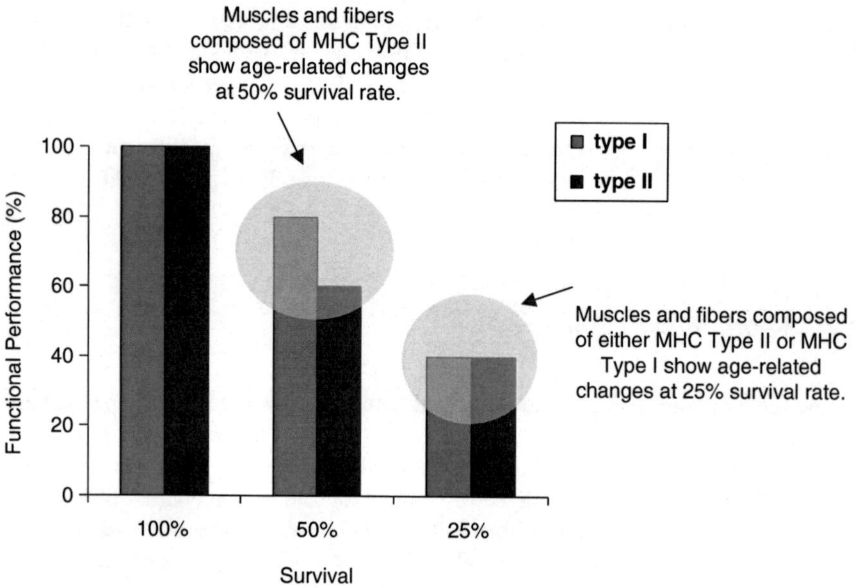

Fig. 5 Skeletal muscle aging phenotype. Contractility or functional performance of muscles and individual skeletal muscles, composed predominantly of type I (*gray*) or type II (*black*) myosin heavy chain (MHC), show characteristic aging phenotypes across the lifespan (defined as percentage of survival). At 50% survival, the contractility (e.g., specific force, velocity of shortening) of muscles and fibers composed of MHC type II show a greater reduction compared to muscle (fibers) composed of MHC type I, demonstrating that type II fibers are very sensitive to the aging process. At 25% survival, contractility is reduced in muscles (fibers) regardless of MHC isoform composition, demonstrating a susceptibility of both MHC isoform types to aging

age-related decline in specific force. In contrast to changes in structural properties as a mechanism to explain force declines with aging, an age-related slowing of myosin ATPase may explain the decline in unloaded shortening velocity.

The hypothesis that age-related deterioration of specific force involves structural changes was suggested by a series of studies showing changes in myosin structure can compromise muscle function. For example, oxidation of cysteine residues (thiols) in myosin decreases force production, velocity, and ATPase activity (Perkins et al. 1997). Because thiols are not involved in the catalytic mechanism of myosin ATPase, this strongly suggests that a structural perturbation in myosin occurs as a result of oxidation, culminating in altered myosin function. Specific structural changes within the myosin molecule have also been implicated in a variety of muscle disorders.

4 Age-Related Structural Changes in Muscle Protein Structure

4.1 EPR Spectroscopy

If age-related structural changes in myosin contribute to the decline in specific force, it is important to be able to determine myosin structural states during contraction. The determination of protein structure requires experimental technology with high resolution and specificity. Electron paramagnetic resonance (EPR) has both high resolution and specificity needed to analyze purified proteins, as well as large protein complexes and intact cells (reviewed in (Thomas et al. 2009). EPR is a spectroscopic technique that detects and quantifies *signals* corresponding to distinct protein structures and motions (dynamics). Special extrinsic probes are used to label proteins because EPR spectroscopy detects unpaired electrons, which are not found on most stable proteins. The special extrinsic probes, termed spin-labels, are stable nitroxide free radicals. The successful use of spin labels in the investigation of protein structure and dynamics usually requires the probe to be both small and site-specific. Particularly, the attachment of the spin-labels to the protein of interest is strategic and selective. Although the probes are attached to the protein, the probes cause very little steric perturbations and do not perturb the function of the protein.

In order for spectroscopic methods to test protein structure certain locations within the protein complex (e.g., actin-myosin) must be probed specifically. Two commonly used spin labels, maleimide and iodoacetamide derivatives, are specific for cysteine residues (Cys). A very small number of Cys residues are reactive in a given protein and Cys is a fairly uncommon amino acid. Recently, site-directed mutagenesis makes it possible to label virtually any amino acid residue by site-directed labeling. Site-directed spin labeling is also achieved by removing reactive Cys residues by mutation and introducing single Cys at the site of choice.

An important advantage of spin-labels is that high-quality EPR data can be obtained with these probes under physiological or near – physiological conditions, on small amounts of protein, and within a time frame of seconds (e.g., Lowe et al. 2001; Zhong et al. 2006). This is obviously advantageous for studying the relationship between protein structural dynamics and physiological functions. Depending on probe choice, sample preparation, labeling procedures, sample orientation, and other conditions of the experiment, spin-label probes report protein orientation (structural state), rotational motion, conformational changes, or information about the local environment of the protein. The EPR spectrum offers the possibility of high resolution, because different probe behaviors give rise to distinct spectral lines that can be resolved and quantified.

In principle, EPR involves the absorption of light by the biological sample (e.g., muscle fibers; Fig. 6; Thompson et al. 2001). The light source is a microwave klystron or diode in an EPR spectrometer. Microwaves (λ) are sent through a waveguide

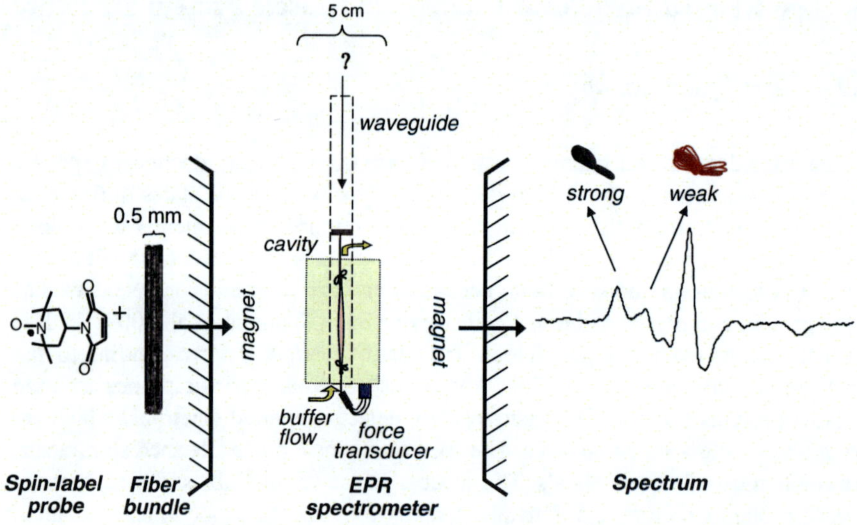

Fig. 6 Schematic of the major components of a muscle fiber EPR experiment. EPR with site-specific spin labeling is the only method that is able to directly determine the fraction of myosin heads in the weak-binding and strong-binding structural states during a muscle contraction. EPR has shown that force is produced when the myosin head changes from a weak-binding structure (*red*), in which the catalytic domain is dynamically disordered, to a strong-binding structure (*black*), in which the head is rigid. A nitroxide spin-label probe (*far left*) is reacted with a small bundle of permeabilized muscle fibers under conditions in which the probe is specific for SH-1 on the myosin head. A capillary tube containing the labeled fibers is fixed in a resonant cavity perpendicular to the magnetic field (located between two magnets), with one end of the fiber bundle attached to a force transducer and the other end stabilized to hold the fibers isometrically. The cavity is custom-designed to allow buffers to flow over the fibers at a designated flow rate that allows for adequate diffusion of substances (e.g., ATP and Ca^{2+}). During muscle rigor, relaxation, or maximal isometric contraction microwave energy (~10 GHz) is delivered into the cavity through a waveguide and is absorbed by the unpaired electrons in the labeled fibers. The result is a high resolution spectrum

into a resonator that contains the sample (cavity). Within the resonant cavity, a strong magnetic field induces a magnetic moment in the unpaired electrons (•) and results in the absorption of microwave radiation, termed "resonance". Next, the magnetic field is scanned and an EPR spectrum is obtained (the derivative of the absorption spectrum is plotted).

4.2 Myosin and EPR

EPR spectroscopy and site-directed spin labeling methods have been used considerably for investigating myosin because EPR is extremely sensitive to myosin's structural changes and EPR's high resolution permits the quantitative analysis of myosin's distinct structural states. These methods have allowed researchers in the field of skeletal muscle physiology to probe previously unexplored domains of muscle proteins, and to test and refine molecular models of muscle structure and function (reviewed in Thomas et al. 2009).

One method of specifically detecting the movements of myosin cross-bridges (structural information about a specific domain with myosin) is by using EPR spectroscopy in combination with site-specific spin-label probes that are introduced directly into the acto-myosin assembly of the sarcomere (to a distinct residue within myosin) in permeabilized muscle fibers (discussed in the previous section). The SH1–SH2 helix was among the first sites labeled covalently with probes on myosin, because SH1 (Cys707) is the most reactive Cys residue in the catalytic domain of the myosin head and can therefore be labeled specifically with a wide range of thiol reagents. Thus, SH1 (Cys707) is the site on the myosin cross-bridge most commonly used for labeling with spectroscopic probes. The mobility of the iodoacetamide spin labels (IASLs), attached to SH1, is sensitive to conformational changes near the myosin active site that occurs with ATP binding, hydrolysis, and P_i release.

The spin-label probes have been used to resolve and quantify distinct structural states of myosin that occur with muscle contraction in skeletal muscle fibers (reviewed in Thomas et al. 2009). The advantage of investigating muscle fibers with intact contractile units is the ability to directly correlate muscle function with protein (myosin) structure. Specifically, EPR has been used to show that the myosin head has two primary structures: a weak-binding (to actin) structural state that is dynamically disordered, in which no force is produced, and a strong-binding (to actin) structural state, in which force is produced (Figs. 4 and 6). The quantitative resolution by EPR of the weak-binding and strong-binding (pre- and post-powerstroke) structural states of myosin in active muscle, coupled with simultaneous measurement of muscle force, enables analysis of the coupling between thermodynamics and structural mechanics within the muscle filament lattice. Thus, the resolved EPR lines are correlated directly with intermediates in the myosin ATPase cycle.

Figure 7 is a representation of the low-field EPR spectra collected during maximal isometric contraction. During an experiment, three different spectra are collected, rigor, relaxation and contraction. Each rigor, relaxation, and contraction spectrum is collected at the same field positions (e.g., 1,024) and over the same gauss range (e.g., 38-gauss). After collection, the 38-gauss low-field EPR spectra are analyzed to determine the fraction of myosin heads in the strong-binding structural state *(x)* during muscle contraction. For each experiment (e.g., fiber bundle), the spectrum obtained during maximal isometric contraction (V_{Con}) are analyzed as a linear combination of the spectra obtained during rigor and relaxation using

$$V_{Con} = xV_{Rig} + (1 - x)V_{Rel} \tag{1}$$

where V_{Rig} (rigor) corresponds to all heads in the strong-binding structural state ($x = 1$) and V_{Rel} (relaxation) corresponds to all heads in the weak-binding structural state ($x = 0$). Thus, for the contraction spectrum, x is solved at each of the field

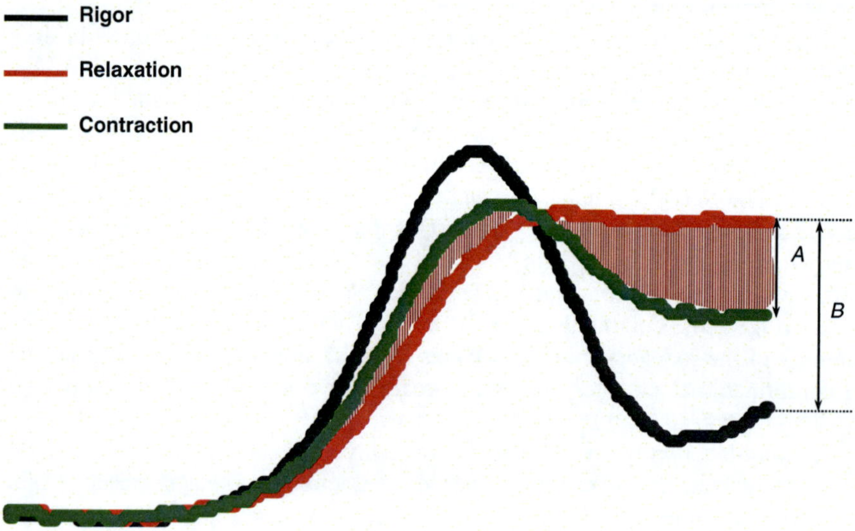

Fig. 7 Portion of the low-field EPR spectra of spin-labeled muscle fibers. The EPR spectra are collected (3,425 G central peak, 38 G sweep width, 5.0 G peak-to-peak modulation amplitude, and 16 mW microwave power) under conditions of rigor in which all heads are in the strong-binding structural state (*black*), relaxation in which all heads are in the weak-binding structural state (*red*), and contraction (*green*). Spectra obtained during maximal isometric contraction are analyzed as a linear combination of the spectra obtained during rigor and relaxation. For each fiber bundle, the spectrum obtained during maximal isometric contraction (V_{Con}) is analyzed as a linear combination of the spectra obtained during rigor and relaxation using $V_{Con} = xV_{Rig} + (1 - x)V_{Rel}$, where V_{Rig} (rigor) corresponds to all heads in the strong-binding structural state ($x = 1$), and V_{Rel} (relaxation) corresponds to all heads in the weak-binding structural state ($x = 0$). Thus for the contraction spectrum, x is solved at each of the 1,024 field positions to determine the fraction of myosin heads in the strong-binding structural state, using $x = (V_{Con} - V_{Rel})/(V_{Rig} - V_{Rel})$. A/B is equal to the mole fraction of x of myosin heads in the strong-binding state, as indicated in Eq. 2

positions (e.g., 1,024) to determine the fraction of myosin heads in the strong-binding structural state as follows

$$x = (V_{Con} - V_{Rel})/(V_{Rig} - V_{Rel}) = A/B$$

and as illustrated in Fig. 7.

4.3 Age-Related Structural Changes in Myosin Protein Structure

Figure 8 summarizes the major findings of one of the first studies to investigate age-related alterations in the distribution of myosin structural states (Lowe et al. 2001). The study hypothesized that the reduced force-generating ability of skeletal muscle fibers from aged animals was due to a decreased population of myosin heads in the strong-binding (force-generating) structural state during muscle contraction relative to skeletal muscle fibers from younger animals. The results clearly demonstrated that the only detected difference between young and aged fibers is in the mole fraction x of myosin heads in the strongly bound structural state. During a maximal isometric contraction (same length contraction), 30% fewer

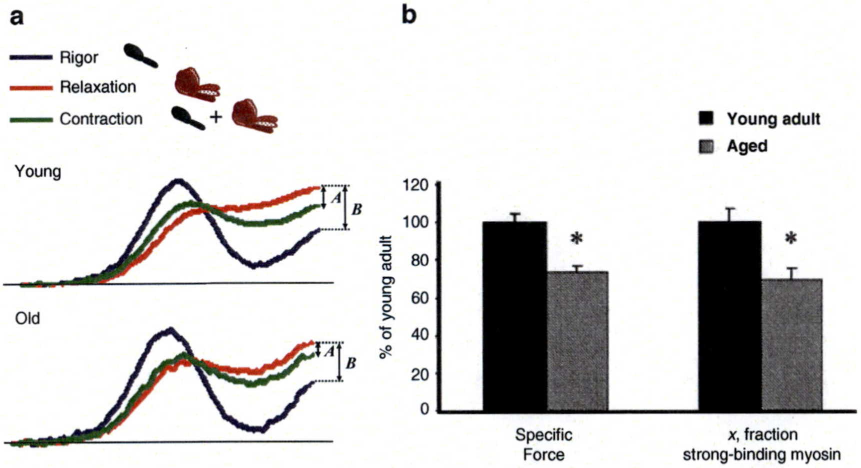

Fig. 8 Differences in fiber-specific force and myosin structural distribution with aging. Panel A – Representative EPR spectra of spin-labeled muscle fibers from young adult and aged rats. For the aged fibers the contraction spectrum (*green*) is closer to the relaxation spectrum (*red*) than is the contraction spectrum (*green*) to its respective relaxation spectrum (*red*) of the young adult fibers; i.e., A/B is less for the aged fibers. Panel B – Deficits of specific force (force/cross-sectional area) and fraction of myosin heads in the strong-binding structural state during maximal isometric contraction. * – Significantly different from young adult. The myosin structural changes can provide a molecular explanation for age-related decline in skeletal muscle force generation (Modified from Lowe et. al. 2001)

myosin heads are in the strong-binding, i.e., force-generating, structural state in fibers from aged animals ($x = 0.22$) than in fibers from younger animals ($x = 0.32$). Because it is usually proposed that the force generated during contraction is directly proportional to x and because the decrease in x (30%) is in good agreement with the decrease in specific force (27%), the study provides direct evidence that the decrement in force-generating capacity of fibers from aged animals is a direct result of a reduced fraction of myosin heads in the strong-binding structural state. This approach has effectively proven the number of force-generating cross-bridges per unit area during a muscle contraction is reduced with age, indicating that age-related changes in myosin structure represent a likely molecular mechanism underlying muscle weakness. Therefore, evidence suggests that the structure and function of myosin are altered with age.

5 ATPase Activity

5.1 Experimental Approaches

ATPase measurements are required for progress in understanding the molecular basis of age-related deterioration of contraction velocity because it is well known that shortening velocity is directly related to, and dependent upon, myosin ATPase activity. Several experimental approaches are used to determine ATPase activity, permeabilized skinned fibers, myofibrils, and purified proteins (actin and myosin). The myofibril preparation is a relatively simplistic experimental system for studying ATPase activity. Myofibrils are functional muscle units comprised of sarcomeres such that the thick-thin filament structure is maintained. Myofibrils in solution containing ATP and Ca^{2+} contract freely corresponding to the condition of an *unloaded* muscle fiber during shortening. The myofibrillar preparation allows quantitative determination of the protein concentration and specific enzymatic activity (ATPase).

In muscle, total protein content is mainly actin (18–22%) and myosin (43–50%) (Ingalls et al. 1998; Yates and Greaser 1983). However, since the interaction between actin and myosin in the muscle filament lattice also depends on other structural and regulatory proteins, a more direct assessment of age-related changes in the interaction between muscle myosin and actin requires specific measurements of biochemical and structural properties of actin and myosin. Purified actin and myosin can be used to determine ATPase activity too, permitting the determination of enzymatic changes in each of these proteins, independently of the other and without interference from other myofibrillar proteins.

Myosin ATPase activity can be determined at high and low salt concentrations, revealing critical information about the proteins (e.g., Bobkova et al. 1999). Myofibrillar ATPase activity at high salt concentration eliminates effects of actin and other proteins, and is sensitive to post-translational changes in myosin, particularly the covalent modification of specific cysteines and lysines, depending on the principal cation present (Ca^{2+} or K^+) and on the affected sites and modifying reagents. Although myofibrillar ATPase activity at high salt concentration provides critical information about myosin, it is not physiological. Physiological ATPase depends on the state of myosin as well as actin and is directly associated with contraction velocity or V_o (e.g., Barany 1967; Marston and Taylor 1980). Physiological ATPase can be determined in myofibrils or in purified myosin and actin (i.e., less than 0.3 M ionic strength, in the presence $MgCl_2$).

In addition to determining myosin ATPase with purified proteins, purified actin and myosin can be directly studied using the in vitro motility assay, a novel method for analyzing the interaction of actin and myosin at the single-molecule level. In this assay, isolated myosin molecules are immobilized on the glass surface, fluorescent actin filaments are added, and sliding movement of these filaments, initiated by addition of ATP, is directly observed under an optical microscope.

5.2 *Age-Related Changes in Myosin ATPase*

5.2.1 Myofibrils

To more directly assess the possibility of low myosin ATPase activity being a mechanism underlying age-reduced shortening velocity, Ca^{2+} – activated myosin ATPase activity is measured in freely contracting myofibrils corresponding to conditions of maximal unloaded shortening velocity. Under these conditions, Ca^{2+}-activated myosin ATPase activity is 16% lower with age and this decrease is similar to the reduced shortening velocity in permeabilized fibers from the same experimental group (Lowe et al. 2004). The results provide important evidence that changing actin-myosin interactions contribute to age-related inhibition of contractility.

Considering the contracting myofibrils and the respective ATPase activity data, it is possible to draw conclusions about how age influences the ATPase cycle. It has been shown that ADP release is the rate-limiting step during an isometric contraction in fibers and in myofibrils that are chemically cross-linked such that they do not shorten during contraction (Dantzig et al. 1992; Lionne et al. 2002). In contrast, both ADP release and P_i release have been implicated as the rate-limiting step during shortening contractions (Lionne et al. 1995, 2002). Therefore, it is probable that P_i or ADP release from myosin during a shortening contraction is reduced with age. Based on the available evidence, it is likely that ADP release is the rate-limiting factor with age because during a maximal isometric contraction there is an increase in the apparent rate constant for myosin detachment from actin with age and ADP release is the critical step controlling detachment (Fig. 4).

5.3 High-Salt ATPase

Age-related molecular changes in myosin are observed in high-salt ATPase activities of myofibrils and purified myosin (Fig. 9b) (Prochniewicz et al. 2005). Like single muscle fibers described previously, the age-induced changes are muscle-specific, strain-dependent and independent of changes in myosin isoform expression (reviewed in Prochniewicz et al. 2007). The observed changes in the ATPase activities provide strong indication of age-related post-translational modifications of myosin. However, high-salt ATPase activities do not provide sufficient information to determine the sites or nature of modification, nor to predict the functional consequences.

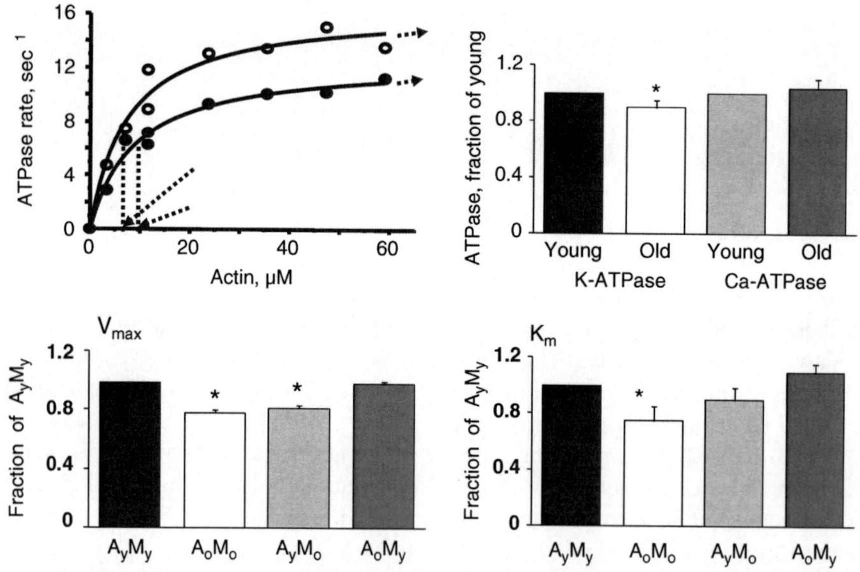

Fig. 9 Age-related molecular changes in myosin are observed in high-salt ATPase activities of purified protein. Panel A – Representative experiment with isolated myosin and actin proteins from young and old rats to determine Vmax (the maximum rate) and Km (actin concentration at half Vmax). The actin-activated myosin ATPase rate was measured at increasing concentrations of actin. The Vmax and Km were determined with O = young myosin, young actin, M_yA_y, Vmax = 16.43 ± 1.00 s^{-1}, Km = 7.44 ± 1.63 μM; • = old myosin, old actin, M_oA_o, Vmax = 12.55 ± 0.62 s^{-1}, Km = 8.98 ± 1.48 μM. Panel B. Age-related changes in myosin high-salt ATPase experiments. Age-related changes in the ATPase activity of myosin. K-ATPase was determined in the presence of 0.6 M KCl 50 mM Tris, and 10 mM EDTA; Ca-ATPase was determined in the presence of 0.6 M KCl, 50 mM Tris, and 10 mM $CaCl_2$. The ATPase rates for old myosin are expressed as fraction of the rates for young myosin. Panels C, D – Age-related changes in the actomyosin function are due primarily to changes in myosin. V_{max} and K_m for actin-activated myosin ATPase are determined as in panel A. Data are normalized to the value for young actin and young myosin, A_yM_y. A_oM_o = old actin and old myosin; A_yM_o = young actin and old myosin; A_oM_y = old actin and young myosin. * – Statistically significant (Modified from Prochniewicz et al. 2005)

5.4 *Purified Protein*

Biochemical experiments on purified actin and myosin provide more direct and detailed evidence about age-related changes in actin-myosin interactions with age, thus contributing to age-related inhibition of contractility (Prochniewicz et al. 2005). The studies show a decrease in two parameters of the actomyosin ATPase, V_{max} (activity extrapolated to infinite actin concentration) and K_m (the concentration of actin at half V_{max}) (Fig. 9a), providing direct support for the role of changes in the contractile proteins in the deterioration of muscle function. Subsequent mixing of actin and myosin from young and old muscle in four combinations shows that the age-related decrease in V_{max} is primarily due to changes in myosin. This decrease in V_{max} is consistent with the findings showing age-related alterations in the structural states of myosin (transitions from weak to strong interactions, Fig. 9c). Yet, the age-related decrease in K_m is due to changes in both proteins, as actin from young muscle attenuates the age-related decrease in K_m for myosin from old muscle (Fig. 9d). The changes in K_m are related to changes in the equilibria between actin and myosin-nucleotide complexes at the final stages of the cycle. The data from these experiments indicate that changes in actin, together with changes in myosin, are involved in the molecular mechanism of age-related deterioration of muscle contractility.

5.5 *In Vitro Motility Assay*

Studies using the in vitro motility assay show direct evidence supporting the role of molecular changes in myosin in age-related deterioration of contractility (D'Antona et al. 2003; Hook and Larsson 2000; Hook et al. 2001). In one study using purified myosin from the human vastus lateralis muscle demonstrates that the observed decrease of sliding speed of actin on myosin from aged muscle is comparable to the age-related decrease in the maximal unloaded shortening velocity V_o reported in single skeletal muscle fibers (D'Antona et al. 2003). Consistent with the human study, a decrease in actin sliding speed on myosin was confirmed in aging studies using purified proteins from muscles from rodents (Hook and Larsson 2000; Hook et al. 2001).

Interestingly, there are differences in the extent of deterioration between experimental preparations. For instance, the 12–25% decrease of actin sliding speed on myosin with age is much less pronounced than the decrease in V_o (47%) in permeabilized fibers (Hook and Larsson 2000; Hook et al. 2001; Li and Larsson 1996). The greater age-effects on fiber contractility than on the sliding velocity of purified actin filament in the in vitro motility reflect age-related changes in myosin as well as in the structural and thin filament proteins within the fiber lattice. Furthermore, the quantitative differences between age-related changes in contractile and enzymatic functions could also result from the complex mechanism of mechanochemical coupling in the actomyosin interaction.

5.6 *Single Permeabilized Skeletal Muscle Fibers/Isometric Contractions*

In contrast to contractions that allow shortening to occur, investigations focused on isometric contractions (muscle contracts without a change in length), in which the myosin ATPase rate is much slower and its kinetics are strongly affected by cross-bridge strain reveal changes in energetic efficiency and myosin cross-bridge kinetics with age (Lowe et al. 2002). Studies using the permeabilized fiber preparation, in which simultaneous measurements of force and ATPase activity were determined, show that fibers generate ~20 lower maximum force without changes in the ATPase activity (Lowe et al. 2002). This result indicates a decrease in the energetic efficiency, a partial uncoupling between ATPase activity and force generation, during isometric contraction in aged muscle. Moreover, the apparent rate constant for the dissociation of strong-binding myosin from actin was ~30% greater in fibers from aged animals, indicating that the lower force produced by fibers from aged animals is due to a greater flux of myosin heads from the strong-binding state to the weak-binding state during contraction (Fig. 10). The changes in cross-bridge kinetics are consistent with the observed structural changes in myosin during contraction with age.

Overall, the observed changes in contractility with age using permeabilized fibers, myofibrils, and purified proteins provide strong indication of age-related post-translational chemical modifications.

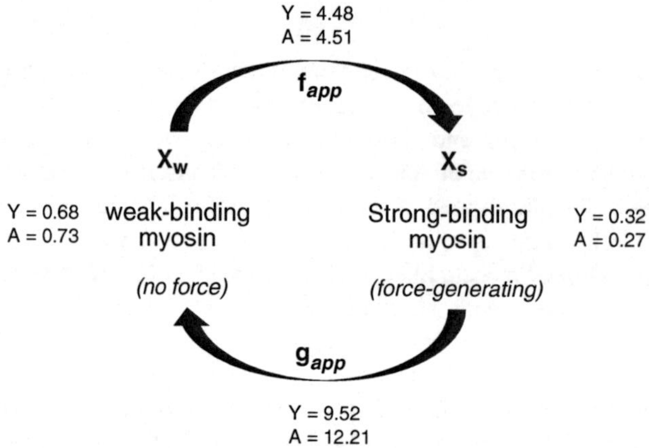

Fig. 10 Myosin structure and kinetics during a maximal isometric contraction are altered with age. Myosin structural data, from electron paramagnetic resonance spectroscopy experiments show that the age-related fractional reduction of myosin in the strong-binding structural state (x_s) during an isometric contraction is proportional to the decline in force in those fibers (i.e., a 16% decline in force generation with age relates to a 16% reduction in x_s). The age-related reduction in strong-binding (force-generating) myosin is due to an increase in the apparent rate constant for myosin detachment from actin (g_{app}) from 9.52 to 12.21. Y, young adult; A, aged; f_{app}, apparent rate constant for myosin attachment to actin; x_w, fraction of myosin heads in weak-binding structural state (Modified from Lowe et al. 2002)

6 Oxidative Stress

Age-related deterioration of muscle function may involve ‘damage’ of muscle proteins by reactive oxygen and nitrogen (ROS and NOS) species. Post-translational chemical modification of proteins affects the protein’s structural and functional integrity and in vitro studies where there is an elevation in ROS and NOS show contractile inhibition (e.g., Callahan et al. 2001; Lamb and Posterino 2003). In this chapter, the term ROS includes not only the oxygen radicals but also non-radical derivatives of O_2, such as H_2O_2. Hence, all oxygen radicals are ROS, but not all ROS are oxygen radicals. ‘Reactive’ is a relative term because superoxide anions ($O_2^{\bullet-}$) and H_2O_2 react fast and are very selective in their reactions, whereas, hydroxyl radical ($HO^{\bullet}$) reacts fast and is very promiscuous.

6.1 ROS Generation

ROS are generated in multiple compartments and by multiple enzymes within the cell. ROS are continually generated as byproducts of normal aerobic metabolism, yet can be produced to a greater extent under stress and pathological conditions, as well as taken up from the external environment. Examples of ROS include unstable oxygen radicals such as superoxide anion ($O_2^{\bullet-}$) and hydroxyl radical ($HO^{\bullet}$), non-radical molecules like hydrogen peroxide (H_2O_2) and peroxynitrite ($ONOO^-$). While nitric oxide ($NO^{\bullet}$) itself is not highly reactive or toxic, the reaction of $NO^{\bullet}$ with other molecules in the cell can produce more toxic species (O_2-derived species leading to the formation of reactive nitrogen oxide species). For instance, the reaction of $NO^{\bullet}$ with $O_2^{\bullet-}$ produces peroxynitrite ($ONOO^-$) which may be the primary mechanism by which $NO^{\bullet}$ causes cellular injury and alterations in function because it is a highly reactive species that can oxidize cellular lipids, proteins, and nucleic acids.

6.2 Role of Mitochondria

Intracellular ROS are primarily generated by the mitochondria. Mitochondria consume ~90% of a cell’s oxygen to support oxidative phosphorylation (OXPHOS) which is the major metabolic system for ATP. Specifically, the process uses the oxidation of NADH or $FADH_2$ to generate a potential energy for protons across the mitochondrial inner membrane. Subsequently, this potential energy for protons is used to phosphorylate ADP. At several sites along the electron transport chain, electrons can directly react with oxygen or other electron acceptors and generate free radicals.

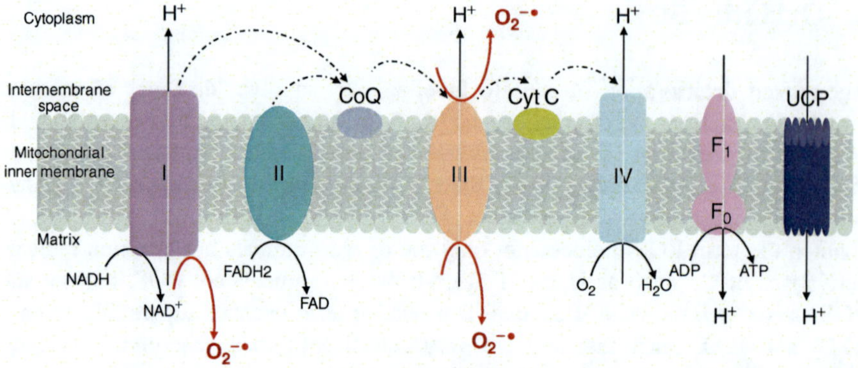

Fig. 11 Diagram showing the relationships of mitochondrial oxidative phosphorylation to energy production (ATP) and reactive oxygen species (ROS) production. Dashed lines indicate the flow of electrons donated from either NADH or FADH2 to complexes I–IV. As a result of electron transport, protons (H+) are translocated into the intermembrane space of the mitochondria creating a proton gradient across the inner mitochondrial membrane. The proton gradient is necessary to drive ATP production via ATP synthase, but superoxide anions are produced ($O_2^{\bullet-}$) at sites I and III. Uncoupling proteins (UCP) reduce the overall mitochondrial proton gradient. The different complexes of oxidative phosporylation, designated I to IV and coded by color, are complex I (NADH: ubiquinone oxidoreductase) encompassing a flavin mononucleotide and six Fe-S centers; complex II (succinate: ubiquinone oxidoreductase) involving a flavin adenine dinucleotide, three Fe-S centers, and a cytochrome b; complex III (ubiquinol: cytochrome c oxidoreductase) encompassing cytochromes b, c1 and the Rieske Fe-S center; complex IV (cytochrome c oxidase) encompassing cytochromes a + a3 and CuA and CuB; and the H+-translocating ATP synthase (F_1 and F_0) (Figure adapted from Balaban et al. 2005; Wolkow and Iser, 2006)

Figure 11 summarizes the mitochondrial electron transport chain (ETC). Carbohydrates (TCA cycle) and fats (β-oxidation), provide the reducing equivalents necessary to initiate electron transport through the mitochondrial ETC, a series of protein complexes that reside in the mitochondrial inner membrane (MIM; Balaban et al. 2005; Wolkow and Iser 2006). Two electrons donated from NADH + H^+ to complex I (NADH dehydrogenase) or from succinate to complex II (succinate dehydrogenase, SDH) are passed sequentially to the membrane-bound electron carrier, ubiquinone (coenzyme Q or CoQ) to give ubisemiquinone (CoQH•) and then ubiquinol (CoQH2). Ubiquinol transfers its electrons to complex III (ubiquinol: cytochrome c oxidoreductase), which transfers them to cytochrome c. From cytochrome c, the electrons flow to complex IV (cytochrome c oxidase, COX), which reduces molecular oxygen to water in the final step. Each of these electron transport chain (ETC) complexes incorporates multiple electron carriers. Complexes I, II, and III encompass several iron-sulfur (Fe-S) centers, whereas complexes III and IV encompass the b + c1 and a + a3 cytochromes, respectively. Electron transfer by complexes I, III and IV is coupled to proton transport across the MIM to the intermembrane space. Thus, electron transport through the ETC is coupled to the export of 2 (via Complex II) or 3 (via Complex I) protons into the mitochondrial intermembrane space.

Superoxide production occurs at two major sites along the electron transport chain, complex I and complex III, because large changes in the potential energy of the electrons, relative to the reduction of oxygen, occur. The relative contributions of complexes I and III to ROS production appear to be dependent on types of tissues, species and experimental conditions. The eight iron sulphur centers and/or the active site flavin mononucleotide are proposed to be the sites of ROS production at complex I (NADH coenzyme Q reductase) (Liu et al. 2002). The production of superoxide at complex I occurs at the matrix side of the inner membrane because the centers are proposed to be at this side of the membrane. The site of superoxide production at complex III is thought to be unstable ubisemiquinone molecules (St-Pierre et al. 2002). Superoxide produced by complex III is released to both the matrix and cytosolic sides of the mitochondria with about 80% going to the matrix and 20% going to the intermembrane space (Cadenas 2004).

ROS production is substrate-, tissue-, cell-, and organelle-specific. Under physiological conditions, about 0.2% of the total oxygen consumption is directed to ROS generation (St Pierre et al. 2002). More importantly, the mitochondrial metabolic state can influence the rate of ROS production (Frisard and Ravussin 2006). When oxygen consumption is low and the potential energy for protons is high (state 4), complexes of the ETC are in reduced states and superoxide production is highest. ROS production is increased when the electron carriers in the initial steps of the ETC harbor excess electrons, i.e., remain reduced, which can result from either inhibition of OXPHOS. Electrons residing in the electron carriers; for example, the unpaired electron of ubisemiquinone bound to the CoQ binding sites of complexes I, II, and III; can be donated directly to O_2 to generate superoxide anion.

6.3 Oxidative Stress/Damage to Macromolecules by ROS

In general, there is a balance between free radical production and the many antioxidant defense mechanisms within the skeletal muscle fibers. Thus, 'oxidative stress' can be viewed as a disturbance in the prooxidant-antioxidant balance in favor of prooxidant, leading to oxidative damage (Fig. 12). Increased levels of ROS can directly or indirectly damage macromolecules such as phospholipids, nucleic acids, and proteins. Since ROS are generated in the mitochondria, they can damage mitochondrial macromolecules either at or near the site of their formation. Mitochondria have two membranes, an inner highly proteinaceous membrane (80% protein) and an outer, porous membrane. Some proteins are attached loosely to the surface of the membranes whereas others are integral parts of the membrane (embedded). ROS damage to proteins as a direct result of oxidative stress or as a consequence of lipid peroxidation can result in protein cross-linking, degradation of proteins and loss of function because of the close physical association of phospholipids and proteins in mitochondrial membranes. ROS can damage other macromolecules, outside the mitochondria, yet within the muscle fiber. Figure 12 is schematic of the sarcomere showing how increased damage to actin and myosin may potentially interrupt actomyosin interactions which could result in skeletal muscle contractility deterioration.

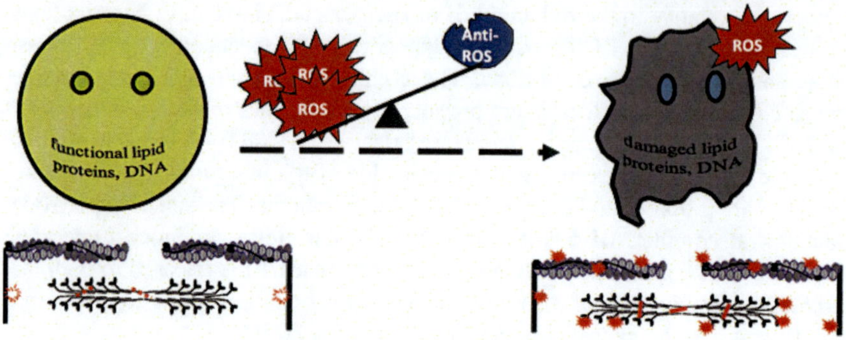

Fig. 12 Oxidative Stress. 'Oxidative stress' is a disturbance in the prooxidant–antioxidant balance in favor of prooxidant, leading to oxidative damage. Increased levels of reactive oxygen species can directly or indirectly damage macromolecules such as phospholipids, nucleic acids, and proteins. In the skeletal muscle sarcomere, increased damage to actin and myosin has potential to interrupt actomyosin interaction resulting in skeletal muscle contractility deterioration

6.4 *ROS Attack: Proteins*

Almost all amino acid residues in a protein can be oxidized by ROS. Oxidative products of amino acid residues include the formation of disulfide bonds at cysteine residues, carbonyl derivatives, and many others oxidized residues, such as methionine sulfoxide. These oxidative modifications lead to functional changes in various types of proteins, which have substantial physiological impact. For instance, oxidative damage to enzymes causes a modification of their activity, while oxidant-derived injury to structural proteins and chaperones produce protein aggregation.

Specifically, the accumulation of damaged proteins is dependent upon the balance between many different processes including: (1) the rate of ROS synthesis by any one of the numerous mechanisms; (2) the ability of various antioxidants to scavenge ROS; (3) the ability to repair nucleic acid damage leading to generation of altered proteins that are highly sensitive to oxidation; (4) the concentrations of proteases that degrade oxidized forms of proteins); (5) the generation of cross-linked proteins that inhibit the proteolytic degradation of oxidized proteins; (6) and the ability to repair oxidation of sulfur-containing amino acid residues of proteins.

6.5 *4-Hydroxy-2-nonenal (HNE)*

Mechanisms of damage and/or cell signaling can be direct, for example through the effects of superoxide, or can be introduced into proteins by reaction with aldehydes formed during lipid peroxidation (e.g. 4-hydroxy-2-nonenal or malondialdehyde that react with the ϵ-amino group). 4-hydroxy-2-nonenal, HNE, is a reactive aldehyde that originates from the peroxidation of membranes and forms a mixture of adduct types on the side-chains of cysteine, lysine, and histidine through a Michael-type nucleophilic addition. The HNE adducts may inhibit protein function. For example, the adenine nucleotide transporter is particularly susceptible, as is the matrix enzyme, aconitase (Yan and Sohal 1998). In both, the degree of damage (measured as protein carbonyls) is correlated with the loss of protein function (Yan and Sohal 1998).

6.6 *3-Nitrotyrosine (3-NT)*

3-nitrotyrosine (3-NT) has been identified as a stable marker of protein oxidative damage. This post-translational chemical modification can alter protein function and is associated with acute and chronic disease states. 3-nitrotyrosine, 3-NT, is formed when tyrosine is nitrated by peroxynitrite, a highly reactive molecule generated by the reaction of nitric oxide with superoxide. During muscle contraction the individual fibers are exposed to periodic fluxes of nitric oxide and superoxide leading to favorable conditions for the formation of peroxynitrite. Tyrosine nitration has been shown to inhibit protein function by altering a protein's conformation, imposing steric restrictions to the catalytic site, and preventing tyrosine phosphorylation (Cassina et al. 2000). Taken together, the functional significance of tyrosine nitration depends on two factors (1) the site of modification and (2) the extent of the protein population containing functionally significant modifications.

6.7 *Oxidative Stress and Muscle Dysfunction*

Skeletal muscle is vulnerable to oxidative stress for several reasons. First, skeletal muscle proteins are exposed to stress during contraction because there is rapid and coordinated changes in energy supply and oxygen flux. Subsequently, there is an increase in electron flux and leakage from the mitochondrial electron transport chain. Second, the high concentration of myoglobin within skeletal muscle also plays a role because the heme-containing protein is known to confer greater sensitivity to free radical-induced damage to surrounding macromolecules by converting hydrogen peroxide to other more highly reactive oxygen species (Ostdal et al. 1997).

Skeletal muscle fiber-type differences in susceptibility to oxidative stress may be mechanistically related to the aging phenotype discussed earlier in the chapter (both fiber types show susceptibility to age-related dysfunction, but the time course of change is fiber type-dependent). There are fundamental metabolic differences between slow-twitch aerobic fibers and fast-twitch glycolytic fibers. In particular, the major energy pathway utilized in type I fibers occurs through oxidative metabolism, whereas the glycolytic pathway is the primary means for generating energy in type II fibers. Thus, type I fibers likely produce greater ROS via mitochondrial oxidative phosphorylation compared with type II fibers. To counter the effects of ROS, type I fibers have higher antioxidant capacities that prevent or attenuate oxidative damage (Ji et al. 1998; Ji 2001, 2002; Reid and Durham 2002). Although type II fibers may generate lower levels of ROS during metabolism than do type I fibers, type II fibers may be more susceptible to oxidative stress because their antioxidant defenses are less robust.

7 Age-Related Post-translational Modifications of Proteins

7.1 Aging

With aging, under basal skeletal muscle conditions oxidant production is increased and the redox state shifts to a more oxidative environment (Ji et al. 1998; Ji 2001). While some antioxidants are increased in aging skeletal muscle, the extent of increase is muscle-specific and not global to all enzymes. Thus, the burden of defending against the increased load of free radicals may be greater than the compensatory change in antioxidants. If the antioxidant system is inadequate and key skeletal muscle proteins are modified, the proteasome must remove damaged proteins (one of the major degradation pathways for damaged proteins in skeletal muscle). Yet, the proteasome function in muscle declines with aging (Husom et al. 2004, 2005). Thus, the fundamental changes in cell redox status and the ability to remove free radical damaged proteins likely contribute to the age-related changes in muscle contractility discussed above.

7.2 Myosin and Actin – Key Contractile Proteins and Post-translational Modifications

As discussed earlier, in aged muscle, there is a reduction in the fraction of myosin heads in the strong-binding structural state, such that there are fewer myosin-actin interactions capable of generating force (Fig. 8). In addition, a significant age-related inhibition of myosin ATPase, critical for generating force, is reported from investigations of isolated proteins (myosin and actin) (Fig. 9). Thus, mechanisms that decrease or interrupt the interaction of myosin and actin are likely to explain the age-related reduction in force-generating capacity.

One mechanism that may play a role in the age-related decline in contractility (e.g., interrupt the interaction of myosin and actin) is an *accumulation* of damage from post-translational chemical modifications (e.g., oxidative damage) to myofibrillar proteins (Fig. 12) because in vitro studies demonstrate that peroxynitrite impairs both energetics and contractility of permeabilized muscle fibers (Callahan et al. 2001). Age-related oxidative damage of myosin and actin are probably accumulating in muscle as a consequence of decreased muscle protein turnover (Balagopal et al. 1997; Ferrington et al. 2005; Husom et al. 2005).

To date, studies of in vivo oxidative modifications of myosin and actin focus on selective markers of oxidative damage such as nitration, formation of HNE adducts, oxidation of cysteines and glycation. In a systematic study of in vivo oxidative modifications of myosin and actin and aging, both nitration and the formation of HNE adducts were evaluated (Thompson et al. 2006). The levels of these two markers of oxidative stress, 3-nitrotyrosine and HNE-adducts, on myosin and actin did not increase with age. This finding suggests that *accumulation* of oxidative damage to these two key myofibrillar proteins does not occur with age.

In contrast to the similar amounts of 3-nitrotyrosine and HNE-adducts on myosin and actin with age, the results of several other investigations suggest that actin and myosin have protein-specific differences in susceptibility to oxidation (Kaldor and Min 1975; Prochniewicz et al. 2005; Srivastava and Kanungo 1982). Studies on purified proteins show an age-related decrease in cysteine content in myosin, but cysteine content of actin is unaffected by age (Prochniewicz et al. 2005). The implication of this finding for muscle contractility depends on the still unknown localization of oxidized sites. Oxidation of one or two reactive myosin cysteines (Cys 707 and Cys 696) could result in significant deterioration of muscle contractility, but myosin contains about 40 cysteines, and the functional role of the majority is not known (Bobkov et al. 1997; Crowder and Cooke 1984).

7.3 SERCA and Post Translational Modifications

The sarco/endoplasmic reticulum Ca-ATPase (SERCA) is a membrane protein responsible for the active transport of calcium from the cytosol into the sarcoplasmic reticulum lumen, thus removing Ca^{2+} from the vicinity of the contractile proteins and causing muscle relaxation. Therefore, changes in the Ca-ATPase function have a direct impact on muscle performance. Ca-ATPase function decreases in an age-dependent manner.

The SERCA protein is probably the most extensively investigated muscle protein, from a biochemical perspective with aging. These investigations focus on what sites are vulnerable to oxidative stress, and how the modification or damage alters protein function with increasing age. Normal aging of skeletal muscle is associated with increased nitration; in particular, specific nitration of the SERCA2a isoform in slow-twitch muscle (Viner et al. 1999). Tyrosine nitration increases by at least threefold in skeletal muscle during normal aging, and correlates with a 40% loss in Ca^{2+}-ATPase activity during normal aging. Mass spectrometry analysis reveals an age-dependent accumulation of 3-NT at positions

294 and 295 of the SERCA2 protein, suggesting that these tyrosines play a critical role in muscle function. In vitro studies also demonstrate that SERCA2a is inherently sensitive to tyrosine nitration with concomitant functional deficits. Because the physiological role of the Ca-ATPase is to mediate muscle relaxation, the consequence of nitration-induced inhibition of SERCA2a most likely explains the slower contraction and relaxation times observed in skeletal muscle with normal aging.

Aging also leads to a partial loss of SERCA1 isoform activity, and a molecular rationale for this phenomenon may be the age-dependent oxidation of specific cysteine residues (Viner 1999a, b). Mapping of the specific cysteine residues reveals nine cysteine residues targeted by age-dependent oxidation in vivo, and six cysteine residues partially lost upon oxidant treatment in vitro. Interestingly, the residues affected in vivo do not completely match those targeted in vitro, suggesting that modification of some residues do not contribute significantly to the loss of SERCA function with age. Taken together, these studies provide some insights about the molecular mechanisms responsible for age-related alterations in calcium regulation in skeletal muscle.

7.4 Aging Skeletal Muscle Phenotype – Nitration and Skeletal Muscle Proteins

One goal of global proteomic experiments in the field of aging is the identification and functional characterization of post-translationally modified proteins in vivo, and to determine whether such modifications are mechanistically related to specific aging phenotypes. The recent development of high resolution separation techniques and mass spectrometry (MS) instrumentation permits the identification of functionally important post-translational protein modifications occurring during aging.

In order to evaluate the role of oxidative stress and the skeletal muscle aging phenotype, comparison of damaged skeletal muscle proteins in two muscles, the soleus and semimembranosus, each composed of different skeletal muscle fiber types (Fugere et al. 2006). Specifically, the soleus muscle is composed of >90% type I fibers, whereas the semimembranosus is composed of >90% type IIB fibers. In these series of experiments, it was hypothesized that with aging the semimembranosus (type II) muscle would accumulate a greater amount of protein tyrosine nitration compared to proteins in the soleus (type I) muscle (Fugere et al. 2006). Previous in vitro studies show impairment in both energetic and contractility when permeabilized skeletal muscle fibers were exposed to peroxynitrite (Callahan et al. 2001). Moreover, the extent of functional decline is consistent with age-induced changes in single fiber contractile properties, suggesting that protein nitration may contribute to underlying mechanism for the age-related functional decrement (Thompson and Brown 1999).

The results of this proteomic study revealed five modified proteins, identified by MALDI-TOF Mass Spectrometry and confirmed with MS/MS and Western immunoblotting included the sarcoplasmic reticulum Ca^{+2}-ATPase (SERCA2a), aconitase, β-enolase, TPI, and carbonic anhydrase III, exhibited an age-dependent increase in 3-NT content in both type I and type II muscles. Confirming the aging phenotype between the two different muscles, significant levels of 3-NT modification were present at an earlier age in the semimembranosus muscle.

The biological function of the identified proteins include energy production (TPI, β enolase, aconitase, carbonic anhydrase III), and calcium homeostasis (SR Ca-ATPase). Previous studies reveal that mitochondrial aconitase is one of the major intracellular targets of nitric oxide, and the decrease in aconitase activity has been attributed to the direct reactions of nitric oxide with the iron-sulfur cluster (Patel et al. 2003). In addition, previous studies demonstrate oxidative modifications of carbonic andydrase III in vivo with a concomitant decrease in catalytic activities in liver tissue. There is increasing evidence that links β-enolase and TPI as targets for nitration in Alzheimer's disease. Taken together, these studies provide some insights about the molecular mechanisms (disturbance in energy metabolism) responsible for the observed phenotypic changes in skeletal muscle.

7.5 Carbonylation

One prominent marker of oxidative stress in aging skeletal muscle is protein carbonylation. Protein carbonylation can occur through metal catalyzed oxidation. In this reaction metals (copper and iron) catalyze the formation of highly-reactive, short-lived hydroxyl radicals that modify nearby amino acids (e.g. proline, arginine, lysine, and threonine). Protein carbonylation can also occur through a reaction of nucleophilic amino acid side chains with lipid oxidation products (e.g., HNE). In this reaction lipid peroxidation leads to the generation of aldehyde-containing byproducts, which covalently modify nucleophilic amino acid side chains on proteins (cysteine, histidine and lysine).

There are several ways to identify carbonylated proteins including (1) immunoassays that are based on derivatization with 2,4-dinitrophenyhydrazine followed by treatment with anti-2,4-dinitrophenol antibodies and secondary peroxidase-labeled antibodies, and (2) biotin hydrazide for derivatization of proteins with carbonyl groups followed by advanced proteomic tools such as two-dimensional gel separation and detection with fluorescently labeled avidin, affinity enrichment with biotin–streptavidin liquid chromatography tandem mass spectrometric (LC-MS/MS) analysis, enrichment using avidin affinity chromatography, followed by LC-MS/MS, and enrichment using avidin affinity chromatography followed by iTRAQ-based quantitative proteomics (Fig. 13). Using enrichment protocols followed by advanced proteomic technology allows for the identification of proteins susceptible to carbonylation.

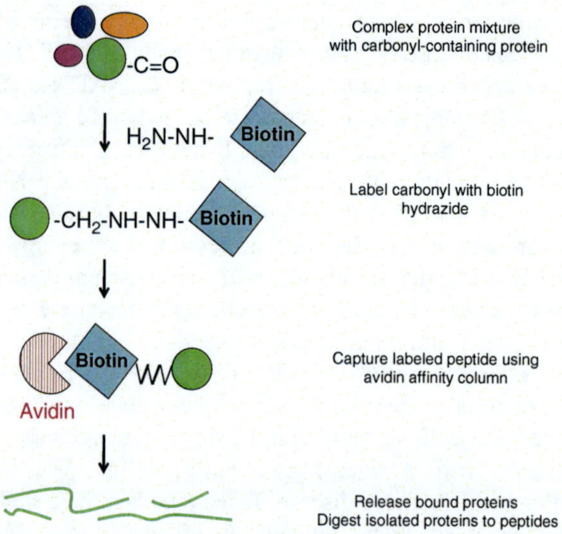

Fig. 13 Enrichment Strategy for the Identification of Carbonylated Proteins. In this strategy the carbonylated proteins are labeled with biotin hydrazide (derivatization of proteins with carbonyl groups) followed by enrichment using avidin affinity chromatography, and ulitimately identified by mass spectrometry

7.6 *Carbonylation: Identification of Susceptible Mitochondrial Proteins in Fast-Twitch and Slow-Twitch Muscle with Aging*

Differences in mitochondrial protein carbonylation may contribute to the age-related changes in muscle phenotype (fast- versus slow-twitch) described earlier in this chapter. Advanced quantitative proteomic profiling to identify proteins susceptible to carbonylation in a muscle type (slow- vs fast-twitch) and age-dependent manner yields very interesting results. With aging, fast-twitch muscle has twice as many carbonylated mitochondrial proteins compared to slow-twitch muscle (78 and 38 carbonylated proteins in the fast-twitch and slow-twitch muscle, respectively; Feng et al. 2008).

Bioinformatic analysis of the set of carbonylated proteins, using Ingenuity Pathway Analysis (IPA) to identify functions and canonical pathways, reveals that the carbonylated proteins belong to pathways and functional classes already known to be impaired in aging skeletal muscle. IPA is a knowledge database generated from peer-reviewed scientific publications that enables discovery of highly represented functions and pathways from large, quantitative data sets. Eight canonical pathways and six biological functions are common to both muscle types (Table 1). The carbonylated proteins unique to fast-twitch muscle map to two distinct pathway (cellular function/maintenance and cell death) and two distinct functions (tryptophan metabolism and synthesis/degradation of ketone bodies) in the IPA environment. In contrast, no significant functions or pathways are assigned to the carbonylated

Table 1 Ingenuity Pathway Analysis (IPA) pathways and functions significantly represented by carbonylated proteins

Canonical pathway	Function
Both muscle types	
Oxidative phosphorylation	Carbohydrate metabolism
Mitochondrial dysfunction	Cell signaling
Butanoate metabolism	Energy production
Fatty acid metabolism	Amino acid metabolism
Valine, leucine, and isoleucine degradation	Lipid metabolism
Citric cycle	Small molecule biochemistry
Fatty acid elongation	
Pyruvate metabolism	
Unique to Fast-twitch muscle	
Tryptophan metabolism	Cellular function and maintenance
Synthesis and degradation of ketone bodies	Cell death

proteins identified only in slow-twitch muscle. The finding of distinct pathways and functions in fast-twitch muscle is potentially significant, given the fact that fast-twitch muscle is known to show more rapid decline with age than slow-twitch muscle does.

7.7 *Age-Dependent Protein Carbonylation and Impaired Biochemical Functions*

Using a two-pronged proteomic strategy, determining changes in carbonylated proteins and changes in protein abundance with age, 20 of the identified susceptible proteins in fast-twitch muscle show significant increases in carbonylation with age. Although it is beyond the scope of this chapter to discuss each protein in detail, several proteins are highlighted. Voltage-dependent anion channel (VDAC) protein and its binding partner ADP/ATP translocase protein show significant increases in carbonylation with aging and map to “Cellular function and maintenance” within the IPA environment. VDAC enables transport of ions, such as calcium ions (Ca^{2+}), across the inner-mitochondrial membrane, critical to mitochondrial function. Interestingly, impaired mitochondrial cycling of Ca^{2+} is associated with aging skeletal muscle. Thus, it is possible to hypothesize that increased carbonyl modification of these proteins critical to mitochondrial inner membrane transport may contribute to this impaired cellular function in aged fast-twitch muscle.

IPA enables identification of biochemical pathways represented by proteins showing changes in carbonylation with age that may not be apparent via visual inspection of the list of proteins. There are 13 canonical pathways and 7 biological functions represented by the proteins that increase in carbonylation with age (Table 2). Although it is beyond the scope of this chapter to discuss each pathway and function, several pathways are highlighted below to demonstrate the valuable tool of IPA. For instance, proteins with enzymatic activity mapping to five of the steps in fatty acid metabolism show increased age-dependent carbonylation. The identification of

Table 2 Significant canonical pathways mapped to protein showing age-dependent quantitative changes by IPA in fast-twitch muscle

Canonical pathway	Function
Oxidative Phosphorylation	Carbohydrate metabolism
Mitochondrial dysfunction	Cell signaling
Fatty acid metabolism	Energy production
Valine, leucine, and isoleucine degradation	Amino acid metabolism
Citric cycle	Lipid metabolism
Fatty acid elongation	Small molecule biochemistry
Pyruvate metabolism	Cell death
Tryptophan metabolism	
Synthesis and degradation of ketone bodies	
Propanoate metabolism	
B-alanine metabolism	
Lysine degradation	
Glutathione metabolism	

proteins showing susceptibility to carbonylation within the fatty acid metabolism pathway is very interesting based on (1) lipid content is known to increase in aging skeletal muscle, and (2) aging skeletal muscle has a decreased ability to oxidize fatty acid for energy generation. Decreased fatty acid metabolism may increase the presence of toxic lipids within skeletal muscle tissue, leading to more carbonylation, setting up a feedback scenario by which carbonylation impairs function and leads to further lipid perioxidation and modification and dysfunction of these proteins.

7.8 *Glycation and Aging Skeletal Muscle*

Protein glycation is another likely explanation for skeletal muscle dysfunction with age. Advanced glycation end products (AGEs) are a diverse class of post-translational modifications stemming from reactive aldehyde reactions. Because of the highly diverse reaction pathways leading to AGE formation, AGEs with a variety of chemical structures have been identified. The accumulation of AGEs is associated in the pathogenesis of many degenerative diseases because AGEs reduces their susceptibility to degradation.

N^{ϵ}-(carboxymethyl)lysine (CML, a 1-carboxyalkyl group is attached to the epsilon amino group of a lysine residue) is the major AGE-product in vivo and is often used as a biomarker of damage and increased oxidative stress. CML is formed by either oxidative breakdown of Amadori products or via adduction of lipid aldehydes generated from peroxidation of membrane (Fig. 14a, b). CML-modified proteins, determined biochemically and immunohistochemically, have extracellular as well as intracellular deposition. They are found in plasma, renal tissues, and retinas of diabetic patients and renal failure patients (Misselwitz et al. 2002; Saxena et al. 1999; Uesugi et al. 2001; Dyer et al. 1993; McCance et al. 1993). The severity of the

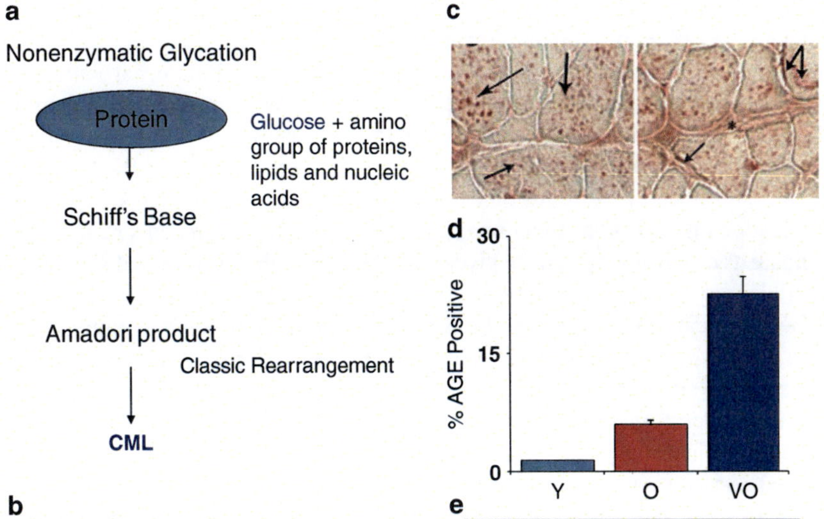

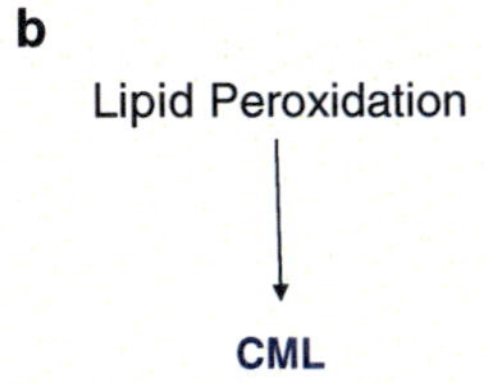

e

	Score	% coverage	Peptides
VDAC	112	54	11
B-enolase	88	37	11
CK	177	39	15
Actin	109	32	10
CAIII	74	46	6

Fig. 14 Glycation and Aging Skeletal Muscle (Snow et al. 2007), N-(carboxymethyl)lysine (CML, a 1-carboxyalkyl group is attached to the epsilon amino group of a lysine residue) is the major AGE-product in vivo and is often used as a biomarker of damage and increased oxidative stress. Panel A, B – CML is formed by either oxidative breakdown of Amadori products or via adduction of lipid aldehydes generated from peroxidation of membrane. Panel C – There are two characteristic patterns of the CML-immunolabeling of individual muscle fibers (intracellular punctuate labeling and labeling at the fiber periphery) in skeletal muscle from very old rats. Panel D – There is a tenfold increase in the percentage of individual fibers containing CML-modified proteins with age. Panel E – Using proteomic technology (mass spectrometry and bioinformatics) to identify the proteins susceptible to CML-modification, the CML-modified proteins are critical enzymes involved in energy production

tissue lesion (e.g., atherosclerosis) correlates with the tissue AGE concentration (Marx et al. 2004). With age, the concentration of CML in tissues increases significantly in cartilage and skin collagen (Verzijl et al. 2000). These findings suggest glycoxidation reactions and oxidative stress may be involved in the development of age-related deterioration of skeletal muscle function. Although the basal level of glycation in muscle protein is small (0.2 mmol/mol lysine) there is a tenfold increase in the percentage of individual fibers containing CML-modified proteins with age (Fig. 14d). There are two characteristic patterns of the CML-immunolabeling of individual muscle fibers (intracellular punctuate labeling and labeling at the fiber periphery (Fig. 14c) suggesting that there are targeted or susceptible proteins.

Using proteomic technology (mass spectrometry and bioinformatics) to identify the proteins susceptible to CML-modification, the CML-modified proteins are critical enzymes involved in energy production (Fig. 14e). Creatine kinase, carbonic anhydrase III, β-enolase, actin, and voltage-dependent anion channel 1 are susceptible to CML-modification, with β-enolase showing an accumulation of CML with age in skeletal muscle. Because lysines are at the exposed surface of β-enolase, the protein may function as a scavenger of CML, sparing other proteins from AGE-modification and potential functional impairment. β-enolase appears to be a good candidate as a scavenger because glycation of this protein has minimal impairment on cellular physiology (glycolytic flux).

The significance of glycation of other skeletal muscle proteins on muscle function is unknown, yet in vitro studies show that glycation decreases myosin and actin interactions (Ramamurthy et al. 2001). The glycation of myosin is detected in the skeletal muscle of aged rats (Syrovy and Hodny 1992). Interestingly, glycation of purified myosin from young rats decreases actin motility and also decreases K^+-activated and actin-activated ATPase activities (1). Thus, modification of lysine-rich nucleotide- and actin-binding regions of the myosin molecule is a possible mechanism for the functional loss.

In summary, the advancement of experimental technologies, quantitative proteomics and bioinformatics, identifies possible underlying mechanism responsible for the aging muscle phenotype. Thus, it will be possible to generate new hypotheses on ROS-induced mechanisms of post-translational chemical modifications (e.g., carbonylation) as well as possible connections between protein modifications and cellular functions already known to be impaired in aging muscle. These numerous hypotheses provide targets for future testing, a step closer to understanding the role of protein post-translational chemical modification in aging muscle decline.

It should be noted that with aging other oxidative modifications might accumulate and/or a site-specific amino acid modification of critical residues on these proteins could adversely affect function and contribute to muscle weakness. Additionally, an important limitation in the characterization of modified proteins from aged tissue is the fact that the data provide only a snapshot of a dynamic process, as proteins are constantly being synthesized and degraded in most tissues. Lastly, current knowledge about post-translational modification, and the techniques available to measure them, may not permit the quantitative analysis of all potential post-translational modifications of a given protein of interest as well as its functional characterization.

8 Age-Related Changes in Protein Expression Levels

8.1 Myosin and Actin

Stoichiometry between myosin and actin is critical for skeletal muscle contractility. Maintenance of the stoichiometry between myosin and actin depends on the balance between the protein synthesis and protein degradation. With aging, there is evidence for decreased myosin heavy chain synthesis rates and a loss in the regulation of the proteasome, the main protease responsible for degrading myofibrillar proteins. Thus, changes in rates of synthesis or degradation could lead to protein-specific declines in either actin or myosin content.

Detailed experiments, in both animal and human, show age-related decreases in myosin but not actin content in muscles composed of MHC type II (D'Antona et al. 2003; Thompson et al. 2006). The reduction in myosin protein expression without a change in actin content alters the optimal stoichiometry, leading to a decrease in the number of active cross-bridges contributing to force generation. In contrast, MHC content was unaffected by age in muscles composed of type I MHC isoform or composed of both type I and type II MHC isoforms indicating muscle-specific molecular changes (Moran et al. 2005; Thompson et al. 2006). Advanced proteomic technology has made possible analysis of age-related changes in the whole muscle proteome, yielding differentially expressed proteins with age (up-regulation and down-regulation). The comparison of results for different muscles shows that changes in the expression levels of contractile proteins are muscle specific. The main consequence of changes in expression levels of myosin and other contractile proteins is a change in stoichiometry. Thus, changes in protein stoichiometry may provide a mechanism for the observed aging muscle phenotypes (i.e., weakness in the fast-twitch muscle compared to the slow-twitch muscle).

Another mechanism that may explain age-related muscle dysfunction is a shift in skeletal muscle protein isoforms. As noted earlier in this chapter, myosin is a hexamer composed of two heavy chains, two regulatory light chains and two essential light chains such that specific protein isoforms confer contractility (e.g., MHC type II fibers contract faster than MHC type I fibers). Single permeabilized fiber experiments evaluating contractility combined with micro-analysis of isoform composition with SDS-PAGE detect age-related shifts in isoforms that are muscle and fiber-dependent, but these results do not explain the total changes in muscle contractility.

8.2 *Muscle Proteome-Protein Expression*

Over the past 4 years there is evidence of age-related changes in the whole skeletal muscle proteome. In two studies, using mass spectrometry to identify proteins, the analyses of the proteomes detect proteins differently expressed with age (Gelfi et al. 2006; Piec et al. 2005). In both studies, the expression levels for all three myosin light chains were down-regulated. Although more studies are needed to draw conclusions about the changes in the whole skeletal muscle proteome with age, a comparison of results for the two identified studies shows that changes in the expression levels of contractile proteins are muscle specific. As noted earlier, the main consequence of changes in expression levels of contractile proteins is a change in the stoichiometry, which could provide one of the explanations of age-related changes in contractile function.

9 Conclusion

Reduced muscle function and its attendant decrease in physical performance with age is a significant public health problem. Oxidative damage to key skeletal muscle proteins may be a contributing factor in sarcopenia. Age-related changes in the interaction between the contractile proteins actin and myosin provide some insights about potential molecular mechanisms responsible for age-related alterations in contractility. However, conclusive results require a more complete determination of the extent and location of oxidized sites, with parallel assessment of functional interactions of the proteins. An important limitation in the characterization of damaged proteins from muscle tissue is the fact that the data provide only a snapshot of a dynamic process, as proteins are constantly being synthesized and degraded in most tissues. Furthermore, current knowledge about post-translational modification due to oxidative stress, and the techniques available to measure them, may not permit the quantitative analysis of all potential modifications of a given protein of interest, as well as its functional characterization. It is likely that the future will see a significant increase in the number of specific modifications of proteins known, and an increase in our ability to associate them with specific aging phenotypes.

References

Balaban, R. S., Nemoto, S., Finkel, T. (2005). Mitochondria, oxidants, and aging. *Cell, 120*, 483–495.

Balagopal, P., Rooyackers, O. E., Adey, D. B., Ades, P. A., Nair, K. S. (1997). Effects of aging on in vivo synthesis of skeletal muscle myosin heavy-chain and sarcoplasmic protein in humans. *American Journal of Physiology. Endocrinology and Metabolism, 273*, E790–E800.

Barany, M. (1967). ATPase activity of myosin correlated with speed of muscle shortening. *The Journal of General Physiology, 50*(Suppl), 197–218.

Baumgartner, R. N., Koehler, K. M., Gallagher, D., Romero, L., Heymsfield, S. B., Ross, R. R., Garry, P.J., Lindeman R. D. (1998). Epidemiology of sarcopenia among the elderly in New Mexico. *American Journal of Epidemiology, 147*, 755–763.

Bobkov, A. A., Bobkova, E. A., Homsher, E., Reisler, E. (1997). Activation of regulated actin by SH1-modified myosin subfragment 1. *Biochemistry, 36*, 7733–7738.

Bobkova, E. A., Bobkov, A. A., Levitsky, D. I., Reisler, E. (1999). Effects of SH1 and SH2 modifications on myosin similarities and differences. *Biophysical Journal, 76*, 1001–1007.

Cadenas, E. (2004). Nitric oxide, mitochondrial function, and cell signaling. *Free Radical Biology and Medicine, 36*, S18–S18.

Callahan, L. A., She, Z. W., Nosek, T. M. (2001). Superoxide, hydroxyl radical, and hydrogen peroxide effects on single-diaphragm fiber contractile apparatus. *Journal of Applied Physiology, 90*, 45–54.

Cassina, A. M., Hodara, R., Souza, J. M., Thomson, L., Castro, L., Ischiropoulos, H., Freeman B.A., Radi, R. (2000). Cytochrome c nitration by peroxynitrite. *Journal of Biological Chemistry, 275*, 21409–21415.

Crowder, M. S. & Cooke, R. (1984). The effect of myosin sulfhydryl modification on the mechanics of fiber contraction. *Journal of Muscle Research and Cell Motility, 5*, 131–146.

D'Antona, G., Pellegrino, M. A., Adami, R., Rossi, R., Carlizzi, C. N., Canepari, M., Saltin, B., Bottinelli, R. (2003). The effect of ageing and immobilization on structure and function of human skeletal muscle fibres. *Journal of Physiology-London, 552*, 499–511.

Dantzig, J. A., Goldman, Y. E., Millar, N. C., Lacktis, J., Homsher, E. (1992). Reversal of the cross-bridge force-generating transition by photogeneration of phosphate in rabbit psoas muscle-fibers. *Journal of Physiology-London, 451*, 247–278.

Dyer, D. G., Dunn, J. A., Thorpe, S. R., Bailie, K. E., Lyons, T. J., McCance, D. R., Baynes, J.W. (1993). Accumulation of Maillard reaction-products in skin collagen in diabetes and aging. *The Journal of Clinical Investigation, 91*, 2463–2469.

Feng, J., Xie, H. W., Meany, D. L., Thompson, L. V., Arriaga, E. A., Griffin, T. J. (2008). Quantitative proteomic profiling of muscle type-dependent and age-dependent protein carbonylation in rat skeletal muscle mitochondria. *The Journals of Gerontology. Series A: Biological Sciences and Medical Sciences, 63*, 1137–1152.

Ferrington, D. A., Husom, A. D., Thompson, L. V. (2005). Altered proteasome structure, function, and oxidation in aged muscle. *Faseb Journal, 19*, 644–646.

Frisard, M. & Ravussin, E. (2006). Energy metabolism and oxidative stress – impact on the metabolic syndrome and the aging process. *Endocrine, 29*, 27–32.

Fugere, N. A., Ferrington, D. A., Thompson, L. V. (2006). Protein nitration with aging in the rat semimembranosus and soleus muscles. *The Journals of Gerontology. Series A: Biological Sciences and Medical Sciences, 61*, 806–812.

Gelfi, C., Vigano, A., Ripamonti, M., Pontoglio, A., Begum, S., Pellegrino, M. A., Grassi, B., Bottinelli, R., Wait, R., Cerretelli, P. (2006). The human muscle proteome in aging. *Journal of Proteome Research, 5*, 1344–1353.

Hook, P. & Larsson, L. (2000). Actomyosin interactions in a novel single muscle fiber in vitro motility assay. *Journal of Muscle Research and Cell Motility, 21*, 357–365.

Hook, P., Sriramoju, V., Larsson, L. (2001). Effects of aging on actin sliding speed on myosin from single skeletal muscle cells of mice, rats, and humans. *American Journal of Physiology. Cell Physiology, 280*, C782–C788.

Husom, A. D., Peters, E. A., Kolling, E. A., Fugere, N. A., Thompson, L. V., Ferrington, D. A. (2004). Altered proteasome function and subunit composition in aged muscle. *Archives of Biochemistry and Biophysics, 421*, 67–76.

Husom, A. D., Ferrington, D. A., Thompson, L. V. (2005). Age-related differences in the adaptive potential of type I skeletal muscle fibers. *Experimental Gerontology, 40*, 227–235.

Ingalls, C. P., Warren, G. L., Armstrong, R. B. (1998). Dissociation of force production from MHC and actin contents in muscles injured by eccentric contractions. *Journal of Muscle Research and Cell Motility, 19*, 215–224.

Ji, L. L. (2001). Exercise at old age: does it increase or alleviate oxidative stress? *Healthy Aging for Functional Longevity, 928*, 236–247.

Ji, L. L. (2002). Exercise-induced modulation of antioxidant defense. *Increasing Healthy Life Span: Conventional Measures and Slowing the Innate Aging Process, 959*, 82–92.

Ji, L. L., Leeuwenburgh, C., Leichtweis, S., Gore, M., Fiebig, R., Hollander, J., Bejma, J. (1998). Oxidative stress and aging – role of exercise and its influences on antioxidant systems. *Towards Prolongation of the Healthy Life Span, 854*, 102–117.

Kaldor, G. & Min, B. K. (1975). Enzymatic studies on skeletal myosin a and actomyosin of aging rats. *Federation Proceedings, 34*, 191–194.

Lamb, G. D. & Posterino, G. S. (2003). Effects of oxidation and reduction on contractile function in skeletal muscle fibres of the rat. *Journal of Physiology-London, 546*, 149–163.

Li, X. P. & Larsson, L. (1996). Maximum shortening velocity and myosin isoforms in single muscle fibers from young and old rats. *American Journal of Physiology – Cell Physiology, 270*, C352–C360.

Lionne, C., Brune, M., Webb, M. R., Travers, F., Barman, T. (1995). Time-resolved measurements show that phosphate release is the rate-limiting step on myofibrillar ATPases. *FEBS Letters, 364*, 59–62.

Lionne, C., Iorga, B., Candau, R., Piroddi, N., Webb, M. R., Belus, A., Travers, F., Barman, T. (2002). Evidence that phosphate release is the rate-limiting step on the overall ATPase of psoas myofibrils prevented from shortening by chemical cross-linking. *Biochemistry, 41*, 13297–13308.

Liu, Y. B., Fiskum, G., Schubert, D. (2002). Generation of reactive oxygen species by the mitochondrial electron transport chain. *Journal of Neurochemistry, 80*, 780–787.

Lowe, D. A., Surek, J. T., Thomas, D. D., Thompson, L. V. (2001). Electron paramagnetic resonance reveals age-related myosin structural changes in rat skeletal muscle fibers. *American Journal of Physiology – Cell Physiology, 280*, C540–C547.

Lowe, D. A., Thomas, D. D., Thompson, L. V. (2002). Force generation, but not myosin ATPase activity, declines with age in rat muscle fibers. *American Journal of Physiology – Cell Physiology, 283*, C187–C192.

Lowe, D. A., Husom, A. D., Ferrington, D. A., Thompson, L. V. (2004). Myofibrillar myosin ATPase activity in hindlimb muscles from young and aged rats. *Mechanisms of Ageing and Development, 125*, 619–627.

Marston, S. B. & Taylor, E. W. (1980). Comparison of the myosin and actomyosin ATPase mechanisms of the 4 types of vertebrate muscles. *Journal of Molecular Biology, 139*, 573–600.

Marx, N., Walcher, D., Ivanova, N., Rautzenberg, K., Jung, A., Friedl, R., Hombach, V., de Caterina, R., Basta, G., Wautier, M.P., Wautiers, J.L. (2004). Thiazolidinediones reduces endothelial expression of receptors for advanced glycation end products. *Diabetes, 53*, 2662–2668.

McCance, D. R., Dyer, D. G., Dunn, J. A., Bailie, K. E., Thorpe, S. R., Baynes, J. W., Lyons, T.J. (1993). Maillard reaction-products and their relation to complications in insulin-dependent diabetes-mellitus. *Journal of Clinical Investigation, 91*, 2470–2478.

Misselwitz, J., Franke, S., Kauf, E., John, U., Stein, G. (2002). Advanced glycation end products in children with chronic renal failure and type 1 diabetes. *Pediatric Nephrology, 17*, 316–321.

Moran, A. L., Warren, G. L., Lowe, D. A. (2005). Soleus and EDL muscle contractility across the lifespan of female C57BL/6 mice. *Experimental Gerontology, 40*, 966–975.

Ostdal, H., Skibsted, L. H., Andersen, H. J. (1997). Formation of long-lived protein radicals in the reaction between H_2O_2-activated metmyoglobin and other proteins. *Free Radical Biology and Medicine, 23*, 754–761.

Patel, H., Li, X., Karan, H. I. (2003). Amperometric glucose sensors based on ferrocene containing polymeric electron transfer systems – a preliminary report. *Biosensors & Bioelectronics, 18*, 1073–1076.

Perkins, W. J., Han, Y. S., Sieck, G. C. (1997). Skeletal muscle force and actomyosin ATPase activity reduced by nitric oxide donor. *Journal of Applied Physiology, 83*, 1326–1332.

Piec, I., Listrat, A., Alliot, J., Chambon, C., Taylor, R.G., Bechet, D. (2005) Differential proteome analysis of aging in rat skeletal muscle. *Faseb Journal, 19*, 1143–1145.

Prochniewicz, E., Walseth, T. F., Thomas, D. D. (2004). Structural dynamics of actin during active interaction with myosin: different effects of weakly and strongly bound myosin heads. *Biochemistry, 43*, 10642–10652.

Prochniewicz, E., Thomas, D. D., Thompson, L. V. (2005). Age-related decline in actomyosin function. *Journals of Gerontology Series A – Biological Sciences and Medical Sciences, 60*, 425–431.

Prochniewicz, E., Thompson, L. V., Thomas, D. D. (2007). Age-related decline in actomyosin structure and function. *Experimental Gerontology, 42*, 931–938.

Ramamurthy, B., Hook, P., Jones, A. D., Larsson, L. (2001). Changes in myosin structure and function in response to glycation. *FASEB Journal, 15*, 2415–2422.

Reid, M. B. & Durham, W. J. (2002). Generation of reactive oxygen and nitrogen species in contracting skeletal muscle – Potential impact on aging. *Increasing Healthy Life Span: Conventional Measures and Slowing the Innate Aging Process, 959*, 108–116.

Roubenoff, R. (2000). Sarcopenia and its implications for the elderly. *European Journal of Clinical Nutrition, 54*, S40–S47.

Saxena, A. K., Saxena, P., Wu, X. L., Obrenovich, M., Weiss, M. F., Monnier, V. M. (1999). Protein aging by carboxymethylation of lysines generates sites for divalent metal and redox active copper binding: Relevance to diseases of glycoxidative stress. *Biochemical and Biophysical Research Communications, 260*, 332–338.

Srivastava, S. K. & Kanungo, M. S. (1982). Aging modulates some properties of skeletal myosin ATPase of rat. *Biochemical Medicine, 28*, 266–272.

St-Pierre, J., Buckingham, J. A., Roebuck, S. J., Brand, M. D. (2002). Topology of superoxide production from different sites in the mitochondrial electron transport chain. *Journal of Biological Chemistry, 277*, 44784–44790.

Syrovy, I. & Hodny, Z. (1992). Nonenzymatic glycosylation of myosin – effects of diabetes and aging. *General Physiology and Biophysics, 11*, 301–307.

Thomas, D. D., Kast, D., Korman, V. L. (2009). Site-directed spectroscopic probes of actomyosin structural dynamics. *Annual Review of Biophysics, 38*, 347–369.

Thompson, L. V. (2009). Age-related muscle dysfunction. *Experimental Gerontology, 44*, 106–111.

Thompson, L. V. & Brown, M. (1999). Age-related changes in contractile properties of single skeletal fibers from the soleus muscle. *Journal of Applied Physiology, 86*, 881–886.

Thompson, L. V., Lowe, D. A., Ferrington, D. A., Thomas, D. D. (2001). Electron paramagnetic resonance: a high-resolution tool for muscle physiology. *Exercise and Sport Sciences Reviews, 29*, 3–6.

Thompson, L. V., Durand, D., Fugere, N. A., Ferrington, D. A. (2006). Myosin and actin expression and oxidation in aging muscle. *Journal of Applied Physiology, 101*, 1581–1587.

Uesugi, N., Sakata, N., Horiuchi, S., Nagai, R., Takeya, M., Meng, J., Saito, T., Takebayashi, S. (2001). Glycoxidation-modified macrophages and lipid peroxidation products are associated with the progression of human diabetic nephropathy. *American Journal Of Kidney Diseases, 38*, 1016–1025.

Verzijl, N., DeGroot, J., Oldehinkel, E., Bank, R. A., Thorpe, S. R., Baynes, J. W., Bayliss, M.T., Bijlsma, J.W.J., Lafeber, F., TeKoppele, J.M. (2000). Age-related accumulation of Maillard reaction products in human articular cartilage collagen. *The Biochemical Journal, 350*, 381–387.

Viner, R. I., Ferrington, D. A., Williams, T. D., Bigelow, D. J., Schoneich, C. (1999). Protein modification during biological aging: selective tyrosine nitration of the SERCA2a isoform of the sarcoplasmic reticulum Ca2+-ATPase in skeletal muscle. *Biochemical Journal, 340*, 657–669.

Viner, R. I., Williams, T. D., Schoneich, C. (1999). Peroxynitrite modification of protein thiols: Oxidation, nitrosylation, and S-glutathiolation of functionally important cysteine residue(s) in the sarcoplasmic reticulum Ca-ATPase. *Biochemistry, 38*, 12408–12415.

Wolkow, C. A. & Iser, W. B. (2006). Uncoupling protein homologs may provide a link between mitochondria, metabolism and lifespan. *Ageing Research Reviews, 5*, 196–208.

Yan, L. J. & Sohal, R. S. (1998). Mitochondrial adenine nucleotide translocase is modified oxidatively during aging. *Proceedings of the National Academy of Sciences of the USA, 95*, 12896–12901.

Yates, L. D. & Greaser, M. L. (1983). Quantitative-determination of myosin and actin in rabbit skeletal-muscle. *Journal of Molecular Biology, 168*, 123–141.

Zhong, S., Lowe, D. A., Thompson, L. V. (2006). Effects of hindlimb unweighting and aging on rat semimembranosus muscle and myosin. *Journal of Applied Physiology, 101*, 873–880.

Excitation-Contraction Coupling Regulation in Aging Skeletal Muscle

Osvaldo Delbono

Abstract Aging is associated with decreasing strength that can lead to impaired performance of daily living activities in the elderly. Functional and structural decline in the neuromuscular system has been recognized as a cause of this impairment and loss of independence, but the age-related loss of strength is greater than the loss of muscle mass in mammals, including humans, and the underlying mechanisms remain only partially understood. This chapter focuses on skeletal muscle excitation-contraction uncoupling (ECU), external calcium-dependent skeletal muscle contraction, the role of JP-45 and other recently discovered molecules of the muscle T-tubule-sarcoplasmic reticulum junction (triad) in excitation-contraction coupling (ECC), the neural influence of skeletal muscle, and the role of trophic factors–particularly insulin-like growth factor-I (IGF-1)–in structural and functional modifications of the motor unit and the neuromuscular junction with aging. A better understanding of the triad proteins involved in muscle ECC and nerve/muscle interactions and their regulation will lead to more rational interventions to delay or prevent muscle weakness with aging.

Keywords Skeletal muscle • Aging • Sarcopenia • Insulin-like growth factor 1 • Denervation

---

O. Delbono (✉)
Departments of Internal Medicine, Section on Gerontology and Geriatric Medicine, Department of Physiology and Pharmacology, Molecular Medicine and Neuroscience Programs, Wake Forest University School of Medicine, Medical Center Boulevard, Winston-Salem, NC 27157, USA
e-mail: odelbono@wfubmc.edu

G.S. Lynch (ed.), *Sarcopenia – Age-Related Muscle Wasting and Weakness*,
DOI 10.1007/978-90-481-9713-2_6,

1 Age-related Decrease in Strength is Greater than the Decrease in Muscle Mass in Humans

Weakness with old age can be partially attributed to a well-recognized decrease in muscle mass. Some studies in humans directly relate this diminished strength to muscle atrophy (Kent-Braun and Ng 1999), while others find that it is greater than the decrease in muscle mass (Lynch et al. 1999). For example, the decline in normalized force (force/muscle mass, Nm/kg) in the knee extensors has been found to follow a curve, starting at about 40 years and declining by about 28% from 40–49 to 70–79 years (Lynch et al. 1999). *In vitro* studies of single human muscle fiber contractility also reveal a decrease in specific force (force/cross-sectional area) with age (Frontera et al. 2000a). Therefore, the intrinsic force-generating capacity of the skeletal muscle per contractile unit may be impaired in aging mammals, including humans. Postulated mechanisms include alterations to the excitation-contraction coupling process (Delbono et al. 1995; Renganathan et al. 1998; Wang et al. 2000) and decreased actin-myosin cross-bridge stability (Lowe et al. 2002). For a review, see (Payne and Delbono 2004).

2 Excitation-Contraction Uncoupling

Skeletal muscle contraction is initiated by action potentials generated in the motor neuron and conducted via its axons, culminating in release of acetylcholine at the motor-end plate. Acetylcholine binds to nicotinic acetylcholine receptors, leading to an increase in sodium and potassium conductance in the end-plate membrane. End-plate potentials at the muscle membrane generate action potentials that are conducted to the sarcolemmal infoldings (T-tubules).

The transduction of changes in sarcolemmal potential to elevated intracellular calcium concentration is a key event that precedes muscle contraction (Dulhunty 2006). Electro-mechanical transduction in muscle cells requires the participation of the dihydropyridine receptor (DHPR) (Schneider and Chandler 1973) located at the sarcolemmal T-tubule. The DHPR is a voltage-gated L-type Ca^{2+} channel (dihydropyridine-sensitive), and its activation evokes Ca^{2+} release from an intracellular store (SR) through ryanodine-sensitive calcium channels (RyR1) into the myoplasm. The functional consequence of the reduced number, function, or interaction of these receptors is reduced intracellular calcium mobilization and force development (Delbono et al. 1997). Calcium binds to troponin C, leading to cross-linkages between actin and myosin and sliding of thin-on-thick filaments to produce force (Loeser and Delbono 2009). Uncoupling of the excitation-contraction machinery is a major factor in age-dependent decline in the force- generating capacity of individual cells (Delbono 2002).

Aging muscle fibers exhibit less specific force than those from young-adult or middle-aged animals but similar endurance and recovery from fatigue (Gonzalez et al. 2000b)

(González and Delbono 2001a, b). Whether excitation-contraction uncoupling results from altered neural control of muscle gene expression is not known. However, a series of studies support this concept. First, denervation results in a significant decrease in DHPR functional expression and alterations in excitation-contraction coupling in skeletal muscle from adult rats (Delbono 1992). Second, nerve crush leads to reduced levels of mRNA-encoding DHPR subunits and RyR1 in muscle (Ray et al. 1995), and studies show that both DHPR and RyR1 expression depend on skeletal muscle innervation (Kyselovic et al. 1994; Pereon et al. 1997b). Third, during development, DHPR mRNA levels change in relation to fiber innervation (Chaudari and Beam 1993). Fourth, myotube depolarization triggers the appearance of (+)-[^{3}H]PN 200–110 binding sites (Pauwels et al. 1987). Finally, exercise and chronic stimulation *in vivo* increase DHPR expression in homogenates of soleus and EDL muscles (Saborido et al. 1995; Pereon et al. 1997a). Thus, fiber-type composition, DHPR and RyR1, and excitation-contraction coupling seem to depend on nerve stimulation and muscle activity.

We are starting to understand how nerve stimulation of muscle activity influences muscle phenotype and the specific sarcolemmal-nuclear signaling pathways involved in muscle gene expression at different ages. Increasing evidence points to a decline in neural influence on skeletal muscle at later ages (Messi and Delbono 2003), leading to changes in muscle composition that result in excitation-contraction uncoupling (Payne and Delbono 2004).

3 IGF-1 Regulates Skeletal Muscle Excitation-contraction Coupling

IGF-1 may affect functional interactions between nerve and muscle by regulating transcription of the DHPRα_{1S} gene (Zheng et al. 2001). Although the DHPRα1 subunit is critical to excitation-contraction coupling, the basic mechanisms regulating its gene expression are unknown. To understand them, we isolated and sequenced the 1.2-kb 5′ flanking-region fragment immediately upstream of the mouse DHPRα_{1S} gene (Zheng et al. 2002). Luciferase reporter constructs driven by different promoter regions of that gene were used for transient transfection assays in muscle C2C12 cells. We found that three regions, corresponding to the CREB, GATA-2, and SOX-5 consensus sequences within this flanking region, are important for DHPRα_{1S} gene transcription, and antisense oligonucleotides against them significantly reduced charge movement in C2C12 cells (Zheng et al. 2002). This study demonstrates that the transcription factors CREB, GATA-2, and SOX-5 play a significant role in the expression of skeletal muscle DHPRα_{1S}.

Whether IGF-1 regulates these transcription factors and subsequent expression of the DHPRα_{1S} gene is not known. Using a approach similar to that described above (Zheng et al. 2002), we investigated the effects of IGF-1 on various promoter deletion/luciferase reporter constructs. They were transfected into C2C12 cells, and IGF-I effects were measured by recording luciferase activity. IGF-I significantly enhanced DHPRα_{1S} transcription, carrying the CREB binding site but not in CREB core binding site mutants. A gel mobility shift assay using a double-stranded oligonucleotide for the CREB site in the promoter region and competition experiments with excess unlabeled or mutated promoter oligonucleotide and unlabeled consensus CREB oligonucleotide indicate that IGF-1 induces CREB binding to the DHPRα_{1S} promoter. We prevented IGF-1 from mediating enhanced charge movement by incubating the cells with antisense but not sense oligonucleotides against CREB. These preliminary results support the conclusion that IGF-1 regulates DHPRα_{1S} transcription in muscle cells by acting on the CREB element of the promoter (Zheng et al. 2001). Confirming these results in skeletal muscle will be important as well as determining whether IGF-1/CREB signaling and the signaling pathway linking IGF-1R to CREB activation is preserved in aging mammals. We hypothesize that these effects are mediated by the direct action of IGF-1 on muscle cells, perhaps via activation of satellite cells (Barton-Davis et al. 1998), but may involve neuronal access to muscle-derived IGF-1.

Muscle IGF-1 is known to have target-derived trophic effects on motor neurons (Messi and Delbono 2003), so its overexpression is effective in delaying or preventing the deleterious effects of aging in both tissues. Since age-related decline in muscle function stems partly from motor neuron loss, we created a tetanus toxin fragment-C (TTC) fusion protein to target IGF-1 to motor neurons. IGF-1-TTC was shown to retain IGF-1 activity as indicated by [^{3}H]thymidine incorporation into L6 myoblasts. Spinal cord motor neurons effectively bound and internalized the IGF-1-TTC *in vitro*. Similarly, IGF-1-TTC injected into skeletal muscles was taken up and transported back to the spinal cord *in vivo*, a process that could be prevented by denervation of the injected muscles. Three monthly IGF-1-TTC injections into muscles of aging mice did not increase muscle weight or fiber size but significantly increased single fiber specific force over aged controls injected with saline, IGF-1, or TTC. None of the injections changed muscle fiber- type composition, but neuromuscular junction postterminals were larger and more complex in muscle fibers injected with IGF-1-TTC compared to the other groups, suggesting preservation of muscle fiber innervation. This work demonstrates that induced overexpression of IGF-1 in spinal cord motor neurons of aging mice prevents muscle fiber specific force decline, a hallmark of aging skeletal muscle (Payne et al. 2006).

4 External Ca^{2+}-Dependent Contraction in Aging Skeletal Muscle and IGF-1

We have shown that a population of fast muscle fibers from aging mice depends on external Ca^{2+} to maintain tetanic force during repeated contractions (Payne et al. 2004). We hypothesized that age-related denervation in muscle fibers plays a role in initiating this contractile deficit and that preventing denervation by IGF-1 overexpression would prevent external Ca^{2+}-dependent contraction in aging mice, which was true. To determine whether IGF-1 overexpression affects muscle or nerve, aging mice were injected with a tetanus toxin fragment-C (TTC) fusion protein that targets IGF-1 to spinal cord motor neurons, and this treatment prevented external Ca^{2+}-dependent contraction. We also showed that injections of the IGF-1-TTC fusion protein prevented age-related alterations to the nerve terminals at the neuromuscular junctions. We conclude that the slow, age-related denervation of fast muscle fibers is responsible for dependence on external Ca^{2+} to maintain tetanic force in a population of muscle fibers from senescent mice (Payne et al. 2007).

More recently, we examined the role of extracellular Ca^{2+}, voltage-induced influx of external Ca^{2+} ions, sarcoplasmic reticulum (SR) Ca^{2+} depletion during repeated contractions, store-operated Ca^{2+} entry (SOCE), SR ultrastructure, SR subdomain localization of the ryanodine receptor, and sarcolemmal excitability in muscle force decline with aging. These experiments demonstrated that external Ca^{2+}, but not Ca^{2+} influx, is needed to maintain fiber force with repeated electrical stimulation. Decline in fiber force is associated with depressed SR Ca^{2+} release. SR Ca^{2+} depletion, SOCE, and the putative segregated Ca^{2+} release store do not play a significant role in external Ca^{2+}-dependent contraction. Note that a significant number of action potentials fail in senescent mouse muscle fibers subjected to a high stimulation frequency. These results indicate that failure to generate action potentials accounts for decreased intracellular Ca^{2+} mobilization and tetanic force in aging muscle exposed to a Ca^{2+}-free medium (Payne et al. 2009).

5 The Sarcoplasmic Reticulum Junctional Face Membrane Protein JP-45 Plays a Role in Skeletal Muscle Excitation-Contraction Uncoupling with Aging

JP-45 has been reported exclusively in skeletal muscle, and its expression decreases with aging. It colocalizes with the Ca^{2+}-release channel (the ryanodine receptor) and interacts with calsequestrin and the skeletal muscle DHPRα1 subunit (Anderson et al. 2006). We identified the JP-45 domains and the $Ca_v1.1$ involved in this interaction and investigated the functional effect of JP-45 on excitation-contraction coupling. Its cytoplasmic domain, comprising residues 1–80, interacts with two distinct and functionally relevant domains of DHPRα1 subunit, the I–II loop and the C-terminal region. Interaction with the I–II loop occurs through the loop's α-interacting domain. A DHPR subunit, β1a, also interacts with the cytosolic domain of JP-45, drastically reducing the interaction between JP-45 and the I–II loop.

The functional effect of JP-45 on DHPRα1 subunit activity was assessed by investigating charge movement in differentiated C2C12 myotubes after overexpressing or depleting JP–45. Overexpression decreased peak charge-movement and shifted VQ1/2 to a more negative potential (−10 mV). Depletion decreased both the amount of DHPRα1subunit and peak charge-movements. These results demonstrated that JP-45 is important for functional expression of voltage-dependent Ca^{2+} channels (Anderson et al. 2006).

Another recent study demonstrates that deleting the gene that encodes JP-45 results in decreased muscle strength in young mice by decreasing functional expression of the DHPRα1 subunit, the molecule that couples membrane depolarization and calcium release from the sarcoplasmic reticulum. These results point to JP-45 as one of the molecules involved in the development or maintenance of skeletal muscle strength (Delbono et al. 2007). Whether JP-45 is modulated by neural activity and/or trophic factors is unknown.

In the last decade, a series of triad proteins have been identified, including mitsugumin-29 (Shimuta et al. 1998; Takeshima et al. 1998), junctophilin (Takeshima et al. 2000), SRP-27/TRIC-A (Yazawa et al. 2007; Bleunven et al. 2008), and junctate/hambug (Treves et al. 2000). However, their role in excitation-contraction coupling is only partially understood (Treves et al. 2009), and nerve-dependent regulation of their expression is unknown.

6 Changes in Skeletal Muscle Innervation with Aging

Muscle weakness in aging mammals may result from primary neural or muscular etiological factors or a combination (Delbono 2003). Experimental muscle denervation leads to loss in absolute and specific force (Finol et al. 1981; Dulhunty and Gage 1985). Although denervation contributes to the functional impairment of skeletal muscle with aging (Larsson and Ansved 1995), its prevalence in human and animal models of aging remains to be determined.

Some studies, particularly in the last decade, have focused on the mechanisms underlying neuromuscular impairments in old age. Several aspects have been investigated: the phenomenon known as excitation-contraction uncoupling (ECU) (Delbono et al. 1995; Wang et al. 2002), which leads to a decline in muscle specific force (force normalized to a cross-sectional area) (Gonzalez et al. 2000a); the loss in muscle mass associated with a decrease in muscle fibers as well as fiber atrophy (Lexell 1995; Dutta 1997); changes in fiber type (Larsson et al. 1991; Frontera et al. 2000b; Messi and Delbono 2003; Lauretani et al. 2006); decreased maximal isometric force and slower sliding speed of actin on myosin (Brooks and Faulkner 1994; Hook et al. 1999); and impaired recovery after eccentric contraction (Faulkner et al. 1993; Rader and Faulkner 2006). Identifying the triggers of these changes remains elusive. Some suggestions include decreased muscle loading (Tseng et al. 1995), oxidative damage (Weindruch 1995; Powers and Jackson 2008), age-dependent decrease in IGF-1 expression or tissue sensitivity (Renganathan et al. 1997; Owino et al. 2001; Shavlakadze et al. 2005), and decline in satellite cell proliferation (Decary et al. 1997).

Interaction between skeletal muscle and neuron is crucial to the capacity of both to survive and function throughout life. Thus, muscle atrophy and weakness may result from primary neural or muscular etiological factors or a combination. Growing evidence supports a role for the nervous system in age-related structural and functional alterations in skeletal muscle (Edstrom et al. 2007). The number of motor neurons in the lumbosacral spinal cord of humans has been shown to decrease after the age of 60, and the number of large and intermediate-sized myelinated axon fibers decreases with age in the ventral roots with no change in small fiber numbers (Ceballos et al. 1999; Verdu et al. 2000; Delbono 2003).

Motor units decrease with motor neurons, as measured with electromyography in humans and *in situ* calculation in rats. As with motor neuron fibers, the loss of motor units seems to be greatest among the largest and fastest. A decline in the number and size of anterior horn cells in the cervical and lumbosacral spinal cord and cytons in motor neuron columns in the lumbar spinal cord in humans with age has been reported (Jacob 1998). These studies found fewer large and intermediate-diameter cytons, which are the largest and fastest motor neurons (Liu et al. 1996; Hashizume et al. 1988). In fact, aged motor units exhibit increased amplitude and duration of action potentials, supporting the idea that those remaining grow larger (Larsson 1995; Larsson and Ansved 1995). Morphological evidence of this process can be found in the muscle. Fiber loss and atrophy with age is greatest among fast type-2 fibers, a finding that agrees with the loss of large and intermediate-sized motor neuron fibers and large motor units. Fiber type "grouping" has been found in human muscle with age, indicating a denervation/re-innervation process (Delbono 2003). More direct evidence of a slow denervation process with aging is provided by the increased prevalence of old muscle fibers staining positive for neural cell adhesion molecule (Urbancheck et al. 2001).

Overall fiber loss and a preferential decrease in type-2 fiber number and size in mixed fiber-type lower limb muscles, such as the vastus lateralis, is observed with aging (for a review see (Delbono 2003)). However, all lower limb muscles may not respond similarly to aging. Numbers of tibialis anterior, a predominantly type-2 muscle, have been shown to decrease, with compensatory hypertrophy in the remaining fibers to maintain overall muscle size (Lexell, unpublished results). Conversely, a recent report documents preferential atrophy of type-2 fibers in biceps brachii, an upper limb muscle, but not reduced numbers. This finding is consistent with clinical studies showing better preservation of upper limb muscle function with age (Payne and Delbono 2004).

Several groups have reported skeletal muscle denervation and reinervation and motor unit remodeling or loss in aging rodents or humans (Hashizume et al. 1988; Kanda and Hashizume 1989; Einsiedel and Luff 1992; Kanda and Hashizume 1992; Doherty et al. 1993; Johnson et al. 1995; Zhang et al. 1996). Motor-unit remodeling leads to changes in fiber-type composition (Pette and Staron 2001). During development, muscle fiber-type phenotype is determined by interactions with subpopulations of ventral spinal cord motor neurons that activate contraction at different rates, ranging from 10 (slow fibers) to 100 (fast-fatigue resistant) or 150 Hz (fast-fatigue sensitive) (Buller et al. 1960a, b; Greensmith and Vrbova 1996). Age-related motor-unit remodeling appears to involve denervation of fast muscle fibers with re-innervation by axonal sprouting from slow fibers (Lexell 1995), (Larsson 1995; Kadhiresan et al. 1996), (Frey et al. 2000). When denervation outpaces re-innervation, a population of muscle fibers becomes atrophic and is functionally excluded. Although denervation contributes to skeletal muscle atrophy and functional impairment with aging (Larsson and Ansved 1995), its time course and prevalence in human and animal models of aging remain to be determined. Urbancheck et al. (2001) analyzed the contribution of denervation to deficits in specific force in skeletal muscle in 27–29-month (old) compared with 3-month (young) rats (Urbancheck et al. 2001). Contraction force recordings together with muscle immunostaining for NCAM protein, a marker of fiber denervation (Andersson et al. 1993; Gosztonyi et al. 2001), showed a significantly higher number of denervated fibers in old rats. The area of denervated fibers detected by positive staining with NCAM antibodies accounts for a significant fraction of the decline in specific force (Urbancheck et al. 2001).

We hypothesized that denervation in aging skeletal muscle is more extensive than predicted by standard functional and structural assays and asked whether it is a fully or partially developed process. To address these two questions, we combined electrophysiological and immunohisto-chemical assays to detect the expression of tetrodotoxin (TTX)-resistant sodium channels ($Na_v1.5$) in flexor digitorum brevis (FDB) muscles from young-adult and senescent mice. The FDB muscle was selected for its fast fiber-type composition (~70% type IIx, 13% IIa, and 17% type I) (González et al. 2003) and because the shortness of the fibers makes them suitable for patch-clamp recordings (Wang et al. 2005).

Two sodium channel isoforms are expressed in skeletal muscle, the TTX-sensitive $Na_v1.4$ and the TTX-resistant $Na_v1.5$. Both were originally isolated from rat skeletal muscle and denominated SkM1 (Trimmer et al. 1989) and SkM2 (Kallen et al. 1990), respectively. To determine the status of denervation of individual fibers from adult and senescent mice, we took advantage of the following properties of the $Na_v1.5$ channel: (1) its expression after denervation but absence in innervated adult muscle; (2) its early increase in expression, recorded 24 h after denervation in hindlimb muscles (Yang et al., 1991); and (3) its relative insensitivity to TTX (Redfern et al. 1970; Pappone, 1980; Kallen et al. 1990; White et al. 1991).

Sodium current density measured with the macropatch cell-attached technique did not show significant differences between FDB fibers from young and old mice. The TTX dose-response curve, using the whole cell voltage-clamp technique, showed three populations of fibers in senescent mice, one similar to fibers from young mice (TTX-sensitive), another similar to fibers from experimentally denervated muscle (TTX-resistance), and a third intermediate group. Partially and fully denervated fibers constituted approximately 50% of the total number of fibers tested, which agrees with the percent of fibers shown to be positive for the $Na_v1.5$ channel by specific immunostaining (Wang et al. 2005). These results confirmed our hypothesis that muscle denervation is more extensive than that reported using more classical techniques.

Recovery from denervation implies nerve sprouting and re-innervation by the same or neighboring motor units. Different methods of inducing transient nerve injury and recovery have been employed with contrasting results. Slower regeneration and re-innervation in aged compared to young motor endplates was recorded in response to crush injury of the peripheral nerve (Kawabuchi et al. 2001; Edstrom et al. 2007). The difference in the time needed to recover was attributed to a transient failure in the spatiotemporal relationship between Schwann cells, axons, and the postsynaptic acetylcholine receptor regions during re-innervation in aged rats (Kawabuchi et al. 2001); that is, nerve/muscle interactions contribute significantly to impaired recovery after nerve injury in the aged.

However, in apparent contrast, a comparable capacity for regeneration has been shown in muscles from very old compared to young rats (Carlson et al. 2001). Effects of age on muscle regeneration were studied by injecting the local anesthetic, bupivacaine, in fast-twitch muscles. It induced similar muscle fiber damage and reduced the mean tetanic tension in fast-twitch muscles from young adult (4-month) and old (32- and 34-month) rats. The same authors investigated muscle regeneration using heterochronic transplantation of nerve-intact extensor digitorum longus (EDL), a fast-twitch muscle. EDL muscles from 4- or 32-month-old rats were cross-transplanted in place of the same muscle in 4-month-old hosts. As a control, contralateral muscles were autotransplanted back into the donors. After 60 days, the old-into-young muscle transplants regenerated as successfully as the young-into-young autotransplants. Lack of nerve damage provided favorable conditions for muscle regeneration, together with an age-related effect of the local environment on the transplants (Carlson et al. 2001).

As evidence of the importance of neural factors in nerve regeneration, the same group reported that when axons are allowed to regenerate in an endoneurial environment, there is no evidence of age-related impairment in muscle re-innervation (Cederna et al. 2001). Therefore, although old muscle can regenerate as successfully as young muscle, an intact nerve supply seems critical to recovery, together with less clearly defined factors associated with the local environment. We believe that one of these factors, vital for the protection of nerve and muscle from age-related degeneration, is IGF-1 secretion and signaling.

7 Age-Dependent Modifications and Plasticity of the Neuromuscular Junction

Neural alterations occur at the ventral spinal cord motor neuron, peripheral nerve, and neuromuscular junction in aging mammals. Age-related changes have been documented in neuronal soma size (Liu et al. 1996; Kanda and Hashizume 1998) and number (Hashizume et al. 1988; Zhang et al. 1996; Jacob 1998) in the spinal cord and in peripheral nerve in tibialis nerves of mice aged 6-33 months (Ceballos et al. 1999), including accumulation of collagen in the perineurium and lipid droplets in the perineurial cells, together with an increase in macrophages and mast cells. From 6 to 12 months, numbers of Schwann cells associated with myelinated fibers (MF) decrease slightly in parallel with an increase in their internodal length, but then increase in older nerves in parallel with a greater incidence of demyelination and remyelination. The reported unmyelinated axon (UA) to myelinated fiber (UA/MF) ratio is about 2 until 12 months, decreasing to 1.6 by 27 months. In older mice, the loss of nerve fibers involves UA (50% loss at 27–33 months) more than MF (35%). In aged nerves, wide incisures and infolded or outfolded myelin loops are frequent, resulting in an increased irregularity in the morphology of fibers along the internodes (Ceballos et al. 1999).

In summary, adult mouse nerves (12–20 month) show several features of progressive degeneration, whereas general nerve disorganization and marked fiber loss occur from 20 months on (Ceballos et al. 1999). The deterioration of myelin sheaths during aging may be due to decreased expression of the major myelin proteins (P0, PMP22, MBP). Axonal atrophy, frequently seen in aged nerves, may be explained by reduced expression and axonal transport of cytoskeletal proteins in the peripheral nerve (Verdu et al. 2000). The incidence and severity of the age-related peripheral nerve changes seem to depend on the animal's genetic background. Thus, histological examination conducted on isolated sciatic nerves and brachial plexuses revealed more pronounced axonal degeneration and remyelination in B6C3F1 and C3H than in C57BL mice (Tabata et al. 2000). Impaired nerve regeneration in animals and humans has been correlated with diminished anterograde and retrograde axonal transport (Kerezoudi and Thomas 1999), and retardation in the slow axonal transport of cytoskeletal elements during maturation and aging has been reported (McQuarrie et al. 1989; Cross et al. 2008). This reduced axonal transport could account for the inability of the motor neuron in old mice to expand the field of innervation in response to partial denervation (Jacob and Robbins 1990).

Alterations of the neuromuscular junction in association with aging have been attributed to its "instability" (Balice-Gordon 1997). The process of neuromuscular synapse formation and activity-dependent editing of neuromuscular synaptic connections is better understood (Personius and Balice-Gordon 2000) than the events leading to denervation in aging mammals. Apparently, after synapse formation, the terminals of the same axon, described as a *cartel*, exhibit heterogeneity in terms of acetylcholine release, which may contribute to nerve terminal selection in the developmental transition from innervation of each muscle fiber by multiple nerve endings to the adult one-on-one pattern. Activity plays a crucial role in synapse elimination during this period (for a review see (Personius and Balice-Gordon 2000)).

These concepts prompt the interesting hypothesis that senescent mammals retain a similar mechanism for eliminating neuromuscular synapse. The level of physical activity among the elderly is highly variable and considered important for successful neuromuscular function. Endurance exercise modulates the neuromuscular junction of C57BL/6NNia aging mice (Fahim 1997). When synaptic terminals occupying motor endplates in adult rats were electrically silenced by the sodium channel blocker tetrodotoxin or the acetylcholine receptor blocker a-bungarotoxin, they were frequently displaced by regenerating axons that were both inactive and synaptically ineffective. This study concludes that neither evoked nor spontaneous activation of acetylcholine receptors is required for competitive re-occupation of neuromuscular synaptic sites by regenerating motor axons in adult rats (Costanzo et al. 2000).

Experimental denervation of skeletal muscle from aging rodents leads to a series of changes, such as re-orientation of costameres (rib-like structures formed by dystrophin and b-dystroglycan) (Bezakova and Lomo 2001), proliferation of triadic membranes (Salvatori et al. 1988), decrease in charge movement (functional expression of the dihydropyridine receptor voltage sensor), and alterations in the sarcoplasmic reticulum calcium-release channel (Delbono 1992; Delbono and Stefani 1993; Delbono and Chu 1995; (Delbono et al. 1997; Wang et al. 2000). The molecular substrate for these alterations is only partially understood. We hypothesize that age-related denervation may induce these structural and functional changes in mammalian, including human, muscle. Costameric proteins transmit mechanical lateral forces and provide structural integrity when mechanically loaded muscle fibers contract (Straub and Campbell 1997). Muscle activity and muscle agrin, two orders of magnitude lower than the effective concentration of neural agrin, regulate the organization of cytoskeletal proteins in skeletal muscle fibers (Bezakova and Lomo 2001). It would be interesting to explore these molecular changes in aging muscle and examine the potential beneficial effect of muscle agrin on costamere structure and force development.

The studies reported above strongly implicate neural alterations in the onset and progression of age-related decline in skeletal muscle function. Interventions focused on spinal cord motor neurons, their axons, and associated nonneuronal cells and the neuromuscular junction slow or even reverse age-related impairments in skeletal muscle.

8 Trophic Factors Regulate Spinal Cord Motor Neuron Structure and Function

Classic neurotrophic theory (Davies 1996) describes a well-established role for target-derived neurotrophic factors, including the neurotrophin, NGF, in regulating survival of developing neurons in the peripheral and central nervous systems (Gibbons et al. 2005). Some other studies point to a continued role for target-derived trophic factors in the plasticity of adult and aged neurons (Cowen and Gavazzi 1998; Orike et al. 2001). A series of studies suggests a role for neurotrophins, at least, in the adult neuromuscular system. Neural activity appears to contribute significantly to the trophic interactions between nerve and muscle at the adult neuromuscular junction. Neurotrophins regulate the development of synaptic function (Lohof et al. 1993), and a formulation of the neurotrophin hypothesis proposes that they participate in activity-induced modification of synaptic transmission (Schinder and Poo 2000). Potentiation of synaptic efficacy by brain-derived neurotrophic factor is facilitated by presynaptic depolarization at developing neuromuscular synapses (Boulanger and Poo 1999; Leßmann and Brigadski 2009). Using a model system of nerve/muscle co-culture in which neurotrophin-4 (NT-4) is overexpressed in a subpopulation of postsynaptic myocytes, presynaptic potentiation was restricted to synapses on myocytes overexpressing NT-4. Nearby synapses formed by the same neuron on control myocytes were not affected (Wang et al. 1998). Furthermore, the production of endogenous NT-4 messenger RNA in rat skeletal muscle was regulated by muscle activity; the amount of NT-4 mRNA decreased after blocking neuromuscular transmission with alpha-bungarotoxin and increased during postnatal development and after electrical stimulation. Finally, NT-4 may mediate the effects of exercise and electrical stimulation on neuromuscular performance (Funakoshi et al. 1995). Thus, muscle-derived NT-4 appears to act as an activity-dependent, muscle-derived neurotrophic signal for the growth and remodeling of the adult neuromuscular junction.

These investigations of the complex role of neural activity in regulating nerve-target interactions have not extended to the aging neuromuscular junction. However, a close correlation between altered ligand-receptor expression(s) and axonal/terminal aberrations in senescence supports a role for neurotrophin signaling in age-related degeneration of cutaneous innervation (Bergman et al. 2000). An age-related decrease in target neurotrophin expression, notably NT3 and NT4, correlated with site-specific loss of sensory terminals combined with aberrant growth of regenerating/sprouting axons into new target fields (Bergman et al. 2000).

The role of IGF-1 and related binding proteins in neural control of aging skeletal muscle excitation-contraction coupling and fiber-type composition in mammals is under investigation. Systemic overexpression of human IGF-1 cDNA in transgenic mice resulted in IGF-1 overexpression in a broad range of visceral organs and increased concentrations in serum (Mathews et al. 1988). These mice exhibited increased body weight and organomegaly but only a modest improvement in muscle mass.

Because of the possible confounding effects of systemic expression, Coleman et al. targeted IGF-1 overexpression specifically to striated muscle (Coleman et al. 1995) using a myogenic expression vector containing regulatory elements from both the 5′- and 3′-flanking regions of the avian skeletal α-actin gene. IGF-1 overexpression in cultured muscle cells causes precocious alignment and fusion of myoblasts into terminally differentiated myotubes and elevated levels of myogenic basic helix-loop-helix factors, intermediate filament, and contractile protein mRNA (Coleman et al. 1995). Transgenic mice carrying a single copy of the hybrid skeletal α-actin/hIGF-1 transgene had hIGF-1 mRNA levels that were approximately half those of the endogenous murine skeletal α-actin gene on a per-allele basis but conferred substantial tissue-specific overexpression without elevating serum levels of IGF-1. This localized, muscle-specific overexpression of human IGF-1 caused significant hypertrophy of myofibers, suggesting that IGF-1 is a more potent myogenic stimulus when derived from sustained autocrine/paracrine release than when administrated exogenously. Similar hypertrophy has been observed in response to simple intramuscular injections of IGF-1 in adult rats (Adams and McCue 1998).

Effects of IGF-1 on muscle in aging animals have also been investigated. In old mice, muscle-specific overexpression of IGF-1 preserves skeletal muscle force and DHPR expression (Renganathan et al. 1998; Musaro et al. 2001), while viral-mediated, muscle-specific expression prevents age-related loss of type-IIB fibers (Barton-Davis et al. 1998). There is evidence that the capacity of IGF-1 to induce muscle hypertrophy declines in adult and senescent mice (Chakravarthy et al. 2001). However, its effects on fiber specific force are sustained until late ages (González and Delbono 2001c), suggesting that the pathways it uses to control fiber size and to generate force diverge. Overexpression of the mIGF-1 isoform, corresponding to the human IGF-1Ea gene, resulted in sustained mouse muscle hypertrophy and regenerative capacity throughout life (Musaro et al. 2001), indicating that this muscle-specific splice variant of the IGF-1 gene plays a different role in muscle molecular composition and function than the other IGF-1 splice variants (see below).

Messi et al. (2003) tested the hypothesis that target-derived IGF-1 prevents alterations in neuromuscular innervation in aging mammals (Messi and Delbono 2003). We used senescent wild-type mice as a model of deficient IGF-1 secretion and signaling and S1S2 transgenic mice to investigate the role sustained IGF-1 overexpression in striated muscle plays in neuromuscular innervation. Analysis of the nerve terminal in EDL muscles from senescent mice showed that sustained overexpression of IGF-1 in skeletal muscle partially or completely reversed the decrease in cholinesterase-stained zones (CSZ) exhibiting nerve terminal branching, number of nerve branches at the CSZ, and nerve branch points. Target-derived IGF-1 also prevented age-related decreases in the postterminal α-bungarotoxin immunostained area. Postsynaptic folds were fewer and longer as shown by electron microscopy.

Overexpression of IGF-1 in skeletal muscle may also prevent the switch in muscle fiber-type composition recorded in senescent mice. The use of the S1S2 IGF-1 transgenic mouse model allowed us to provide morphological evidence for the role of target-derived IGF-1 in spinal cord motor neurons in senescent mice. The main conclusion of this study was that muscle IGF-1 prevents age-dependent changes in nerve terminal and neuromuscular junction, influencing muscle fiber-type composition and, potentially, muscle function (Barton-Davis et al. 1998) (Musaro et al. 2001; Delbono 2002).

9 *Effects of IGF-1 on Neurons*

The role of IGF-1 in motor neuron survival has been examined during embryonic or postnatal life (Neff et al. 1993) as well as in spinal cord pathology (Rind and von Bartheld 2002; Dobrowolny et al. 2005; Messi et al. 2007). For example, in young rodents, IGF-1 expression is upregulated in Schwann cells and astrocytes following spinal cord and peripheral nerve injury, while IGF-binding protein 6 is strongly upregulated in the injured motor neurons (Hammarberg et al. 1998). In regions of muscle enriched with neuromuscular junctions, IGF-II was strongly upregulated in satellite and possibly glial cells during recovery from sciatic nerve crush (Pu et al. 1999) while IGF-1 showed less significant changes. In young animals, systemic administration of IGF-1 decreases lesion-induced motor neuron cell death and promotes muscle re-innervation (Vergani et al. 1998). It also promotes neurogenesis and synaptogenesis in diverse areas of the central nervous system, such as the hypocampal dentate gyrus during postnatal development (O'Kusky et al. 2000), and increases proliferation of granule cell progenitors (Ye et al. 1996).

These studies suggest that IGF-1 might have beneficial effects on spinal cord motor neurons from senescent mammals. However, transgenic overexpression of IGF-1 in the central nervous system does not improve excitation-contraction coupling or neuromuscular performance in the mouse (Ye et al. 1996; Moreno et al. 2006). In contrast to localized motor neuron expression, widespread IGF-1 may be deleterious for neuronal function or muscle innervation (Moreno et al. 2006).

During embryonic and postnatal development, specific sets of CNS neurons show high levels of IGF-1 receptor gene expression combined with IGF-1 expression, while in hippocampal and cortical neurons, receptor and IGF-1 expression are localized in different cell groups (Bondy et al. 1992). These expression patterns suggest that IGF-1 exerts autocrine and paracrine effects in the CNS in addition to its previously described paracrine (muscle-derived) actions on spinal cord motor neurons. While these mechanisms contribute undoubtedly to the development of the appropriate neuronal phenotype and probably to its maintenance in adulthood, its involvement in aging processes remains substantially untested. Despite these uncertainties, an age-related decline in neuronal as well as muscle-derived IGF-1 combined with altered IGF-1 resistance through reduced expression or sensitivity of the receptor may contribute to the atrophy or death of motor and other CNS neurons in aging mammals. Through the previously described mechanisms, these changes may trigger a cascade of events leading to decreased skeletal muscle gene transcription.

10 Concluding Remarks

Age-related decline in the neuromuscular system is a recognized cause of impaired physical performance and loss of independence in the elderly. Epidemiological data associate these changes with increased risk of morbidity, disability, and mortality in the elderly (Winograd et al. 1991; Baumgartner et al. 1998; Ryall et al. 2008).

We argue for the importance of neural factors in age-related impairment of mammalian skeletal muscle structure and function. Decreased local production of IGF-1 and/or neurotrophins and tissue resistance to these factors through altered receptor expression or responsiveness may result in loss and atrophy of spinal cord motor neurons. In fact, declining motor neuron function may be more extensive than that predicted by structural assays. Preliminary data support the concept that reduced IGF-1 synthesis may cause the failure of an IGF-mediated pathway to decrease CREB phosphorylation. In turn, reduced CREB phosphorylation may result in reduced DHPRα_{1S} transcription, excitation-contraction uncoupling, and decreased muscle force.

The characterization of a number of triad proteins is shedding light on the molecular signaling involved in excitation-gene expression and excitation-contraction coupling (Carrasco et al. 2004). The role of neural factors in regulating the expression and function of these newly identified triad proteins is a necessary focus of research in the coming years. We hypothesize that neural factors (autocrine trophic factors, nerve activity and connectivity) play a vital role in preventing age-related excitation-contraction uncoupling. Based on this hypothesis, we predict that interventions aimed at counteracting nerve loss will play an important part in ameliorating the loss of force exhibited in animal models of aging as well as in elderly humans.

Acknowledgments Results reported in this article were obtained with the support of the National Institutes of Health/National Institute on Aging (AG15820, AG13934, and AG033385) and Muscular Dystrophy Association of America's grants to Osvaldo Delbono and the Wake Forest University Claude D. Pepper Older Americans Independence Center (P30-AG21332).

References

Adams, G. R. & Mccue, S. A. (1998). Localized infusion of IGF-I results in skeletal muscle hypertrophy in rats. *J Appl Physiol, 84*, 1716–22.

Andersson, A. M., Olsen, M., Zhernosekov, D., Gaardsvoll, H., Krog, L., Linnemann, D., Bock, E. (1993). Age-related changes in expression of the neural cell adhesion molecule in skeletal muscle: a comparative study of newborn, adult and aged rats. *Biochem J, 290*(Pt 3), 641–8.

Anderson, A. A., Altafaj, X., Zheng, Z., Wang, Z. M., Delbono, O., Ronjat, M., Treves, S., Zorzato, F. (2006). The junctional SR protein JP-45 affects the functional expression of the voltage-dependent Ca2+ channel Cav1.1. *J Cell Sci, 119*, 2145–55.

Balice-Gordon, R. J. (1997). Age-related changes in neuromuscular innervation. *Muscle & Nerve, S5*, S83–S87.

Barton-Davis, E. R., Shoturma, D. I., Musaro, A., Rosenthal, N., Sweeney, H. L. (1998). Viral mediated expression of insulin-like growth factor I blocks the aging-related loss of skeletal muslce function. *Proceedings of the National Academy of Science, 95*, 15603–15607.

Baumgartner, R. N., Koehler, K. M., Gallagher, D. (1998). Epidemiology of sarcopenia among the elderly in New Mexico. *American Journal of Epidemiology, 147*, 755–763.

Bergman, E., Ulfhake, B., Fundin, B. T. (2000). Regulation of NGF-family ligands and receptors in adulthood and senescence: correlation to degenerative and regenerative changes in cutaneous innervation. *Eur J Neurosci, 12*, 2694–706.

Bezakova, G. & Lomo, T. (2001). Muscle activity and muscle agrin regulate the organization of cytoskeletal proteins and attached acetylcholine receptor (AchR) aggregates in skeletal muscle fibers. *J Cell Biol, 153*, 1453–63.

Bleunven, C., Treves, S., Jinyu, X., Leo, E., Ronjat, M., DE waard, M., Kern, G., Flucher, B. E., Zorzato, F. (2008). SRP-27 is a novel component of the supramolecular signalling complex involved in skeletal muscle excitation-contraction coupling. *Biochem J, 411*, 343–9.

Bondy, C., Werner, H., JR Roberts, C. T., Leroith, D. (1992). Cellular pattern of type-I insulin-like growth factor receptor gene expression during maturation of the rat brain: comparison with insulin-like growth factors I and II. *Neuroscience, 46*, 909–23.

Boulanger, L. & Poo, M. M. (1999). Presynaptic depolarization facilitates neurotrophin-induced synaptic potentiation. *Nat Neurosci, 2*, 346–51.

Brooks, S. V. & Faulkner, J. A. (1994). Skeletal muscle weakness in old age: underlying mechanisms. *Medicine & Science in Sports & Exercise, 26*, 432–9.

Buller, A. J., Eccles, J. C., Eccles, R. M. (1960a). Differentiation of fast and slow muscles in the cat hind limb. *Journal of Physiology, 150*, 399–416.

Buller, A. J., Eccles, J. C., Eccles, R. M. (1960b). Interactions between motoneurones and muscles in respect of the characteristic speeds of their responses. *Journal of Physiology, 150*, 417–439.

Carlson, B. M., Dedkov, E. I., Borisov, A. B., Faulkner, J. A. (2001). Skeletal muscle regeneration in very old rats. *J Gerontol A Biol Sci Med Sci, 56*, B224–33.

Carrasco, M., Jaimovich, E., Kemmerling, U., Hidalgo, C. (2004). Signal transduction and gene expression regulated by calcium release from internal stores in excitable cells. *Biological Research, 37*, 701–712.

Ceballos, D., Cuadras, J., Verdu, E., Navarro, X. (1999). Morphometric and ultrastructural changes with ageing in mouse peripheral nerve. *J Anat, 195*(Pt 4), 563–76.

Cederna, P. S., Asato, H., Gu, X., Van Der Meulen, J., Jr Kuzon, W. M., Carlson, B. M., Faulkner, J. A. (2001). Motor unit properties of nerve-intact extensor digitorum longus muscle grafts in young and old rats. *J Gerontol A Biol Sci Med Sci, 56*, B254–8.

Chakravarthy, M. V., Fiorotto, M. L., Schwartz, R. J., Booth, F. W. (2001). Long-term insulin-like growth factor-1 expression in skeletal muscles attenuates the enhanced in vitro proliferation ability of the resident satellite cells in transgenic mice. *Mechanisms of Ageing and Development, 122*, 1303–1320.

Chaudari, N. & Beam, K. G. (1993). mRNA for cardiac calcium channel is expressed during development of skeletal muscle. *Developmental Biology, 155*, 507–515.

Coleman, M. E., Demayo, F., Yin, K. C., Lee, H. M., Geske, R., Montgomery, C., Schwartz, R. J. (1995). Myogenic vector expression of insulin-like growth factor I stimulates muscle cell differentiation and myofiber hypertrophy in transgenic mice. *Journal of Biological Chemistry, 270*, 12109–16.

Costanzo, E. M., Barry, J. A., Ribchester, R. R. (2000). Competition at silent synapses in reinnervated skeletal muscle. *Nat Neurosci, 3*, 694–700.

Cowen, T. & Gavazzi, I. (1998). Plasticity in adult and ageing sympathetic neurons. *Prog Neurobiol, 54*, 249–88.

Cross, D. J., Flexman, J. A., Anzai, Y., Maravilla, K. R., Minoshima, S. (2008). Age-related decrease in axonal transport measured by MR imaging in vivo. *NeuroImage, 39*, 915.

Davies, A. M. (1996). The neurotrophic hypothesis: where does it stand? *Philos Trans R Soc Lond B Biol Sci, 351*, 389–94.

Decary, S., Mouly, V., Hamida, C. B., Sautet, A., Barbet, J. P., Butler-Browne, G. S. (1997). Replicative potential and telomere length in human skeletal muscle: implications for satellite cell-mediated gene therapy. *Hum Gene Ther, 8*, 1429–38.

Delbono, O. (1992). Calcium current activation and charge movement in denervated mammalian skeletal muscle fibres. *Journal of Physiology, 451*, 187–203.

Delbono, O. (2002). Molecular Mechanisms and Therapeutics of the Deficit in Specific Force in Ageing Skeletal Muscle. *Biogerontology, 3*, 265–270.

Delbono, O. (2003). Neural Control of Aging Skeletal Muscle. *Aging Cell, 2*, 21–29.

Delbono, O. & CHU, A. (1995). Ca2+ release channels in rat denervated skeletal muscles. *Experimental Physiology, 80*, 561–74.

Delbono, O. & Stefani, E. (1993). Calcium current inactivation in denervated rat skeletal muscle fibres. *Journal of Physiology, 460*, 173–83.

Delbono, O., O'rourke, K. S., Ettinger, W. H. (1995). Excitation-calcium release uncoupling in aged single human skeletal muscle fibers. *Journal of Membrane Biology, 148*, 211–22.

Delbono, O., Renganathan, M., Messi, M. L. (1997). Excitation-Ca2+ release-contraction coupling in single aged human skeletal muscle fiber. *Muscle and Nerve Supplement, 5*, S88–92.

Delbono, O., Xia, J., Treves, S., Wang, Z. M., Jimenez-Moreno, R., Payne, A. M., Messi, M. L., Briguet, A., Schaerer, F., Nishi, M., Takeshima, H., Zorzato, F. (2007). Loss of skeletal muscle strength by ablation of the sarcoplasmic reticulum protein JP45. *Proc Natl Acad Sci U S A, 104*, 20108–20113.

Dobrowolny, G., Giacinti, C., Pelosi, L., Nicoletti, C., Winn, N., Barberi, L., Molinaro, M., Rosenthal, N., Musaro, A. (2005). Muscle expression of a local Igf-1 isoform protects motor neurons in an ALS mouse model. *J Cell Biol, 168*, 193–9.

Doherty, T. J., Vandervoort, A. A., Taylor, A. W., Brown, W. F. (1993). Effects of motor unit losses on strength in older men and women. *Journal of Applied Physiology, 74*, 868–74.

Dulhunty, A. F. (2006). Excitation-contraction coupling from the 1950s into the new millennium. *Clin Exp Pharmacol Physiol, 33*, 763–72.

Dulhunty, A. F. & Gage, P. W. (1985). Excitation-contraction coupling and charge movement in denervated rat extensor digitorum longus and soleus muscles. *J Physiol, 358*, 75–89.

Dutta, C. (1997). Significance of sarcopenia in the elderly. *Journal of Nutrition, 127*, 992S–993S.

Edstrom, E., Altun, M., Bergman, E., Johnson, H., Kullberg, S., Ramirez-Leon, V., Ulfhake, B. (2007). Factors Contributing to neuromuscular impairment and sarcopenia during aging. *Physiol Behav, 92*, 129–35.

Einsiedel, L. J. & Luff, A. R. (1992). Effect of partial denervation on motor units in the ageing rat medial gastrocnemius. *Journal of the Neurological Sciences, 112*, 178–184.

Fahim, M. A. (1997). Endurance exercise modulates neuromuscular junction of C57BL/6NNia aging mice. *J Appl Physiol, 83*, 59–66.

Faulkner, J. A., Brooks, S. V., Opiteck, J. A. (1993). Injury to skeletal muscle fibers during contractions: conditions of occurrence and prevention. *Physical Therapy, 73*, 911–921.

Finol, H. J., Lewis, D. M., Owens, R. (1981). The effects of denervation on contractile properties or rat skeletal muscle. *J Physiol, 319*, 81–92.

Frey, D., Schneider, C., Xu, L., Borg, J., Spooren, W., Caroni, P. (2000). Early and selective loss of neuromuscular synapse subtypes with low sprouting competence in motoneuron diseases. *Journal of Neuroscience, 20*, 2534–2542.

Frontera, W. R., Hughes, V. A., Fielding, R. A., Fiatarone, M. A., Evans, W. J., Roubenoff, R. (2000a). Aging of skeletal muscle: a 12-yr longitudinal study. *J Appl Physiol, 88*, 1321–6.

Frontera, W. R., Suh, D., Krivickas, L. S., Hughes, V. A., Goldstein, R., Roubenoff, R. (2000b). Skeletal muscle fiber quality in older men and women. *Am J Physiol Cell Physiol, 279*, C611–8.

Funakoshi, H., Belluardo, N., Arenas, E., Yamamoto, Y., Casabona, A., Persson, H., Ibanez, C. F. (1995). Muscle-derived neurotrophin-4 as an activity-dependent trophic signal for adult motor neurons. *Science, 268*, 1495–9.

Gibbons, A., Wreford, N., Pankhurst, J., Bailey, K. (2005). Continuous supply of the neurotrophins BDNF and NT-3 improve chick motor neuron survival in vivo. *International Journal of Developmental Neuroscience, 23*, 389.

González, E. & Delbono, O. (2001a). Age-dependent fatigue in single intact fast- and slow-fibers from mouse EDL and soleus skeletal muscles. *Mechanisms of Ageing and Development, 122*, 1019–1032.

González, E. & Delbono, O. (2001b). Recovery from fatigue in fast and slow single intact skeletal muscle fibers from aging mouse. *Muscle Nerve, 24*, 1219–1224.

González, E. & Delbono, O. (2001c) Sustained overexpression of IGF-1 prevents age-dependent decline in skeletal muscle fiber force and intracellular calcium. *Society for Neuroscience*, 31st Annual Meeting, 519.13.

Gonzalez, A., Kirsch, W. G., Shirokova, N., Pizarro, G., Brum, G., Pessah, I. N., Stern, M. D., Cheng, H., Rios, E. (2000a). Involvement of multiple intracellular release channels in calcium sparks of skeletal muscle. *Proc Natl Acad Sci U S A, 97*, 4380–5.

Gonzalez, E., Messi, M. L., Delbono, O. (2000b). The specific force of single intact extensor digitorum longus and soleus mouse muscle fibers declines with aging. *J Membr Biol, 178*, 175–83.

González, E., Messi, M. L., Zheng, Z., Delbono, O. (2003). Insulin-like growth factor-1 prevents age-related decrease in specific force and intracellular Ca2+ in single intact muscle fibres from transgenic mice. *J Physiol, 552*, 833–44.

Gosztonyi, G., Naschold, U., Grozdanovic, Z., Stoltenburg-Didinger, G., Gossrau, R. (2001). Expression of Leu-19 (CD56, N-CAM) and nitric oxide synthase (NOS) I in denervated and reinnervated human skeletal muscle. *Microsc Res Tech, 55*, 187–97.

Greensmith, L. & Vrbova, G. (1996). Motoneurone survival: a functional approach. *Trends Neurosci, 19*, 450–5.

Hammarberg, H., Risling, M., Hokfelt, T., Cullheim, S., Piehl, F. (1998). Expression of insulin-like growth factors and corresponding binding proteins (IGFBP 1-6) in rat spinal cord and peripheral nerve after axonal injuries. *Journal of Comparative Neurology, 400*, 57–72.

Hashizume, K., Kanda, K., Burke, R. (1988). Medial gastrocnemius motor nucleus in the rat: Age-related changes in the number and size of motoneurons. *The Journal of Comparative Neurology, 269*, 425–430.

Hook, P., Li, X., Sleep, J., Hughes, S., Larsson, L. (1999). In vitro motility speed of slow myosin extracted from single soleus fibres from young and old rats. *Journal of Physiology, 520*, 463–71.

Jacob, J. M. (1998). Lumbar motor neuron size and number is affected by age in male F344 rats. *Mechanisms of Aging and Development, 106*, 205–216.

Jacob, J. M. & Robbins, N. (1990). Age differences in morphology of reinnervation of partially denervated mouse muscle. *J Neurosci, 10*, 1530–40.

Johnson, H., Mossberg, K., Arvidsson, U., Piehl, F., Hokfelt, T., Ulfhake, B. (1995). Increase in alpha-CGRP and GAP-43 in aged motoneurons: A study of peptides, growth factors, and ChAT mRNA in the lumbar spinal cord of senescent rats with symptoms of hindlimb incapacities. *The Journal of Comparative Neurology, 359*, 69–89.

Kadhiresan, V. A., Hassett, C. A., Faulkner, J. A. (1996). Properties of single motor units in medial gastrocnemius muscles of adult and old rats. *Journal of Physiology, 493*, 543–52.

Kallen, R. G., Sheng, Z. H., Yang, J., Chen, L. Q., Rogart, R. B., Barchi, R. L. (1990). Primary structure and expression of a sodium channel characteristic of denervated and immature rat skeletal muscle. *Neuron, 4*, 233–42.

Kanda, K. & Hashizume, K. (1989). Changes in properties of the medial gastrocnemius motor units in aging rats. *Journal of Neurophysiology, 1989*, 737–746.

Kanda, K. & Hashizume, K. (1992). Factors causing difference in force output among motor units in the rat medial gastrocnemius muscle. *Journal of Physiology, 448*, 677–695.

Kanda, K. & Hashizume, K. (1998). Effects of long-term physical exercise on age-related changes of spinal motoneurons and peripheral nerves in rats. *Neurosci Res, 31*, 69–75.

Kawabuchi, M., Zhou, C. J., Wang, S., Nakamura, K., Liu, W. T., Hirata, K. (2001). The spatiotemporal relationship among Schwann cells, axons and postsynaptic acetylcholine receptor regions during muscle reinnervation in aged rats. *Anat Rec, 264*, 183–202.

Kent-Braun, J. A. & Ng, A. V. (1999). Specific strength and voluntary muscle activation in young and elderly women and men. *Journal of Applied Physiology, 87*, 22–9.

Kerezoudi, E. & Thomas, P. K. (1999). Influence of age on regeneration in the peripheral nervous system. *Gerontology, 45*, 301–6.

Kyselovic, J., Leddy, J. J., Ray, A., Wigle, J., Tuana, B. S. (1994). Temporal differences in the induction of dihydropyridine receptor subunits and ryanodine receptors during skeletal muscle development. *Journal of Biological Chemistry, 269*, 21770–21777.

Larsson, L. (1995). Motor units: remodeling in aged animals. *Journals of Gerontology. Series A, Biological Sciences Medical Sciences, 50*, 91–5.

Larsson, L. & Ansved, T. (1995). Effects of ageing on the motor unit. *Progress in Neurobiology, 45*, 397–458.

Larsson, L., Ansved, T., Edstrom, L., Gorza, L., Schiaffino, S. (1991). Effects of age on physiological, immunohistochemical and biochemical properties of fast-twitch single motor units in the rat. *Journal of Physiology, 443*, 257–275.

Lauretani, F., Bandinelli, S., Bartali, B., Di Iorio, A., Giacomini, V., Corsi, A. M., Guralnik, J. M., Ferrucci, L. (2006). Axonal degeneration affects muscle density in older men and women. *Neurobiol Aging, 27*, 1145–54.

Leßmann, V. & Brigadski, T. (2009) Mechanisms, locations, and kinetics of synaptic BDNF secretion: An update. *Neurosci Res, 65*, 316–317.

Lexell, J. (1995). Human aging, muscle mass, and fiber type composition. *Journals of Gerontology. Series A, Biological Sciences Medical Sciences, 50*, 11–16.

Liu, R.-H., Bertolotto, C., Engelhardt, J. K., Chase, M. H. (1996). Age-related changes in soma size of neurons in the spinal cord motor column of the cat. *Neuroscience Letters, 211*, 163–166.

Loeser, R. F. & Delbono, O. (2009). The musculoskeletal and joint system. In Halter, J. B., Ouslander, J. G., Tinetti, M. E., Studenski, S., High, K. P., Asthana, S. (Eds.), *Hazzard's Geriatric Medicine and Gerontology* (6th ed.). New York: McGraw-Hill.

Lohof, A. M., Ip, N. Y., Poo, M. M. (1993). Potentiation of developing neuromuscular synapses by the neurotrophins NT-3 and BDNF. *Nature, 363*, 350–3.

Lowe, D. A., Thomas, D. D., Thompson, L. V. (2002). Force generation, but not myosin ATPase activity, declines with age in rat muscle fibers. *Am J Physiol Cell Physiol, 283*, C187–92.

Lynch, N. A., Metter, E. J., Lindle, R. S., Fozard, J. L., Tobin, J. D., Roy, T. A., Fleg, J. L., Hurley, B. F. (1999). Muscle quality I. Age-associated differences between arm and leg muscle groups. *Journal of Applied Physiology, 86*, 188–94.

Mathews, L. S., Hammer, R. E., Behringer, R. R., D'ercole, A. J., Bell, G. I., Brinster, R. L., Palmiter, R. D. (1988). Growth enhancement of transgenic mice expressing human insulin-like growth factor I. *Endocrinology, 123*, 2827–2833.

Mcquarrie, I. G., Brady, S. T., Lasek, R. J. (1989). Retardation in the slow axonal transport of cytoskeletal elements during maturation and aging. *Neurobiol Aging, 10*, 359–65.

Messi, M. L. & Delbono, O. (2003). Target-derived trophic effect on skeletal muscle innervation in senescent mice. *Journal of Neuroscience, 23*, 1351–1359.

Messi, M. L., Clark, H. M., Prevette, D. M., Oppenheim, R. W., Delbono, O. (2007). The lack of effect of specific overexpression of IGF-1 in the central nervous system or skeletal muscle on pathophysiology in the G93A SOD-1 mouse model of ALS. *Exp Neurol, 207*, 52–63.

Moreno, R., Messi, M., Zheng, Z., Wang, Z.-M., Ye, P., J, D. E. Delbono, O. (2006) Role of sustained overexpression of central nervous system IGF-1 in the age-dependent decline of mouse excitation-contraction coupling. *Journal of Membrane Biology,* 212**,** 147–161.

Musaro, A., Mccullagh, K. J., Paul, A., Houghton, L., Dobrowolny, G., Molinaro, M., Barton-Davis, E. R., Sweeney, H. L., Rosenthal, N. (2001). Localized IGF-1 transgene expression sustains hypertrophy and regeneration in senescent skeletal muscle. *Nature Genetics, 27*, 195–200.

Neff, N. T., Prevette, D. M., Houenou, L. J., Lewis, M. E., Glicksman, M. A., Yin, Q.-W., Oppenheim, R. W. (1993). Insulin-like growth factors: Putative muscle-derived trophic agents that promote motoneuron survival. *Journal of Neurobiology, 24*, 1578–1588.

O'kusky, J. R., Ye, P., D'ercole, J. (2000). Insulin-Like growth factor-1 promotes neurogenesis and synaptogenesis in the hippocampal dentate gyrus during postnatal development. *The Journal of Neuroscience, 15*, 8435–8442.

Orike, N., Thrasivoulou, C., Wrigley, A., Cowen, T. (2001). Differential regulation of survival and growth in adult sympathetic neurons: an in vitro study of neurotrophin responsiveness. *J Neurobiol, 47*, 295–305.

Owino, W., Yang, S. Y., Goldspink, G. (2001). Age-related loss of skeletal muscle function and the inability to express the autocrine form of insulin-like growth factor-1 (MGF) in response to mechanical overload. *FEBS Letters, 505*, 259–263.

Pappone, P. A. (1980). Voltage-clamp experiments in normal and denervated mammalian skeletal muscle fibres. *J Physiol, 306*, 377–410.

Pauwels, P. J., Van Assouw, H. P., Leysen, J. E. (1987). Depolarization of chick myotubes triggers the appearance of (+)-[3H]PN 200-110-Binding Sites. *Molecular Pharmacology, 32*, 785–791.

Payne, A. M. & Delbono, O. (2004). Neurogenesis of excitation-contraction uncoupling in aging skeletal muscle. *Exerc Sport Sci Rev, 32*, 36–40.

Payne, A. M., Zheng, Z., Gonzalez, E., Wang, Z. M., Messi, M. L., Delbono, O. (2004). External Ca2+-dependent excitation-contraction coupling in a population of ageing mouse skeletal muscle fibres. *Journal of Physiology, 560*(1), 137–157.

Payne, A. M., Messi, M. L., Zheng, Z., Delbono, O. (2006). Motor neuron targeting of IGF-1 attenuates age-related external Ca2+-dependent skeletal muscle contraction in senescent mice. *Journal of Physiology, 570*, 283–294.

Payne, A. M., Messi, M. L., Zheng, Z., Delbono, O. (2007). Motor neuron targeting of IGF-1 attenuates age-related external Ca(2+)-dependent skeletal muscle contraction in senescent mice. *Exp Gerontol, 42*, 309–19.

Payne, A. M., Jimenez-Moreno, R., Wang, Z. M., Messi, M. L., Delbono, O. (2009). Role of Ca2+, Membrane Excitability, and Ca2+ Stores in Failing Muscle Contraction with Aging. *Experimental Gerontology, 44*, 261–273.

Pereon, Y., Navarro, J., Hamilton, M., Booth, F. W., Palade, P. (1997a). Chronic stimulation differentially modulates expression of mRNA for dihydropyridine receptor isoforms in rat fast twitch skeletal muscle. *Biochemical Biophysical Research Communications, 235*, 217–222.

Pereon, Y., Sorrentino, V., Dettbarn, C., Noireaud, J., Palade, P. (1997b). Dihydropyridine receptor and ryanodine receptor gene expression in long-term denervated rat muscles. *Biochemical and Biophysical Research Communications, 240*, 612–617.

Personius, K. E. & Balice-Gordon, R. J. (2000). Activity-dependent editing of neuromuscular synaptic connections. *Brain Res Bull, 53*, 513–22.

Pette, D. & Staron, R. S. (2001). Transitions of muscle fiber phenotypic profiles. *Histochem Cell Biol, 115*, 359–72.

Powers, S. K. & Jackson, M. J. (2008). Exercise-Induced Oxidative Stress: Cellular Mechanisms and Impact on Muscle Force Production. *Physiol. Rev., 88*, 1243–1276.

Pu, S. F., Zhuang, H. X., Marsh, D. J., Ishii, D. N. (1999). Time-dependent alteration of insulin-like growth factor gene expression during nerve regeneration in regions of muscle enriched with neuromuscular junctions. *Brain Research. Molecular Brain Research, 63*, 207–16.

Rader, E. P. & Faulkner, J. A. (2006). Effect of aging on the recovery following contraction-induced injury in muscles of female mice. *J Appl Physiol, 101*, 887–892.

Ray, A., Kyselovic, J., Leddy, J. J., Wigle, J., Jasmin, B. J., Tuana, B. S. (1995). Regulation of dihydropyridine and ryanodine receptor gene expression in skeletal muscle. Role of nerve, protein kinase C and cAMP pathways. *The Journal of Biological Chemistry, 270*, 25837–25844.

Redfern, P., Lundh, H., Thesleff, S. (1970). Tetrodotoxin resistant action potentials in denervated rat skeletal muscle. *Eur J Pharmacol, 11*, 263–5.

Renganathan, M., Sonntag, W. E., Delbono, O. (1997). L-type Ca2+ channel-insulin-like growth factor-1 receptor signaling impairment in aging rat skeletal muscle. *Biochemical Biophysical Research Communications, 235*, 784–9.

Renganathan, M., Messi, M. L., Delbono, O. (1998). Overexpression of IGF-1 exclusively in skeletal muscle prevents age-related decline in the number of dihydropyridine receptors. *Journal of Biological Chemistry, 273*, 28845–28851.

Rind, H. B. & Von Bartheld, C. S. (2002). Target-derived cardiotrophin-1 and insulin-like growth factor-I promote neurite growth and survival of developing oculomotor neurons. *Mol Cell Neurosci, 19*, 58–71.

Ryall, J., Schertzer, J., Lynch, G. (2008). Cellular and molecular mechanisms underlying age-related skeletal muscle wasting and weakness. *Biogerontology, 9*, 213.

Saborido, A., Molano, F., Moro, G., Megias, A. (1995). Regulation of dihydropyridine receptor levels in skeletal and cardiac muscle by exercise training. *Pflugers Archives- European Journal of Physiology, 429*, 364–369.

Salvatori, S., Damiani, E., Zorzato, F., Volpe, P., Pierobon, S., Quaglino, D., Salviati, G., Margreth, A. (1988). Denervation-induced proliferative changes of triads in rabbit skeletal muscle. *Muscle Nerve, 11*, 1246–1259.

Schinder, A. F. & Poo, M. (2000). The neurotrophin hypothesis for synaptic plasticity. *Trends Neurosci, 23*, 639–45.

Schneider, M. F. & Chandler, W. K. (1973). Voltage dependent charge movement of skeletal muscle: a possible step in excitation-contraction coupling. *Nature, 242*, 244–6.

Shavlakadze, T., Winn, N., Rosenthal, N., Grounds, M. D. (2005). Reconciling data from transgenic mice that overexpress IGF-I specifically in skeletal muscle. *Growth Horm IGF Res, 15*, 4–18.

Shimuta, M., Komazaki, S., Nishi, M., Iino, M., Nakagawara, K., Takeshima, H. (1998). Structure and expression of mitsugumin29 gene. *FEBS Lett, 431*, 263–7.

Straub, V. & Campbell, K. P. (1997). Muscular dystrophies and the dystrophin-glycoprotein complex. *Curr Opin Neurol, 10*, 168–75.

Tabata, H., Ikegami, H., Kariya, K. (2000). A parallel comparison of age-related peripheral nerve changes in three different strains of mice. *Exp Anim, 49*, 295–9.

Takeshima, H., Shimuta, M., Komazaki, S., Ohmi, K., Nishi, M., Iino, M., Miyata, A., Kangawa, K. (1998). Mitsugumin29, a novel synaptophysin family member from the triad junction in skeletal muscle. *Biochem J, 331*(Pt 1), 317–22.

Takeshima, H., Komazaki, S., Nishi, M., Iino, M., Kangawa, K. (2000). Junctophilins: a novel family of junctional membrane complex proteins. *Mol Cell, 6*, 11–22.

Treves, S., Feriotto, G., Moccagatta, L., Gambari, R., Zorzato, F. (2000). Molecular cloning, expression, functional characterization, chromosomal localization, and gene structure of junctate, a novel integral calcium binding protein of sarco(endo)plasmic reticulum membrane. *J Biol Chem, 275*, 39555–68.

Treves, S., Vukcevic, M., Maj, M., Thurnheer, R., Mosca, B. Zorzato, F. (2009) Minor sarcoplasmic reticulum membrane components that modulate excitation-contraction coupling in striated muscles. *J Physiol, 587*, 3071–3079.

Trimmer, J. S., Cooperman, S. S., Tomiko, S. A., Zhou, J. Y., Crean, S. M., Boyle, M. B., Kallen, R. G., Sheng Z. H., Barchi, R. L., Sigworth, F. J., Goodman, R. H., Agnew, W. S., Mandel, G. (1989). Primary structure and functional expression of a mammalian skeletal muscle sodium channel. *Neuron, 3*, 33–49.

Tseng, B. S., Marsh, D. R., Hamilton, M. T., Booth, F. W. (1995). Strength and aerobic training attenuate muscle wasting and improve resistance to the development of disability with aging. *J Gerontol A Biol Sci Med Sci, 50 Spec No*, 113–9.

Urbancheck, M. G., Picken, E. B., Kalliainen, L. K., Kuzon, W. M. (2001). Specific force deficit in skeletal muscles of old rats is partially explained by the existence of denervated muscle fibers. *Journal of Gerontology: Biological Sciences, 56A*, B191–B198.

Verdu, E., Ceballos, D., Vilches, J. J., Navarro, X. (2000). Influence of aging on peripheral nerve function and regeneration. *J Peripher Nerv Syst, 5*, 191–208.

Vergani, L., Di Giulio, A. M., Losa, M., Rossoni, G., Muller, E. E., Gorio, A. (1998). Systemic administration of insulin-like growth factor decreases motor neuron cell death and promotes muscle reinnervation. *Journal of Neuroscience Research, 54*, 840–7.

Wang, X., Berninger, B., Poo, M. (1998). Localized synaptic actions of neurotrophin-4. *J Neurosci, 18*, 4985–92.

Wang, Z.-M., Messi, M. L., Delbono, O. (2000). L-type Ca2+ channel charge movement and intracellular Ca2+ in skeletal muscle fibers from aging mice. *Biophysical Journal, 78*, 1947–1954.

Wang, Z.-M., Messi, M. L., Delbono, O. (2002). Sustained overexpression of IGF-1 prevents age-dependent decrease in charge movement and intracellular calcium in mouse skeletal muscle. *Biophysical Journal, 82*, 1338–1344.

Wang, Z. M., Zheng, Z., Messi, M. L., Delbono, O. (2005). Extension and magnitude of denervation in skeletal muscle from ageing mice. *J Physiol, 565*, 757–64.

Weindruch, R. (1995). Interventions based on the possibility that oxidative stress contributes to sarcopenia. *J Gerontol A Biol Sci Med Sci, 50*, 157–61.

White, M. M., Chen, L. Q., Kleinfield, R., Kallen, R. G., Barchi, R. L. (1991). SkM2, a Na+ channel cDNA clone from denervated skeletal muscle, encodes a tetrodotoxin-insensitive Na+ channel. *Mol Pharmacol, 39*, 604–8.

Winograd, C. H., Gerety, M. B., Chung, M., Goldstein, M. K., Dominguez, F. J., Vallone, R. (1991). Screening for frailty: criteria and prefictors of outcomes. *Journal of the American Geriatrics Society, 39*, 778–784.

Yang, J. S., Sladky, J. T., Kallen, R. G., Barchi, R. L. (1991). TTX-sensitive and TTX-insensitive sodium channel mRNA transcripts are independently regulated in adult skeletal muscle after denervation. *Neuron, 7*, 421–7.

Yazawa, M., Ferrante, C., Feng, J., Mio, K., Ogura, T., Zhang, M., Lin, P. H., Pan, Z., Komazaki, S., Kato, K., Nishi, M., Zhao, X., Weisleder, N., Sato, C., Ma, J., Takeshima, H. (2007). TRIC channels are essential for Ca2+ handling in intracellular stores. *Nature, 448*, 78–82.

Ye, P., Xing, Y., Dai, Z., D'ercole, J. (1996). In vivo actions of insulin-like growth factor-I (IGF-1) on cerebellum development in transgenic mice: evidence that IGF-1 increases proliferation of granule cells progenitors. *Developmental Brain Research, 95*, 44–54.

Zhang, C., Goto, N., Suzuki, M., Ke, M. (1996). Age-related reductions in number and size of anterior horn cells at C6 level of the human spinal cord. *Okajimas Folia Anatomica Japonica, 73*, 171–7.

Zheng, Z., Messi, M. L., Delbono, O. (2001). Age-dependent IGF-1 regulation of gene transcription of Ca^{2+} channels in skeletal muscle. *Mechanisms of Aging and Development, 122*, 373–384.

Zheng, Z., Wang, Z.-M., Delbono, O. (2002). Charge movement and transcription regulation of L-type calcium channel alpha-1S in skeletal muscle cells. *Journal of Physiology, 540*, 397–409.

Alterations in Mitochondria and Their Impact in Aging Skeletal Muscle

Russell T. Hepple

Abstract There is an abundance of studies examining the involvement of mitochondria in aging, including their role in the functional and structural deterioration of skeletal muscle with aging. Despite years of study, the precise involvement of mitochondria in the aging of skeletal muscle remains to be fully understood. This chapter provides some context for the current knowledge in this area and areas that will be refined through further study. It will examine the issue of "mitochondrial dysfunction" in aging; why it occurs and the functional consequences. The potential impact of three important age-related changes in mitochondria will be considered here: a reduced capacity for generating cellular energy in the form of adenosine triphosphate (ATP); an increased susceptibility to apoptosis; and an increase in reactive oxygen species (ROS) production with aging. The chapter considers the extent to which the mitochondrial content may be up-regulated in response to muscle activity as a means of assessing the malleability of the age-related impairments in mitochondria. Given the central importance of mitochondrial biology to so many facets of normal cell function, particularly in tissues with a wide metabolic scope like skeletal muscle, new discoveries about the significance of changes in mitochondria for aging skeletal muscles, and their potential remedy through lifestyle modification (e.g., exercise training, diet) and/ or medical intervention (e.g., pharmaceuticals, gene therapy), will remain at the forefront of our quest to promote healthy aging.

Keywords Apoptosis • Denervation • Exercise • Mitochondria • Mitochondrial biogenesis • Mitochondrial dysfunction • Plasticity • Reactive oxygen species

---

R.T. Hepple (✉)
Faculty of Kinesiology and Faculty of Medicine
University of Calgary, Calgary, Canada
e-mail: hepple@ucalgary.ca

G.S. Lynch (ed.), *Sarcopenia – Age-Related Muscle Wasting and Weakness*,
DOI 10.1007/978-90-481-9713-2_7,

1 Introduction

Aging is associated with myriad changes in physiological function. Amongst the most visible of these changes is a progressive loss of skeletal muscle mass and function, known as sarcopenia, a process that begins in approximately the 5th to 6th decade of life (Lexell et al. 1988; Hepple 2003). There is an abundance of studies examining the involvement of mitochondria in aging, including their role in the functional and structural deterioration of skeletal muscle with aging. Despite years of study, the precise involvement of mitochondria in the aging of skeletal muscle remains to be fully understood. On the other hand, the appeal of a central involvement of mitochondria in age-related changes in skeletal muscle is that this could provide a unifying explanation for both the loss of skeletal muscle mass (e.g., by increasing the incidence of apoptosis and increasing ROS-induced activation of the proteasome) and the decline in skeletal muscle contractile function (e.g., by reducing muscle aerobic capacity and oxidizing proteins involved in muscle contractile responses) with aging.

There is a multitude of ways that mitochondria might be involved in sarcopenia. The potential impact of three important age-related changes in mitochondria will be considered here: (1) a reduced capacity for generating cellular energy in the form of adenosine triphosphate (ATP) (Conley et al. 2000; Drew et al. 2003; Tonkonogi et al. 2003), (2) an increased susceptibility to apoptosis (Chabi et al. 2008; Seo et al. 2008), (3) and an increase in reactive oxygen species (ROS) production with aging (Capel et al. 2005; Mansouri et al. 2006; Chabi et al. 2008). In the context of explaining age-related muscle atrophy, mitochondria have been implicated in: (i) fiber loss, atrophy and breakage (Lee et al. 1998; Wanagat et al. 2001; Bua et al. 2002); (ii) an increase in apoptosis (Dirks and Leeuwenburgh 2002; Marzetti et al. 2008; Seo et al. 2008); and (iii) activation of protein degradation pathways via increased reactive oxygen species (ROS) generation (Muller et al. 2007; Hepple et al. 2008). In the context of explaining impaired muscle contractile function with aging, mitochondria have been implicated in: (i) the decline of aerobic contractile function secondary to reduced muscle oxidative capacity (Hepple et al. 2003; Hagen et al. 2004) and reduced muscle ATP generating capacity (Hepple et al. 2004a); (ii) impaired cross-bridge function secondary to oxidative damage to contractile proteins (Lowe et al. 2001; Prochniewicz et al. 2005; Thompson et al. 2006); and (iii) impaired Ca^{2+} handling secondary to oxidative damage to the Ca^{2+} handling apparatus (Fano et al. 2001; Boncompagni et al. 2006; Fugere et al. 2006; Thomas et al. 2009).

This chapter will examine the issue of mitochondrial dysfunction in aging muscles. The first point of examination will be to determine the extent to which mitochondrial dysfunction occurs (and what is meant by "mitochondrial dysfunction"). We will then examine why mitochondrial dysfunction occurs, and the functional consequences of mitochondrial dysfunction in aging muscles. Finally, we will consider the extent to which the mitochondrial content may be up-regulated in response to muscle activity as a means of assessing the malleability of the age-related impairments in mitochondria.

2 Age-Related Changes in Mitochondrial Function

Because of the complexity of mitochondrial structural and biochemical organization and the many roles that mitochondria serve within the cell, the ways in which mitochondrial function may be altered, and the consequences thereof, is vast. Underscoring this point, despite the fact that the mitochondrion was discovered more than a century ago (Altmann 1890), new insights into the scope of mitochondrial functional alteration, both in a physiological and pathological context, continue to this day. Perhaps reflecting a limited appreciation of the normal scope of mitochondrial function, although the term "mitochondrial dysfunction" is used extensively in the literature, the criteria used in making this qualification are often vague or inaccurate. In the interest of keeping things simple and given the central importance of mitochondria to cellular energy provision, one criterion that will be considered here in specifying a decline in mitochondrial function with aging is whether the capacity for energy provision per unit of mitochondrial volume is reduced. This criterion is distinct from a reduction in mitochondrial volume per se because a reduced skeletal muscle mitochondrial volume could occur in response to reduced physical activity with aging and reduce muscle oxidative capacity without impacting the ability of individual mitochondria to generate energy.

2.1 *Evidence for Reduced Oxidative Capacity Per Mitochondrion*

Although many studies have demonstrated a reduced mitochondrial oxidative capacity with aging at the level of whole muscle (e.g., enzyme assays using whole muscle homogenates) (Essen-Gustavsson and Borges 1986; Coggan et al. 1992; Sugiyama et al. 1993), these studies do not reveal the extent to which these declines might reflect a lower mitochondrial content due to a more sedentary lifestyle with aging versus changes intrinsic to the aged mitochondria themselves. Conley and colleagues provided the first in vivo estimation of mitochondrial function in aging skeletal muscle. Their study showed that there was a greater decline in the oxidative capacity of human vastus lateralis muscle (inferred from phosphocreatine recovery following knee extensor exercise) of aged subjects than could be accounted for by the reduction in mitochondrial volume density (measured by electron microscopy in muscle cross sections taken from biopsy samples), revealing a reduced oxidative capacity per volume of mitochondria in aged human skeletal muscle (Conley et al. 2000) (Fig. 1). Others have examined the function of mitochondria ex vivo using mitochondria isolated from muscles of aged individuals or organisms and the results have been mixed, with some groups finding reduced oxidative capacity or ATP production per unit of mitochondria in aged rodents (Desai et al. 1996; Drew et al. 2003; Mansouri et al. 2006) and aged humans (Short et al. 2005), and others finding no change in mitochondria isolated from skeletal muscles of older humans relative to younger adults (Rasmussen et al. 2003; Hutter et al. 2007). In addition to this, it has been

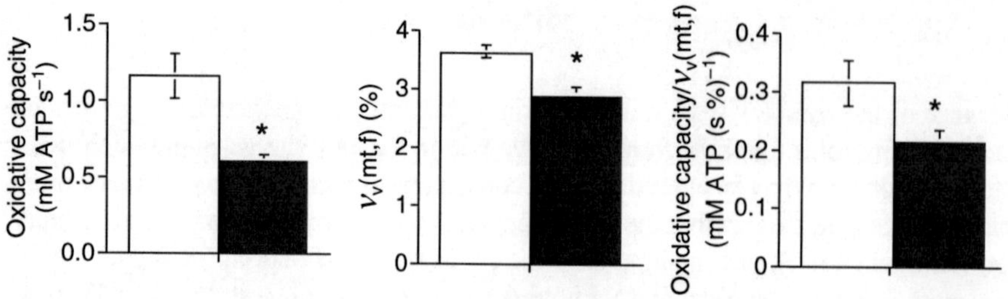

Fig. 1 The rate of muscle phosphocreatine resynthesis following muscle contractions, used as a surrogate of muscle oxidative capacity, was slower in muscle of elderly human adults (*black bar*) versus young adults (*open bar*) (*left panel*). Although the mitochondrial volume density (Vv[mt,f]%) was also lower in the elderly subjects (*middle panel*), taking the quotient of oxidative capacity and Vv[mt,f]% revealed a lower oxidative capacity per mitochondrion in the muscle of elderly humans (*right panel*) (Figure reproduced from Conley et al. [2000], with permission from The Physiological Society)

shown that whereas mitochondrial volume density does not decline between adulthood and senescence in rat fast- or slow-twitch muscles (Mathieu-Costello et al. 2005) (Fig. 2, panel A), there is a significant reduction in mitochondrial electron transport chain enzyme activities across this age range (Hagen et al. 2004; Hepple et al. 2006) (Fig. 2, panel B), indicating a reduced oxidative power per mitochondrion in aged skeletal muscles.

One of the factors suggested to account for inconsistency in some of the findings is that isolating mitochondria may underestimate the potential for mitochondrial dysfunction with aging by selectively harvesting the healthiest mitochondria (Tonkonogi et al. 2003). Although this has not been rigorously tested experimentally, it has been hypothesized that due to increasing fragility of some mitochondria with aging (Terman and Brunk 2004), this would result in selective harvest of the healthiest mitochondria in the aged muscles, thereby leading to an underestimate of the extent of mitochondrial dysfunction in isolated mitochondrial fractions (Tonkonogi et al. 2003). Furthermore, as mitochondria in skeletal muscle exist in varying degrees of a reticulum (Bakeeva et al. 1978; Kayar et al. 1988; Ogata and Yamasaki 1997), experimental isolation of mitochondria would disrupt this structural arrangement, which could also obscure important changes in mitochondrial function that would be evident in vivo.

Two other factors relating to the human literature may also contribute to inconsistency in observing mitochondrial dysfunction in studies of human subjects. Firstly, the screening measures required for human studies often results in loss of the least healthy subjects (Stathokostas et al. 2004), and it would be expected that this would bias the measures against identifying mitochondrial dysfunction. Secondly, mitochondrial function measurements in humans have not so far included subjects who are amongst the oldest of old (>75 years). Since the progression of sarcopenia exhibits a marked acceleration both in terms of declining muscle mass (Lexell et al. 1988; Hagen et al. 2004; Baker and Hepple 2006) and impaired

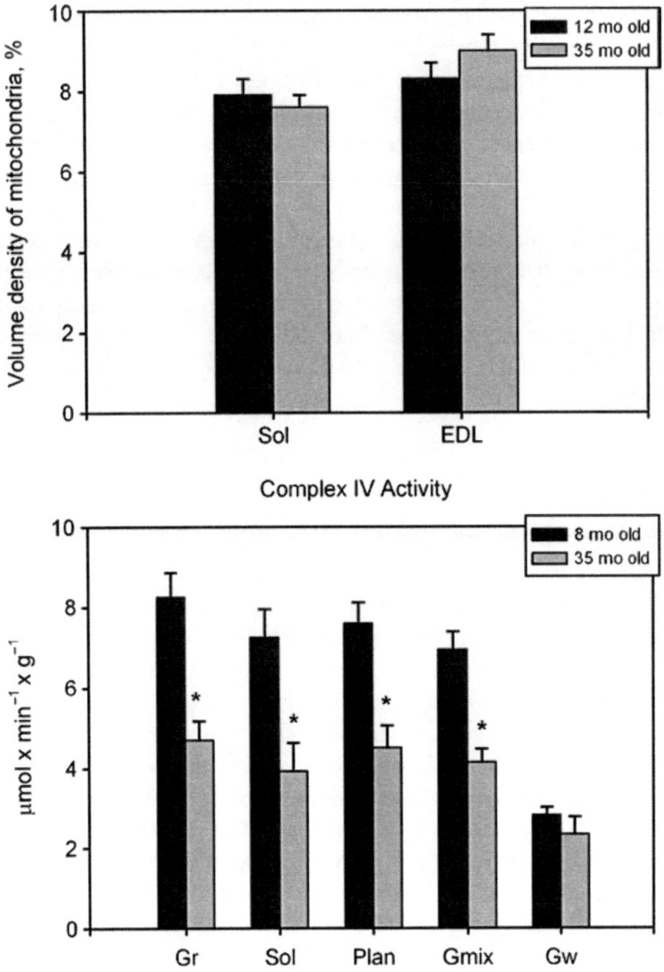

Fig. 2 Whereas aging is not associated with a reduction in mitochondrial volume density in either the slow-twitch soleus (Sol) muscle or the fast-twitch extensor digitorum longus (EDL) muscle between adulthood (12 month) and senescence (35 month old) in rats (*top panel*), there is a significant reduction in complex IV activity of the electron transport chain in the slow-twitch soleus muscle and mixed fast-twitch muscles like the red region of gastrocnemius (Gr), the mixed region of gastrocnemius (Gmix) and the plantaris (Plan) muscle (The top panel was adapted with permission from the American Physiological Society from data provided in Mathieu-Costello [2005]. The bottom panel was adapted with permission from Mary Ann Liebert, Inc. from a figure appearing in Hepple et al. [2006])

muscle function (Hagen et al. 2004; Hepple et al. 2004a) between late middle age and senescence, study of very old human subjects may reveal changes in mitochondrial function that have not been previously identified. These important issues remain to be adequately addressed in the literature.

One means by which the oxidative capacity per mitochondrion could be reduced with aging is via a selective loss of electron transport chain function. As the activities of different mitochondrial enzymes normally scale proportionally across a wide range of muscle oxidative capacity (Davies et al. 1981), this kind of alteration could be revealed by examining the activity of electron transport chain enzymes relative to other mitochondrial enzyme pathways. In this context, complex IV of the electron transport chain often exhibits a disproportionately lower activity with aging relative to other mitochondrial enzymes (Navarro and Boveris 2007). This has also been seen in aged skeletal muscles (Hepple et al. 2005, 2006). The reasons for a greater decline in complex IV activity remain to be agreed upon, but strong candidates include the accumulation of oxidative damage (Navarro and Boveris 2007; Choksi et al. 2008) and/or incorrect assembly of the subunit proteins.

2.2 *Aged Mitochondria Exhibit Greater ROS Generation*

Another indication of impaired mitochondrial function with aging is an increase in mitochondrial ROS generation. Although some ROS production is a normal part of mitochondrial physiology (Droge 2002) and is considered essential to facilitate adaptations in skeletal muscle (Gomez-Cabrera et al. 2005), excessive ROS production can lead to adverse consequences for skeletal myocytes. There are several studies showing that mitochondria isolated from skeletal muscles of aged humans (Capel et al. 2005) or rodent models (Bejma and Ji 1999; Capel et al. 2004; Mansouri et al. 2006; Vasilaki et al. 2006; Chabi et al. 2008) emit higher levels of ROS. On the basis of experiments using rotenone to inhibit complex I, it was suggested that the majority of the increase in mitochondrial ROS emission with aging was from complex I due to reverse electron transfer between complex II and complex I (Capel et al. 2005) (Fig. 3).

In summarizing age-related changes in mitochondrial function, although the findings are not uniformly in agreement, several lines of evidence suggest that aging is associated with a reduction in skeletal muscle mitochondrial oxidative capacity which exceeds that explainable by a reduction in muscle mitochondrial content. Furthermore, mitochondria from aged muscles pump out higher levels of ROS, which contributes to the greater accumulation of oxidative damage with aging, and likely plays a key role in impaired muscle function with aging and its greater vulnerability to apoptosis and excessive protein degradation. Therefore, while physical inactivity may be contributing to a declining muscle oxidative capacity with aging, the basis of mitochondrial functional alterations with aging likely includes aging-specific changes that are not reversible by restoring physical activity alone.

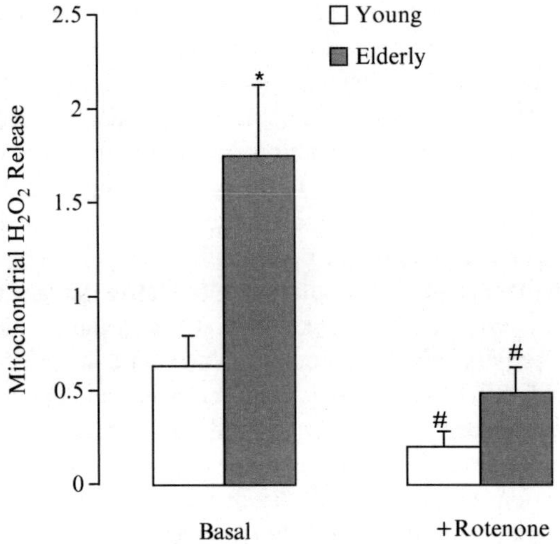

Fig. 3 Reactive oxygen species (ROS) production, as determined by hydrogen peroxide (H_2O_2) release, increases with aging in mitochondria isolated from human skeletal muscle. Furthermore, inhibition of Complex I ROS generation with Rotenone markedly reduces the elevated ROS generation with aging, suggesting that most of the increase in ROS with aging is due to reverse electron transfer from complex II to complex I (Figure reproduced from Capel et al. [2005], with permission from Elsevier)

3 Factors Accounting for Mitochondrial Dysfunction in Aged Muscles

Given the aforementioned evidence for mitochondrial dysfunction in aged muscles, an important question is why this occurs. Many different ideas are currently being explored, with some gaining experimental support. These include a reduced mitochondrial turnover, which leads to accumulation of poorly functioning mitochondria, and denervation which by some mechanism yet to be fully identified, leads to increased ROS generation and also low mitochondrial content in afflicted fibers.

3.1 Evidence for Decreased Mitochondrial Turnover with Aging

Mitochondrial protein exhibits a continual turnover, with the enzymes having a half-life of approximately 7 days (Booth and Holloszy 1977), although recent evidence from murine liver suggests mitochondrial turnover may be much more rapid, on the order of 2 days (Miwa et al. 2008). One reason for this high rate of turnover is that mitochondria normally produce some ROS, which even at physiological levels may oxidatively damage the mitochondrial proteins and mitochondrial DNA, leading to impaired enzyme function. This impaired enzyme function, particularly if it were to occur in the electron transport chain, could elevate ROS production and lead to a downward spiral in mitochondrial function. Thus, continual renewal of mitochondrial proteins is thought to be essential to the proper function of the mitochondria. It follows that changes in the rate at which mitochondria are turned over with aging can contribute to age-related cellular impairment.

Consistent with the idea that accumulation of oxidative damage can impair mitochondrial enzyme activity, elevating oxidative stress in aging muscle can reduce aconitase enzyme activity without reducing its protein content (Bota et al. 2002). The significance of this observation is that aconitase has an iron-sulfur center, which renders it particularly susceptible to oxidative damage, and thus it provides a useful biomarker of oxidative damage in mitochondria. In accounting for impaired mitochondrial function in aged skeletal muscles it is relevant that a major enzyme involved in the degradation of oxidatively damaged mitochondrial proteins (Lon protease) declines with aging (Bota et al. 2002), and mitochondrial protein synthesis rate declines in aged muscle (Rooyackers et al. 1996). Further to this latter point, there is evidence that the reduced mitochondrial protein synthesis may occur secondary to a reduced drive on mitochondrial biogenesis, based upon the decreased expression of peroxisome proliferator activated receptor coactivator gamma 1 alpha (PGC-1α) in aged skeletal muscle (Baker et al. 2006; Chabi et al. 2008). Finally, mitochondrial autophagy, whereby whole organelles are engulfed and enzymatically degraded in lysosomes, is thought to be impaired in aging muscles (Terman and Brunk 2004). Collectively, these changes lead to a reduced mitochondrial protein turnover with aging, due to the combined effects of reduced mitochondrial protein synthesis, impaired removal of oxidatively damaged mitochondrial proteins, and reduced mitochondrial autophagy. As implied above, the expected impact of this reduced mitochondrial turnover would not only be manifest as a reduced oxidative capacity per unit of mitochondrial volume because the longer mitochondrial protein dwell-time would exacerbate accumulation of oxidative damage, but also an increase in mitochondrial ROS generation secondary to, for example, a relatively greater reduction in complex IV activity (by allowing oxygen to accumulate to higher levels this favors production of ROS). This is consistent with the above-mentioned increase in mitochondrial ROS generation with aging in both rodent (Mansouri et al. 2006) and human (Capel et al. 2005) skeletal muscles.

3.2 *Role of Denervation in Mitochondrial Dysfunction*

The mechanistic basis for an increase in mitochondrial ROS production with aging may be due in part to a decreased mitochondrial renewal and resultant accumulation of 'aged' mitochondria, as described above (Section 3.1). In addition to this, recent evidence suggests that denervation may also be a predisposing factor. For example, skeletal mitochondrial ROS generation was shown to increase following surgical denervation (Adhihetty et al. 2007; Muller et al. 2007), and in disease models where there is loss of skeletal muscle α-motor neurons (Muller et al. 2007). Denervation is thought to affect muscle fibers and the mitochondria therein in several important ways. Perhaps the most important is the removal of neurotrophic influences that affect the drive on mitochondrial biogenesis. This is consistent with evidence showing a decreased expression of factors involved in driving mitochondrial biogenesis (e.g., PGC-1α, PGC-1β, mitochondrial transcription factor A) in skeletal muscle following denervation (Raffaello et al. 2006; Adhihetty et al. 2007; Sacheck et al. 2007) and that the pattern of their decline mirrors a decline in mitochondrial enzyme activities (Adhihetty et al. 2007). In addition, denervation is thought to increase phospholipase A signaling, resulting in hydrolysis of the mitochondrial membrane phospholipids and subsequent release of mitochondrial membrane-derived hydroperoxides (Bhattacharya et al. 2009). Finally, denervation also leads to an increase in pro-apoptotic factors, particularly those involving mitochondrial-driven apoptosis (Adhihetty et al. 2007). Collectively, therefore, denervation can have several important effects on mitochondria that may contribute to the increase in ROS generation observed in aging muscles.

4 Role of Mitochondria in Age-Related Muscle Deterioration

As noted in the Introduction, the appeal of a role for mitochondria in sarcopenia is that it may provide a unifying explanation for both the reduction of muscle mass and the impairment in contractile function in aging muscles. To this end, the following section will address the evidence that mitochondria are involved in both the mass and functional declines in aging skeletal muscles.

4.1 Involvement of Mitochondria in Age-Related Muscle Atrophy

As noted above, mitochondria in aging skeletal muscle exhibit numerous changes and several of these could have important implications in the context of age-related muscle atrophy. Firstly, the age-related increase in mitochodrial ROS generation is thought to induce protein degradation via NF-κB-induced activation of the proteasome (Jackman and Kandarian 2004; Powers et al. 2005). Although direct evidence of how this might be involved in sarcopenia remains to be provided, this idea is consistent with evidence that proteasome activity increases in aging skeletal muscle in a manner that is similar to the trajectory of age-related muscle atrophy (Hepple et al. 2008) (Fig. 4). This view is also consistent with observations in mice showing that muscle atrophy with (i) aging, (ii) superoxide dismutase 1 knockout, and (iii) experimental models of amyotrophic lateral sclerosis, correlates with the amount of muscle mitochondrial ROS production (Muller et al. 2007). Furthermore,

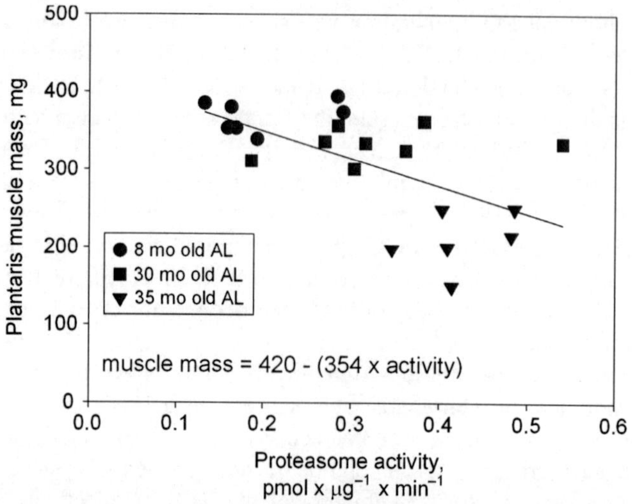

Fig. 4 Plantaris muscle mass versus the chymotrypsin-like activity of the proteasome in young adult (8 month old), late middle aged (30 month old) and senescent (35 month old) rats (Figure reproduced from Hepple et al. [2008], with permission from The American Physiological Society)

an increase in mitochondrial ROS generation following surgical denervation in skeletal muscle precedes muscle atrophy by several days (Muller et al. 2007). Despite the appeal of denervation being a cause of muscle atrophy in aged muscle, it is important to note that it is currently not known whether death of α-motor neurons is the cause versus the effect of myofiber atrophy and/or death in aged muscles. Interestingly, recent experiments in transgenic mice have examined a muscle-specific over-expression of uncoupling protein 1 (the isoform normally found in brown adipose tissue) by using a muscle creatine kinase promotor to limit expression to the myocytes, and these mice exhibit deterioration of neuromuscular junctions and retrograde α-motor neuron degeneration (Dupuis et al. 2009), showing that mitochondrial dysfunction within myocytes can be a cause of denervation. In addition, these animals exhibited a progressive loss of muscle mass (Dupuis et al. 2009). As such, these latter experiments show that abnormalities in mitochondrial metabolism within skeletal muscle fibers can be an initiating event in denervation. Therefore, the extent to which denervation is the initiating event in muscle atrophy with aging versus denervation occurring secondary to mitochondrial dysfunction in aging myocytes requires further study.

Some of the most compelling data examining the role of mitochondrial dysfunction in age-related muscle atrophy has been the studies examining the co-localization of mitochondrial dysfunction and mitochondrial DNA (mtDNA) damage with focal regions of fiber atrophy and breakage along the length of individual muscle fibers in aged muscle (Lee et al. 1998; Wanagat et al. 2001; Bua et al. 2002). The hypothesis most frequently cited to explain the significance of the aforementioned co-localization phenomenon is that mtDNA damage occurs segmentally along the length of individual muscle fibers (due to the accumulated effects of ROS) and that this damage is propagated by clonal expansion of damaged/mutated mtDNA within this region, leading to synthesis of mitochondria containing faulty electron transport chain enzymes (specifically those containing mtDNA-encoded subunits), which in turn is eventually manifest as a complex IV deficient fiber segment. This focal mitochondrial dysfunction is thought to have numerous consequences, including insufficient ATP supply, impaired protein synthesis, increased susceptibility to apoptosis, and increased mitochondrial ROS production, all of which may contribute to fiber atrophy and/or death (Wanagat et al. 2001).

Despite the elegance of experiments supporting this hypothesis, and the logical appeal of the explanation, the significance of this phenomenon for sarcopenia should be carefully scrutinized. Firstly, the only study to have examined this phenomenon in skeletal muscles from aging humans (Bua et al. 2006) found that although muscle fiber segments exhibiting complex IV deficiency co-localized with regions having a large burden of mtDNA damage, these fiber segments were not atrophied relative to regions with normal complex IV activity (Bua et al. 2006) (Fig. 5). Secondly, patients with so-called mtDNA disease exhibit much higher

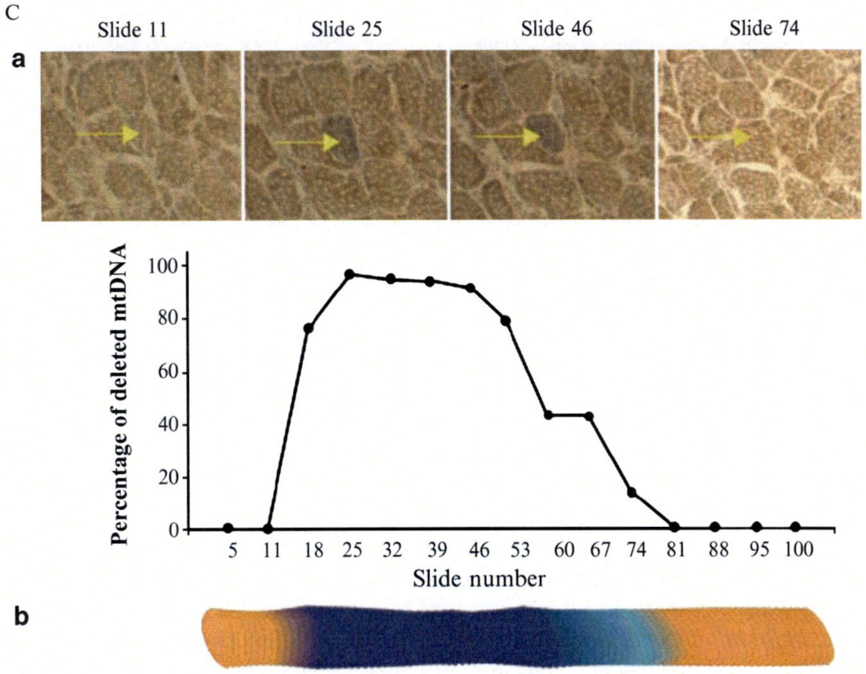

Fig. 5 Serial cross-sections of human skeletal muscle doubly-stained for succinate dehydrogenase and complex IV activity (*top panels*) depicting a fiber with a lack of complex IV activity (blue fiber indicated by *arrow*). Although fiber segments with complex IV deficiency (depicted as the blue region in the reconstructed fiber, *bottom panel*) exhibited high levels of deleted mitochondrial DNA (*middle panel*), these regions did not exhibit atrophy relative to fibers with normal complex IV activity (depicted as orange regions in the reconstructed fiber, bottom panel) (Reproduced from Bua et al. [2006], with permission from The American Society of Human Genetics)

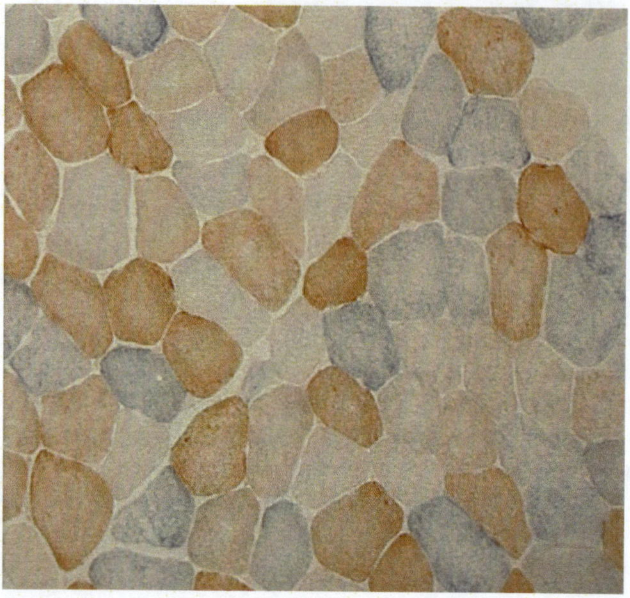

Fig. 6 Succinate dehydrogenase and complex IV doubly-stained cross-section of muscle from a patient with heteroplasmic mtDNA mutation. Note that the complex IV deficient fibers (blue fibers) are no different in size than fibers with normal complex IV activity (brown-orange fibers) (Reproduced from Taivassalo and Haller [2005], with permission from The American College of Sports Medicine)

burdens of mtDNA damage at the whole muscle level and very much higher fractions of muscle fibers exhibiting complex IV enzyme activity deficiency, and yet in these patients neither individual muscle fibers lacking complex IV activity (Fig. 6) nor their muscles as a whole are grossly atrophied relative to healthy individuals of the same age (Jacobs 2003). As such, the degree to which this phenomenon might contribute to sarcopenia remains an important area of investigation.

As suggested above, one specific manner in which mitochondria are proposed to be involved in sarcopenia involves apoptosis (Pollack and Leeuwenburgh 2001; Chabi et al. 2008; Seo et al. 2008). Mitochondria play a key role in regulating apoptosis, via the mitochondrial permeability transition pore (mPTP) which regulates the release of cytochrome c into the cytoplasm. A variety of stimuli, such as high Ca^{2+} and high ROS exposure, can lead to opening of the mPTP, allowing cytochrome c to leak out of the mitochondria and into the cytoplasm. Once released into the cytoplasm, cytochrome c binds with Apaf-1 and caspase 9, leading to the formation of an apoptosome, activation of caspase 9 and subsequent commitment of the apoptotic pathway via activation of caspase 3. In support of a role for apoptosis in age-related muscle atrophy, many studies have reported an increase in pro-apoptotic signaling in aged muscles (Alway et al. 2002; Dirks and Leeuwenburgh 2002; Giresi et al. 2005; Baker and Hepple 2006; Rice and Blough 2006; Chabi et al. 2008). On the other hand, differences in the degree of muscle atrophy between

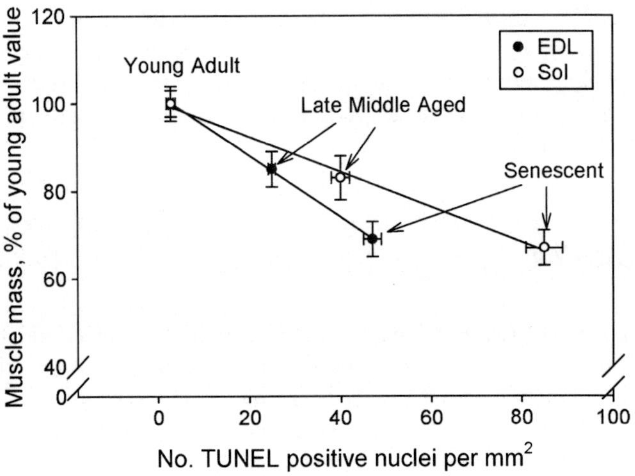

Fig. 7 Muscle mass in the fast-twitch extensor digitorum longus (EDL) muscle and slow-twitch soleus muscle (Sol) versus the density of TUNEL-positive nuclei (a marker of apoptotic nuclei) as sarcopenia progresses with aging (Data reproduced from Rice et al. [2006])

individuals in senescent animals do not track well with differences in expression of pro-apoptotic transcripts (Baker and Hepple 2006). In addition, although the progression of muscle atrophy with aging correlates generally with an increase in number of apoptotic nuclei in both fast-twitch and slow-twitch muscles, it is striking that there are markedly more apoptotic nuclei in the slow-twitch soleus muscle than the fast-twitch extensor digitorum longus muscle, despite very similar amounts of atrophy (Fig. 7; data taken from (Rice and Blough 2006)). This difference may relate to the fact that muscle fibers are multi-nucleated and, therefore, apoptotic loss of a nucleus within a given myocyte does not need to result in loss of the myocyte entirely. As such, a difference in the incidence of apoptotic nuclei between muscles having the same amount of atrophy could reflect differences in the ability of these muscles to regenerate and repair, e.g., via recruitment of satellite cells. Whether this or another explanation applies awaits further investigation.

Notwithstanding some uncertainty about the degree to which apoptosis directly explains the degree of muscle atrophy with aging, recent data suggests that accumulation of non-heme iron in skeletal muscle mitochondria may be one mechanism leading to an increased incidence of mitochondrial-mediated apoptosis in aged skeletal muscle. Specifically, accumulation of non-heme iron with aging is hypothesized to exacerbate mitochondrial ROS generation (and thus oxidative damage) via the Fenton reaction, wherein the increased mitochondrial damage leads to an increased probability of mPTP opening (Seo et al. 2008). This notion is consistent with the aforementioned increase in mitochondrial ROS generation in aged skeletal muscles (Section 2.2), and observations indicating greater accumulation of non-heme iron in mitochondria isolated from aged skeletal muscle (Seo et al. 2008). In addition, mitochondria from aged muscles exhibit a greater release of cytochrome c in response to ROS-induced stress (Chabi et al. 2008), which may in part explain the increased susceptibility to mitochondrial-driven apoptosis in aging muscle. Thus, collectively, there is substantial evidence that apoptosis increases in aged muscles and that age-related changes in mitochondria are likely to be involved.

4.2 *Involvement of Mitochondria in Age-Related Muscle Dysfunction*

In addition to the potential involvement of mitochondria in the age-related loss of muscle mass, there is considerable support for the involvement of mitochondria in impaired muscle function with aging. For example, there is a progressive decline in skeletal muscle aerobic function with aging that is not due to loss of capillaries (Hepple and Vogell 2004; Mathieu-Costello et al. 2005), but rather correlates with a progressive loss of mitochondrial oxidative capacity in aging muscles (Hagen et al. 2004) (Fig. 8). As noted in Section 3, a decline in muscle mitochondrial oxidative capacity may be caused by a reduction in the expression of PGC-1α in aged muscles (Baker et al. 2006; Chabi et al. 2008). In this context, it is important to note that aged muscles, particularly in senescence, are characterized by an accumulation of very small muscle fibers. Although this area requires further study, it seems likely that a large proportion of these small fibers are denervated (Hepple et al. 2004b) and that a sub-fraction of these may be attempting to regenerate. The reason this is relevant here is that these small fibers have lower levels of markers of

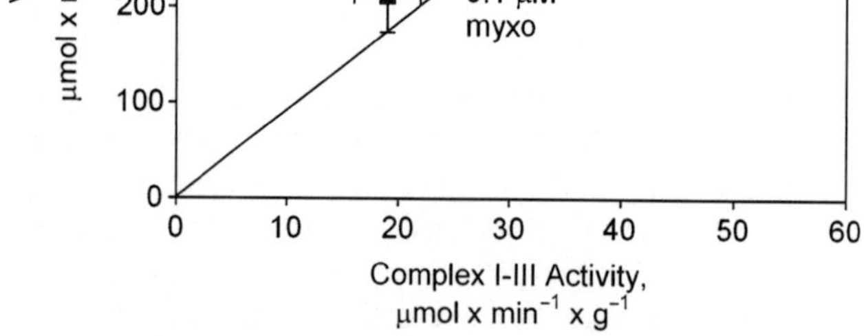

Fig. 8 Muscle maximal oxygen uptake (VO_{2max}) in pump-perfused rat hindlimb versus the flux capacity of complex I–III in homogenates of gastrocnemius muscle. The figure shows that the age-related decline in VO_{2max} parallels the decline in flux capacity through a key part of the mitochondrial electron transport chain (Reproduced from Hagen et al. [2004], with permission from The Gerontological Society of America)

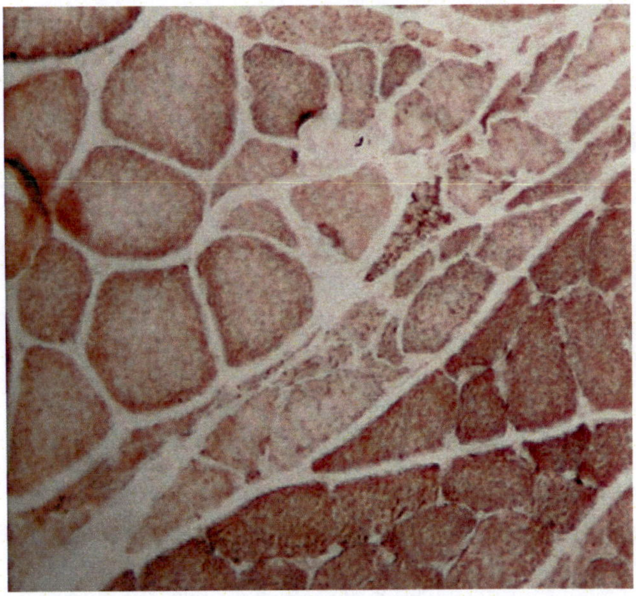

Fig. 9 Senescent rat gastrocnemius muscle cross-section stained for complex IV activity. Note that the very small fibers have a lower complex IV activity than the larger fibers, showing that the accumulation of these very small fibers in aged muscle, particularly in senescence, contributes to the overall decline in muscle oxidative capacity with aging (R.T. Hepple [unpublished])

mitochondrial content (e.g., complex IV activity) (Fig. 9), and because of this they contribute significantly to the lower muscle oxidative capacity. Furthermore, denervation, or perhaps failure to reinnervate, may be constraining the mitochondrial content of these fibers, secondary to the aforementioned reduction in drive on mitochondrial biogenesis that occurs in denervated muscle (Adhihetty et al. 2007) (Section 3.1). Thus, the reduction of muscle mitochondrial oxidative capacity with aging may have an important neurological involvement. This point needs further consideration in the experimental literature.

As noted in Section 3.2, aged muscles are also characterized by mitochondria that emit higher levels of ROS. This increase in mitochondrial ROS generation in aging skeletal muscles can exacerbate oxidative damage to proteins, which has been shown to inhibit the biological activity of enzymes, particularly those containing iron-sulfur centers (Bota et al. 2002; Ma et al. 2009). In addition, several proteins involved in muscle contraction are known to be specifically targeted by oxidative stress, and thus, likely contribute to the impairment in muscle contractile function with aging. Prochniewicz et al. (2005) previously showed using in vitro motility assays that although actin function was unaltered with aging, the catalytically active portion of myosin (heavy meromyosin) was impaired in muscles of aged versus young adult rats. In addition, this difference in actin versus myosin function with aging corresponded to differences in the susceptibility of actin versus myosin to accumulate oxidative damage to cysteine molecules (Prochniewicz et al. 2005). Similarly, there is an increase in oxidative damage, particularly nitrotyrosine damage, to the sarcoplasmic reticulum ATPase in aged muscles (Fugere et al. 2006; Thomas et al. 2009), and this is thought to contribute to decreases in maximal SERCA activity in aged muscle (Thomas et al. 2009). As such, the collective evidence suggests that oxidative damage to various proteins within skeletal muscle, and the mitochondria therein, can lead to functional deterioration in aging skeletal muscle.

5 Plasticity of Mitochondria in Aging Muscles

Given the above evidence of reduced mitochondrial oxidative capacity and increased ROS generation with aging, both of which have been attributed in part to accumulation of damaged mitochondria secondary to reduced mitochondrial renewal, an obvious question is whether aged muscle simply loses the capacity to increase its mitochondrial content. The majority of what we know about this question has been obtained from experiments examining changes in muscle mitochondrial oxidative capacity in response to exercise training or chronic electrical stimulation. Significantly, an emerging concept is that the capacity for mitochondrial biogenesis in response to muscle activation, while relatively preserved in the younger of the old, becomes severely impaired in the oldest old.

There are many studies showing that aged muscles can respond favorably by increasing markers of mitochondrial content in response to endurance exercise training in both the human (Orlander and Aniansson 1980; Hagberg et al. 1989; Meredith et al. 1989; Short et al. 2003) and animal model (Cartee and Farrar 1987; Rossiter et al. 2005; Betik et al. 2008) literature. However, it is important to realize that these prior studies have not considered potential differences in the endurance training responses between late middle age versus the senescent period (i.e., when survival rates drop below 50%), and it is the senescent period when the consequences of aging for skeletal muscle become most severe. To address this issue, we recently examined the effect of aging on the responses of the skeletal muscle aerobic machinery to endurance training in rat skeletal muscles. Interestingly, whereas skeletal muscle aerobic function (in situ maximal oxygen consumption) and mitochondrial enzyme activities increased significantly when endurance exercise training was imposed in late middle age and continued for 7 weeks (Betik et al. 2008) (Fig. 10), the skeletal muscles completely lost this positive adaptation when the training was continued for 7 months into the senescent period (Betik et al. 2009) (Fig. 11). Further to this, the normally robust response of PGC-1α expression to endurance exercise training seen in studies of rodents (Baar et al. 2002; Terada et al. 2002) and young adult humans (Norrbom et al. 2004) was abolished in senescent rat skeletal muscles following 7 months of endurance exercise training in both the slow-twitch soleus muscle and the fast-twitch plantaris muscle (Fig. 12) (Betik et al. 2009). On the basis of these results, therefore, it appears that senescent

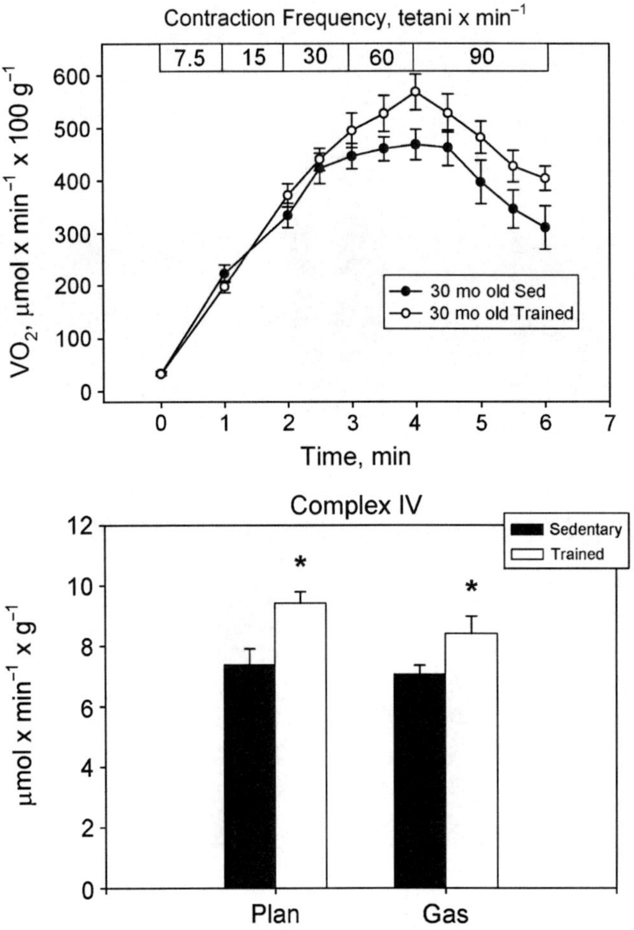

Fig. 10 Muscle oxygen uptake during incremental muscle contractions in distal rat hindlimb muscles pump-perfused in situ (*top*) and the activity of complex IV in homogenates of plantaris (Plan) and gastrocnemius (Gas) muscle (*bottom*) in sedentary late middle aged rats and late middle aged rats exercise-trained for 7 weeks (Reproduced from Betik et al. [2008], with permission from The Physiological Society [London])

muscle in particular has a markedly diminished capacity to increase mitochondrial biogenesis in response to an endurance training stimulus, and that this is due in part to an impaired ability to up-regulate PGC-1α. This finding of reduced adaptability with endurance training in senescence is consistent with studies demonstrating that skeletal muscle from the oldest old also has a diminished plasticity in response to resistance exercise training (Slivka et al. 2008; Raue et al. 2009) and functional overload (Blough and Linderman 2000).

The aforementioned results indicate that senescent skeletal muscle loses its ability to generate new mitochondria in advanced age, suggesting that the reduced

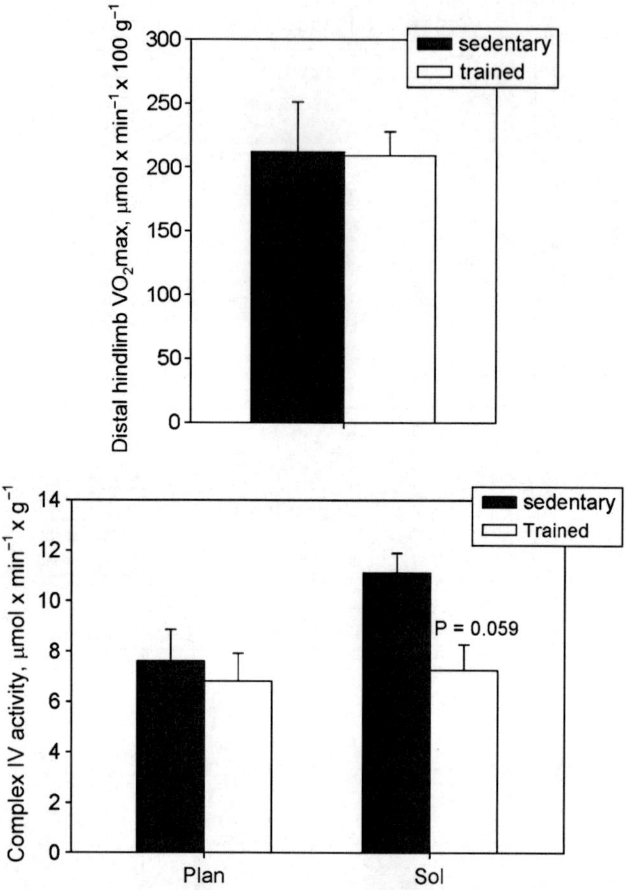

Fig. 11 Muscle maximal oxygen uptake (VO_2max) during incremental muscle contractions in distal rat hindlimb muscles pump-perfused in situ (*top*) and the activity of complex IV in homogenates of plantaris (Plan) and Soleus (Sol) muscle (*bottom*) in sedentary senescent rats and senescent rats trained for 7 months beginning in late middle age (Data reproduced from Betik et al. [2009])

mitochondrial turnover rate with aging is secondary to this diminished capacity to make new mitochondria. However, an important question remains: is it that senescent muscle loses its adaptive plasticity per se, or is the limitation the result of the much lower exercise stimulus that can be sustained in very old age. To help address this issue, a recent study examined the response of young adult versus senescent skeletal muscle to an acute bout of low frequency electrical stimulation. Interestingly, these experiments revealed that whereas the cell signaling pathway, including molecules involved in driving mitochondrial biogenesis (e.g., adenosine monophosphate protein kinase [AMPK] activation), was relatively intact in the highly oxidative region of the tibialis anterior muscle, there was a blunted response

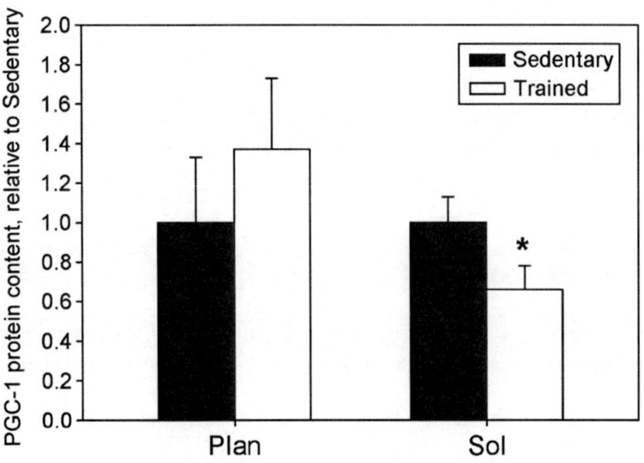

Fig. 12 PGC-1 protein expression in plantaris (Plan) and soleus (Sol) muscles of sedentary senescent rats and senescent rats trained for 7 months beginning in late middle age. *$P<0.05$ versus Sedentary group (Data reproduced from Betik et al. [2009])

in the highly glycolytic region of this muscle in senescence (Ljubicic and Hood 2009). These data are generally consistent with another study showing that AMPK activation is markedly blunted in aged muscles following either pharamacological stimuli or an acute exercise bout (Reznick et al. 2007). What is not yet clear, however, is the degree to which an attenuated mitochondrial biogenesis response is a general property of all muscle fibers in an aged muscle, versus there being an increasing proportion of muscle fibers which cannot contribute to the whole muscle mitochondrial biogenesis response (e.g., those that have become denervated and/or which are undergoing regeneration). Irrespective of this point, the growing consensus is that aging muscle, particularly in senescence, displays an impaired ability to up-regulate mitochondrial biogenesis and this in turn plays an important role in the attenuated benefits of endurance exercise training for skeletal muscle aerobic capacity in senescence. Future studies need to address whether this loss of adaptive plasticity is an immutable consequence of aging, or if other interventions yet to be identified can help restore the adaptive response to increased muscle use.

6 Conclusions

Mitochondrial changes in aging skeletal muscles, and the implications these have for the decline in both muscle mass and its function with aging, have constituted an intensive area of study. The aforementioned chapter provides some context for the current knowledge in this area and areas that will be refined through further study. Given the central importance of mitochondrial biology to so many facets of normal cell function, particularly in tissues with a wide metabolic scope like skeletal muscle, new discoveries about the significance of changes in mitochondria for aging skeletal muscles, and their potential remedy through lifestyle modification (e.g., exercise training, diet) and/or medical intervention (e.g., pharmaceuticals, gene therapy), will remain at the forefront of our quest to promote healthy aging.

References

Adhihetty, P. J., O'Leary, M. F., Chabi, B., Wicks, K. L., Hood, D. A. (2007). Effect of denervation on mitochondrially mediated apoptosis in skeletal muscle. *Journal of Applied Physiology, 102*, 1143–1151.

Altmann, R. (1890). Die Elementarorganismen und ihre Beziehungen zu den Zellen. *Verlag von Veit & Comp Leipzig.*

Alway, S. E., Degens, H., Krishnamurthy, G., Smith, C. A. (2002). Potential role for Id myogenic repressors in apoptosis and attenuation of hypertrophy in muscles of aged rats. *AJP – Cell Physiology, 283*, C66–C76.

Baar, K., Wende, A. R., Jones, T. E., Marison, M., Nolte, L. A., Chen, M., Kelly, D. P., Holloszy, J. O. (2002). Adaptations of skeletal muscle to exercise: rapid increase in the transcriptional coactivator PGC-1. *The FASEB Journal, 16*, 1879–1886.

Bakeeva, L. E., Chentsov, Y. S., Skulachev, V. P. (1978). Mitochondrial framework (reticulum mitochondriale) in rat diaphragm muscle. *Biochim Biophys Acta, 501*, 349–369.

Baker, D. J. & Hepple, R. T. (2006). Elevated caspase and AIF signaling correlates with the progression of sarcopenia during aging in male F344BN rats. *Experimental Gerontology, 41*, 1149–1156.

Baker, D. J., Betik, A. C., Krause, D. J., Hepple, R. T. (2006). No decline in skeletal muscle oxidative capacity with aging in long-term caloric restricted rats: effects are independent of mtDNA integrity. *The Journals of Gerontology. Series A: Biological Sciences and Medical Sciences, 61A*, 675–684.

Bejma, J. & Ji, L. L. (1999). Aging and acute exercise enhance free radical generation in rat skeletal muscle. *Journal of Applied Physiology, 87*, 465–470.

Betik, A. C., Baker, D. J., Krause, D. J., McConkey, M. J., Hepple, R. T. (2008). Exercise training in late middle-aged male Fischer 344 x Brown Norway F1-hybrid rats improves skeletal muscle aerobic function. *Experimental Physiology, 93*, 863–871.

Betik, A. C., Thomas, M. M., Wright, K. J., Riel, C. D., Hepple, R. T. (2009). Exercise training from late middle age until senescence does not attenuate the declines in skeletal muscle aerobic function. *American Journal of Physiology: Regulatory, Integrative and Comparative Physiology, 297*, R744–R755.

Bhattacharya, A., Muller, F. L., Liu, Y., Sabia, M., Liang, H., Song, W., Jang, Y. C., Ran, Q., Van Remmen, H. (2009). Denervation induces cytosolic phospholipase A2-mediated fatty acid hydroperoxide generation by muscle mitochondria. *The Journal of Biological Chemistry, 284*, 46–55.

Blough, E. R. & Linderman, J. K. (2000). Lack of skeletal muscle hypertrophy in very aged male Fischer 344 X Brown Norway rats. *Journal of Applied Physiology, 88*, 1265–1270.

Boncompagni, S., d'Amelio, L., Fulle, S., Fano, G., Protasi, F. (2006). Progressive disorganization of the excitation-contraction coupling apparatus in aging human skeletal muscle as revealed by electron microscopy: A possible role in the decline of muscle performance. *The Journals of Gerontology. Series A: Biological Sciences and Medical Sciences, 61*, 995–1008.

Booth, F. W. & Holloszy, J. O. (1977). Cytochrome c turnover in rat skeletal muscles. *The Journal of Biological Chemistry, 252*, 416–419.

Bota, D. A., Van Remmen, H., Davies, K. J. (2002). Modulation of Lon protease activity and aconitase turnover during aging and oxidative stress. *FEBS Letters, 532*, 103–106.

Bua, E., Johnson, J., Herbst, A., Delong, B., McKenzie, D., Salamat, S., Aiken, J. M. (2006). Mitochondrial DNA-deletion mutations accumulate intracellularly to detrimental levels in aged human skeletal muscle fibers. *American Journal of Human Genetics, 79*, 469–480.

Bua, E. A., McKiernan, S. H., Wanagat, J., McKenzie, D., Aiken, J. M. (2002). Mitochondrial abnormalities are more frequent in muscles undergoing sarcopenia. *Journal of Applied Physiology, 92*, 2617–2624.

Capel, F., Buffiere, C., Patureau, M. P., Mosoni, L. (2004). Differential variation of mitochondrial H2O2 release during aging in oxidative and glycolytic muscles in rats. *Mechanisms of Ageing and Development, 125*, 367–373.

Capel, F., Rimbert, V., Lioger, D., Diot, A., Rousset, P., Mirand, P. P., Boirie, Y., Morio, B., Mosoni, L. (2005). Due to reverse electron transfer, mitochondrial H2O2 release increases with age in human vastus lateralis muscle although oxidative capacity is preserved. *Mechanisms of Ageing and Development, 126*, 505–511.

Cartee, G. D. & Farrar, R. P. (1987). Muscle respiratory capacity and VO 2max in identically trained young and old rats. *Journal of Applied Physiology, 63*, 257–261.

Chabi, B., Ljubicic, V., Menzies, K. J., Huang, J. H., Saleem, A., Hood, D. A. (2008). Mitochondrial function and apoptotic susceptibility in aging skeletal muscle. *Aging Cell, 7*, 2–12.

Choksi, K. B., Nuss, J. E., Deford, J. H., Papaconstantinou, J. (2008). Age-related alterations in oxidatively damaged proteins of mouse skeletal muscle mitochondrial electron transport chain complexes. *Free Radical Biology & Medicine, 45*, 826–838.

Coggan, A. R., Spina, R. J., King, D. S., Rogers, M. A., Brown, M., Nemeth, P. M., Holloszy, J. O. (1992). Histochemical and enzymatic comparison of the gastrocnemius muscle of young and elderly men and women. *Journal of Gerontology, 47*, B71–B76.

Conley, K. E., Jubrias, S. A., Esselman, P. C. (2000). Oxidative capacity and aging in human muscle. *The Journal of Physiology, 526.1*, 203–210.

Davies, K. J., Packer, L., Brooks, G. A. (1981). Biochemical adaptation of mitochondria, muscle, and whole-animal respiration to endurance training. *Archives of Biochemistry and Biophysics, 209*, 539–554.

Desai, V. G., Weindruch, R., Hart, R. W., Feuers, R. J. (1996). Influences of age and dietary restriction on gastrocnemius electron transport system activities in mice. *Archives of Biochemistry and Biophysics, 333*, 145–151.

Dirks, A. & Leeuwenburgh, C. (2002). Apoptosis in skeletal muscle with aging. *AJP - Regulatory, Integrative and Comparative Physiology, 282*, R519–R527.

Drew, B., Phaneuf, S., Dirks, A., Selman, C., Gredilla, R., Lezza, A., Barja, G., Leeuwenburgh, C. (2003). Effects of aging and caloric restriction on mitochondrial energy production in gastrocnemius muscle and heart. *American Journal of Physiology: Regulatory, Integrative and Comparative Physiology, 284*, R474–R480.

Droge, W. (2002). Free radicals in the physiological control of cell function. *Physiological Reviews, 82*, 47–95.

Dupuis, L., Gonzalez de Aguilar, J. L., Echaniz-Laguna, A., Eschbach, J., Rene, F., Oudart, H., Halter, B., Huze, C., Schaeffer, L., Bouillaud, F., Loeffler, J. P. (2009). Muscle mitochondrial uncoupling dismantles neuromuscular junction and triggers distal degeneration of motor neurons. *PLoS ONE, 4*, e5390.

Essen-Gustavsson, B. & Borges, O. (1986). Histochemical and metabolic characteristics of human skeletal muscle in relation to age. *Acta Physiologica Scandinavica, 126*, 107–114.

Fano, G., Mecocci, P., Vecchiet, J., Belia, S., Fulle, S., Polidori, M. C., Felzani, G., Senin, U., Vecchiet, L., Beal, M. F. (2001). Age and sex influence on oxidative damage and functional status in human skeletal muscle. *Journal of Muscle Research and Cell Motility, 22*, 345–351.

Fugere, N. A., Ferrington, D. A., Thompson, L. V. (2006). Protein nitration with aging in the rat semimembranosus and soleus muscles. *The Journals of Gerontology. Series A: Biological Sciences and Medical Sciences, 61*, 806–812.

Giresi, P. G., Stevenson, E. J., Theilhaber, J., Koncarevic, A., Parkington, J., Fielding, R. A., Kandarian, S. C. (2005). Identification of a molecular signature of sarcopenia. *Physiological Genomics, 21*, 253–263.

Gomez-Cabrera, M. C., Borras, C., Pallardo, F. V., Sastre, J., Ji, L. L., Vina, J. (2005). Decreasing xanthine oxidase-mediated oxidative stress prevents useful cellular adaptations to exercise in rats. *The Journal of Physiology Online, 567*, 113–120.

Hagberg, J. M., Graves, J. E., Limacher, M. C., Woods, D. R., Legget, S. H., Cononie, C. C., Gruber, J. J., Pollock, M. L. (1989). Cardiovascular responses of 70- to 79-yr-old men and women to exercise training. *Journal of Applied Physiology, 66*, 2589–2594.

Hagen, J. L., Krause, D. J., Baker, D. J., Fu, M., Tarnopolsky, M. A., Hepple, R. T. (2004). Skeletal muscle aging in F344BN F1-hybrid rats: I. Mitochondrial dysfunction contributes to the age-associated reduction in VO2max. *The Journals of Gerontology. Series A: Biological Sciences and Medical Sciences, 59A*, 1099–1110.

Hepple, R. T. (2003). Sarcopenia – A critical perspective. *Science of Aging Knowledge Environment, 2003*, pe31.

Hepple, R. T., Hagen, J. L., Krause, D. J., Jackson, C. C. (2003). Aerobic power declines with aging in rat skeletal muscles perfused at matched convective O2 delivery. *Journal of Applied Physiology, 94*, 744–751.

Hepple, R. T., Hagen, J. L., Krause, D. J., Baker, D. J. (2004a). Skeletal muscle aging in F344BN F1-hybrid rats: II. Improved contractile economy in senescence helps compensate for reduced ATP generating capacity. *The Journals of Gerontology Series A: Biological Sciences and Medical Sciences, 59A*, 1111–1119.

Hepple, R. T., Ross, K. D., Rempfer, A. B. (2004b). Fiber Atrophy and Hypertrophy in Skeletal Muscles of Late Middle-Aged Fischer 344 x Brown Norway F1-Hybrid Rats. *The Journals of Gerontology Series A: Biological Sciences and Medical Sciences, 59*, B108–B117.

Hepple, R. T., Baker, D. J., Kaczor, J. J., Krause, D. J. (2005). Long-term caloric restriction abrogates the age-related decline in skeletal muscle aerobic function. *The FASEB Journal, 19*, 1320–1322.

Hepple, R. T., Baker, D. J., McConkey, M., Murynka, T., Norris, R. (2006). Caloric restriction protects mitochondrial function with aging in skeletal and cardiac muscles. *Rejuvenation Research, 9*, 219–222.

Hepple, R. T., Qin, M., Nakamoto, H., Goto, S. (2008). Caloric restriction optimizes the proteasome pathway with aging in rat plantaris muscle: implications for sarcopenia. *American Journal of Physiology: Regulatory, Integrative and Comparative Physiology, 295*, R1231–1237.

Hepple, R. T. & Vogell, J. E. (2004). Anatomic capillarization is maintained in relative excess of fiber oxidative capacity in some skeletal muscles of late middle aged rats. *Journal of Applied Physiology, 96*, 2257–2264.

Hutter, E., Skovbro, M., Lener, B., Prats, C., Rabol, R., Dela, F., Jansen-Durr, P. (2007). Oxidative stress and mitochondrial impairment can be separated from lipofuscin accumulation in aged human skeletal muscle. *Aging Cell, 6*, 245–256.

Jackman, R. W., Kandarian, S. C. (2004). The molecular basis of skeletal muscle atrophy. *AJP – Cell Physiology, 287*, C834–C843.

Jacobs, H. T. (2003). The mitochondrial theory of aging: dead or alive? *Aging Cell, 2*, 11–17.

Kayar, S. R., Hoppeler, H., Mermod, L., Weibel, E. R. (1988). Mitochondrial size and shape in equine skeletal muscle: a three-dimensional reconstruction study. *Anatomical Record, 222*, 333–339.

Lee, C. M., Lopez, M. E., Weindruch, R., Aiken, J. M. (1998). Association of age-related mitochondrial abnormalities with skeletal muscle fiber atrophy. *Free Radical Biology and Medicine, 25*, 964–972.

Lexell, J., Taylor, C. C., Sjostrom, M. (1988). What is the cause of the ageing atrophy? Total number, size and proportion of different fiber types studied in whole vastus lateralis muscle from 15- to 83-year-old men. *Journal of the Neurological Sciences, 84*, 275–294.

Ljubicic, V. & Hood, D. A. (2009). Diminished contraction-induced intracellular signaling towards mitochondrial biogenesis in aged skeletal muscle. *Aging Cell, 8*, 394–404.

Lowe, D. A., Surek, J. T., Thomas, D. D., Thompson, L. V. (2001). Electron paramagnetic resonance reveals age-related myosin structural changes in rat skeletal muscle fibers. *American Journal of Physiology – Cell Physiology, 280*, C540–C547.

Ma, Y. S., Wu, S. B., Lee, W. Y., Cheng, J. S., Wei, Y. H. (2009). Response to the increase of oxidative stress and mutation of mitochondrial DNA in aging. *Biochimica et Biophysica Acta, 1790*, 1021–1029.

Mansouri, A., Muller, F. L., Liu, Y., Ng, R., Faulkner, J., Hamilton, M., Richardson, A., Huang, T. T., Epstein, C. J., Van, R. H. (2006). Alterations in mitochondrial function, hydrogen peroxide release and oxidative damage in mouse hind-limb skeletal muscle during aging. *Mechanisms of Ageing and Development, 127*, 298–306.

Marzetti, E., Wohlgemuth, S. E., Lees, H. A., Chung, H. Y., Giovannini, S., Leeuwenburgh, C. (2008). Age-related activation of mitochondrial caspase-independent apoptotic signaling in rat gastrocnemius muscle. *Mechanisms of Ageing and Development, 129*, 542–549.

Mathieu-Costello, O., Ju, Y., Trejo-Morales, M., Cui, L. (2005). Greater capillary-fiber interface per fiber mitochondrial volume in skeletal muscles of old rats. *Journal of Applied Physiology, 99*, 281–289.

Meredith, C. N., Frontera, W. R., Fisher, E. C., Hughes, V. A., Herland, J. C., Edwards, J., Evans, W. J. (1989). Peripheral effects of endurance training in young and old subjects. *Journal of Applied Physiology, 66*, 2844–2849.

Miwa, S., Lawless, C., von Zglinicki, T. (2008). Mitochondrial turnover in liver is fast in vivo and is accelerated by dietary restriction: application of a simple dynamic model. *Aging Cell, 7*, 920–923.

Muller, F. L., Song, W., Jang, Y. C., Liu, Y., Sabia, M., Richardson, A., Van Remmen, H. (2007). Denervation-induced skeletal muscle atrophy is associated with increased mitochondrial ROS production. *American Journal of Physiology: Regulatory, Integrative and Comparative Physiology, 293*, R1159–1168.

Navarro, A. & Boveris, A. (2007). The mitochondrial energy transduction system and the aging process. *AJP – Cell Physiology, 292*, C670–C686.

Norrbom, J., Sundberg, C. J., Ameln, H., Kraus, W. E., Jansson, E., Gustafsson, T. (2004). PGC-1{α} mRNA expression is influenced by metabolic perturbation in exercising human skeletal muscle. *Journal of Applied Physiology, 96*, 189–194.

Ogata, T. & Yamasaki, Y. (1997). Ultra-high-resolution scanning electron microscopy of mitochondria and sarcoplasmic reticulum arrangement in human red, white, and intermediate muscle fibers. *Anatomical Record, 248*, 214–223.

Orlander, J. & Aniansson, A. (1980). Effects of physical training on skeletal muscle metabolism and ultrastructure in 70 to 75-year-old men. *Acta Physiologica Scandinavica, 109*, 149–154.

Pollack, M. & Leeuwenburgh, C. (2001). Apoptosis and Aging: Role of the Mitochondria. *The Journals of Gerontology. Series A: Biological Sciences and Medical Sciences, 56*, B475–B482.

Powers, S. K., Kavazis, A. N., DeRuisseau, K. C. (2005). Mechanisms of disuse muscle atrophy: role of oxidative stress. *AJP – Regulatory, Integrative and Comparative Physiology, 288*, R337–R344.

Prochniewicz, E., Thomas, D. D., Thompson, L. V. (2005). Age-related decline in actomyosin function. *The Journals of Gerontology. Series A: Biological Sciences and Medical Sciences, 60*, 425–431.

Raffaello, A., Laveder, P., Romualdi, C., Bean, C., Toniolo, L., Germinario, E., Megighian, A., Danieli-Betto, D., Reggiani, C., Lanfranchi, G. (2006). Denervation in murine fast-twitch muscle: short-term physiological changes and temporal expression profiling. *Physiological Genomics, 25*, 60–74.

Rasmussen, U. F., Krustrup, P., Kjaer, M., Rasmussen, H. N. (2003). Human skeletal muscle mitochondrial metabolism in youth and senescence: no signs of functional changes of ATP formation and mitochondrial capacity. *Pflugers Archiv, 446*, 270–278.

Raue, U., Slivka, D., Minchev, K., Trappe, S. (2009). Improvements in whole muscle and myocellular function are limited with high-intensity resistance training in octogenarian women. *Journal of Applied Physiology, 106*, 1611–1617.

Reznick, R. M., Zong, H., Li, J., Morino, K., Moore, I. K., Yu, H. J., Liu, Z. X., Dong, J., Mustard, K. J., Hawley, S. A., Befroy, D., Pypaert, M., Hardie, D. G., Young, L. H., Shulman, G. I. (2007). Aging-associated reductions in AMP-activated protein kinase activity and mitochondrial biogenesis. *Cell Metabolism, 5*, 151–156.

Rice, K. M. & Blough, E. R. (2006). Sarcopenia-related apoptosis is regulated differently in fast- and slow-twitch muscles of the aging F344/NxBN rat model. *Mechanisms of Ageing and Development, 127*, 670–679.

Rooyackers, O. E., Adey, D. B., Ades, P. A., Nair, K. S. (1996). Effect of age on in vivo rates of mitochondrial protein synthesis in human skeletal muscle. *Proceedings of the National Academy of Sciences of the United States of America, 93*, 15364–15369.

Rossiter, H. B., Howlett, R. A., Holcombe, H. H., Entin, P. L., Wagner, H. E., Wagner, P. D. (2005). Age is no barrier to muscle structural, biochemical and angiogenic adaptations to training up to 24 months in female rats. *Journal de Physiologie, 565*, 993–1005.

Sacheck, J. M., Hyatt, J. P., Raffaello, A., Jagoe, R. T., Roy, R. R., Edgerton, V. R., Lecker, S. H., Goldberg, A. L. (2007). Rapid disuse and denervation atrophy involve transcriptional changes similar to those of muscle wasting during systemic diseases. *The FASEB Journal, 21*, 140–155.

Seo, A. Y., Xu, J., Servais, S., Hofer, T., Marzetti, E., Wohlgemuth, S. E., Knutson, M. D., Chung, H. Y., Leeuwenburgh, C. (2008). Mitochondrial iron accumulation with age and functional consequences. *Aging Cell, 7*, 706–716.

Short, K. R., Vittone, J. L., Bigelow, M. L., Proctor, D. N., Rizza, R. A., Coenen-Schimke, J. M., Nair, K. S. (2003). Impact of aerobic exercise training on age-related changes in insulin sensitivity and muscle oxidative capacity. *Diabetes, 52*, 1888–1896.

Short, K. R., Bigelow, M. L., Kahl, J., Singh, R., Coenen-Schimke, J., Raghavakaimal, S., Nair, K. S. (2005). Decline in skeletal muscle mitochondrial function with aging in humans. *Proceedings of the National Academy of Sciences of the United States of America, 102*, 5618–5623.

Slivka, D., Raue, U., Hollon, C., Minchev, K., Trappe, S. (2008). Single muscle fiber adaptations to resistance training in old (>80 yr) men: Evidence for limited skeletal muscle plasticity. *American Journal of Physiology: Regulatory, Integrative and Comparative Physiology, 295*, R273–R280.

Stathokostas, L., Jacob-Johnson, S., Petrella, R. J., Paterson, D. H. (2004). Longitudinal changes in aerobic power in older men and women. *Journal of Applied Physiology, 97*, 781–789.

Sugiyama, S., Takasawa, M., Hayakawa, M., Ozawa, T. (1993). Changes in skeletal muscle, heart and liver mitochondrial electron transport activities in rats and dogs of various ages. *Biochemistry and Molecular Biology International, 30*, 937–944.

Terada, S., Goto, M., Kato, M., Kawanaka, K., Shimokawa, T., Tabata, I. (2002). Effects of low-intensity prolonged exercise on PGC-1 mRNA expression in rat epitrochlearis muscle. *Biochemical and Biophysical Research Communications, 296*, 350–354.

Terman, A. & Brunk, U. T. (2004). Myocyte aging and mitochondrial turnover. *Experimental Gerontology, 39*, 701–705.

Thomas, M. M., Vigna, C., Betik, A. C., Tupling, A. R., Hepple, R. T. (2009). Initiating treadmill exercise training in late middle age offers modest adaptations in Ca2+ handling but enhances protein oxidative damage in senescent rat skeletal muscle. *American Journal of Physiology: Regulatory, Integrative and Comparative Physiology, 298*, R1269–R1278.

Thompson, L. V., Durand, D., Fugere, N. A., Ferrington, D. A. (2006). Myosin and actin expression and oxidation in aging muscle. *Journal of Applied Physiology, 101*(6), 1581–1587.

Tonkonogi, M., Fernstrom, M., Walsh, B., Ji, L. L., Rooyackers, O., Hammarqvist, F., Wernerman, J., Sahlin, K. (2003). Reduced oxidative power but unchanged antioxidative capacity in skeletal muscle from aged humans. *Pflugers Archiv, 446*, 261–269.

Vasilaki, A., Mansouri, A., Remmen, H., der Meulen, J. H., Larkin, L., Richardson, A. G., McArdle, A., Faulkner, J. A., Jackson, M. J. (2006). Free radical generation by skeletal muscle of adult and old mice: effect of contractile activity. *Aging Cell, 5*, 109–117.

Wanagat, J., Cao, Z., Pathare, P., Aiken, J. M. (2001). Mitochondrial DNA deletion mutations colocalize with segmental electron transport system abnormalities, muscle fiber atrophy, fiber splitting, and oxidative damage in sarcopenia. *The FASEB Journal, 15*, 322–332.

Skeletal Muscle Collagen: Age, Injury and Disease

Luc E. Gosselin

Abstract Collagen is the most common protein of the extracellular matrix and has several important functions in skeletal muscle, including the provision of both tensile strength and elasticity, the transmission of muscular forces to the bones, the regulation of cell attachment and differentiation, and mechanical and ionic filtration by the basal lamina. Aging is associated with significant changes in the connective tissue compartment of skeletal muscle. This chapter describes the effect of aging on skeletal muscle collagen, how injury affects collagen metabolism, how collagen is remodeled with advancing age and in severe muscle diseases like Duchenne muscular dystrophy. The regulation of collagen metabolism in normal and damaged skeletal muscle is complex and likely involves the interaction of several cell types and growth factors. Muscles with different activation patterns exhibit marked differences in collagen mRNA levels as well as collagen characteristics, indicating that mechanical load mediates collagen biosynthesis. Injured skeletal muscle contains elevated levels of inflammatory cells, which are known to secrete pro- and anti-inflammatory cytokines. Chronic inflammation plays a key role in the development of fibrosis in dystrophic muscle, although the mechanisms that regulate this process are not well understood. Both neutrophils and macrophages play important roles in the regulation of collagen remodeling post-injury by releasing various cytokines that mediate the behavior of inflammatory cells, fibroblasts and satellite cells. The behavior of these cells can be affected by extrinsic factors such as basal levels of growth hormone, which also changes with advancing age.

Keywords Aging • Collagen • Fibrosis • Force transmission • Inflammation • Growth factors • Mechanical loading • Muscle architecture • Muscular dystrophy • Tissue remodeling

---

L.E. Gosselin (✉)
Department of Exercise and Nutrition Sciences, University at Buffalo,
211 Kimball Tower, Buffalo, NY 14214-8028, USA
e-mail: gosselin@buffalo.edu

G.S. Lynch (ed.), *Sarcopenia – Age-Related Muscle Wasting and Weakness*,
DOI 10.1007/978-90-481-9713-2_8, © Springer Science+Business Media B.V. 2011

1 Overview of Collagen in Skeletal Muscle

Collagen is the most common protein of the extracellular matrix (ECM) (Laurent 1987) and has several important functions in skeletal muscle, including: (1) provision of both tensile strength and elasticity; (2) transmission of muscular forces to the bones; (3) regulation of cell attachment and differentiation; and (4) mechanical and ionic filtration by the basal lamina (Minor 1980; Nimni and Harkness 1988; Hay 1991). From the collagen family of proteins, fibrillar collagen type I and type III, the basement membrane collagen type IV, and some of the minor types (e.g. V, VI, VII, XV, XVIII) have been characterized in skeletal muscle (Duance et al. 1977; Light and Champion 1984; Kovanen et al. 1988; Hurme et al. 1991). The epimysium is composed primarily of type I collagen whereas the perimysium contains both type I and III (with type I predominating) (Light and Champion 1984). On the basis of their structural properties type I collagen is suggested to confer tensile strength and rigidity (Mays et al. 1988) whereas type III collagen confers compliance (Burgeson 1987) to intramuscular connective tissue. Fibroblasts synthesize the fibrillar collagen types in muscle (Hurme et al. 1991), although skeletal muscle cells are known to produce mRNA for types I and III collagen (Takala and Virtanen 2000).

Collagen is unique because the protein undergoes extensive post-translational modification both in the intra- and extracellular space. Prolyl-4-hydroxylase (P4H) is an intracellular posttranslational enzyme involved in the hydroxylation of prolyl residues necessary for the formation of the stable collagen triple-helix (Kovanen 2002). Molecular maturation of collagen (i.e., formation of reducible and nonreducible cross-links) is an essential extracellular post-translational process that affords tensile strength to the protein (Viidik 1968; Eyre et al. 1984). The rate-limiting step involves the extracellular oxidation of lysine and hydroxylysine residues by the enzyme lysyl oxidase, thus forming semialdehydes that can undergo further chemical transformations throughout the life of the protein (Eyre et al. 1984; Reiser et al. 1992). The maturation of collagen alters its mechanical and biochemical properties, leading to increased tensile strength (Viidik 1968; Eyre et al. 1984), decreased solubility (Ricard-Blum and Ville 1989) and enhanced resistance to some proteases (Cheung and Nimni 1982).

Collagen concentration in the extracellular space can be controlled either intracellularly prior to secretion or extracellularly following secretion. Intracellular procollagen turnover may be influenced by altering synthesis and/or degradation rate (Bienkowski et al. 1978; Laurent et al. 1985; Laurent 1987; McAnulty and Laurent 1987). As much as 90% of procollagen may be degraded intracellularly within minutes of synthesis (Laurent 1987). Two pathways for this intracellular degradation are proposed: Golgi apparatus and the lysosomes (Laurent 1987). In the extracellular space, the newly synthesized forms of collagen are degraded more quickly than the mature, cross-linked collagen (Laurent 1987). Matrix metalloproteinases (MMPs), also known as collagenases, are the enzymes responsible for the initiation of the extracellular degradation of the collagen triple-helix (Stetler-Stevenson 1996). Fibrillar collagens (I, II, III) are degraded by MMP-1, MMP-8, and MMP-13, whereas the gelatinases MMP-2 and MMP-9 degrade type IV collagen and gelatin (Birkedal-Hansen et al. 1993). Tissue inhibitors of matrix metalloproteinases (TIMP-1,-2,-3, and -4) regulate the activity of MMPs by binding either the active or latent forms of MMPs (Edwards et al. 1996). In skeletal muscle, MMP-2 is constitutively expressed, whereas MMP-9 appears following acute skeletal muscle damage (Kherif et al. 1999). In vivo, fibroblasts, polymorphonuclear leukocytes, neutrophils, and macrophages are responsible for the secretion of MMPs as well as the growth factors involved in the regulation of the expression of the MMPs and TIMPs (Birkedal-Hansen et al. 1993).

2 Effect of Aging on Skeletal Muscle Collagen

Aging is associated with significant changes in the connective tissue compartment of skeletal muscle. The relative distribution of type I collagen increases from birth to senescence, whereas the relative distribution of type III collagen decreases during the same period (Kovanen and Suominen 1989). The concentration of type IV collagen also increases in skeletal muscle with age (Kovanen et al. 1988). In addition to these changes, both concentration of collagen and extent of nonreducible cross-linking significantly increase in senescent skeletal muscle (Zimmerman et al. 1993; Gosselin et al. 1994, 1998) and cardiac tissue (Thomas et al. 1992). The age-related increase in skeletal muscle collagen content occurs without any changes in the activities of P4H or galactosylhydroxylysysl glucosyltransferase (Kovanen and Suominen 1989), two post-translational modification enzymes whose activities reflect collagen synthesis rate. Moreover, Mays et al. (1988) reported that the fractional synthesis rate of collagen in rat skeletal muscle decreases approximately tenfold from 1- to 24-months of age. These results suggest that increases in collagen concentration in senescent skeletal muscle are a result of a decreased rate of resorption out of proportion to the reduced biosynthetic activity. Biopsies from the vastus lateralis muscles of young and old sedentary men and women revealed that intramuscular endomysial collagen and collagen cross-linking (hydroxylsylpyridoline) were unchanged with aging but that the advanced glycation end product, pentosidine, was increased by ~200% (Haus et al. 2007). These data suggested that the synthesis and degradation of contractile proteins (actin and myosin) and proteins involved in the transfer of muscle forces (collagen and pyridinoline cross-links), were tightly regulated during aging and that changes in the glycation-related cross-linking of intramuscular connective tissue possibly contributes to the age-related changes in force transmission and overall muscle function (Haus et al. 2007).

Endurance exercise training can lower the extent of collagen cross-linking in senescent cardiac (Thomas et al. 1992) and skeletal (Zimmerman et al. 1993; Gosselin et al. 1998) muscle, suggestive that collagen turnover is increased during periods of altered use. The impact of increased collagen concentration and cross-linking on repair of injured senescent skeletal muscle is unknown. Increased cross-linking increases collagen's resistance to proteolytic degradation (Cheung and Nimni 1982), allowing slower collagen degradation in senescent skeletal muscle. Whether or not this affects muscle repair is unknown. It is also possible that increased collagen concentration may impair the migration of satellite cells in cases where the basement membrane is destroyed in the damaged area, though this remains speculative.

3 Effect of Injury on Skeletal Muscle Collagen Metabolism

Despite positive benefits achieved from exercise training, some studies have indicated that skeletal muscles of older adults are more susceptible to injury during exercise than muscles of younger adults (Zerba et al. 1990; Brooks and Faulkner 1994; Faulkner et al. 1995). Senescent skeletal muscles can be further compromised since repair occurs more slowly compared to young muscle (Brooks and Faulkner 1990), and because of a limited potential for satellite cell activation (Schultz and Lipton 1982). The slowed response time for repair may be partially attributed to decreases in protein synthesis observed with aging (Welle et al. 1993). Thus, any beneficial gains from exercise may be lost during a prolonged period of muscle repair due to inactivity.

Although exercises involving lengthening or 'eccentric' contractions, appear to cause more injury (Armstrong et al. 1983; McCully and Faulkner 1986) than shortening contractions, muscle injury has also been reported to occur with the latter (McCormick and Thomas 1992). Muscle injury is typically manifested by a decrement in maximal specific force (force/cross sectional area), and morphologically by alterations in Z-line pattern (i.e., Z-line streaming) (Friden et al. 1983) and infiltration by inflammatory cells (Tidball; see Chapter 16). Catabolism of damaged intra- and extracellular proteins is a necessary step in the injury/repair process and involves the activity of calpains (Tidball and Spencer 2000). Additionally, satellite cells and muscle fibroblasts are activated (Tidball 1995), presumably from local growth factors such as fibroblast growth factor (FGF) and insulin-like growth factor I (IGF-I). Participation by these cells as well as inflammatory cells is essential for the repair of the damaged muscle fibers. Thus, repair of the muscle involves the coordinated processes from several cell types, each of which having separate and distinct roles. Successful repair of skeletal muscle depends not only on remodeling the damaged intracellular (contractile, cytoskeletal) proteins, but also the surrounding extracellular matrix, including collagen.

Extensive evidence indicates that the extracellular matrix is remodeled during muscle repair. Following acute exercise-induced muscle damage, the mRNA level of type IV collagen increases within 6 h after inducement of damage (Han et al. 1999). The level of mRNA for types I and III collagen subsequently increase coordinately with mRNA of P4H α- and β- subunits and lysyl oxidase, in addition to the P4H activity. As determined by immunohistochemistry, a qualitative transitory increase in the expression of type III collagen has been noted in mouse skeletal muscle following exercise-induced injury (Myllyla et al. 1986). It is known that collagen metabolism is down-regulated with aging (Mays et al. 1988), and that accumulation of intramuscular connective tissue occurs (Kovanen and Suominen 1989; Zimmerman et al. 1993; Gosselin et al. 1994, 1998) together with altered functional properties (Kovanen et al. 1984; Gosselin et al. 1994, 1998). However, there is a dearth of information regarding how collagen expression is regulated in aged skeletal muscle following muscle injury.

4 Do Extrinsic Factors Affect Collagen Remodeling in Aged Damaged Muscle?

Growth hormone (GH) has pronounced effects on organ and tissue growth. Body growth of hypophysectomized rats and Lewis dwarf rats deficient in GH is markedly reduced but can be reversed by GH supplementation (Guler et al. 1988; Gosteli-Peter et al. 1994; Martinez et al. 1996). During aging, myofibrillar protein synthesis decreases (Welle et al. 1993) as do the circulating levels of serum GH (Florini et al. 1985). However, when old rats are supplemented with GH, protein synthesis is increased to levels similar to that observed in young rats (Sonntag et al. 1985). It was reported recently that increased GH availability stimulates matrix collagen synthesis in skeletal muscle and tendon, but with no effect on myofibrillar protein synthesis, indicating that GH might be more important in strengthening the matrix tissue than for skeletal muscle hypertrophy in adult human musculotendinous tissue (Doessing et al. 2010).

GH is thought to function indirectly on skeletal muscle via the action of insulin-like growth factor I (IGF-I), a growth promoting peptide factor (Schwander et al. 1983). When physiological concentrations of IGF-I are applied to myoblasts grown in tissue culture, cell mitotic activity and protein synthesis significantly increases (Florini 1987; Johnson and Allen 1990). The target of IGF-I not only includes myoblasts but other cell types as well. For example, cultured fibroblasts exposed to physiological concentrations of IGF-I increase collagen synthesis (Goldstein et al. 1989; Gillery et al. 1992), whereas addition of an antibody specific to the IGF-I receptor (αIR-3) inhibits fibroblast collagen synthesis (Goldstein et al. 1989). Although the liver produces the majority of IGF-I (Sonntag et al. 1985), other tissues, including skeletal muscle, can also produce IGF-I (Sonntag et al. 1985; Jennische and Hansson 1987; Jennische and Olivecrona 1987; Yan et al. 1993). The action of IGF-I on muscle is dependent not only upon the local concentration of IGF-I, but also on the pattern of growth factor receptor expression (Rubin and Baserga 1995). Whether or not aging alters IGF-I receptor density in skeletal muscle, and what impact this may have during muscle repair is unclear.

5 Duchenne Muscular Dystrophy: Collagen Metabolism Run Amok

DMD is an X-chromosome linked disorder resulting in the loss of the muscle protein dystrophin (Hoffman et al. 1987), a large protein localized to the inner surface of the muscle cell membrane (Watkins et al. 1988). Dystrophin-deficient muscle is damaged to a greater degree given the same recruitment history due to its innate membrane fragility (Petrof et al. 1993; Petrof 1998). Consequently, the muscles undergo cycles of injury and repair that result in progressive muscle fiber loss, weakness, and extensive fibrosis. The diaphragm is particularly affected, and humans typically suffer from respiratory failure early in life (Inkley et al. 1974).

The *mdx* mouse shares a genetic and biochemical homology with human muscular dystrophy and is commonly used to study DMD. Although limb skeletal muscles from *mdx* mice are capable of significant regeneration, the diaphragm muscle exhibits progressive degeneration similar to that observed in skeletal muscle from patients with DMD (Stedman et al. 1991). The mechanisms responsible for this divergent response are not known, but may be due to differences in inflammation secondary to muscle activation pattern.

Data indicates that the process of diaphragm fibrosis has commenced by 6 weeks of age in *mdx* mice (Gosselin et al. 2004), and that the extent of diaphragm fibrosis increases progressively thereafter such that by 16 months of age, hydroxyproline concentration in *mdx* diaphragm is elevated ~sevenfold (Stedman et al. 1991). These biochemical changes are associated with a significant increase in diaphragm stiffness (Stedman et al. 1991). Collagen is also involved in the regulation of cell attachment and differentiation, and mechanical and ionic filtration by the basal lamina (Minor 1980; Nimni and Harkness 1988; Hay 1991). Hence, excessive collagen may therefore serve as a barrier for targeted drug or gene therapy. In spite of these important physiological functions, there is a dearth of information regarding the mechanisms that regulate collagen metabolism in damaged and dystrophic skeletal muscle.

Collagen accretion in the extracellular space is a function of both synthesis and degradation. Significant increases in type I collagen mRNA (Goldspink et al. 1994; Gosselin and Martinez 2004; Gosselin et al. 2004; Gosselin and Williams 2006) have been observed in *mdx* diaphragm. Interestingly, the level of type I collagen mRNA, expressed per μg RNA, is similar in diaphragm and gastrocnemius muscle from 9-week-old *mdx* mice, despite the fact that the diaphragm accumulates significantly more collagen (Gosselin and Williams 2006). RNA concentration in *mdx* diaphragm is ~80% higher than in *mdx* gastrocnemius (Gosselin and Williams 2006), suggestive that a hypercellular environment exists in *mdx* diaphragm. Assuming a constant mRNA to RNA ratio in both muscles, the diaphragm muscle contains approximately 80% more type I collagen mRNA per unit weight. This difference could theoretically result in significantly greater collagen synthesis and accretion in the diaphragm. Whether or not fibroblast proliferation occurs in vivo in dystrophic diaphragm muscle and contributes to the hypercellularity remains to be determined. Such a finding however would be of significant biological consequence, even in the absence of elevated levels of pro-fibrotic cytokines.

Matrix metalloproteinases (MMPs) are a group of zinc-dependent enzymes that initiate the extracellular degradation of collagen (Hay 1991; Nagase et al. 2006). Of the 20 or so different MMPs (Nagase et al. 2006), MMP-9 and MMP-2 have been the most studied in mammalian skeletal muscle. MMP-2 is constitutively expressed in normal skeletal muscle whereas MMP-9 is absent (Kherif et al. 1999). However, in response to various forms of injury, such as that induced by cardiotoxin (Kherif et al. 1999) or ischemia-reperfusion (Muhs et al. 2003), MMP-9 mRNA and activity significantly increase within 24 h post-injury and appears to be expressed primarily by neutrophils (Kherif et al. 1999; Muhs et al. 2003). In contrast, the active form of MMP-2 does not begin to increase until ~72 h post-injury, and increases further at 7 days, suggestive that these two MMPs have unique roles in the remodeling of the ECM. Interestingly, MMP-9 and MMP-2 are elevated in skeletal muscle from 3-month-old *mdx* mice (Kherif et al. 1999), findings that are paradoxical to the development of fibrosis in dystrophic skeletal muscle.

MMP-9 has been shown to be involved in the recruitment of inflammatory cells in the post-ischemic liver model (Khandoga et al. 2006). In other models of injury and fibrosis, MMP-9 blockade significantly decreases the extent of inflammation and fibrosis (Corbel et al. 2001a, b; Tan et al. 2006), suggestive that MMP-9 may either directly or indirectly mediate the behavior of inflammatory cells or fibroblasts. The basal lamina, which contains type IV collagen, is known to bind a number of growth factors, including bFGF (DiMario et al. 1989; Yamada et al. 1989). Given the rapidity of MMP-9 up-regulation following muscle damage and of its action on type IV collagen, MMP-9 may play a crucial role in the pathogenesis of fibrosis in *mdx* muscle, either through stimulating the inflammatory response or through its action on the basal lamina (i.e. growth factor release/activation). Indeed, when *mdx* mice were administered with Batimastat, an inhibitor of MMP's, resulted in reduced muscle necrosis and infiltration with inflammatory cells (Kumar et al. 2010). Additionally, MMP-9 gene deletion in *mdx* mice significantly reduced the extent of skeletal muscle injury and inflammation (Li et al. 2009).

An interesting feature of dystrophin-deficiency across species is the expression of grouped and segmental necrosis (Cazzato 1968; Anderson et al. 1988; Cox et al. 1993; D'Amore et al. 1994). Grouped fiber necrosis is more typical of extracellular rather than intracellular events (Bridges 1986). As a consequence of muscle activation, the sarcolemma accumulates transient breaks, which allow the release of factors that initiate wound healing (McNeil and Khakee 1992). DNA microarray analysis of adult *mdx* limb muscle revealed that approximately 30% of all differentially regulated genes were associated with inflammation (Porter et al. 2002), and that several of the inflammatory genes identified in the muscle from *mdx* mouse were also found to be upregulated in muscle from DMD patients (Chen et al. 2000). The leakage of material from dystrophin-deficient muscle results in the accumulation of inflammatory cells in both endomysial and perimysial connective tissue (Tanabe et al. 1986; Carnwath and Shotton 1987; McDouall et al. 1990; Spencer et al. 2000). Dystrophin-deficient muscle is damaged to a greater degree given the same recruitment history due to its innate membrane fragility (Menke and Jockusch 1991; Petrof et al. 1993). Therefore, the same factors released fleetingly by normal muscle to promote wound healing are present chronically in dystrophic muscle and may have pathologic consequences.

The presence of inflammatory cells is increased in skeletal muscle from patients with DMD and in *mdx* mice. The major infiltrating cell types in dystrophin-deficient muscle are macrophages (Engel and Arahata 1986; Spencer et al. 1997), T cells (Engel and Arahata 1986; Spencer et al. 1997), and eosinophils (Cai et al. 2000). Nguyen and Tidball (Nguyen and Tidball 2003) demonstrated that macrophages caused significant myotube lysis when co-cultured together. Furthermore, Wehling et al. (2001) reported that macrophage depletion from *mdx* muscles significantly reduced the concentration of regenerative muscle fibers. These findings support the hypothesis that macrophage accumulation secondary to inflammation can promote muscle injury. Given the persistent inflammatory response in dystrophic muscle, it is possible that an altered extracellular environment exists that promotes muscle fibrosis. Both TNF and TGF-β1 are produced by macrophages and are known to stimulate collagen metabolism. Moreover, their levels have been reported to be increased in muscular dystrophy (Bernasconi et al. 1995; Iannaccone et al. 1995; Lundberg et al. 1995; Tews and Goebel 1996; Murakami et al. 1999; Porreca et al. 1999; Hartel et al. 2001; Andreetta et al. 2006; Zhou et al. 2006). Given that the extracellular environment contains increased levels of and these cytokines, and because of their biologic actions observed in vitro, these cytokines may have prominent yet unknown in vivo roles in the pathogenesis of fibrosis in DMD.

6 Summary

Regulation of collagen metabolism in normal and damaged skeletal muscle is complex and likely involves the interaction of several cell types and growth factors. Moreover, within a given organism, muscles with different activation patterns exhibit marked differences in collagen mRNA levels as well as collagen characteristics – indicative that mechanical load mediates collagen biosynthesis. Injured skeletal muscle contains elevated levels of inflammatory cells, which are known to secrete pro- and anti-inflammatory cytokines such as TNF-α and TGF-β1. Moreover, the expression of bFGF is also up-regulated in damaged and/or dystrophic skeletal muscle. Significant evidence exists to suggest chronic inflammation plays a key role in the development of fibrosis in dystrophic muscle, though the mechanisms that regulate this process are not well understood. Both neutrophils and macrophages play important roles in the regulation of collagen remodeling post-injury by releasing various cytokines that mediate the behavior of inflammatory cells, fibroblasts and satellite cells. Moreover, the behavior of these cells can be affected by extrinsic factors such as basal levels of growth hormone, which changes with age.

References

Anderson, J. E., Bressler, B. H., Ovalle, W. K. (1988). Functional regeneration in the hindlimb skeletal muscle of the mdx mouse. *Journal of Muscle Research and Cell Motility, 9*, 499–515.

Andreetta, F., Bernasconi, P., Baggi, F., Ferro, P., Oliva, L., Arnoldi, E., Cornelio, F., Mantegazza, R., Confalonieri, P. (2006). Immunomodulation of TGF-beta 1 in mdx mouse inhibits connective tissue proliferation in diaphragm but increases inflammatory response: implications for antifibrotic therapy. *Journal of Neuroimmunology, 175*, 77–86.

Armstrong, R. B., Ogilvie, R. W., Schwane, J. A. (1983). Eccentric exercise-induced injury to rat skeletal muscle. *Journal of Applied Physiology, 54*, 80–93.

Bernasconi, P., Torchiana, E., Confalonieri, P., Brugnoni, R., Barresi, R., Mora, M., Cornelio, F., Morandi, L., Mantegazza, R. (1995). Expression of transforming growth factor-beta 1 in dystrophic patient muscles correlates with fibrosis. Pathogenetic role of a fibrogenic cytokine. *Journal of Clinical Investigation, 96*, 1137–1144.

Bienkowski, R. S., Baum, B. J., Crystal, R. G. (1978). Fibroblasts degrade newly synthesized collagen within the cell before secretion. *Nature, 276*, 413–416.

Birkedal-Hansen, H., Moore, W. G., Bodden, M. K., Windsor, L. J., Birkedal-Hansen, B., DeCarlo, A., Engler, J. A. (1993). Matrix metalloproteinases: a review. *Critical Reviews in Oral Biology and Medicine, 4*, 197–250.

Bridges, L. R. (1986). The association of cardiac muscle necrosis and inflammation with the degenerative and persistent myopathy of MDX mice. *Journal of the Neurological Sciences, 72*, 147–157.

Brooks, S. V. & Faulkner, J. A. (1990). Contraction-induced injury: recovery of skeletal muscles in young and old mice. *The American Journal of Physiology, 258*, C436–C442.

Brooks, S. V. & Faulkner, J. A. (1994). Skeletal muscle weakness in old age: underlying mechanisms. *Medicine and Science in Sports and Exercise, 26*, 432–439.

Burgeson, R. E. (1987). The collagens of skin. *Current Problems in Dermatology, 17*, 61–75.

Cai, B., Spencer, M. J., Nakamura, G., Tseng-Ong, L., Tidball, J. G. (2000). Eosinophilia of dystrophin-deficient muscle is promoted by perforin- mediated cytotoxicity by T cell effectors. *The American Journal of Pathology, 156*, 1789–1796.

Carnwath, J. W. & Shotton, D. M. (1987). Muscular dystrophy in the mdx mouse: histopathology of the soleus and extensor digitorum longus muscles. *Journal of the Neurological Sciences, 80*, 39–54.

Cazzato, G. (1968). Considerations about a possible role played by connective tissue proliferation and vascular disturbances in the pathogenesis of progressive muscular dystrophy. *European Neurology, 1*, 158–179.

Chen, Y. W., Zhao, P., Borup, R., Hoffman, E. P. (2000). Expression profiling in the muscular dystrophies: identification of novel aspects of molecular pathophysiology. *The Journal of Cell Biology, 151*, 1321–1336.

Cheung, D. T. & Nimni, M. E. (1982). Mechanism of crosslinking of proteins by glutaraldehyde II. Reaction with monomeric and polymeric collagen. *Connective Tissue Research, 10*, 201–216.

Corbel, M., Caulet-Maugendre, S., Germain, N., Molet, S., Lagente, V., Boichot, E. (2001a). Inhibition of bleomycin-induced pulmonary fibrosis in mice by the matrix metalloproteinase inhibitor batimastat. *The Journal of Pathology, 193*, 538–545.

Corbel, M., Lanchou, J., Germain, N., Malledant, Y., Boichot, E., Lagente, V. (2001b). Modulation of airway remodeling-associated mediators by the antifibrotic compound, pirfenidone, and the matrix metalloproteinase inhibitor, batimastat, during acute lung injury in mice. *European Journal of Pharmacology, 426*, 113–121.

Cox, G. A., Cole, N. M., Matsumura, K., Phelps, S. F., Hauschka, S. D., Campbell, K. P., Faulkner, J. A., Chamberlain, J. S. (1993). Overexpression of dystrophin in transgenic mdx mice eliminates dystrophic symptoms without toxicity. *Nature, 364*, 725–729.

D'Amore, P. A., Brown, R. H., Jr., Ku, P. T., Hoffman, E. P., Watanabe, H., Arahata, K., Ishihara, T., Folkman, J. (1994). Elevated basic fibroblast growth factor in the serum of patients with Duchenne muscular dystrophy. *Annals of Neurology, 35*, 362–365.

DiMario, J., Buffinger, N., Yamada, S., Strohman, R. C. (1989). Fibroblast growth factor in the extracellular matrix of dystrophic (mdx) mouse muscle. *Science, 244*, 688–690.

Doessing, S., Heinemeier, K. M., Holm, L., Mackey, A. L., Schjerling, P., Rennie, M., Smith, K., Reitelseder, S., Kappelgaard, A. M., Rasmussen, M. H., Flyvbjerg, A., Kjaer, M. (2010). Growth hormone stimulates the collagen synthesis in human tendon and skeletal muscle without affecting myofibrillar protein synthesis. *Journal de Physiologie, 588*, 341–351.

Duance, V. C., Restall, D. J., Beard, H., Bourne, F. J., Bailey, A. J. (1977). The location of three collagen types in skeletal muscle. *FEBS Letters, 79*, 248–252.

Edwards, D. R., Beaudry, P. P., Laing, T. D., Kowal, V., Leco, K. J., Leco, P. A., Lim, M. S. (1996). The roles of tissue inhibitors of metalloproteinases in tissue remodelling and cell growth. *International Journal of Obesity and Related Metabolic Disorders, 20*(Suppl 3), S9–S15.

Engel, A. G. & Arahata, K. (1986). Mononuclear cells in myopathies: quantitation of functionally distinct subsets, recognition of antigen-specific cell-mediated cytotoxicity in some diseases, and implications for the pathogenesis of the different inflammatory myopathies. *Human Pathology, 17*, 704–721.

Eyre, D. R., Paz, M. A., Gallop, P. M. (1984). Cross-linking in collagen and elastin. *Annual Review of Biochemistry, 53*, 717–748.

Faulkner, J. A., Brooks, S. V., Zerba, E. (1995). Muscle atrophy and weakness with aging: contraction-induced injury as an underlying mechanism. *The Journals of Gerontology. Series A: Biological Sciences and Medical Sciences, 50*(Spec No), 124–129.

Florini, J. R. (1987). Hormonal control of muscle growth. *Muscle & Nerve, 10*, 577–598.

Florini, J. R., Prinz, P. N., Vitiello, M. V., Hintz, R. L. (1985). Somatomedin-C levels in healthy young and old men: relationship to peak and 24-hour integrated levels of growth hormone. *Journal of Gerontology, 40*, 2–7.

Friden, J., Sjostrom, M., Ekblom, B. (1983). Myofibrillar damage following intense eccentric exercise in man. *International Journal of Sports Medicine, 4*, 170–176.

Gillery, P., Leperre, A., Maquart, F. X., Borel, J. P. (1992). Insulin-like growth factor-I (IGF-I) stimulates protein synthesis and collagen gene expression in monolayer and lattice cultures of fibroblasts. *Journal of Cellular Physiology, 152*, 389–396.

Goldspink, G., Fernandes, K., Williams, P. E., Wells, D. J. (1994). Age-related changes in collagen gene expression in the muscles of mdx dystrophic and normal mice. *Neuromuscular Disorders, 4*, 183–191.

Goldstein, R. H., Poliks, C. F., Pilch, P. F., Smith, B. D., Fine, A. (1989). Stimulation of collagen formation by insulin and insulin-like growth factor I in cultures of human lung fibroblasts. *Endocrinology, 124*, 964–970.

Gosselin, L. E. & Martinez, D. A. (2004). Impact of TNF-alpha blockade on TGF-beta1 and type I collagen mRNA expression in dystrophic muscle. *Muscle & Nerve, 30*, 244–246.

Gosselin, L. E. & Williams, J. E. (2006). Pentoxifylline fails to attenuate fibrosis in dystrophic (mdx) diaphragm muscle. *Muscle & Nerve, 33*, 820–823.

Gosselin, L. E., Martinez, D. A., Vailas, A. C., Sieck, G. C. (1994). Passive length-force properties of senescent diaphragm: relationship with collagen characteristics. *Journal of Applied Physiology, 76*, 2680–2685.

Gosselin, L. E., Adams, C., Cotter, T. A., McCormick, R. J., Thomas, D. P. (1998). Effect of exercise training on passive stiffness in locomotor skeletal muscle: role of extracellular matrix. *Journal of Applied Physiology, 85*, 1011–1016.

Gosselin, L. E., Williams, J. E., Deering, M., Brazeau, D., Koury, S., Martinez, D. A. (2004). Localization and early time course of TGF-beta 1 mRNA expression in dystrophic muscle. *Muscle & Nerve, 30*, 645–653.

Gosteli-Peter, M. A., Winterhalter, K. H., Schmid, C., Froesch, E. R., Zapf, J. (1994). Expression and regulation of insulin-like growth factor-I (IGF-I) and IGF-binding protein messenger ribonucleic acid levels in tissues of hypophysectomized rats infused with IGF-I and growth hormone. *Endocrinology, 135*, 2558–2567.

Guler, H. P., Zapf, J., Scheiwiller, E., Froesch, E. R. (1988). Recombinant human insulin-like growth factor I stimulates growth and has distinct effects on organ size in hypophysectomized rats. *Proceedings of the National Academy of Sciences of the United States of America, 85*, 4889–4893.

Han, X. Y., Wang, W., Komulainen, J., Koskinen, S. O., Kovanen, V., Vihko, V., Trackman, P. C., Takala, T. E. (1999). Increased mRNAs for procollagens and key regulating enzymes in rat skeletal muscle following downhill running. *Pflugers Archiv, 437*, 857–864.

Hartel, J. V., Granchelli, J. A., Hudecki, M. S., Pollina, C. M., Gosselin, L. E. (2001). Impact of prednisone on TGF-beta1 and collagen in diaphragm muscle from mdx mice. *Muscle & Nerve, 24*, 428–432.

Haus, J. M., Carrithers, J. A., Trappe, S. W., Trappe, T. A. (2007). Collagen, cross-linking, and advanced glycation end products in aging human skeletal muscle. *Journal of Applied Physiology, 103*, 2068–2076.

Hay, E. D. (1991). *Cell biology of the extracellular matrix*. New York: Plenum Press.

Hoffman, E. P., Brown, R. H., Jr., Kunkel, L. M. (1987). Dystrophin: the protein product of the Duchenne muscular dystrophy locus. *Cell, 51*, 919–928.

Hurme, T., Kalimo, H., Sandberg, M., Lehto, M., Vuorio, E. (1991). Localization of type I and III collagen and fibronectin production in injured gastrocnemius muscle. *Laboratory Investigation, 64*, 76–84.

Iannaccone, S., Quattrini, A., Smirne, S., Sessa, M., de Rino, F., Ferini-Strambi, L., Nemni, R. (1995). Connective tissue proliferation and growth factors in animal models of Duchenne muscular dystrophy. *Journal of the Neurological Sciences, 128*, 36–44.

Inkley, S. R., Oldenburg, F. C., Vignos, P. J., Jr. (1974). Pulmonary function in Duchenne muscular dystrophy related to stage of disease. *The American Journal of Medicine, 56*, 297–306.

Jennische, E. & Hansson, H. A. (1987). Regenerating skeletal muscle cells express insulin-like growth factor I. *Acta Physiologica Scandinavica, 130*, 327–332.

Jennische, E. & Olivecrona, H. (1987). Transient expression of insulin-like growth factor I immunoreactivity in skeletal muscle cells during postnatal development in the rat. *Acta Physiologica Scandinavica, 131*, 619–622.

Johnson, S. E. & Allen, R. E. (1990). The effects of bFGF, IGF-I, and TGF-beta on RMo skeletal muscle cell proliferation and differentiation. *Experimental Cell Research, 187*, 250–254.

Khandoga, A., Kessler, J. S., Hanschen, M., Khandoga, A. G., Burggraf, D., Reichel, C., Hamann, G. F., Enders, G., Krombach, F. (2006). Matrix metalloproteinase-9 promotes neutrophil and T cell recruitment and migration in the postischemic liver. *Journal of Leukocyte Biology, 79*, 1295–1305.

Kherif, S., Lafuma, C., Dehaupas, M., Lachkar, S., Fournier, J. G., Verdiere-Sahuque, M., Fardeau, M., Alameddine, H. S. (1999). Expression of matrix metalloproteinases 2 and 9 in regenerating skeletal muscle: a study in experimentally injured and mdx muscles. *Developmental Biology, 205*, 158–178.

Kovanen, V. (2002). Intramuscular extracellular matrix: complex environment of muscle cells. *Exercise and Sport Sciences Reviews, 30*, 20–25.

Kovanen, V. & Suominen, H. (1989). Age- and training-related changes in the collagen metabolism of rat skeletal muscle. *European Journal of Applied Physiology and Occupational Physiology, 58*, 765–771.

Kovanen, V., Suominen, H., Heikkinen, E. (1984). Mechanical properties of fast and slow skeletal muscle with special reference to collagen and endurance training. *Journal of Biomechanics, 17*, 725–735.

Kovanen, V., Suominen, H., Risteli, J., Risteli, L. (1988). Type IV collagen and laminin in slow and fast skeletal muscle in rats–effects of age and life-time endurance training. *Collagen and Related Research, 8*, 145–153.

Kumar, A., Bhatnagar, S., Kumar, A. (2010). Matrix Metalloproteinase Inhibitor Batimastat Alleviates Pathology and Improves Skeletal Muscle Function in Dystrophin-Deficient mdx Mice. *American Journal of Pathology, 177*(1), 248–260.

Laurent, G. (1987). Dynamic state of collagen: pathways of collagen degradation in vivo and their possible role in regulation of collagen mass. *American Journal of Physiology (Cell Physiology), 252*, C1–C9.

Laurent, G. J., McAnulty, R. J., Gibson, J. (1985). Changes in collagen synthesis and degradation during skeletal muscle growth. *American Journal of Physiology (Cell Physiology), 249*, C352–C355.

Li, H., Mittal, A., Makonchuk, D. Y., Bhatnagar, S., Kumar, A. (2009). Matrix metalloproteinase-9 inhibition ameliorates pathogenesis and improves skeletal muscle regeneration in muscular dystrophy. *Human Molecular Genetics, 18*, 2584–2598.

Light, N. & Champion, A. E. (1984). Characterization of muscle epimysium, perimysium and endomysium collagens. *The Biochemical Journal, 219*, 1017–1026.

Lundberg, I., Brengman, J. M., Engel, A. G. (1995). Analysis of cytokine expression in muscle in inflammatory myopathies, Duchenne dystrophy, and non-weak controls. *Journal of Neuroimmunology, 63*, 9–16.

Martinez, D. A., Orth, M. W., Carr, K. E., Vanderby, R., Jr., Vailas, A. C. (1996). Cortical bone growth and maturational changes in dwarf rats induced by recombinant human growth hormone. *The American Journal of Physiology, 270*, E51–E59.

Mays, P. K., Bishop, J. E., Laurent, G. J. (1988). Age-related changes in the proportion of types I and III collagen. *Mechanisms of Ageing and Development, 45*, 203–212.

McAnulty, R. & Laurent, G. J. (1987). Collagen synthesis and degradation in vivo. Evidence for rapid rates of collagen turnover with extensive degradation of newly synthesized collagen in tissues of the adult rat. *Collagen Related Research, 7*, 93–104.

McCormick, K. M. & Thomas, D. P. (1992). Exercise-induced satellite cell activation in senescent soleus muscle. *Journal of Applied Physiology, 72*, 888–893.

McCully, K. K. & Faulkner, J. A. (1986). Characteristics of lengthening contractions associated with injury to skeletal muscle fibers. *Journal of Applied Physiology, 61*, 293–299.

McDouall, R. M., Dunn, M. J., Dubowitz, V. (1990). Nature of the mononuclear infiltrate and the mechanism of muscle damage in juvenile dermatomyositis and Duchenne muscular dystrophy. *Journal of the Neurological Sciences, 99*, 199–217.

McNeil, P. L. & Khakee, R. (1992). Disruptions of muscle fiber plasma membranes. Role in exercise-induced damage. *The American Journal of Pathology, 140*, 1097–1109.

Menke, A., Jockusch, H. (1991). Decreased osmotic stability of dystrophin-less muscle cells from the mdx mouse. *Nature, 349*, 69–71.

Minor, R. R. (1980). Collagen metabolism: a comparison of diseases of collagen and diseases affecting collagen. *The American Journal of Pathology, 98*, 225–280.

Muhs, B. E., Plitas, G., Delgado, Y., Ianus, I., Shaw, J. P., Adelman, M. A., Lamparello, P., Shamamian, P., Gagne, P. (2003). Temporal expression and activation of matrix metalloproteinases-2, -9, and membrane type 1-matrix metalloproteinase following acute hindlimb ischemia. *The Journal of Surgical Research, 111*, 8–15.

Murakami, N., McLennan, I. S., Nonaka, I., Koishi, K., Baker, C., Hammond-Tooke, G. (1999). Transforming growth factor-beta2 is elevated in skeletal muscle disorders. *Muscle & Nerve, 22*, 889–898.

Myllyla, R., Salminen, A., Peltonen, L., Takala, T. E., Vihko, V. (1986). Collagen metabolism of mouse skeletal muscle during the repair of exercise injuries. *Pflugers Archiv, 407*, 64–70.

Nagase, H., Visse, R., Murphy, G. (2006). Structure and function of matrix metalloproteinases and TIMPs. *Cardiovascular Research, 69*, 562–573.

Nguyen, H. X. & Tidball, J. G. (2003). Interactions between neutrophils and macrophages promote macrophage killing of rat muscle cells in vitro. *Journal de Physiologie, 547*, 125–132.

Nimni, M. & Harkness, R. (1988). *Molecular structures and functions of collagen* (Vol. 1). Boca Raton: CRC.

Petrof, B. J. (1998). The molecular basis of activity-induced muscle injury in Duchenne muscular dystrophy. *Molecular and Cellular Biochemistry, 179*, 111–123.

Petrof, B. J., Shrager, J. B., Stedman, H. H., Kelly, A. M., Sweeney, H. L. (1993). Dystrophin protects the sarcolemma from stresses developed during muscle contraction. *Proceedings of the National Academy of Sciences of the United States of America, 90*, 3710–3714.

Porreca, E., Guglielmi, M. D., Uncini, A., Di Gregorio, P., Angelini, A., Di Febbo, C., Pierdomenico, S. D., Baccante, G., Cuccurullo, F. (1999). Haemostatic abnormalities, cardiac involvement and serum tumor necrosis factor levels in X-linked dystrophic patients. *Thrombosis and Haemostasis, 81*, 543–546.

Porter, J. D., Khanna, S., Kaminski, H. J., Rao, J. S., Merriam, A. P., Richmonds, C. R., Leahy, P., Li, J., Guo, W., Andrade, F. H. (2002). A chronic inflammatory response dominates the skeletal muscle molecular signature in dystrophin-deficient mdx mice. *Human Molecular Genetics, 11*, 263–272.

Reiser, K., McCormick, R. J., Rucker, R. B. (1992). Enzymatic and nonenzymatic cross-linking of collagen and elastin. *The FASEB Journal, 6*, 2439–2449.

Ricard-Blum, S., Ville, G. (1989). Collagen cross-linking. *The International Journal of Biochemistry, 21*, 1185–1189.

Rubin, R. & Baserga, R. (1995). Insulin-like growth factor-I receptor. Its role in cell proliferation, apoptosis, and tumorigenicity. *Laboratory Investigation, 73*, 311–331.

Schultz, E. & Lipton, B. H. (1982). Skeletal muscle satellite cells: changes in proliferation potential as a function of age. *Mechanisms of Ageing and Development, 20*, 377–383.

Schwander, J. C., Hauri, C., Zapf, J., Froesch, E. R. (1983). Synthesis and secretion of insulin-like growth factor and its binding protein by the perfused rat liver: dependence on growth hormone status. *Endocrinology, 113*, 297–305.

Sonntag, W. E., Hylka, V. W., Meites, J. (1985). Growth hormone restores protein synthesis in skeletal muscle of old male rats. *Journal of Gerontology, 40*, 689–694.

Spencer, M. J., Walsh, C. M., Dorshkind, K. A., Rodriguez, E. M., Tidball, J. G. (1997). Myonuclear apoptosis in dystrophic mdx muscle occurs by perforin- mediated cytotoxicity. *Journal of Clinical Investigation, 99*, 2745–2751.

Spencer, M. J., Marino, M. W., Winckler, W. M. (2000). Altered pathological progression of diaphragm and quadriceps muscle in TNF-deficient, dystrophin-deficient mice. *Neuromuscular Disorders, 10*, 612–619.

Stedman, H. H., Sweeney, H. L., Shrager, J. B., Maguire, H. C., Panettieri, R. A., Petrof, B., Narusawa, M, Leferovich, J. M., Sladky, J. T., Kelly, A. M. (1991). The mdx mouse diaphragm reproduces the degenerative changes of Duchenne muscular dystrophy. *Nature, 352*, 536–539.

Stetler-Stevenson, W. G. (1996). Dynamics of matrix turnover during pathologic remodeling of the extracellular matrix. *The American Journal of Pathology, 148*, 1345–1350.

Takala, T. E. & Virtanen, P. (2000). Biochemical composition of muscle extracellular matrix: the effect of loading. *Scandinavian Journal of Medicine & Science in Sports, 10*, 321–325.

Tan, R. J., Fattman, C. L., Niehouse, L. M., Tobolewski, J. M., Hanford, L. E., Li, Q., Monzon, F. A., Parks, W. C., Oury, T. D. (2006). Matrix metalloproteinases promote inflammation and fibrosis in asbestos-induced lung injury in mice. *American Journal of Respiratory Cell and Molecular Biology, 35*, 289–297.

Tanabe, Y., Esaki, K., Nomura, T. (1986). Skeletal muscle pathology in X chromosome-linked muscular dystrophy (mdx) mouse. *Acta Neuropathologica, 69*, 91–95.

Tews, D. S. & Goebel, H. H. (1996). Cytokine expression profile in idiopathic inflammatory myopathies. *Journal of Neuropathology and Experimental Neurology, 55*, 342–347.

Thomas, D. P., McCormick, R. J., Zimmerman, S. D., Vadlamudi, R. K., Gosselin, L. E. (1992). Aging- and training-induced alterations in collagen characteristics of rat left ventricle and papillary muscle. *The American Journal of Physiology, 263*, H778–H783.

Tidball, J. G. (1995). Inflammatory cell response to acute muscle injury. *Medicine and Science in Sports and Exercise, 27*, 1022–1032.

Tidball, J. G. & Spencer, M. J. (2000). Calpains and muscular dystrophies. *The International Journal of Biochemistry & Cell Biology, 32*, 1–5.

Viidik, A. (1968). Elasticity and tensile strength of the anterior cruciate ligament in rabbits as influenced by training. *Acta Physiologica Scandinavica, 74*, 372–380.

Watkins, S. C., Hoffman, E. P., Slayter, H. S., Kunkel, L. M. (1988). Immunoelectron microscopic localization of dystrophin in myofibres. *Nature, 333*, 863–866.

Wehling, M., Spencer, M. J., Tidball, J. G. (2001). A nitric oxide synthase transgene ameliorates muscular dystrophy in mdx mice. *The Journal of Cell Biology, 155*, 123–131.

Welle, S., Thornton, C., Jozefowicz, R., Statt, M. (1993). Myofibrillar protein synthesis in young and old men. *The American Journal of Physiology, 264*, E693–E698.

Yamada, S., Buffinger, N., DiMario, J., Strohman, R. C. (1989). Fibroblast growth factor is stored in fiber extracellular matrix and plays a role in regulating muscle hypertrophy. *Medicine and Science in Sports and Exercise, 21*, S173–S180.

Yan, Z., Biggs, R. B., Booth, F. W. (1993). Insulin-like growth factor immunoreactivity increases in muscle after acute eccentric contractions. *Journal of Applied Physiology, 74*, 410–414.

Zerba, E., Komorowski, T. E., Faulkner, J. A. (1990). Free radical injury to skeletal muscles of young, adult, and old mice. *The American Journal of Physiology, 258*, C429–C435.

Zhou, L., Porter, J. D., Cheng, G., Gong, B., Hatala, D. A., Merriam, A. P., Zhou, X., Rafael, J. A., Kaminski, H. J. (2006). Temporal and spatial mRNA expression patterns of TGF-beta1, 2, 3 and TbetaRI, II, III in skeletal muscles of mdx mice. *Neuromuscular Disorders, 16*, 32–38.

Zimmerman, S. D. M. R., Vadlamudi, R. K., Thomas, D. P. (1993). Age and training alter collagen characteristics in fast- and slow-twitch rat limb muscle. *Journal of Applied Physiology, 75*, 1670–1674.

Nuclear Apoptosis and Sarcopenia

Stephen E. Alway and Parco M. Siu

Abstract Apoptosis is a well-conserve cellular disassembly process, which has been implicated in a variety of diseases. Unlike cells with a single nucleus, apoptotic signaling can target individual nuclei in multi-nucleated skeletal muscle cells without necessarily eliminating the entire cell (muscle fiber). This targeted apoptosis or "nuclear apoptosis" appears to have a role in regulating aging-induced muscle loss (sarcopenia) by reducing the myofiber volume (i.e. cytoplasm) that can be supported in a single muscle fibre. Recent investigations indicate that apoptotic signaling in aged skeletal muscles occurs through three apoptotic pathways. The intrinsic or mitochondria apoptotic pathway has been most widely studied in muscle. Mitochondria dysfunction and increased mitochondria permeability lead to activation of *c*ysteine-*asp*artic acid prote*ases* (caspases) and eventually DNA fragmentation in sarcopenia. The death receptor (extrinsic) apoptotic pathway has been strongly implicated in sarcopenia and other conditions of muscle loss with aging or disuse. TNF-α is thought to initiate apoptotic signaling via the death receptor, and this can proceed to activate the effort proteases (e.g., caspase 3) independent from mitochondria signaling. Nevertheless, there is some cross-talk between the intrinsic and the extrinsic apoptotic pathways. Finally, a few studies have shown data to suggest that the endoplasmic reticulum-stress apoptotic pathway may also have a role in sarcopenia, although the importance of this pathway relative to the other two pathways is less clear. Both myonuclei and satellite cells appear to be susceptible to nuclear apoptosis in sarcopenia.

S.E. Alway (✉)
Department of Exercise Physiology, and Center for Cardiovascular and Respiratory Sciences, West Virginia University School of Medicine, Robert C Byrd Health Sciences Center, 1 Medical Center Drive, Morgantown, WV 26506, USA
e-mail: salway@hsc.wvu.edu

P.M. Siu
Department of Health Technology and Informatics, The Hong Kong Polytechnic University, Hung Hom, Kowloon, Hong Kong, China
e-mail: htpsiu@inet.polyu.edu.hk

G.S. Lynch (ed.), *Sarcopenia – Age-Related Muscle Wasting and Weakness*,
DOI 10.1007/978-90-481-9713-2_9,

Keywords Nuclear cell death • Apoptosis • Skeletal myofiber • Satellite cell • Mitochondria • Muscle atrophy

1 Apoptosis

Apoptosis is a fundamental biological process that is highly conserved among species ranging from worm to human (Ellis et al. 1991; Yuan 1996) for elimination of cells from tissues in an energy dependent manner. The term "apoptosis" originates from Greek (apo – from; ptosis – falling) which means "falling off". The phenomenon of apoptosis was first systematically described in nematode *Caenorhabditis elegans* by Kerr and colleagues (Kerr et al. 1972). The distinctive morphological characteristics of apoptosis include cell shrinkage, cell membrane blebbing, chromatin condensation, internucleosomal degradation of chromosomal DNA, and formation of membrane-bound fragments called apoptotic bodies (Kerr et al. 1972). The morphological and biochemical characteristics of apoptosis are unique and clearly distinguish it from necrotic cell death. Homologous apoptotic regulatory death genes have been identified in a variety of organisms including mammals and humans (Sulston and Horvitz 1977).

In the past several decades, there has been a better understanding of the biological role and the regulatory mechanisms of apoptosis in life science and disease and aging. Apoptosis is necessary for the elimination of damaged, aberrant, or harmful cells. Apoptosis also participates in normal embryonic development, tissue turnover, and immunological function (Thompson 1995). Apoptosis coordinates the balance among cell proliferation, differentiation, and cell death in multicellular organisms. Therefore, it is reasonable to conclude that health would be threatened if apoptosis is not adequately maintained or if it is disrupted. In fact, aberrant regulation of apoptosis has been demonstrated to contribute to the pathogenesis of severe diseases including viral infections, cancers, autoimmune diseases (e.g., systemic lupus erythematosus and rheumatoid arthritis), loss of pancreatic beta-cell in diabetes mellitus, toxin-induced liver disease, and acquired immune deficiency syndrome (AIDS), myocardial and cerebral ischemic injuries and neurodegenerative diseases and muscle loss associated with aging such as Alzheimer's and Parkinson's diseases (Williams 1991; Thompson 1995; Duke et al. 1996; Yuan and Yankner 2000; Lee and Pervaiz 2007; McMullen et al. 2009; Cacciapaglia et al. 2009; Campisi and Sedivy 2009).

2 Muscle Specific Apoptotic Signalling – Nuclear Apoptosis

Apoptosis was initially described as a process that was responsible for elimination of entire cells, and this was essential for maintaining the homeostasis of cell growth and death especially in cells with a high proliferative rate. In the context of single cells, the term apoptosis has a clearly defined process leading to elimination of the nucleus and therefore the cell. However, the better term to describe this same process in multinucleated post mitotic cell populations including cardiomyocytes and skeletal myofibres is “nuclear apoptosis”. This is because elimination of a single nucleus can occur without the death of the entire (multinucleated) muscle cell although this may result in smaller cells. We propose that the process of apoptotic loss of myonuclei in skeletal muscle should be best described as “nuclear apoptosis”. Nuclear apoptosis can occur without inflammation or disturbing adjacent proteins or organelles.

The concept of “nuclear apoptosis” (i.e., death of a nucleus without death of the entire cell) is intriguing and exciting. By definition, nuclear apoptosis involves cell signalling that is so precise that specific individual nuclei can be targeted for elimination in a multinucleated skeletal myofiber without targeting other nuclei. Thus, nuclear apoptosis requires amazingly precise targeting of some nuclei but not others within a single muscle fibre.

Evidence accumulated over the last several years has shown that apoptosis is a significant contributor to muscle degeneration (Primeau et al. 2002; Adhihetty and Hood 2003; Dirks and Leeuwenburgh 2005; Tews 2005; Siu and Alway 2005a, 2006b; Siu et al. 2006; Pistilli et al. 2006b; Adhihetty et al. 2008, 2009; Marzetti et al. 2008c, 2009b; Lees et al. 2009; Smith et al. 2009). However, apoptosis in skeletal muscle is unique for several reasons. First, skeletal muscle is multi-nucleated. Thus, the removal of one myonucleus by apoptosis will not produce “wholesale” muscle cell death, but it does result in a loss of gene expression within the local myonuclear domain, potentially leading to cellular atrophy. Second, muscle contains two morphologically and biochemically distinct subfractions of mitochondria (subsarcolemmal, SS and intermyofibrillar, IMF) that exist in different regions of the fibre could produce regional differences in the sensitivity to apoptotic stimuli within the cell (Adhihetty et al. 2007a, 2008, 2009). Third, skeletal muscle is a malleable tissue capable of changing its mitochondrial content and/or composition in response to chronic alterations in muscle use or disuse. Such variations in mitochondrial content and/or composition can undoubtedly influence the degree of organelle-directed apoptotic signalling in skeletal muscle.

Evidence that not all myonuclei in a single myofiber become apoptotic during muscle loss has been observed in experimental denervation and denervation-associated disease (e.g., infantile spinal muscular atrophy). This further supports the hypothesis of “nuclear apoptosis” in modulating the myofiber volume by controlling the successive myofiber segments. The hypothesis of nuclear apoptosis is consistent with the proposed “nuclear domain hypothesis” which explains the phenomenon of cell size remodelling of myofiber by adding or subtracting nuclei because each nucleus controls a specifically defined cytoplasmic area (Fig. 1). The skeletal myofiber is a differentiated but highly plastic cell type which adapts to loading and unloading. The nuclear domain hypothesis predicts that a nucleus controls a defined volume of cellular territory in each myofiber. Therefore, addition of extra nuclei (from satellite cells) into the myofiber is required to support the increment of cell size in order to achieve muscle hypertrophy and removal of the myonuclei is needed to allow the muscle to atrophy. If fewer nuclei are available, less cytoplasmic area could be supported. Generally, there is a tight relationship between nuclear number and muscle fibre cross-sectional area and volume. Nevertheless, this relationship is not perfect, because the nuclear domain increases

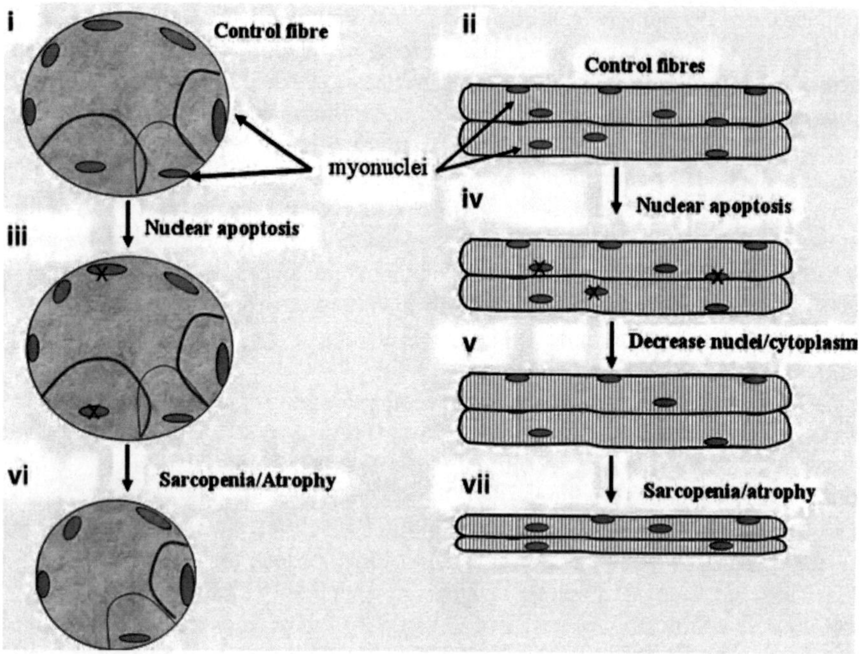

Fig. 1 Muscle fibres are illustrated in cross section (**a, c, f**) or longitudinally (**b, d, e, g**). Myonuclei in muscle fibres control a fixed cytoplasmic domain (**c**)). Nuclear are targeted for elimination by apoptosis (**red; c, d**). Fewer nuclei are unable to maintain the cytoplasmic area (**e**) and these results in fibre atrophy and ultimately sarcopenia (**f** and **g**)

slightly with age (i.e., less nuclei/cytoplasm area). With age there is a loss of satellite cells or muscle precussor cells (MPCs), which reduces the muscle's ability to replace nuclei (Brack et al. 2005, 2007; Bruusgaard et al. 2006; Brack and Rando 2007). This results in a somewhat transient increase in the nuclear domain with aging, but the excessive domain size triggers fibre atrophy (Brack et al. 2005) which in turn restores the original nuclear domain size, but also contributes to sarcopenia (Fig. 1).

3 Apoptosis Signaling Pathways in Muscle

In single nucleated cell populations, apoptosis functions to destroy and eliminate the entire cell through a cascade of cellular suicide steps. One of the distinctive characteristics of apoptosis is that it allows the execution of cells in the absence of inflammation and therefore it does not disturb neighboring cells. This characteristic of apoptosis permits highly selective dismissal of targeted individual cells among the whole cell population. Apoptosis-induced myonuclear debris removal likely involves the ubiquitin-proteasome pathway, as well as autophagy in many cell types (Yang et al. 2009; Korolchuk et al. 2009) including skeletal muscle (Attaix et al. 2005; Combaret et al. 2009). Literature relating to how the ubiquitin-proteasome and autophagy pathways are associated with apoptosis in muscle is currently scarce and further investigation in this area is warranted.

Three primary apoptotic pathways have been identified in mediating cellular signalling transduction leading to the implementation of apoptosis in muscle cells (Fig. 2). These apoptotic pathways include mitochondria-dependent (intrinsic), death receptor-mediated (extrinsic), and endoplasmic reticulum-calcium stress-induced pathways (Li et al. 1998; Gorman et al. 2000; Nakagawa et al. 2000; Phaneuf and Leeuwenburgh 2002; Green and Kroemer 2004; Spierings et al. 2005). These apoptotic pathways are named based on the origin of stimulus and the subcellular site that carries out the signalling events. Various gene products play a role in regulating the process of apoptosis. These proteins include B-cell leukaemia/lymphoma-2 (BCL-2) family proteins, caspases, inhibitors of apoptosis proteins (IAPs), caspase-independent apoptotic factors including apoptosis inducing factor (AIF), endonuclease G (EndoG) and heat requirement A2 protein

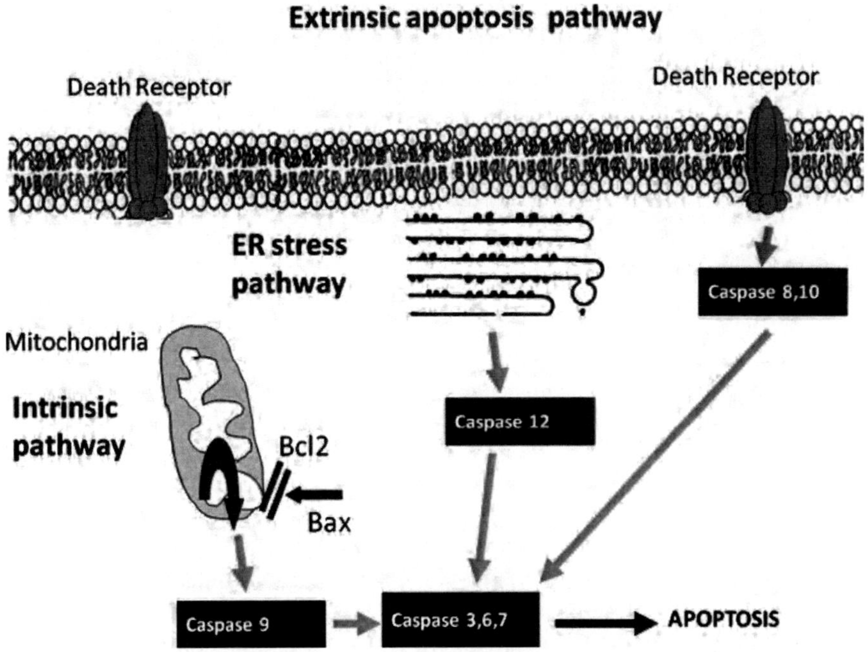

Fig. 2 Three apoptotic pathways have been identified in sarcopenia. These include the intrinsic (mitochondria pathway) which involves mitochondria dysfunction and increased mitochondria permeability. A series of downstream signalling events results in activation of initiator caspases (e.g., caspase 9) and effector caspases (e.g., caspase 3) and finally apoptosis. The endoplasmic reticulum (ER)-calcium stress pathway activates initiator caspases (e.g., caspase 12) then effector caspases (e.g., caspase 3 or 7). The extrinsic pathway is activated by a ligand (e.g., TNF-α) and activates initiator caspases (e.g., caspase 8) and the effector caspases (e.g., caspase 3) and through this to apoptosis

(HtrA2/Omi), and other apoptosis-related proteins like cytochrome *c*, apoptosis protease activating factor-1 (Apaf-1), apoptosis repressor with caspase recruitment domain (ARC), Smac/DIABLO, p53, heat shock proteins (HSPs) and others. The participation of these apoptotic factors are selective in nature and are largely dependent on the apoptotic pathway being invoked. For example, initiator caspases-8, -9, and -12 are activated when cells are exposed to an appropriate stress stimulus. When apoptosis is stimulated by TNF-α and FasL which subsequently activate the death receptor apoptotic pathway, caspase 8 is the initiator caspase being triggered and responsible for the mediation of the corresponding subsequent signalling transduction (Li et al. 1998; Sun et al. 1999). Smac/Diablo is also thought to mediate the pro-apoptotic function of TNF-α- regulated PUMA (Yu et al. 2007). Apoptotic signalling initiated by intracellular calcium disturbance and endoplasmic reticulum stress is attributed to initial activation of caspase-12 (Nakagawa et al. 2000; Nakagawa and Yuan 2000) whereas caspase 9 mediates the mitochondria-dependent apoptosis through the interaction of procaspase 9 with Apaf-1, dATP/ATP, and cytochrome c. Although different initiator caspases (caspase 8, -12, and -9) are responsible for the initial signalling transduction in different apoptotic pathways, the signals eventually converge on the activation of common effector caspases-3, -6, or -7, which function to progress to the final dismissal of the target cell.

4 Intrinsic Apoptotic Pathway

4.1 Role of Mitochondria in the Intrinsic Apoptosis Pathway in Muscle

An accumulating body of evidence suggests that disruptions in mitochondrial function precedes the initiation of the intrinsic apoptotic pathway in sarcopenia of aging (Siu et al. 2005b; Pistilli et al. 2006b; Chabi et al. 2008; Seo et al. 2008) as well as disuse-associated muscle atrophy (Siu and Alway 2005b; Adhihetty et al. 2007b).

Mitochondria play a critical role in maintaining cellular integrity through the regulation of apoptosis (Fig. 3). When mitochondria localized proteins are released to the cytosol, they can initiate a cascade of proteolytic events that converge on the nucleus leading to the fragmentation of DNA and elimination of the nucleus. This compromises muscle cell viability and ultimately leads to cell death (Bernardi 1999) in non-muscle cells. The release of these apoptotic proteins, include cytochrome *c*, endonuclease G (EndoG), Smac/Diablo and apoptosis-inducing factor (AIF), through either the mitochondrial permeability transition pore (mtPTP) (Kroemer and Reed 2000; Precht et al. 2005; Forte and Bernardi 2006; Rasola and Bernardi 2007; Knudson and Brown 2008) or the homooligomeric Bax mitochondria apoptotic channels (MAC) in the outer mitochondria membrane, occurs in response

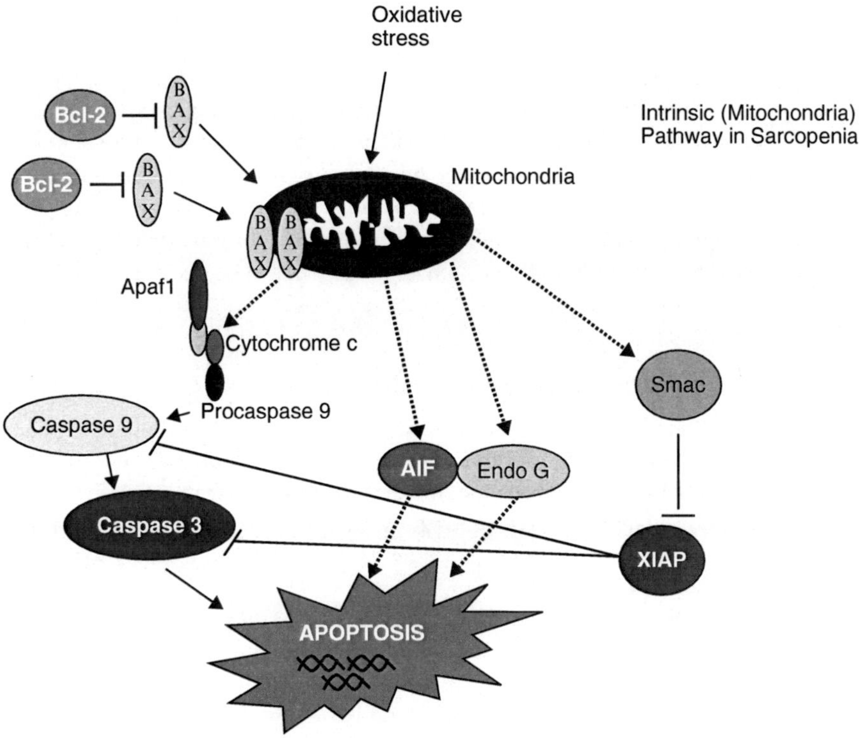

Fig. 3 The intrinsic (mitochondria) pathway is activated in sarcopenia. Pro-apoptotic factors (e.g., Bax) heterodimerise to form a mitochondria channel which releases caspase dependent (e.g., cytochrome c) or caspase independent (e.g., AIF, Endo G, Smac/Diablo) pro-apoptotic factors and result in DNA fragmentation and nuclear apoptosis in muscle

to cellular stressors including ROS (Dejean et al. 2006a, b; Martin et al. 2007). Putative components of the MAC channel are Bax and Bak, whereas Bcl2 acts as a negative regulator of this channel (Dejean et al. 2005, 2006a, b). Thus, this intimate connection between mitochondrial function and the viability of skeletal muscle suggests that this organelle likely plays a significant role in the progression of aging and sarcopenia. Indeed, it is evident that in skeletal muscle of aged individuals, the induction of apoptosis is greater when compared to younger subjects. Oxidative stress is greater in muscles of old vs. young adult animals (Ryan et al. 2008) and may have a role in regulating increased apoptotic signalling (Siu et al. 2008). Furthermore, increased oxidative stress may increase peroxidation of the mitochondrial lipid cardiolipin, Bax mobilization and release of cytochrome c (Huang et al. 2008). The increase in cytochrome *c* and EndoG release from the mitochondria of aged individuals (Tamilselvan et al. 2007) is paralleled by an increase in cleavage and activation of caspase 3 (Alway et al. 2003b; Siu et al. 2005b; Chabi et al. 2008), and p53 mediated apoptosis (Tamilselvan et al. 2007). A consequence of apoptosis is a loss in myonuclear number, resulting in a reduction in myofiber diameter to maintain a constant myonuclear domain size (Dirks and Leeuwenburgh 2005; Pistilli et al. 2006b; Wang et al. 2008; Alway and Siu 2008; Pistilli and Alway 2008). This decrease in fibre area results in whole muscle atrophy, especially in fast muscles which have a high percentage of type II myosin heavy chain. This suggests that there is a significant mitochondrial involvement in the progression of sarcopenia. Greater mitochondrial dysfunction is also evident in muscles with higher type II muscle fibre composition, and this may be key to the preferential loss of type II fibres found in the elderly (Conley et al. 2007a).

4.2 Oxidative Stress and Mitochondria

The free radical theory of aging first proposed by Harman more than five decades ago (Harman 1956), suggests that mitochondria dysfunction from oxidative damage to mitochondria DNA (mtDNA) caused by reactive oxygen species (ROS) is a central factor contributing to aging (Harman 1992, 2003, 2006, 2009; Malinska et al. 2009; Kadenbach et al. 2009).

The mitochondrion is the main cellular site for ROS; however, it is not the only site for ROS production. Nevertheless, it is reasonable to expect that mitochondrial components will be susceptible to oxidative damage. In particular, mtDNA in muscle is particularly susceptible to oxidative damage (Hagen et al. 2004; Murray et al. 2007; Ricci et al. 2007; Meissner 2007; Meissner et al. 2008) due to its proximity to the electron transport chain (ETC), the lack of protective histones and an inefficient repair system compared to nuclear DNA (Wei and Lee 2002; Lee and Wei 2007; Ma et al. 2009). Mutations in mtDNA can lead to the synthesis of defective respiratory chain elements, which may impair oxidative phosphorylation, increase ROS production or decrease ATP availability (Harman 2006; Malinska et al. 2009; Kadenbach et al. 2009; Ma et al. 2009). Several lines of evidence support the idea that mtDNA damage and mutations contribute to aging in muscle (reviewed in (Dirks et al. 2006; Dirks Naylor and Leeuwenburgh 2008; Marzetti et al. 2009b). For example, mice expressing a mutated mtDNA polymerase accumulate mtDNA mutations and display a premature aging phenotype, which includes extensive sarcopenia, compared to wild-type littermates (Kujoth et al. 2005, 2006).

Oxidative stress is greater in muscles of old vs. young adult animals (Ryan et al. 2008) and may have a role in regulating increased apoptotic signalling (Siu et al. 2008). Furthermore, increased oxidative stress may function to elevate peroxidation of the mitochondrial lipid cardiolipin, as well as Bax mobilization and release of cytochrome c (Huang et al. 2008). The increase in cytochrome *c* and EndoG release from the mitochondria of aged individuals (Tamilselvan et al. 2007) is paralleled by an increase in cleavage and activation of caspase 3 (Alway et al. 2003b; Siu et al. 2005b; Chabi et al. 2008), and p53 mediated apoptosis (Tamilselvan et al. 2007).

4.3 BCL2 Protein Family

The BCL-2 family serves as an important upstream intracellular checkpoint which plays a crucial role in the coordination of the apoptotic signalling (Danial and Korsmeyer 2004). BCL-2 family members share homology within four conserved sequence motifs which are: BH1, BH2, BH3, and BH4 family proteins. In general, the BCL-2 family consists of three subclasses: (a) anti-apoptotic proteins (e.g., Bcl-2, Bcl-XL, Bcl-W, A1, and Mcl-1), (b) multidomain pro-apoptotic proteins (Bax, Bak, and Bok), and (c) BH3-only pro-apoptotic proteins (Bid, Bad, Bim, Bik, Dp5/Hrk, Noxa, and Puma) (Chao and Korsmeyer 1998; Mikhailov et al. 2001, 2003; Danial and Korsmeyer 2004). All pro-apoptotic members and most anti-apoptotic members contain the BH3 domain and this domain is believed to be essential for the interactions among the family members (Korsmeyer 1995; Chao and Korsmeyer 1998; Korsmeyer 1999). The BH3 sequence motif has a hydrophobic α-helix which is favourable for protein interaction, and this is the putative region responsible for the association among the BCL-2 family members through homo- or hetero-oligomerisation (Chao and Korsmeyer 1998; Mikhailov et al. 2001, 2003; Er et al. 2007). The strict control which balances cell survival and apoptotic cell death is believed to be primarily regulated by the relative ratio of pro- and anti-apoptotic BCL-2 members (Chao et al. 1995; Korsmeyer et al. 1995; Chao and Korsmeyer 1998; Danial and Korsmeyer 2004). Among the BCL-2 family members, pro-apoptotic Bax and anti-apoptotic Bcl-2 have been well-studied. These proteins are thought to constitute the main components in the regulation of mitochondria apoptotic channels or pores. Essentially, Bcl-2 forms a homodimer with Bax and prevents its translocation to the mitochondria in non-apoptotic conditions. However, an apoptotic stimulus translocates Bax to mitochondria and phosphorylates it. Bax undergoes conformational change to expose its N-terminus (Hsu et al. 1997; Wolter et al. 1997; Basanez et al. 1999; Desagher and Martinou 2000; Cartron et al. 2002) to allow the Bax–Bax-oligomerisation and insertion of Bax into the outer mitochondrial membrane (Zha et al. 1996), which mediates the subsequent release of the apoptogenic factors (e.g., cytochrome, EndoG, AIF etc.) from the mitochondrial intermembrane space (Narita et al. 1998; Reed et al. 1998; Shimizu et al. 1999; Tsujimoto and Shimizu 2000; Tsujimoto et al. 2006; Kroemer et al. 2007). Bcl-2 functions to prevent the Bax–Bax-oligomerisation and therefore opposes the pro-apoptotic activity of Bax (Yin et al. 1994; Korsmeyer 1995, 1999; Korsmeyer et al. 1995; Reed 1997, 2006; Reed et al. 1998; Antonsson et al. 2000; Kroemer et al. 2007; Lalier et al. 2007).

4.4 Caspase (Cysteine-dependent Aspartic Acid Specific Protease) Dependent Signalling

The involvement of the pro-apoptotic role of cysteine-dependent aspartate proteases (caspases) has been extensively studied and several members appear to have a critical role in apoptotic signaling transduction (Earnshaw et al. 1999; Chang and Yang 2000; Grutter 2000; Degterev et al. 2003). Although caspase 9 is an exception (Stennicke and Salvesen 1999; Stennicke et al. 1999), other caspases are synthesized as inactive zymogens (i.e., procaspases). When procaspases undergo cleavage or oligomerisation-mediated self-/auto-activation by an apoptotic signal, they are converted from their inactive procaspases to the active protease (Earnshaw et al. 1999; Deveraux et al. 1999; Stennicke and Salvesen 1999; Stennicke et al. 1999; Deveraux and Reed 1999; Chang and Yang 2000; Grutter 2000). Caspase 9 is an initiator caspase which has been shown to mediate the signalling of mitochondria-mediated apoptosis. Caspase 9 participates in a protein complex, the apoptosome. The interaction of procaspase 9 with Apaf-1, cytochrome c (which is released from the mitochondria), and ATP/dATP in the cytosol activates caspase 9 which cleaves procaspase 3 and activates it (Chang and Yang 2000; Shiozaki et al. 2002; Acehan et al. 2002; Shi 2002a, b, 2004). Caspase 3 is a common downstream effector (executer) caspase for initiating DNA destruction. Cellular substrates for caspase 3 cleavage include proteins responsible for cell cycle regulation (e.g., $p21^{Cip1/Waf1}$), apoptotic cell death (e.g., Bcl-2 and IAP), DNA repair (e.g., poly(ADP-ribose) polymerase (PARP) and inhibitor of caspase-activated DNase (ICAD), cell signal transduction (e.g., Akt/PKB), and cytoskeletal structural scaffold (e.g., gelsolin), etc. (Chang and Yang 2000).

4.5 Caspase-independent Apoptotic Signalling

Mitochondria-housed proteins including apoptosis-inducing factor (AIF), endonuclease G (EndoG) and high temperature requirement protein A2 (HtrA2/Omi) have been shown to be able to induce apoptosis without the involvement of caspases (Joza et al. 2001; Li et al. 2001; Blink et al. 2004). AIF is a mitochondrial flavoprotein that has both oxidoreductase and apoptosis-inducing activities (Joza et al. 2001, 2005; Cande et al. 2002a, b). Although the full physiologic importance of AIF is not yet completely known, it is clear that AIF has an important role in mitochondrial-mediated apoptosis. The apoptotic function of AIF may be the result of a putative DNA binding site which results in chromatin condensation and DNA fragmentation (Lipton and Bossy-Wetzel 2002; Ye et al. 2002). EndoG is an well-conserved nuclear-encoded endonuclease, which can induce chromosomal DNA cleavage in a caspase-independent manner (Li et al. 2001). In contrast, the apoptotic properties of a serine protease HtrA2/Omi are less well defined. It has been thought that HtrA2/Omi induces apoptosis via the mechanism similar to Smac/DIABLO, in which the apoptosis-suppressing activities of IAPs are removed through a caspase-regulated process (Hegde et al. 2002; Shi 2004; Shiozaki and Shi 2004). However, it has also been shown that the apoptosis-inducing ability of HtrA2/Omi can function via its proteolytic activity in the absence of caspase activation (Blink et al. 2004; Suzuki et al. 2004). These caspase-independent proteins are normally housed in the mitochondrial intermembrane space, but they are released into cytosol once in response to an apoptotic stimulus (Joza et al. 2001; Li et al. 2001; Cande et al. 2002b; Blink et al. 2004).

It is known that cytosolic and nuclear levels of AIF and EndoG are elevated in skeletal muscles of old and senescent animals (Leeuwenburgh et al. 2005; Siu and Alway 2006a; Marzetti et al. 2008c). This confirms a central role for apoptosis in sarcopenia, but the extent to which caspase-dependent vs. caspase-independent signalling dominates apoptotic elimination of nuclei has not yet been established. Although a role for HtrA2/Omi has been suggested in response to myocardial injury or heart failure (Siu et al. 2007; Bhuiyan and Fukunaga 2007), it has not been established that HtrA2/Omi is elevated in sarcopenia.

4.6 Mitochondria-associated Apoptotic Suppressors

A group of mitochondrially stored endogenous proteins have been shown to function in suppressing pro-apoptotic signaling. Members of this Inhibitor of Apoptosis (IAP) family include X-linked inhibitor of apoptosis (XIAP), apoptosis repressor with caspases recruitment domain protein (ARC), and Fas-associated death domain protein-like interleukin 1α-converting enzyme-like inhibitory protein (FLIP). XIAP is a fundamental conserved gene product among many species (Deveraux et al. 1998; Shi 2002b). The anti-apoptotic ability of XIAP is attributed to the conserved baculovirus inhibitor of apoptosis repeat (BIR) motif which is the essential part for the inhibition on initiator as well as effector caspases and all protein members in IAP family are found to carry at least one of this BIR motif (Deveraux et al. 1998, 1999; Salvesen and Duckett 2002; Sanna et al. 2002; Chowdhury et al. 2008). ARC and FLIP are two endogenous apoptosis-suppressing proteins with high expression levels in muscle tissue (Irmler et al. 1997; Koseki et al. 1998). It is possible that the high resistance of mature muscle tissues to apoptosis is related to the abundant expressions of these two apoptotic suppressors, although this has not been definitively shown. The apoptotic suppressive effects of ARC and FLIP are thought to be due to their inhibiting interactions with selective caspases, in particular, caspase 8 which is the initiator caspase in the death receptor-mediated apoptosis (Irmler et al. 1997; Koseki et al. 1998; Abmayr et al. 2004; Heikaus et al. 2008; Yu et al. 2009b). Additional observations indicate that ARC is able to interact with pro-apoptotic Bax protein and so exhibits the apoptosis suppressive effect by influencing the mitochondria-mediated apoptotic signaling (Gustafsson et al. 2004).

Regulation of the extrinsic pathway is very complex, with some proteins appearing to have dual roles. For example, c-FLIP (L) is widely regarded as an inhibitor of initiator caspase 8 activation and cell death in the extrinsic pathway; however, it is also capable of enhancing procaspase 8 activation through heterodimerisation of their respective protease domains. Cleavage of the inter-subunit linker of c-FLIP(L) by procaspase 8 potentiates the activation process by enhancing heterodimerisation between the two proteins and elevates the proteolytic activity of unprocessed caspase-(C)8 (Yu et al. 2009b). FLIP's role in regulation of apoptosis may be in part related to the individual splice variants (i.e., protein isoforms). For example, $FLIP_S$ versus $FLIP_L$ or $FLIP_c$. For example, disruption of NF-Kappa B regulation of FLIPc has been implicated in muscle wasting diseases such as Limb-girdle muscular dystrophy type 2A (Benayoun et al. 2008) although it is not known if similar deregulations occur in aging muscles.

4.7 *Sarcopenia-associated Mitochondria Mediated Signalling in Apoptosis*

Sarcopenia is a complex pathology which is not fully understood. Several factors are thought to contribute to sarcopenia including increases in inflammation and oxidative stress, loss of systemically or locally generated growth signals, neural factors and reduced muscle progenitor stem cell function. Not only do post-mitotic myocytes exhibit apoptosis during atrophy induced by denervation and unloading (Allen et al. 1997; Jin et al. 2001; Jejurikar et al. 2002; Alway et al. 2003a, b; Siu and Alway 2005a; Siu et al. 2005c), but apoptosis is thought to have an important role in the aging associated loss of muscle mass or sarcopenia (Dirks and Leeuwenburgh 2002; Pollack et al. 2002; Alway et al. 2002a, b; Leeuwenburgh 2003; Dirks and Leeuwenburgh 2004; Siu et al. 2005c). Evidence for myonuclei undergoing apoptosis via the intrinsic pathway in aging has been shown by increases in TUNEL positive nuclei, increases in the frequency of nuclei with DNA strand breaks and in the expression of pro-apoptotic genes and proteins including Bax, caspase 3, apoptosis-inducing factor (AIF) and apoptotic protease- activating factor (Apaf1) in aged and atrophied muscles in mammals and non-mammals including birds, worms and flies (Alway et al. 2002a, b; Senoo-Matsuda et al. 2003; Siu et al. 2004, 2005c; Zheng et al. 2005; Siu and Alway 2005a, 2006a, b; Dirks and Leeuwenburgh 2006; Pistilli et al. 2006b; li-Youcef et al. 2007; Dirks Naylor and Leeuwenburgh 2008).

5 Extrinsic Apoptotic Signalling in Skeletal Muscle

One potential mechanism contributing to the onset of sarcopenia may be the increase in circulating cytokines which activates the extrinsic apoptotic pathway. The circulating concentrations of specific cytokines have been shown to be elevated in the serum as a result of aging. In humans, serum levels of catabolic cytokines, such as TNF-α (Sandmand et al. 2003; Schaap et al. 2009) and IL-6 (Bruunsgaard 2002; Forsey et al. 2003; Pedersen et al. 2003; Schaap et al. 2009), are increased in healthy elderly compared to young adults. Serum concentrations of TNF-α have been proposed as a prognostic marker of all cause-mortality in centenarians (Bruunsgaard et al. 2003b) and with age-associated pathology and mortality in 80-year old adults (Bruunsgaard et al. 2003a).

5.1 *Tumour Necrosis Factor-α (TNF-α) and Death Receptor Signalling*

Several studies have also drawn associations between the increases in circulating cytokines and the sarcopenic process (Visser et al. 2002; Pedersen et al. 2003; Schaap et al. 2006, 2009). Specifically, elevated circulating levels of TNF-α are associated with lower appendicular skeletal muscle mass (Pedersen et al. 2003) and reduced knee extensor and grip strength (Visser et al. 2002).

Tumour necrosis factor-α (TNF-α) is a pleiotropic cytokine that has an important role in many different physiological and pathological processes including immune and inflammatory responses (Wajant et al. 2003; Wajant 2009). TNF-α-induced apoptosis is mediated by its interactions with cell-surface receptors such as extrinsic signalling through TNF receptor 1 (TNFR1) and TNF receptor 2 (TNFR2) (Wajant et al. 2003; Wajant 2009). The extrinsic death ligand associated apoptotic pathway in sarcopenia is thought to be activated by ligands such as TNF-α. Ligand binding induces trimerisation of death receptors, activation of caspase 8 and subsequently executioner caspases, such as caspase 3 (Ricci et al. 2007). The contribution of the extrinsic apoptotic pathway to skeletal muscle mass losses, especially during aging, has been less well studied than the intrinsic pathway (Phillips and Leeuwenburgh 2005). However, activation of this pathway does appear to play a role in aging associated muscle loss (Fig. 4).

The increase in circulating concentrations TNF-α in aged animals may initiate pro-apoptotic signalling upon binding to the type I TNF receptor (TNFR). Upon binding, a death inducing signalling complex (DISC) is formed at the cytoplasmic portion of the TNFR, composed of adaptor proteins such as Fas associated death domain protein (FADD), TNFR associated death domain protein (TRADD) and procaspase 8 (reviewed in Sprick and Walczak (2004)). Formation of the DISC stimulates cleavage of procaspase 8 into the functional initiator caspase 8. Once cleaved, caspase 8 stimulates cleavage and activation of the executioner caspase 3, which is directly linked to pro-apoptotic changes. Thus, this pathway represents an extrinsic pathway of apoptosis activated by binding of a ligand (TNF-α) to a cell surface death receptor (type-I TNFR).

Nuclear factor-kB (NF-kB) is the best-known mediator of TNF-α-associated cellular responses. NF-kB is a group of dimeric transcription factors which are members of the NF-kB/Rel family, including p50, p52, p65 (Rel-A), Rel-B, and c-Rel (Shih et al. 2009; Kearns and Hoffmann 2009). The activity of NF-kB is normally regulated by the IkB family of inhibitors, which bind to and sequester NF-kB in the cytoplasm (Shih et al. 2009). Activation of NF-kB is triggered by IkB phosphorylation by IKK kinases and subsequent proteasomal degradation, which allows NF-kB to translocate to the nucleus, where it binds to the kB consensus sequences and modulates specific target genes (Kearns and Hoffmann 2009; Vallabhapurapu and Karin 2009). NF-kB is thought to provide a protective role in TNF-α-induced apoptosis. This is because NF-kB is a transcriptional activator of anti-apoptotic proteins including c-FLIP, Bcl-2 and Bcl-XL (Vallabhapurapu and

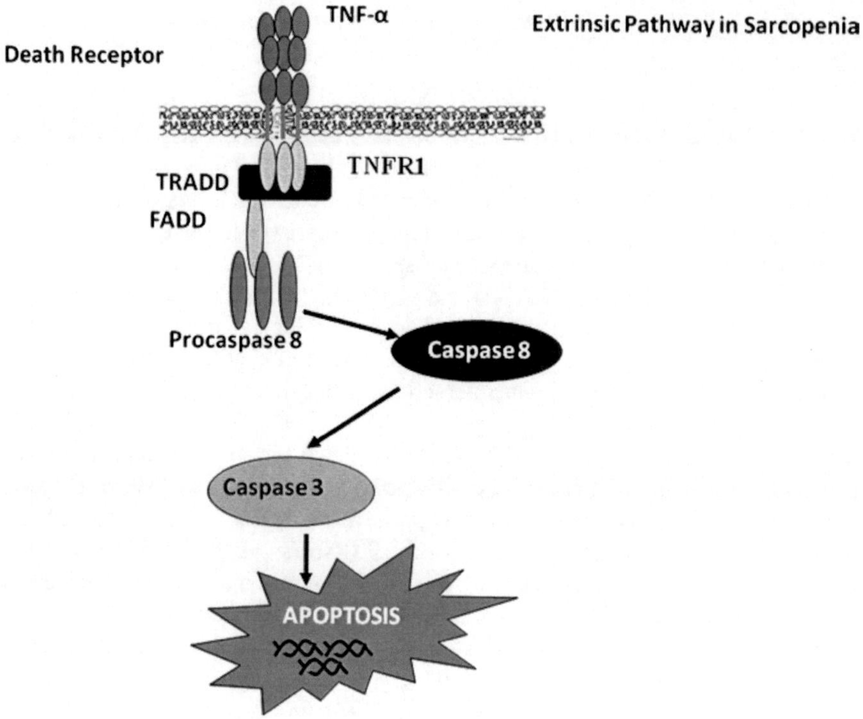

Fig. 4 The extrinsic (death receptor) pathway is activated in aging and contributes to sarcopenia. A ligand (e.g., TNF-α) binds to the death receptor and TNFR1, activates procaspase 8 and caspase 8 which in turn activates caspase 3 and DNA fragmentation

Karin 2009). However, NF-kB can also promote apoptosis when activated by pro-apoptotic proteins including p53, Fas and Fas ligand (Burstein and Duckett 2003; Dutta et al. 2006; Fan et al. 2008).

p53 upregulated modulator of apoptosis (PUMA) is a downstream target of p53 and a BH3-only Bcl-2 family member(Lee et al. 2009; Chipuk and Green 2009; Ghosh et al. 2009b). It is induced by p53 following exposure to DNA-damaging agents, such as gamma-irradiation and commonly used chemotherapeutic drugs or oxidative stress (Lee et al. 2009; Chipuk and Green 2009; Ghosh et al. 2009a). It is also activated by a variety of nongenotoxic stimuli independent of p53, such as serum starvation, kinase inhibitors, glucocorticoids, endoplasmic reticulum stress, and ischemia/reperfusion (Nickson et al. 2007; Yu and Zhang 2008). The pro-apoptotic function of PUMA is mediated by its interactions with anti-apoptotic Bcl-2 family members such as Bcl-2 and Bcl-XL which lead to Bax/Bak-dependent mitochondrial dysfunction mitochondria permeability and caspase activation (Chipuk and Green 2009). In addition, PUMA is directly activated by NF-kB and contributes to TNF-α-induced apoptosis (Wang et al. 2009).

Based on the well-documented increase in circulating TNF-α levels with aging (Bruunsgaard et al. 1999, 2001, 2003a, b; Bruunsgaard 2002; Visser et al. 2002; Pedersen et al. 2003; Sandmand et al. 2003; Schaap et al. 2006, 2009) and increases in apoptosis of myonuclei in aged skeletal muscles (Allen et al. 1997; Siu et al. 2005c; Pistilli et al. 2006b), we examined whether apoptotic signalling via the extrinsic pathway contributed to sarcopenia. Our data show that pro- and anti-apoptotic proteins in the extrinsic apoptotic pathway are affected by aging in fast (plantaris) and slow (soleus) skeletal muscles of rats (Pistilli et al. 2006b). Similarly, Marzetti et al. (2009a, b) report elevated TNF-α and TNF-receptor 1 in muscles of old rodents. Together, these data suggest that TNF-α mediated signalling may be an important element triggering the extrinsic apoptotic pathway in and leading to sarcopenia in aging muscles.

Muscles from aged rats are significantly smaller and exhibit a larger incidence in fragmented DNA. This suggests that there is a higher level of nuclear apoptosis in muscles from aged animals. In addition, muscles from aged rodents have higher TNFR and FADD mRNA content (measured by semi-quantitative RT-PCR) and protein contents for FADD, Bid, and FLIP, and enzymatic activities of caspase 8 and caspase 3, when compared to muscles from young adult rodents. Although there is an increase in mRNA expression for the TNFR as measured by the semi-quantitative approach, the protein content for the TNFR remains unchanged (Pistilli et al. 2006a, b). This may be explained by the fact that the TNFR antibody utilized in western immunoblots recognizes the soluble form of the receptor. Thus, the changes in the membrane bound form of the receptor, measured by PCR, and the amount of the soluble TNFR may not be equivalent. While fast contracting muscles are generally more susceptible to apoptosis and sarcopenic muscle loss, the pro-apoptotic changes have been reported to be expressed in a similar fashion in both plantaris and soleus muscles; however strong relationships were observed between markers of apoptosis and muscle loss in the fast plantaris muscle that were not observed in the soleus (Pistilli et al. 2006a). These data extend the previous demonstration that type II fibres are preferentially affected by aging and suggest that type II fibre containing skeletal muscles may be more susceptible to muscle mass loses via the extrinsic apoptotic pathway (Pistilli et al. 2006b).

We have found activation of the extrinsic apoptotic signalling pathway in muscles of old rats (Pistilli et al. 2006a, 2007; Siu et al. 2008), and therefore we speculate that circulating TNF-α may be the initiator of this pathway in skeletal muscle. Nevertheless, we cannot rule out the possibility that other pathways that we did not examine may have been activated by circulating TNF-α in aging muscle. For example, TNF-α has been shown to directly promote protein degradation (Garcia-Martinez, et al. 1993a, b; Llovera et al. 1997, 1998) and apoptosis within skeletal muscle (Carbo et al. 2002; Figueras et al. 2005). Furthermore, intravenous injection of recombinant TNF-α increases protein degradation in rat skeletal muscles and this is associated with the increased activity of the ubiquitin-dependent proteolytic pathway (Garcia-Martinez et al. 1993a, 1995; Llovera et al. 1997, 1998). In addition, elevated TNF-α concentrations in cell culture for 24–48 h increases apoptosis in skeletal myoblasts as determined by DNA fragmentation (Meadows et al. 2000;

Foulstone et al. 2001). A reduction of procaspase 8 occurs within 6h of incubating myoblasts in vitro with recombinant TNF-α, suggesting a TNF-α mediated cleavage and activation of this initiator caspase in myoblast cultures (Stewart et al. 2004).

Lees and co-workers (Lees et al. 2009) have recently shown that satellite cells (i.e., MPCs) isolated from hindlimb muscles of old rats have increased TNF-α-induced nuclear factor-kappa B (NF-kB) activation and expression of mRNA levels for TRAF2 and the cell death-inducing receptor, Fas (CD95), in response to prolonged (24 h) TNF-α treatment compared to in MPCs isolated from muscles of young animals. These findings suggest that age-related differences may exist in the regulatory mechanisms responsible for NF-kB inactivation, which may in turn have an effect on TNF-α-induced apoptotic signalling. Systemic and muscle levels of TNF-α increase with aging, and this should have an even more profound increase in activation of apoptotic gene targets through the extrinsic pathway, as compared to MPCs in muscles of young adult rats (Krajnak et al. 2006; Lees et al. 2009).

The effects of TNF-α on apoptosis are not limited to in vitro conditions, because a systemic elevation of TNF-α in vivo increases DNA fragmentation within rodent skeletal muscle (Carbo et al. 2002). Based on the observation that TNF-α mRNA was not different between muscles from young adult and aged rats, it is reasonable to assume that muscle-derived TNF-α does not act in an autocrine manner to stimulate the pro-apoptotic signalling observed in this study. Data from Pistilli and co-workers (Pistilli et al. 2006b) are consistent with the hypothesis that the well-documented systemic elevation of TNF-α with age, may increase the likelihood of ligand binding to the TNFR and stimulate apoptotic signalling of the extrinsic pathway downstream of the TNFR and contribute to sarcopenia in skeletal muscle of old rats.

5.2 Cross-talk Between Extrinsic and Intrinsic Apoptotic Signalling

Cross-talk between extrinsic and intrinsic apoptotic pathways was recently reviewed (Sprick and Walczak 2004). Cross-talk between these pathways is the result of the cleavage of the pro-apoptotic BCL-2 family member Bid. Cleaved and activated caspase 8 cannot only serve to activate caspase 3, which is the executioner caspase, but also cleave full-length Bid into a truncated version (tBid) (Tang et al. 2000). tBid then interacts with pro-apoptotic Bax, to stimulate apoptotic signalling from the mitochondria (Grinberg et al. 2005). As has been previously shown, apoptotic signalling from the mitochondria stimulates cleavage of procaspase 9, which then serves to activate caspase 3 (Johnson and Jarvis 2004). Thus, both the extrinsic and intrinsic apoptotic pathways converge on caspase 3, which then fully engages pro-apoptotic signalling. Skeletal muscles from aged rodents contained a greater protein expression of full-length Bid, which raises the possibility that cross talk between the extrinsic pathway and the intrinsic pathway may occur in aged skeletal muscles (Fig. 5).

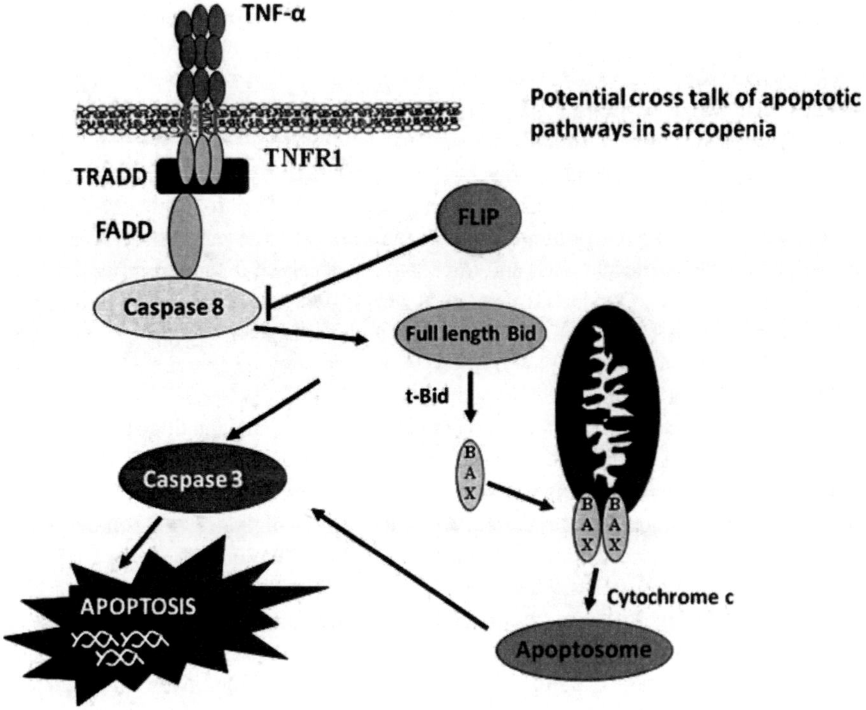

Fig. 5 The potential cross talk between the extrinsic and intrinsic apoptotic signalling pathways are shown

6 Exercise Modulation of Apoptosis in Sarcopenia

Various perturbations have been used to determine if aging increases the sensitivity of skeletal muscle to apoptosis and apoptosis signalling cascades. These include increases in muscle loading, loading followed by a period of unloading, disuse, denervation or muscle unloading, and aerobic exercise.

6.1 Interventions by Muscle Loading

The evidence presented above indicates that mitochondrial dysfunction is a major contributing factor to the path physiology of aging including sarcopenia. While muscle disuse decreases mitochondria function leading to apoptosis (Adhihetty et al. 2003; Siu and Alway 2005a; Bourdel-Marchasson et al. 2007), chronic exercise improves mitochondria function (Daussin et al. 2008; Lanza et al. 2008) and reduces apoptotic signalling (Siu et al. 2004).

Adaptation to chronic loading has been shown to improve anti-apoptotic proteins in skeletal muscle including XIAP (Siu et al. 2005d), Bcl2 (Song et al. 2006), and reduce DNA fragmentation (Siu and Alway 2006a) (Song et al. 2006) and lower pro-apoptotic proteins including Bax (Song et al. 2006), ARC (Siu and Alway 2006a), AIF (Siu and Alway 2006a). In contrast, models of muscle unloading show most of the appositive apoptotic signalling such as elevations in Bax, Apaf1, AIF (Pistilli et al. 2006b), cytosolic levels of Id2 and p53 (Siu et al. 2006) and the Bax/Bcl2 ratio (Song et al. 2006).

Reduced levels of pro-apoptotic proteins may provide one mechanism to explain the improvements in muscle mass and force that are observed in humans after a period of resistance exercise. Our lab (Roman et al. 1993; Ferketich et al. 1998) and others (Charette et al. 1991; Welle et al. 1995; Parise and Yarasheski 2000; Deschenes and Kraemer 2002; Mayhew et al. 2009) have shown that resistance exercise is an effective tool to reduce but not eliminate sarcopenia in aging humans. Although aging has generally been shown to attenuate the absolute extent of muscle adaptations that are possible with increased loading (Alway et al. 2002a; Degens and Alway 2003; Degens 2007; Degens et al. 2007), it is not known how much of this might be the result of increased nuclear apoptosis in skeletal muscle. Interestingly, several studies have reported unexpected improvements in mitochondrial function in both young adult and aged subjects as a result of resistance exercise training. For example, the mitochondrial capacity for ATP synthesis increases after resistance training (Jubrias et al. 2001; Conley et al. 2007b; Tarnopolsky 2009). Resistance exercise also increases antioxidant enzymes and decreases oxidative stress (Parise et al. 2005; Johnston et al. 2008). Furthermore, 26 weeks of whole body resistance exercise was shown to reverse the gene expression of mitochondrial proteins that were associated with normal aging, to that observed in young subjects (Melov et al. 2007). Although we have found that resistance training did not increase the relative volume of mitochondria in muscle fibres of young adults, resistance exercise stimulated mitochondria biogenesis to maintain the myofibrillar to mitochondria volume (Alway et al. 1989; Alway 1991). In addition, aging attenuates the adaptive response to improve the muscle's ability to buffer pro-oxidants in response to chronic muscle loading (Ryan et al. 2008). Nevertheless, there is some improvement in antioxidant enzymes and the ability to buffer oxidative stress in response to loading conditions (Ryan et al. 2008). Therefore, it is possible that, resistance training could also improve mitochondria function and stimulate mitochondrial biogenesis in aged individuals. If muscle loading improves not only antioxidant enzymes levels but it also reduces Bax accumulation in mitochondria, we would expect that apoptosis signalling should be decreased. This would lead to improved muscle recovery following disuse and reduce sarcopenia.

6.2 *Apoptotic Elimination of MPCs Reduces Muscle Hypertrophic Adaptation to Loading*

It is thought that myonuclei maintain a constant cytoplasm to nuclei ratio, (i.e. "nuclear domain", see Fig. 1), and that hypertrophy requires that fibres add new nuclei (Schultz 1989, 1996; Schultz and McCormick 1994). Because myonuclei are post mitotic (Schultz 1989, 1996; Schultz and McCormick 1994), satellite cells/MPCs provide the only important source for adding new nuclei to initiate muscle regeneration, muscle hypertrophy, and postnatal muscle growth in muscles of both young and aged animals (Rosenblatt et al. 1994; Phelan and Gonyea 1997; McCall et al. 1998; Allen et al. 1999; Hawke and Garry 2001; Adams et al. 2002). MPCs are critical for muscle growth because muscle hypertrophy is markedly reduced or eliminated completely after irradiation to prevent MPC activation (Rosenblatt et al. 1994; Hawke and Garry 2001). Growth of adult skeletal muscle requires activation and differentiation of satellite cells/MPCs and increased protein synthesis and accumulation of proteins, and this necessitates increased transcription of muscle genes (Dirks and Leeuwenburgh 2002; Pollack et al. 2002; Alway et al. 2002b; Leeuwenburgh 2003; Dirks and Leeuwenburgh 2004; Siu et al. 2005c). Thus, there is little doubt that MPC activation and differentiation are critical components in determining muscle adaptation and growth.

If MPCs are activated normally, but they either do not differentiate or do not survive to participate in increased protein synthesis, then muscle adaptation would be compromised. Elevation of apoptosis (lower MPC survival) in muscles from aged animals (Renault et al. 2002; Siu et al. 2005c) could explain the poorer adaptation to repetitive loading in aging. We have shown that the most recently activated satellite cells/MPCs during loading are also the most susceptible to apoptosis (Pollack et al. 2002; Alway et al. 2002a, b; Leeuwenburgh 2003; Dirks and Leeuwenburgh 2004). Based on these data, we hypothesize that MPC contribution to chronic loading-induced adaptation (hypertrophy) is lower in muscles of old animals because apoptosis is higher (Degens and Alway 2003), and fewer MPCs survive to contribute to muscle adaptation (Chakravarthy et al. 2001).

6.3 Regulation of Apoptotic Signalling by Aerobic Exercise

Although acute endurance exercise has been shown to increase apoptotic signalling under some conditions including dystrophies and other pathologies (Sandri et al. 1997; Podhorska-Okolow et al. 1998, 1999) long-term adaptation to endurance exercise has been shown to lower mitochondria-associated apoptosis in heart and skeletal muscle of rodents (Siu et al. 2004; Kwak et al. 2006; Song et al. 2006; Peterson et al. 2008); however, it does not improve muscle mass or act as a countermeasure to sarcopenia (Alway et al. 1996; Marzetti et al. 2008a). This might be in part due to aerobically-induced pathways that are generally inhibitory to muscle growth (e.g., AMPK).

Apoptosis has been shown to occur in cardiac (Dalla et al. 2001; Hu et al. 2008; Molina et al. 2009) and skeletal muscles (Dalla et al. 2001; Vescovo and Dalla 2006; Libera et al. 2009) of experimental models of chronic heart failure. Apoptosis in skeletal muscle has been linked to elevated circulating levels of TNF-α (Adams et al. 1999; Vescovo et al. 2000). Although nuclear apoptosis has been detected in muscles of humans with severe chronic heart failure (Conraads et al. 2009), it does not appear to be a large component of muscle loss associated when the disease is less severe (Dirks and Jones 2006; Yu et al. 2009a). Complicating the treatment of heart failure and related cardiovascular diseases is the likelihood that drugs including statins which are routinely prescribed to reduce hypercholesterolemia, may themselves have a pro-apoptotic role in skeletal muscle (Adams et al. 2008). Such increases in apoptosis are likely to have devastating effects when statins are combined with sarcopenia, where muscle loss is already high. Although aerobic exercise appears to reduce several skeletal muscle problems of persons suffering from severe chronic heart failure (Linke et al. 2005) and an exercise-induced improvement in antioxidant enzymes is correlated to reduced apoptosis in muscles of patients with chronic heart failure (Siu et al. 2004, 2005a; Song et al. 2006), currently there are no data to definitively address if aerobic exercise reduces apoptosis in heart failure patients. The role or aerobic exercise on nuclear apoptosis of skeletal muscle has not been well-studied but limited data suggest that apoptosis signalling is reduced by aerobic exercise in cardiac and skeletal muscle of young, diseased and aged animals (Siu et al. 2004; Kwak et al. 2006; Song et al. 2006; Peterson et al. 2008; Marzetti et al. 2008a, b).

7 Summary and Conclusions

Sarcopenia involves complex of several cellular mechanisms which together contribute to muscle loss during aging. Among them, nuclear apoptosis has recently emerged as an important factor involved in the pathophysiology of sarcopenia. Several lines of evidence support the hypothesis that mitochondrial (intrinsic), extrinsic (death receptor) and endoplasmic reticulum-calcium stress activated apoptotic signalling, occurs in skeletal muscles of old mammals. Nevertheless, it has not been determined to what extent sarcopenia would be reduced, if apoptotic signalling could be fully blocked. Although there is evidence that reducing Bax markedly reduces apoptosis associated muscle loss with denervation (Siu and Alway 2006b), it is not known if this is also the case with aging. We cannot rule the possibility that the apoptotic signalling events may occur to simply eliminate dysfunctional nuclei and/or damaged muscle fibres, whose perseverance would be detrimental for organ function.

Even though a cause and effect relationship between apoptosis and sarcopenia has not been unequivocally determined, evidence that muscle loss is reduced in Bax null mice (Siu and Alway 2006b), and experimental interventions to accelerate muscle loss in aged animals also elevates apoptosis (Siu and Alway 2005a; Siu et al. 2005b, c, d, 2006, 2008; Pistilli et al. 2007) strongly suggests that a causal relationship likely exists between nuclear apoptosis and muscle loss, and this may also extend to aging associated muscle loss. Furthermore, activation of mitochondrial apoptotic signalling during the early phases of disuse muscle atrophy (Siu and Alway 2005b; Siu and Alway 2009) suggests that this may exist to balance muscle size and the metabolic or functional needs of the animal. If this is true, nuclear apoptosis may be a fundamentally important mechanism that regulates myonuclei number and, therefore controls the extent of muscle growth (or atrophy) in aging. Apoptotic signalling may be modified by loading and aerobic forms of exercise, but it remains to be seen how effective exercise might be in slowing or preventing apoptosis in sarcopenia. Clearly further research is required to better understand the complex cellular mechanisms underlying muscle atrophy that occurs in sarcopenia, and the importance of apoptosis in this process. Unravelling the regulatory factors in the apoptotic pathways will be a necessary step prior to having the ability to design effective interventions and countermeasures for sarcopenia.

References

Abmayr, S., Crawford, R. W., Chamberlain, J. S. (2004). Characterization of ARC, apoptosis repressor interacting with CARD, in normal and dystrophin-deficient skeletal muscle. *Human Molecular Genetics, 13*, 213–221.

Acehan, D., Jiang, X., Morgan, D. G., Heuser, J. E., Wang, X., Akey, C. W. (2002). Three-dimensional structure of the apoptosome: implications for assembly, procaspase-9 binding, and activation. *Molecular Cell, 9*, 423–432.

Adams, G. R., Caiozzo, V. J., Haddad, F., Baldwin, K. M. (2002). Cellular and molecular responses to increased skeletal muscle loading after irradiation. *American Journal of Physiology. Cell Physiology, 283*, C1182–C1195.

Adams, V., Jiang, H., Yu, J., Mobius-Winkler, S., Fiehn, E., Linke, A., Weigl, C., Schuler, G., Hambrecht, R. (1999). Apoptosis in skeletal myocytes of patients with chronic heart failure is associated with exercise intolerance. *Journal of the American College of Cardiology, 33*, 959–965.

Adams, V., Doring, C., Schuler, G. (2008). Impact of physical exercise on alterations in the skeletal muscle in patients with chronic heart failure. *Frontiers in Bioscience, 13*, 302–311.

Adhihetty, P. J., Hood, D. A. (2003). Mechanisms of apoptosis in skeletal muscle. *Basic and Applied Myology, 13*, 171–179.

Adhihetty, P. J., Irrcher, I., Joseph, A. M., Ljubicic, V., Hood, D. A. (2003). Plasticity of skeletal muscle mitochondria in response to contractile activity. *Experimental Physiology, 88*, 99–107.

Adhihetty, P. J., Ljubicic, V., Hood, D. A. (2007a). Effect of chronic contractile activity on SS and IMF mitochondrial apoptotic susceptibility in skeletal muscle. *American Journal of Physiology. Endocrinology and Metabolism, 292*, E748–E755.

Adhihetty, P. J., O'Leary, M. F., Chabi, B., Wicks, K. L., Hood, D. A. (2007b). Effect of denervation on mitochondrially mediated apoptosis in skeletal muscle. *Journal of Applied Physiology, 102*, 1143–1151.

Adhihetty, P. J., O'Leary, M. F., Hood, D. A. (2008). Mitochondria in skeletal muscle: adaptable rheostats of apoptotic susceptibility. *Exercise and Sport Sciences Reviews, 36*, 116–121.

Adhihetty, P. J., Uguccioni, G., Leick, L., Hidalgo, J., Pilegaard, H., Hood, D. A. (2009). The role of PGC-1{alpha} on mitochondrial function and apoptotic susceptibility in muscle. *American Journal of Physiology. Cell Physiology, 297*(1), C217–C225.

Allen, D. L., Linderman, J. K., Roy, R. R., Bigbee, A. J., Grindeland, R. E., Mukku, V., Edgerton, V. R. (1997). Apoptosis: a mechanism contributing to remodeling of skeletal muscle in response to hindlimb unweighting. *The American Journal of Physiology, 273*, C579–C587.

Allen, D. L., Roy, R. R., Edgerton, V. R. (1999). Myonuclear domains in muscle adaptation and disease. *Muscle & Nerve, 22*, 1350–1360.

Alway, S. E. (1991). Is fiber mitochondrial volume density a good indicator of muscle fatigability to isometric exercise? *Journal of Applied Physiology, 70*, 2111–2119.

Alway, S. E., Siu, P. M. (2008). Nuclear apoptosis contributes to sarcopenia. *Exercise and Sport Sciences Reviews, 36*, 51–57.

Alway, S. E., MacDougall, J. D., Sale, D. G. (1989). Contractile adaptations in the human triceps surae after isometric exercise. *Journal of Applied Physiology, 66*, 2725–2732.

Alway, S. E., Coggan, A. R., Sproul, M. S., Abduljalil, A. M., Robitaille, P. M. (1996). Muscle torque in young and older untrained and endurance-trained men. *The Journals of Gerontology. Series A: Biological Sciences and Medical Sciences, 51*, B195–B201.

Alway, S. E., Degens, H., Krishnamurthy, G., Smith, C. A. (2002a). Potential role for Id myogenic repressors in apoptosis and attenuation of hypertrophy in muscles of aged rats. *American Journal of Physiology. Cell Physiology, 283*, C66–C76.

Alway, S. E., Degens, H., Lowe, D. A., Krishnamurthy, G. (2002b). Increased myogenic repressor Id mRNA and protein levels in hindlimb muscles of aged rats. *American Journal of Physiology: Regulatory, Integrative and Comparative Physiology, 282*, R411–R422.

Alway, S. E., Degens, H., Krishnamurthy, G., Chaudhrai, A. (2003a). Denervation stimulates apoptosis but not Id2 expression in hindlimb muscles of aged rats. *The Journals of Gerontology. Series A: Biological Sciences and Medical Sciences, 58*, 687–697.

Alway, S. E., Martyn, J. K., Ouyang, J., Chaudhrai, A., Murlasits, Z. S. (2003b). Id2 expression during apoptosis and satellite cell activation in unloaded and loaded quail skeletal muscles. *American Journal of Physiology: Regulatory, Integrative and Comparative Physiology, 284*, R540–R549.

Antonsson, B., Montessuit, S., Lauper, S., Eskes, R., Martinou, J. C. (2000). Bax oligomerization is required for channel-forming activity in liposomes and to trigger cytochrome c release from mitochondria. *The Biochemical Journal, 345*(Pt 2), 271–278.

Attaix, D., Ventadour, S., Codran, A., Bechet, D., Taillandier, D., Combaret, L. (2005). The ubiquitin-proteasome system and skeletal muscle wasting. *Essays in Biochemistry, 41*, 173–186.

Basanez, G., Nechushtan, A., Drozhinin, O., Chanturiya, A., Choe, E., Tutt, S., Wood, K. A., Hsu, Y., Zimmerberg J., Youle R.J. (1999). Bax, but not Bcl-xL, decreases the lifetime of planar phospholipid bilayer membranes at subnanomolar concentrations. *Proceedings of the National Academy of Sciences of the United States of America, 96*, 5492–5497.

Benayoun, B., Baghdiguian, S., Lajmanovich, A., Bartoli, M., Daniele, N., Gicquel, E., Bourg, N., Raynaud, F., Pasquier, M. A., Suel, L., Lochmuller, H., Lefranc, G., Richard, I. (2008). NF-kappaB-dependent expression of the antiapoptotic factor c-FLIP is regulated by calpain 3, the protein involved in limb-girdle muscular dystrophy type 2A. *The FASEB Journal, 22*, 1521–1529.

Bernardi, P. (1999). Mitochondria in muscle cell death. *Italian Journal of Neurological Sciences, 20*, 395–400.

Bhuiyan, M. S. & Fukunaga, K. (2007). Inhibition of HtrA2/Omi ameliorates heart dysfunction following ischemia/reperfusion injury in rat heart in vivo. *European Journal of Pharmacology, 557*, 168–177.

Blink, E., Maianski, N. A., Alnemri, E. S., Zervos, A. S., Roos, D., Kuijpers, T. W. (2004). Intramitochondrial serine protease activity of Omi/HtrA2 is required for caspase-independent cell death of human neutrophils. *Cell Death and Differentiation, 11*, 937–939.

Bourdel-Marchasson, I., Biran, M., Dehail, P., Traissac, T., Muller, F., Jenn, J., Raffard, G., Franconi, J. M., Thiaudiere, E. (2007). Muscle phosphocreatine post-exercise recovery rate is related to functional evaluation in hospitalized and community-living older people. *The Journal of Nutrition, Health & Aging, 11*, 215–221.

Brack, A. S. & Rando, T. A. (2007). Intrinsic changes and extrinsic influences of myogenic stem cell function during aging. *Stem Cell Reviews, 3*, 226–237.

Brack, A. S., Bildsoe, H., Hughes, S. M. (2005). Evidence that satellite cell decrement contributes to preferential decline in nuclear number from large fibres during murine age-related muscle atrophy. *Journal of Cell Science, 118*, 4813–4821.

Brack, A. S., Conboy, M. J., Roy, S., Lee, M., Kuo, C. J., Keller, C., Rando, T. A. (2007). Increased Wnt signaling during aging alters muscle stem cell fate and increases fibrosis. *Science, 317*, 807–810.

Bruunsgaard, H. (2002). Effects of tumor necrosis factor-alpha and interleukin-6 in elderly populations. *European Cytokine Network, 13*, 389–391.

Bruunsgaard, H., Andersen-Ranberg, K., Jeune, B., Pedersen, A. N., Skinhoj, P., Pedersen, B. K. (1999). A high plasma concentration of TNF-alpha is associated with dementia in centenarians. *The Journals of Gerontology. Series A: Biological Sciences and Medical Sciences, 54*, M357–M364.

Bruunsgaard, H., Pedersen, M., Pedersen, B. K. (2001). Aging and proinflammatory cytokines. *Current Opinion in Hematology, 8*, 131–136.

Bruunsgaard, H., Ladelund, S., Pedersen, A. N., Schroll, M., Jorgensen, T., Pedersen, B. K. (2003a). Predicting death from tumour necrosis factor-alpha and interleukin-6 in 80-year-old people. *Clinical and Experimental Immunology, 132*, 24–31.

Bruunsgaard, H., Andersen-Ranberg, K., Hjelmborg, J. B., Pedersen, B. K., Jeune, B. (2003b). Elevated levels of tumor necrosis factor alpha and mortality in centenarians. *The American Journal of Medicine, 115*, 278–283.

Bruusgaard, J. C., Liestol, K., Gundersen, K. (2006). Distribution of myonuclei and microtubules in live muscle fibers of young, middle-aged, and old mice. *Journal of Applied Physiology, 100*, 2024–2030.

Burstein, E. & Duckett, C. S. (2003). Dying for NF-kappaB? Control of cell death by transcriptional regulation of the apoptotic machinery. *Current Opinion in Cell Biology, 15*, 732–737.

Cacciapaglia, F., Spadaccio, C., Chello, M., Gigante, A., Coccia, R., Afeltra, A., Amoroso, A. (2009). Apoptotic molecular mechanisms implicated in autoimmune diseases. *European Review for Medical and Pharmacological Sciences, 13*, 23–40.

Campisi, J. & Sedivy, J. (2009). How does proliferative homeostasis change with age? What causes it and how does it contribute to aging? *The Journals of Gerontology. Series A: Biological Sciences and Medical Sciences, 64*, 164–166.

Cande, C., Cecconi, F., Dessen, P., Kroemer, G. (2002a). Apoptosis-inducing factor (AIF): key to the conserved caspase-independent pathways of cell death? *Journal of Cell Science, 115*, 4727–4734.

Cande, C., Cohen, I., Daugas, E., Ravagnan, L., Larochette, N., Zamzami, N., Kroemer, G. (2002b). Apoptosis-inducing factor (AIF): a novel caspase-independent death effector released from mitochondria. *Biochimie, 84*, 215–222.

Carbo, N., Busquets, S., van, R. M., Alvarez, B., Lopez-Soriano, F. J., Argiles, J. M. (2002). TNF-alpha is involved in activating DNA fragmentation in skeletal muscle. *British Journal of Cancer, 86*, 1012–1016.

Cartron, P. F., Moreau, C., Oliver, L., Mayat, E., Meflah, K., Vallette, F. M. (2002). Involvement of the N-terminus of Bax in its intracellular localization and function. *FEBS Letters, 512*, 95–100.

Chabi, B., Ljubicic, V., Menzies, K. J., Huang, J. H., Saleem, A., Hood, D. A. (2008). Mitochondrial function and apoptotic susceptibility in aging skeletal muscle. *Aging Cell, 7*, 2–12.

Chakravarthy, M. V., Spangenburg, E. E., Booth, F. W. (2001). Culture in low levels of oxygen enhances in vitro proliferation potential of satellite cells from old skeletal muscles. *Cellular and Molecular Life Sciences, 58*, 1150–1158.

Chang, H. Y. & Yang, X. (2000). Proteases for cell suicide: functions and regulation of caspases. *Microbiology and Molecular Biology Reviews, 64*, 821–846.

Chao, D. T., Korsmeyer, S. J. (1998). BCL-2 family: regulators of cell death. *Annual Review of Immunology, 16*, 395–419.

Chao, D. T., Linette, G. P., Boise, L. H., White, L. S., Thompson, C. B., Korsmeyer, S. J. (1995). Bcl-XL and Bcl-2 repress a common pathway of cell death. *The Journal of Experimental Medicine, 182*, 821–828.

Charette, S. L., McEvoy, L., Pyka, G., Snow-Harter, C., Guido, D., Wiswell, R. A., Marcus, R. (1991). Muscle hypertrophy response to resistance training in older women. *Journal of Applied Physiology, 70*, 1912–1916.

Chipuk, J. E. & Green, D. R. (2009). PUMA cooperates with direct activator proteins to promote mitochondrial outer membrane permeabilization and apoptosis. *Cell Cycle, 8*(17), 2692–2696.

Chowdhury, I., Tharakan, B., Bhat, G. K. (2008). Caspases – an update. *Comparative Biochemistry and Physiology. Part B: Biochemistry & Molecular Biology, 151*, 10–27.

Combaret, L., Dardevet, D., Bechet, D., Taillandier, D., Mosoni, L., Attaix, D. (2009). Skeletal muscle proteolysis in aging. *Current Opinion in Clinical Nutrition and Metabolic Care, 12*, 37–41.

Conley, K. E., Amara, C. E., Jubrias, S. A., Marcinek, D. J. (2007a). Mitochondrial function, fibre types and ageing: new insights from human muscle in vivo. *Experimental Physiology, 92*, 333–339.

Conley, K. E., Jubrias, S. A., Amara, C. E., Marcinek, D. J. (2007b). Mitochondrial dysfunction: impact on exercise performance and cellular aging. *Exercise and Sport Sciences Reviews, 35*, 43–49.

Conraads, V. M., Hoymans, V. Y., Vermeulen, T., Beckers, P., Possemiers, N., Maeseneer, M. D., Vrints, C., Martinet, W. (2009). Exercise capacity in chronic heart failure patients is related to active gene transcription in skeletal muscle and not apoptosis. *European Journal of Cardiovascular Prevention and Rehabilitation, 16*, 325–332.

Dalla, L. L., Sabbadini, R., Renken, C., Ravara, B., Sandri, M., Betto, R., Angelini, A., Vescovo, G. (2001). Apoptosis in the skeletal muscle of rats with heart failure is associated with increased serum levels of TNF-alpha and sphingosine. *Journal of Molecular and Cellular Cardiology, 33*, 1871–1878.

Dalla, L., Ravara, B., Gobbo, V., Betto, D.D., Germinario, E., Angelini, A., Evangelista, S., Vescovo, G. (2009). Skeletal muscle proteins oxidation in chronic right heart failure in rats: can different beta-blockers prevent it to the same degree? *International Journal of Cardiology, 143*, 192–199.

Danial, N. N. & Korsmeyer, S. J. (2004). Cell death: critical control points. *Cell, 116*, 205–219.

Daussin, F. N., Zoll, J., Ponsot, E., Dufour, S. P., Doutreleau, S., Lonsdorfer, E., Ventura-Clapier, R., Mettauer, B., Piquard, F., Geny, B., Richard, R. (2008). Training at high exercise intensity promotes qualitative adaptations of mitochondrial function in human skeletal muscle. *Journal of Applied Physiology, 104*, 1436–1441.

Degens, H. (2007). Age-related skeletal muscle dysfunction: causes and mechanisms. *Journal of Musculoskeletal & Neuronal Interactions, 7*, 246–252.

Degens, H. & Alway, S. E. (2003). Skeletal muscle function and hypertrophy are diminished in old age. *Muscle & Nerve, 27*, 339–347.

Degens, H., Swisher, A. K., Heijdra, Y. F., Siu, P. M., Dekhuijzen, P. N., Alway, S. E. (2007). Apoptosis and Id2 expression in diaphragm and soleus muscle from the emphysematous hamster. *American Journal of Physiology: Regulatory, Integrative and Comparative Physiology, 293*, R135–R144.

Degterev, A., Boyce, M., Yuan, J. (2003). A decade of caspases. *Oncogene, 22*, 8543–8567.

Dejean, L. M., Martinez-Caballero, S., Guo, L., Hughes, C., Teijido, O., Ducret, T., Ichas, F., Korsmeyer, S. J., Antonsson, B., Jonas, E. A., Kinnally, K. W. (2005). Oligomeric Bax is a component of the putative cytochrome c release channel MAC, mitochondrial apoptosis-induced channel. *Molecular Biology of the Cell, 16*, 2424–2432.

Dejean, L. M., Martinez-Caballero, S., Kinnally, K. W. (2006a). Is MAC the knife that cuts cytochrome c from mitochondria during apoptosis? *Cell Death and Differentiation, 13*, 1387–1395.

Dejean, L. M., Martinez-Caballero, S., Manon, S., Kinnally, K. W. (2006b). Regulation of the mitochondrial apoptosis-induced channel, MAC, by BCL-2 family proteins. *Biochimica et Biophysica Acta, 1762*, 191–201.

Desagher, S. & Martinou, J. C. (2000). Mitochondria as the central control point of apoptosis. *Trends in Cell Biology, 10*, 369–377.

Deschenes, M. R. & Kraemer, W. J. (2002). Performance and physiologic adaptations to resistance training. *American Journal of Physical Medicine & Rehabilitation, 81*, S3–16.

Deveraux, Q. L. & Reed, J. C. (1999). IAP family proteins–suppressors of apoptosis. *Genes Development, 13*, 239–252.
Deveraux, Q. L., Roy, N., Stennicke, H. R., Van, A. T., Zhou, Q., Srinivasula, S. M., Alnemri, E. S., Salvesen, G. S., Reed, J. C. (1998). IAPs block apoptotic events induced by caspase 8 and cytochrome c by direct inhibition of distinct caspases. *The EMBO Journal, 17*, 2215–2223.
Deveraux, Q. L., Stennicke, H. R., Salvesen, G. S., Reed, J. C. (1999). Endogenous inhibitors of caspases. *Journal of Clinical Immunology, 19*, 388–398.
Dirks, A. J. & Jones, K. M. (2006). Statin-induced apoptosis and skeletal myopathy. *American Journal of Physiology. Cell Physiology, 291*, C1208–C1212.
Dirks, A. & Leeuwenburgh, C. (2002). Apoptosis in skeletal muscle with aging. *American Journal of Physiology: Regulatory, Integrative and Comparative Physiology, 282*, R519–R527.
Dirks, A. J. & Leeuwenburgh, C. (2004). Aging and lifelong calorie restriction result in adaptations of skeletal muscle apoptosis repressor, apoptosis-inducing factor, X-linked inhibitor of apoptosis, caspase 3, and caspase-12. *Free Radical Biology & Medicine, 36*, 27–39.
Dirks, A. J. & Leeuwenburgh, C. (2005). The role of apoptosis in age-related skeletal muscle atrophy. *Sports Medicine, 35*, 473–483.
Dirks, A. J. & Leeuwenburgh, C. (2006). Tumor necrosis factor alpha signaling in skeletal muscle: effects of age and caloric restriction. *The Journal of Nutritional Biochemistry, 17*, 501–508.
Dirks Naylor, A. J. & Leeuwenburgh, C. (2008). Sarcopenia: the role of apoptosis and modulation by caloric restriction. *Exercise and Sport Sciences Reviews, 36*, 19–24.
Dirks, A. J., Hofer, T., Marzetti, E., Pahor, M., Leeuwenburgh, C. (2006). Mitochondrial DNA mutations, energy metabolism and apoptosis in aging muscle. *Ageing Research Reviews, 5*, 179–195.
Duke, R. C., Ojcius, D. M., Young, J. D. (1996). Cell suicide in health and disease. *Scientific American, 275*, 80–87.
Dutta, J., Fan, Y., Gupta, N., Fan, G., Gelinas, C. (2006). Current insights into the regulation of programmed cell death by NF-kappaB. *Oncogene, 25*, 6800–6816.
Earnshaw, W. C., Martins, L. M., Kaufmann, S. H. (1999). Mammalian caspases: structure, activation, substrates, and functions during apoptosis. *Annual Review of Biochemistry, 68*, 383–424.
Ellis, R. E., Yuan, J. Y., Horvitz, H. R. (1991). Mechanisms and functions of cell death. *Annual Review of Cell Biology, 7*, 663–698.
Er, E., Lalier, L., Cartron, P. F., Oliver, L., Vallette, F. M. (2007). Control of Bax homo-dimerization by its carboxy-terminal. *The Journal of Biological Chemistry, 282*(34), 24938–24947.
Fan, Y., Dutta, J., Gupta, N., Fan, G., Gelinas, C. (2008). Regulation of programmed cell death by NF-kappaB and its role in tumorigenesis and therapy. *Advances in Experimental Medicine and Biology, 615*, 223–250.
Ferketich, A. K., Kirby, T. E., Alway, S. E. (1998). Cardiovascular and muscular adaptations to combined endurance and strength training in elderly women. *Acta Physiologica Scandinavica, 164*, 259–267.
Figueras, M., Busquets, S., Carbo, N., Almendro, V., Argiles, J. M., Lopez-Soriano, F. J. (2005). Cancer cachexia results in an increase in TNF-alpha receptor gene expression in both skeletal muscle and adipose tissue. *International Journal of Oncology, 27*, 855–860.
Forsey, R. J., Thompson, J. M., Ernerudh, J., Hurst, T. L., Strindhall, J., Johansson, B., Nilsson, B. O., Wikby, A. (2003). Plasma cytokine profiles in elderly humans. *Mechanisms of Ageing and Development, 124*, 487–493.
Forte, M. & Bernardi, P. (2006). The permeability transition and BCL-2 family proteins in apoptosis: co-conspirators or independent agents? *Cell Death and Differentiation, 13*, 1287–1290.
Foulstone, E. J., Meadows, K. A., Holly, J. M., Stewart, C. E. (2001). Insulin-like growth factors (IGF-I and IGF-II) inhibit C2 skeletal myoblast differentiation and enhance TNF alpha-induced apoptosis. *Journal of Cellular Physiology, 189*, 207–215.
Garcia-Martinez, C., Agell, N., Llovera, M., Lopez-Soriano, F. J., Argiles, J. M. (1993a). Tumour necrosis factor-alpha increases the ubiquitinization of rat skeletal muscle proteins. *FEBS Letters, 323*, 211–214.

Garcia-Martinez, C., Lopez-Soriano, F. J., Argiles, J. M. (1993b). Acute treatment with tumour necrosis factor-alpha induces changes in protein metabolism in rat skeletal muscle. *Molecular and Cellular Biochemistry, 125*, 11–18.

Garcia-Martinez, C., Llovera, M., Agell, N., Lopez-Soriano, F. J., Argiles, J. M. (1995). Ubiquitin gene expression in skeletal muscle is increased during sepsis: involvement of TNF-alpha but not IL-1. *Biochemical and Biophysical Research Communications, 217*, 839–844.

Ghosh, A. P., Walls, K. C., Klocke, B. J., Toms, R., Strasser, A., Roth, K. A. (2009). The proapoptotic BH3-only, Bcl-2 family member, Puma is critical for acute ethanol-induced neuronal apoptosis. *Journal of Neuropathology and Experimental Neurology, 68*, 747–756.

Gorman, A. M., Ceccatelli, S., Orrenius, S. (2000). Role of mitochondria in neuronal apoptosis. *Developmental Neuroscience, 22*, 348–358.

Green, D. R., Kroemer, G. (2004). The pathophysiology of mitochondrial cell death. *Science, 305*, 626–629.

Grinberg, M., Schwarz, M., Zaltsman, Y., Eini, T., Niv, H., Pietrokovski, S., Gross, A. (2005). Mitochondrial carrier homolog 2 is a target of tBID in cells signaled to die by tumor necrosis factor alpha. *Molecular and Cellular Biology, 25*, 4579–4590.

Grutter, M. G. (2000). Caspases: key players in programmed cell death. *Current Opinion in Structural Biology, 10*, 649–655.

Gustafsson, A. B., Tsai, J. G., Logue, S. E., Crow, M. T., Gottlieb, R. A. (2004). Apoptosis repressor with caspase recruitment domain protects against cell death by interfering with Bax activation. *The Journal of Biological Chemistry, 279*, 21233–21238.

Hagen, J. L., Krause, D. J., Baker, D. J., Fu, M. H., Tarnopolsky, M. A., Hepple, R. T. (2004). Skeletal muscle aging in F344BN F1-hybrid rats: I. Mitochondrial dysfunction contributes to the age-associated reduction in VO2max. *The Journals of Gerontology. Series A: Biological Sciences and Medical Sciences, 59*, 1099–1110.

Harman, D. (1956). Aging: a theory based on free radical and radiation chemistry. *Journal of Gerontology, 11*, 298–300.

Harman, D. (1992). Role of free radicals in aging and disease. *Annals of the New York Academy of Sciences, 673*, 126–141.

Harman, D. (2003). The free radical theory of aging. *Antioxidants Redox Signaling, 5*, 557–561.

Harman, D. (2006). Free radical theory of aging: an update: increasing the functional life span. *Annals of the New York Academy of Sciences, 1067*, 10–21.

Harman, D. (2009). Origin and evolution of the free radical theory of aging: a brief personal history, 1954–2009. *Biogerontology, 10*, 773–781.

Hawke, T. J. & Garry, D. J. (2001). Myogenic satellite cells: physiology to molecular biology. *Journal of Applied Physiology, 91*, 534–551.

Hegde, R., Srinivasula, S. M., Zhang, Z., Wassell, R., Mukattash, R., Cilenti, L., DuBois, G., Lazebnik, Y., Zervos, A. S., Fernandes-Alnemri, T., Alnemri, E. S. (2002). Identification of Omi/HtrA2 as a mitochondrial apoptotic serine protease that disrupts inhibitor of apoptosis protein-caspase interaction. *The Journal of Biological Chemistry, 277*, 432–438.

Heikaus, S., Kempf, T., Mahotka, C., Gabbert, H. E., Ramp, U. (2008). Caspase 8 and its inhibitors in RCCs in vivo: the prominent role of ARC. *Apoptosis, 13*, 938–949.

Hsu, Y. T., Wolter, K. G., Youle, R. J. (1997). Cytosol-to-membrane redistribution of Bax and Bcl-X(L) during apoptosis. *Proceedings of the National Academy of Sciences of the United States of America, 94*, 3668–3672.

Hu, A., Jiao, X., Gao, E., Li, Y., Sharifi-Azad, S., Grunwald, Z., Ma, X. L., Sun, J. Z. (2008). Tonic beta-adrenergic drive provokes proinflammatory and proapoptotic changes in aging mouse heart. *Rejuvenation Research, 11*, 215–226.

Huang, Z., Jiang, J., Tyurin, V. A., Zhao, Q., Mnuskin, A., Ren, J., Belikova, N. A., Feng, W., Kurnikov, I. V., Kagan, V. E. (2008). Cardiolipin deficiency leads to decreased cardiolipin peroxidation and increased resistance of cells to apoptosis. *Free Radical Biology Medicine, 44*, 1935–1944.

Irmler, M., Thome, M., Hahne, M., Schneider, P., Hofmann, K., Steiner, V., Bodmer, J. L., Schroter, M., Burns, K., Mattmann, C., Rimoldi, D., French, L. E., Tschopp, J. (1997). Inhibition of death receptor signals by cellular FLIP. *Nature, 388*, 190–195.

Jejurikar, S. S., Marcelo, C. L., Kuzon, W. M., Jr. (2002). Skeletal muscle denervation increases satellite cell susceptibility to apoptosis. *Plastic and Reconstructive Surgery, 110*, 160–168.

Jin, H., Wu, Z., Tian, T., Gu, Y. (2001). Apoptosis in atrophic skeletal muscle induced by brachial plexus injury in rats. *The Journal of Trauma, 50*, 31–35.

Johnson, C. R. & Jarvis, W. D. (2004). Caspase-9 regulation: an update. *Apoptosis, 9*, 423–427.

Johnston, A. P., De, L. M., Parise, G. (2008). Resistance training, sarcopenia, and the mitochondrial theory of aging. *Applied Physiology, Nutrition, and Metabolism, 33*, 191–199.

Joza, N., Susin, S. A., Daugas, E., Stanford, W. L., Cho, S. K., Li, C. Y., Sasaki, T., Elia, A. J., Cheng, H. Y., Ravagnan, L., Ferri, K. F., Zamzami, N., Wakeham, A., Hakem, R., Yoshida, H., Kong, Y. Y., Mak, T. W., Zuniga-Pflucker, J. C., Kroemer, G., Penninger, J. M. (2001). Essential role of the mitochondrial apoptosis-inducing factor in programmed cell death. *Nature, 410*, 549–554.

Joza, N., Oudit, G. Y., Brown, D., Benit, P., Kassiri, Z., Vahsen, N., Benoit, L., Patel, M. M., Nowikovsky, K., Vassault, A., Backx, P. H., Wada, T., Kroemer, G., Rustin, P., Penninger, J. M. (2005). Muscle-specific loss of apoptosis-inducing factor leads to mitochondrial dysfunction, skeletal muscle atrophy, and dilated cardiomyopathy. *Molecular and Cellular Biology, 25*, 10261–10272.

Jubrias, S. A., Esselman, P. C., Price, L. B., Cress, M. E., Conley, K. E. (2001). Large energetic adaptations of elderly muscle to resistance and endurance training. *Journal of Applied Physiology, 90*, 1663–1670.

Kadenbach, B., Ramzan, R., Vogt, S. (2009). Degenerative diseases, oxidative stress and cytochrome c oxidase function. *Trends in Molecular Medicine, 15*, 139–147.

Kearns, J. D. & Hoffmann, A. (2009). Integrating computational and biochemical studies to explore mechanisms in NF-{kappa}B signaling. *The Journal of Biological Chemistry, 284*, 5439–5443.

Kerr, J. F., Wyllie, A. H., Currie, A. R. (1972). Apoptosis: a basic biological phenomenon with wide-ranging implications in tissue kinetics. *British Journal of Cancer, 26*, 239–257.

Knudson, C. M. & Brown, N. M. (2008). Mitochondria potential, bax "activation", and programmed cell death. *Methods in Molecular Biology, 414*, 95–108.

Korolchuk, V. I., Mansilla, A., Menzies, F. M., Rubinsztein, D. C. (2009). Autophagy inhibition compromises degradation of ubiquitin-proteasome pathway substrates. *Molecular Cell, 33*, 517–527.

Korsmeyer, S. J. (1995). Regulators of cell death. *Trends in Genetics, 11*, 101–105.

Korsmeyer, S. J. (1999). BCL-2 gene family and the regulation of programmed cell death. *Cancer Research, 59*, 1693s–1700s.

Korsmeyer, S. J., Yin, X. M., Oltvai, Z. N., Veis-Novack, D. J., Linette, G. P. (1995). Reactive oxygen species and the regulation of cell death by the Bcl-2 gene family. *Biochimica et Biophysica Acta, 1271*, 63–66.

Koseki, T., Inohara, N., Chen, S., Nunez, G. (1998). ARC, an inhibitor of apoptosis expressed in skeletal muscle and heart that interacts selectively with caspases. *Proceedings of the National Academy of Sciences of the United States of America, 95*, 5156–5160.

Krajnak, K., Waugh, S., Miller, R., Baker, B., Geronilla, K., Alway, S. E., Cutlip, R. G. (2006). Proapoptotic factor Bax is increased in satellite cells in the tibialis anterior muscles of old rats. *Muscle & Nerve, 34*, 720–730.

Kroemer, G. & Reed, J. C. (2000). Mitochondrial control of cell death. *Natural Medicines, 6*, 513–519.

Kroemer, G., Galluzzi, L., Brenner, C. (2007). Mitochondrial membrane permeabilization in cell death. *Physiological Reviews, 87*, 99–163.

Kujoth, G. C., Hiona, A., Pugh, T. D., Someya, S., Panzer, K., Wohlgemuth, S. E., Hofer, T., Seo, A.Y., Sullivan, R., Jobling, W. A., Morrow, J. D., Van, R. H., Sedivy, J. M., Yamasoba, T.,

Tanokura, M., Weindruch, R., Leeuwenburgh, C., Prolla, T. A. (2005). Mitochondrial DNA mutations, oxidative stress, and apoptosis in mammalian aging. *Science, 309*, 481–484.

Kujoth, G. C., Leeuwenburgh, C., Prolla, T. A. (2006). Mitochondrial DNA mutations and apoptosis in mammalian aging. *Cancer Research, 66*, 7386–7389.

Kwak, H. B., Song, W., Lawler, J. M. (2006). Exercise training attenuates age-induced elevation in Bax/Bcl-2 ratio, apoptosis, and remodeling in the rat heart. *The FASEB Journal, 20*, 791–793.

Lalier, L., Cartron, P. F., Juin, P., Nedelkina, S., Manon, S., Bechinger, B., Vallette, F. M. (2007). Bax activation and mitochondrial insertion during apoptosis. *Apoptosis, 12*, 887–896.

Lanza, I. R., Short, D. K., Short, K. R., Raghavakaimal, S., Basu, R., Joyner, M. J., McConnell, J. P., Nair, K. S. (2008). Endurance exercise as a countermeasure for aging. *Diabetes, 57*(11), 2933–42.

Lee, S. C. & Pervaiz, S. (2007). Apoptosis in the pathophysiology of diabetes mellitus. *The International Journal of Biochemistry & Cell Biology, 39*, 497–504.

Lee, H. C. & Wei, Y. H. (2007). Oxidative stress, mitochondrial DNA mutation, and apoptosis in aging. *Experimental Biology and Medicine (Maywood), 232*, 592–606.

Lee, D. H., Rhee, J. G., Lee, Y. J. (2009). Reactive oxygen species up-regulate p53 and Puma; a possible mechanism for apoptosis during combined treatment with TRAIL and wogonin. *British Journal of Pharmacology, 157*, 1189–1202.

Lees, S. J., Zwetsloot, K. A., Booth, F. W. (2009). Muscle precursor cells isolated from aged rats exhibit an increased tumor necrosis factor- alpha response. *Aging Cell, 8*, 26–35.

Leeuwenburgh, C. (2003). Role of apoptosis in sarcopenia. *The Journals of Gerontology. Series A: Biological Sciences and Medical Sciences, 58*, 999–1001.

Leeuwenburgh, C., Gurley, C. M., Strotman, B. A., Dupont-Versteegden, E. E. (2005). Age-related differences in apoptosis with disuse atrophy in soleus muscle. *American Journal of Physiology: Regulatory, Integrative and Comparative Physiology, 288*, R1288–R1296.

Li, H., Zhu, H., Xu, C. J., Yuan, J. (1998). Cleavage of BID by caspase 8 mediates the mitochondrial damage in the Fas pathway of apoptosis. *Cell, 94*, 491–501.

Li, L. Y., Luo, X., Wang, X. (2001). Endonuclease G is an apoptotic DNase when released from mitochondria. *Nature, 412*, 95–99.

Linke, A., Adams, V., Schulze, P. C., Erbs, S., Gielen, S., Fiehn, E., Mobius-Winkler, S., Schubert, A., Schuler, G., Hambrecht, R. (2005). Antioxidative effects of exercise training in patients with chronic heart failure: increase in radical scavenger enzyme activity in skeletal muscle. *Circulation, 111*, 1763–1770.

Lipton, S. A. & Bossy-Wetzel, E. (2002). Dueling activities of AIF in cell death versus survival: DNA binding and redox activity. *Cell, 111*, 147–150.

li-Youcef, N., Lagouge, M., Froelich, S., Koehl, C., Schoonjans, K., Auwerx, J. (2007). Sirtuins: the 'magnificent seven', function, metabolism and longevity. *Annali Medici, 39*, 335–345.

Llovera, M., Garcia-Martinez, C., Agell, N., Lopez-Soriano, F. J., Argiles, J. M. (1997). TNF can directly induce the expression of ubiquitin-dependent proteolytic system in rat soleus muscles. *Biochemical and Biophysical Research Communications, 230*, 238–241.

Llovera, M., Garcia-Martinez, C., Agell, N., Lopez-Soriano, F. J., Authier, F. J., Gherardi, R. K., Argiles, J. M. (1998). Ubiquitin and proteasome gene expression is increased in skeletal muscle of slim AIDS patients. *International Journal of Molecular Medicine, 2*, 69–73.

Ma, Y. S., Wu, S. B., Lee, W. Y., Cheng, J. S., Wei, Y. H. (2009). Response to the increase of oxidative stress and mutation of mitochondrial DNA in aging. *Biochimica et Biophysica Acta, 1790*(10), 1021–9.

Malinska, D., Kudin, A. P., bska-Vielhaber, G., Vielhaber, S., Kunz, W. S. (2009). Chapter 23 Quantification of superoxide production by mouse brain and skeletal muscle mitochondria. *Methods in Enzymology, 456*, 419–437.

Martin, C., Dubouchaud, H., Mosoni, L., Chardigny, J. M., Oudot, A., Fontaine, E., Vergely, C., Keriel, C., Rochette, L., Leverve, X., Demaison, L. (2007). Abnormalities of mitochondrial functioning can partly explain the metabolic disorders encountered in sarcopenic gastrocnemius. *Aging Cell, 6*, 165–177.

Marzetti, E., Groban, L., Wohlgemuth, S. E., Lees, H. A., Lin, M., Jobe, H., Giovannini, S., Leeuwenburgh, C., Carter, C. S. (2008a). Effects of short-term GH supplementation and treadmill exercise training on physical performance and skeletal muscle apoptosis in old rats. *American Journal of Physiology: Regulatory, Integrative and Comparative Physiology, 294*, R558–R567.

Marzetti, E., Lawler, J. M., Hiona, A., Manini, T., Seo, A. Y., Leeuwenburgh, C. (2008b). Modulation of age-induced apoptotic signaling and cellular remodeling by exercise and calorie restriction in skeletal muscle. *Free Radical Biology & Medicine, 44*, 160–168.

Marzetti, E., Wohlgemuth, S. E., Lees, H. A., Chung, H. Y., Giovannini, S., Leeuwenburgh, C. (2008c). Age-related activation of mitochondrial caspase-independent apoptotic signaling in rat gastrocnemius muscle. *Mechanisms of Ageing and Development, 129*, 542–549.

Marzetti, E., Carter, C. S., Wohlgemuth, S. E., Lees, H. A., Giovannini, S., Anderson, B., Quinn, L. S., Leeuwenburgh, C. (2009a). Changes in IL-15 expression and death-receptor apoptotic signaling in rat gastrocnemius muscle with aging and life-long calorie restriction. *Mechanisms of Ageing and Development, 130*, 272–280.

Marzetti, E., Hwang, J. C., Lees, H. A., Wohlgemuth, S. E., Dupont-Versteegden, E. E., Carter, C. S., Bernabei, R., Leeuwenburgh, C. (2009b). Mitochondrial death effectors: relevance to sarcopenia and disuse muscle atrophy. *Biochimimica et Biophysica Acta, 1800*(3), 235–244.

Mayhew, D. L., Kim, J. S., Cross, J. M., Ferrando, A. A., Bamman, M. M. (2009). Translational signaling responses preceding resistance training-mediated myofiber hypertrophy in young and old humans. *Journal of Applied Physiology*, In Press.

McCall, G. E., Allen, D. L., Linderman, J. K., Grindeland, R. E., Roy, R. R., Mukku, V. R., Edgerton, V. R. (1998). Maintenance of myonuclear domain size in rat soleus after overload and growth hormone/IGF-I treatment. *Journal of Applied Physiology, 84*, 1407–1412.

McMullen, C. A., Ferry, A. L., Gamboa, J. L., Andrade, F. H., Dupont-Versteegden, E. E. (2009). Age-related changes of cell death pathways in rat extraocular muscle. *Experimental Gerontology, 44*(6–7), 420–425.

Meadows, K. A., Holly, J. M., Stewart, C. E. (2000). Tumor necrosis factor-alpha-induced apoptosis is associated with suppression of insulin-like growth factor binding protein-5 secretion in differentiating murine skeletal myoblasts. *Journal of Cellular Physiology, 183*, 330–337.

Meissner, C. (2007). Mutations of mitochondrial. *Zeitschrift für Gerontologie und Geriatrie, 40*, 325–333.

Meissner, C., Bruse, P., Mohamed, S. A., Schulz, A., Warnk, H., Storm, T., Oehmichen, M. (2008). The 4977 bp deletion of mitochondrial DNA in human skeletal muscle, heart and different areas of the brain: a useful biomarker or more? *Experimental Gerontology, 43*, 645–652.

Melov, S., Tarnopolsky, M. A., Beckman, K., Felkey, K., Hubbard, A. (2007). Resistance exercise reverses aging in human skeletal muscle. *PLoS ONE, 2*, e465.

Mikhailov, V., Mikhailova, M., Pulkrabek, D. J., Dong, Z., Venkatachalam, M. A., Saikumar, P. (2001). Bcl-2 prevents Bax oligomerization in the mitochondrial outer membrane. *The Journal of Biological Chemistry, 276*, 18361–18374.

Mikhailov, V., Mikhailova, M., Degenhardt, K., Venkatachalam, M. A., White, E., Saikumar, P. (2003). Association of Bax and Bak homo-oligomers in mitochondria. Bax requirement for Bak reorganization and cytochrome c release. *The Journal of Biological Chemistry, 278*, 5367–5376.

Molina, E. J., Gupta, D., Palma, J., Torres, D., Gaughan, J. P., Houser, S., Macha, M. (2009). Novel experimental model of pressure overload hypertrophy in rats. *The Journal of Surgical Research, 153*, 287–294.

Murray, A. J., Edwards, L. M., Clarke, K. (2007). Mitochondria and heart failure. *Current Opinion in Clinical Nutrition and Metabolic Care, 10*, 704–711.

Nakagawa, T. & Yuan, J. (2000). Cross-talk between two cysteine protease families. Activation of caspase-12 by calpain in apoptosis. *The Journal of Cell Biology, 150*, 887–894.

Nakagawa, T., Zhu, H., Morishima, N., Li, E., Xu, J., Yankner, B. A., Yuan, J. (2000). Caspase-12 mediates endoplasmic-reticulum-specific apoptosis and cytotoxicity by amyloid-beta. *Nature, 403*, 98–103.

Narita, M., Shimizu, S., Ito, T., Chittenden, T., Lutz, R. J., Matsuda, H., Tsujimoto, Y. (1998). Bax interacts with the permeability transition pore to induce permeability transition and cytochrome c release in isolated mitochondria. *Proceedings of the National Academy of Sciences of the United States of America, 95*, 14681–14686.

Nickson, P., Toth, A., Erhardt, P. (2007). PUMA is critical for neonatal cardiomyocyte apoptosis induced by endoplasmic reticulum stress. *Cardiovascular Research, 73*, 48–56.

Parise, G. & Yarasheski, K. E. (2000). The utility of resistance exercise training and amino acid supplementation for reversing age-associated decrements in muscle protein mass and function. *Current Opinion in Clinical Nutrition and Metabolic Care, 3*, 489–495.

Parise, G., Phillips, S. M., Kaczor, J. J., Tarnopolsky, M. A. (2005). Antioxidant enzyme activity is up-regulated after unilateral resistance exercise training in older adults. *Free Radical Biology & Medicine, 39*, 289–295.

Pedersen, M., Bruunsgaard, H., Weis, N., Hendel, H. W., Andreassen, B. U., Eldrup, E., Dela, F., Pedersen, B. K. (2003). Circulating levels of TNF-alpha and IL-6-relation to truncal fat mass and muscle mass in healthy elderly individuals and in patients with type-2 diabetes. *Mechanisms of Ageing and Development, 124*, 495–502.

Peterson, J. M., Bryner, R. W., Sindler, A., Frisbee, J. C., Alway, S. E. (2008). Mitochondrial apoptotic signaling is elevated in cardiac but not skeletal muscle in the obese Zucker rat and is reduced with aerobic exercise. *Journal of Applied Physiology, 105*, 1934–1943.

Phaneuf, S. & Leeuwenburgh, C. (2002). Cytochrome c release from mitochondria in the aging heart: a possible mechanism for apoptosis with age. *American Journal of Physiology: Regulatory, Integrative and Comparative Physiology, 282*, R423–R430.

Phelan, J. N. & Gonyea, W. J. (1997). Effect of radiation on satellite cell activity and protein expression in overloaded mammalian skeletal muscle. *The Anatomical Record, 247*, 179–188.

Phillips, T. & Leeuwenburgh, C. (2005). Muscle fiber specific apoptosis and TNF-alpha signaling in sarcopenia are attenuated by life-long calorie restriction. *The FASEB Journal, 19*, 668–670.

Pistilli, E. E. & Alway, S. E. (2008). Systemic elevation of interleukin-15 in vivo promotes apoptosis in skeletal muscles of young adult and aged rats. *Biochemical and Biophysical Research Communications, 373*, 20–24.

Pistilli, E. E., Jackson, J. R., Alway, S. E. (2006a). Death receptor-associated pro-apoptotic signaling in aged skeletal muscle. *Apoptosis, 11*, 2115–2126.

Pistilli, E. E., Siu, P. M., Alway, S. E. (2006b). Molecular regulation of apoptosis in fast plantaris muscles of aged rats. *The Journals of Gerontology. Series A: Biological Sciences and Medical Sciences, 61*, 245–255.

Pistilli, E. E., Siu, P. M., Alway, S. E. (2007). Interleukin-15 responses to aging and unloading-induced skeletal muscle atrophy. *American Journal of Physiology. Cell Physiology, 292*, C1298–C1304.

Podhorska-Okolow, M., Sandri, M., Zampieri, S., Brun, B., Rossini, K., Carraro, U. (1998). Apoptosis of myofibres and satellite cells: exercise-induced damage in skeletal muscle of the mouse. *Neuropathology and Applied Neurobiology, 24*, 518–531.

Podhorska-Okolow, M., Krajewska, B., Carraro, U., Zabel, M. (1999). Apoptosis in mouse skeletal muscles after physical exercise. *Folia Histochemica et Cytobiologica, 37*, 127–128.

Pollack, M., Phaneuf, S., Dirks, A., Leeuwenburgh, C. (2002). The role of apoptosis in the normal aging brain, skeletal muscle, and heart. *Annals of the New York Academy of Sciences, 959*, 93–107.

Precht, T. A., Phelps, R. A., Linseman, D. A., Butts, B. D., Le, S. S., Laessig, T. A., Bouchard, R.J., Heidenreich, K. A. (2005). The permeability transition pore triggers Bax translocation to mitochondria during neuronal apoptosis. *Cell Death and Differentiation, 12*, 255–265.

Primeau, A. J., Adhihetty, P. J., Hood, D. A. (2002). Apoptosis in heart and skeletal muscle. *Canadian Journal of Applied Physiology, 27*, 349–395.

Rasola, A. & Bernardi, P. (2007). The mitochondrial permeability transition pore and its involvement in cell death and in disease pathogenesis. *Apoptosis, 12*, 815–833.

Reed, J. C. (1997). Double identity for proteins of the Bcl-2 family. *Nature, 387*, 773–776.
Reed, J. C. (2006). Proapoptotic multidomain Bcl-2/Bax-family proteins: mechanisms, physiological roles, and therapeutic opportunities. *Cell Death and Differentiation, 13*, 1378–1386.
Reed, J. C., Jurgensmeier, J. M., Matsuyama, S. (1998). Bcl-2 family proteins and mitochondria. *Biochimica et Biophysica Acta, 1366*, 127–137.
Renault, V., Thornell, L. E., Butler-Browne, G., Mouly, V. (2002). Human skeletal muscle satellite cells: aging, oxidative stress and the mitotic clock. *Experimental Gerontology, 37*, 1229–1236.
Ricci, C., Pastukh, V., Mozaffari, M., Schaffer, S. W. (2007). Insulin withdrawal induces apoptosis via a free radical-mediated mechanism. *Canadian Journal of Physiology and Pharmacology, 85*, 455–464.
Roman, W. J., Fleckenstein, J., Stray-Gundersen, J., Alway, S. E., Peshock, R., Gonyea, W. J. (1993). Adaptations in the elbow flexors of elderly males after heavy- resistance training. *Journal of Applied Physiology, 74*, 750–754.
Rosenblatt, J. D., Yong, D., Parry, D. J. (1994). Satellite cell activity is required for hypertrophy of overloaded adult rat muscle. *Muscle & Nerve, 17*, 608–613.
Ryan, M. J., Dudash, H. J., Docherty, M., Geronilla, K. B., Baker, B. A., Haff, G. G., Cutlip, R. G., Alway, S. E. (2008). Aging-dependent regulation of antioxidant enzymes and redox status in chronically loaded rat dorsiflexor muscles. *The Journals of Gerontology. Series A: Biological Sciences and Medical Sciences, 63*, 1015–1026.
Salvesen, G. S. & Duckett, C. S. (2002). IAP proteins: blocking the road to death's door. *Nature Reviews. Molecular Cell Biology, 3*, 401–410.
Sandmand, M., Bruunsgaard, H., Kemp, K., Andersen-Ranberg, K., Schroll, M., Jeune, B. (2003). High circulating levels of tumor necrosis factor-alpha in centenarians are not associated with increased production in T lymphocytes. *Gerontology, 49*, 155–160.
Sandri, M., Podhorska-Okolow, M., Geromel, V., Rizzi, C., Arslan, P., Franceschi, C., Carraro, U. (1997). Exercise induces myonuclear ubiquitination and apoptosis in dystrophin-deficient muscle of mice. *Journal of Neuropathology and Experimental Neurology, 56*, 45–57.
Sanna, M. G., da Silva, C. J., Ducrey, O., Lee, J., Nomoto, K., Schrantz, N., Deveraux, Q. L., Ulevitch, R. J. (2002). IAP suppression of apoptosis involves distinct mechanisms: the TAK1/JNK1 signaling cascade and caspase inhibition. *Molecular and Cellular Biology, 22*, 1754–1766.
Schaap, L. A., Pluijm, S. M., Deeg, D. J., Visser, M. (2006). Inflammatory markers and loss of muscle mass (sarcopenia) and strength. *The American Journal of Medicine, 119*, 526–17.
Schaap, L. A., Pluijm, S. M., Deeg, D. J., Harris, T. B., Kritchevsky, S. B., Newman, A. B., Colbert, L. H., Pahor, M., Rubin, S. M., Tylavsky, F. A., Visser, M. (2009). Higher Inflammatory Marker Levels in Older Persons: Associations With 5-Year Change in Muscle Mass and Muscle Strength. *The Journals of Gerontology. Series A: Biological Sciences and Medical Sciences, 64A*(11), 1183–1189.
Schultz, E. (1989). Satellite cell behavior during skeletal muscle growth and regeneration. *Medicine and Science in Sports and Exercise, 21*, S181–S186.
Schultz, E. (1996). Satellite cell proliferative compartments in growing skeletal muscles. *Developmental Biology, 175*, 84–94.
Schultz, E., McCormick, K. M. (1994). Skeletal muscle satellite cells. *Reviews of Physiology Biochemistry and Pharmacology, 123*, 213–257.
Senoo-Matsuda, N., Hartman, P. S., Akatsuka, A., Yoshimura, S., Ishii, N. (2003). A complex II defect affects mitochondrial structure, leading to ced-3- and ced-4-dependent apoptosis and aging. *The Journal of Biological Chemistry, 278*, 22031–22036.
Seo, A. Y., Xu, J., Servais, S., Hofer, T., Marzetti, E., Wohlgemuth, S. E., Knutson, M. D., Chung, H. Y., Leeuwenburgh, C. (2008). Mitochondrial iron accumulation with age and functional consequences. *Aging Cell, 7*, 706–716.
Shi, Y. (2002a). Apoptosome: the cellular engine for the activation of caspase-9. *Structure, 10*, 285–288.
Shi, Y. (2002b). Mechanisms of caspase activation and inhibition during apoptosis. *Molecular Cell, 9*, 459–470.

Shi, Y. (2004). Caspase activation, inhibition, and reactivation: a mechanistic view. *Protein Science, 13*, 1979–1987.

Shih, V. F., Kearns, J. D., Basak, S., Savinova, O. V., Ghosh, G., Hoffmann, A. (2009). Kinetic control of negative feedback regulators of NF-kappaB/RelA determines their pathogen- and cytokine-receptor signaling specificity. *Proceedings of the National Academy of Sciences of the United States of America, 106*, 9619–9624.

Shimizu, S., Narita, M., Tsujimoto, Y. (1999). Bcl-2 family proteins regulate the release of apoptogenic cytochrome c by the mitochondrial channel VDAC. *Nature, 399*, 483–487.

Shiozaki, E. N., Chai, J., Shi, Y. (2002). Oligomerization and activation of caspase-9, induced by Apaf-1 CARD. *Proceedings of the National Academy of Sciences of the United States of America, 99*, 4197–4202.

Shiozaki, E. N. & Shi, Y. (2004). Caspases, IAPs and Smac/DIABLO: mechanisms from structural biology. *Trends in Biochemical Sciences, 29*, 486–494.

Siu, P. M. & Alway, S. E. (2005a). Id2 and p53 participate in apoptosis during unloading-induced muscle atrophy. *American Journal of Physiology. Cell Physiology, 288*, C1058–C1073.

Siu, P. M. & Alway, S. E. (2005b). Mitochondria-associated apoptotic signalling in denervated rat skeletal muscle. *Journal de Physiologie, 565*, 309–323.

Siu, P. M. & Alway, S. E. (2006a). Aging alters the reduction of pro-apoptotic signaling in response to loading-induced hypertrophy. *Experimental Gerontology, 41*, 175–188.

Siu, P. M. & Alway, S. E. (2006b). Deficiency of the Bax gene attenuates denervation-induced apoptosis. *Apoptosis, 11*, 967–981.

Siu, P. M. & Alway, S. E. (2009). Response and adaptation of skeletal muscle to denervation stress: the role of apoptosis in muscle loss. *Frontiers in Bioscience, 14*, 432–452.

Siu, P. M., Bryner, R. W., Martyn, J. K., Alway, S. E. (2004). Apoptotic adaptations from exercise training in skeletal and cardiac muscles. *The FASEB Journal, 18*, 1150–1152.

Siu, P. M., Bryner, R. W., Murlasits, Z., Alway, S. E. (2005a). Response of XIAP, ARC, and FLIP apoptotic suppressors to 8 wk of treadmill running in rat heart and skeletal muscle. *Journal of Applied Physiology, 99*, 204–209.

Siu, P. M., Pistilli, E. E., Alway, S. E. (2005b). Apoptotic responses to hindlimb suspension in gastrocnemius muscles from young adult and aged rats. *American Journal of Physiology: Regulatory, Integrative and Comparative Physiology, 289*, R1015–R1026.

Siu, P. M., Pistilli, E. E., Butler, D. C., Alway, S. E. (2005c). Aging influences cellular and molecular responses of apoptosis to skeletal muscle unloading. *American Journal of Physiology. Cell Physiology, 288*, C338–C349.

Siu, P. M., Pistilli, E. E., Ryan, M. J., Alway, S. E. (2005d). Aging sustains the hypertrophy-associated elevation of apoptotic suppressor X-linked inhibitor of apoptosis protein (XIAP) in skeletal muscle during unloading. *The Journals of Gerontology. Series A: Biological Sciences and Medical Sciences, 60*, 976–983.

Siu, P. M., Pistilli, E. E., Murlasits, Z., Alway, S. E. (2006). Hindlimb unloading increases muscle content of cytosolic but not nuclear Id2 and p53 proteins in young adult and aged rats. *Journal of Applied Physiology, 100*, 907–916.

Siu, P. M., Bae, S., Bodyak, N., Rigor, D. L., Kang, P. M. (2007). Response of caspase-independent apoptotic factors to high salt diet-induced heart failure. *Journal of Molecular and Cellular Cardiology, 42*, 678–686.

Siu, P. M., Pistilli, E. E., Alway, S. E. (2008). Age-dependent increase in oxidative stress in gastrocnemius muscle with unloading. *Journal of Applied Physiology, 105*, 1695–1705.

Smith, M. I., Huang, Y. Y., Deshmukh, M. (2009). Skeletal muscle differentiation evokes endogenous XIAP to restrict the apoptotic pathway. *PLoS ONE, 4*, e5097.

Song, W., Kwak, H. B., Lawler, J. M. (2006). Exercise training attenuates age-induced changes in apoptotic signaling in rat skeletal muscle. *Antioxidants Redox Signaling, 8*, 517–528.

Spierings, D., McStay, G., Saleh, M., Bender, C., Chipuk, J., Maurer, U., Green, D. R. (2005). Connected to death: the (unexpurgated) mitochondrial pathway of apoptosis. *Science, 310*, 66–67.

Sprick, M. R. & Walczak, H. (2004). The interplay between the Bcl-2 family and death receptor-mediated apoptosis. *Biochimica et Biophysica Acta, 1644*, 125–132.

Stennicke, H. R. & Salvesen, G. S. (1999). Catalytic properties of the caspases. *Cell Death and Differentiation, 6*, 1054–1059.
Stennicke, H. R., Deveraux, Q. L., Humke, E. W., Reed, J. C., Dixit, V. M., Salvesen, G. S. (1999). Caspase-9 can be activated without proteolytic processing. *The Journal of Biological Chemistry, 274*, 8359–8362.
Stewart, C. E., Newcomb, P. V., Holly, J. M. (2004). Multifaceted roles of TNF-alpha in myoblast destruction: a multitude of signal transduction pathways. *Journal of Cellular Physiology, 198*, 237–247.
Sulston, J. E. & Horvitz, H. R. (1977). Post-embryonic cell lineages of the nematode, Caenorhabditis elegans. *Developmental Biology, 56*, 110–156.
Sun, X. M., MacFarlane, M., Zhuang, J., Wolf, B. B., Green, D. R., Cohen, G. M. (1999). Distinct caspase cascades are initiated in receptor-mediated and chemical-induced apoptosis. *The Journal of Biological Chemistry, 274*, 5053–5060.
Suzuki, Y., Takahashi-Niki, K., Akagi, T., Hashikawa, T., Takahashi, R. (2004). Mitochondrial protease Omi/HtrA2 enhances caspase activation through multiple pathways. *Cell Death and Differentiation, 11*, 208–216.
Tamilselvan, J., Jayaraman, G., Sivarajan, K., Panneerselvam, C. (2007). Age-dependent upregulation of p53 and cytochrome c release and susceptibility to apoptosis in skeletal muscle fiber of aged rats: role of carnitine and lipoic acid. *Free Radical Biology & Medicine, 43*, 1656–1669.
Tang, D., Lahti, J. M., Kidd, V. J. (2000). Caspase 8 activation and bid cleavage contribute to MCF7 cellular execution in a caspase 3-dependent manner during staurosporine-mediated apoptosis. *The Journal of Biological Chemistry, 275*, 9303–9307.
Tarnopolsky, M. A. (2009). Mitochondrial DNA shifting in older adults following resistance exercise training. *Applied Physiology, Nutrition, and Metabolism, 34*, 348–354.
Tews, D. S. (2005). Muscle-fiber apoptosis in neuromuscular diseases. *Muscle & Nerve, 32*, 443–458.
Thompson, C. B. (1995). Apoptosis in the pathogenesis and treatment of disease. *Science, 267*, 1456–1462.
Tsujimoto, Y. & Shimizu, S. (2000). VDAC regulation by the Bcl-2 family of proteins. *Cell Death and Differentiation, 7*, 1174–1181.
Tsujimoto, Y., Nakagawa, T., Shimizu, S. (2006). Mitochondrial membrane permeability transition and cell death. *Biochimica et Biophysica Acta, 1757*, 1297–1300.
Vallabhapurapu, S. & Karin, M. (2009). Regulation and function of NF-kappaB transcription factors in the immune system. *Annual Review of Immunology, 27*, 693–733.
Vescovo, G. & Dalla, L. L. (2006). Skeletal muscle apoptosis in experimental heart failure: the only link between inflammation and skeletal muscle wastage? *Current Opinion in Clinical Nutrition and Metabolic Care, 9*, 416–422.
Vescovo, G., Volterrani, M., Zennaro, R., Sandri, M., Ceconi, C., Lorusso, R., Ferrari, R., Ambrosio, G. B., Dalla, L. L. (2000). Apoptosis in the skeletal muscle of patients with heart failure: investigation of clinical and biochemical changes. *Heart, 84*, 431–437.
Visser, M., Pahor, M., Taaffe, D. R., Goodpaster, B. H., Simonsick, E. M., Newman, A. B., Nevitt, M., Harris, T. B. (2002). Relationship of interleukin-6 and tumor necrosis factor-alpha with muscle mass and muscle strength in elderly men and women: the Health ABC Study. *The Journals of Gerontology. Series A: Biological Sciences and Medical Sciences, 57*, M326–M332.
Wajant, H. (2009). The role of TNF in cancer. *Results and Problems in Cell Differentiation, 49*, 1–15.
Wajant, H., Pfizenmaier, K., Scheurich, P. (2003). Tumor necrosis factor signaling. *Cell Death and Differentiation, 10*, 45–65.
Wang, W., Fang, H., Groom, L., Cheng, A., Zhang, W., Liu, J., Wang, X., Li, K., Han, P., Zheng, M., Yin, J., Wang, W., Mattson, M. P., Kao, J. P., Lakatta, E. G., Sheu, S. S., Ouyang, K., Chen, J., Dirksen, R. T., Cheng, H. (2008). Superoxide flashes in single mitochondria. *Cell, 134*, 279–290.

Wang, P., Qiu, W., Dudgeon, C., Liu, H., Huang, C., Zambetti, G. P., Yu, J., Zhang, L. (2009). PUMA is directly activated by NF-kappaB and contributes to TNF-alpha-induced apoptosis. *Cell Death and Differentiation, 16*, 1192–1202.

Wei, Y. H., Lee, H. C. (2002). Oxidative stress, mitochondrial DNA mutation, and impairment of antioxidant enzymes in aging. *Exp Biol Med (Maywood), 227*, 671–682.

Welle, S., Thornton, C., Statt, M. (1995). Myofibrillar protein synthesis in young and old human subjects after three months of resistance training. *The American Journal of Physiology, 268*, E422–E427.

Williams, G. T. (1991). Programmed cell death: apoptosis and oncogenesis. *Cell, 65*, 1097–1098.

Wolter, K. G., Hsu, Y. T., Smith, C. L., Nechushtan, A., Xi, X. G., Youle, R. J. (1997). Movement of Bax from the cytosol to mitochondria during apoptosis. *The Journal of Cell Biology, 139*, 1281–1292.

Yang, F., Yang, Y. P., Mao, C. J., Cao, B. Y., Cai, Z. L., Shi, J. J., Huang, J. Z., Zhang, P., Liu, C. F. (2009). Role of autophagy and proteasome degradation pathways in apoptosis of PC12 cells overexpressing human alpha-synuclein. *Neuroscience Letters, 454*, 203–208.

Ye, H., Cande, C., Stephanou, N. C., Jiang, S., Gurbuxani, S., Larochette, N., Daugas, E., Garrido, C., Kroemer, G., Wu, H. (2002). DNA binding is required for the apoptogenic action of apoptosis inducing factor. *Nature Structural Biology, 9*, 680–684.

Yin, X. M., Oltvai, Z. N., Korsmeyer, S. J. (1994). BH1 and BH2 domains of Bcl-2 are required for inhibition of apoptosis and heterodimerization with Bax. *Nature, 369*, 321–323.

Yu, J. & Zhang, L. (2008). PUMA, a potent killer with or without p53. *Oncogene, 27*(Suppl 1), S71–S83.

Yu, J., Wang, P., Ming, L., Wood, M. A., Zhang, L. (2007). SMAC/Diablo mediates the proapoptotic function of PUMA by regulating PUMA-induced mitochondrial events. *Oncogene, 26*, 4189–4198.

Yu, J. W., Jeffrey, P. D., Shi, Y. (2009a). Mechanism of procaspase 8 activation by c-FLIPL. *Proceedings of the National Academy of Sciences of the United States of America, 106*, 8169–8174.

Yu, J. G., Sewright, K., Hubal, M. J., Liu, J. X., Schwartz, L. M., Hoffman, E. P., Clarkson, P. M. (2009b). Investigation of gene expression in C(2)C(12) myotubes following simvastatin application and mechanical strain. *Journal of Atherosclerosis and Thrombosis, 16*, 21–29.

Yuan, J. (1996). Evolutionary conservation of a genetic pathway of programmed cell death. *Journal of Cellular Biochemistry, 60*, 4–11.

Yuan, J. & Yankner, B. A. (2000). Apoptosis in the nervous system. *Nature, 407*, 802–809.

Zha, H., ime-Sempe, C., Sato, T., Reed, J. C. (1996). Proapoptotic protein Bax heterodimerizes with Bcl-2 and homodimerizes with Bax via a novel domain (BH3) distinct from BH1 and BH2. *The Journal of Biological Chemistry, 271*, 7440–7444.

Zheng, J., Edelman, S. W., Tharmarajah, G., Walker, D. W., Pletcher, S. D., Seroude, L. (2005). Differential patterns of apoptosis in response to aging in Drosophila. *Proceedings of the National Academy of Sciences of the United States of America, 102*, 12083–12088.

Age-Related Changes in the Molecular Regulation of Skeletal Muscle Mass

Aaron P. Russell and Bertrand Lèger

Abstract Maintaining skeletal muscle mass and function throughout the entire lifespan is a prerequisite for good health and independent living. While skeletal muscle has an amazing ability for self-renewal and regeneration, its capacity to perform these tasks declines with age. The age-related loss in skeletal muscle mass and function, known as sarcopenia, is a major contributor to the increase in falls and fractures in the elderly. As such, it impacts dramatically upon the quality of life and independence of our aged community and places considerable stain on healthcare systems. At present there are no treatments which stop sarcopenia. Considerable research has focused on identifying the molecular signals which regulate skeletal muscle protein synthesis, degradation and regeneration and how these signals may be perturbed during the ageing process. Regulation of signalling hormones including growth hormone (GH) and insulin-like growth factor -1 (IGF-1), as well as the Akt (protein kinase B) and serum response factor (SRF) signalling pathways have been implicated in age-related changes in muscle protein synthesis and degradation. These factors, as well as those governing muscle stem cell renewal, are presently considered as potential therapeutic targets to combat age-related muscle wasting. This chapter will provide an overview of the age-related regulation of these molecular targets in skeletal muscle.

Keywords Akt signalling • Muscle protein synthesis • Myogenesis • Sarcopenia

---

A.P. Russell (✉)
School of Exercise and Nutrition Sciences, Centre for Physical Activity and Nutrition, Deakin University, Burwood 3125, Australia
e-mail: aaron.russell@deakin.edu.au

B. Lèger
Basic and Clinical Myology Laboratory, Department of Physiology, The University of Melbourne, Parkville 3010, Australia
and
Institut de Recherche en Réadaptation et Réinsertion, 1950 Sion, Switzerland
e-mail: Bertrand.Leger@crr-suva.ch

G.S. Lynch (ed.), *Sarcopenia – Age-Related Muscle Wasting and Weakness*,
DOI 10.1007/978-90-481-9713-2_10,

1 Introduction

Skeletal muscle comprises about 40% of body mass and plays vital roles in regulating metabolism, notably via insulin stimulated glucose uptake (~80%), maintaining posture and controlling movement. Significant reductions in the quantity and quality of skeletal muscle increase the risk of disease including diabetes and heart disease. It also compromises the level of physical independence which results in a reduced quality of life. These muscle related complications are most notably observed in our ageing population (Lexell 1995; Mahoney et al. 1994). The loss of skeletal muscle mass and function with age, also known as sarcopenia, is a major contributor to falls and fractures in the elderly (Mahoney et al. 1994). Sarcopenia significantly reduces the quality of independent living, is the fifth leading cause of death in our aged population and places significant socio-economic pressure on family members and health-care systems (Mahoney et al. 1994). It is well known that the maintenance of skeletal muscle mass is tightly regulated by processes controlling muscle protein synthesis (MPS) and muscle protein breakdown (MPB). Recently our understanding of the molecular signaling factors which detect external environmental cues, transmit these signals within the cells of the body and stimulate the synthesis or breakdown of muscle proteins has improved (Glass 2003). However, what is not well understood are the age-related changes in molecular signaling proteins which contribute to the inability to maintain skeletal muscle mass as we age. This chapter will discuss the age-related changes in key molecular targets which influence skeletal muscle hypertrophy, atrophy and regeneration.

2 Molecular Factors Controlling Muscle Hypertrophy and Atrophy

Maintaining skeletal muscle mass is dependent upon tightly regulated processes governing protein synthesis, protein degradation as well as muscle cell regeneration. Recently, significant advances have been made in understanding the factors controlling skeletal muscle hypertrophy and atrophy using pharmacological and genetic manipulation in cellular and rodent models (Bodine et al. 2001a, b; Pallafacchina et al. 2002; Rommel et al. 2001). Combined, these studies have underlined Akt-1 (also called PKB; Protein Kinase B), a serine/threonine kinase, as a pivotal point in the hypertrophy, and more recently, in the atrophy signalling pathways (Stitt et al. 2004; Latres et al. 2005).

2.1 Akt-1 Signalling and Muscle Hypertrophy

Akt-1 is activated following a series of intracellular signalling cascades involving insulin-like growth factor 1 (IGF-1) and phosphatidylinositol 3-kinase (PI3K) (Datta et al. 1999; Rommel et al. 2001; Vivanco and Sawyers 2002). A downstream target of Akt-1 is glycogen synthase kinase-3β (GSK-3β). The phosphorylation of GSK-3β by Akt-1 (Jefferson et al. 1999; Welsh et al. 1997) releases its inhibition of the translation initiation factor eIF2B (Rhoads 1999). Akt-1 also phosphorylates and activates the mammalian target of rapamycin (mTOR) (Pallafacchina et al. 2002), with the latter phosphorylating and activating $p70^{S6K}$ as well as phosphorylating and releasing the inhibitory effect of PHAS-1/4E-BP1 (Rhoads 1999). Phosphorylation of both $p70^{S6K}$ and PHAS-1/4E-BP1 leads to the activation of pathways promoting protein synthesis and translation initiation, respectively. Hence the Akt-1/GSK-3β and Akt-1/mTOR pathways are important for muscle hypertrophy.

In ageing skeletal muscle there appears to be perturbations in the Akt-1-muscle growth stimulating pathway which may be initiated up-stream due to reductions in insulin-like growth factor-1 (IGF-1). In aging human skeletal muscle a reduction in IGF mRNA has been observed (Leger et al. 2008; Welle 2002). IGF-1 is an important determinant of skeletal muscle growth and repair via its activation of Akt-1 signaling (Rommel et al. 2001). With reduced levels of IGF-1 in the elderly, the capacity to phosphorylate and active Akt-1 at rest or in response to anabolic stimuli, such as following a meal or exercise, would be compromised. Recently, our group has observed that in muscle biopsy samples from elderly males, when compared with young male subjects, there is an elevated level of total Akt-1 protein. However this was not matched by an elevated increase in phosphorylated Akt-1. The inability of the older skeletal muscle to phosphorylate more of the available Akt-1 pool demonstrates a reduced efficiency of Akt-1 phosphorylation. This observation supports those made in older rat skeletal muscle (Haddad and Adams 2006) and suggests an age-related reduction in the efficiency to phosphorylate skeletal muscle Akt-1.

The downstream GSK-3β and mTOR pathways, two axis independently stimulated by Akt-1, regulate muscle growth and have also been measured and compared in muscle biopsies from elderly and younger subjects (Cuthbertson et al. 2005; Leger et al. 2008). Increased levels of total and phosphorylated GSK-3β have been observed in older subjects (Leger et al. 2008). The increased pool of GSK-3β protein may be a result of increased protein translation or protein stability, aimed at providing the cell with a source to maintain protein synthesis. These observations suggest the existence of a mechanism which is able to phosphorylate GSK-3β, independently of Akt-1; an observation not without precedent (Hornberger et al. 2004). In line with this is the recent suggestion that muscle protein synthesis rates may be increased in a futile attempt to maintain muscle mass, however increased levels of protein degradation could be the determining factor during age-related muscle wasting (Kimball et al. 2004), at least in rodents. The total and phosphorylated protein levels of mTOR and its downstream targets $p70^{S6k}$ (Cuthbertson et al. 2005), but not 4E-BP1 (Leger et al. 2008), were shown to be reduced in elderly, when compared with younger muscle.

2.2 *Akt-1 Signalling and Muscle Atrophy*

The activation of Akt-1 has been shown to be important for reducing the activity of pathways involved in muscle protein breakdown (Stitt et al. 2004). The ubiquitin proteasome pathway (UPP) is a major player is skeletal muscle protein breakdown (Lecker et al. 1999). The recent identification of two muscle specific members of the UPP, atrogin-1/MAFbx and MuRF1 (Bodine et al. 2001a; Gomes et al. 2001), has resulted in numerous investigations into the role and regulation in skeletal muscle loss (Glass 2005; Russell 2009). Atrogin-1/MAFbx and MuRF1 are seen as important markers of skeletal muscle atrophy and appear to be regulated Akt-1/forkhead-O F-box (FoXO) signaling (Sandri et al. 2006; Stitt et al. 2004). Akt-1 is able to phosphorylate the FoXO family of transcription factors. When phosphorylated the FoXO proteins are sequestered to the cytoplasm (Brunet et al. 1999). As the FoXO transcription factors have been shown to increase gene transcription of atrogin-1 and MuRF1 (Sandri et al. 2004; Stitt et al. 2004) their translocation to the cytoplasm inhibits their ability to transcribe these genes (Stitt et al. 2004; Latres et al. 2005)

In aged rats Akt-1 activity is decreased with a concomitant increase in atrogin-1 and MuRF1 mRNA levels in the fast-twitch tibialis anterior muscle (Clavel et al. 2006). In contrast, atrogin-1 and MuRF1 mRNA levels are reduced in the mixed-fibre gastrocnemius muscle in rats (Edstrom et al. 2006). These contradicting results suggest that atrogin-1 and MuRF1 regulation might be muscle fibre type specific, at least during age-related muscle wasting. In human studies however, altered Akt/FoXO signaling does not seem to control atrogin-1 and MuRF-1 levels which appear not to be influenced by age; at least in elderly men (Leger et al. 2008; Welle et al. 2003; Whitman et al. 2005). Contrary to this, elevated MuRF1 mRNA levels have been found in skeletal muscle of women aged 85 years compared to 23 year old women (Raue et al. 2007). As this is the only study to compare atrogin-1 or MuRF1 levels in elderly women, as distinct from men, there may be a gender bias favouring increased protein degradation in elderly women. Whether this is a consequence of hormonal and signalling changes occurring with menopause, or merely a factor of the particularly advanced age of the subjects (Raue et al. 2007) compared with other sarcopenia studies (Leger et al. 2008; Welle et al. 2003; Whitman et al. 2005) remains to be explored.

The issue of altered protein synthesis or degradation as the principle regulator of age-related muscle wasting has recently been discussed, with comparisons between rodent and human studies highlighted (Rennie et al. 2009). Muscle wasting in aged rats does not appear to be due to reduced protein synthesis which suggests protein degradation is elevated (Kimball et al. 2004). In contrast, studies in healthy elderly humans do not show a reduction in basal protein synthesis or degradation rates (Volpi et al. 2001; Cuthbertson et al. 2005). Therefore, attention has been given to the protein synthetic response to anabolic stimuli such as feeding and exercise. Recent data suggests an attenuated anabolic response to amino acids as well as exercise in the elderly when compared with younger subjects (Cuthbertson et al. 2005; Katsanos et al. 2005; Wilkes et al. 2009). This anabolic resistance with age is also associated with attenuated increase in mTOR, a key signaling protein in the Akt pathway (Cuthbertson et al. 2005). Furthermore, ageing muscle also has a reduced capacity to blunt proteolysis in response to insulin; an effect potentially mediated through blunted Akt activation (Wilkes et al. 2009). It is evident that considerable perturbations along the Akt-1 signaling pathway occur during the ageing process. These perturbations may negatively influence the ability of the elderly muscle to increase protein synthesis in response to an anabolic stimuli, or maintain protein synthesis, when faced with a catabolic insult such as illness or injury. Identifying factors which might be contributing to the perturbation of this pathway may lead to the identification of therapeutic targets.

3 Molecular Factors Regulating IGF-1 Levels and Akt-1 Activation in Elderly Muscle

3.1 Growth Hormone

Growth hormone (GH) plays a significant role in muscle development (Herrington and Carter-Su 2001) with much of its anabolic effects mediated via insulin-like growth factor-1 (IGF-1). In fact, IGF-1 gene transcription is controlled by GH via a Janus kinase-2 (JAK2)/signal transducer and activator of transcription-5b (STAT5b) signaling pathway (Lupu et al. 2001; Tollet-Egnell et al. 1999; Woelfle and Rotwein 2004). The reduced IGF levels in aged muscle may be linked to reduced circulating levels of GH (Zadik et al. 1985), GH-receptor content (Leger et al. 2008) or GH sensitivity (Corpas et al. 1993). The precise mechanisms regulating GH and IGF levels in aged skeletal muscle are unknown. A possible mechanism may be the catabolic cytokine, tumor necrosis factor-α (TNFα) which is known to decrease IGF-1 mRNA in C2C12 myotubes (Frost et al. 2003) and is increased in aging skeletal muscle (Greiwe et al. 2001; Leger et al. 2008). TNFα is known to regulate the transcription of suppressor of cytokine signaling-3 (SOCS3) (Emanuelli et al. 2001), with the latter able to inhibit GH signaling to JAK2 and STAT5b (Hansen et al. 1999; Ram and Waxman 1999). We have recently shown that SOCS3 levels are increased in humans although this was not associated with reduced STATb phosphorylation (Leger et al. 2008). This suggests that the age-related reduction in IGF-1 mRNA may be influenced by a GH/SOCS3 pathway but independnt of STAT5b transcriptional pertubation.

3.2 Striated Activator of Rho Signaling (STARS)/Serum Response Factor (SRF) Signaling

STARS is a muscle specific actin-binding protein which binds to the I-band of the sarcomere and to actin filaments (Arai et al. 2002; Mahadeva et al. 2002). STARS stimulates the binding of free G-actin to F-actin filaments; a process increasing actin polymerization and reducing the pool of G-actin (Arai et al. 2002). The reduction in the pool of free G-actin removes its inhibition of the transcriptional co-activator myocardin-related transcription factor-A (MRTF-A) (Sotiropoulos et al. 1999). This permits the nuclear translocation of MRTF-A where it increases the transcriptional activity of serum response factor (SRF) (Miralles et al. 2003). The activation of this signaling pathway has been shown to increase cardiac hypertrophy in mice (Kuwahara et al. 2005). Work from our laboratory has recently shown that STARS, as well as members of its signalling pathway, may play a role in human skeletal muscle hypertrophy and atrophy (Lamon et al. 2009). Following 8 weeks of hypertrophy-stimulating resistance training, STARS, MRTF-A, MRTF-B and SRF mRNA as well as RhoA and nuclear SRF protein levels were all increased. This was associated with increases in several SRF target genes; the structural protein α-actin (Carson et al. 1996), the motor protein myosin heavy chain type IIa (MHC IIa) (Allen et al. 2001), and the insulin-like growth factor-1 (IGF-1) (Charvet et al. 2006). Importantly, following 8 weeks of de-training and concomitant muscle atrophy, the increases in the STARS signaling pathway, as well the SRF target genes, returned to base-line.

Recently, STARS, MRTF-A and SRF have been shown to be reduced in skeletal muscles from aged 24-month-old mice (Sakuma et al. 2008). In another study SRF protein levels were also reduced in mice at 15 months of age with an associated decrease in the SRF target gene, α-actin (Lahoute et al. 2008). Of further interest was the report of reduced SRF protein levels in muscle biopsies form elderly subjects (Lahoute et al. 2008). Combined, these results suggest that the loss of members of the STARS signaling pathway, in particular SRF, may contribute to age-related muscle wasting. As IGF-1, a transcriptional target of SRF, is also reduced in aged muscle it is tempting to speculate that a compromised STARS/SRF signaling pathway may be responsible, in part, for reduced IGF-1 levels and associated age-related muscle wasting. The importance of the SRF/IGF-1 axis may not be isolated to skeletal muscle. SRF activity is reduced in aged liver (Supakar and Roy 1996) and the transgenic disruption of hepatic SRF results in impaired liver function and IGF-1 production (Sun et al. 2009). An ageing-associated decline in SRF activityed may well play a vital role in reduced circulating IGF-1 and therefore perturb the pathways involved in muscle growth and regeneration.

3.3 Myostatin – a Negative Regulator of Muscle Mass

Myostatin, also called growth and differentiation factor-8 (GDF-8), is a regulatory factor primarily expressed in skeletal muscle lineage throughout embryonic development as well as in adult animals. It is a member of the transforming growth factor-β family which is known to regulate cell proliferation, differentiation, apoptosis, gene expression and inhibits muscle development (Gonzalez-Cadavid et al. 1998; McPherron et al. 1997). Myostatin is known to activate the activin type IIB receptor which regulates the SMAD 3 signaling pathway to inhibit MyoD and decrease the movement of myogenic stem cells from the G to the S phase (Thomas et al. 2000; McFarlane et al. 2006; Langley et al. 2002). Mutation in the myostatin gene results in exaggerated muscle hypertrophy in animals (Grobet et al. 1997) and one case has also been observed in humans (Schuelke et al. 2004).

Alterations in myostatin mRNA expression with ageing are inconsistent and its involvement in age-related muscle wasting is controversial. Studies in rodents have shown an increase (Baumann et al. 2003), no change (Kawada et al. 2001) or a decrease (Nishimura et al. 2007) in myostatin mRNA levels with age. However, it has recently been reported that myostain knock-out mice, when compared with wild-type mice, have an increase in quadriceps muscle mass when measured at 4–5 months of age. This preservation of muscle mass remained in the myostain knock-out mice aged 37–30 months, supporting a potential role of myostatin in sarcopenia (Morissette et al. 2009b).

In humans, such discrepancy has also been observed as myostatin mRNA levels in elderly skeletal muscle has been shown to be either increased (Raue et al. 2006; Leger et al. 2008) or unchanged (Welle et al. 2002). Myostatin protein levels have been shown to be increased in skeletal muscle of older when compared to younger subjects (Leger et al. 2008). Myostatin levels are known to be inhibited by GH (Liu et al. 2003). Therefore, a perturbation in GH levels or GH activity with age may result in increased myostatin levels. In elderly muscle the increase in myostatin levels and a reduced efficiency of Akt-1 phosphorylation suggests an potential inhibitory effect by myostatin (Leger et al. 2008). In support of this it has been shown in C2C12 muscle cells that myostatin reduces the activity of Akt-1 (Morissette et al. 2009a). Additionally, overexpression of myostatin in the tibialis anterior muscle of Sprague Dawley male rats by electrotransfer attenuated the phosphorylation of Akt-1, tuberous sclerosis complex 2, ribosomal protein S6 and 4E-BP1, demonstrating that myostatin can act as a negative regulator of Akt-1/ mTOR pathway *in vivo* (Amirouche et al. 2009).

4 Satellite Cells and Muscle Regeneration

Quiescent skeletal muscle precursor cells of satellite cells reside between the basal lamina and plasma membrane of muscle fibres (Hawke and Garry 2001). In response to stress induced by weight bearing activities and/or trauma these cells are activated whereby the exit their quiescent state, proliferate and eventually terminally differentiate to repair the muscle (Hawke and Garry 2001). The ability of SC to be activated and proliferate under anabolic stimuli has been suggested to contribute to the development of sarcopenia (Conboy et al. 2003). Additionally, reduced SC population has also been proposed as a mechanism responsible for the loss of muscle mass during ageing (Verdijk et al. 2007; Renault et al. 2002). Several studies in rodents have shown that SC numbers decrease with advancing age (Brack et al. 2005; Dedkov et al. 2003; Gibson and Schultz 1983; Shefer et al. 2006) while others do not (Nnodim 2000; Schafer et al. 2005). Similarly, human studies have also demonstrated such conflicting results, with some studies reporting a decrease in the number of SC in older subjects (Sajko et al. 2004; Verdijk et al. 2007;

Renault et al. 2002; Kadi et al. 2004) and others observing no age-related changes (Dreyer et al. 2006; Petrella et al. 2006; Roth et al. 2000; Verney et al. 2008). Such discrepancy may be attributed to the different age categories of the subjects included in these studies or to the specific muscle group examined. Recently, it was shown that in human skeletal muscle the population of SC is maintained until at least the seventh decade of life, but quickly declines thereafter (Snijders et al. 2009). If the pool of SC is sufficient to effectively repair muscle during most of the adult life then a limiting factor may be the functionality of the SC. Indeed, SC function is largely controlled by extrinsic cell factors which have been shown to be impaired during muscle regeneration with age (Brack and Rando 2007).

4.1 *Microvasculature and Hormonal Regulation*

The systemic environment has a major influence on every tissue in the body and is responsible for the efficient circulation and delivery of key paracrine and endocrine factors. During ageing the capillary network and the capillary-myofiber contacts are reduced which is associated with a corresponding decrease in the secretion of endothelial-derived growth factor (EGF) (Ryan et al. 2006). As SC activation depends upon the action of a broad range of paracrine as well as endocrine factors such as IGF-I, FGF and HGF (Kadi et al. 2005) it would appear that SC activity is closely associated with the microvasculature (Brack and Rando 2007; Christov et al. 2007). Recently, it has been shown that endothelial cells, or multipotent stem cells derived from blood vessels such as pericytes and mesangioblasts, secrete soluble growth factors including IGF-1, HGF, bFGF, PDGF-BB and VEGF which directly influence SC proliferation (Christov et al. 2007). Therefore, it would be expected that changes in the microvasculature would directly influence SC function with increasing age. The profound influence of the systemic component on SC activation has been demonstrated by heterochronic parabiotic pairings. In that experiment, young and old mice shared the same circulatory system exposing old mice to factors present in young serum (Conboy et al. 2005). Under these conditions, activation and regeneration potential of SC from old mice was fully restored.

4.2 *Notch Signalling*

Satellite cell (SC) activation, proliferation and cell linage determination has been shown to be regulated by the Notch signaling pathway (Conboy and Rando 2002). It has been established that aberrant Notch signaling occurs in aged muscle and plays a major role in the reduced capacity of muscle regeneration in aged muscle. SC from adult and aged muscle has similar expression of Notch as well as the Notch ligand and inhibitor, Delta-1 and Numb (Artavanis-Tsakonas et al. 1995,

1999). However following injury SC from adult, but not aged muscle are able to upregulate the Notch ligand Delta-1 (Conboy et al. 2003). The upregulation of Delta-1 is associated with a reduction in the Notch inhibitor Numb as well as an increase in SC proliferation (Conboy et al. 2003). The inability of aged muscle to upregulate Delta-1 does not result in a reduction in the levels of Notch, but rather a reduction in activated Notch. Following muscle injury adult mice, but not aged mice, are able to increase the expression of Delta-1; a response associated with increased SC activation (Conboy et al. 2003). In aged mice, the activation of Notch results in an improved capacity for muscle regeneration, similar to that observed in adult mice. These results demonstrate that the age-related decline in muscle regeneration is linked to insufficient Notch activation via Delta-1. Developing effective strategies to stimulate Notch activation, with the aim of enhancing SC proliferation and differentiation, is a key goal in maintaining skeletal muscle regeneration and reducing muscle wasting in the elderly population.

5 Conclusion

Age-related muscle wasting is a relatively slow, yet relentless process which has debilitating consequences for our elderly community. The maintenance of healthy skeletal muscle mass throughout the lifespan requires the precise coordination of processes controlling protein synthesis and degradation as well as activation of quiescent satellite cells for regeneration. The stimulation of these pathways in response to extracellular anabolic stress such as diet and exercise or catabolic stress such as trauma and injury depends on the ability of molecular targets to detect and transmit these stress signals to the appropriate pathways. These pathways are often interrelated so that a small perturbation in one facet can have numerous consequences on several muscle-related functions. As our society ages and we demand higher living standards and quality of life, understanding how muscle loss occurs with age will remain a key priority for medical research.

References

Allen, D. L., Sartorius, C. A., Sycuro, L. K., Leinwand, L. A. (2001). Different pathways regulate expression of the skeletal myosin heavy chain genes. *The Journal of Biological Chemistry, 276*, 43524–43533.

Amirouche, A., Durieux, A. C., Banzet, S., Koulmann, N., Bonnefoy, R., Mouret, C., Bigard, X., Peinnequin, A., Freyssenet, D. (2009). Down-regulation of Akt/mammalian target of rapamycin signaling pathway in response to myostatin overexpression in skeletal muscle. *Endocrinology, 150*, 286–294.

Arai, A., Spencer, J. A., Olson, E. N. (2002). STARS, a striated muscle activator of Rho signaling and serum response factor-dependent transcription. *The Journal of Biological Chemistry, 277*, 24453–24459.

Artavanis-Tsakonas, S., Matsuno, K., Fortini, M. E. (1995). Notch signaling. *Science, 268*, 225–232.

Artavanis-Tsakonas, S., Rand, M. D., Lake, R. J. (1999). Notch signaling: Cell fate control and signal integration in development. *Science, 284*, 770–776.

Baumann, A. P., Ibebunjo, C., Grasser, W. A., Paralkar, V. M. (2003). Myostatin expression in age and denervation-induced skeletal muscle atrophy. *Journal of Musculoskeletal & Neuronal Interactions, 3*, 8–16.

Bodine, S. C., Latres, E., Baumhueter, S., Lai, V. K., Nunez, L., Clarke, B. A., Poueymirou, W. T., Panaro, F. J., Na, E., Dharmarajan, K., Pan, Z. Q., Valenzuela, D. M., Dechiara, T. M., Stitt, T. N., Yancopoulos, G. D., Glass, D. J. (2001). Identification of ubiquitin ligases required for skeletal muscle atrophy. *Science, 294*, 1704–1708.

Bodine, S. C., Stitt, T. N., Gonzalez, M., Kline, W. O., Stover, G. L., Bauerlein, R., Zlotchenko, E., Scrimgeour, A., Lawrence, J. C., Glass, D. J., Yancopoulos, G. D. (2001). Akt/mTOR pathway is a crucial regulator of skeletal muscle hypertrophy and can prevent muscle atrophy in vivo. *Nature Cell Biology, 3*, 1014–1019.

Brack, A. S. & Rando, T. A. (2007). Intrinsic changes and extrinsic influences of myogenic stem cell function during aging. *Stem Cell Reviews, 3*, 226–237.

Brack, A. S., Bildsoe, H., Hughes, S. M. (2005). Evidence that satellite cell decrement contributes to preferential decline in nuclear number from large fibres during murine age-related muscle atrophy. *Journal of Cell Science, 118*, 4813–4821.

Brunet, A., Bonni, A., Zigmond, M. J., Lin, M. Z., Juo, P., Hu, L. S., Anderson, M. J., Arden, K. C., Blenis, J., Greenberg, M. E. (1999). Akt promotes cell survival by phosphorylating and inhibiting a Forkhead transcription factor. *Cell, 96*, 857–868.

Carson, J. A., Schwartz, R. J., Booth, F. W. (1996). SRF and TEF–1 control of chicken skeletal alpha-actin gene during slow-muscle hypertrophy. *The American Journal of Physiology, 270*, C1624–C1633.

Charvet, C., Houbron, C., Parlakian, A., Giordani, J., Lahoute, C., Bertrand, A., Sotiropoulos, A., Renou, l., Schmitt, A., Melki, J., Li, Z., Daegelen, D., Tuil, D. (2006). New role for serum response factor in postnatal skeletal muscle growth and regeneration via the interleukin 4 and insulin-like growth factor 1 pathways. *Molecular and Cellular Biology, 26*, 6664–6674.

Christov, C., Chretien, F., Abou-Khalil, R., Bassez, G., Vallet, G., Authier, F. J., Bassaglia, Y., Shinin, V., Tajbakhsh, S., Chazaud, B., Gherardi, R. K. (2007). Muscle satellite cells and endothelial cells: Close neighbors and privileged partners. *Molecular Biology of the Cell, 18*, 1397–1409.

Clavel, S., Coldefy, A. S., Kurkdjian, E., Salles, J., Margaritis, I., Derijard, B. (2006). Atrophy-related ubiquitin ligases, atrogin-1 and MuRF1 are up-regulated in aged rat Tibialis Anterior muscle. *Mechanisms of Ageing and Development, 127*, 794–801.

Conboy, I. M. & Rando, T. A. (2002). The regulation of Notch signaling controls satellite cell activation and cell fate determination in postnatal myogenesis. *Developmental Cell, 3*, 397–409.

Conboy, I. M., Conboy, M. J., Smythe, G. M., Rando, T. A. (2003). Notch-mediated restoration of regenerative potential to aged muscle. *Science, 302*, 1575–1577.

Conboy, I. M., Conboy, M. J., Wagers, A. J., Girma, E. R., Weissman, I. L., Rando, T. A. (2005). Rejuvenation of aged progenitor cells by exposure to a young systemic environment. *Nature, 433*, 760–764.

Corpas, E., Harman, S. M., Blackman, M. R. (1993). Human growth hormone and human aging. *Endocrine Reviews, 14*, 20–39.

Cuthbertson, D., Smith, K., Babraj, J., Leese, G., Waddell, T., Atherton, P., Wackerhage, H., Taylor, P. M., Rennie, M. J. (2005). Anabolic signaling deficits underlie amino acid resistance of wasting, aging muscle. *The Faseb Journal, 19*, 422–424.

Datta, S. R., Brunet, A., Greenberg, M. E. (1999). Cellular survival: A play in three Akts. *Genes & Development, 13*, 2905–2927.

Dedkov, E. I., Borisov, A. B., Wernig, A., Carlson, B. M. (2003). Aging of skeletal muscle does not affect the response of satellite cells to denervation. *The Journal of Histochemistry and Cytochemistry, 51*, 853–863.

Dreyer, H. C., Blanco, C. E., Sattler, F. R., Schroeder, E. T., Wiswell, R. A. (2006). Satellite cell numbers in young and older men 24 hours after eccentric exercise. *Muscle & Nerve, 33*, 242–253.

Edstrom, E., Altun, M., Hagglund, M., Ulfhake, B. (2006). Atrogin-1/MAFbx and MuRF1 are downregulated in aging-related loss of skeletal muscle. *The Journals of Gerontology. Series A: Biological Sciences and Medical Sciences, 61*, 663–674.

Emanuelli, B., Peraldi, P., Filloux, C., Chavey, C., Freidinger, K., Hilton, D. J., Hotamisligil, G. S., Van Obberghen, E. (2001). SOCS-3 inhibits insulin signaling and is up-regulated in response to tumor necrosis factor-alpha in the adipose tissue of obese mice. *The Journal of Biological Chemistry, 276*, 47944–47949.

Frost, R. A., Nystrom, G. J., Lang, C. H. (2003). Tumor necrosis factor-alpha decreases insulin-like growth factor-I messenger ribonucleic acid expression in C2C12 myoblasts via a Jun N-terminal kinase pathway. *Endocrinology, 144*, 1770–1779.

Gibson, M. C. & Schultz, E. (1983). Age-related differences in absolute numbers of skeletal muscle satellite cells. *Muscle & Nerve, 6*, 574–580.

Glass, D. J. (2003). Signalling pathways that mediate skeletal muscle hypertrophy and atrophy. *Nature Cell Biology, 5*, 87–90.

Glass, D. J. (2005). Skeletal muscle hypertrophy and atrophy signaling pathways. *The International Journal of Biochemistry & Cell Biology, 37*, 1974–1984.

Gomes, M. D., Lecker, S. H., Jagoe, R. T., Navon, A., Goldberg, A. L. (2001). Atrogin-1, a muscle-specific F-box protein highly expressed during muscle atrophy. *Proceedings of the National Academy of Sciences of the United States of America, 98*, 14440–14445.

Gonzalez-Cadavid, N. F., Taylor, W. E., Yarasheski, K., Sinha-Hikim, I., MA, K., Ezzat, S., Shen, R., Lalani, R., Asa, S., Mamita, M., Nair, G., Arver, S., Bhasin, S. (1998). Organization of the human myostatin gene and expression in healthy men and HIV-infected men with muscle wasting. *Proceedings of the National Academy of Sciences of the United States of America, 95*, 14938–14943.

Greiwe, J. S., Cheng, B., Rubin, D. C., Yarasheski, K. E., Semenkovich, C. F. (2001). Resistance exercise decreases skeletal muscle tumor necrosis factor alpha in frail elderly humans. *The FASEB Journal, 15*, 475–482.

Grobet, L., Martin, L. J., Poncelet, D., Pirottin, D., Brouwers, B., Riquet, J., Schoeberlein, A., Dunner, S., Menissier, F., Massabanda, J., Fries, R., Hanset, R., Georges, M. (1997). A deletion in the bovine myostatin gene causes the double-muscled phenotype in cattle. *Nature Genetics, 17*, 71–74.

Haddad, F. & Adams, G. R. (2006). Aging-sensitive cellular and molecular mechanisms associated with skeletal muscle hypertrophy. *Journal of Applied Physiology, 100*, 1188–1203.

Hansen, J. A., Lindberg, K., Hilton, D. J., Nielsen, J. H., Billestrup, N. (1999). Mechanism of inhibition of growth hormone receptor signaling by suppressor of cytokine signaling proteins. *Molecular Endocrinology, 13*, 1832–1843.

Hawke, T. J. & Garry, D. J. (2001). Myogenic satellite cells: Physiology to molecular biology. *Journal of Applied Physiology, 91*, 534–551.

Herrington, J. & Carter-Su, C. (2001). Signaling pathways activated by the growth hormone receptor. *Trends in Endocrinology and Metabolism, 12*, 252–257.

Hornberger, T. A., Stuppard, R., Conley, K. E., Fedele, M. J., Fiorotto, M. L., Chin, E. R., Esser, K. A. (2004). Mechanical stimuli regulate rapamycin-sensitive signalling by a phosphoinositide 3-kinase-, protein kinase B- and growth factor-independent mechanism. *The Biochemical Journal, 380*, 795–804.

Jefferson, L. S., Fabian, J. R., Kimball, S. R. (1999). Glycogen synthase kinase-3 is the predominant insulin-regulated eukaryotic initiation factor 2B kinase in skeletal muscle. *The International Journal of Biochemistry & Cell Biology, 31*, 191–200.

Kadi, F., Charifi, N., Denis, C., Lexell, J. (2004). Satellite cells and myonuclei in young and elderly women and men. *Muscle & Nerve, 29*, 120–127.

Kadi, F., Charifi, N., Denis, C., Lexell, J., Andersen, J. L., Schjerling, P., Olsen, S., Kjaer, M. (2005). The behaviour of satellite cells in response to exercise: What have we learned from human studies? *Pflugers Archives, 451*, 319–327.

Katsanos, C. S., Kobayashi, H., Sheffield-Moore, M., Aarsland, A., Wolfe, R. R. (2005). Aging is associated with diminished accretion of muscle proteins after the ingestion of a small bolus of essential amino acids. *The American Journal of Clinical Nutrition, 82*, 1065–1073.

Kawada, S., Tachi, C., Ishii, N. (2001). Content and localization of myostatin in mouse skeletal muscles during aging, mechanical unloading and reloading. *Journal of Muscle Research and Cell Motility, 22*, 627–633.

Kimball, S. R., O'malley, J. P., Anthony, J. C., Crozier, S. J., Jefferson, L. S. (2004). Assessment of biomarkers of protein anabolism in skeletal muscle during the life span of the rat: Sarcopenia despite elevated protein synthesis. *American Journal of Physiology. Endocrinology and Metabolism, 287*, E772–E780.

Kuwahara, K., Barrientos, T., Pipes, G. C., LI, S., Olson, E. N. (2005). Muscle-specific signaling mechanism that links actin dynamics to serum response factor. *Molecular and Cellular Biology, 25*, 3173–3181.

Lahoute, C., Sotiropoulos, A., Favier, M., Guillet-Deniau, I., Charvet, C., Ferry, A., Butler-Browne, G., Metzger, D., Tuil, D., Daegelen, D. (2008). Premature aging in skeletal muscle lacking serum response factor. *PLoS ONE, 3*, e3910.

Lamon, S., Wallace, M. A., Leger, B., Russell, A. P. (2009). Regulation of STARS and its downstream targets suggest a novel pathway involved in human skeletal muscle hypertrophy and atrophy. *Journal de Physiologie, 587*, 1795–1803.

Langley, B., Thomas, M., Bishop, A., Sharma, M., Gilmour, S., Kambadur, R. (2002). Myostatin inhibits myoblast differentiation by down-regulating MyoD expression. *The Journal of Biological Chemistry, 277*, 49831–49840.

Latres, E., Amini, A. R., Amini, A. A., Griffiths, J., Martin, F. J., Wei, Y., Lin, H. C., Yancopoulos, G. D., Glass, D. J. (2005). Insulin-like growth factor-1 (IGF-1) inversely regulates atrophy-induced genes via the phosphatidylinositol 3-kinase/Akt/mammalian target of rapamycin (PI3K/Akt/mTOR) pathway. *The Journal of Biological Chemistry, 280*, 2737–2744.

Lecker, S. H., Solomon, V., Mitch, W. E., Goldberg, A. L. (1999). Muscle protein breakdown and the critical role of the ubiquitin-proteasome pathway in normal and disease states. *The Journal of Nutrition, 129*, 227S–237S.

Leger, B., Derave, W., De Bock, K., Hespel, P., Russell, A. P. (2008). Human sarcopenia reveals an increase in SOCS-3 and myostatin and a reduced efficiency of Akt phosphorylation. *Rejuvenation Research, 11*, 163–175B.

Lexell, J. (1995). Human aging, muscle mass, and fiber type composition. *The Journals of Gerontology. Series A: Biological Sciences and Medical Sciences, 50 Spec No*, 11–16.

Liu, W., Thomas, S. G., Asa, S. L., Gonzalez-Cadavid, N., Bhasin, S., Ezzat, S. (2003). Myostatin is a skeletal muscle target of growth hormone anabolic action. *The Journal of Clinical Endocrinology and Metabolism, 88*, 5490–5496.

Lupu, F., Terwilliger, J. D., Lee, K., Segre, G. V., Efstratiadis, A. (2001). Roles of growth hormone and insulin-like growth factor 1 in mouse postnatal growth. *Developmental Biology, 229*, 141–162.

Mahadeva, H., Brooks, G., Lodwick, D., Chong, N. W., Samani, N. J. (2002). ms1, a novel stress-responsive, muscle-specific gene that is up-regulated in the early stages of pressure overload-induced left ventricular hypertrophy. *FEBS Letters, 521*, 100–104.

Mahoney, J., Sager, M., Dunham, N. C., Johnson, J. (1994). Risk of falls after hospital discharge. *Journal of the American Geriatrics Society, 42*, 269–274.

McFarlane, C., Plummer, E., Thomas, M., Hennebry, A., Ashby, M., Ling, N., Smith, H., Sharma, M., Kambadur, R. (2006). Myostatin induces cachexia by activating the ubiquitin proteolytic system through an NF-kappaB-independent, FoxO1-dependent mechanism. *The Journal of Cell Physiology, 209*(2), 501–514.

McPherron, A. C., Lawler, A. M., Lee, S. J. (1997). Regulation of skeletal muscle mass in mice by a new TGF-beta superfamily member. *Nature, 387*, 83–90.

Miralles, F., Posern, G., Zaromytidou, A. I., Treisman, R. (2003). Actin dynamics control SRF activity by regulation of its coactivator Mal. *Cell, 113*, 329–342.

Morissette, M., Cook, S., Buranasombati, C., Rosenberg, M., Rosenzweig, A. (2009). Myostatin inhibits IGF-I induced myotube hypertrophy through Akt. *American Journal of Physiology. Cell Physiology, 297*(5), 1124–1132.

Morissette, M. R., Stricker, J. C., Rosenberg, M. A., Buranasombati, C., Levitan, E. B., Mittleman, M. A., Rosenzweig, A. (2009). Effects of myostatin deletion in aging mice. *Aging Cell, 8*, 573–583.

Nishimura, T., Oyama, K., Kishioka, Y., Wakamatsu, J., Hattori, A. (2007). Spatiotemporal expression of decorin and myostatin during rat skeletal muscle development. *Biochemical and Biophysical Research Communications, 361*, 896–902.

Nnodim, J. O. (2000). Satellite cell numbers in senile rat levator ani muscle. *Mechanisms of Ageing and Development, 112*, 99–111.

Pallafacchina, G., Calabria, E., Serrano, A. L., Kalhovde, J. M., Schiaffino, S. (2002). A protein kinase B-dependent and rapamycin-sensitive pathway controls skeletal muscle growth but not fiber type specification. *Proceedings of the National Academy of Sciences of the United States of America, 99*, 9213–9218.

Petrella, J. K., Kim, J. S., Cross, J. M., Kosek, D. J., Bamman, M. M. (2006). Efficacy of myonuclear addition may explain differential myofiber growth among resistance-trained young and older men and women. *American Journal of Physiology. Endocrinology and Metabolism, 291*, E937–E946.

Ram, P. A. & Waxman, D. J. (1999). SOCS/CIS protein inhibition of growth hormone-stimulated STAT5 signaling by multiple mechanisms. *The Journal of Biological Chemistry, 274*, 35553–35561.

Raue, U., Slivka, D., Jemiolo, B., Hollon, C., Trappe, S. (2006). Myogenic gene expression at rest and after a bout of resistance exercise in young (18–30 yr) and old (80–89 yr) women. *Journal of Applied Physiology, 101*, 53–59.

Raue, U., Slivka, D., Jemiolo, B., Hollon, C., Trappe, S. (2007). Proteolytic gene expression differs at rest and after resistance exercise between young and old women. *The Journals of Gerontology. Series A: Biological Sciences and Medical Sciences, 62*, 1407–1412.

Renault, V., Thornell, L. E., Eriksson, P. O., Butler-Browne, G., Mouly, V. (2002). Regenerative potential of human skeletal muscle during aging. *Aging Cell, 1*, 132–139.

Rennie, M. J., Selby, A., Atherton, P., Smith, K., Kumar, V., Glover, E. L., Philips, S. M. (2009). Facts, noise and wishful thinking: Muscle protein turnover in aging and human disuse atrophy. *Scandinavian Journal of Medicine & Science in Sports, 20*(1), 5–9.

Rhoads, R. E. (1999). Signal transduction pathways that regulate eukaryotic protein synthesis. *The Journal of Biological Chemistry, 274*, 30337–30340.

Rommel, C., Bodine, S. C., Clarke, B. A., Rossman, R., Nunez, L., Stitt, T. N., Yancopoulos, G. D., Glass, D. J. (2001). Mediation of IGF-1-induced skeletal myotube hypertrophy by PI(3)K/Akt/mTOR and PI(3)K/Akt/GSK3 pathways. *Nature Cell Biology, 3*, 1009–1013.

Roth, S. M., Martel, G. F., Ivey, F. M., Lemmer, J. T., Metter, E. J., Hurley, B. F., Rogers, M. A. (2000). Skeletal muscle satellite cell populations in healthy young and older men and women. *The Anatomical Record, 260*, 351–358.

Russell, A. P. (2009). The molecular regulation of skeletal muscle mass. *Clinical and Experimental Pharmacology & Physiology, 37*(3), 378–384.

Ryan, N. A., Zwetsloot, K. A., Westerkamp, L. M., Hickner, R. C., Pofahl, W. E., Gavin, T. P. (2006). Lower skeletal muscle capillarization and VEGF expression in aged vs. young men. *Journal of Applied Physiology, 100*, 178–185.

Sajko, S., Kubinova, L., Cvetko, E., Kreft, M., Wernig, A., Erzen, I. (2004). Frequency of M-cadherin-stained satellite cells declines in human muscles during aging. *The Journal of Histochemistry and Cytochemistry, 52*, 179–185.

Sakuma, K., Akiho, M., Nakashima, H., Akima, H., Yasuhara, M. (2008). Age-related reductions in expression of serum response factor and myocardin-related transcription factor A in mouse skeletal muscles. *Biochimica et Biophysica Acta, 1782*, 453–461.

Sandri, M., Sandri, C., Gilbert, A., Skurk, C., Calabria, E., Picard, A., Walsh, K., Schiaffino, S., Lecker, S. H., Goldberg, A. L. (2004). Foxo transcription factors induce the atrophy-related ubiquitin ligase atrogin-1 and cause skeletal muscle atrophy. *Cell, 117*, 399–412.

Sandri, M., Lin, J., Handschin, C., Yang, W., Arany, Z. P., Lecker, S. H., Goldberg, A. L., Spiegelman, B. M. (2006). PGC-1alpha protects skeletal muscle from atrophy by suppressing

FoxO3 action and atrophy-specific gene transcription. *Proceedings of the National Academy of Sciences of the United States of America, 103*, 16260–16265.
Schafer, R., Zweyer, M., Knauf, U., Mundegar, R. R., Wernig, A. (2005). The ontogeny of soleus muscles in mdx and wild type mice. *Neuromuscular Disorders, 15*, 57–64.
Schuelke, M., Wagner, K. R., Stolz, L. E., Hubner, C., Riebel, T., Komen, W., Braun, T., Tobin, J. F., Lee, S. J. (2004). Myostatin mutation associated with gross muscle hypertrophy in a child. *The New England Journal of Medicine, 350*, 2682–2688.
Shefer, G., Vande Mark, D. P., Richardson, J. B., Yablonka-Reuveni, Z. (2006). Satellite-cell pool size does matter: Defining the myogenic potency of aging skeletal muscle. *Developmental Biology, 294*, 50–66.
Snijders, T., Verdijk, L. B., Van Loon, L. J. (2009). The impact of sarcopenia and exercise training on skeletal muscle satellite cells. *Ageing Research Reviews, 8*(4), 328–338.
Sotiropoulos, A., Gineitis, D., Copeland, J., Treisman, R. (1999). Signal-regulated activation of serum response factor is mediated by changes in actin dynamics. *Cell, 98*, 159–169.
Stitt, T. N., Drujan, D., Clarke, B. A., Panaro, F., Timofeyva, Y., Kline, W. O., Gonzalez, M., Yancopoulos, G. D., Glass, D. J. (2004). The Igf-1/PI3K/Akt pathway prevents expression of muscle atrophy-induced ubiquitin ligases by inhibiting Foxo transcription factors. *Molecular Cell, 14*, 395–403.
Sun, K., Battle, M. A., Misra, R. P., Duncan, S. A. (2009). Hepatocyte expression of serum response factor is essential for liver function, hepatocyte proliferation and survival, and post-natal body growth in mice. *Hepatology, 49*, 1645–1654.
Supakar, P. C. & Roy, A. K. (1996). Role of transcription factors in the age-dependent regulation of the androgen receptor gene in rat liver. *Biological Signals, 5*, 170–179.
Thomas, M., Langley, B., Berry, C., Sharma, M., Kirk, S., Bass, J., Kambadur, R. (2000). Myostatin, a negative regulator of muscle growth, functions by inhibiting myoblast proliferation. *The Journal of Biological Chemistry, 275*, 40235–40243.
Tollet-Egnell, P., Flores-Morales, A., Stavreus-Evers, A., Sahlin, L., Norstedt, G. (1999). Growth hormone regulation of SOCS-2, SOCS-3, and CIS messenger ribonucleic acid expression in the rat. *Endocrinology, 140*, 3693–3704.
Verdijk, L. B., Koopman, R., Schaart, G., Meijer, K., Savelberg, H. H., Van Loon, L. J. (2007). Satellite cell content is specifically reduced in type II skeletal muscle fibers in the elderly. *American Journal of Physiology. Endocrinology and Metabolism, 292*, E151–E157.
Verney, J., Kadi, F., Charifi, N., Feasson, L., Saafi, M. A., Castells, J., Piehl-Aulin, K., Denis, C. (2008). Effects of combined lower body endurance and upper body resistance training on the satellite cell pool in elderly subjects. *Muscle & Nerve, 38*, 1147–1154.
Vivanco, I. & Sawyers, C. L. (2002). The phosphatidylinositol 3-Kinase Akt pathway in human cancer. *Nature Reviews. Cancer, 2*, 489–501.
Volpi, E., Sheffield-Moore, M., Rasmussen, B. B., Wolfe, R. R. (2001). Basal muscle amino acid kinetics and protein synthesis in healthy young and older men. *JAMA, 286*, 1206–1212.
Welle, S. (2002). Cellular and molecular basis of age-related sarcopenia. *Canadian Journal of Applied Physiology, 27*, 19–41.
Welle, S., Bhatt, K., Shah, B., Thornton, C. (2002). Insulin-like growth factor-1 and myostatin mRNA expression in muscle: Comparison between 62–77 and 21–31 yr old men. *Experimental Gerontology, 37*, 833–839.
Welle, S., Brooks, A. I., Delehanty, J. M., Needler, N., Thornton, C. A. (2003). Gene expression profile of aging in human muscle. *Physiological Genomics, 14*, 149–159.
Welsh, G. I., Stokes, C. M., Wang, X., Sakaue, H., Ogawa, W., Kasuga, M., Proud, C. G. (1997). Activation of translation initiation factor eIF2B by insulin requires phosphatidyl inositol 3-kinase. *Febs Letters, 410*, 418–422.
Whitman, S. A., Wacker, M. J., Richmond, S. R., Godard, M. P. (2005). Contributions of the ubiquitin-proteasome pathway and apoptosis to human skeletal muscle wasting with age. *Pflugers Archives, 450*, 437–446.
Wilkes, E. A., Selby, A. L., Atherton, P. J., Patel, R., Rankin, D., Smith, K., Rennie, M. J. (2009). Blunting of insulin inhibition of proteolysis in legs of older subjects may contribute to age-related sarcopenia. *The American Journal of Clinical Nutrition, 90*(5), 1343–1350.

Woelfle, J. & Rotwein, P. (2004). In vivo regulation of growth hormone-stimulated gene transcription by STAT5b. *American Journal of Physiology. Endocrinology and Metabolism, 286*, E393–E401.

Zadik, Z., Chalew, S. A., Mccarter, R. J., JR Meistas, M., Kowarski, A. A. (1985). The influence of age on the 24-hour integrated concentration of growth hormone in normal individuals. *The Journal of Clinical Endocrinology and Metabolism, 60*, 513–516.

Genetic Variation and Skeletal Muscle Traits: Implications for Sarcopenia

Stephen M. Roth

Abstract Skeletal muscle is one of the most heritable quantitative traits studied to date, with heritability estimates ranging from 30% to 85% for muscle strength measures and 50–80% for lean mass measures. The strong genetic contribution to skeletal muscle traits indicates the possibility of using genetic approaches to individualize treatment approaches for sarcopenia or even aid in prevention strategies through the use of genetic screening prior to functional limitations. While these possibilities provide the rationale and motivation for genetic studies of skeletal muscle traits, few genes have been identified to date that appear to contribute to variation in either skeletal muscle strength or mass phenotypes, let alone sarcopenia itself. The *ACE*, *ACTN3*, *CNTF*, and *VDR* genes have been associated with skeletal muscle strength in two or more papers each, while the *AR*, *TRHR*, and *VDR* genes have been similarly associated with muscle mass. Only the *VDR* gene has been significantly associated with sarcopenia itself as an endpoint phenotype but replication of this initial finding has not yet been performed. Large-scale clinical studies relying on advanced genome-wide association techniques are needed to provide further insights into potentially clinically relevant genes that contribute to skeletal muscle traits, with identified genes then explored functionally to determine the likelihood that genetic screening can assist in the prevention and treatment of sarcopenia.

Keywords Genotype • Heritability • Muscle mass • Muscle strength • Polymorphism

S.M. Roth (✉)
Department of Kinesiology, School of Public Health, University of Maryland, College Park, MD 20742, USA
e-mail: sroth1@umd.edu

G.S. Lynch (ed.), *Sarcopenia – Age-Related Muscle Wasting and Weakness*,
DOI 10.1007/978-90-481-9713-2_11, © Springer Science+Business Media B.V. 2011

1 Introduction

Aging is associated with a decline in skeletal muscle mass, strength, power and physical functioning, generally termed sarcopenia (Dutta and Hadley 1995). These well-documented losses of muscle strength, mass, and muscle quality (limb strength/limb muscle mass) with age (Lindle et al. 1997; Baumgartner et al. 1998; Janssen et al. 2002; Lauretani et al. 2003; Sowers et al. 2005; Ploutz-Snyder et al. 2002) have important health consequences, because this deterioration in muscle structure and function is associated with an increased risk of falls, hip fractures, and functional decline (Schultz et al. 1997; Aniansson et al. 1984; Janssen et al. 2004; Newman et al. 2003a; Lauretani et al. 2003; Sowers et al. 2005). Muscle strength is independently associated with functional ability in the elderly (Hyatt et al. 1990; Visser et al. 2000a; Kwon et al. 2001; Purser et al. 2003; Rantanen et al. 1998; Foldvari et al. 2000; Lauretani et al. 2003; Pendergast et al. 1993) and may explain up to 25% of the variance in overall functional ability (Buchner and deLateur 1991). Furthermore, sarcopenia is related to a reduction in the performance of activities of daily living (Nybo et al. 2001), which may lead to further declines in muscle mass and strength and greater reductions in the performance of those activities. The net effect of this cycle can result in marked disablement, predisposing older individuals to falls, injuries and disability (Rantanen et al. 2000).

Although the loss of muscle mass is associated with the decline in strength in older adults, the strength decline is much more rapid than the concomitant loss of muscle mass, suggesting a decline in muscle quality (Goodpaster et al. 2006). The loss of muscle strength is an independent predictor of mortality in the elderly, more so than loss of muscle mass (Metter et al. 2002; Rantanen et al. 2000, 2003; Fujita et al. 1995; Laukkanen et al. 1995; Newman et al. 2003b). Thus, the relationship of muscle mass and strength to mortality may rest in the higher functional capacity associated with having more muscle strength and mass, and an inverse association with functional limitations and disability. Sex differences have been shown, with women showing an earlier age of onset of sarcopenia (Lauretani et al. 2003; Janssen et al. 2002), and a greater prevalence of functional impairment at any age in comparison to men (Lauretani et al. 2003; Rantanen and Avela 1997; Ostchega et al. 2000; Dunlap et al. 2002; Visser et al. 2000b), most likely owing to their lower muscle mass and strength levels compared to men throughout the adult age span (Frontera et al. 1991; Lindle et al. 1997; Rantanen and Avela 1997; Lauretani et al. 2003). The consequences of sarcopenia-related disability are significant both in terms of personal quality of life and to the overall economy, with healthcare costs related to sarcopenia in the United States estimated to be $18.5 billion dollars for adults 60 years and older for the year 2000 (Janssen et al. 2004).

Though the losses of muscle mass and strength begin on average between 40 and 50 years of age, losses for any particular individual are quite variable. For example, investigators from our laboratories at the University of Maryland have reported substantial age-related declines in strength and muscle quality in men and women from the Baltimore Longitudinal Study of Aging (BLSA) (Lindle et al. 1997;

Lynch et al. 1999). However, we've observed enormous inter-individual variability in muscle strength within each age group that could not be explained by previous muscular activity levels. For example, the highest strength values for 80–96 year old men and women were two to four times higher than the lowest strength values in 20–39 year old men and women (Table 1). Furthermore, at least 15% of the men and women >60 year had strength values that were above the average values for 20 year old subjects. Similar inter-individual variations existed for leg muscle mass (Lindle et al. 1997) and for muscle quality in both older men and women (Lynch et al. 1999). Sarcopenia has been reported in community-dwelling men and women below the age of 50 year (Melton et al. 2000; Tanko et al. 2002; Janssen et al. 2002; Lauretani et al. 2003), and recently, sarcopenia associated with compromised physical functioning was shown to occur in nearly one in ten women aged 34–58 year (mid-life) (Sowers et al. 2005), providing further support for the variable onset of muscle strength losses and an indication of susceptibility to sarcopenia in some individuals. Various research groups are currently exploring the possibility that a portion of this inter-individual variability and susceptibility to early muscle losses is due to genetic factors, which could someday be used to identify susceptible men and women and individualize their prevention and treatment interventions.

This review discusses the genetic aspects of skeletal muscle traits with an emphasis on sarcopenia, including examination of heritability, linkage analysis, and specific genes associated with relevant traits. While skeletal muscle remains one of the most heritable health-related quantitative phenotypes studied to date, the identification of specific contributing genes remains at the early stages and much work remains to determine the future clinical importance of genetic contributions to sarcopenia risk. This review will not address the potential role of mitochondrial DNA mutations in the development of sarcopenia (Hiona and Leeuwenburgh 2008), as these genetic variations represent age-related, sporadic modifications of DNA sequence rather than stable, genome-wide genetic variants present since birth in all somatic cells.

2 Heritability of Skeletal Muscle Traits

Variation in skeletal muscle traits among individuals can be attributed to environmental factors, genetic factors, or the interaction of both. While the influence of environmental factors such as physical activity and diet have been broadly investigated, only recently have studies begun to address the specific genetic influences on skeletal muscle traits that may explain the inter-individual variability noted above. The earliest of these studies examined familial aggregation of body composition traits in twins, especially exploiting the slight but important differences between monozygotic and dizygotic twin pairs. Monozygotic twins share not only 100% of genetic variation in their DNA sequence, but also share the intrauterine environment and very likely a similar environment through adolescence. Dizygotic

Table 1 Lowest and highest concentric knee extension strength in each decade of the adult life span in 1,283 men and women from the Baltimore Longitudinal Study of Aging (Shock et al. 1984; Lindle et al. 1997; Lynch et al. 1999; Ferrucci 2008)

Age Range (year)	20–29	30–39	40–49	50–59	60–69	70–79	80–96
Men (N=661)	101–248 (N=21)	57–317 (N=60)	37–411 (N=102)	55–205 (N=156)	38–330 (N=114)	19–178 (N=117)	16–239 (N=90)
Women (N=622)	28–126 (N=22)	29–151 (N=73)	27–134 (N=102)	20–240 (N=168)	11–136 (N=125)	17–140 (N=83)	12–117 (N=49)

Data are isokinetic peak torque values (Nm) at 180 deg/s.

twins on the other hand similarly share the intrauterine and external environment through young adulthood, but share only approximately 50% of their genetic variation. Thus, correlations performed between monozygotic and dizygotic twin pairs can be compared and estimates of genetic contribution, termed heritability, can be determined (Bouchard et al. 1985, 1997; Roth 2007). When traits exhibit closer correlation in monozygotic compared to dizygotic twins, the assumption is that genetic factors are contributing to the closer correlation in monozygotic twins and heritability can be calculated from the extent of difference observed in the correlation values.

Clark reported one of the first heritability studies with relevance to skeletal muscle in 1956 (Clark 1956). In that report, a series of anthropometric traits were compared in monozygotic and dizygotic twins, including measures of arm and calf circumference both of which were greater than 60% heritable. Later studies provided more direct measures of skeletal muscle traits. For example, the heritability of grip strength was estimated between 30% and 50% in several early studies (Montoye et al. 1975; Venerando and Milani-Comparetti 1970; Kovar 1975). In a study of older twins, genetic factors accounted for 65% of the variance in grip strength in 260 mono- and dizygotic twins (59–69 year), even after adjusting for body weight, height and age (Reed et al. 1991). More recently, twin studies have revealed heritability values for muscle strength phenotypes ranging from 30% to 85% depending on the conditions of the strength measure (e.g., limb, contraction angle, velocity, and type) (Thomis et al. 1998a, 2004; Perusse et al. 1987a, b; Huygens et al. 2004a; Karlsson et al. 1979; Reed et al. 1991; Thomis et al. 1998a; Arden and Spector 1997; Zhai et al. 2004; Ropponen et al. 2004). Skeletal muscle fiber type composition has also been shown to be a heritable trait (Komi et al. 1977; Simoneau and Bouchard 1995), though variability in the biopsy technique and heterogeneity of fiber type distribution within skeletal muscle make these estimates remarkably challenging. The hypothesis that genetic factors may influence muscular strength is also supported by data from rats in which a 1.5- to 5.2-fold divergence between the muscular strength of 11 different strains with the lowest and highest strength levels has been reported (Biesiadecki et al. 1998).

With regard to skeletal muscle mass, evidence for significant heritability has been identified across a number of traits, with the first studies reporting heritability of limb circumferences (Clark 1956, Huygens et al. 2004b; Loos et al. 1997; Susanne 1977; Thomis et al. 1997). The first direct study of lean body mass (LBM) was performed by Bouchard et al. (1985) who reported 80% heritability of LBM by hydrodensitometry in twin pairs. Later Forbes et al. (1995) reported 70% heritability of LBM by the potassium 40 counting method, and Seeman et al. (1996) and Arden et al. (1997) provided the first estimates (50–80%) using dual energy x-ray absorptiometry (DXA). Other studies have reported similar findings (Nguyen et al. 1998; Loos et al. 1997; Thomis et al. 1998b; Livshits et al. 2007; Karasik et al. 2009) and recently Prior and colleagues (2007) reported significant heritability of lean mass and calf cross-sectional area (CSA) in families of African-descent, providing the first evidence of heritability values in this race group, which is known to higher muscle mass and strength traits compared to subjects of European descent (Aloia et al. 2000; Gallagher et al. 1997; Jones et al. 2002; Visser et al. 2000a; Newman et al. 2006). Across these studies, heritability estimates greater than 50% are not uncommon for muscle mass measurements.

Perhaps most relevant for this discussion are the various studies that have examined heritability within older subjects. In addition to the study of grip strength by Reed and colleagues (1991) discussed above, several other reports have demonstrated significant heritability values for muscle strength in older individuals (Frederiksen et al. 2002, 2003; Tiainen et al. 2004, 2005, 2009; Zhai et al. 2004, 2005). For example, Frederiksen and colleagues (2002) showed heritability of grip strength at 50% across several age groups from 46 to 96 year. The change in muscle strength with advancing age has also been found to be heritable (Carmelli et al. 2000; Zhai et al. 2004), though some studies indicate that the contribution of environmental factors appears to increase at older ages (Carmelli and Reed 2000; Tiainen et al. 2004). With regard to the more general trait of functional performance, the results are more mixed with moderate heritability for lower-extremity function in older male twins (Carmelli et al. 2000), low heritability reported for age-related functional impairment in male twins (Gurland et al. 2004), and low but significant heritability for older female twins in the rate of change of physical function with age, with a non-significant genetic component in older male twins (Christensen et al. 2002, 2003). These findings are consistent with the idea that more general, multi-component traits are likely to be influenced by a wider range of environmental factors, especially in older individuals (Tiainen et al. 2005; Harris et al. 1992). Overall, genetic variation explains a significant fraction of the inter-individual variability in skeletal muscle phenotypes, including muscle traits in older individuals. While there is strong evidence for a heritable component to muscle phenotypes, the genetic analysis of muscle architecture is in its infancy.

3 Linkage Analysis and Skeletal Muscle Traits

After the familial aggregation and heritability of a trait is firmly established, until recently the next step in genetic analysis was to perform linkage analysis studies in families. The goal of linkage analysis was to rely on the shared genetic variation with families to identify chromosome locations that harbor genes and gene variants that contribute to trait variation. By determining several hundred genotypes spread across the genome in each of the individuals of several families, linkage analysis would identify those regions most closely correlated with the trait of interest. Significantly correlated regions are assumed to harbor genetic variation relevant to the trait of interest, though these identified regions are often quite extensive, with many potential genes. Thus, linkage analysis is useful for narrowing the potential list of candidate genes from many thousand to several hundred, but considerable work remains even after a linkage study to confidently determine the specific contributing genes.

In the first genome-wide linkage analysis for genes related to muscle mass, Chagnon et al. (2000) examined microsatellite markers in the Quebec Family Study, which consisted of 748 subjects from 194 families. Fat-free mass (FFM) was calculated from percent body fat determined by hydrostatic weighing. Significant linkages were observed for a CA-repeat within the insulin-like growth factor 1 receptor (*IGF1R*) on 15q25-26, and at two markers at 18q12; moderate linkage was noted on 7p15.3, with the authors noting possible candidate genes of neuropeptide Y (*NPY*) and growth hormone-releasing hormone (*GHRH*) receptor in that location. A second study by Chagnon et al. (2001) examined body composition in 364 sibling pairs from 99 families from the HERITAGE Family Study before and after 20 weeks of aerobic exercise training. In that analysis, no significant loci were identified for baseline FFM, though change in FFM in response to aerobic exercise training was linked to loci at *IGF1*, 1q22, and 8q24.12. Livshits and colleagues (2007) reported significant linkage with LBM in 3180 female twin pairs at chromosomes 12q24.3 and 14q22.3. Most recently, Karasik et al. (2009) reported significant linkage in 1346 adults from 327 families from the Framingham study for leg lean mass measured by DXA. Two loci (12p12.3-12p13.2 and 14q21.3-22.1) were identified as having bivariate linkage with both leg lean mass and bone phenotypes.

Two studies have examined strength-related phenotypes in family-based linkage analysis. De Mars and colleagues (2008a) reported significant linkage signals for torque-velocity ratios of the knee flexors and extensors (strongest signal at 15q23), as well as for the torque-velocity slope of the knee extensors. The same group reported significant linkage for the torque-length relationship of the knee flexors (strongest signal at 14q24.3) and isometric knee torque in 283 male siblings from 105 families (De Mars et al. 2008b).

A few linkage studies have been performed in a more focused manner, isolating a small number of regions in order to better identify potential candidate genes. In the HERITAGE Family Study, Sun et al. (1999) performed a focused linkage analysis around a microsatellite marker in the *IGF1* locus. In 502 individuals from 99 families, the *IGF1* locus was not significantly linked with baseline FFM, though was significantly associated with the change in FFM after aerobic exercise training, consistent with the genome-wide linkage results of Chagnon and colleagues (2000) described above. Huygens et al. (2004c) performed a gene-specific linkage analysis for the *RB1* locus in 329 young Caucasian male siblings from 146 families for trunk strength and identified multiple linkage peaks for trunk flexion measures with no evidence of linkage for trunk extension measures. In a second study, Huygens and colleagues (2004c) performed a gene-targeted single-point (one marker per gene) linkage analysis in the myostatin pathway (across 10 genes) in the same young male cohort for various measures of muscle mass and strength. Significant linkage was reported with markers D2S118, D6S1051, and D11S4138 for knee extension and flexion peak torque measures. These markers are in the *MSTN* (myostatin, formerly *GDF8*), *CDKN1A*, and *MYOD1* genes, respectively. Huygens et al. (2005) then performed an expanded multi-point (multiple markers per gene) linkage analysis in 367 young Caucasian male siblings from 145 families with nine genes involved in the myostatin signaling pathway and various measures of muscle strength. Significant linkages were reported on four chromosomal regions with knee muscle strength measures: chromosome 13q21 (D13S1303), chromosome 12p12-p11 (D12S1042), chromosome 12q12-q13.1 (D12S85), and chromosome 12q23.3-q24.1 (D12S78).

Only one linkage study has targeted older individuals in particular. In 2008, Tiainen et al. (2009) examined 397 microsatellite markers in 217 female twin pairs aged 66 to 75 years from the Finnish Twin Study on Aging. Significant linkages were reported for knee extensor isometric strength on chromosome 15q14, for leg extensor power on chromosome 8q24.23, and for calf muscle CSA on chromosomes 20q13.31 and 9q34.3. Importantly, the linkage noted at 9q34 was similarly observed by Chagnon and colleagues (2001) for change in FFM in response to exercise training, providing some of the first evidence of replication of a locus related to skeletal muscle mass across different linkage studies.

Recently, linkage analysis studies have given way to genome-wide association studies that can be used to identify specific gene regions in unrelated individuals by use of high-density single nucleotide polymorphism microarrays, which allow as many as 1 million genotypes to be determined and used in association analyses. These studies have been successful at identifying a clinically relevant candidate gene for age-related macular degeneration (Klein et al. 2005), and have provided important novel targets for other health-related traits (Lindgren et al. 2009; Graham et al. 2009). Only one such study has been performed for skeletal muscle traits to date. In 2009, Liu and colleagues examined 379,319 polymorphisms across the genome in nearly 1,000 unrelated U.S. whites for association with LBM measured by DXA. In the initial genome-wide analysis, two polymorphisms were identified as statistically significant (with Bonferroni corrected p values at 7×10^{-8}) and another 146 polymorphisms approaching statistical significance. The two significant polymorphisms are both located in the *TRHR* gene, which encodes the thyrotropin-releasing hormone receptor. These two polymorphisms were then genotyped in three replication cohorts consisting of over 6000 total white and Chinese subjects and consistent significant associations were observed in those analyses. Because of the importance of thyroid hormone in skeletal muscle development (Larsson et al. 1994; Norenberg et al. 1996; Soukup and Jirmanova 2000), the *TRHR* gene is thus recognized as an important candidate gene for future investigation. Though currently unpublished, other research groups have genome-wide association data available and additional findings are expected before the end of 2010.

4 Genetic Variation and Skeletal Muscle Traits

The ultimate goal of linkage and genome-wide association studies is the identification of specific genes and gene variants with clinically relevant influences on skeletal muscle traits important to physical function. The advent of genome-wide association studies provides an important technical improvement in the ability to identify specific loci for in-depth investigation, though as mentioned above, only one has been published to date for skeletal muscle traits. As loci are replicated across studies, specific gene variants will be identified and their clinical relevance determined. In the next sections, specific genes and gene sequence variants that have been associated with skeletal muscle phenotypes will be discussed, including those associated with muscle strength, muscle mass, and sarcopenia in particular. Genes related to skeletal muscle adaptation will only be discussed briefly as this is not a focus of this chapter. While the reference lists for these sections will be comprehensive, only those genes examined in multiple investigations or otherwise shown to be functional in some way will be discussed in detail. Replication of genetic associations, especially those of generally weak genetic influence, is generally considered the gold standard for considering a gene important to a trait, though other approaches exist (Khoury et al. 2005, 2007).

4.1 Genetic Variation and Skeletal Muscle Strength

The identification of genetic factors important to skeletal muscle strength is remarkably difficult owing to the fact that multiple strength variables are commonly measured in different studies, including different muscle groups (forearm, knee extensor, leg), contraction types (isometric, isotonic, isokinetic), and measurement instruments. Moreover, different genes are likely to contribute to different aspects of strength that may not be reflected across the different measurement types. Additionally, some studies have included measurements of muscle quality or muscle power given their importance to physical function, especially for the elderly (Dutta et al. 1997; Bassey et al. 1992). All this means that for a particular gene or genotype of interest, the chances of finding replication across multiple studies for the same trait are small. This has both positive and negative implications: though few studies demonstrate replication and thus few studies have found evidence of the importance of any one gene, when genes are found to be important across multiple, different strength measurements the likelihood the gene is truly important to muscle strength improves.

Table 2 summarizes the genes that have been studied in relation to skeletal muscle strength measurements, focusing on genes associated with baseline strength values; genes related to muscle strength adaptation to exercise training are discussed in a later section. Genes that have been studied in only one paper or that have not been replicated in some way and are not discussed here in detail include: *COL1A1* (Van Pottelbergh et al. 2001, 2002); *BDKRB2* (Hopkinson et al. 2006); *DIO1* (Peeters et al. 2005); *MYLK* (Clarkson et al. 2005b); *IL6* (Walston et al. 2005); *TNF* (Liu et al. 2008a); *NR3C1* (van Rossum et al. 2004; Peeters et al. 2008); *AR* (Walsh et al. 2005); and *IL15* and *IL15RA* (Pistilli et al. 2008).

Angiotensin Converting Enzyme (ACE) *ACE* and its insertion/deletion (I/D) polymorphism is arguably the most studied of genes with regard to exercise

Table 2 Genes and gene sequence variants associated with skeletal muscle strength phenotypes in multiple studies

Gene	References	Variants Examined	Subjects	Skeletal Muscle Strength Measurements
ACE	Woods et al. (2001)	I/D polymorphism	83 postmenopausal women	Change in isometric strength of adductor pollicis in response to HRT
	Hopkinson et al. (2004)	I/D polymorphism	103 COPD patients	Quadriceps isometric strength
	Williams et al. (2005)	I/D polymorphism	81 young men	Quadriceps isometric strength
	Moran et al. (2006)	*ACE* I/D and haplotype	1,027 adolescents	Handgrip strength and vertical jump in females
	Wagner et al. (2006)	I/D polymorphism	62 young men and women	Contraction velocity and isometric force in multiple muscles
	Yoshihara et al. (2009)	I/D polymorphism	431 older Japanese men and women	Hand grip strength and walking speed
ACTN3	Clarkson et al. (2005)	R577X	602 young men and women	Biceps isometric strength in females
	Delmonico et al. (2007)	R577X	157 older men and women	Knee extensor peak power in women
	Vincent et al. (2007)	R577X	90 young men	Isokinetic knee extensor strength
	Delmonico et al. (2008)	R577X	1,367 older men and women	Physical limitation and walk performance
	Walsh et al. (2008)	R577X	848 men and women	Isokinetic knee extensor strength in women
CNTF	Roth et al. (2001)	rs1800169	494 men and women	Isokinetic knee extensor and flexor strength and muscle quality
	Arking et al. (2006)	rs1800169 and *CNTF* haplotype	363 older women	Hand grip strength

CNTFR	Roth et al. (2003)	C-1703T, T1069A, C174T	465 men and women	Isokinetic knee extensor and flexor strength
	De Mars et al. (2007)	C-1703T, T1069A, C174T, and others	493 older men and women	Knee flexor and extensor strength measurements
IGF2	Sayer et al. (2002)	ApaI	693 older men and women	Hand grip strength in men
	Schrager et al. (2004)	ApaI	596 men and women	Isokinetic strength of multiple muscle groups
MSTN	Seibert et al. (2001)	K153R	286 older women	Composite isometric strength score
	Corsi et al. (2002)	K153R	450 older men and women	Composite isometric strength score
	Kostek et al. (2009)	K153R, A55T	23 young African Americans	Isometric biceps strength
VDR	Geusens et al. (1997)	BsmI	501 older women	Isometric quadriceps and handgrip strength
	Grundberg et al. (2004)	BsmI, poly A repeat	175 young women	Isokinetic knee flexor strength
	Roth et al. (2004)	FokI	302 older men	Isometric knee extensor strength
	Wang et al. (2006)	ApaI, BsmI, TaqI	109 young Chinese women	Multiple knee and elbow strength measures
	Windelinckx et al. (2007)	BsmI, TaqI, and FokI	493 older men and women	Multiple quadriceps strength measures
	Hopkinson et al. (2008)	FokI and BsmI	107 COPD patients; 104 control men and women	Isometric quadriceps strength

performance phenotypes (Jones et al. 2004) and several investigations have targeted skeletal muscle traits in particular. Folland and coworkers (2000) first reported no significant association between *ACE* genotype and quadriceps isometric strength in 33 young males, though differences in muscle strength response to strength training were observed. Woods et al. (2001) found that the rate of change in muscle force in response to hormone replacement therapy (HRT) was stronger in I/I compared to D/D genotype carriers in a study of 83 older postmenopausal women. Thomis and colleagues (1998b) found that the *ACE* I/D polymorphism was not significantly associated with elbow flexor strength in a study of 57 young male twins. Hopkinson et al. (2006) reported significantly higher knee extensor maximal strength in chronic obstructive pulmonary disease (COPD) patients carrying the D-allele compared to I/I patients, though the association was not observed in 101 age-matched healthy controls. Williams et al. (2005) examined quadriceps muscle strength in 81 young Caucasian men and reported that baseline isometric strength was significantly associated with *ACE* genotype, with I-allele homozygotes showing the lowest strength values. Moran and colleagues (2006) examined handgrip strength and vertical jump in 1,027 Greek adolescents and reported higher handgrip strength and vertical jump scores in females carrying the I/I genotype. No significant associations were observed in males. The authors performed haplotype analysis of the *ACE* gene region using three polymorphisms and determined that the I/D polymorphism explained the bulk of the explained genetic variance. Pescatello and co-workers (2006) studied the I/D genotype in relation to elbow flexor strength in 631 young men and women and reported no association with muscle strength in either arm. Wagner et al. (2006) examined leg press strength variables in 62 young men and women. They showed that no single muscle phenotype was consistently associated with *ACE* I/D genotype, but that combinations of traits including contraction velocity, isometric force, and optimum contraction velocity differed among the three genotype groups in both men and women with I/I genotype carriers exhibiting lower maximum and optimum contraction velocity compared to I/D and D/D carriers. McCauley and colleagues (2009) did not observe any associations between *ACE* I/D genotype and knee extensor isometric or isokinetic torques in 79 young males, though serum ACE activity was associated with *ACE* genotype as expected. Charbonneau et al. (2008) reported higher quadriceps muscle volume in D/D carriers in a study of 225 older men and women, but no genotype differences were observed for muscle strength (1RM). Finally, Yoshihara et al. (2009) recently reported that the I/D polymorphism was associated with physical function in 431 elderly Japanese subjects, with higher hand grip and 10 m maximum walking speed in D/D carriers. In summary, *ACE* genotype has been associated with muscle strength variables in a number of studies, but those associations are balanced by several studies showing no association or inconsistencies among findings. There is little evidence to suggest that *ACE* genotype is a strong contributor to inter-individual variation in skeletal muscle strength.

Alpha Actinin 3 (ACTN3) The *ACTN3* gene and its nonsense R577X polymorphism has generated considerable attention following a number of cross-sectional investigations in elite athletes that pointed to a considerable disadvantage for X/X carriers in sprint and power related activities (Yang et al. 2003; Niemi and Majamaa 2005; Roth et al. 2008). Several groups then moved to examine quantitative traits to determine the underlying phenotype impacted by the alpha actinin 3 protein deficiency resulting from the X/X genotype. Clarkson and colleagues (2005a) reported that X/X women had lower baseline isometric strength than the R/R women in a study of 602 young men and women. No association was observed in men. Delmonico and coworkers (2007) examined knee extensor concentric peak power in 157 older men and women. Contrary to expectation, women X/X carriers exhibited greater relative peak power than both R/X and R/R genotypes. In men, no genotype differences were observed. Both men and women participated in a strength training program that indicated a stronger adaptation for R/R carriers compared to X/X carriers. Vincent and colleagues (2007) studied the R577X polymorphism in relation to isometric and isokinetic knee extensor strength in 90 young men and reported lower concentric peak torque at 300 deg/s in X/X compared to R/R homozygotes. The authors also reported a lower proportion of type IIx muscle fibers in X/X vs R/R homozygotes. In a study of 1,367 older adults (70–79 year), Delmonico et al. (2008) reported greater losses of 400 m walk time performance over 5 years in male X/X vs R-allele carriers, while X/X women had a 35% greater risk of lower extremity physical limitation compared to R/R women. Walsh et al. (2008) examined knee extensor shortening and lengthening peak torque values in 848 adults (22–90 year) and reported that X/X women displayed lower knee extensor strength values compared with R/X + R/R women. No genotype-related differences were observed in men. Women X/X homozygotes also displayed lower levels of FFM, as described in the next section. Some studies have not been able to confirm these genotype differences. For example, Norman and colleagues (2009) reported no significant associations with muscle power or torque-velocity relationships among *ACTN3* genotypes in a study of 120 moderately to well-trained men and women. They were also unable to confirm the difference in fiber type proportion reported by Vincent and colleagues (2007). Similarly, McCauley and colleagues (2009) did not observe any associations between *ACTN3* genotype and knee extensor isometric or isokinetic torques in 79 young males. The general consensus among these studies is that *ACTN3* X/X carriers may have modestly lower skeletal muscle strength and power in comparison to R-allele carriers, with the work of Delmonico and colleagues (2008) indicating potential clinical importance for the X/X genotype in older men and women.

Ciliary Neurotrophic Factor (CNTF) Three studies have examined genetic variation in the *CNTF* gene and/or its receptor, *CNTFR*. Roth and colleagues (2001) first reported that a null mutation (rs1800169; A/G: A = null allele) in the *CNTF* gene was associated with muscle strength and muscle quality in 494 men and women across the adult age span. Homozygotes of the rare null allele (A/A) had lower strength while heterozygotes had higher strength than G/G carriers across multiple muscle strength and muscle quality measurements. Arking et al. (2006) examined eight polymorphisms surrounding the *CNTF* locus, including the rare rs1800169 nonsense polymorphism in 363 older Caucasian women. Haplotype analysis revealed a significant association with handgrip strength that was completely explained by the rs1800169 A-allele, such that A/A individuals exhibited lower handgrip strength compared to G-allele carriers. In a follow-up study, Roth et al. (2008) examined multiple polymorphisms in the *CNTFR* gene in association with strength variables in 465 men and women (20–90 year). For the C174T polymorphism, T-allele carriers exhibited significantly higher quadriceps and hamstrings concentric and eccentric isokinetic strength at both 30 and 180 deg/s compared to C/C carriers, but these differences were not significant after adjustment for lower limb lean mass. No differences were observed for polymorphisms in the promoter region or elsewhere in the gene. De Mars and coworkers (2007) examined polymorphisms in both the *CNTF* and the *CNTFR* genes in 493 middle-aged and older men and women with measures of knee flexor and extensor strength. T-allele carriers of the C-1703T polymorphism in *CNTFR* exhibited higher strength levels for multiple measures compared to C/C homozygotes, including all knee flexor torque values. In middle-aged women, A-allele carriers at the T1069A locus in *CNTFR* exhibited lower concentric knee flexor isokinetic and isometric torque compared to T/T homozygotes. The *CNTF* null allele was not associated with any strength measures, nor were any *CNTF***CNTFR* interactions observed. These findings indicate the potential for significant influences of *CNTF* and *CNTFR* gene variants on skeletal muscle strength, though inconsistencies have been noted for *CNTFR*. The frequency of the rare A/A genotype in *CNTF* is so low that, despite some consistent findings of lower muscle strength, public health significance is uncertain, though clinical importance may be had for those particular individuals.

Estrogen Receptor (ESR1) The estrogen receptor alpha is expressed in skeletal muscle, indicating a potential sensitivity to estrogen signaling (Wiik et al. 2009). While several studies have examined genetic variation in the *ESR1* gene in relation to muscle strength measures, none have confirmed any association. Salmen et al. (2002) examined 331 early postmenopausal women during a 5-year hormone replacement therapy trial for associations with the *ESR1* gene. Neither baseline nor 5-year grip strength values were associated with *ESR1* genotype. Vandevyver and colleagues (1999) examined 313 postmenopausal Caucasian women with measures of grip and quadriceps strength and reported no associations with *ESR1* genotype. Grundberg et al. (2005) reported no association between a TA-repeat polymorphism in the *ESR1* gene and several muscle strength measures in 175 Swedish women (20–39 year). Ronkainen and co-workers (2008) examined *ESR1* genotype in 434 older women (63–76 year) and found no significant association with hand grip or knee extension strength or leg extension power.

Insulin-like Growth Factor 2 (IGF2) Two studies have examined the *IGF2* gene in relation to strength phenotypes. Sayer et al. (2002) performed grip strength analysis in 693 older men and women and examined association with the *IGF2* ApaI polymorphism. *IGF2* genotype was associated with grip strength in men but not women, with G/G genotype having lower strength compared to A/A genotype carriers. Interestingly, an independent but additive effect of birth weight on grip strength values was also noted in men. Schrager and colleagues (2004) examined the same ApaI polymorphism in relation to muscle strength and power phenotypes in 485 men and women. They reported significantly lower arm and leg isokinetic strength measures in A/A women compared to G/G women, differences that were not observed in men. *IGF2* is imprinted in mammals such that only the paternal allele is transcribed (Zemel et al. 1992), thus analyses in these studies focused on comparing homozygote groups rather than heterozygotes. The results of these studies stand in direct contrast to each other, and indicate that any influence of *IGF2* genotype on strength-related traits is going to be minor or the result of interaction with other yet-to-be identified factors.

Myostatin-Related Genes After myostatin's discovery in the late 1990s, it emerged as a potential target of gene association studies and multiple polymorphisms were identified in the human gene (*MSTN*) (Ferrell et al. 1999). Initial investigations reported associations with skeletal muscle strength, but the sample sizes were very small owing in part to low allele frequencies of the common polymorphisms. Seibert et al. (2001) reported lower strength in older African American women (70–79 year) with the R-allele compared to K/K genotype at the *MSTN* K153R polymorphism, but the sample size was quite low ($n=55$). Corsi et al. (2002) reported lower isometric muscle strength (averaged across eight muscle groups) in R-allele carriers of the K153R polymorphism in 450 older men and women. Though consistent with the findings of Seibert (2001), the sample size of R-allele carriers was only seven making the findings inconclusive. Because the common polymorphisms have rare allele frequencies, the clinical significance of *MSTN* genetic variation is unlikely. Two groups have recently examined genes within the myostatin signaling pathway, including the myostatin receptor (activin-type II receptor B; *ACVR2B*) and follistatin (*FST*), a myostatin inhibitor. Walsh et al. (2007) examined the genetic association of *ACVR2B* and *FST* with muscle strength in 593 men and women across the adult age span. In women but not men, *ACVR2B* haplotype was significantly associated with knee extensor concentric peak torque. *FST* haplotype was not associated with muscle strength. Kostek et al. (2005) reported significant associations with the *MSTN* gene in 23 African Americans for biceps isometric strength. The *FST* gene was also associated with baseline one-repetition maximum strength levels. Again, the sample sizes of the genotype groups with significant findings were small making the clinical relevance of these findings uncertain but generally not compelling.

Vitamin D Receptor (VDR) Vitamin D deficiency has been consistently associated with lower muscle strength (Ceglia 2008) and has been discussed as a potential mechanism of sarcopenia (Montero-Odasso and Duque 2005). In one of the first gene associations for skeletal muscle traits, Geusens et al. (1997) demonstrated a significant relationship between the *VDR* BsmI polymorphism and both isometric quadriceps and hand grip strength in 501 elderly, healthy women, with 23% higher quadriceps strength and 7% higher grip strength in the b/b compared to B/B genotype carriers. These findings were subsequently supported in a subgroup of these same women (Vandevyver et al. 1999). In contrast, Grundberg et al. (2005) examined two polymorphisms (poly A repeat and BsmI) within *VDR* in relation to muscle strength in 175 women aged 20–39 year.

They found greater hamstrings isokinetic muscle strength in women homozygous for the shorter poly A repeat (ss) compared to women homozygous for the long poly A repeat (LL). No associations were reported with quadriceps or grip strength. Similar findings were reported for the BsmI variant (b and B alleles) given the significant linkage disequilibrium between the s and B alleles. Thus, the B/B genotype group exhibited higher hamstrings strength in contrast to the Geusens et al. findings. Roth and colleagues (2008) reported significant associations with the *VDR* FokI polymorphism (f and F alleles) and knee extensor isometric strength in 302 older Caucasian men (f/f higher than F/F), but these associations were no longer significant once leg FFM was accounted for in the models, suggesting that the genotype-strength associations were explained by differences in muscle mass. Wang et al. (2006) examined the ApaI, BsmI, and TaqI *VDR* polymorphisms in 109 young Chinese women in relation to knee and elbow torque measures. At the ApaI locus, A/A women exhibited lower elbow flexor concentric peak torque and lower knee extensor eccentric peak torque compared to either A/a or a/a carriers. For the BsmI locus, the b/b carriers demonstrated lower knee flexor concentric peak torque than the B-allele carriers. No associations were observed for the TaqI locus. Windelinckx and colleagues (2007) examined the BsmI, TaqI, and FokI *VDR* polymorphisms in 493 middle-aged and older men and women for association with various muscle strength phenotypes, with BsmI and TaqI combined in a haplotype analysis. In women, the FokI polymorphism was associated with quadriceps isometric and concentric strength, with higher levels in f/f homozygotes compared to F-allele carriers. In men, the BsmI/TaqI haplotype was associated with quadriceps isometric strength with Bt/Bt homozygotes exhibiting greater strength than bT haplotype carriers. In a study involving 107 COPD patients and 104 healthy controls, Hopkinson et al. (2006) reported Fok1 F/F carriers had lower quadriceps isometric strength than f-allele carriers. The b-allele of the Bsm1 polymorphism was associated with greater strength compared to B-allele carriers in COPD patients but not in controls. In summary, VDR genetic variation has been associated with muscle strength variables in numerous studies, though inconsistencies have been noted. Studies having examined the BsmI locus are mixed with regard to their findings and future studies need to incorporate the haplotype of BsmI and TaqI rather than looking at either site independently. The *VDR* FokI site is considered functional (Arai et al. 1997; Jurutka et al. 2000) and two studies reported higher strength in f/f compared to F/F carriers, so this site should be investigated more thoroughly for possible clinical significance.

In summary, several genes have been associated with skeletal muscle strength phenotypes in multiple studies. While none of these genes can yet be tagged as conclusively contributing to inter-individual variation in strength phenotypes, their consistency across multiple studies is encouraging. These genes will require additional validation and clarification as to their specific roles in modifying strength-related traits, with the eventual goal to determine their clinical importance to sarcopenia.

4.2 Genetic Variation and Skeletal Muscle Mass

Table 3 summarizes the genes that have been studied in relation to skeletal muscle mass measurements, focusing on genes associated with baseline muscle mass values; genes related to muscle mass adaptation to exercise training are discussed in a later section. Genes that have been studied in only one paper or that have not been replicated in some way and are not discussed here include: *MTHFR* (Liu et al. 2008b); *CNTF* and *CNTFR* (Roth et al. 2000, 2008); *COL1A1* (Van Pottelbergh et al. 2001); *TNF* (Liu et al. 2008a); *IL15* and *IL15RA* (Pistilli et al. 2008); *COMT* (Ronkainen et al. 2008); *ESR1* (Ronkainen et al. 2008); *NR3C1* (Peeters et al. 2008); and *IGF2* (Schrager et al. 2004).

Angiotensin Converting Enzyme (ACE) The majority of papers examining the *ACE* I/D polymorphism have been focused on muscle strength rather than muscle mass phenotypes, though some studies have examined both. Most have shown no significant association (Thomis et al. 1998a; Pescatello et al. 2006), though Charbonneau et al. (2008) reported higher quadriceps muscle volume in D/D compared to I/I carriers in a study of 225 older men and women (50–85 year). Thus, it appears unlikely that *ACE* genotype contributes significantly to muscle mass phenotypes, which is similar to the conclusion for muscle strength traits.

Alpha Actinin 3 (ACTN3) As discussed above, several studies have examined the potential for the *ACTN3* R577X polymorphism to explain variability in muscle strength measures. Many of those same papers have also examined muscle mass variables, though the results are less consistent. Vincent and colleagues (2007) did not observe any genotype difference in FFM determined by bioelectrical impedance in their study of 90 young men. Norman et al. (2009) reported no significant genotype associations with FFM determined by skinfold measurements in 120 young men and women. Delmonico et al. (2008) reported no significant genotype associations with DXA-measured FFM in their study of 1,367 older adults (70–79 year). Walsh et al. (2008) examined 848 adult men and women (22–90 year) and found that X/X women displayed lower levels of both total body FFM and lower limb FFM compared with R/X+R/R women. Concomitant differences were noted for muscle strength that were explained by the FFM differences, as discussed in the previous section. No genotype-related differences were observed in men. Thus, only Walsh et al. (2008) have found evidence of an association between muscle mass and the *ACTN3* null allele, indicating at best a minor role for this polymorphism in explaining inter-individual variability in this trait.

Androgen Receptor (AR) Walsh and colleagues (2005) examined the association between the *AR* CAG-repeat polymorphism with muscle strength and mass variables in two cohorts of older men and women. Though they found no association between muscle strength and *AR* genotype, significant genotype associations with FFM were observed in the men of both cohorts. The androgen receptor is a nuclear transcription factor, for which testosterone is an important ligand. The CAG-repeat sequence in exon 1 of the *AR* gene appears to modulate receptor transcriptional activity (Chamberlain et al. 1994). Subjects were grouped according to

Table 3 Genes and gene sequence variants associated with skeletal muscle mass phenotypes in multiple studies

Gene	References	Variants Examined	Subjects	Skeletal Muscle Mass Measurements
AR	Walsh et al. (2005)	CAG repeat	295 men (cohort 1) and 202 men and women (cohort 2)	FFM (DXA) in men in both cohorts
FST	Walsh et al. (2007)	Haplotype analysis	593 men and women	FFM (DXA) in men
	Kostek et al. (2009)	A-5003T	23 young African American	Biceps cross-sectional area
TRHR	Liu et al. (2009)	rs16892496, rs7832552	1,000 men women (cohort 1); 1,488 men and women (cohort 2); 2,955 Chinese men and women (cohort 3); 1,972 men and women from 593 families (cohort 4)	LBM (DXA) in all four cohorts
VDR	Van Pottelbergh et al. (2002)	TaqI, ApaI, FokI	271 older men	FFM (DXA)
	Roth et al. (2004)	FokI, BsmI	302 older men	FFM (DXA)

FFM, fat-free mass; LBM, lean body mass; DXA, dual-energy X-ray absorptiometry. Gene abbreviations are defined in the text.

the length of the CAG repeat, with subjects grouped for short and long fragments. Men in both cohorts with the long fragment lengths demonstrated significantly greater appendicular skeletal muscle mass and higher relative total lean mass. The results could not be explained by genotype-based differences in either bioavailable or total testosterone. Additional work is required to determine the extent to which the *AR* CAG-repeat polymorphism contributes to muscle mass variation, though these consistent findings in two cohorts is encouraging.

Myostatin-Related Genes Despite the strong physiological evidence behind myostatin as a candidate gene for muscle mass traits, genetic variation in the *MSTN* gene has not been associated with muscle mass (Ivey et al. 2000; Kostek et al. 2005). Kostek et al. (2009) did report strength differences for *MSTN* in a small number of African American subjects, as noted above. Two studies have examined myostatin-related genes in relation to muscle mass phenotypes. In 593 men and women across the adult age span, Walsh et al. (2007) reported significant associations between follistatin (*FST*) haplotype and leg FFM in men but not women, but no association with FFM and haplotype structure in the myostatin receptor, *ACVR2B*. Strength differences were discussed in the previous section. Kostek et al. (2005) also examined the *FST* gene and found that African Americans carriers of the *FST* T-allele had greater biceps CSA than A/A genotype carriers for the A-5003T polymorphism, but sample sizes were small. There is little compelling evidence that *MSTN* or myostatin-related genes are major contributors to skeletal muscle mass, though minor contributions are indicated.

Thyrotropin-Releasing Hormone Receptor (TRHR) As described above, Liu and colleagues (2008a) identified *TRHR* as a potential candidate gene for skeletal muscle mass from the first genome-wide association study for this trait. After the initial genome-wide analysis that identified two polymorphisms in the *TRHR* locus, the authors performed separate replication studies in three cohorts consisting of over 6,000 total white and Chinese subjects and consistent significant associations with LBM were observed in those analyses. Importantly, interactions between *TRHR* and genes in the growth hormone/insulin-like growth factor (GH/IGF1) pathway were explored and tentative connections were indicated. Though only a single paper, the multiple replications pointing to *TRHR* provide strength for this as a potentially important candidate gene for muscle mass variation.

Vitamin D Receptor (VDR) *VDR* genetic variation has been studied fairly extensively for muscle strength phenotypes, as described above, but fewer studies have focused on skeletal muscle mass. Van Pottelbergh and colleagues (2001) reported associations between the TaqI (T and t alleles)/ApaI (A and a alleles) haplotypes and lean mass in 271 older men (>70 year). The highest lean mass was observed in the At-At haplotype group, which differed most from haplotypes containing T-allele homozygosity (e.g., aT-aT, AT-aT, and AT-AT haplotypes). This relationship was not observed, however, in a group of younger men from the same study. Roth et al. (2008) reported significant associations with the *VDR* FokI polymorphism (f and F alleles) and leg FFM in 302 older Caucasian men, with concomitant differences in muscle strength as noted above. No significant differences were associated with the *VDR* BsmI site. This study is described in more detail in the section on genes specifically associated with sarcopenia. Thus, only two studies have examined *VDR* genotype in relation to skeletal muscle mass phenotypes, but the results provide some evidence for positive association.

In summary, remarkably few studies have provided evidence of genetic association of specific candidate genes with muscle mass phenotypes despite the strong heritability of the trait. The strongest findings are perhaps those with the least evidence, as *TRHR* and *AR* have at least been replicated, but only one research group has contributed to each of those studies. Presumably the advent of genome-wide association studies will provide a greater push for identifying potential candidate genes with relevance to skeletal muscle mass.

4.3 Genetic Variation and Sarcopenia

While a number of studies have addressed specific genes and genetic variants in relation to skeletal muscle strength and mass phenotypes, only one study to date has specifically targeted a measure of sarcopenia per se. Roth and colleagues (2004) analyzed the influence of the *VDR* BsmI and FokI variants on muscle strength and mass in a cohort of 302 older (58–93 year) Caucasian men with measures of FFM by DXA. *VDR* FokI genotype was significantly associated with total lean mass, appendicular lean mass, and normalized appendicular lean mass (all $P<0.05$), with the F/F group demonstrating significantly lower mass than the F/f and f/f groups. In addition, the group categorized the men as normal or sarcopenic based on the definition of Baumgartner et al. (1998), which relies on a cutoff value based on appendicular FFM relative to body weight (kg/m^2). Logistic regression revealed a significant 2-fold higher risk for sarcopenia in *VDR* Fok I F/F homozygotes than carriers of the f-allele (OR = 2.17; 95%CI = 1.19–3.85; $P=0.03$). Quadriceps muscle strength was also significantly lower in the F/F group compared to the F/f and f/f groups, but this association was eliminated when the analysis controlled for differences in total body lean mass. No significant differences were associated with the *VDR* BsmI site. Thus, *VDR* FokI genotype was significantly associated with lean mass and sarcopenia in this cohort of older Caucasian men, with concomitant differences in muscle strength. Vitamin D deficiency has been consistently associated with lower muscle strength (Ceglia 2008), and appears to be related to type II fiber atrophy (Pfeifer et al. 2002), thus making it an important potential mechanism in the etiology of sarcopenia in some individuals (Montero-Odasso and Duque 2005). The FokI polymorphism in the *VDR* gene affects the translational start site of the gene (Arai et al. 1997; Jurutka et al. 2000) thus making it a potentially functional polymorphism, though other variants in the *VDR* gene may interact in a more complex haplotype (Uitterlinden et al. 2004). Obviously, considerable work remains to be done to take the many genes outlined above and address the clinical relevance of sarcopenia in particular.

4.4 Genetic Variation and Skeletal Muscle Adaptation to Training

Though not an emphasis of this chapter, several studies have examined the role of genetic variation in the adaptation of skeletal muscle to exercise training, especially strength or resistance training. The adaptation of skeletal muscle to strength training is a heritable trait in itself (Thomis et al. 1998b) and linkage studies have been successfully performed using such traits as outcome variables, as described above (Chagnon et al. 2000; Sun et al. 1999). Moreover, specific genes have been studied and specific gene variants identified as being potentially important to skeletal muscle adaptation. The bulk of these studies have been described most recently in the updated Human Gene Map for Performance and Health-Related Fitness Phenotypes (Bray et al. 2009).

Genetic variation important to skeletal muscle adaptation has relevance for sarcopenia in multiple contexts. First, the identification of particular genes that contribute to inter-individual variation in skeletal muscle adaptation provide insights into the basic biology of skeletal muscle, which could be exploited in multiple ways to facilitate new or improved intervention techniques for muscle disorders and sarcopenia in particular. Second, the possibility exists that the same gene variants important to skeletal muscle adaptation could also be important to skeletal muscle development and thus baseline phenotypes, though the case can equally be made that different genetic contributions can be expected for these two different traits. Finally, because exercise training in general and strength training in particular are considered some of the most important interventions for the prevention and treatment of sarcopenia (Roth et al. 2000), understanding the genetic contributions to muscle adaptation, especially in older men and women, will allow improved application of such interventions via genetic screening.

A number of genes have been identified as potentially important for skeletal muscle adaptation, though arguably none have emerged as clinically meaningful as of this writing. Similar to the situation with baseline skeletal muscle phenotypes, the bulk of these genes remain unreplicated or replicated across different training stimuli or measurement methods, making traditional genetic replication analysis challenging. In fact, the variations on exercise training interventions are arguably more numerous than those related to measurement of skeletal muscle strength, and variations on both of these are often seen across different gene association studies related to muscle strength adaptation. Genes studied in relation to skeletal muscle adaptation include: *PPARD* with muscle volume response to lifestyle intervention (Thamer et al. 2008); *IGF1*, *IGFBP3*, and *PPP3R1* (calcineurin) with muscle strength and volume responses to strength training (Kostek et al. 2005, Hand et al. 2007); *RST* with upper arm muscle strength and muscle CSA responses to strength training (Pistilli et al. 2007); *TNF*, *TNFR1*, *TNFR2*, and *IL6* with measures of physical function before and after exercise training (Nicklas et al. 2005); *IGF2*, *ACTN3*, and *MYLK* in different studies with muscle damage in response to a damaging exercise protocol (Devaney et al. 2007; Clarkson et al. 2005b); *ACE* with muscle strength and mass responses to various exercise training protocols (Folland et al. 2000; Charbonneau et al. 2008; Thomis et al. 1998a; Williams et al. 2005; Pescatello et al. 2006; Frederiksen et al. 2003); *IL15* and *IL15RA* (IL-15 receptor) with muscle strength and size responses to strength training (Riechman et al. 2004, Pistilli et al. 2008); *MSTN* and *FST* polymorphisms with muscle strength and size traits in response to strength training (Thomis et al. 1998b; Kostek et al. 2009; Ivey et al. 2000); *ACTN3* with muscle strength and size responses to strength training (Clarkson et al. 2005a; Delmonico et al. 2007); and *BMP2* with muscle size response to strength training (Devaney et al. 2009).

5 Conclusions and Future Directions

Despite remarkably high heritability values, only modest progress has been made in identifying the specific genetic contributors to skeletal muscle strength and mass phenotypes. Only seven genes have been positively associated with strength-related traits in multiple cohorts (Table 2), and the findings are not always consistent within the replication analyses. Similarly, only four such genes have been identified for muscle mass and two of those genes were internally replicated rather than being confirmed in a second paper (Table 3). No genes have been replicated for association with sarcopenia per se, though *VDR* has been associated with sarcopenia in one study and associated with muscle mass and strength phenotypes in multiple studies.

Not only have few genes been identified, but their contribution to genetic variation is also generally quite small. None of the genes identified in the present chapter have been shown to conclusively contribute more than 5% of the inter-individual variation to their respective traits, and most are on the order of 1–3%. These results mirror what has recently been found for other highly heritable traits: genome-wide association studies are finding genes with relatively small influence that in no way explain the overall genetic influence predicted by heritability estimates (Maher 2008). This could reflect the major limitation of genome-wide association studies and most genetic association studies to date in that these have focused almost exclusively on single nucleotide polymorphisms, which though important are not the only DNA-related components that contribute to genetic influence. In addition to typical polymorphisms, copy number variation (CNV; multiple copies of the same gene), epistasis (multiple genes coordinated in a pathway), complex gene*environment interactions, and epigenetic factors are also contributing to the genetic component of inter-individual variability (Altshuler et al. 2008) and these more complex phenomena are just beginning to be studied in large-scale investigations.

An important contributor to inter-individual variation in age-related muscle traits will likely be epigenetic factors, which have already been shown to be important to aging tissues in general (Kahn and Fraga 2009). Epigenetics generally refers to chemical modifiers to DNA and histone proteins that alter DNA regulation without a direct change to the DNA sequence itself with consequences for normal development and disease risk (Hirst and Marra 2009). DNA methylation has been shown to decline with aging in several species including humans (Bollati et al. 2009) and DNA methylation has important consequences for gene expression. Importantly, modification of epigenetic factors appears to be related to environmental conditions (Foley et al. 2009; Baccarelli et al. 2009). So, both age and environment are likely to contribute to epigenetic changes in skeletal muscle tissue that will alter gene regulation and contribute to age-related losses in strength and mass, thus affecting physical function. How environmental conditions will alter epigenetic factors in a way meaningful for skeletal muscle traits and sarcopenia risk is as yet unclear, but certainly this represents another avenue of exploration for future studies.

An underlying theme when considering the genetic aspects of skeletal muscle traits generally and sarcopenia in particular is that of a "threshold" level for these traits below which physical function (e.g., activities of daily living) is impaired. Once a person's strength falls below a certain threshold, physical function becomes impaired. Such a threshold would surely be defined differently for each individual, but within reason we can expect clinically meaningful thresholds to be established across various physical characteristics, especially sex, age, height, weight, and body composition. This threshold concept has been discussed by a number of groups (Ferrucci et al. 1997; Walston and Fried 1999; Visser et al. 2005; McNeil et al. 2005).

Because genetic variation (including epigenetics) will tend to have subtle influences on skeletal muscle and sarcopenia-related traits, the general hypothesis is that genetic variation will tend to push trait values closer to or farther away from this threshold, thus altering an individual's risk for impaired physical function. Thus, identifying individuals with genetic susceptibility to lower levels of skeletal muscle strength or mass who are closer to their likely threshold for physical limitation will allow for early, targeted interventions to help prevent early losses. This is the concept behind personalized or genetic medicine. Early identification for individuals genetically susceptible to sarcopenia could result in a dramatic improvement in health care costs, by introducing interventions prior to the onset of associated infirmities. Of course, finding these genes and developing the individualized interventions will take many years if the last decade provides any clue to future progress. One potential approach to speed discovery will be to examine genes related to bone structure and mass, which may have a pleiotropic influence on skeletal muscle traits (Karasik and Kiel 2008). The development of more sophisticated genome-wide association studies that include copy number variants may also aid in this search. Even if genes of only minor effect are identified that don't lend themselves to genetic screening and personalized medicine, those genes will point to potential physiological pathways that can be manipulated through more typical means and lend insight into the underlying etiology of sarcopenia in different individuals (Khoury et al. 2007; Burke 2003).

References

Aloia, J. F., Vaswani, A., Feuerman, M., Mikhail, M., Ma, R. (2000). Differences in skeletal and muscle mass with aging in black and white women. *American Journal of Physiology. Endocrinology and Metabolism, 278*, E1153–E1157.

Altshuler, D., Daly, M. J., Lander, E. S. (2008). Genetic mapping in human disease. *Science, 322*, 881–888.

Aniansson, A., Zetterberg, C., Hadberg, M., Henriksson, K. G. (1984). Impaired muscle function with aging. A background factor in the incidence of fractures of the proximal end of the femur. *Clinical Orthopaedics and Related Research, 191*, 193–201.

Arai, H., Myiyamoto, K., Taketani, Y., Yamamoto, H., Iemori, Y., Morita, K., Tonai, T., Nishisho, T., Mori, S., Takeda, E. (1997). A vitamin D receptor gene polymorphism in the translation initiation codon: effect on protein activity and relation to bone mineral density in Japanese women. *Journal of Bone and Mineral Research, 12*, 915–921.

Arden, N. K. & Spector, T. D. (1997). Genetic influences on muscle strength, lean body mass, and bone mineral density: a twin study. *Journal of Bone and Mineral Research, 12*, 2076–2081.

Arking, D. E., Fallin, D. M., Fried, L. P., Li, T., Beamer, B. A., Xue, Q. L., Chakravarti, A., Walston, J. (2006). Variation in the ciliary neurotrophic factor gene and muscle strength in older Caucasian women. *Journal of the American Geriatrics Society, 54*, 823–826.

Baccarelli, A., Wright, R. O., Bollati, V., Tarantini, L., Litonjua, A. A., Suh, H. H., Zanobetti, A., Sparrow, D., Vokonas, P. S., Schwartz, J. (2009). Rapid DNA methylation changes after exposure to traffic particles. *American Journal of Respiratory and Critical Care Medicine, 179*, 572–578.

Bassey, E. J., Fiatarone, M. A., O'Neill, E. F., Kelly, M., Evans, W. J., Lipsitz, L. A. (1992). Leg extensor power and functional performance in very old men and women. *Clinical Science, 82*, 321–327.

Baumgartner, R. N., Koehler, K. M., Gallagher, D., Romero, L., Heymsfield, S. B., Ross, R. R., Garry, P. J., Lindeman, R. D. (1998). Epidemiology of sarcopenia among the elderly in New Mexico. *American Journal of Epidemiology, 147*, 755–763.

Biesiadecki, B. J., Brand, P. H., Metting, P. J., Koch, L. G., Britton, S. L. (1998). Phenotypic variation in strength among eleven inbred strains of rats. *Proceedings of the Society for Experimental Biology and Medicine, 219*, 126–131.

Bollati, V., Schwartz, J., Wright, R., Litonjua, A., Tarantini, L., Suh, H., Sparrow, D., Vokonas, P., Baccarelli, A. (2009). Decline in genomic DNA methylation through aging in a cohort of elderly subjects. *Mechanisms of Ageing and Development, 130*, 234–239.

Bouchard, C., Savard, R., Despres, J. P., Tremblay, A., LeBlanc, C. (1985). Body composition in adopted and biological siblings. *Human Biology, 57*, 61–75.

Bouchard, C., Malina, R. M., Perusse, L. (1997). *Genetics of fitness and physical performance.* Champaign, IL: Human Kinetics.

Bray, M. S., Hagberg, J. M., Perusse, L., Rankinen, T., Roth, S. M., Wolfarth, B., Bouchard, C. (2009). The human gene map for performance and health-related fitness phenotypes: the 2006-2007 update. *Medicine and Science in Sports and Exercise, 41*, 35–73.

Buchner, D. M. & deLateur, B. (1991). The importance of skeletal muscle strength to physical function in older adults. *Annals of Behavioral Medicine, 13*, 91–98.

Burke, W. (2003). Genomics as a probe for disease biology. *The New England Journal of Medicine, 349*, 969–974.

Carmelli, D. & Reed, T. (2000). Stability and change in genetic and environmental influences on hand-grip strength in older male twins. *Journal of Applied Physiology, 89*, 1879–1883.

Carmelli, D., Kelly-Hayes, M., Wolf, P. A., Swan, G. E., Jack, L. M., Reed, T., Guralnik, J. M. (2000). The contribution of genetic influences to measures of lower-extremity function in older male twins. *The Journals of Gerontology. Series A: Biological Sciences and Medical Sciences, 55A*, B49–B53.

Ceglia, L. (2008). Vitamin D and skeletal muscle tissue and function. *Molecular Aspects of Medicine, 29*, 407–414.

Chagnon, Y. C., Borecki, I., Perusse, L., Roy, S., Lacaille, M., Chagnon, M., Ho-Kim, M. A., Rice, T., Province, M. A., Rao, D. C., Bouchard, C. (2000). Genome-wide search for genes related to the fat-free body mass in the Quebec Family Study. *Metabolism, 49*, 203–207.

Chagnon, Y. C., Rice, T., Perusse, L., Borecki, I., Ho-Kim, M. A., Lacaille, M., Pare, C., Bouchard, L., Gagnon, J., Leon, A. S., Skinner, J. S., Wilmore, J. H., Rao, D. C., Bouchard, C. (2001). Genomic scan for genes affecting body composition before and after training in Caucasians from HERITAGE. *Journal of Applied Physiology, 90*, 1777–1787.

Chamberlain, N. L., Driver, E. D., Miesfeld, R. L. (1994). The length and location of CAG trinucleotide repeats in the androgen receptor N-terminal domain affect transactivation function. *Nucleic Acids Research, 22*, 3181–3186.

Charbonneau, D. E., Hanson, E. D., Ludlow, A. T., Delmonico, M. J., Hurley, B. F., Roth, S. M. (2008). ACE genotype and the muscle hypertrophic and strength responses to strength training. *Medicine and Science in Sports and Exercise, 40*, 677–683.

Christensen, K., Gaist, D., Vaupel, J. W., McGue, M. (2002). Genetic contribution to rate of change in functional abilities among Danish twins aged 75 years or more. *American Journal of Epidemiology, 155* 132–139.

Christensen, K., Frederiksen, H., Vaupel, J. W., McGue, M. (2003). Age trajectories of genetic variance in physical functioning: a longitudinal study of Danish twins aged 70 years and older. *Behavior Genetics, 33*, 125–136.

Clark, P. J. (1956). The heritability of certain anthropometric characters of ascertained from measurements of twins. *American Journal of Human Genetics, 8*, 49–54.

Clarkson, P. M., Devaney, J. M., Gordish-Dressman, H., Thompson, P. D., Hubal, M. J., Urso, M., Price, T. B., Angelopoulos, T. J., Gordon, P. M., Moyna, N. M., Pescatello, L. S., Visich, P. S., Zoeller, R. F., Seip, R. L., Hoffman, E. P. (2005a). ACTN3 genotype is associated with increases in muscle strength in response to resistance training in women. *Journal of Applied Physiology, 99*, 154–163.

Clarkson, P. M., Hoffman, E. P., Zambraski, E., Gordish-Dressman, H., Kearns, A., Hubal, M., Harmon, B., Devaney, J. M. (2005b). ACTN3 and MLCK genotype associations with exertional muscle damage. *Journal of Applied Physiology, 99*, 564–569.

Corsi, A.M., Ferrucci, L., Gozzini, A., Tanini, A., Brandi, M.L. (2002). Myostatin polymorphisms and age-related sarcopenia in the Italian population. Journal of the American Geriatrics Society, *50*, 1463.

De Mars, G., Windelinckx, A., Beunen, G., Delecluse, C., Lefevre, J., Thomis, M. A. (2007). Polymorphisms in the CNTF and CNTF receptor genes are associated with muscle strength in men and women. *Journal of Applied Physiology, 102*, 1824–1831.

De Mars, G., Windelinckx, A., Huygens, W., Peeters, M. W., Beunen, G. P., Aerssens, J., Vlietinck, R., Thomis, M. A. (2008a). Genome-wide linkage scan for contraction velocity characteristics of knee musculature in the Leuven Genes for Muscular Strength Study. *Physiological Genomics, 35*, 36–44.

De Mars, G., Windelinckx, A., Huygens, W., Peeters, M. W., Beunen, G. P., Aerssens, J., Vlietinck, R., Thomis, M. A. (2008b). Genome-wide linkage scan for maximum and length-dependent knee muscle strength in young men: significant evidence for linkage at chromosome 14q24.3. *Journal of Medical Genetics, 45*, 275–283.

Delmonico, M. J., Kostek, M. A., Doldo, N. A., Hand, B. D., Walsh, S., Conway, J. M., Carignan, C., Roth, S. M., Hurley, B. F. (2007). The alpha-actinin-3 (ACTN3) R577X polymorphism influences knee extensor peak power response to strength training in older men and women. *The Journals of Gerontology. Series A: Biological Sciences and Medical Sciences, 62*, 206–212.

Delmonico, M. J., Zmuda, J. M., Taylor, B. C., Cauley, J. A., Harris, T. B., Manini, T. M., Schwartz, A., Li, R., Roth, S. M., Hurley, B. F., Bauer, D. C., Ferrell, R. E., Newman, A. B. (2008). Association of the ACTN3 genotype and physical functioning with age in older adults. *The Journals of Gerontology. Series A: Biological Sciences and Medical Sciences, 63*, 1227–1234.

Devaney, J. M., Hoffman, E. P., Gordish-Dressman, H., Kearns, A., Zambraski, E., Clarkson, P. M. (2007). IGF-II gene region polymorphisms related to exertional muscle damage. *Journal of Applied Physiology, 102*, 1815–1823.

Devaney, J. M., Tosi, L. L., Fritz, D. T., Gordish-Dressman, H. A., Jiang, S., Orkunoglu-Suer, F. E., Gordon, A. H., Harmon, B. T., Thompson, P. D., Clarkson, P. M., Angelopoulos, T. J., Gordon, P. M., Moyna, N. M., Pescatello, L. S., Visich, P. S., Zoeller, R. F., Brandoli, C., Hoffman, E. P., Rogers, M. B. (2009). Differences in fat and muscle mass associated with a functional human polymorphism in a post-transcriptional BMP2 gene regulatory element. *Journal of Cellular Biochemistry, 107*, 1073–1082.

Dunlap, D. D., Manheim, L. M., Sohn, M. W., Liu, X., Chang, R. W. (2002). Incidence of functional limitation in older adults: the impact of gender race, and chronic conditions. *Archives of Physical Medicine and Rehabilitation, 83*, 964–971.

Dutta, C. & Hadley, E. C. (1995). The significance of sarcopenia in old age. *The Journals of Gerontology. Series A: Biological Sciences and Medical Sciences, 50A*, 1–4.

Dutta, C., Hadley, E. C., Lexell, J. (1997). Sarcopenia and physical performance in old age: overview. *Muscle & Nerve, 5 (Supplementary)*, S5–S9.

Ferrell, R. E., Conte, V., Lawrence, E. C., Roth, S. M., Hagberg, J. M., Hurley, B. F. (1999). Frequent sequence variation in the human myostatin (GDF8) gene as a marker for analysis of muscle-related phenotypes. *Genomics, 62*, 203–207.

Ferrucci, L. (2008). The Baltimore Longitudinal Study of Aging (BLSA): a 50-year-long journey and plans for the future. *The Journals of Gerontology. Series A: Biological Sciences and Medical Sciences, 63*, 1416–1419.

Ferrucci, L., Guralnik, J. M., Buchner, D., Kasper, J., Lamb, S. E., Simonsick, E. M., Corti, M. C., Bandeen-Roche, K., Fried, L. P. (1997). Departures from linearity in the relationship between measures of muscular strength and physical performance of the lower extremities: the Women's Health and Aging Study. *The Journals of Gerontology. Series A: Biological Sciences and Medical Sciences, 52*, M275–M285.

Foldvari, M., Clark, M., Laviolette, L. C., Bernstein, M. A., Kaliton, D., Castaneda, C., Pu, C. T., Hausdorff, M. J., Fielding, R. A., Singh, M. A. F. (2000). Association of muscle power with functional status in community-dwelling elderly women. *The Journals of Gerontology. Series A: Biological Sciences and Medical Sciences, 55A*, M192–M199.

Foley, D. L., Craig, J. M., Morley, R., Olsson, C. J., Dwyer, T., Smith, K., Saffery, R. (2009). Prospects for epigenetic epidemiology. *American Journal of Epidemiology, 169*, 389–400.

Folland, J., Leach, B., Little, t, Hawker, K., Myerson, S., Montgomery, H., Jones, D. (2000). Angiotensin-converting enzyme genotype affects the response of human skeletal muscle to functional overload. *Experimental Physiology, 85*, 575–579.

Forbes, G. B., Sauer, E. P., Weitkamp, L. R. (1995). Lean body mass in twins. *Metabolism, 44*, 1442–1446.

Frederiksen, H., Gaist, D., Petersen, H. C., Hjelmborg, J., McGue, M., Vaupel, J. W., Christensen, K. (2002). Hand grip strength: a phenotype suitable for identifying genetic variants affecting mid- and late-life physical functioning. *Genetic Epidemiology, 23*, 110–122.

Frederiksen, H., Bathum, L., Worm, C., Christensen, K., Puggaard, L. (2003). ACE genotype and physical training effects: a randomized study among elderly Danes. *Aging Clinical and Experimental Research, 15*, 284–291.

Frontera, W. R., Hughes, V. A., Lutz, K. J., Evans, W. J. (1991). A cross-sectional study of muscle strength and mass in 45- to 78-yr-old men and women. *Journal of Applied Physiology, 71*, 644–650.

Fujita, Y., Nakamura, Y., Hiraoka, J., Kobayashi, K., Sakata, K., Nagai, Y., Yanagawa, H (1995). Physical-strength tests and mortality among visitors to health-promotion centers in Japan. *Journal of Clinical Epidemiology, 48*,1349–1359.

Gallagher, D., Visser, M., De Meersman, R. E., Sepulveda, D., Baumgartner, R. N., Pierson, R. N., Harris, T., Heymsfield, S. B. (1997). Appendicular skeletal muscle mass: effects of age gender, and ethnicity. *Journal of Applied Physiology, 83*, 229–239.

Geusens, P., Vandevyver, C., VanHoof, J., Cassiman, J. J., Boonen, S., Rous, J. (1997). Quadriceps and grip strength are related to vitamin D receptor genotype in elderly nonobese women. *Journal of Bone and Mineral Research, 12*, 2082–2088.

Goodpaster, B. H., Park, S. W., Harris, T. B., Kritchevsky, S. B., Nevitt, M., Schwartz, A. V., Simonsick, E. M., Tylavsky, F. A., Visser, M., Newman, A. B. (2006). The loss of skeletal muscle strength, mass, and quality in older adults: the health, aging and body composition study. *The Journals of Gerontology. Series A: Biological Sciences and Medical Sciences, 61*, 1059–1064.

Graham, R. R., Hom, G., Ortmann, W., Behrens, T. W. (2009). Review of recent genome-wide association scans in lupus. *Journal of Internal Medicine, 265*, 680–688.

Grundberg, E., Brandstrom, H., Ribom, E. L., Ljunggren, O., Mallmin, H., Kindmark, A. (2004). Genetic variation in the human vitamin D receptor is associated with muscle strength fat mass and body weight in Swedish women. *European Journal of Endocrinology, 150*, 323–328.

Grundberg, E., Ribom, E. L., Brandstrom, H., Ljunggren, O., Mallmin, H., Kindmark, A. (2005). A TA-repeat polymorphism in the gene for the estrogen receptor alpha does not correlate with muscle strength or body composition in young adult Swedish women. *Maturitas, 50*, 153–160.

Gurland, B. J., Page, W. F., Plassman, B. L. (2004). A twin study of the genetic contribution to age-related functional impairment. *The Journals of Gerontology. Series A: Biological Sciences and Medical Sciences, 59A*, 859–863.

Hand, B. D., Kostek, M. C., Ferrell, R. E., Delmonico, M. J., Douglass, L. W., Roth, S. M., Hagberg, J. M., Hurley, B. F. (2007). Influence of promoter region variants of insulin-like growth factor pathway genes on the strength-training response of muscle phenotypes in older adults. *Journal of Applied Physiology, 103*, 1678–1687.

Harris, J. R., Pedersen, N. L., McClearn, G. E., Plomin, R., Nesselroade, J. R. (1992). Age differences in genetic and environmental influences for health from the Swedish Adoption/Twin Study of Aging. *Journal of Gerontology, 47*, P213–P220.

Hiona, A. & Leeuwenburgh, C. (2008). The role of mitochondrial DNA mutations in aging and sarcopenia: implications for the mitochondrial vicious cycle theory of aging. *Experimental Gerontology, 43*, 24–33.

Hirst, M. & Marra, M. A. (2009). Epigenetics and human disease. *The International Journal of Biochemistry & Cell Biology, 41*, 136–146.

Hopkinson, N. S., Nickol, A. H., Payne, J., Hawe, E., Man, W. D., Moxham, J., Polkey, M. I. (2004). Angiotensin converting enzyme genotype and strength in chronic obstructive pulmonary disease. *American Journal of Respiratory and Critical Care Medicine, 170*, 395–399.

Hopkinson, N. S., Eleftheriou, K. I., Payne, J., Nickol, A. H., Hawe, E., Moxham, J., Montgomery, H., Polkey, M. I. (2006). +9/+9 Homozygosity of the bradykinin receptor gene polymorphism is associated with reduced fat-free mass in chronic obstructive pulmonary disease. *The American Journal of Clinical Nutrition, 83*, 912–917.

Hopkinson, N. S., Li, K. W., Kehoe, A., Humphries, S. E., Roughton, M., Moxham, J., Montgomery, H., Polkey, M. I. (2008). Vitamin D receptor genotypes influence quadriceps strength in chronic obstructive pulmonary disease. *The American Journal of Clinical Nutrition, 87*, 385–390.

Huygens, W., Thomis, M. A., Peeters, M. W., Aerssens, J., Janssen, R., Vlietinck, R. F., Beunen, G. (2004a). Linkage of myostatin pathway genes with knee strength in men. *Physiological Genomics, 17*, 264–270.

Huygens, W., Thomis, M. A., Peeters, M. W., Aerssens, J., Janssen, R. G. J. H., Vlietinck, R. F., Beunen, G. (2004b). A quantitative trait locus on 13q14.2 for trunk strength. *Twin Research, 7*, 603–606.

Huygens, W., Thomis, M. A., Peeters, M. W., Vlietinck, R. F., Beunen, G. P. (2004c). Determinants and upper-limit heritabilities of skeletal muscle mass and strength. *Canadian Journal of Applied Physiology, 29*, 186–200.

Huygens, W., Thomis, M. A., Peeters, M. W., Aerssens, J., Vlietinck, R., Beunen, G. P. (2005). Quantitative trait loci for human muscle strength: linkage analysis of myostatin pathway genes. *Physiological Genomics, 22*, 390–397.

Hyatt, R. H., Whitelaw, M. N., Bhat, A., Scott, S., Maxwell, J. D. (1990). Association of muscle strength with functional status of elderly people. *Age and Ageing, 19*, 330–336.

Ivey, F. M., Roth, S. M., Ferrell, R. E., Tracy, B. L., Lemmer, J. T., Hurlbut, D. E., Martel, G. F., Siegel, E. L., Fozard, J. L., Jeffrey, M. E., Fleg, J. L., Hurley, B. F. (2000). Effects of age, gender, and myostatin genotype on the hypertrophic response to heavy resistance strength training. *The Journals of Gerontology. Series A: Biological Sciences and Medical Sciences, 55*, M641–M648.

Janssen, I., Heymsfield, S. B., Ross, R. (2002). Low relative skeletal muscle mass (sarcopenia) in older persons is associated with functional impairment and physical disability. *Journal of the American Geriatrics Society, 50*, 889–896.

Janssen, I., Shepard, D. S., Katzmarzyk, P. T., Roubenoff, R. (2004). The healthcare costs of sarcopenia in the United States. *Journal of the American Geriatrics Society, 52*, 80–85.

Jones, A., Montgomery, H. E., Woods, D. R. (2002). Human performance: a role for the ACE genotype? *Exercise and Sport Sciences Reviews, 30*, 184–190.

Jones, A., Shen, W., St Onge, M. P., Gallagher, D., Heshka, S., Wang, Z., Heymsfield, S. B. (2004). Body-composition differences between African American and white women: relation to resting energy requirements. *The American Journal of Clinical Nutrition, 79*, 780–786.

Jurutka, P. W., Remus, L. S., Whitfield, G. K., Thompson, P. D., SHsieh, J-c, Zitzer, H., Tavakkoli, P., Galligan, M. A., Dang, H. T. L., Haussler, C. A., Haussler, M. R. (2000). The polymorphic N terminus in human vitamin D receptor isoforms influences transcriptional activity by modulating interaction with transcription factor IIB. *Molecular Endocrinology, 14*, 401–420.

Kahn, A. & Fraga, M. F. (2009). Epigenetics and aging: status, challenges, and needs for the future. *The Journals of Gerontology. Series A: Biological Sciences and Medical Sciences, 64*, 195–198.

Karasik, D. & Kiel, D. P. (2008). Genetics of the musculoskeletal system: a pleiotropic approach. *Journal of Bone and Mineral Research, 23*, 788–802.

Karasik, D., Zhou, Y., Cupples, L. A., Hannan, M. T., Kiel, D. P., Demissie, S. (2009). Bivariate genome-wide linkage analysis of femoral bone traits and leg lean mass: Framingham study. *Journal of Bone and Mineral Research, 24*, 710–718.

Karlsson, J., Komi, P. V., Viitasalo, J. H. T. (1979). Muscle strength and muscle characteristics in monozygous and dizygous twins. *Acta Physiologica Scandinavica, 106*, 319–325.

Khoury, M. J., Davis, R., Gwinn, M., Lindegren, M. L., Yoon, P. (2005). Do we need genomic research for the prevention of common diseases with environmental causes? *American Journal of Epidemiology, 161*, 799–805.

Khoury, M. J., Little, J., Gwinn, M., Ioannidis, J. P. (2007). On the synthesis and interpretation of consistent but weak gene-disease associations in the era of genome-wide association studies. *International Journal of Epidemiology, 36*, 439–445.

Klein, R. J., Zeiss, C., Chew, E. Y., Tsai, J. Y., Sackler, R. S., Haynes, C., Henning, A. K., Sangiovanni, J. P., Mane, S. M., Mayne, S. T., Bracken, M. B., Ferris, F. L., Ott, J., Barnstable, C., Hoh, J. (2005). Complement factor H polymorphism in age-related macular degeneration. *Science, 308*, 385–389.

Komi, P. V., Viitasalo, J. H. T., Havu, M., Thorstensson, A., Sjodin, B., Karlsson, J. (1977). Skeletal muscle fibres and muscle enzyme activities in monozyogous and dizygous twins of both sexes. *Acta Physiologica Scandinavica, 100*, 385–392.

Kostek, M. C., Delmonico, M. J., Reichel, J. B., Roth, S. M., Douglass, L., Ferrell, R. E., Hurley, B. F. (2005). Muscle strength response to strength training is influenced by insulin-like growth factor 1 genotype in older adults. *Journal of Applied Physiology, 98*, 2147–2154.

Kostek, M. A., Angelopoulos, T. J., Clarkson, P. M., Gordon, P. M., Moyna, N. M., Visich, P. S., Zoeller, R. F., Price, T. B., Seip, R. L., Thompson, P. D., Devaney, J. M., Gordish-Dressman, H., Hoffman, E. P., Pescatello, L. S. (2009). Myostatin and follistatin polymorphisms interact with muscle phenotypes and ethnicity. *Medicine and Science in Sports and Exercise, 41*, 1063–1071.

Kovar, R. (1975). Letter: Motor performances in twins. *Acta Geneticae Medicae et Gemellologiae (Roma), 24*, 174.

Kwon, I. S., Oldaker, S., Schrager, M. A., Talbot, L. A., Fozard, J. L., Metter, E. J. (2001). Relationship between muscle strength and the time taken to complete a standardized walk-turn-walk test.

The Journals of Gerontology. Series A: Biological Sciences and Medical Sciences, 56A, B398–B404.

Larsson, L., Li, X., Teresi, A., Salviati, G. (1994). Effects of thyroid hormone on fast- and slow-twitch skeletal muscles in young and old rats. *Journal de Physiologie, 481*, 149–161.

Laukkanen, P., Heikkinen, E., Kauppinen, M. (1995). Muscle strength and mobility as predictors of survival in 75–84-year-old people. *Age and Ageing, 24*, 468–473.

Lauretani, F., Russo, C. R., Bandinelli, S., Bartali, B., Cavazzini, C., Di Iorio, A., Corsi, A. M., Rantanen, T., Guralnik, J. M., Ferrucci, L. (2003). Age-associated changes in skeletal muscles and their effect on mobility: an operational diagnosis of sarcopenia. *Journal of Applied Physiology, 95*, 1851–1860.

Lindgren CM, Heid IM, Randall JC, Lamina C, Steinthorsdottir V, Qi L, Speliotes EK, Thorleifsson G, Willer CJ, Herrera BM, Jackson AU, Lim N, Scheet P, Soranzo N, Amin N, Aulchenko YS, Chambers JC, Drong A, Luan J, Lyon HN, Rivadeneira F, Sanna S, Timpson NJ, Zillikens MC, Zhao JH, Almgren P, Bandinelli S, Bennett AJ, Bergman RN, Bonnycastle LL, Bumpstead SJ, Chanock SJ, Cherkas L, Chines P, Coin L, Cooper C, Crawford G, Doering A, Dominiczak A, Doney AS, Ebrahim S, Elliott P, Erdos MR, Estrada K, Ferrucci L, Fischer G, Forouhi NG, Gieger C, Grallert H, Groves CJ, Grundy S, Guiducci C, Hadley D, Hamsten A, Havulinna AS, Hofman A, Holle R, Holloway JW, Illig T, Isomaa B, Jacobs LC, Jameson K, Jousilahti P, Karpe F, Kuusisto J, Laitinen J, Lathrop GM, Lawlor DA, Mangino M, McArdle WL, Meitinger T, Morken MA, Morris AP, Munroe P, Narisu N, Nordström A, Nordström P, Oostra BA, Palmer CN, Payne F, Peden JF, Prokopenko I, Renström F, Ruokonen A, Salomaa V, Sandhu MS, Scott LJ, Scuteri A, Silander K, Song K, Yuan X, Stringham HM, Swift AJ, Tuomi T, Uda M, Vollenweider P, Waeber G, Wallace C, Walters GB, Weedon MN; Wellcome Trust Case Control Consortium, Witteman JC, Zhang C, Zhang W, Caulfield MJ, Collins FS, Davey Smith G, Day IN, Franks PW, Hattersley AT, Hu FB, Jarvelin MR, Kong A, Kooner JS, Laakso M, Lakatta E, Mooser V, Morris AD, Peltonen L, Samani NJ, Spector TD, Strachan DP, Tanaka T, Tuomilehto J, Uitterlinden AG, van Duijn CM, Wareham NJ, Hugh Watkins; Procardis Consortia, Waterworth DM, Boehnke M, Deloukas P, Groop L, Hunter DJ, Thorsteinsdottir U, Schlessinger D, Wichmann HE, Frayling TM, Abecasis GR, Hirschhorn JN, Loos RJ, Stefansson K, Mohlke KL, Barroso I, McCarthy MI; Giant Consortium. (2009). Genome-wide association scan meta-analysis identifies three Loci influencing adiposity and fat distribution. *PLoS Genetics, 5*, e1000508.

Lindle, R. S., Metter, E. J., Lynch, N. A., Fleg, J. L., Fozard, J. L., Tobin, J. D., Roy, T. A., Hurley, B. F. (1997). Age and gender comparisons of muscle strength in 654 women and men aged 20-93 yr. *Journal of Applied Physiology, 83*, 1581–1587.

Liu, D., Metter, E. J., Ferrucci, L., Roth, S. M. (2008a). TNF promoter polymorphisms associated with muscle phenotypes in humans. *Journal of Applied Physiology, 105*, 859–867.

Liu, X., Zhao, L. J., Liu, Y. J., Xiong, D. H., Recker, R. R., Deng, H. W. (2008b). The MTHFR gene polymorphism is associated with lean body mass but not fat body mass. *Human Genetics, 123*, 189–196.

Liu, X. G., Tan, L. J., Lei, S. F., Liu, Y. J., Shen, H., Wang, L., Yan, H., Guo, Y. F., Xiong, D. H., Chen, X. D., Pan, F., Yang, T. L., Zhang, Y. P., Guo, Y., Tang, N. L., Zhu, X. Z., Deng, H. Y., Levy, S., Recker, R. R., Papasian, C. J., Deng, H. W. (2009). Genome-wide association and replication studies identified TRHR as an important gene for lean body mass. *American Journal of Human Genetics, 84*, 418–423.

Livshits, G., Kato, B. S., Wilson, S. G., Spector, T. D. (2007). Linkage of genes to total lean body mass in normal women. *The Journal of Clinical Endocrinology and Metabolism, 92*, 3171–3176.

Loos, R., Thomis, M. A. I., Maes, H. H., Beunen, G. P., Claessens, A. L., Derom, C., Legius, E., Derom, R., Vlietinck, R. F. (1997). Gender-specific regional changes in genetic structure of muscularity in early adolescence. *Journal of Applied Physiology, 82*, 1802–1804.

Lynch, N. A., Metter, E. J., Lindle, R. S., Fozard, J. L., Tobin, J. D., Roy, T. A., Fleg, J. L., Hurley, B. F. (1999). Muscle quality. I. Age-associated differences between arm and leg muscle groups. *Journal of Applied Physiology, 86*, 188–194.

Maher, B. (2008). Personal genomes: the case of the missing heritability. *Nature, 456*, 18–21.
McCauley, T., Mastana, S. S., Hossack, J., Macdonald, M., Folland, J. P. (2009). Human angiotensin-converting enzyme I/D and alpha-actinin 3 R577X genotypes and muscle functional and contractile properties. *Experimental Physiology, 94*, 81–89.
McNeil, C. J., Doherty, T. J., Stashuk, D. W., Rice, C. L. (2005). Motor unit number estimates in the tibialis anterior muscle of young, old, and very old men. *Muscle & Nerve, 31*, 461–467.
Melton, L. J., Khosla, S., Crowson, C. S., O'Connor, M. K., O'Fallon, W. M., Riggs, B. L. (2000). Epidemiology of sarcopenia. *Journal of the American Geriatrics Society, 48*, 625–630.
Metter, E. J., Talbot, L. A., Schrager, M. A., Conwit, R. (2002). Skeletal muscle strength as a predictor of all-cause mortality in healthy men. *The Journals of Gerontology. Series A: Biological Sciences and Medical Sciences, 57*, B359–B365.
Montero-Odasso, M., Duque, G. (2005). Vitamin D in the aging musculoskeletal system: an authentic strength preserving hormone. *Molecular Aspects of Medicine, 26*, 203–219.
Montoye, H. J., Metzner, H. L., Keller, J. B. (1975). Familial aggregation of strength and heart rate response to exercise. *Human Biology, 47*, 17–36.
Moran, C. N., Vassilopoulos, C., Tsiokanos, A., Jamurtas, A. Z., Bailey, M. E., Montgomery, H. E., Wilson, R. H., Pitsiladis, Y. P. (2006). The associations of ACE polymorphisms with physical, physiological and skill parameters in adolescents. *European Journal of Human Genetics, 14*, 332–339.
Newman, A. B., Haggerty, C. L., Goodpaster, B. H., Harris, T., Kritchevsky, S., Nevitt, M., Miles, T. P., Visser, M. (2003a). Strength and muscle quality in a well-functioning cohort of older adults: the HealthAging and Body Composition Study. *Journal of the American Geriatrics Society, 51*, 323–330.
Newman, A. B., Kupelian, V., Visser, M., Simonsick, E., Goodpaster, B. H., Nevitt, M., Kritchevsky, S., Tylavsky, F., Rubin, S. M., Harris, T. B. (2003b). Sarcopenia: alternative definitions and associations with lower extremity function. *Journal of the American Geriatrics Society, 51*, 1602–1609.
Newman, A. B., Kupelian, V., Visser, M., Simonsick, E. M., Goodpaster, B. H., Kritchevsky, S. B., Kritchevsky, S., Tylavsky, F., Rubin, S. M., Harris, T. B. (2006). Strength, but not muscle mass, is associated with mortality in the health, aging and body composition study cohort. *The Journals of Gerontology. Series A: Biological Sciences and Medical Sciences, 61A*, 72–77.
Nguyen, T. V., Howard, G. M., Kelly, P. J., Eisman, J. A. (1998). Bone mass lean mass, and fat mass: same genes or environments? *American Journal of Epidemiology, 147*, 3–16.
Nicklas, B. J., Mychaleckyj, J., Kritchevsky, S., Palla, S., Lange, L. A., Lange, E. M., Messier, S. P., Bowden, D., Pahor, M. (2005). Physical function and its response to exercise: associations with cytokine gene variation in older adults with knee osteoarthritis. *The Journals of Gerontology. Series A: Biological Sciences and Medical Sciences, 60A*, 1292–1298.
Niemi, A. K. & Majamaa, K. (2005). Mitochondrial DNA and ACTN3 genotypes in Finnish elite endurance and sprint athletes. *European Journal of Human Genetics, 13*, 965–969.
Norenberg, K. M., Herb, R. A., Dodd, S. L., Powers, S. K. (1996). The effects of hypothyroidism on single fibers of the rat soleus muscle. *Canadian Journal of Physiology and Pharmacology, 74*, 362–367.
Norman, B., Esbjornsson, M., Rundqvist, H., Osterlund, T., von Walden, F., Tesch, P. A. (2009). Strength, power, fiber types, and mRNA expression in trained men and women with different ACTN3 R577X genotypes. *Journal of Applied Physiology, 106*, 959–965.
Nybo, H., Gaist, D., Jeune, B., McGue, M., Vaupel, J. W., Christensen, K. (2001). Functional status and self-rated health in 2 262 nonagenarians: the Danish 1905 Cohort Survey. *Journal of the American Geriatrics Society, 49*, 601–609.
Ostchega, Y., Harris, T. B., Hirsch, R., Parsons, V. L., Kington, R. (2000). The prevalence of functional limitations and disability in older persons in the US: data from the National Health and Nutrition Examination Survey III. *Journal of the American Geriatrics Society, 48*, 1132–1135.
Peeters, R. P., van den Beld, A. W., van Toor, H., Uitterlinden, A. G., Janssen, J. A. M. J. L., Lamberts, S. W. J., Visser, T. J. (2005). A polymorphism in type I deiodinase is associated with

circulating free insulin-like growth factor I levels and body composition in humans. *The Journal of Clinical Endocrinology and Metabolism, 90*, 256–263.

Peeters, G. M., van Schoor, N. M., van Rossum, E. F., Visser, M., Lips, P. (2008). The relationship between cortisol, muscle mass and muscle strength in older persons and the role of genetic variations in the glucocorticoid receptor. *Clinical Endocrinology (Oxf), 69*, 673–682.

Pendergast, D. R., Fisher, N. M., Calkins, E. (1993). Cardiovascular, neuromuscular, and metabolic alterations with age leading to frailty. *Journal of Gerontology, 48*(Spec), 61–67.

Perusse, L., LeBlanc, C., Tremblay, A., Allard, C., Theriault, G., Landry, F., Talbot, J., Bouchard, C. (1987a). Familial aggregation in physical fitness, coronary heart disease risk factors, and pulmonary function measurements. *Preventive Medicine, 16*, 607–615.

Perusse, L., Lortie, G., LeBlanc, C., Tremblay, A., Theriault, G., Bouchard, C. (1987b). Genetic and environmental sources of variation in physical fitness. *Annals of Human Biology, 14*, 425–434.

Pescatello, L. S., Kostek, M. A., Gordish-Dressman, H., Thompson, P. D., Seip, R. L., Price, T. B., Angelopoulos, T. J., Clarkson, P. M., Gordon, P. M., Moyna, N. M., Visich, P. S., Zoeller, R. F., Devaney, J. M., Hoffman, E. P. (2006). ACE ID genotype and the muscle strength and size response to unilateral resistance training. *Medicine and Science in Sports and Exercise, 38*, 1074–1081.

Pfeifer, M., Begerow, B., Minne, H. W. (2002). Vitamin D and muscle function. *Osteoporosis International, 13*, 187–194.

Pistilli, E. E., Gordish-Dressman, H., Seip, R. L., Devaney, J. M., Thompson, P. D., Price, T. B., Angelopoulos, T. J., Clarkson, P. M., Moyna, N. M., Pescatello, L. S., Visich, P. S., Zoeller. R. F., Hoffman, E. P., Gordon, P. M. (2007). Resistin polymorphisms are associated with muscle, bone, and fat phenotypes in white men and women. *Obesity (Silver Spring), 15*, 392–402.

Pistilli, E. E., Devaney, J. M., Gordish-Dressman, H., Bradbury, M. K., Seip, R. L., Thompson, P. D., Angelopoulos, T. J., Clarkson, P. M., Moyna, N. M., Pescatello, L. S., Visich, P. S., Zoeller, R. F., Gordon, P. M., Hoffman, E. P. (2008). Interleukin-15 and interleukin-15R alpha SNPs and associations with muscle, bone, and predictors of the metabolic syndrome. *Cytokine, 43*, 45–53.

Ploutz-Snyder, L. L., Manini, T., Ploutz-Snyder, R. J., Wolf, D. A. (2002). Functionally relevant thresholds of quadriceps femoris strength. *The Journals of Gerontology. Series A: Biological Sciences and Medical Sciences, 57*, B144–B152.

Prior, S. J., Roth, S. M., Wang, X., Kammerer, C., Miljkovic-Gacic, I., Bunker, C. H., Wheeler, V. W., Patrick, A. L., Zmuda, J. M. (2007). Genetic and environmental influences on skeletal muscle phenotypes as a function of age and sex in large, multigenerational families of African heritage. *Journal of Applied Physiology, 103*, 1121–1127.

Purser, J. L., Pieper, C. F., Poole, C., Morey, M. (2003). Trajectories of leg strength and gait speed among sedentary older adults: longitudinal pattern of dose response. *The Journals of Gerontology. Series A: Biological Sciences and Medical Sciences, 58A*, M1125–M1134.

Rantanen, T., Avela, J. (1997). Leg extension power and walking speed in very old people living independently. *The Journals of Gerontology. Series A: Biological Sciences and Medical Sciences, 52A*, M225–M231.

Rantanen, T., Guralnik, J. M., Izmirlian, G., Williamson, J. D., Simonsick, E. M., Ferrucci, L., Fried, L. P. (1998). Association of muscle strength with maximum walking speed in disabled older women. *American Journal of Physical Medicine & Rehabilitation, 77*, 299–305.

Rantanen, T., Harris, T., Leveille, S. G., Visser, M., Foley, D., Masaki, K., Guralnik, J. M. (2000). Muscle strength and body mass index as long-term predictors of mortality in initially healthy men. *The Journals of Gerontology. Series A: Biological Sciences and Medical Sciences, 55A*, M168–M173.

Rantanen, T., Volpato, S., Ferrucci, L., Heikkinen, E., Fried, L. P., Guralnik, J. M. (2003). Handgrip strength and cause-specific and total mortality in older disabled women: exploring the mechanism. *Journal of the American Geriatrics Society, 51*, 636–641.

Reed, T., Fabsitz, R. R., Selby, J. V., Carmelli, D. (1991). Genetic influences and grip strength norms in the NHLBI twin study males aged 59-69. *Annals of Human Biology, 18*, 425–432.

Riechman, S. E., Balasekaran, G., Roth, S. M., Ferrell, R. E. (2004). Association of interleukin-15 protein and interleukin-15 receptor genetic variation with resistance exercise training responses. *Journal of Applied Physiology, 97*, 2214–2219.

Ronkainen, P.H., Pollanen, E., Tormakangas, T., Tiainen, K., Koskenvuo, M., Kaprio, J., Rantanen, T., Sipila, S., Kovanen, V. (2008). Catechol-o-methyltransferase gene polymorphism is associated with skeletal muscle properties in older women alone and together with physical activity. PLoS One, *3*, e1819.

Ropponen, A., Levalahti, E., Videman, T., Kaprio, J., Battie, M. C. (2004). The role of genetics and environment in lifting force and isometric trunk extensor endurance. *Physical Therapy, 84*, 608–621.

Roth, S. M. (2007). *Genetics primer for exercise science and health*. Champaign, IL: Human Kinetics.

Roth, S. M., Ferrell, R. E., Hurley, B. F. (2000). Strength training for the prevention and treatment of sarcopenia. *The Journal of Nutrition, Health & Aging, 4*, 143–155.

Roth, S. M., Schrager, M. A., Ferrell, R. E., Riechman, S. E., Metter, E. J., Lynch, N. A., Lindle, R. S., Hurley, B. F. (2001). CNTF genotype is associated with muscular strength and quality in humans across the adult age span. *Journal of Applied Physiology, 90*, 1205–1210.

Roth, S. M., Metter, E. J., Lee, M. R., Hurley, B. F., Ferrell, R. E. (2003). C174T polymorphism in the CNTF receptor gene is associated with fat-free mass in men and women. *Journal of Applied Physiology, 95*, 1425–1430.

Roth, S. M., Zmuda, J. M., Cauley, J. A., Shea, P. R., Ferrell, R. E. (2004). Vitamin D receptor genotype is associated with fat-free mass and sarcopenia in elderly men. *The Journals of Gerontology. Series A: Biological Sciences and Medical Sciences, 59A*, 10–15.

Roth, S. M., Walsh, S., Liu, D., Metter, E. J., Ferrucci, L., Hurley, B. F. (2008). The ACTN3 R577X nonsense allele is under-represented in elite-level strength athletes. *European Journal of Human Genetics, 16*, 391–394.

Salmen, T., Heikkinen, A. M., Mahonen, A., Kroger, H., Komulainen, M., Saarikoski, S., Honkanen, R., Partanen, J., Maenpaa, P. H. (2002). Relation of estrogen receptor-alpha gene polymorphism and hormone replacement therapy to fall risk and muscle strength in early postmenopausal women. *Annali Medici, 34*, 64–72.

Sayer, A. A., Syddall, H. E., O'Dell, S. D., Chen, X. H., Briggs, P. J., Briggs, R., Day, I. N. M., Cooper, C. (2002). Polymorphism of the IGF2 gene birth weight and grip strength in adult men. *Age and Ageing, 31*, 468–470.

Schrager, M. A., Roth, S. M., Ferrell, R. E., Metter, E. J., Russek-Cohen, E., Lynch, N. A., Lindle, R. S., Hurley, B. F. (2004). Insulin-like growth factor-2 genotype fat-free mass, and muscle performance across the adult life span. *Journal of Applied Physiology, 97*, 2176–2183.

Schultz, A. B., Ashton-Miller, J. A., Alexander, N. B. (1997). What leads to age and gender differences in balance maintenance and recovery? *Muscle & Nerve, 20*(5 supplementary), S60–S64.

Seeman, E., Hopper, J. L., Young, N. R., Formica, C., Goss, P., Tsalamandris, C. (1996). Do genetic factors explain associations between muscle strength lean mass, and bone density? A twin study. *American Journal of Physiology, 270*, E320–E327.

Seibert, M. J., Xue, Q.-L., Fried, L. P., Walston, J. D. (2001). Polymorphic variation in the human myostatin (GDF-8) gene and association with strength measures in the Women's Health and Aging Study II cohort. *Journal of the American Geriatrics Society, 49*, 1093–1096.

Shock, N. W., Gruelich, R. C., Andres, R. A., Arenberg, D., Costa, P. T., Lakatta, E. G., Tobin, J. D. (1984). *Normal human aging. The Baltimore Longitudinal Study of Aging*. Washington, DC: U.S. Government Printing Office.

Simoneau, J. A. & Bouchard, C. (1995). Genetic determinism of fiber type proportion in human skeletal muscle. *The FASEB Journal, 9*, 1091–1095.

Soukup, T. & Jirmanova, I. (2000). Regulation of myosin expression in developing and regenerating extrafusal and intrafusal muscle fibers with special emphasis on the role of thyroid hormones. *Physiological Research, 49*, 617–633.

Sowers, M. R., Crutchfild, M., Richards, K., Wilkin, M. K., Furniss, A., Jannausch, M., Zhang, D., Gross, M. (2005). Sarcopenia is related to physical functioning and leg strength in middle-aged women. *The Journals of Gerontology. Series A: Biological Sciences and Medical Sciences, 60A*, 486–490.

Sun, G., Gagnon, J., Chagnon, Y. C., Perusse, L., Despres, J. P., Leon, A. S., Wilmore, J. H., Skinner, J. S., Borecki, I., Rao, D. C., Bouchard, C. (1999). Association and linkage between an insulin-like growth factor-1 gene polymorphism and fat free mass in the HERITAGE Family Study. *International Journal of Obesity, 23*, 929–935.

Susanne, C. (1977). Heritability of anthropological characters. *Human Biology, 49*, 573–580.

Tanko, L. B., Movsesyan, L., Mouritzen, U., Christiansen, C., Svendsen, O. L. (2002). Appendicular lean tissue mass and the prevalence of sarcopenia among healthy women. *Metabolism, 51*, 69–74.

Thamer, C., Machann, J., Stefan, N., Schafer, S. A., Machicao, F., Staiger, H., Laakso, M., Bottcher, M., Claussen, C., Schick, F., Fritsche, A., Haring, H. U. (2008). Variations in PPARD determine the change in body composition during lifestyle intervention: a whole-body magnetic resonance study. *The Journal of Clinical Endocrinology and Metabolism, 93*, 1497–1500.

Thomis, M. A. I., Van Leemputte, M., Maes, H. H., Blimkie, C. J., Claessens, A. L., Marchal, G., Willems, E., Vlietinck, R. F., Beunen, G. P. (1997). Multivariate genetic analysis of maximal isometric muscle force at different elbow angles. *Journal of Applied Physiology, 82*, 959–967.

Thomis, M. A. I., Beunen, G. P., Maes, H. H., Blimkie, C. J., Van Leemputte, M., Claessens, A. L., Marchal, G., Willems, E., Vlietinck, R. F. (1998a). Strength training: importance of genetic factors. *Medicine and Science in Sports and Exercise, 30*, 724–731.

Thomis, M. A. I., Beunen, G. P., Van Leemputte, M., Maes, H. H., Blimkie, C. J., Claessens, A. L., Marchal, G., Willems, E., Vlietinck, R. F. (1998b). Inheritance of static and dynamic arm strength and some of its determinants. *Acta Physiologica Scandinavica, 163*, 59–71.

Thomis, M. A., Huygens, W., Heuninckx, S., Chagnon, M., Maes, H. H. M., Claessens, A. L., Vlietinck, R., Bouchard, C., Beunen, G. P. (2004). Exploration of myostatin polymorphisms and the angiotensin-converting enzyme insertion/deletion genotype in responses of human muscle to strength training. *European Journal of Applied Physiology, 92*, 267–274.

Tiainen, K., Sipila, S., Alen, M., Heikkinen, E., Kaprio, J., Koskenvuo, M., Tolvanen, A., Pajala, S., Rantanen, T. (2004). Heritability of maximal isometric muscle strength in older female twins. *Journal of Applied Physiology, 96*, 173–180.

Tiainen, K., Sipila, S., Alen, M., Heikkinen, E., Kaprio, J., Koskenvuo, M., Tolvanen, A., Pajala, S., Rantanen, T. (2005). Shared genetic and environmental effects on strength and power in older female twins. *Medicine and Science in Sports and Exercise, 37*, 72–78.

Tiainen, K., Sipila, S., Kauppinen, M., Kaprio, J., Rantanen, T. (2009). Genetic and environmental effects on isometric muscle strength and leg extensor power followed up for three years among older female twins. *Journal of Applied Physiology, 106*, 1604–1610.

Uitterlinden, A. G., Fang, Y., van Meurs, J. B., Pols, H. A., Van Leeuwen, J. P. (2004). Genetics and biology of vitamin D receptor polymorphisms. *Gene, 338*, 143–156.

Van Pottelbergh, I., Goemaere, S., Nuytinck, L., De Paepe, A., Kaufman, J. M. (2001). Association of the type I collagen alpha1 sp1 polymorphism bone density and upper limb muscle strength in community-dwelling elderly men. *Osteoporosis International, 12*, 895–901.

Van Pottelbergh, I., Goemaere, S., De Bacquer, D., De Paepe, A., Kaufman, J. M. (2002). Vitamin D receptor gene allelic variants bone density, and bone turnover in community-dwelling men. *Bone, 31*, 631–637.

van Rossum, E. F., Voorhoeve, P. G., Te Velde, S. J., Koper, J. W., Delemarre-van de Waal, H. A., Kemper, H. C., Lamberts, S. W. (2004). The ER22/23EK polymorphism in the glucocorticoid receptor gene is associated with a beneficial body composition and muscle strength in young adults. *The Journal of Clinical Endocrinology and Metabolism, 89*, 4004–4009.

Vandevyver, C., VanHoof, J., Declerck, K., Stinissen, P., Vandervorst, C., Michiels, L., Cassiman, J. J., Boonen, S., Raus, J., Geusens, P. (1999). Lack of association between estrogen receptor genotypes and bone mineral density fracture history, or muscle strength in elderly women. *Journal of Bone and Mineral Research, 14*, 1576–1582.

Venerando, A., Milani-Comparetti, M. (1970). Twin studies in sport and physical performance. *Acta Geneticae Medicae et Gemellologiae (Roma), 19*, 80–82.

Vincent, B., De Bock, K., Ramaekers, M., Van den Eede, E., Van Leemputte, M., Hespel, P., Thomis, M. A. (2007). ACTN3 (R577X) genotype is associated with fiber type distribution. *Physiological Genomics, 32*, 58–63.

Visser, M., Deeg, D. J. H., Lips, P., Harris, T. B., Bouter, L. M. (2000a). Skeletal muscle mass and muscle strength in relation to lower-extremity performance in older men and women. *Journal of the American Geriatrics Society, 48*, 381–386.

Visser, M., Newman, A. B., Nevitt, M. C., Kritchevsky, S. B., Stamm, E. B., Goodpaster, B. H., Harris, T. B. (2000b). Reexamining the sarcopenia hypothesis Muscle mass versus muscle strength. Health, Aging, and Body Composition Study Research Group. *Annals of the New York Academy of Sciences, 904*, 456–461.

Visser, M., Goodpaster, B. H., Kritchevsky, S., Newman, A. B., Nevitt, M., Rubin, S. M., Simonsick, E., Harris, T. B. (2005). Muscle mass muscle strength, and muscle fat infiltration as predictors of incident mobility limitations in well-functioning older persons. *The Journals of Gerontology. Series A: Biological Sciences and Medical Sciences, 60A*, 324–333.

Wagner, H., Thaller, S., Dahse, R., Sust, M. (2006). Biomechanical muscle properties and angiotensin-converting enzyme gene polymorphism: a model-based study. *European Journal of Applied Physiology, 98*, 507–515.

Walsh, S., Zmuda, J. M., Cauley, J. A., Shea, P. R., Metter, E. J., Hurley, B. F., Ferrell, R. E., Roth, S. M. (2005). Androgen receptor CAG repeat polymorphism is associated with fat-free mass in men. *Journal of Applied Physiology, 98*, 132–137.

Walsh, S., Metter, E. J., Ferrucci, L., Roth, S. M. (2007). Activin-type II receptor B (ACVR2B) and follistatin haplotype associations with muscle mass and strength in humans. *Journal of Applied Physiology, 102*, 2142–2148.

Walsh, S., Liu, D., Metter, E. J., Ferrucci, L., Roth, S. M. (2008). ACTN3 genotype is associated with muscle phenotypes in women across the adult age span. *Journal of Applied Physiology, 105*, 1486–1491.

Walston, J. & Fried, L. P. (1999). Frailty and the older man. *Medical Clinics of North America, 83*, 1173–1194.

Walston, J., Arking, D. E., Fallin, D., Li, T., Beamer, B., Xue, Q., Ferrucci, L., Fried, L. P., Chakravarti, A. (2005). IL-6 gene variation is not associated with increased serum levels of IL-6 muscle, weakness, or frailty in older women. *Experimental Gerontology, 40*, 344–352.

Wang, P., Ma, L. H., Wang, H. Y., Zhang, W., Tian, Q., Cao, D. N., Zheng, G. X., Sun, Y. L. (2006). Association between polymorphisms of vitamin D receptor gene ApaI, BsmI and TaqI and muscular strength in young Chinese women. *International Journal of Sports Medicine, 27*, 182–186.

Wiik, A., Ekman, M., Johansson, O., Jansson, E., Esbjornsson, M. (2009). Expression of both oestrogen receptor alpha and beta in human skeletal muscle tissue. *Histochemistry and Cell Biology, 131*, 181–189.

Williams, A. G., Day, S. H., Folland, J. P., Gohlke, P., Dhamrait, S., Montgomery, H. E. (2005). Circulating angiotensin converting enzyme activity is correlated with muscle strength. *Medicine and Science in Sports and Exercise, 37*, 944–948.

Windelinckx, A., De Mars, G., Beunen, G., Aerssens, J., Delecluse, C., Lefevre, J., Thomis, M. A. (2007). Polymorphisms in the vitamin D receptor gene are associated with muscle strength in men and women. *Osteoporosis International, 18*, 1235–1242.

Woods, D., Onambele, G., Woledge, R., Skelton, D., Bruce, S., Humphries, S. E., Montgomery, H. (2001). Angiotensin-I converting enzyme genotype-dependent benefit from hormone replacement therapy in isometric muscle strength and bone mineral density. *The Journal of Clinical Endocrinology and Metabolism, 86*, 2200–2204.

Yang, N., MacArthur, D. G., Gulbin, J. P., Hahn, A. G., Beggs, A. H., Easteal, S., North, K. N. (2003). ACTN3 genotype is associated with human elite athletic performance. *American Journal of Human Genetics, 73*, 627–631.

Yoshihara, A., Tobina, T., Yamaga, T., Ayabe, M., Yoshitake, Y., Kimura, Y., Shimada, M., Nishimuta, M., Nakagawa, N., Ohashi, M., Hanada, N., Tanaka, H., Kiyonaga, A., Miyazaki, H. (2009), Physical function is weakly associated with angiotensin-converting enzyme gene I/D polymorphism in elderly Japanese subjects. *Gerontology, 55*, 387–392.

Zemel, S., Boartolomei, M.S., Tilghman, S.M. (1992). Physical linkage of two mammalian imprinted genes, H19 and insulin-like growth factor 2. Nature Genetics, *2*, 61–65.

Zhai, G., Stankovich, J., Ding, C., Scott, F., Cicuttini, F., Jones, G. (2004). The genetic contribution to muscle strength, knee pain, cartilage volume, bone size, and radiographic osteoarthritis: a sibpair study. *Arthritis and Rheumatism, 50*, 805–810.

Zhai, G., Ding, C., Stankovich, J., Cicuttini, F., Jones, G. (2005). The genetic contribution to longitudinal changes in knee structure and muscle strength: a sibpair study. *Arthritis and Rheumatism, 52*, 2830–2834.

Proteomic and Biochemical Profiling of Aged Skeletal Muscle

Kathleen O'Connell, Philip Doran, Joan Gannon, Pamela Donoghue, and Kay Ohlendieck

Abstract Muscle proteomics is concerned with the large-scale profiling of the protein complement from contractile tissues in order to enhance our biochemical knowledge of fundamental physiological processes, as well as the pathophysiological mechanisms that underlie neuromuscular disorders. Since the loss of skeletal muscle mass and strength is one of the most striking features of the senescent body, a large number of proteomic studies have recently attempted the global analysis of age-related fibre degeneration. Although the large size of the muscle proteome and its broad range of expression levels complicates a comprehensive cataloguing of the entire muscle protein complement, mass spectrometry-based proteomic studies have succeeded in the identification of many novel sarcopenia-specific markers. Changes in the expression of affected muscle proteins, as well as altered post-translational modifications, can now be used to establish a reliable biomarker signature of age-dependent fibre wasting. Muscle proteins that are changed during aging belong to the regulatory and contractile elements of the actomyosin apparatus, key bioenergetic pathways, the myofibrillar remodeling machinery and the cellular stress response. The proteomic profiling of crude muscle extracts and distinct subcellular fractions agrees with the notion that sarcopenia of old age is due to a multi-factorial pathology. Changes in muscle markers of the contractile apparatus and energy metabolism strongly indicate a fast-to-slow fibre transition process and a shift to more aerobic-oxidative metabolism during aging. In the long-term, newly established biomarkers of sarcopenia might be useful for the design of improved diagnostic procedures and the identification of new therapeutic targets.

Keywords Mass spectrometry • Muscle aging • Muscle proteome • Muscle proteomics • Sarcopenia

K. O'Connell, J. Gannon, and K. Ohlendieck (✉)
Department of Biology, National University of Ireland, Maynooth, Co. Kildare, Ireland
e-mail: kay.ohlendieck@nuim.ie

P. Doran
Department of Biological Chemistry, University of California, Los Angeles, CA, USA

P. Donoghue
Conway Institute, University College Dublin, Belfield, Ireland

G.S. Lynch (ed.), *Sarcopenia – Age-Related Muscle Wasting and Weakness*,
DOI 10.1007/978-90-481-9713-2_12,

1 Introduction

Since skeletal muscle fibres represent the most abundant type of tissue in mammalians, primary pathological changes in the neuromuscular system have profound secondary effects on overall body homeostasis and bioenergetic requirements. It is therefore not surprising that patients suffering from inherited muscular dystrophies and related muscle wasting disorders have also functional impairments in other organ systems (Emery and Muntoni 2003). However, loss in skeletal muscle mass and associated contractile weakness may also occur as a critical co-morbidity in human disease. Secondary muscular dysfunction is seen in common disorders such as diabetes mellitus (Phielix and Mensink 2008), the metabolic syndrome (Wells et al. 2008), congestive heart disease (Dalla Libera et al. 2008), cancer-associated cachexia (Melstrom et al. 2007), sepsis (Smith et al. 2008), renal failure (Adams and Vaziri 2006) and chronic obstructive pulmonary disease (Wuest and Degens 2007). Importantly, during the natural aging process, a gradual reduction in muscle mass and a progressive decline in contractile strength is seen in all humans to a varying degree (Thompson 2009). It is not well understood whether muscle degeneration during aging is primarily due to abnormalities in the contractile tissue itself or a secondary consequence of severely impaired innervation patterns (Carlson 2004). The results from a large number of cross-sectional and longitudinal studies do not agree on the exact extent of age-dependent muscle degeneration (Forbes and Reina 1970; Baumgartner et al. 1995; Lindle et al. 1997; Proctor et al. 1999; Melton et al. 2000; Janssen et al. 2002) and how individual muscles are differentially affected during aging (Frontera et al. 2008), but concur that human aging is clearly associated with a severely impaired structure and function of the cells comprising the musculoskeletal system (Vandervoort 2002).

Progressive muscular dysfunction may prevent elderly patients from living an independent life and may require outside help despite the lack of other medical ailments (Rolland et al. 2008; Thompson 2009). The vastly improved availability of high-quality nutrients, enhanced hygiene, superior medical care and hugely improved pharmacological interventions have achieved an unprecedented extension of human longevity over the last few decades. It is now imperative to acquire the scientific basis of evidence to aid the development of new therapeutic strategies for the promotion of healthy aging (Lynch et al. 2007). In this respect, it is crucial to elucidate the molecular and cellular mechanisms that render the aged neuromuscular system more susceptible to degeneration (Doherty 2003). High-throughput and large-scale approaches used in the emerging biomedical fields of genomics, proteomics and metabolomics suggest themselves as ideal tools for the identification of novel markers of sarcopenia (Doran et al. 2007a). Currently, both proper diagnostic criteria to fully describe the different stages of skeletal muscle aging and suitable treatment options to reverse sarcopenia are lacking. The establishment of a disease- and stage-specific biomarker signature of sarcopenia would therefore greatly aid in the development of better diagnostic tools and the identification of novel therapeutic targets to treat age-dependent fibre degeneration.

2 Skeletal Muscle Proteomics

In the post-genomic era, skeletal muscle proteomics attempts the global profiling of voluntary contractile tissues in order to identify and catalogue the entire fibre protein complement and determine alterations in the abundance, post-translational modifications and oligomeric status of muscle proteins in development, differentiation, disease and aging (Isfort 2002). This includes the proteomic profiling of motor units, distinct muscles, individual classes of muscle fibres and defined subcellular fractions such as mitochondria, the contractile apparatus or the sarcoplasmic reticulum. Muscle proteomics employs standardized biochemical methodology to efficiently separate, unequivocally identify and comprehensively characterise muscle-associated protein species. The techniques of choice are mass spectrometric peptide fingerprinting for routine high-throughput analyses, and peptide fragmentation analysis and chemical peptide sequencing for targeted proteomics (Aebersold and Mann 2003). The long-term goal of muscle proteomics is to decisively improve our biochemical knowledge of fundamental physiological processes related to the many cellular functions of contractile tissues, as well as the elucidation of the molecular mechanisms that underlie neuromuscular pathology.

2.1 Mass Spectrometry-Based Proteomics

In contrast to the traditional reductionist approach focusing on specific proteins, complexes or pathways, modern proteomics attempts to carry out large-scale high-throughput analyses of entire cellular protein complements (de Hoog and Mann 2004). The combination of highly accurate mass spectrometric methods and optimized electrophoretic and chromatographic separation technology has provided an unprecedented capability for the swift qualitative and quantitative analysis of large numbers of proteins (Ferguson and Smith 2003). Since mass spectrometric peptide fingerprinting or peptide fragmentation techniques are dependent on the existence of suitable protein- or DNA-based databanks for sequence comparisons, the information generated by the human genome project and related sequencing projects for other species form an integral part of any proteomic workflow. Modern proteomics can identify individual protein isoforms and determine potential changes in their concentration or post-translational modifications from extremely small amounts of biological material. Especially the introduction of differential fluorescent tagging approaches has improved the simultaneous analysis of several proteomes (Viswanathan et al. 2006). Muscle proteomics in particular is concerned with the global identification, cataloguing and comparative analysis of the protein complement present in distinct subcellular fibre fractions, differing muscle fibres and subtypes of muscles (Isfort 2002).

Optimized biochemical methods are used for the comprehensive and reproducible separation of the accessible muscle proteome or subproteomes. Subsequently the individual constituents of mixtures of peptides, proteins and supramolecular complexes are rapidly identified and characterized by a variety of mass spectrometric techniques (Domon and Aebersold 2006). In muscle biology, the majority of proteomic profiling exercises have been carried out with gel electrophoretic separation methods, as reviewed by Doran et al. (2007b). See the flow chart of Fig. 1 for an outline of a typical proteomic profiling exercise that employs fluorescent tagging technology. Unlu and co-workers (1997) first described this powerful comparative method and Tonge et al. (2001) have evaluated the capabilities of its 2D software analysis program. Fluorescent difference in-gel electrophoresis, usually abbreviated as DIGE analysis, represents a highly accurate quantitative technique that enables the separation of multiple proteomes on the same two-dimensional gel, thereby greatly reducing the introduction of potential artifacts due to gel-to-gel variations (Marouga et al. 2005). Although all gel-based separation techniques have their limitations, two-dimensional methods with isoelectric focusing in the first dimension and ionic detergent-based slab gel electrophoresis in the second dimension are still the method of choice for most proteomic pilot studies (Gorg et al. 2004; Wittmann-Liebold et al. 2006). Two-dimensional gel electrophoresis underestimates the number of integral membrane proteins present in a crude tissue extract and does not properly separate or account for protein species with extreme p*I*-values, very large molecular masses, low abundance and/or extensive post-translational modifications. It is important to keep these technical restrictions in mind when analysing skeletal muscle fibres. Recently, the application of detergent extraction procedures and the careful application of subcellular fractionation procedures has improved the scope of proteomic investigations and has included many integral components in subproteomic approaches (Sadowski et al. 2008; Zheng and Foster 2009). Thus, crucial proteins involved in the regulation of mitochondria, plasmalemma, endoplasmic reticulum, nucleus and cytosol are now routinely included in the subproteomic screening of normal and pathological tissue preparations (Tan et al. 2008).

The proteomic identification of proteins of interest is usually accomplished by standardized biochemical techniques, such as mass spectrometric peptide fingerprinting, peptide fragmentation analysis, chemical peptide sequencing, the comparison of the relative electrophoretic mobility using two-dimensional gel databanks, immunoblotting surveys employing monoclonal antibody libraries and large-scale microscopical screening. The core technique of most proteomic studies is represented by mass spectrometry whereby a variety of instruments are commonly employed for the identification and characterization of biomolecules. Mass spectrometers produce and separate ions according to their mass-to-charge ratio (m/z). The suitability of mass spectrometric instruments is defined by their resolving power, i.e. the analytical ability to differentiate between two ions of similar mass, and most importantly by their mass accuracy (Domon and Aebersold 2006). Electromagnetic fields are used to separate ions derived from biomolecules under vacuum conditions. Mass spectrometers consist of a sample introduction device, an ionization source, a mass analyzer, a detector and a digitizer. Hence, the core functions of these components are ion generation, ion separation, ion detection and the

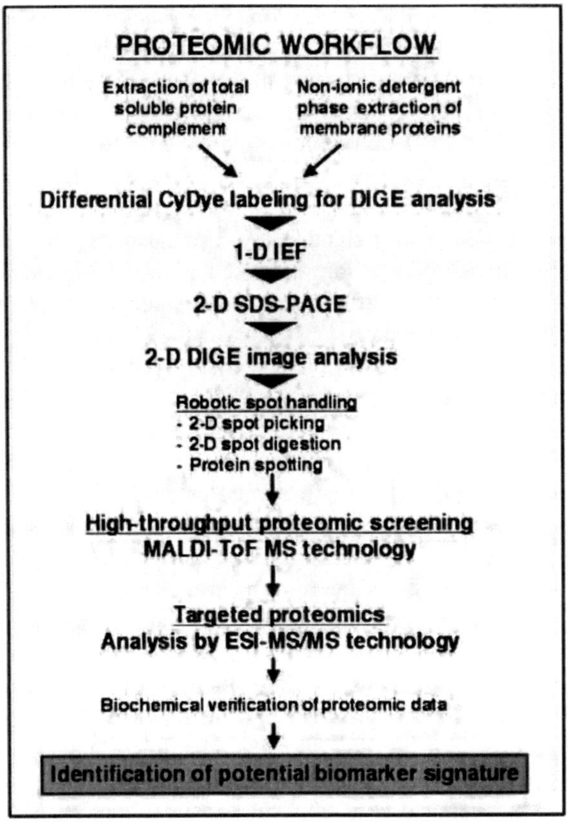

Fig. 1 Overview of proteomic difference in-gel electrophoretic analysis. Shown is the routine proteomic workflow employed for the standardized identification of novel protein biomarkers. The constituents of proteomes or subproteomes are fluorescently tagged and then separated by two-dimensional gel electrophoresis, using isoelectric focusing (IEF) in the first dimension and sodium dodecyl sulfate polyacrylamide gel electrophoresis (SDS-PAGE) in the second dimension. Following fluorescent difference in-gel electrophoresis (DIGE), proteins are identified by matrix-assisted laser desorption/ionization time-of-flight (MALDI-ToF) or electrospray ionisation (ESI) mass spectrometry (MS)

recording of a mass spectrum (Canas et al. 2006). The development of two key methods, matrix-assisted laser desorption/ionization (MALDI) and electrospray ionisation (ESI), has improved the large-scale analysis of complex protein mixtures to an unprecedented extent (Fenn et al. 1989; Zaluzec et al. 1995). These mass spectrometric techniques can therefore be considered the key facilitators of protein biochemistry that have actually enabled the establishment of modern proteomics.

Mass spectrometric peptide fingerprinting relies on the assumption that the controlled digestion of a protein results in the generation of a unique set of peptides that exhibit a highly reproducible combination of molecular masses (Webster and Oxley 2005). The comparison of the determined molecular masses of a sub-set of trypsin-generated peptides with theoretical in silico generated peptide masses leads to the identification of a specific protein species. A certain degree of proteolytic miscleavage has to be taking into account during the bioinfornatic analysis. In the case of muscle proteins, the exhaustive digestion with sequencing-grade trypsin usually produces a distinct peptide population ranging in molecular mass from approximately 500–2,500 kDa (Doran et al. 2007b). MALDI-based Time-of-Flight (ToF) mass spectrometry involves the irradiation of a co-precipitate, consisting of trypsin-generated peptides and a suitable UV-light absorbing matrix, by a nano-second laser pulse. Since different ions traverse a constant electric field according to their mass-to-charge ratio, a differential signal is generated for individual ions when they reach the detector, which transforms analogue signals into digital signals and records a mass spectrum. MALDI-ToF mass spectrometry is an extremely robust, rapid and cost-effective system for the high-throughput identification of unknown proteins (Webster and Oxley 2005). However, for targeted proteomics and the generation of large data sets of peptide sequences and the evaluation of post-translational modifications, ESI is the preferred method of choice. The ESI technique is based on the fact that high voltage triggers an electric spray in a liquid flowing through a narrow capillary. Charged small droplets are formed in a solution of peptides and suspended in a gaseous atmosphere. During an evaporation process, charged peptide analytes escape from micro-drops and are then analyzed by mass spectrometry (Fenn et al. 1989). See Table 1 for an example of the proteomic identification of typical muscle biomarkers. Shown are the primary sequences of peptides generated from mitochondrial ATP synthase and pyruvate dehydrogenase from aged skeletal muscle using ESI-MS/MS technology. The application of ESI- and MALDI-based methodology for studying complex mixtures of biomolecules has revolutionized biochemical research. With respect to muscle biology, the application of state-of-the-art genomic, proteomic and metabolomic approaches has at least partially overcome the problems associated with the traditional reductionist approach investigating individual genes or single proteins. In the future, it is hoped that high-throughput methodology will enable a detailed molecular understanding of biological problems at the systems level (Aggarwal and Lee 2003), including sarcopenia of old age.

2.2 *Proteomic Profiling of Skeletal Muscle*

A motor unit consists of a single α-motor neuron and all its innervated contractile fibres (Chan et al. 2001). The hierarchy of biological organization within a functional motor unit is represented in ascending order by the genome of the nerve and its corresponding muscle fibres, their transcriptomes, subproteomes and lastly the total neuromuscular proteome (Doran et al. 2007b). Although a recent study on muscle aging has attempted the simultaneous proteomic profiling of both rat sciatic nerve and gastrocnemius muscle (Capitanio et al. 2009), most proteomic studies on skeletal muscle have focused on the fibre population without its neuronal elements

Table 1 Proteomic identification of mitochondrial markers in aged rat skeletal muscle using ESI-MS/MS technology

Name of protein	Peptide sequence	Accession no.	Isolectric point (p*I*)	Molecular mass (kDa)	Peptides matched	Mascot score	% coverage
Mitochondrial ATP Synthase D Chain	KAIGNALKS	ATP5H_RAT	6.2	18.7	13	598	86
	KIPVPEDKY						
	KYTALVDAEEKE						
	KSWNETFHTRL						
	KNCAQFVTGSQARV						
	KYNALKIPVPEDKY						
	KYTALVDAEEKEDVKN						
	RANVDKPGLVDDFKNKY						
	RKYPYWPHQPIENL						
	KTIDWVSFVEIMPQNQKA						
	RLASLSEKPPAIDWAYYRA						
	KNMIPFDQMTIDDLNEVFPETKL						
	KIKNMIPFDQMTIDDLNEVFPETKL						
Pyruvate dehydrogenase	KDIIFAIKK	Q6AY95_RAT	6.2	39.3	15	584	24
	KDFLIPIGKA						
	KDIIFAIKKT						
	KVVSPWNSEDAKG						
	RVTGADVPMPYAKI						
	KILEDNSIPQVKD						
	RVTGADVPMPYAKI						
	KEGIECEVINLRT						
	REAINQGMDEELERD						
	RIMEGPAFNFLDAPAVRV						
	KVFLLGEEVAQYDGAYKV						
	RTIRPMDIEAIEASVMKT						
	KTYYMSAGLQPVPIVFRG						
	REAINQGMDEELERDEKV						
	KSAIRDDNPVVMLENELMYGVAFELPTEAQSKD						

Name of protein	Peptide sequence	Accession no.	Isolectric point (p*I*)	Molecular mass (kDa)	Peptides matched	Mascot score	% coverage
H^+-transporting two-sector ATPase alpha chain	RLTELLKQ KLELAQYRE RVLSIGDGIARV KAVDSLVPIGRG RGYLDKLEPSKI KTSIAIDTIINQKR KGIRPAINVGLSVSRV REAYPGDVFYLHSRL RILGADTSVDLEETGRV KLKEIVTNFLAGFEP RTGAIVDVPVGDELLGRV KQGQYSPMAIEEQVAVIYAGVRG REVAAFAQFGSDLDAATQQLLSRG	A35730	7.2	58.9	13	564	31

(Piec et al. 2005; Gelfi et al. 2006a; O'Connell et al. 2007; Doran et al. 2008; Feng et al. 2008; Lombardi et al. 2009). It is, however, important to stress that motor neurons form an integral part of the physiological units that regulate and maintain excitation–contraction coupling and muscle relaxation. In contrast to the relatively stable skeletal muscle genome, the fibre proteome does not exist as a distinct cohort of biomolecules. For obvious biological reasons, any tissue-specific protein complement is constantly changing and adapting to altered physiological and pathological demands. This phenomena is even more pronounced in the case of the muscle proteome, since skeletal muscles belong to the class of excessively plastic and adaptable tissues (Pette 2001; Flueck and Hoppeler 2003). The heterogeneous character of individual muscles and the inescapable influence of neuromuscular activity on fibre distribution make the proteomic profiling of diseased or aged muscles more complex as compared to many other tissues. Besides biological considerations, another major hurdle for the comprehensive cataloging and differential analysis of muscle proteomes is the concentration range of proteins. It is currently difficult to accurately determine differences in skeletal muscle protein density.

However, proteomic studies have determined the dynamic range of plasma protein concentrations and predict that at least nine orders of magnitude separate one of the most abundant elements of this body fluid, albumin, and the rarest protein in this body fluid, interleukin-6 (Pieper et al. 2003). The concentration range of plasma proteins involved in immune defense, coagulation and metabolite transportation has been estimated from pg/ml-values at the low abundance end to mg/ml-values at the high abundance end (Anderson and Anderson 2002). A similar dynamic range in protein concentration probably also exists in contractile tissues. If one takes into account the fact that the human genome consists of approximate 30,000 genes which in turn produce several 100,000 individual proteins, it is safe to assume that the number of protein isoforms in the skeletal muscle proteome exceeds the number of muscle-specific genes. Therefore, for both technical and biological reasons, the current mass spectrometric recording of the electrophoretically or chromatographically separated muscle protein complement can only represent a partial documentation of the entire fibre proteome. Even the most sophisticated approaches for the simultaneous visualization of the soluble components derived from a specific proteome, such as fluorescence difference in-gel electrophoresis (Viswanathan et al. 2006), can only separate a few thousand proteins (Doran et al. 2006). Thus, even proteomic studies of tissues with a relatively low number of individual classes of proteins and a considerably narrower range of protein concentrations as observed in plasma, can only determine the near-to-total proteome.

Over the last few years, muscle proteomics has identified the most abundant components of contractile fibres from various species, including humans and the most important animal species used for biomedical research. Most studies have focused on the total soluble protein complement, but more discriminatory approaches covering low-abundance elements from distinct subcellular fractions and membrane-associated proteins are emerging. The cataloguing of total muscle proteomes has included tissues derived from mouse (Raddatz et al. 2008), rat (Yan et al. 2001), rabbit (Donoghue et al. 2007), chicken (Doherty et al. 2004), sheep (Hamelin et al. 2007), pig (Kim et al. 2004), cow (Bouley et al. 2005) and human (Gelfi et al. 2003). Subproteomic profiles have been reported for the cytosolic, microsomal, nuclear and mitochondrial fraction (Forner et al. 2006; Vitorino et al. 2007). Muscle protein expression levels were determined under developmental, physiological, pathological and aging conditions. Comparative studies have included the proteomic characterization of myoblast differentiation (Kislinger et al. 2005), muscle transformation (Donoghue et al. 2005), the effect of endurance exercise (Burniston 2008), muscular hypertrophy (Hamelin et al. 2006), disuse fibre atrophy (Isfort et al. 2000), adaptation to hypobaric hypoxia (Vigano et al. 2008), sepsis-related muscle damage (Duan et al. 2006), hypoxia-associated metabolic modulations (De Palma et al. 2007), neonatal muscle fibre necrosis of postural muscles (Le Bihan et al. 2006), denervation–reinnervation cycles (Sun et al. 2006), x-linked muscular dystrophy (Doran et al. 2006), dysferlionpathy (De Palma et al. 2006) and aging (Doran et al. 2008). Post mortem changes in the fibre proteome have been profiled for agriculturally important animal muscles, i.e. bovine and porcine meat (Lametsch and Bendixen 2001; Jia et al. 2006). Since post-translational modifications (PTM) play a crucial role in protein function and are responsible for much of the heterogeneity in muscle proteins, the establishment of proteomic maps based on common PTMs has been initiated. This includes the identification of critical glycosylation, phosphorylation, nitration and carbonylation sites and their role in health and disease (Kanski et al. 2005; Meany et al. 2007; Gannon et al. 2008; O'Connell et al. 2008a; Feng et al. 2008).

3 Proteomics of Muscle Aging

To better understand aging of the neuromuscular system, numerous proteomic studies have been carried out over the last few years. Major studies are listed in Table 2. The usual workflow of gel electrophoresis-based proteomic studies of muscle aging and the subsequent biochemical and cell biological characterization of novel mass spectrometry-identified protein markers is illustrated in Fig. 2. The general trend of altered protein expression patterns agrees with the findings from previous physiological, biochemical, cell biological and genomic studies (Piec et al. 2005; Gelfi et al. 2006a; Dencher et al. 2006, 2007; O'Connell et al. 2007; Doran et al. 2007c, 2008; Lombardi et al. 2009; Capitanio et al. 2009). However, certain results from transcriptomic analyses of muscle aging do not concur with proteomic investigations (Welle et al. 2001; Giresi et al. 2005; Dennis et al. 2008). Several genes that encode mitochondrial enzymes are down-regulated in aged fibres (Kayo et al. 2001), while the protein ratio between mitochondrial and glycolytic muscle proteins was shown to be increased (Piec et al. 2005; Gelfi et al. 2006a; Doran et al. 2008). Possibly, age-related alterations at the transcriptional and proteomic level do not correspond for all classes of proteins. Transcriptomic investigations have demonstrated that the age-related up-regulation of genes

Table 2 Proteomic profiling studies of sarcopenia of old age

Mass spectrometry-based proteomic analysis	Species	Skeletal muscle type or fraction	References
Profiling of total soluble proteome	Human	Vastus lateralis	Gelfi et al. 2006a
Profiling of total soluble proteome	Rat	Gastrocnemius	Piec et al. 2005 O'Connell et al. 2007 Doran et al. 2008 Lombardi et al. 2009
Profiling of motor unit	Rat	Sciatic nerve and gastrocnemius	Capitanio et al. 2009
Profiling of small heat shock proteins	Rat	Gastrocnemius	Doran et al. 2007c
PTM analysis of protein glycosylation	Rat	Gastrocnemius	O'Connell et al. 2007
PTM analysis of protein nitration	Rat	Gastrocnemius	Kanski et al. 2005
PTM analysis of protein carbonylation	Rat	Mitochondria	Feng et al. 2008
PTM analysis of protein phosphorylation	Rat	Gastrocnemius	Gannon et al. 2008
Subproteomic analysis	Rat	Mitochondria	Dencher et al. 2006, 2007

includes factors involved in stress response, apoptosis, inflammation, proteolysis and neuronal regulation (Roth et al. 2002). In contrast, aging is associated with a down-regulation of genes that encode muscle proteins engaged in fibre remodeling, the regulation of energy metabolism and muscle growth (Dennis et al. 2008). These results indicate that progressive muscle weakness in the elderly is a highly complex process. The large-scale proteomic profiling of aging muscles might throw new light on the multi-factorial etiology of sarcopenia and determine the pathobiochemical hierarchy in the many pathways that lead to contractile dysfunction.

Previous biomedical studies have established that the loss in skeletal muscle mass and function during aging is associated with a large variety of molecular and cellular abnormalities (Faulkner et al. 2007; Edstrom et al. 2007). This includes a shift to a slower-twitching fibre population (Prochniewicz et al. 2007), decreased protein synthesis of myofibrillar components (Balagopal et al. 1997), disturbed ion handling (Schoneich et al. 1999), a blunted stress response (Kayani et al. 2008), progressive denervation (Carlson 2004), decreased capillarisation (Degens 1998), excitation–contraction uncoupling (Delbono et al. 1995), oxidative stress (Squier and Bigelow 2000), mitochondrial dysfunction (Figueiredo et al. 2008), increased susceptibility to apoptosis (Dirks and Leeuwenburgh 2002), a metabolic disequilibrium (Vandervoort and Symons 2001), progressive decline in energy intake (Roberts 1995), a reduced regenerative potential (Renault et al. 2002) and inadequate levels of essential growth factors and hormones indispensable for the maintenance of the excitation–contraction–relaxation cycle (Lee et al. 2007). Proteomics promises to unearth what primary changes within this complex molecular pathogenesis cause detrimental down-stream alterations. The large-scale protein biochemical analysis of muscle aging may also elucidate what compensatory adaptation processes, repair mechanisms and stress responses are initiated to limit age-dependent fibre degeneration.

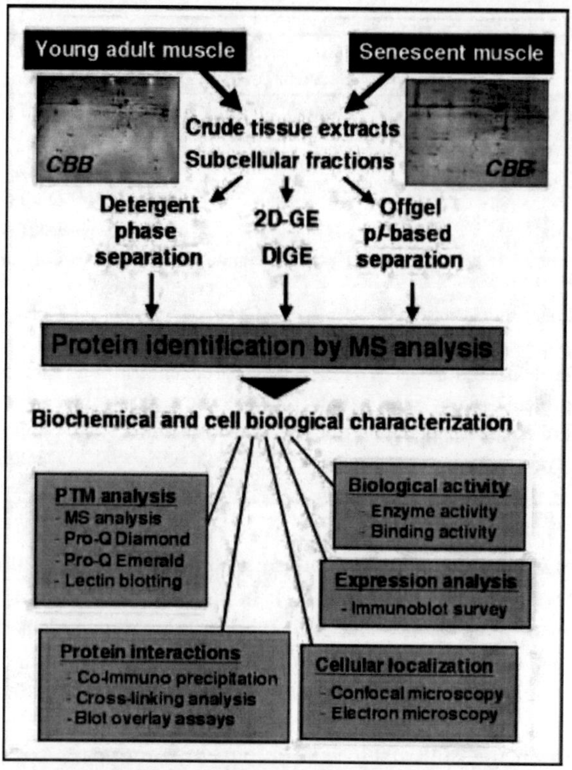

Fig. 2 Proteomic workflow for the identification and characterization of novel biomarkers of skeletal muscle aging. Shown is the gel-electrophoresis (GE)-based separation of young adult versus senescent muscle extracts. Two-dimensional gels were stained with colloidal Coomassie Blue (CCB) dye (O'Connell et al. 2007). Difference in-gel electrophoresis (DIGE) is routinely used for the fluorescent tagging and separation of muscle proteomes and mass spectrometric technology is usually employed to unequivocally identify proteins that exhibit a changed abundance during fibre aging. Potential alterations in the biological activity, oligomeric status, expression level, subcellular localization and/or post-translational modifications of newly identified skeletal muscle proteins are then determined by standard biochemical and cell biological methods

3.1 Remodeling of the Contractile Apparatus during Aging

Major physiological and cell biological differences exist between type-I, type-IIa and type-IIb fibres. Differences in motor neuron size, capillary density, myoglobin content, mitochondrial density and metabolite content closely relate to the biochemical composition of the contractile apparatus (Pette and Staron 1990; Punkt 2002; Spangenburg and Booth 2003). A major interest in muscle aging research is to understand what exact changes on the protein level cause a loss of contractile strength in both slow and fast muscles (Prochniewicz et al. 2007). The proteomic analysis of aged muscle has revealed a generally perturbed protein expression pattern in senescent muscle (Piec et al. 2005; O'Connell et al. 2007; Doran et al. 2008;

Lombardi et al. 2009; Capitanio et al. 2009), including many of the proteins belonging to the contractile apparatus that makes up approximately 50% of the total muscle protein complement. The supramolecular protein assemblies forming the thick and thin filaments of the basic contractile units exist in a great variety of fibre type-specific isoforms (Pette and Staron 1990). In the presence of ATP, a highly complex and cyclic coupling process between actin filaments and myosin head structures provides the molecular basis for the sliding of thin filaments past thick filaments causing distinct increments of sarcomere shortening (Gordon et al. 2000; Fitts 2008). The contractile status is controlled by the cytoplasmic Ca^{2+}-concentration whereby the troponin complex and tropomyosin strands directly regulate and enable actomyosin interactions for force generation (Swartz et al. 2006; Kreutziger et al. 2007). In skeletal muscles, a close relationship exists between isoform expression patterns of contractile proteins and metabolic fibre properties (Pette and Staron 2001). Thus, to understand the molecular mechanisms that underlie age-related fibre type shifting, a special interest focuses on potential alterations in the isoforms of myosin, actin, troponin or tropomyosin. Myosins consist of a hexameric structure consisting of 2 MHC heavy chains and various MLC light chains (Clark et al. 2002; Bozzo et al. 2005). A recent study by Capitanio et al. (2009) has shown a clear age-dependent transformation process within the pool of myosin heavy chain isoforms, i.e. a transition from fast MHC-IIb to MHC-IIa to slow MHC-I in 8-month versus 22-month old rat gastrocnemius muscle. This pattern of MHC changes is in line with observed adaptive processes in chronic electro-stimulated fast muscle (Pette 2001) and exercised muscles (Sullivan et al. 1995).

Fast-to-slow transformation is evidently associated with a shift to more oxidative metabolism and a concomitant change in the aged contractile apparatus to slower kinetics. The proteomic profiling of fast muscles following chronic low-frequency stimulation has shown that light and heavy chains of myosin undergo a stepwise replacement from fast to slow isoforms (Donoghue et al. 2005, 2007). Previous biochemical studies have shown similar effects of the neuromuscular activity on the expression of individual subunits of troponin (Pette and Staron 2001). In analogy, a comparable process appears to occur during muscle aging causing a drastic increase in the abundance of slow isoforms of key contractile elements in senescent fibres (Gelfi et al. 2006a; Doran et al. 2008; Capitanio et al. 2009). A comprehensive proteomic study of rat muscle aging, using the highly discriminatory fluorescent difference in-gel electrophoresis technique, has identified the slow myosin light chain isoform MLC-2 as one of the most drastically altered muscle proteins in this animal model of sarcopenia (Doran et al. 2008). Thus, both myosin light chains and heavy chains seem to shift towards slower isoforms. Application of the phospho-specific fluorescent dye ProQ-Diamond demonstrated that the abundance of the slow MLC-2 protein is not only drastically increased, but that its phosphorylation levels are even more enhanced in senescent gastrocnemius fibres (Gannon et al. 2008). This supports the idea of an age-related shift to a slower-twitching fibre population and suggests changed expression levels and altered post-translational modifications in myosin components as novel candidates for establishing a biomarker signature of muscle aging.

3.2 *Metabolic Adaptations in Aged Skeletal Muscle*

Findings from the proteomic analysis of bioenergetic adaptations in aged muscle agree with previous physiological and biochemical studies of fibre aging. The results from different proteomic studies of muscle aging have demonstrated that a general shift occurs in major metabolic pathways towards a more oxidative muscle metabolism (Doran et al. 2009a). However, species-specific differences appear to exist with respect to the degree of modifications in distinct rate-limiting enzymes and metabolite transporters, as well as in the complexity of these changes in particular pathways (Piec et al. 2005; Gelfi et al. 2006a; Doran et al. 2008). When studying the effects of physiological or pathological factors on contractile function, it is crucial to take into account the influence of patterns of innervation and activity on the metabolic and bioenergetic properties of skeletal muscles. In the case of diseased and aged muscles, it has clearly been documented that long-term inactivity inevitably results in disuse atrophy which results in a drastic reduction in tissue mass and contractile strength (Kandarian and Jackman 2006). Proteomic studies have to take into account the heterogeneity of skeletal muscles and build on the previous biochemical and physiological knowledge on fibre type characteristics and how they relate to specific marker proteins. Distinct protein expression signatures can be conveniently employed to differentiate between type I and type II fibres. The abundance and or isform expression pattern of many metabolic enzymes, excitation–contraction coupling elements, ion-handling proteins and contractile components can be used to determine fibre type distributions. The proteomic profiling of fast-twitching fibres agrees with a predominantly glycolytic metabolism, a high recruitment frequency, an easily fatigable phenotype and a high maximum power output (Okumura et al. 2005; Gelfi et al. 2006b). On the other hand, the protein complement of slower fibres is perfectly adapted to oxidative metabolism, a low recruitment frequency, resistance to fatigue and a low maximum power output (Okumura et al. 2005; Gelfi et al. 2006b). Especially striking is the difference in the density of myosin isoforms, glycolytic enzymes, citric acid cycle enzymes, oxidative phosphorylation elements, the oxygen carrier myoglobin and the fatty acid binding protein FABP. In addition, the abundance of Ca^{2+}-dependent binding proteins, pumps, channels and exchangers differs considerably between fast and slow muscles (Froemming et al. 2000). These established fibre type-specific markers could now be used for the interpretation of proteomic profiles generated by mass spectrometry-based muscle aging studies.

Major age-dependent alterations in the expression of catabolic enzymes and rate-limiting transporter molecules have been demonstrated by proteomics (Piec et al. 2005; Doran et al. 2008). As an example, Fig. 3 illustrates the age-related increase in the enzyme adenylate kinase. The expression of the soluble AK1 isoform was shown to be increased using both fluorescent difference in-gel electrophoresis and two-dimensional immunoblotting. Adenylate kinase, in conjunction with creatine kinase, maintaines a major nucleotide pathway in skeletal muscle. Increased levels of the AK1 isoform suggest adaptive processes that regulate

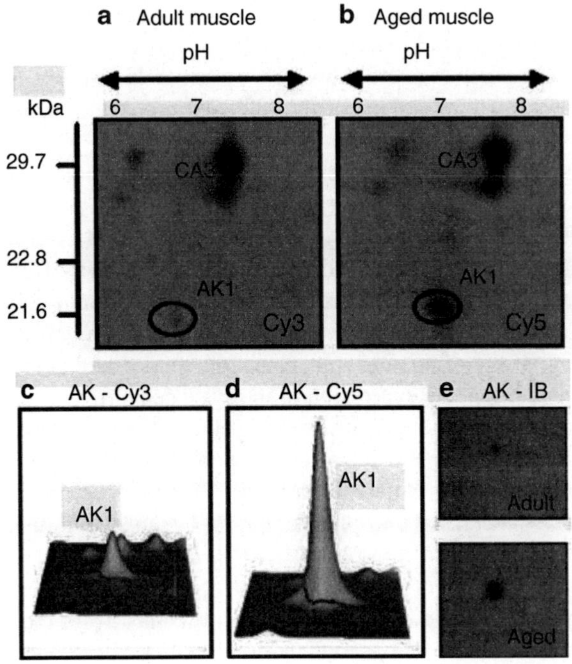

Fig. 3 Proteomic profiling of adenylate kinase isoform AK1 in senescent skeletal muscle. Shown is an expanded view of fluorescently tagged two-dimensional gels of the young adult muscle proteome versus the aged muscle proteome. Preparations from differently aged rat gastrocnemius muscles were labelled with the CyDyes Cy3 (**a**) and Cy5 (**b**). In panels (**c**) and (**d**) are shown the comparative graphic representation of the AK1 spot in young adult versus aged fibres, respectively. A major two-dimensional protein spot of approximately 30 kDa represents the abundant muscle enzyme carbonic anhydrase (CA3). The portion of the two-dimensional gel illustrated covers the range of approximately pH 7 to pH 8 in the first dimension and a molecular mass range of approximately 20–30 kDa in the second dimension. While the CA3 spot exhibits comparable levels between adult and aged muscle, the AK1 protein is clearly increased in aged muscle. The elevated expression level of adenylate kinase was confirmed by two-dimensional immunoblot (IB) analysis (**e**). Standard methods were employed for fluorescent difference in-gel electrophoresis and immunoblotting (Doran et al. 2006)

nucleotide ratios in aging fibres. Other skeletal muscle proteins that exhibit an age-related change in concentration are involved in the transportation of oxygen, the provision of fatty acids and the removal of carbon dioxide, as well as the maintenance of glycolysis, the citric acid cycle and oxidative phosphorylation (Piec et al. 2005; Gelfi et al. 2006a; Doran et al. 2008). Muscle aging is associated with a reduced glycolytic flux due to a drastic reduction in key glycolytic enzymes, such as pyruvate kinase, phosphofructokinase and enolase. The reduction of the key regulatory enzyme pyruvate kinase was shown by Deep Purple staining (O'Connell et al. 2007), a DIGE-based study (Doran et al. 2008) and PTM analysis (O'Connell et al. 2008a). Pyruvate kinase facilitates the final oxidoreduction-phosphorylation reaction during glycolysis that converts phosphoenolpyruvate to

ATP and pyruvate (Munoz and Ponce 2003). The decreased expression of the PK-M1 isoform of pyruvate kinase agrees with a shift to more aerobic-oxidative metabolism in senescent muscle. Although pyruvate kinase levels are reduced during aging, the remaining cohort of this glycolytic enzyme exhibits drastically increased levels of both N-glycosylation (O'Connell et al. 2008a) and tyrosine nitration (Kanski et al. 2005). Abnormal post-translational modifications in metabolic enzymes are believed to negatively affect the biological activity of glycolytic enzymes, which was shown to be true in the case of the PK-M1 isoform. Senescent muscle are characterized by a reduced pyruvate kinase activity (O'Connell et al. 2008a). Enhanced N-glycosylation probably influences protein stability, cellular targeting, inter- and intra-molecular interactions, and coupling efficiency between substrates and active site of this enzyme, causing a diminished glycolytic flux rate in aged fibres. In addition, the expression of pyruvate dehydrogenase, the metabolic linker between glycolysis and the citric acid cycle, is lower in aged fibres (Doran et al. 2008). Consequently, the transformation of pyruvate into acetyl-CoA is reduced in sarcopenia. The proteomic analysis of the phosphoprotein cohort of aged muscle showed increased phosphorylation for lactate dehydrogenase, albumin and aconitase, and decreased phosphorylation in cytochrome-c-oxidase, creatine kinase and enolase (Gannon et al. 2008). Hence, age-related changes in the muscle phosphoproteome are associated with metabolic enzymes from the cytosolic and mitochondrial compartment. This agrees with the idea that sarcopenia is a highly complex muscle disease that causes drastic alterations in the expression and molecular structure of important metabolic regulators.

The biochemical analysis of the fast-to-slow transformation process in chronic electro-stimulated fast muscles strongly suggests that the two most crucial limiting factors of oxidative metabolism are represented by the availability of oxygen and the rate of fatty acid transportation (Kaufmann et al. 1989). Since in senescent muscles an up-regulation of both the fatty acid transporter FABP and the oxygen-carrier myoglobin has been demonstrated by proteomic analysis (Doran et al. 2008), these alterations in biomarkers suggest that senescent fibres switch to a more aerobic-oxidative metabolism. In agreement with this major metabolic adaptation is the increased expression of citric acid cycle enzymes such as succinate dehydrogenase, isocitrate dehydrogenase and malate dehydrogenase in senescent muscles (Piec et al. 2005; Gelfi et al. 2006a; Doran et al. 2008). Recently Lombardi et al. (2009) have determined both the transcriptomic and proteomic profile of aged rat muscle employing a combination of DNA array and native blue PAGE technology. Aging seems to differentially affect the abundance, supramolecular organization and activity of the various mitochondrial complexes associated with the oxidative phosphorylation pathway. Although aging is generally associated with a shift to more oxidative muscle metabolism, senescent human muscles showed a more pronounced transition from predominantly glycolytic to mitochondrial energy generation (Gelfi et al. 2006a) as compared to small mammalians such as rats (Piec et al. 2005). These species-specific differences should be taken into account in animal model studies. The extrapolation of results from aging rat muscle to the human aging process should be undertaken with caution.

3.3 Cellular Stress Response in Aged Skeletal Muscle

The fast and efficient up-regulation of stress proteins is an essential cellular survival mechanism that prevents excess protein degradation and deleterious protein aggregation during tissue injury (Ellis and van der Vies 1991). In healthy adult muscle fibres, the natural response to stressful conditions involves a diverse array of molecular chaperones, mostly belonging to the very large family of heat shock proteins. During fibre adaptation or cellular regeneration phases, molecular chaperones stabilize denatured muscle proteins and facilitate the correct folding and conformational maturation in nascent peptides (McArdle and Jackson 2000). Muscle chaperones protect fibres during extensive contractile activity, traumatic injury, hyperthermia, hypoxic insult, ischemic damage and neuromuscular pathology (Nishimura and Sharp 2005). A common feature of chaperoning heat shock proteins is a promotor region that contains a consensus-binding sequence for HSF1 (Amin et al. 1988), the heat shock transcription factor that is associated with the response of cells following exposure to acute stressors (Anckar and Sistonen 2007). Heat shock proteins are classified according to their relative molecular mass. Besides the widely distributed Hsp60s, Hsp70s, Hsp90s and Hsp100s, some low-molecular-mass heat shock proteins are specifically induced during muscle injury (Golenhofen et al. 2004). These small members of the cytoprotective chaperone complement of skeletal muscles are characterized by a α-crystallin domain, a conserved 90-residue carboxy-terminal sequence (van Montfort et al. 2001). A major function of muscle-specific small heat shock proteins is the prevention of deleterious protein aggregation, and they are especially involved in the modulation of intermediate filament assembly (Nicholl and Quinlan 1994).

Heat shock proteins are relatively soluble and abundant, making them ideal candidates for proteomic investigations. Over the last few years, a large number of proteomic studies have identified cellular chaperones in muscle tissues. Most mass spectrometry-based analyses showed increased levels of heat shock proteins in the neuromuscular system following exposure to physiological or pathological stressors. This included the large-scale screening of fibre transformation following chronic electro-stimulation (Donoghue et al. 2005, 2007), moderate intensity endurance exercise (Burniston 2008), myoblast differentiation (Gonnet et al. 2008; Tannu et al. 2004), muscular hypertrophy (Hamelin et al. 2006), nerve crush-induced denervation (Sun et al. 2006), experimental muscular atrophy following hindlimb suspension (Seo et al. 2006), dystrophinopathy-associated necrosis (Doran et al. 2006), experimental exon-skipping therapy of muscular dystrophy (Doran et al. 2009b), dysferlin-related myopathy (De Palma et al. 2006), burn sepsis-induced stress (Duan et al. 2006), hypoxia-related stress (Bosworth et al. 2005) and post mortem changes in muscle fibres (Jia et al. 2006), as well as age-dependent muscle degeneration (Piec et al. 2005; O'Connell et al. 2007; Doran et al. 2007c; Feng et al. 2008; Lombardi et al. 2009; Capitanio et al. 2009). Interestingly, mass spectrometry-based proteomics of aged muscles has shown increased levels of distinct small chaperones, especially the cardiovascular heat shock protein cvHsp (Doran et al. 2007c).

The chaperone cvHsp appears to counter-act deleterious protein aggregation in the cytosol, sarcolemma and actomyosin apparatus of aged muscle (Doran et al. 2007c). In addition, increased concentrations of the ubiquitous small heat shock protein αB-crystallin were also detected by the proteomic profiling of senescent fibres (Doran et al. 2008). The family of small heat shock proteins quickly responds during stressful conditions and facilitates the disintegration of polydisperse assemblies into smaller subunits. This process is ATP-independent whereby small chaperone subunits bind to unfolding substrate and then reform into larger complexes (Stamler et al. 2005). The age-dependent activation of the cytoprotective protein complement of skeletal muscles seems to counter-act increased levels of denatured proteins in senescent fibres, especially abundant elements such as non-functional myosins, actins, troponins and tropomysoins (Vandervoort 2002; Prochniewicz et al. 2007). Increased chaperone levels represent an essential cellular rescue mechanism for eliminating the potentially destructive accumulation of inactive muscle protein aggregates. During aging, adaptive fibre transformation occurs in skeletal muscles. The fast-to-slow transition process encompasses major cellular remodeling. This includes the degeneration of the fastest-twitching fibre population, the activation of the satellite pool of muscle precursor cells and a certain degree of phenotypic fibre shifting within a contractile unit. However, since senescent muscles have a reduced regenerative capacity, adaptive fibre modulation probably triggers excessive detrimental protein aggregation as compared to healthy adult tissues. This in turn requires a massive cellular stress response to prevent contractile dysfunction. Therefore, in the context of a blunted stress response involving large heat shock proteins in aged muscle (Kayani et al. 2008), the drastic up-regulation of low-molecular-mass chaperones probably represents a compensatory mechanism that mostly supports filament remodeling (Doran et al. 2007c).

Continuous contractile activity clearly influences the expression of heat shock proteins (Neufer et al. 1998). Key chaperones containing the α-crystallin domain are up-regulated following chronic contraction patterns (Donoghue et al. 2007). In analogy to chronic neuromuscular activity, similar fibre transition processes occur in aged muscle. The concomitant damage to the actomyosin apparatus and associated cytoskeletal network may therefore trigger an increased synthesis of small heat shock proteins (Doran et al. 2009a). In contrast, cellular stress does not generate a sufficient response by larger heat shock proteins, such as those encoded by the Hsp70 gene (Liu et al. 2006). The up-regulation of Hsp70 and related chaperones is usually part of a highly coordinated stress response that prevents extensive muscular atrophy by limiting the stress-induced rate of cellular degeneration (Chung and Ng 2006). High levels of Hsp70 are essential for the stabilization of metabolic pathways, the prevention of high rates of apoptosis and the facilitation of physiological adaptation to changed functional demands. An age-related impairment of the Hsp70 response is believed to play a key role in contractile deficits (McArdle et al. 2004). It is therefore not surprising that skeletal muscles of aged transgenic mice with over-expressed levels of Hsp70 are partially protected against fibre degeneration (Broome et al. 2006).

This suggests that a well-designed pharmacological approach to enhance the natural stress response could potentially eliminate excessive fibre damage in aged muscle. In other areas of biomedicine, the drug-induced modulation of the cellular stress response has already gained considerable attention, as reviewed by Soti et al. (2005). Various inducers, co-inducers and inhibitors of specific heat shock proteins are currently evaluated as emerging therapeutic vehicles for the treatment of heart disease, diabetes, cancer and neurodegenerative disorders (Calderwood et al. 2006; Shamaei-Tousi et al. 2007). Since the up-regulation of small heat shock proteins, such as αB-crystallin or cvHsp, may represent an auto-protective mechanism in senescent muscle, a further increase in their expression levels may have therapeutic benefits. Hence, a pharmacologically mediated increase in essential muscle chaperones may be a realistic treatment option for eliminating certain neuromuscular impairments and could decisively improve the survival rate of stressed motor units in the senescent body.

3.4 Excitation–Contraction Uncoupling in Aged Muscle

Ca^{2+}-fluxes represent one of the most crucial second messenger system in contractile tissues (Berchtold et al. 2000). Alterations in Ca^{2+}-levels do not only affect protein activity and key physiological processes, but also gene expression patterns in skeletal muscle. Changes in the cytosolic Ca^{2+}-concentration play a key role in myogenesis, differentiation, fibre transformation, metabolic regulation, excitation–contraction coupling and muscle relaxation. Importantly, cyclic alterations in cytosolic Ca^{2+}-levels determine the contractile status of skeletal muscle fibres. The regulation of Ca^{2+}-homoeostasis and the mediation of the excitation–contraction–relaxation cycle depend on a finely tuned interplay between voltage-sensing receptors in the transverse tubules, Ca^{2+}-release channel units in the junctional sarcoplasmic reticulum, luminal and cytosolic Ca^{2+}-binding proteins, and Ca^{2+}-pumps of the sarcoplasmic reticulum, as well as minor structural components of the triad junction and sarcolemmal ion-regulatory elements such as ion exchangers and ion pumps (Murray et al. 1998). It is therefore not surprising that abnormal Ca^{2+}-handling is involved in a variety of muscle pathologies (MacLennan 2000; Froemming and Ohlendieck 2001), including sarcopenia of old age (Renganathan et al. 1997; O'Connell et al. 2008b).

The physical coupling between the voltage-sensing α_{1S}-subunit of the transverse-tubular dihydropyridine receptor and the ryanodine receptor Ca^{2+}-release channel of the junctional sarcoplasmic reticulum forms the central signal transduction unit during excitation–contraction coupling in mature skeletal muscles (MacLennan et al. 2002). The dihydropyridine receptor from skeletal muscle consists of a α_{1S}-α_2/δ-β-γ configuration. The α_{1S}-subunit represents the principal ion channel pore with three cytoplasmic loops between four repeat segments, whereby the II-III loop domain interacts directly with the junctional calcium release channel. During muscle aging, a drastically lowered supply of Ca^{2+}-ions to contractile proteins occurs due to uncoupling between the two main triad receptors (Renganathan et al. 1997). Excitation–contraction uncoupling appears to be due to a larger number of ryanodine receptors being uncoupled to the voltage-sensing dihydropyridine receptor units as compared to mature fibres. A pathophysiological disconnection between sarcolemmal excitation and muscle contraction may result in alterations in the voltage-gated Ca^{2+}-release mechanism, decreases in myoplasmic Ca^{2+}-elevation in response to surface depolarisation, reduced Ca^{2+}-supply to the actomyosin apparatus and reduced contractile strength. Thus, abnormal Ca^{2+}-handling may account for a significant proportion of the decay in skeletal muscle force during aging (Delbono et al. 1995). A recent immunoblotting and immunofluorescence survey has confirmed the excitation–contraction coupling hypothesis. The Ca^{2+}-binding protein named sarcalumenin, which represents a major mediator of ion shuttling within the longitudinal sarcoplasmic reticulum, was shown to be greatly reduced in aged rat gastrocnemius muscle as compared to young adult specimens (O'Connell et al. 2008b). In addition, key elements of the plasmalemma-associated Ca^{2+}-extrusion system, i.e. the calmodulin-dependent Ca^{2+}-ATPase and the Na^+-Ca^{2+}-exchanger, were also found to be diminished in aged muscle.

Figure 4 summarizes the findings of the immunoblotting survey of essential physiological regulators of Ca^{2+}-homeostasis and how their dysregulation may affect the excitation–contraction–relaxation cycle during aging. The overall protein band pattern of electrophoretically separated crude tissue extracts from 3-month versus 30-month old rat gastrocnemius muscle was very comparable between young adult versus senescent fibres. The previously reported senescence-related decrease in the α_{1S}-subunit of the dihydropryridine receptor, but not its auxiliary α_2-subunit, was confirmed. Immunoblotting of the sarcoplasmic reticulum proteins that mediate Ca^{2+}-buffering and Ca^{2+}-removal, i.e. fast and slow calsequestrins and the Ca^{2+}-pumping ATPase isoforms SERCA1 and SERCA2, suggested a shift to a slower phenotype, but these findings are not statistically significant. In contrast, the reduced expression of the 160 kDa Ca^{2+}-binding protein sarcalumenin and its related glycoprotein product of 53 kDa, as well as the Na^+-Ca^{2+}-exchanger and the PMCA-type Ca^{2+}-ATPase was shown to be significant in aged muscle. Thus, downstream from the coupling defect between the dihydropyridine receptor and the junctional Ca^{2+}-release channel, additional age-dependent changes appear to exist in Ca^{2+}-regulatory elements. Reduced levels of sarcalumenin and the two sarcolemmal Ca^{2+}-extrusion proteins may cause abnormal luminal Ca^{2+}-binding and impaired Ca^{2+}-removal (O'Connell et al. 2008b). This in turn could exacerbate disturbed ion fluxes and diminished triad signaling in senescent muscle and thereby contribute to contractile weakness.

4 Conclusion

Natural aging is a fundamental biological process. The functional decline of skeletal muscle fibres and the loss of total muscle mass are crucial factors that render the human body more susceptible to a metabolic disequilibrium and physical

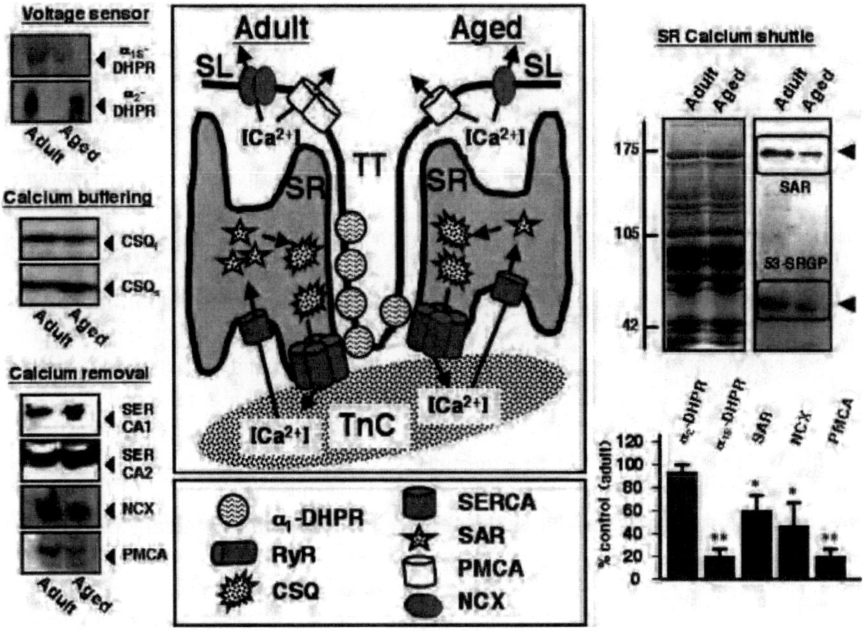

Fig. 4 Overview of the excitation–contraction uncoupling hypothesis of skeletal muscle aging and comparative immunoblot analysis of key Ca^{2+}-handling proteins in young adult versus senescent muscle. Shown is a Coomassie-stained gel and immunoblots of young adult versus aged rat gastrocnemius preparations. Immunoblots were labeled with antibodies to key proteins of the sarcolemma (SL), transverse tubules (TT) and sarcoplasmic reticulum (SR), including sarcalumenin (SAR) and its alternative splice product, the 53 kDa sarcoplasmic reticulum glycoprotein (53-SRGP), fast and slow calsequsetrin (fast CSQf; slow CSQs), fast and slow sarcoplasmic reticulum Ca^{2+}-ATPase (fast SERCA1; slow SERCA2), the Na^{+}-Ca^{2+}-exchanger (NCX), the plasmalemmal Ca^{2+}-ATPase (PMCA), and the α_{1S}- and α_2-subunit of the dihydropryridine receptor (DHPR). Molecular mass standards (in kDa) are indicated on the left of the Coomassie-stained gel panel. The comparative blotting was statistically evaluated using an unpaired Student's *t*-test (n = 6; $*p < 0.05$; $**p < 0.01$. Standard methods were employed for muscle preparations from crude tissue extracts, one-dimensional gel electrophoresis and immunoblot analysis of Ca^{2+}-handling proteins (O'Connell et al. 2007). The central panel outlines the dysregulation of Ca^{2+}-fluxes in senescent fibres and how this may affect the excitation–contraction–relaxation cycle during skeletal muscle aging. Besides the Ca^{2+}-handling proteins that have been analysed by immunoblotting, other key elements of ion homeostasis and muscle regulation are included in this diagram, i.e. the ryanodine receptor (RyR) Ca^{2+}-release channel of the sarcoplasmic reticulum and the troponin subunit TnC

weakness. Besides studying the histological and anatomical effects of muscle aging on frailty and fragility, it is also crucial to determine the molecular mechanisms that underlie age-dependent alterations at the cellular level. The application of modern proteomic methodology for analysing age-related impairments in contractile tissues promises to elucidate the pathobiochemical processes that lead to sarcopenia of old age. Mass spectrometry represents an unrivalled technique for the swift and reliable identification of protein factors involved in pathological pathways or compensatory mechanisms involved in aging. Over the last few years, mass spectrometry-based proteomics has identified a large number of relatively sarcopenia-specific biomarkers. Skeletal muscle proteins that exhibit altered expression levels or changed post-translational modifications during aging include regulatory proteins, contractile elements, metabolic enzymes and cellular stress proteins. The complexity of the observed changes in the senescent muscle proteome confirm the idea that sarcopenia is probably based on a multi-factorial etiology, rather than alterations in just one class of protein factors, regulatory mechanisms or aging-inducing gene clusters. Proteomic profiling studies have established distinct switches in fibre type-specific isoforms of contractile and metabolic proteins during aging, demonstrating an age-related transformation to slower-twitching muscles. The fast-to-slow transition process is accompanied by bioenergetic adaptation mechanisms. The comparative proteomic analysis of adult versus senescent muscles has clearly revealed a drastic shift to more aerobic-oxidative metabolism during aging. The proteomic identification of new sarcopenic biomarkers and their detailed cell biological, physiological and biochemical characterzation will hopefully lead to the prompt development of superior diagnostic tools and the improved design of pharmacological strategies to counter-act the age-induced loss of contractile tissue. Since alterations in the neuromuscular system are of central importance for comprehending the overall pathogenesis of the aging process in humans, the recent findings from proteomic studies will be crucial for improving our general biomedical knowledge on the mechanisms of aging.

Acknowledgements Research in the author's laboratory was supported by a principal investigator grant from Science Foundation Ireland (SFI-04/IN3/B614) and equipment grants from the Irish Health Research Board (HRB-EQ/2003/3, HRB-EQ/2004/2) and the Higher Education Authority (HEA-RERGS-07-NUIM). The authors thank Dr. Marina Lynch (Trinity College Dublin) for her generous help obtaining aged rat muscle, and Ms. Caroline Batchlor (NUI Maynooth) for assistance with mass spectrometry.

References

Adams, G. R. & Vaziri, N. D. (2006). Skeletal muscle dysfunction in chronic renal failure: effects of exercise. *The American Journal of Physiology, 290*, F753–F761.

Aebersold, R. & Mann, M. (2003). Mass spectrometry-based proteomics. *Nature, 422*, 198–207.

Aggarwal, K. & Lee, K. H. (2003). Functional genomics and proteomics as a foundation for systems biology. *Briefings in Functional Genomics & Proteomics, 2*, 175–184.

Amin, J., Ananthan, J., Voellmy, R. (1988). Key features of heat shock regulatory elements. *Molecular and Cellular Biology, 8*, 3761–3769.

Anckar, J. & Sistonen, L. (2007). Heat shock factor 1 as a coordinator of stress and developmental pathways. *Advances in Experimental Medicine and Biology, 594*, 78–88.

Anderson, N. L. & Anderson, N. G. (2002). The human plasma proteome: history, character, and diagnostic prospects. *Molecular & Cellular Proteomics, 1*, 845–867.

Balagopal, P., Rooyackers, O. E., Adey, D. B., Ades, P. A., Nair, K. S. (1997). Effects of aging on in vivo synthesis of skeletal muscle myosin heavy-chain and sarcoplasmic protein in humans. *The American Journal of Physiology, 273*, E790–E800.

Baumgartner, R. N., Stauber, P. M., McHugh, D., Koehler, K. M., Garry, P. J. (1995). Cross-sectional age differences in body composition in persons 60+ years of age. *Journal of Gerontology A Biological Sciences and Medical Sciences, 50*, M307–M316.

Berchtold, M. W., Brinkmeier, H., Muntener, M. (2000). Calcium ion in skeletal muscle: its crucial role for muscle function, plasticity, and disease. *Physiological Reviews, 80*, 1215–1265.

Bosworth, C. A., Chou, C. W., Cole, R. B., Rees, B. B. (2005). Protein expression patterns in zebrafish skeletal muscle: initial characterization and the effects of hypoxic exposure. *Proteomics, 5*, 1362–1371.

Bouley, J., Meunier, B., Chambon, C., De Smet, S., Hocquette, J. F., Picard, B. (2005). Proteomic analysis of bovine skeletal muscle hypertrophy. *Proteomics, 5*, 490–500.

Bozzo, C., Spolaore, B., Toniolo, L., Stevens, L., Bastide, B., Cieniewski-Bernard, C., Fontana, A., Mounier, Y., Reggiani, C. (2005). Nerve influence on myosin light chain phosphorylation in slow and fast skeletal muscles. *The FEBS Journal, 272*, 5771–5785.

Broome, C. S., Kayani, A. C., Palomero, J., Dillmann, W. H., Mestril, R., Jackson, M. J., McArdle, A. (2006). Effect of lifelong overexpression of HSP70 in skeletal muscle on age-related oxidative stress and adaptation after nondamaging contractile activity. *The FASEB Journal, 20*, 1549–1551.

Burniston, J. G. (2008). Changes in the rat skeletal muscle proteome induced by moderate-intensity endurance exercise. *Biochimica et Biophysica Acta, 1784*, 1077–1086.

Calderwood, S. K., Khaleque, M. A., Sawyer, D. B., Ciocca, D. R. (2006). Heat shock proteins in cancer: chaperones of tumorigenesis. *Trends in Biochemical Sciences, 31*, 164–172.

Canas, B., Lopez-Ferrer, D., Ramos-Fernandez, A., Camafeita, E., Calvo, E. (2006). Mass spectrometry technologies for proteomics. *Briefings in Functional Genomics & Proteomics, 4*, 295–320.

Capitanio, D., Vasso, M., Fania, C., Moriggi, M., Vigano, A., Procacci, P., Magnaghi, V., Gelfi, C. (2009). Comparative proteomic profile of rat sciatic nerve and gastrocnemius muscle tissues in ageing by 2-D DIGE. *Proteomics, 9*, 2004–2020.

Carlson, B. M. (2004). Denervation and the aging of skeletal muscle. *Basic and Applied Myology, 14*, 135–140.

Chan, K. M., Doherty, T. J., Brown, W. F. (2001). Contractile properties of human motor units in health, aging, and disease. *Muscle & Nerve, 24*, 1113–1133.

Chung, L. & Ng, Y. C. (2006). Age-related alterations in expression of apoptosis regulatory proteins and heat shock proteins in rat skeletal muscle. *Biochimica et Biophysica Acta, 1762*, 103–109.

Clark, K. A., McElhinny, A. S., Beckerle, M. C., Gregorio, C. C. (2002). Striated muscle cytoarchitecture: an intricate web of form and function. *Annual Review of Cell and Developmental Biology, 18*, 637–706.

Dalla Libera, L., Vescovo, G., Volterrani, M. (2008). Physiological basis for contractile dysfunction in heart failure. *Current Pharmaceutical Design, 14*, 2572–2581.

Degens, H. (1998). Age-related changes in the microcirculation of skeletal muscle. *Advances in Experimental Medicine and Biology, 454*, 343–348.

de Hoog, C. L. & Mann, M. (2004). Proteomics. *Annual Review of Genomics and Human Genetics, 5*, 267–293.

Delbono, O., O'Rourke, K. S., Ettinger, W. H. (1995). Excitation-calcium release uncoupling in aged single human skeletal muscle fibers. *The Journal of Membrane Biology, 148*, 211–222.

Dencher, N. A., Frenzel, M., Reifschneider, N. H., Sugawa, M., Krause, F. (2007). Proteome alterations in rat mitochondria caused by aging. *Annals of the New York Academy of Sciences, 1100*, 291–298.

Dencher, N. A., Goto, S., Reifschneider, N. H., Sugawa, M., Krause, F. (2006). Unraveling age-dependent variation of the mitochondrial proteome. *Annals of the New York Academy of Sciences, 1067*, 116–119.

Dennis, R. A., Przybyla, B., Gurley, C., Kortebein, P. M., Simpson, P., Sullivan, D. H., Peterson, C. A. (2008). Aging alters gene expression of growth and remodeling factors in human skeletal muscle both at rest and in response to acute resistance exercise. *Physiological Genomics, 32*, 393–400.

De Palma, S., Morandi, L., Mariani, E., Begum, S., Cerretelli, P., Wait, R., Gelfi, C. (2006). Proteomic investigation of the molecular pathophysiology of dysferlinopathy. *Proteomics, 6*, 379–385.

De Palma, S., Ripamonti, M., Vigano, A., Moriggi, M., Capitanio, D., Samaja, M., Milano, G., Cerretelli, P., Wait, R., Gelfi, C. (2007). Metabolic modulation induced by chronic hypoxia in rats using a comparative proteomic analysis of skeletal muscle tissue. *Journal of Proteome Research, 6*, 1974–1984.

Dirks, A. & Leeuwenburgh, C. (2002). Apoptosis in skeletal muscle with aging. *The American Journal of Physiology, 282*, R519–R527.

Doherty, T. J. (2003). Aging and sarcopenia. *Journal of Applied Physiology, 95*, 1717–1727.

Doherty, M. K., McLean, L., Hayter, J. R., Pratt, J. M., Robertson, D. H., El, S. A. (2004). The proteome of chicken skeletal muscle: changes in soluble protein expression during growth in a layer strain. *Proteomics, 4*, 2082–2093.

Domon, B. & Aebersold, R. (2006). Mass spectrometry and protein analysis. *Science, 312*, 212–217.

Donoghue, P., Doran, P., Dowling, P., Ohlendieck, K. (2005). Differential expression of the fast skeletal muscle proteome following chronic low-frequency stimulation. *Biochimica et Biophysica Acta, 1752*, 166–176.

Donoghue, P., Doran, P., Wynne, K., Pedersen, K., Dunn, M. J., Ohlendieck, K. (2007). Proteomic profiling of chronic low-frequency stimulated fast muscle. *Proteomics, 7*, 3417–3430.

Doran, P., Martin, G., Dowling, P., Jockusch, H., Ohlendieck, K. (2006). Proteome analysis of the dystrophin-deficient MDX diaphragm reveals a drastic increase in the heat shock protein cvHSP. *Proteomics, 6*, 4610–4621.

Doran, P., Donoghue, P., O'Connell, K., Gannon, J., Ohlendieck, K. (2007a). Proteomic profiling of pathological and aged skeletal muscle fibres by peptide mass fingerprinting. *International Journal of Molecular Medicine, 19*, 547–564.

Doran, P., Gannon, J., O'Connell, K., Ohlendieck, K. (2007b). Proteomic profiling of animal models mimicking skeletal muscle disorders. *Proteomics: Clinical Applications, 1*, 1169–1184.

Doran, P., Gannon, J., O'Connell, K., Ohlendieck, K. (2007c). Aging skeletal muscle shows a drastic increase in the small heat shock proteins αB-crystallin/HspB5 and cvHsp/HspB7. *European Journal of Cell Biology, 86*, 629–640.

Doran, P., O'Connell, K., Gannon, J., Kavanagh, M., Ohlendieck, K. (2008). Opposite pathobiochemical fate of pyruvate kinase and adenylate kinase in aged rat skeletal muscle as revealed by proteomic DIGE analysis. *Proteomics, 8*, 364–377.

Doran, P., Donoghue, P., O'Connell, K., Gannon, J., Ohlendieck, K. (2009a). Proteomics of skeletal muscle aging. *Proteomics, 9*, 989–1003.

Doran, P., Wilton, S. D., Fletcher, S., Ohlendieck, K. (2009b). Proteomic profiling of antisense-induced exon skipping reveals reversal of pathobiochemical abnormalities in dystrophic mdx diaphragm. *Proteomics, 9*, 671–685.

Duan, X., Berthiaume, F., Yarmush, D., Yarmush, M. L. (2006). Proteomic analysis of altered protein expression in skeletal muscle of rats in a hypermetabolic state induced by burn sepsis. *The Biochemical Journal, 397*, 149–158.

Edstrom, E., Altun, M., Bergman, E., Johnson, H., Kullberg, S., Ramirez-Leon, V., Ulfhake, B. (2007). Factors contributing to neuromuscular impairment and sarcopenia during aging. *Physiology & Behavior, 92*, 129–135.

Ellis, R. J. & van der Vies, S. M. (1991). Molecular chaperones. *Annual Review of Biochemistry, 60*, 321–347.

Emery, A. & Muntoni, F. (2003). *Duchenne muscular dystrophy* (3rd ed., pp. 1–270). Oxford, UK: Oxford University Press.

Faulkner, J. A., Larkin, L. M., Claflin, D. R., Brooks, S. V. (2007). Age-related changes in the structure and function of skeletal muscles. *Clinical and Experimental Pharmacology & Physiology, 34*, 1091–1096.

Feng, J., Xie, H., Meany, D. L., Thompson, L. V., Arriaga, E. A., Griffin, T. J. (2008). Quantitative proteomic profiling of muscle type-dependent and age-dependent protein carbonylation in rat skeletal muscle mitochondria. *Journal of Gerontology A Biological Sciences and Medical Sciences, 63*, 1137–1152.

Fenn, J. B., Mann, M., Meng, C. K., Wong, S. F., Whitehouse, C. M. (1989). Electrospray ionization for mass spectrometry of large biomolecules. *Science, 246*, 64–71.

Ferguson, P. L. & Smith, R. D. (2003). Proteome analysis by mass spectrometry. *Annual Review of Biophysics and Biomolecular Structure, 32*, 399–424.

Figueiredo, P. A., Mota, M. P., Appell, H. J., Duarte, J. A. (2008). The role of mitochondria in aging of skeletal muscle. *Biogerontology, 9*, 67–84.

Fitts, R. H. (2008). The cross-bridge cycle and skeletal muscle fatigue. *Journal of Applied Physiology, 104*, 551–558.

Flueck, M. & Hoppeler, H. (2003). Molecular basis of skeletal muscle plasticity: from gene to form and function. *Reviews of Physiology Biochemistry and Pharmacology, 146*, 159–216.

Forbes, G. B. & Reina, J. C. (1970). Adult lean body mass declines with age: some longitudinal observations. *Metabolism, 19*, 653–663.

Forner, F., Foster, L. J., Campanaro, S., Valle, G., Mann, M. (2006). Quantitative proteomic comparison of rat mitochondria from muscle, heart, and liver. *Molecular & Cellular Proteomics, 5*, 608–619.

Froemming, G. R. & Ohlendieck, K. (2001). Role of ion-regulatory membrane proteins in inherited muscle diseases. *Frontiers in Bioscience, 6*, D65–D74.

Froemming, G. R., Murray, B. E., Harmon, S., Pette, D., Ohlendieck, K. (2000). Comparative analysis of the isoform expression pattern of Ca^{2+}-regulatory membrane proteins in fast-twitch, slow-twitch, cardiac, neonatal and chronic low-frequency stimulated muscle fibres. *Biochimica et Biophysica Acta, 1466*, 151–168.

Frontera, W. R., Reid, K. F., Phillips, E. M., Krivickas, L. S., Hughes, V. A., Roubenoff, R., Fielding, R. A. (2008). Muscle fiber size and function in elderly humans: a longitudinal study. *Journal of Applied Physiology, 105*, 637–642.

Gannon, J., Staunton, L., O'Connell, K., Doran, P., Ohlendieck, K. (2008). Phosphoproteomic analysis of aged skeletal muscle. *International Journal of Molecular Medicine, 22*, 33–42.

Gelfi, C., De Palma, S., Cerretelli, P., Begum, S., Wait, R. (2003). Two-dimensional protein map of human vastus lateralis muscle. *Electrophoresis, 24*, 286–295.

Gelfi, C., Vigano, A., De Palma, S., Ripamonti, M., Begum, S., Cerretelli, P., Wait, R. (2006a). 2-D protein maps of rat gastrocnemius and soleus muscles: a tool for muscle plasticity assessment. *Proteomics, 6*, 321–340.

Gelfi, C., Vigano, A., Ripamonti, M., Pontoglio, A., Begum, S., Pellegrino, M. A., Grassi, B., Bottinelli, R., Wait, R., Cerretelli, P. (2006b). The human muscle proteome in aging. *Journal of Proteome Research, 5*, 1344–1353.

Giresi, P. G., Stevenson, E. J., Theilhaber, J., Koncarevic, A., Parkington, J., Fielding, R. A., Kandarian, S. C. (2005). Identification of a molecular signature of sarcopenia. *Physiological Genomics, 21*, 253–263.

Golenhofen, N., Perng, M. D., Quinlan, R. A., Drenckhahn, D. (2004). Comparison of the small heat shock proteins alphaB-crystallin, MKBP, HSP25, HSP20, and cvHSP in heart and skeletal muscle. *Histochemistry and Cell Biology, 122*, 415–425.

Gonnet, F., Bouazza, B., Millot, G. A., Ziaei, S., Garcia, L., Butler-Browne, G. S., Mouly, V., Tortajada, J., Danos, O., Svinartchouk, F. (2008). Proteome analysis of differentiating human myoblasts by dialysis-assisted two-dimensional gel electrophoresis (DAGE). *Proteomics, 8*, 264–278.

Gordon, A. M., Homsher, E., Regnier, M. (2000). Regulation of contraction in striated muscle. *Physiological Reviews, 80*, 853–924.

Gorg, A., Weiss, W., Dunn, M. J. (2004). Current two-dimensional electrophoresis technology for proteomics. *Proteomics, 4*, 3665–3685.

Hamelin, M., Sayd, T., Chambon, C., Bouix, J., Bibé, B., Milenkovic, D., Leveziel, H., Georges, M., Clop, A., Marinova, P., Laville, E. (2006). Proteomic analysis of ovine muscle hypertrophy. *Journal of Animal Science, 84*, 3266–3276.

Hamelin, M., Sayd, T., Chambon, C., Bouix, J., Bibe, B., Milenkovic, D., Leveziel, H., Georges, M., Clop, A., Marinova, P., Laville, E. (2007). Differential expression of sarcoplasmic proteins in four heterogeneous ovine skeletal muscles. *Proteomics, 7*, 271–280.

Isfort, R. J. (2002). Proteomic analysis of striated muscle. *Journal of Chromatography, B771*, 155–165.

Isfort, R. J., Hinkle, R. T., Jones, M. B., Wang, F., Greis, K. D., Sun, Y., Keough, T. W., Anderson, N. L., Sheldon, R. J. (2000). Proteomic analysis of the atrophying rat soleus muscle following denervation. *Electrophoresis, 21*, 2228–2234.

Janssen, I., Heymsfield, S. B., Ross, R. (2002). Low Relative Skeletal Muscle Mass (Sarcopenia) in Older Persons Is Associated with Functional Impairment and Physical Disability. *Journal of the American Geriatrics Society, 50*, 889–896.

Jia, X., Hildrum, K. I., Westad, F., Kummen, E., Aass, L., Hollung, K. (2006). Changes in enzymes associated with energy metabolism during the early post mortem period in longissimus thoracis bovine muscle analyzed by proteomics. *Journal of Proteome Research, 5*, 1763–1769.

Kandarian, S. C. & Jackman, R. W. (2006). Intracellular signaling during skeletal muscle atrophy. *Muscle & Nerve, 33*, 155–65.

Kanski, J., Hong, S. J., Schoneich, C. (2005). Proteomic analysis of protein nitration in aging skeletal muscle and identification of nitrotyrosine-containing sequences in vivo by nanoelectrospray ionization tandem mass spectrometry. *The Journal of Biological Chemistry, 280*, 24261–24266.

Kaufmann, M., Simoneau, J. A., Veerkamp, J. H., Pette, D. (1989). Electrostimulation-induced increases in fatty acid-binding protein and myoglobin in rat fast-twitch muscle and comparison with tissue levels in heart. *FEBS Letters, 245*, 181–184.

Kayani, A. C., Morton, J. P., McArdle, A. (2008). The exercise-induced stress response in skeletal muscle: failure during aging. *Applied Physiology, Nutrition, and Metabolism, 33*, 1033–1041.

Kayo, T., Allison, D. B., Weinbruch, R., Prolla, T. A. (2001). Influences of aging and caloric restriction on the transcriptional profile of skeletal muscle from rhesus monkeys. *Proceedings of the National Academy of Science USA, 98*, 5093–5098.

Kim, N. K., Joh, J. H., Park, H. R., Kim, O. H., Park, B. Y., Lee, C. S. (2004). Differential expression profiling of the proteomes and their mRNAs in porcine white and red skeletal muscles. *Proteomics, 4*, 3422–3428.

Kislinger, T., Gramolini, A. O., Pan, Y., Rahman, K., MacLennan, D. H., Emili, A. (2005). Proteome dynamics during C2C12 myoblast differentiation. *Molecular & Cellular Proteomics, 4*, 887–901.

Kreutziger, K. L., Gillis, T. E., Davis, J. P., Tikunova, S. B., Regnier, M. (2007). Influence of enhanced troponin C Ca2+-binding affinity on cooperative thin filament activation in rabbit skeletal muscle. *The Journal of Physiology, 583*, 337–350.

Lametsch, R., Bendixen, E. (2001). Proteome analysis applied to meat science: characterizing postmortem changes in porcine muscle. *Journal of Agriculture and Food Chemistry, 49*, 4531–4537.

Le Bihan, M. C., Hou, Y., Harris, N., Tarelli, E., Coulton, G. R. (2006). Proteomic analysis of fast and slow muscles from normal and kyphoscoliotic mice using protein arrays, 2-DE and MS. *Proteomics, 6*, 4646–4661.

Lee, C. E., McArdle, A., Griffiths, R. D. (2007). The role of hormones, cytokines and heat shock proteins during age-related muscle loss. *Clinical Nutrition, 26*, 524–534.

Lindle, R. S., Metter, E. J., Lynch, N. A., Fleg, J. L., Fozard, J. L., Tobin, L. (1997). Age and gender comparisons of muscle strength in 654 women and men aged 20-93 yr. *Journal of Applied Physiology, 83*, 1581–1587.

Liu, Y., Gampert, L., Nething, K., Steinacker, J. M. (2006). Response and function of skeletal muscle heat shock protein 70. *Frontiers in Bioscience, 11*, 2802–2827.

Lombardi, A., Silvestri, E., Cioffi, F., Senese, R., Lanni, A., Goglia, F., de Lange, P., Moreno, M. (2009). Defining the transcriptomic and proteomic profiles of rat ageing skeletal muscle by the use of a cDNA array, 2D- and Blue native-PAGE approach. *Journal of Proteomics, 72*, 708–721.

Lynch, G. S., Schertzer, J. D., Ryall, J. G. (2007). Therapeutic approaches for muscle wasting disorders. *Pharmacology & Therapeutics, 113*, 461–487.

MacLennan, D. H. (2000). Ca^{2+} signalling and muscle disease. *European Journal of Biochemistry, 267*, 5291–5297.

MacLennan, D. H., Abu-Abed, M., Kang, C. (2002). Structure-function relationships in Ca(2+) cycling proteins. *Journal of Molecular and Cellular Cardiology, 34*, 897–918.

Marouga, R., David, S., Hawkins, E. (2005). The development of the DIGE system: 2D fluorescence difference gel analysis technology. *Analytical and Bioanalytical Chemistry, 382*, 669–678.

McArdle, A. & Jackson, M. J. (2000). Exercise, oxidative stress and ageing. *Journal of Anatomy, 197*, 539–541.

McArdle, A., Dillmann, W. H., Mestril, R., Faulkner, J. A., Jackson, M. J. (2004). Overexpression of HSP70 in mouse skeletal muscle protects against muscle damage and age-related muscle dysfunction. *The FASEB Journal, 18*, 355–357.

Meany, D. L., Xie, H., Thompson, L. V., Arriaga, E. A., Griffin, T. J. (2007). Identification of carbonylated proteins from enriched rat skeletal muscle mitochondria using affinity chromatography-stable isotope labeling and tandem mass spectrometry. *Proteomics, 7*, 1150–1163.

Melstrom, L. G., Melstrom, K. A., Ding, X. Z., Adrian, T. E. (2007). Mechanisms of skeletal muscle degradation and its therapy in cancer cachexia. *Histology and Histopathology, 22*, 805–814.

Melton, L. J., Khosla, S., Crowson, C. S., O'Connor, M. K., O'Fallon, W. M., Riggs, B. L. (2000). Epidemiology of sarcopenia. *Journal of the American Geriatrics Society, 48*, 625–630.

Munoz, M. E. & Ponce, E. (2003). Pyruvate kinase: current status of regulatory and functional properties. *Comparative Biochemistry and Physiology. Part B: Biochemistry & Molecular Biology, 135*, 197–218.

Murray, B. E., Froemming, G. R., Maguire, P. B., Ohlendieck, K. (1998). Excitation–contraction–relaxation cycle: role of Ca^{2+}-regulatory membrane proteins in normal, stimulated and pathological skeletal muscle (review). *International Journal of Molecular Medicine, 1*, 677–687.

Neufer, P. D., Ordway, G. A., Williams, R. S. (1998). Transient regulation of c-fos, alpha B-crystallin, and hsp70 in muscle during recovery from contractile activity. *The American Journal of Physiology, 274*, C341–C346.

Nicholl, I. D. & Quinlan, R. A. (1994). Chaperone activity of alpha-crystallins modulates intermediate filament assembly. *The EMBO Journal, 13*, 945–953.

Nishimura, R. N. & Sharp, F. R. (2005). Heat shock proteins and neuromuscular disease. *Muscle & Nerve, 32*, 693–709.

O'Connell, K., Gannon, J., Doran, P., Ohlendieck, K. (2007). Proteomic profiling reveals a severely perturbed protein expression pattern in aged skeletal muscle. *International Journal of Molecular Medicine, 20*, 145–153.

O'Connell, K., Doran, P., Gannon, J., Ohlendieck, K. (2008a). Lectin-based proteomic profiling of aged skeletal muscle: Decreased pyruvate kinase isozyme M1 exhibits drastically increased levels of N-glycosylation. *European Journal of Cell Biology, 87*, 793–805.

O'Connell, K., Gannon, J., Doran, P., Ohlendieck, K. (2008b). Reduced expression of sarcalumenin and related Ca2+ -regulatory proteins in aged rat skeletal muscle. *Experimental Gerontology, 43*, 958–961.

Okumura, N., Hashida-Okumura, A., Kita, K., Matsubae, M., Matsubara, T., Takao, T., Nagai, K. (2005). Proteomic analysis of slow- and fast-twitch skeletal muscles. *Proteomics, 5*, 2896–2906.

Pette, D. (2001). Historical Perspectives: plasticity of mammalian skeletal muscle. *Journal of Applied Physiology, 90*, 1119–1124.

Pette, D. & Staron, R. S. (1990). Cellular and molecular diversities of mammalian skeletal muscle fibers. *Reviews of Physiology Biochemistry and Pharmacology, 116*, 1–76.

Pette, D. & Staron, R. S. (2001). Transitions of muscle fiber phenotypic profiles. *Histochemistry and Cell Biology, 115*, 359–372.
Phielix, E. & Mensink, M. (2008). Type 2 diabetes mellitus and skeletal muscle metabolic function. *Physiology & Behavior, 94*, 252–258.
Piec, I., Listrat, A., Alliot, J., Chambon, C., Taylor, R. G., Bechet, D. (2005). Differential proteome analysis of aging in rat skeletal muscle. *The FASEB Journal, 19*, 1143–1145.
Pieper, R., Gatlin, C. L., Makusky, A. J., Russo, P. S., Schatz, C. R., Miller, S. S., Su, Q., McGrath, A. M., Estock, M. A., Parmar, P. P., Zhao, M., Huang, S. T., Zhou, J., Wang, F., Esquer-Blasco, R., Anderson, N. L., Taylor, J., Steiner, S. (2003). The human serum proteome: display of nearly 3700 chromatographically separated protein spots on two-dimensional electrophoresis gels and identification of 325 distinct proteins. *Proteomics, 3*, 1345–1364.
Prochniewicz, E., Thompson, L. V., Thomas, D. D. (2007). Age-related decline in actomyosin structure and function. *Experimental Gerontology, 42*, 931–938.
Proctor, D. N., O'Brien, P. C., Atkinson, E. J., Nair, K. S. (1999). Comparison of techniques to estimate total body skeletal muscle mass in people of different age groups. *The American Journal of Physiology, 277*, E489–E495.
Punkt, K. (2002). Fibre types in skeletal muscles. *Advances in Anatomy, Embryology and Cell Biology, 162*, 1–109.
Raddatz, K., Albrecht, D., Hochgrafe, F., Hecker, M., Gotthardt, M. (2008). A proteome map of murine heart and skeletal muscle. *Proteomics, 8*, 1885–1897.
Renault, V., Thornell, L. E., Eriksson, P. O., Butler-Browne, G., Mouly, V. (2002). Regenerative potential of human skeletal muscle during aging. *Aging Cell, 1*, 132–139.
Renganathan, M., Messi, M. L., Delbono, O. (1997). Dihydropyridine receptor-ryanodine receptor uncoupling in aged skeletal muscle. *The Journal of Membrane Biology, 157*, 247–253.
Roberts, S. B. (1995). Effects of aging on energy requirements and the control of food intake in men. *Journal of Gerontology, 50A*, 101–106.
Rolland, Y., Czerwinski, S., Abellan Van Kan, G., Morley, J. E., Cesari, M., Onder, G., Woo, J., Baumgartner, R., Pillard, F., Boirie, Y., Chumlea, W. M., Vellas, B. (2008). Sarcopenia: its assessment, etiology, pathogenesis, consequences and future perspectives. *The Journal of Nutrition, Health Aging, 12*, 433–450.
Roth, S. M., Ferrell, R. E., Peters, D. G., Metter, E. J., Hurley, B. F., Rogers, M. A. (2002). Influence of age, sex, and strength training on human muscle gene expression determined by microarray. *Physiological Genomics, 10*, 181–190.
Sadowski, P. G., Groen, A. J., Dupree, P., Lilley, K. S. (2008). Sub-cellular localization of membrane proteins. *Proteomics, 8*, 3991–4011.
Schoneich, C., Viner, R. I., Ferrington, D. A., Bigelow, D. J. (1999). Age-related chemical modification of the skeletal muscle sarcoplasmic reticulum Ca^{2+}-ATPase of the rat. *Mechanisms of Ageing and Development, 107*, 221–231.
Seo, Y., Lee, K., Park, K., Bae, K., Choi, I. (2006). A proteomic assessment of muscle contractile alterations during unloading and reloading. *Journal of Biochemistry, 139*, 71–80.
Shamaei-Tousi, A., Halcox, J. P., Henderson, B. (2007). Stressing the obvious? Cell stress and cell stress proteins in cardiovascular disease. *Cardiovascular Research, 74*, 19–28.
Smith, I. J., Lecker, S. H., Hasselgren, P. O. (2008). Calpain activity and muscle wasting in sepsis. *The American Journal of Physiology, 295*, E762–771.
Soti, C., Nagy, E., Giricz, Z., Vigh, L., Csermely, P., Ferdinandy, P. (2005). Heat shock proteins as emerging therapeutic targets. *British Journal of Pharmacology, 146*, 769–780.
Spangenburg, E. E. & Booth, F. W. (2003). Molecular regulation of individual skeletal muscle fibre types. *Acta Physiologica Scandinavia, 178*, 413–424.
Squier, T. C. & Bigelow, D. J. (2000). Protein oxidation and age-dependent alterations in calcium homeostasis. *Frontiers in Bioscience, 5*, D504–D526.
Stamler, R., Kappe, G., Boelens, W., Slingsby, C. (2005). Wrapping the alpha-crystallin domain fold in a chaperone assembly. *Journal of Molecular Biology, 353*, 68–79.
Sullivan, V. K., Powers, S. K., Criswell, D. S., Tumer, N., Larochelle, J. S., Lowenthal, D. (1995). Myosin heavy chain composition in young and old rat skeletal muscle: effects of endurance exercise. *Journal of Applied Physiology, 78*, 2115–2120.

Sun, H., Liu, J., Ding, F., Wang, X., Liu, M., Gu, X. (2006). Investigation of differentially expressed proteins in rat gastrocnemius muscle during denervation–reinnervation. *Journal of Muscle Research and Cell Motility, 27*, 241–250.

Swartz, D. R., Yang, Z., Sen, A., Tikunova, S. B., Davis, J. P. (2006). Myofibrillar troponin exists in three states and there is signal transduction along skeletal myofibrillar thin filaments. *Journal of Molecular Biology, 361*, 420–435.

Tan, S., Tan, H. T., Chung, M. C. (2008). Membrane proteins and membrane proteomics. *Proteomics, 8*, 3924–3932.

Tannu, N. S., Rao, V. K., Chaudhary, R. M., Giorgianni, F., Saeed, A. E., Gao, Y., Raghow, R. (2004). Comparative proteomes of the proliferating C(2)C(12) myoblasts and fully differentiated myotubes reveal the complexity of the skeletal muscle differentiation program. *Molecular & Cellular Proteomics, 3*, 1065–1082.

Thompson, L. V. (2009). Age-related muscle dysfunction. *Experimental Gerontology, 44*, 106–111.

Tonge, R., Shaw, J., Middleton, B., Rowlinson, R., Rayner, S., Young, J., Pognan, F., Hawkins, E., Currie, I., Davison, M. (2001). Validation and development of fluorescence two-dimensional differential gel electrophoresis proteomics technology. *Proteomics, 1*, 377–396.

Unlu, M., Morgan, M. E., Minden, J. S. (1997). Difference gel electrophoresis: a single gel method for detecting changes in protein extracts. *Electrophoresis, 18*, 2071–2077.

Vandervoort, A. A. (2002). Aging of the human neuromuscular system. *Muscle & Nerve, 25*, 17–25.

Vandervoort, A. A. & Symons, T. B. (2001). Functional and metabolic consequences of sarcopenia. *Canadian Journal of Applied Physiology, 26*, 90–101.

van Montfort, R., Slingsby, C., Vierling, E. (2001). Structure and function of the small heat shock protein/alpha-crystallin family of molecular chaperones. *Advances in Protein Chemistry, 59*, 105–156.

Vigano, A., Ripamonti, M., De Palma, S., Capitanio, D., Vasso, M., Wait, R., Lundby, C., Cerretelli, P., Gelfi, C. (2008). Proteins modulation in human skeletal muscle in the early phase of adaptation to hypobaric hypoxia. *Proteomics, 8*, 4668–4679.

Viswanathan, S., Unlu, M., Minden, J. S. (2006). Two-dimensional difference gel electrophoresis. *Nature Protocols, 1*, 1351–1358.

Vitorino, R., Ferreira, R., Neuparth, M., Guedes, S., Williams, J., Tomer, K. B., Domingues, P. M., Appell, H. J., Duarte, J. A., Amado, F. M. (2007). Subcellular proteomics of mice gastrocnemius and soleus muscles. *Analytical Biochemistry, 366*, 156–169.

Webster, J. & Oxley, D. (2005). Peptide mass fingerprinting: protein identification using MALDI-TOF mass spectrometry. *Methods in Molecular Biology, 310*, 227–240.

Welle, S., Brooks, A., Thornton, C. A. (2001). Senescence-related changes in gene expression in muscle: similarities and differences between mice and men. *Physiological Genomics, 5*, 67–73.

Wells, G. D., Noseworthy, M. D., Hamilton, J., Tarnopolski, M., Tein, I. (2008). Skeletal muscle metabolic dysfunction in obesity and metabolic syndrome. *The Canadian Journal of Neurological Sciences, 35*, 31–40.

Wittmann-Liebold, B., Graack, H. R., Pohl, T. (2006). Two-dimensional gel electrophoresis as tool for proteomics studies in combination with protein identification by mass spectrometry. *Proteomics, 6*, 4688–4703.

Wuest, R. C. & Degens, H. (2007). Factors contributing to muscle wasting and dysfunction in COPD patients. *International Journal of Chronic Obstructive Pulmonary Disease, 2*, 289–300.

Yan, J. X., Harry, R. A., Wait, R., Welson, S. Y., Emery, P. W., Preedy, V. R., Dunn, M. J. (2001). Separation and identification of rat skeletal muscle proteins using two-dimensional gel electrophoresis and mass spectrometry. *Proteomics, 1*, 424–434.

Zaluzec, E. J., Gage, D. A., Watson, J. T. (1995). Matrix-assisted laser desorption ionization mass spectrometry: applications in peptide and protein characterization. *Protein Expression and Purification, 6*, 109–123.

Zheng, Y. Z. & Foster, L. J. (2009). Biochemical and proteomic approaches for the study of membrane microdomains. *Journal of Proteomics, 72*, 12–22.

Exercise and Nutritional Interventions to Combat Age-Related Muscle Loss

René Koopman, Lex B. Verdijk, and Luc J.C. van Loon

Abstract Aging is accompanied by a progressive loss of skeletal muscle mass and strength, leading to the loss of functional capacity and an increased risk of developing chronic metabolic diseases such as diabetes. The age-related loss of skeletal muscle mass must be due to a chronic disruption in the balance between muscle protein synthesis and degradation. In addition, it has been suggested that a decline in the number of satellite cells (SC) and/or their ability to become activated can contribute to the development of sarcopenia.

In healthy active older individuals, there does not seem to be a disturbance in muscle protein metabolism in the fasted (basal) state. Consequently, it has been proposed that older muscle has a deficit in the ability to regulate the protein synthetic response to anabolic stimuli, such as food intake and physical activity. Indeed, recent data suggest that the dose-response relationship between myofibrillar protein synthesis and the availability of essential amino acids and/or resistance exercise intensity is shifted down and to the right in elderly humans. This so-called anabolic resistance is now believed to represent a key factor responsible for the age-related decline in skeletal muscle mass. Although physical activity and/or exercise stimulate muscle protein synthesis in both the young and elderly, the hypertrophic response largely depends on the timed administration of amino acids and/or protein prior to, during, and/or after exercise. However, prolonged resistance type exercise training has been shown to be effective as a therapeutic strategy to augment skeletal muscle mass, increase muscle SC content, and

L.J.C. van Loon (✉)
Department of Human Movement Sciences, Maastricht University Medical Centre,
6200 MD, Maastricht, The Netherlands
e-mail: L.vanLoon@HB.unimaas.nl

R. Koopman
Basic and Clinical Myology Laboratory, Department of Physiology,
The University of Melbourne, Australia

L.B. Verdijk
Department of Human Movement Sciences, Nutrition and Toxicology Research Institute
Maastricht (NUTRIM), Maastricht University Medical Centre, Maastricht, The Netherlands

G.S. Lynch (ed.), *Sarcopenia – Age-Related Muscle Wasting and Weakness*,
DOI 10.1007/978-90-481-9713-2_13, © Springer Science+Business Media B.V. 2011

improve functional performance in the elderly. The latter shows that the ability to increase muscle mass is preserved up to very old age. More research is warranted to elucidate the interaction between nutrition, exercise and the skeletal muscle adaptive response. The latter is needed to define more effective strategies that will maximize the therapeutic benefits of lifestyle intervention in the elderly.

Keywords Sarcopenia • Nutrition • Exercise training • Muscle hypertrophy

1 Introduction

Demographics show that the world's population aged 60 years and over will triple within 50 years, from 600 million in the year 2000 to more than two billion by 2050. Two thirds of the elderly people are presently living in the developed world, and this will continue to rise up to 75%. Due to greater longevity, the subpopulation of elderly people aged 80 years and over is presently the fastest growing subpopulation in the developed world (WHO 2008). This global aging will have a major impact on our healthcare system due to increased morbidity and greater need for hospitalization and/or institutionalization. Good health is essential for older people to remain independent and to continue to actively take part in family and community life. Life-long health promotion is warranted to prevent or delay the onset of non-communicable and chronic (metabolic) diseases, like heart disease, stroke, cancer, and diabetes.

Healthy aging depends on a wide range of factors, but the preservation of muscle function is among the most important, allowing the continuation of an independent lifestyle with undiminished activities of daily living. One of the factors that play an important role in the loss of functional performance is the progressive loss of skeletal muscle mass with aging, or sarcopenia (Baumgartner et al. 1998; Forbes and Reina 1970; Melton et al. 2000). Lean muscle mass generally contributes up to ~50% of total bodyweight in young adults but declines with aging to 25% when reaching an age of 75–80 years (Short and Nair 2000; Short et al. 2004). The loss of muscle mass is most notable in the lower limb muscle groups, with the cross-sectional area of the *vastus lateralis* being reduced by as much as 40% between the age of 20 and 80 years (Lexell 1995). As muscle strength is proportionate to muscle mass and cross-sectional area (Fig. 1), the loss of skeletal muscle mass is accompanied by the loss of muscle strength, a decline in functional capacity (Bassey et al. 1992; Brown et al. 1995; Frontera et al. 1991; Landers et al. 2001; Larsson and Karlsson 1978; Lindle et al. 1997; Petrella et al. 2005; Wolfson et al. 1995), and a reduction in whole-body and skeletal muscle oxidative capacity (Nair 1995, 2005; Short and Nair 2000). The absolute decline in muscle mass and muscle oxidative capacity, in combination with a greater fat mass, contributes to the greater risk of developing insulin resistance and/or type 2 diabetes due to the reduced capacity for blood glucose disposal and a greater likelihood of excess lipid deposition in liver and skeletal

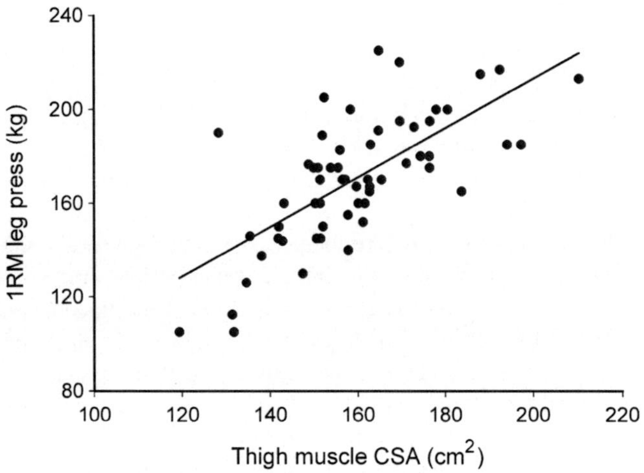

Fig. 1 Correlation between total thigh muscle cross-sectional area (CSA) measured by CT scan and one-repetition maximum (1 RM) leg press strength ($n = 60$, $r = 0.70$; Verdijk and colleagues, unpublished observations)

muscle tissue. The latter will also lead to hyperlipidemia, hypertension, and cardiovascular co-morbidities. Therefore, it is evident that preventing, attenuating, and/or reversing the decline in skeletal muscle mass should form a main target in interventional strategies to promote healthy aging.

2 Aging and Skeletal Muscle

The age-related loss of skeletal muscle mass is facilitated by a combination of factors, which include a less than optimal diet (Campbell and Evans 1996; Campbell and Leidy 2007; Campbell et al. 2001) and a sedentary lifestyle (Nair 2005). The progressive muscle wasting during aging must be due to a disruption in the regulation of skeletal muscle protein turnover, leading to a structural imbalance between muscle protein synthesis and degradation. As satellite cells (SC, undifferentiated myogenic precursor cells) play a key role in the maintenance, growth and repair of myofibers (Snijders et al. 2009; Kadi et al. 2004a), a decline in the number of SC and/or their ability to become activated might also contribute to the development of sarcopenia.

2.1 *Protein Turnover in Senescent Muscle*

Many research groups have assessed basal muscle protein synthesis and/or protein breakdown rates in both young and elderly subjects in an attempt to unravel the proposed impairments in muscle protein metabolism in the elderly (Balagopal et al. 1997; Hasten et al. 2000; Rooyackers et al. 1996; Short et al. 2004; Welle et al. 1993, 1995; Yarasheski et al. 1993, 2002; Cuthbertson et al. 2005; Katsanos et al. 2005, 2006; Paddon-Jones et al. 2004; Volpi et al. 1999, 2000, 2001). Although it was originally reported that healthy old subjects show decreased rates of basal muscle protein synthesis (Balagopal et al. 1997; Hasten et al. 2000; Rooyackers et al. 1996; Short et al. 2004; Welle et al. 1993, 1995; Yarasheski et al. 1993, 2002), more recent studies have failed to reproduce these findings and generally show little or no differences in basal muscle protein synthesis rates between the young and old (Cuthbertson et al. 2005; Katsanos et al. 2005, 2006; Paddon-Jones et al. 2004; Volpi et al. 1999, 2000, 2001). The apparent discrepancy in the reported basal muscle protein synthesis rates can be attributed to differences in health status, habitual physical activity and/or dietary habits between the selected young and elderly subjects and to the applied methodology to assess muscle protein synthesis. It should be noted that the assessment of fractional muscle protein synthetic rate in vivo in humans has its methodological limitations. The sensitivity of the measurement and large inter-subject variance in basal muscle protein synthesis rates limit the ability to detect small, but potentially physiologically relevant differences between groups. In addition, a 30–40% lower basal protein synthesis rate in the elderly, as observed by the earlier studies (Balagopal et al. 1997; Hasten et al. 2000; Rooyackers et al. 1996; Short et al. 2004; Welle et al. 1993, 1995; Yarasheski et al. 1993, 2002), is unlikely representative of a normal physiological condition. Without a similar, concomitant decline in muscle protein breakdown rate, such protein synthesis rates would be accompanied by rapid muscle wasting. In contrast, the rate of muscle loss that is typically observed during aging is relatively small (<2%/year), which must mean that the mismatch between average diurnal rate of muscle protein synthesis and breakdown is small. Therefore, the hypothesis that basal fasting protein synthesis and/or breakdown rates are not (substantially) impaired with aging generally receives more support (Volpi et al. 2001; Rennie 2009). Thus, as basal (fasting) muscle protein synthesis rates do not seem to differ substantially between the young and elderly, many research groups have since re-focused on the proposed anabolic resistance in the elderly to the main anabolic stimuli, i.e. food intake and physical activity.

2.2 *Myofiber Properties of Senescent Muscle*

The progressive decline in skeletal muscle mass with aging is associated with reductions in fiber number, motor units and a relative loss of type II fibers, with a preservation of type I fibers (Larsson et al. 1978). As a result, aging in humans has been shown to be associated with a shift in muscle fiber type distribution towards a higher percentage type I (slow-twitch) versus type II (fast-twitch) muscle fibers (Charifi et al. 2003; Frontera et al. 2000; Hameed et al. 2003;

Verdijk et al. 2007; Kim et al. 2005b; Larsson 1978; Dreyer et al. 2006). In addition, type II muscle fiber size has been shown to decline with age, whereas type I muscle fiber size tends to remain rather constant (Charette et al. 1991, Dreyer et al. 2006; Kim et al. 2005b; Larsson et al. 1978; Lexell et al. 1988; Martel et al. 2006; Verdijk et al. 2007).

In recent years, there has been an increased interest in the potential role that satellite cells (SC) might play in muscle fiber atrophy and the age-related loss of skeletal muscle mass and function. Satellite cells are undifferentiated myogenic precursor cells (or muscle stem cells) that were named after their location, residing in a quiescent state between the basal lamina and the sarcolemma of a muscle fiber (Mauro 1961; Moss and Leblond 1970, 1971). Upon stimulation (e.g. through injury or mechanical loading), SC become activated and start proliferating (Fig. 2). These proliferated cells can then differentiate and fuse together to form new myofibers, or fuse with existing myofibers to donate their nucleus to the sarcoplasm of the myofiber (Hawke and Garry 2001; Mauro 1961; Moss and Leblond 1970, 1971). Alternatively, part of the proliferated cells will return to quiescence as part of a 'self-renewal' process to maintain the satellite cell pool. Normal myonuclei are post-mitotic and SC are the only known source to provide new myonuclei to muscle tissue in vivo. As such, SC play an essential role in myofiber maintenance, repair, and growth, and any changes in SC content and/or function with aging might affect muscle mass and function in the elderly (Snijders et al. 2009).

Although several studies have shown a decline in SC content with aging (Kadi et al. 2004a; Renault et al. 2002; Sajko et al. 2004; Verdijk et al. 2007, 2009a; Verney et al. 2008), some reports did not observe any differences (Dreyer et al. 2006; Hikida et al. 1998; Petrella et al. 2006; Roth et al. 2000; Verney et al. 2008). Recent data from our laboratory indicate that this inconsistency is probably due to the lack of fiber type specific data. We recently reported that type II muscle fiber atrophy with aging is accompanied by a fiber type specific decline in SC content (Verdijk et al. 2007). These findings have been confirmed by other studies showing a specifically lower SC content in the type II muscle fibers with aging (Verdijk et al. 2009a; Verney et al. 2008). However, despite this age-related reduction in SC content, senescent muscle still seems to respond to long-term exercise intervention by increasing muscle mass and strength (Koopman and van Loon 2009). In fact, both the specific atrophy and reduced satellite cell content in type II muscle fibers can be reversed with 12 weeks of resistance type exercise training in the elderly (Verdijk et al. 2009a). Yet, it remains to be determined to what extent the skeletal muscle adaptive response might be attenuated in elderly subjects when compared with young, healthy adults.

The observation that muscle fiber SC content is altered in the elderly has initiated a series of investigations looking at concomitant changes in the activation status of these SC (reviewed in detail elsewhere (Snijders et al. 2009)). The activation and/or suppression of myogenic regulatory factors (MRFs) is required for SC to enter and/or complete their cell cycle. Interestingly, the magnitude of upregulation of MRF mRNA expression appears to be proportional to the degree

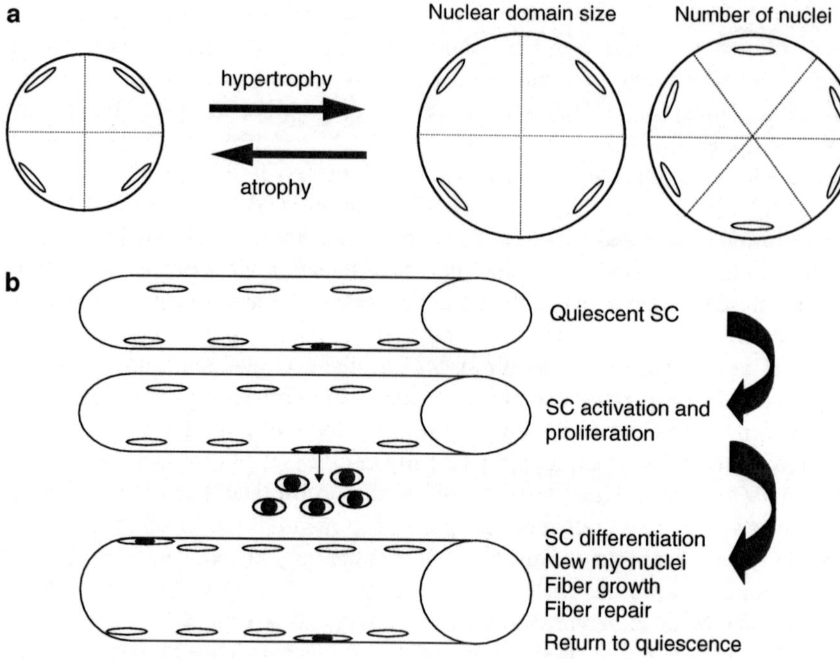

Fig. 2 Myonuclei and satellite cells in the skeletal muscle adaptive response. (**a**) Cross-section of a muscle fiber containing several myonuclei. Muscle hypertrophy is associated with an increase in myonuclear domain size (due to increased protein synthesis induced by the existing myonuclei) and/or an increase in the number of myonuclei. In contrast, atrophy is associated with a reduced myonuclear number and/or domain size. (**b**) Longitudinal section of a muscle fiber containing several myonuclei and a satellite cell. Skeletal muscle satellite cells normally lie quiescent between the basal lamina and the plasma membrane of the associated myofiber. Upon injury or mechanical loading, satellite cells are activated and start to proliferate. Following differentiation, new myonuclei will be incorporated into the myofiber to achieve hypertrophy. In addition, part of the proliferated satellite cells will return to quiescence to replenish the satellite cell pool. Alternatively, differentiated cells can fuse together to generate new myofibers in case of significant myofiber damage (not shown) (Reprinted from Snijders et al. 2009, © (2009). With permission from Elsevier)

of sarcopenia in rodent muscle (Edstrom and Ulfhake 2005). This seems to suggest that senescent muscle is in a state of failing regenerative effort (Kosek et al. 2006; Musaro et al. 1995). In contrast to the upregulated basal MRF mRNA expression, Notch signaling has been reported to decline at a more advanced age. Notch activation plays a pivotal role in the regulation of SC activation, proliferation and differentiation (Conboy and Rando 2002; Nofziger et al. 1999). Attenuated Notch activation in older mice has been implicated in a blunted regenerative response to injury (Conboy and Rando 2002; Schubert 2004). In humans, Notch mRNA expression has been shown to increase following 12 weeks of resistance type exercise training in both young and older subjects (Carey et al. 2007). However, the same study reported that young subjects showed an absolute higher

Notch gene expression at all time points when compared with the elderly. In short, both SC number and their activation status appear to be altered in senescent muscle. Therefore, future studies should not only focus to restore the SC pool but also to define effective strategies to stimulate SC activation. Both strategies may be of vital importance to prevent, delay, and/or even reverse the loss of muscle mass with aging.

3 Food Intake and Muscle Protein Turnover

It has been well established that protein turnover in skeletal muscle tissue is highly responsive to nutrient intake (Rennie et al. 1982). Ingestion of amino acids (AA) and/or protein strongly stimulates muscle protein synthesis and inhibits protein breakdown, resulting in a positive net protein balance in both the young and elderly (Paddon-Jones et al. 2004, 2006; Rennie et al. 1982; Volpi et al. 1998, 1999). Interestingly, data from recent studies suggest that the muscle protein synthetic response to the ingestion of a small amount of (essential) AA (Cuthbertson et al. 2005; Katsanos et al. 2005) is attenuated in the elderly, which is now believed to represent one of the key factors responsible for the age-related decline in skeletal muscle mass. The so-called anabolic resistance in elderly humans has been demonstrated by a rightward and downward shift of the dose-response relationship between myofibrillar protein synthesis and the availability of leucine in the plasma (Cuthbertson et al. 2005). As Cuthbertson et al. (2005) showed that even a very large (40 g) dose of EAA is not able to bring the curve back to values seen in young subjects, some are skeptical of claims that supplementation of meals with extra amino acids or feeding more protein will restore the rate of muscle protein synthesis in older people, relative to those found in the young (Rennie 2009).

The mechanisms that might be responsible for the proposed anabolic resistance to protein and/or amino acid administration in the elderly remain to be elucidated. In addition, it is unclear whether the blunted muscle protein synthetic response to food intake is also accompanied by an attenuated post-prandial decline in muscle protein breakdown in the elderly (Guillet et al. 2004b). Cuthbertson et al. (2005) reported decrements in amounts of signaling protein in the protein kinase B (PKB)-mammalian target of rapamycin (mTOR) pathway in senescent muscle and showed an attenuated rise in the activation of key signaling proteins in the mTOR pathway after ingesting 10 g essential amino acids (EAA) in the elderly versus the young (Cuthbertson et al. 2005). These findings seem to be in line with previous observations by Guillet et al. (Guillet et al. 2004a) showing reduced p70-S6 Kinase 1 phosphorylation following combined AA and glucose infusions in the elderly. Combined, these data suggest that an anabolic signal might not be sensed and/or transduced as well in muscle tissue of elderly compared with younger subjects (Bohe et al. 2003; Cuthbertson et al. 2005). The EAA (Tipton et al. 1999b; Volpi et al. 2003), and leucine in particular (Smith et al. 1992; Norton and Layman 2006), seem to represent the main anabolic signals responsible for the post-prandial increase in muscle protein synthesis. In accordance, recent studies demonstrate that the attenuated muscle protein synthetic response to food intake in the elderly can, at least partly, be compensated for by increasing the leucine content of a meal (Katsanos et al. 2006; Rieu et al. 2006).

Even in the absence of a concomitant increase in plasma insulin, EAA show a dose-dependent stimulation of muscle protein synthesis (Bohe et al. 2003), and as a result, some propose that insulin is rather permissive instead of modulatory (Greenhaff et al. 2008; Rennie 2009; Bohe et al. 2003). Recent data indicate that insulin in the range of ~30–150 μU/ml does not further stimulate muscle protein synthesis (Greenhaff et al. 2008). Interestingly, it seems that muscle protein breakdown is very responsive to changes in insulin concentrations. Data from the Rennie laboratory (Rennie 2009) suggest that insulin levels of 15 μU/ml can almost maximally reduce muscle protein breakdown and there seems to be no further inhibition above 30 μU/ml (Greenhaff et al. 2008). These data suggest that protein breakdown can be already maximally reduced by the intake of a light breakfast in healthy young men. Resting leg protein breakdown has been reported to be either similar (Wilkes et al. 2008) or just slightly increased (Volpi et al. 2001) in older men compared with young controls. However, recent data suggest that muscle protein breakdown is not strongly inhibited by insulin in the elderly (Wilkes et al. 2008). These observations seem to be in line with previous reports suggesting that muscle protein synthesis is resistant to the anabolic action of insulin in the elderly (Rasmussen et al. 2006; Volpi et al. 2000), which may be attributed to a less responsive impact of physiological hyperinsulinemia on the increase in skeletal muscle blood flow and subsequent amino acid availability in aged muscle (Fujita et al. 2006; Rasmussen et al. 2006). The latter would also agree with a reduced activation of the PI-3 kinase/Akt/mTOR signaling pathway and with the lesser increase in the muscle protein synthetic rate following amino acid/protein ingestion in the elderly (Cuthbertson et al. 2005).

As recent data clearly show that the digestion rate of protein is an independent regulating factor of post-prandial protein anabolism (Dangin et al. 2001), any impairments in protein digestion and/or absorption will attenuate and/or reduce the appearance rate of dietary AA in the circulation, thereby lowering the post-prandial muscle protein synthetic rate. Evidence to support the existence of differences in digestion and absorption kinetics and the subsequent muscle protein synthetic response to dietary protein intake between young and elderly humans remains lacking. The latter is largely due to the restrictions set by the methodology that has been used to assess the appearance rate of AA from the gut into the circulation. As free AA and protein-derived AA exhibit a different timing and efficiency of intestinal absorption (Boirie et al. 1996), simply adding labeled free AA to a protein containing drink does not provide an accurate measure of the digestion and absorption kinetics of the ingested dietary protein (Boirie et al. 1995). To accurately assess the appearance rate of AA derived from dietary protein, the labeled AA need to be incorporated in the dietary protein source (Beaufrere et al. 2000; Boirie et al. 1996; Dangin et al. 2002). A series of studies that have applied specifically produced, intrinsically labeled protein have been instrumental in the development of the fast versus slow protein concept (Beaufrere et al. 2000; Boirie et al. 1997a; Dangin et al. 2001, 2002, 2003). These studies show that ingestion of a slowly digested protein (casein) leads to a more positive whole-body protein balance when compared to the ingestion of a fast digestible protein (whey) or a mixture of free AA in healthy, young subjects (Dangin et al. 2001). In contrast, ingestion of a fast protein was shown to result in greater net protein retention when compared with a slow protein when provided in healthy, elderly men (Beaufrere et al. 2000; Boirie et al. 1997a; Dangin et al. 2002, 2003). The latter might be attributed to the proposed anabolic resistance of the muscle protein synthetic machinery to become activated in elderly muscle. However, in a recent study we did not observe any differences in the appearance rate of dietary phenylalanine in the circulation, and plasma amino acid availability between young and elderly men following the intake of 35 g (Koopman et al. 2009b) of intact casein protein. Clearly more research is warranted to determine to what extent anabolic resistance to food (i.e. intact protein) intake exists in elderly humans. Interestingly, previous studies have suggested that amino acid utilization in the splanchnic area is elevated in the elderly (Boirie et al. 1997b; Volpi et al. 1999), which would imply that less of the ingested AA are available for muscle protein synthesis (Boirie et al. 1997a). Our recently obtained data clearly shows that this is not the case following the ingestion of a relatively large bolus of intact casein (Koopman et al. 2009b). In addition, we also recently tested the hypothesis that the ingestion of a protein hydrolysate, i.e. an enzymatically predigested protein, would enhance protein digestion and the absorption rate in elderly men (Koopman et al. 2009a). The latter should theoretically result in a greater increase in plasma AA availability and might improve the post-prandial muscle protein synthetic response. Elderly men were provided with a single bolus of specifically produced intrinsically L-[1-^{13}C]phenylalanine–labeled intact casein or casein hydrolysate. This is the first study to show that ingestion of a casein hydrolysate, as opposed to its intact protein, accelerates the appearance rate of dietary phenylalanine in the circulation, lowers splanchnic phenylalanine extraction, increases post-prandial plasma amino acid availability, and tends to augment the subsequent muscle protein synthetic response in vivo in humans (Koopman et al. 2009a). These findings may indicate that part of the proposed anabolic resistance in the elderly might be compensated for in part by enhancing amino acid availability during the post-prandial period. In accordance, it has been reported that protein pulse feeding (providing up to 80% of daily protein intake in one meal) leads to greater protein retention than ingesting the same amount of protein provided over four meals throughout the day (spread-feeding) in elderly women (Arnal et al. 1999, 2000). In agreement, pulse feeding did not lead to greater protein retention than spread feeding when applied in young females (Arnal et al. 2000).

4 Exercise and Muscle Protein Turnover

Physical activity, in particular resistance type exercise, is a powerful stimulus to promote net muscle protein anabolism, resulting in specific metabolic and morphological adaptations in skeletal muscle tissue. Resistance type exercise training can effectively increase muscle strength, muscle mass and, as such, improve physical performance and functional capacity (Evans 1995). Following a single bout of resistance type exercise, specific signaling pathways are activated, which result in a temporally increase in muscle IGF-1 gene expression (Chesley et al. 1992), whereas myostatin expression is reduced (Raue et al. 2006). As a result, mRNA translation is enhanced (Rommel et al. 2001) and DNA transcription is increased via activation of transcription factors like MyoD and Myogenin (Willoughby and Nelson 2002).

A single bout of resistance exercise rapidly (within 2–4 h, (Phillips et al. 1997)) stimulates muscle protein synthesis, and increased protein synthesis rates, in particular myofibrillar protein synthesis (Welle et al. 1993; Yarasheski et al. 1993; Wilkinson et al. 2008), have been reported to persist for up to 16 h in trained (Tang et al. 2008) and 24–48 h in untrained individuals (Chesley et al. 1992; Phillips et al. 1997; Tang et al. 2008). Muscle protein breakdown is also stimulated following exercise, albeit to a lesser extent than protein synthesis (Biolo et al. 1995; Phillips et al. 1997). The latter results in an improved net muscle protein balance that persists up to 48 h in untrained individuals (Phillips et al. 1997). Net muscle protein balance remains negative in the absence of nutrient intake (Phillips et al. 1997), and as such, both exercise and nutrition are required to obtain a positive muscle net amino acid balance and, as such, allow muscle hypertrophy. Carbohydrate and protein/amino acid ingestion during post-exercise recovery forms an effective strategy to enable net muscle protein accretion. Ingestion of carbohydrate following exercise has been shown to improve net leg amino acid balance without affecting muscle protein synthesis rates (Borsheim et al. 2004). Ingestion of carbohydrate effectively increases plasma insulin levels and stimulates muscle protein anabolism, primarily by reducing muscle protein degradation (Gelfand and Barrett 1987; Hillier et al. 1998). As explained above, insulin should not be regarded as a primary regulator of muscle protein synthesis as insulin exerts only a modest effect on muscle protein synthesis in the absence of elevated amino acid concentrations (Cuthbertson et al. 2005). In rodent models, it has been reported that an increase in circulating plasma insulin concentrations does not further enhance mRNA translation initiation during post-exercise recovery (Fedele et al. 2000; Gautsch et al. 1998; Kimball et al. 2002). In a recent attempt to assess whether carbohydrate co-ingestion is required to maximize post-exercise muscle protein synthesis, we observed no surplus effect of carbohydrate co-ingestion on post-exercise muscle protein synthesis under conditions where ample protein is ingested (Koopman et al. 2007a). Although carbohydrate co-ingestion does not seem to be required to maximize post-exercise muscle protein synthesis rates, it is likely that carbohydrate co-ingestion can further inhibit the post-exercise increase in muscle protein breakdown (Borsheim et al. 2004), thereby improving net protein balance (Borsheim et al. 2004; Roy et al. 1997).

There is a substantial amount of evidence showing that protein/amino acid administration effectively stimulates muscle protein synthesis in a dose-dependent manner. Hyperaminoacidemia, following intravenous amino acid infusion, increases post-exercise muscle protein synthesis rates and prevents the exercise-induced increase in protein degradation (Biolo et al. 1997). In a more practical, physiological setting, oral administration of a large single bolus (Tipton et al. 1999a) or repeated boluses (Koopman et al. 2005, 2006) of a protein and/or amino acid mixture ingested following resistance type exercise also substantially increases muscle protein synthesis rates. Moreover, ingestion of smaller amounts of EAA or intact protein with and without carbohydrate have all been shown to augment post-exercise protein synthesis rates and improve net protein balance (Borsheim et al. 2002; Dreyer et al. 2008; Drummond et al. 2008a; Miller et al. 2003; Rasmussen et al. 2000; Tang et al. 2008; Wilkinson et al. 2007). In short, it has been well established that post-exercise amino acid/protein ingestion represents an effective strategy to augment the anabolic response to exercise and ample amino acid supply to the muscle is crucial to allow hypertrophy following resistance exercise training.

It has been suggested that the timing of amino acid/protein intake is instrumental to further optimize the anabolic response to exercise (Beelen et al. 2008a; Esmarck et al. 2001; Tipton et al. 2001). As a result, several research groups have studied the efficacy of protein/amino acid ingestion prior to and/or during exercise to further augment muscle protein synthesis. Recently, we reported that protein ingestion prior to and during endurance (Koopman et al. 2004) and resistance (Beelen et al. 2008a) type exercise stimulate whole-body (Beelen et al. 2008a; Koopman et al. 2004) and mixed muscle protein synthesis (Beelen et al. 2008a) during exercise. In line with these findings, we have reported that protein intake prior to exercise augments activation of the PI-3 kinase/mTOR-pathway during subsequent post-exercise recovery (Koopman et al. 2007b). In addition, protein ingestion prior to and/or during exercise may further enhance muscle protein anabolism by blunting the exercise induced increase in protein breakdown. Interestingly, a recent study by Fujita et al. showed no additional benefits of the ingestion of small amounts of EAA prior to resistance type exercise on post-exercise muscle protein synthesis rates, despite significantly elevated phosphorylation of S6K1 and 4E-BP1 (Fujita et al. 2008). In addition, a recent study from our lab showed no effect of protein ingestion prior to, during, and after exercise on muscle protein synthesis measured during subsequent overnight recovery (Beelen et al. 2008b). The latter might be attributed to the fact that subjects were studied in the fed state, performing exercise in the evening after receiving a standardized diet throughout the day. Clearly, more research is warranted to assess the impact of timing of food intake on the skeletal muscle adaptive response to exercise.

As discussed previously, the increase in extracellular leucine concentration has been proposed to represent an important nutritional signal that drives the postprandial increase in muscle protein synthesis (Kimball and Jefferson 2004).

The dose-dependent relationship between myofibrillar protein synthesis and the availability of leucine in the plasma (Cuthbertson et al. 2005) has provided a strong foundation for the hypothesis that the ingestion of additional leucine during post-exercise recovery could further accelerate post-exercise muscle protein synthesis rates. Recently, Dreyer et al. reported that ingestion of a leucine-enriched EAA and carbohydrate mixture following resistance type exercise enhances mTOR signaling and muscle protein synthesis in vivo in humans (Dreyer et al. 2008). However, previous observations in our lab showed no surplus value of additional leucine supplementation in either young (Koopman et al. 2005) or old subjects (Koopman et al. 2008) when a substantial amount of protein was being ingested during post-exercise recovery.

5 Aging and the Anabolic Response to Exercise

Muscle protein synthesis is responsive to exercise in both the young and elderly. In studies performed in young and elderly individuals, resistance and endurance type exercise have been shown to stimulate mixed muscle protein synthesis (Drummond et al. 2008b; Fujita et al. 2007; Kumar et al. 2009; Sheffield-Moore et al. 2004; Welle et al. 1995; Yarasheski et al. 1993). Furthermore, SC content has been shown to increase following a single bout of exercise and following more prolonged resistance type exercise training in both the young and the elderly. However, the increase in muscle SC content following eccentric contractions is greater in young compared with elderly humans, suggesting that SC recruitment in response to exercise is blunted in the elderly (Dreyer et al. 2006). SC proliferation and/or differentiation are controlled by the sequential activation and/or suppression of MRFs (i.e. Myf5, MyoD, Myogenin, and Mrf4). Interestingly, MRF mRNA expression appears to increase following resistance type exercise in both young and older adults (Costa et al. 2007; Kim et al. 2005b; Kosek et al. 2006; Raue et al. 2006; McKay et al. 2008; Psilander et al. 2003; Yang et al. 2005). In addition, the impaired Notch signaling in the elderly has been reported to be modulated by strenuous exercise (Carey et al. 2007). In contrast with the upregulation of MRFs, myostatin mRNA expression is found to be down-regulated in response to exercise training in both the young and elderly (Costa et al. 2007; Hulmi et al. 2008; Kim et al. 2005a; Raue et al. 2006; Roth et al. 2003; Walker et al. 2004). Thus, although some studies have reported subtle differences in changes in gene expression and anabolic signaling (Hameed et al. 2003), early studies indicate that the protein synthetic response to resistance type exercise does not differ considerably between the young and elderly (Hasten et al. 2000; Yarasheski et al. 1993). In contrast, a more recent study shows anabolic resistance of anabolic signaling (i.e. 4E-BP1 and S6K1) and muscle protein synthesis to resistance type exercise in elderly men when compared to young controls, with measurements being performed in the post-absorptive state (Kumar et al. 2009). This study is the first to demonstrate that the sigmoidal response of muscle protein synthesis to resistance exercise of different (increasing) intensities is shifted downward in older men compared with younger men (Kumar et al. 2009). In addition, it has recently been suggested that mRNA expression of proteolytic regulators, such as Atrogin-1, are elevated in senescent compared with young muscle at rest and these levels increased even further in response to resistance type exercise in the elderly. These findings from Raue et al. (2007) suggest that the regulation of ubiquitin proteasome-related genes involved with muscle atrophy might be altered in the elderly and protein breakdown may be increased in elderly humans (Raue et al. 2007). However, there is a paucity of data regarding the measurement of muscle protein breakdown in response to exercise in the elderly and it is clear that more work is needed to assess the impact of exercise and specific exercise modalities on post-exercise muscle protein synthesis and breakdown rates and associated myocellular signaling in young and elderly humans.

We have previously shown that muscle protein synthesis rates are lower in elderly (~75 years) compared with young controls under conditions in which resistance type exercise is followed by food intake (Koopman et al. 2006). However, combined ingestion of carbohydrate and protein during recovery from physical activity resulted in similar increases in mixed muscle protein synthesis rates in young and elderly men (Koopman et al. 2006). In line with these findings, Drummond et al. recently reported similar post-exercise muscle protein synthesis rates over a 5 h recovery period in young versus elderly subjects following ingestion of carbohydrate with an EAA mixture (Drummond et al. 2008b). However, their data indicated that the anabolic response to exercise and food intake was delayed in the elderly. During the first 3 h of post-exercise recovery the young subjects showed a substantial increase in muscle protein synthesis rate, which was not observed in the elderly. The latter may be attributed to a more pronounced activation of AMPK and/or reduced ERK1/2 activation during exercise, which seems to be in line with the recently reported attenuated rise in 4E-BP1 phosphorylation following resistance type exercise in the elderly (Kumar et al. 2009). The mechanisms responsible for the delayed intramyocellular activation of the mTOR pathway remain unclear, but might include differences in muscle recruitment, muscle fiber type composition, capacity and/or sensitivity of the muscle protein synthetic machinery, the presence of an inflammatory state, and/or the impact of stress on the cellular energy status of the cell between the young and the elderly.

6 Exercise Training in the Elderly

The clinical relevance of nutritional and/or exercise intervention in the elderly naturally resides in the long-term impact on skeletal muscle mass and strength, and the implications for functional capacity and the risk of developing chronic metabolic disease. In accordance with the previously discussed work it has been well established that the ability of the muscle protein synthetic machinery to respond to anabolic stimuli is preserved, albeit maybe to a lesser extent (Rennie 2009), until very old age (Fiatarone et al. 1990; Frontera et al. 1988). Resistance type exercise interventions have been shown effective in augmenting skeletal muscle mass, increasing muscle strength, and/or improving functional capacity in the elderly (Ades et al. 1996; Bamman et al. 2003; Fiatarone et al. 1990, 1994; Frontera et al. 1988, 1990, 2003; Lexell et al. 1995; Vincent et al. 2002; Brose et al. 2003; Ferri et al. 2003; Godard et al. 2002; Iglay et al. 2007; Kosek et al. 2006; Martel et al. 2006; Verdijk et al. 2009a; Haub et al. 2002). In addition, endurance (Ades et al. 1996; Bamman et al. 2003; Fiatarone et al. 1990, 1994; Frontera et al. 1988, 1990, 2003; Lexell et al. 1995; Vincent et al. 2002) type exercise activities have been shown to enhance skeletal muscle oxidative capacity, resulting in greater endurance capacity (Short et al. 2003, 2004).

The muscle regenerative capacity seems to decline at a more advanced age (i.e. decline in SC number and/or activation status). However, it is obvious that a reduced SC pool size does not prevent the capacity to allow extensive muscle hypertrophy even at an advanced age (Dedkov et al. 2003; Shefer et al. 2006; Thornell et al. 2003). Moreover, resistance type exercise training has been shown to increase muscle fiber size with a concurrent increase in SC content (Kadi et al. 2004b; Kadi and Thornell 2000; Petrella et al. 2006; Olsen et al. 2006). Some (Mackey et al. 2007; Roth et al. 2001; Verdijk et al. 2009a; Verney et al. 2008) but not all (Hikida et al. 1998; Petrella et al. 2006) studies report a substantial increase in SC content following 9–16 weeks of resistance type exercise training in older adults. Recently, we assessed the effects of 12 weeks resistance type exercise training on fiber type specific hypertrophy and SC content in healthy, elderly men (Verdijk et al. 2009a). Elderly men show a reduced type II muscle fiber size and SC content when compared with the type I muscle fibers. Interestingly, prolonged exercise training resulted in a 28% increase in type II muscle fiber size and a concomitant 76% increase in type II muscle fiber SC content in elderly males (Verdijk et al. 2009a). The apparent differences in fiber size and/or SC content between type I and type II muscle fibers prior to intervention were no longer evident after 12 weeks of training. Overall, these findings suggest that SC are instrumental in the generation of new myonuclei to facilitate muscle fiber hypertrophy.

Numerous studies have highlighted the need for protein/amino acid ingestion before, during, and/or after exercise to stimulate muscle protein synthesis and reduce muscle protein breakdown. Remarkably, little evidence exists that dietary co-interventions can further augment the adaptive response to prolonged exercise training in the elderly. Even the proposed importance of ample dietary protein intake in the long-term adaptive response to resistance training in the elderly has been a topic of intense debate (Campbell and Evans 1996; Morais et al. 2006; Campbell and Leidy 2007). The current Recommended Dietary Allowance (RDA) for habitual protein intake of 0.8 g/kg/day (Rand et al. 2003; Trumbo et al. 2002) has been suggested to be marginal to allow lean mass accretion following resistance exercise training in the elderly (Campbell et al. 2002). Moreover, it has been suggested that the RDA is even insufficient for long term maintenance of skeletal muscle mass in sedentary elderly (Campbell et al. 2001). However, more recent work by the same research group indicates that dietary protein requirements do not increase with age, and that a dietary protein allowance of 0.85 g/kg/day is adequate (Campbell et al. 2008).

When habitual dietary protein intake is standardized at 0.9 g/kg/day, exercise induced increases in muscle mass become apparent and a further increase in protein intake does not seem to have any additional effect (Iglay et al. 2007). In addition, data from a recent study by Walrand et al. (2008) indicated that although increased protein intake in the elderly further improved nitrogen balance (by increasing amino acid oxidation), no beneficial effects on muscle protein synthesis and muscle function were observed in that study(Walrand et al. 2008). These observations might explain why most studies fail to observe any additional benefit of nutritional co-intervention on the skeletal muscle adaptive response to prolonged resistance type exercise training in the elderly (Campbell et al. 1995; Fiatarone et al. 1990, 1994; Freyssenet et al. 1996; Frontera et al. 1988; Godard et al. 2002; Haub et al. 2002; Iglay et al. 2007; Meredith et al. 1992; Verdijk et al. 2009b; Welle and Thornton 1998). However, the absence of any benefits of nutritional co-intervention may be attributed to a less than optimal timing of amino acid and/or protein supplementation. Esmarck et al. (2001) concluded that an early intake of a protein supplement immediately after each bout of resistance type exercise, as opposed to 2 h later, is required for skeletal muscle hypertrophy to occur following 12 weeks of intervention in the elderly. However, the absence of any hypertrophy in the control group receiving the same supplement 2 h after cessation of each exercise bout seems to be in conflict with previous studies that show muscle hypertrophy following resistance training without any dietary intervention (Esmarck et al. 2001). Nevertheless, the proposed importance of nutrient timing is supported by more recent studies investigating the impact of amino acid or protein co-ingestion prior to, during, and/or after exercise on the acute muscle protein synthetic response (Beelen et al. 2008a; Tipton et al. 2001). To study the proposed impact of timed protein supplementation during prolonged exercise intervention, we recently compared increases in skeletal muscle mass and strength following 3 months of resistance type exercise training with or without protein ingestion prior to and immediately after each exercise session in elderly males (Verdijk et al. 2009b). However, timed protein supplementation prior to and after each exercise bout did not further increase skeletal muscle hypertrophy in these healthy, elderly men who habitually consumed ~1.0 g protein/kg/day.

Altogether, the available data suggest that sufficient habitual protein intake (~0.9 g/kg/day) combined with a normal meal pattern (i.e. providing ample protein three times per day) will allow substantial gains in muscle mass and strength following resistance type exercise training in the elderly. Additional protein supplementation does not seem to provide large surplus benefits to exercise intervention in healthy, elderly males. Additional protein intake may reduce subsequent voluntary food consumption in the elderly (Fiatarone Singh et al. 2000) and as a consequence some have suggested that supplementation with EAA would be more efficient (Timmerman and Volpi 2008). Clearly, acute studies have shown benefits of timed supplementation with small (~7–15 g) amounts of EAA on muscle protein synthesis (Katsanos et al. 2005; Paddon-Jones et al. 2004, 2006). However, well designed, double-blind, placebo-controlled long-term studies to investigate beneficial and adverse effects of long-term EAA supplementation in the elderly have yet to be performed (Henderson et al. 2009).

7 Future Research

Over the last ~30 years our understanding about the regulation of muscle protein synthesis and degradation and their response to exercise and nutrition has increased tremendously. However, despite significant technical evolution during this time, the sensitivity of the measurement and large inter-subject variance in (basal) muscle protein synthesis rates limit the ability to detect small, but potentially physiologically relevant differences between groups. It should be noted that even minor differences in, for example basal muscle protein synthesis and/or breakdown rate (<10%) would be clinically relevant when calculating their impact over one or more decades before sarcopenia becomes evident. Therefore, more sensitive methods should be developed to assess both muscle protein synthesis and breakdown rates in vivo in humans. In particular, more work is needed to develop valid tracer-models to assess muscle protein breakdown rates in various settings to complement currently used measurements of proteasome and calpain activity.

Even though it has been demonstrated that satellite cell (SC) content is reduced in the elderly, little is known about changes in activation status that occur when we get older. Moreover, the molecular mechanisms controlling SC activation, proliferation, differentiation and self-renewal in vivo in humans remain to be established. The identification of molecular key signatures of quiescent and activated SC may help to determine the precise signaling pathways leading to SC activation. Discovery of these key-regulatory proteins can potentially result in the identification of new targets for nutritional and pharmacological strategies to improve skeletal muscle development in pathological conditions.

Most of the data provided in this chapter agree with the concept that the post-prandial muscle protein synthetic response is set-off by a specific nutritional signal, most likely the post-prandial rise in plasma availability of one or more specific EAA and/or the concomitant insulin response allowing the AA to reach the extracellular matrix of the target tissue, and that the sensitivity and/or capacity of this signaling process is impaired with aging. Much effort is presently being directed toward the discovery of such an extracellular amino acid sensing mechanism in skeletal muscle tissue. The latter will further increase our understanding of the proposed impact of the anabolic resistance to food intake in the etiology of sarcopenia.

Nutrient availability throughout day and night likely plays an important role in the differential response to acute versus long-term exercise intervention. We speculate that potential benefits of (timed) protein and/or amino acid supplementation in the elderly might be restricted to specific elderly subpopulations, e.g. malnourished or frail elderly, and various patient populations. So far, it is evident that the combination of resistance type exercise training with or without post-exercise protein administration represents a feasible and effective strategy to improve muscle mass, strength, and functional performance in the elderly. More research is necessary to study the interaction between exercise and nutrition in the elderly, and the implications for the acute and long-term adaptive response to intervention.

8 Conclusions

The loss of skeletal muscle mass with aging is associated with reduced muscle strength, the loss of functional capacity, and an increased risk of developing chronic metabolic disease. The progressive loss of skeletal muscle mass does not seem to be attributed to age-related changes in basal muscle protein synthesis and/or breakdown rates. Recent work suggests that the muscle protein synthetic response to the main anabolic stimuli, i.e. food intake and/or physical activity, is blunted in the elderly. Despite this proposed anabolic resistance to food intake and/or physical activity, resistance type exercise substantially stimulates net muscle protein accretion when protein is ingested prior to, during, and/or following exercise in both the young and the elderly. In accordance, prolonged resistance type exercise training has proven an effective interventional strategy to prevent and/or treat the loss of muscle mass and strength in the elderly. Research is warranted to provide more insight in the interaction between nutrition, exercise and the skeletal muscle adaptive response. The latter is needed to define more effective nutritional, exercise, and/or pharmaceutical interventional strategies to prevent and/or treat sarcopenia.

Acknowledgements Dr. Koopman was supported by a Rubicon Fellowship from the Netherlands Organisation for Scientific Research (NWO). Dr. Koopman is a C.R. Roper Senior Research Fellow of the Faculty of Medicine, Dentistry and Health Sciences at the University of Melbourne (Victoria, Australia).

References

Ades, P. A., Ballor, D. L., Ashikaga, T., Utton, J. L., Nair, K. S. (1996). Weight training improves walking endurance in healthy elderly persons. *Annals of Internal Medicine, 124*, 568–572.

Arnal, M. A., Mosoni, L., Boirie, Y., Houlier, M. L., Morin, L., Verdier, E., Ritz, P., Antoine, J. M., Prugnaud, J., Beaufrere, B., Mirand, P. P. (1999). Protein pulse feeding improves protein retention in elderly women. *The American Journal of Clinical Nutrition, 69*, 1202–1208.

Arnal, M. A., Mosoni, L., Boirie, Y., Houlier, M. L., Morin, L., Verdier, E., Ritz, P., Antoine, J. M., Prugnaud, J., Beaufrere, B., Mirand, P. P. (2000). Protein feeding pattern does not affect protein retention in young women. *The Journal of Nutrition, 130*, 1700–1704.

Balagopal, P., Rooyackers, O. E., Adey, D. B., Ades, P. A., Nair, K. S. (1997). Effects of aging on in vivo synthesis of skeletal muscle myosin heavy-chain and sarcoplasmic protein in humans. *The American Journal of Physiology, 273*, E790–E800.

Bamman, M. M., Hill, V. J., Adams, G. R., Haddad, F., Wetzstein, C. J., Gower, B. A., Ahmed, A., Hunter, G. R. (2003). Gender differences in resistance-training-induced myofiber hypertrophy among older adults. *The Journals of Gerontology. Series A: Biological Sciences and Medical Sciences, 58*, 108–116.

Bassey, E. J., Fiatarone, M. A., O'Neill, E. F., Kelly, M., Evans, W. J., Lipsitz, L. A. (1992). Leg extensor power and functional performance in very old men and women. *Clinical Science (London), 82*, 321–327.

Baumgartner, R. N., Koehler, K. M., Gallagher, D., Romero, L., Heymsfield, S. B., Ross, R. R., Garry, P. J., Lindeman, R. D. (1998). Epidemiology of sarcopenia among the elderly in New Mexico. *American Journal of Epidemiology, 147*, 755–763.

Beaufrere, B., Dangin, M., Boirie, Y. (2000). The 'fast' and 'slow' protein concept. *Nestlé Nutrition Institute Workshop Series: Clinical & Performance Program, 3*, 121–131. discussion 131–133.

Beelen, M., Koopman, R., Gijsen, A. P., Vandereyt, H., Kies, A. K., Kuipers, H., Saris, W. H., van Loon, L. J. (2008a). Protein coingestion stimulates muscle protein synthesis during resistance-type exercise. *American Journal of Physiology. Endocrinology and Metabolism, 295*, E70–E77.

Beelen, M., Tieland, M., Gijsen, A. P., Vandereyt, H., Kies, A. K., Kuipers, H., Saris, W. H., Koopman, R., van Loon, L. J. (2008b). Coingestion of carbohydrate and protein hydrolysate stimulates muscle protein synthesis during exercise in young men, with no further increase during subsequent overnight recovery. *The Journal of Nutrition, 138*, 2198–2204.

Biolo, G., Maggi, S. P., Williams, B. D., Tipton, K. D., Wolfe, R. R. (1995). Increased rates of muscle protein turnover and amino acid transport after resistance exercise in humans. *The American Journal of Physiology, 268*, E514–E520.

Biolo, G., Tipton, K. D., Klein, S., Wolfe, R. R. (1997). An abundant supply of amino acids enhances the metabolic effect of exercise on muscle protein. *The American Journal of Physiology, 273*, E122–E129.

Bohe, J., Low, A., Wolfe, R. R., Rennie, M. J. (2003). Human muscle protein synthesis is modulated by extracellular, not intramuscular amino acid availability: a dose-response study. *Journal de Physiologie, 552*, 315–324.

Boirie, Y., Fauquant, J., Rulquin, H., Maubois, J. L., Beaufrere, B. (1995). Production of large amounts of [13C] leucine-enriched milk proteins by lactating cows. *The Journal of Nutrition, 125*, 92–98.

Boirie, Y., Gachon, P., Corny, S., Fauquant, J., Maubois, J. L., Beaufrere, B. (1996). Acute postprandial changes in leucine metabolism as assessed with an intrinsically labeled milk protein. *The American Journal of Physiology, 271*, E1083–E1091.

Boirie, Y., Dangin, M., Gachon, P., Vasson, M. P., Maubois, J. L., Beaufrere, B. (1997a). Slow and fast dietary proteins differently modulate postprandial protein accretion. *Proceedings of the National Academy of Sciences of the United States of America, 94*, 14930–14935.

Boirie, Y., Gachon, P., Beaufrere, B. (1997b). Splanchnic and whole-body leucine kinetics in young and elderly men. *The American Journal of Clinical Nutrition, 65*, 489–495.

Borsheim, E., Tipton, K. D., Wolf, S. E., Wolfe, R. R. (2002). Essential amino acids and muscle protein recovery from resistance exercise. *American Journal of Physiology. Endocrinology and Metabolism, 283*, E648–E657.

Borsheim, E., Cree, M. G., Tipton, K. D., Elliott, T. A., Aarsland, A., Wolfe, R. R. (2004). Effect of carbohydrate intake on net muscle protein synthesis during recovery from resistance exercise. *Journal of Applied Physiology, 96*, 674–678.

Brose, A., Parise, G., Tarnopolsky, M. A. (2003). Creatine supplementation enhances isometric strength and body composition improvements following strength exercise training in older adults. *The Journals of Gerontology. Series A: Biological Sciences and Medical Sciences, 58*, 11–19.

Brown, M., Sinacore, D. R., Host, H. H. (1995). The relationship of strength to function in the older adult. *The Journals of Gerontology. Series A: Biological Sciences and Medical Sciences, 50*(Spec No), 55–59.

Campbell, W. W. & Evans, W. J. (1996). Protein requirements of elderly people. *European Journal of Clinical Nutrition, 50*(Suppl 1), S180–S183. discussion S183–185.

Campbell, W. W. & Leidy, H. J. (2007). Dietary protein and resistance training effects on muscle and body composition in older persons. *Journal of the American College of Nutrition, 26*, 696S–703S.

Campbell, W. W., Crim, M. C., Young, V. R., Joseph, L. J., Evans, W. J. (1995). Effects of resistance training and dietary protein intake on protein metabolism in older adults. *The American Journal of Physiology, 268*, E1143–E1153.

Campbell, W. W., Trappe, T. A., Wolfe, R. R., Evans, W. J. (2001). The recommended dietary allowance for protein may not be adequate for older people to maintain skeletal muscle. *The Journals of Gerontology. Series A: Biological Sciences and Medical Sciences, 56*, M373–M380.

Campbell, W. W., Trappe, T. A., Jozsi, A. C., Kruskall, L. J., Wolfe, R. R., Evans, W. J. (2002). Dietary protein adequacy and lower body versus whole body resistive training in older humans. *Journal de Physiologie, 542*, 631–642.

Campbell, W. W., Johnson, C. A., Mccabe, G. P., Carnell, N. S. (2008). Dietary protein requirements of younger and older adults. *The American Journal of Clinical Nutrition, 88*, 1322–1329.

Carey, K. A., Farnfield, M. M., Tarquinio, S. D., Cameron-Smith, D. (2007). Impaired expression of Notch signaling genes in aged human skeletal muscle. *The Journals of Gerontology. Series A: Biological Sciences and Medical Sciences, 62*, 9–17.

Charette, S. L., Mcevoy, L., Pyka, G., Snow-Harter, C., Guido, D., Wiswell, R. A., Marcus, R. (1991). Muscle hypertrophy response to resistance training in older women. *Journal of Applied Physiology, 70*, 1912–1916.

Charifi, N., Kadi, F., Feasson, L., Denis, C. (2003). Effects of endurance training on satellite cell frequency in skeletal muscle of old men. *Muscle and Nerve, 28*, 87–92.

Chesley, A., Macdougall, J. D., Tarnopolsky, M. A., Atkinson, S. A., Smith, K. (1992). Changes in human muscle protein synthesis after resistance exercise. *Journal of Applied Physiology, 73*, 1383–1388.

Conboy, I. M. & Rando, T. A. (2002). The regulation of Notch signaling controls satellite cell activation and cell fate determination in postnatal myogenesis. *Developmental Cell, 3*, 397–409.

Costa, A., Dalloul, H., Hegyesi, H., Apor, P., Csende, Z., Racz, L., Vaczi, M., Tihanyi, J. (2007). Impact of repeated bouts of eccentric exercise on myogenic gene expression. *European Journal of Applied Physiology, 101*, 427–436.

Cuthbertson, D., Smith, K., Babraj, J., Leese, G., Waddell, T., Atherton, P., Wackerhage, H., Taylor, P. M., Rennie, M. J. (2005). Anabolic signaling deficits underlie amino acid resistance of wasting, aging muscle. *The FASEB Journal, 19*, 422–424.

Dangin, M., Boirie, Y., Garcia-Rodenas, C., Gachon, P., Fauquant, J., Callier, P., Ballevre, O., Beaufrere, B. (2001). The digestion rate of protein is an independent regulating factor of postprandial protein retention. *American Journal of Physiology. Endocrinology and Metabolism, 280*, E340–E348.

Dangin, M., Boirie, Y., Guillet, C., Beaufrere, B. (2002). Influence of the protein digestion rate on protein turnover in young and elderly subjects. *The Journal of Nutrition, 132*, 3228S–3233S.

Dangin, M., Guillet, C., Garcia-Rodenas, C., Gachon, P., Bouteloup-Demange, C., Reiffers-Magnani, K., Fauquant, J., Ballevre, O., Beaufrere, B. (2003). The rate of protein digestion affects protein gain differently during aging in humans. *Journal de Physiologie, 549*, 635–644.

Dedkov, E. I., Borisov, A. B., Wernig, A., Carlson, B. M. (2003). Aging of skeletal muscle does not affect the response of satellite cells to denervation. *The Journal of Histochemistry and Cytochemistry, 51*, 853–863.

Dreyer, H. C., Blanco, C. E., Sattler, F. R., Schroeder, E. T., Wiswell, R. A. (2006). Satellite cell numbers in young and older men 24 hours after eccentric exercise. *Muscle and Nerve, 33*, 242–253.

Dreyer, H. C., Drummond, M. J., Pennings, B., Fujita, S., Glynn, E. L., Chinkes, D. L., Dhanani, S., Volpi, E., Rasmussen, B. B. (2008). Leucine-enriched essential amino acid and carbohydrate ingestion following resistance exercise enhances mTOR signaling and protein synthesis in human muscle. *American Journal of Physiology. Endocrinology and Metabolism, 294*, E392–E400.

Drummond, M. J., Bell, J. A., Fujita, S., Dreyer, H. C., Glynn, E. L., Volpi, E., Rasmussen, B. B. (2008a). Amino acids are necessary for the insulin-induced activation of mTOR/S6K1 signaling and protein synthesis in healthy and insulin resistant human skeletal muscle. *Clinical Nutrition, 27*, 447–456.

Drummond, M. J., Dreyer, H. C., Pennings, B., Fry, C. S., Dhanani, S., Dillon, E. L., Sheffield-Moore, M., Volpi, E., Rasmussen, B. B. (2008b). Skeletal muscle protein anabolic response to resistance exercise and essential amino acids is delayed with aging. *Journal of Applied Physiology, 104*, 1452–1461.

Edstrom, E., Ulfhake, B. (2005). Sarcopenia is not due to lack of regenerative drive in senescent skeletal muscle. *Aging Cell, 4*, 65–77.

Esmarck, B., Andersen, J. L., Olsen, S., Richter, E. A., Mizuno, M., Kjaer, M. (2001). Timing of postexercise protein intake is important for muscle hypertrophy with resistance training in elderly humans. *Journal de Physiologie, 535*, 301–311.

Evans, W. J. (1995). Effects of exercise on body composition and functional capacity of the elderly. *The Journals of Gerontology. Series A: Biological Sciences and Medical Sciences, 50*(Spec No), 147–150.

Fedele, M. J., Hernandez, J. M., Lang, C. H., Vary, T. C., Kimball, S. R., Jefferson, L. S., Farrell, P. A. (2000). Severe diabetes prohibits elevations in muscle protein synthesis after acute resistance exercise in rats. *Journal of Applied Physiology, 88*, 102–108.

Ferri, A., Scaglioni, G., Pousson, M., Capodaglio, P., van Hoecke, J., Narici, M. V. (2003). Strength and power changes of the human plantar flexors and knee extensors in response to resistance training in old age. *Acta Physiologica Scandinavica, 177*, 69–78.

Fiatarone, M. A., Marks, E. C., Ryan, N. D., Meredith, C. N., Lipsitz, L. A., Evans, W. J. (1990). High-intensity strength training in nonagenarians. Effects on skeletal muscle. *Jama, 263*, 3029–3034.

Fiatarone, M. A., O'Neill, E. F., Ryan, N. D., Clements, K. M., Solares, G. R., Nelson, M. E., Roberts, S. B., Kehayias, J. J., Lipsitz, L. A., Evans, W. J. (1994). Exercise training and nutritional supplementation for physical frailty in very elderly people. *The New England Journal of Medicine, 330*, 1769–1775.

Fiatarone Singh, M. A., Bernstein, M. A., Ryan, A. D., O'Neill, E. F., Clements, K. M., Evans, W. J. (2000). The effect of oral nutritional supplements on habitual dietary quality and quantity in frail elders. *The Journal of Nutrition, Health and Aging, 4*, 5–12.

Forbes, G. B. & Reina, J. C. (1970). Adult lean body mass declines with age: some longitudinal observations. *Metabolism, 19*, 653–663.

Freyssenet, D., Berthon, P., Denis, C., Barthelemy, J. C., Guezennec, C. Y., Chatard, J. C. (1996). Effect of a 6-week endurance training programme and branched-chain amino acid supplementation on histomorphometric characteristics of aged human muscle. *Archives of Physiology and Biochemistry, 104*, 157–162.

Frontera, W. R., Meredith, C. N., O'Reilly, K. P., Knuttgen, H. G., Evans, W. J. (1988). Strength conditioning in older men: skeletal muscle hypertrophy and improved function. *Journal of Applied Physiology, 64*, 1038–1044.

Frontera, W. R., Meredith, C. N., O'Reilly, K. P., Evans, W. J. (1990). Strength training and determinants of VO2max in older men. *Journal of Applied Physiology, 68*, 329–333.

Frontera, W. R., Hughes, V. A., Lutz, K. J., Evans, W. J. (1991). A cross-sectional study of muscle strength and mass in 45- to 78-yr-old men and women. *Journal of Applied Physiology, 71*, 644–650.

Frontera, W. R., Hughes, V. A., Fielding, R. A., Fiatarone, M. A., Evans, W. J., Roubenoff, R. (2000). Aging of skeletal muscle: a 12-yr longitudinal study. *Journal of Applied Physiology, 88*, 1321–1326.

Frontera, W. R., Hughes, V. A., Krivickas, L. S., Kim, S. K., Foldvari, M., Roubenoff, R. (2003). Strength training in older women: early and late changes in whole muscle and single cells. *Muscle and Nerve, 28*, 601–608.

Fujita, S., Rasmussen, B. B., Cadenas, J. G., Grady, J. J., Volpi, E. (2006). Effect of insulin on human skeletal muscle protein synthesis is modulated by insulin-induced changes in muscle blood flow and amino acid availability. *American Journal of Physiology. Endocrinology and Metabolism, 291*, E745–E754.

Fujita, S., Rasmussen, B. B., Cadenas, J. G., Drummond, M. J., Glynn, E. L., Sattler, F. R., Volpi, E. (2007). Aerobic exercise overcomes the age-related insulin resistance of muscle protein metabolism by improving endothelial function and Akt/mammalian target of rapamycin signaling. *Diabetes, 56*, 1615–1622.

Fujita, S., Dreyer, H.C., Drummond, M.J., Glynn, E.L., Volpi, E., Rasmussen, B.B. (2008). Essential amino acid and carbohydrate ingestion prior to resistance exercise does not enhance post-exercise muscle protein synthesis. *Journal of Applied Physiology, 106*(5), 1730–1739.

Gautsch, T. A., Anthony, J. C., Kimball, S. R., Paul, G. L., Layman, D. K., Jefferson, L. S. (1998). Availability of eIF4E regulates skeletal muscle protein synthesis during recovery from exercise. *The American Journal of Physiology, 274*, C406–C414.

Gelfand, R. A. & Barrett, E. J. (1987). Effect of physiologic hyperinsulinemia on skeletal muscle protein synthesis and breakdown in man. *Journal of Clinical Investigation, 80*, 1–6.

Godard, M. P., Williamson, D. L., Trappe, S. W. (2002). Oral amino-acid provision does not affect muscle strength or size gains in older men. *Medicine and Science in Sports and Exercise, 34*, 1126–1131.

Greenhaff, P. L., Karagounis, L. G., Peirce, N., Simpson, E. J., Hazell, M., Layfield, R., Wackerhage, H., Smith, K., Atherton, P., Selby, A., Rennie, M. J. (2008). Disassociation between the effects of amino acids and insulin on signaling, ubiquitin ligases, and protein turnover in human muscle. *American Journal of Physiology. Endocrinology and Metabolism, 295*, E595–E604.

Guillet, C., Prod'Homme, M., Balage, M., Gachon, P., Giraudet, C., Morin, L., Grizard, J., Boirie, Y. (2004a). Impaired anabolic response of muscle protein synthesis is associated with S6K1 dysregulation in elderly humans. *The FASEB Journal, 18*, 1586–1587.

Guillet, C., Zangarelli, A., Gachon, P., Morio, B., Giraudet, C., Rousset, P., Boirie, Y. (2004b). Whole body protein breakdown is less inhibited by insulin, but still responsive to amino acid, in nondiabetic elderly subjects. *The Journal of Clinical Endocrinology and Metabolism, 89*, 6017–6024.

Hameed, M., Orrell, R. W., Cobbold, M., Goldspink, G., Harridge, S. D. (2003). Expression of IGF-I splice variants in young and old human skeletal muscle after high resistance exercise. *Journal de Physiologie, 547*, 247–254.

Hasten, D. L., Pak-Loduca, J., Obert, K. A., Yarasheski, K. E. (2000). Resistance exercise acutely increases MHC and mixed muscle protein synthesis rates in 78–84 and 23–32 yr olds. *American Journal of Physiology. Endocrinology and Metabolism, 278*, E620–E626.

Haub, M. D., Wells, A. M., Tarnopolsky, M. A., Campbell, W. W. (2002). Effect of protein source on resistive-training-induced changes in body composition and muscle size in older men. *The American Journal of Clinical Nutrition, 76*, 511–517.

Hawke, T. J. & Garry, D. J. (2001). Myogenic satellite cells: physiology to molecular biology. *Journal of Applied Physiology, 91*, 534–551.

Henderson, G. C., Irving, B. A., Nair, K. S. (2009). Potential application of essential amino Acid supplementation to treat sarcopenia in elderly people. *The Journal of Clinical Endocrinology and Metabolism, 94*, 1524–1526.

Hikida, S., Takeuchi, M., Hata, H., Yamana, H., Fujita, H., Shirouzu, K., Matsuno, K., Tanaka, T., Kawaguchi, C., Akiyoshi, K., Tsuru, T., Tanaka, Y., Mizote, H. (1998). Free jejunal graft autotransplantation should be revascularized within 3 hours. *Transplantation Proceedings, 30*, 3446–3448.

Hillier, T. A., Fryburg, D. A., Jahn, L. A., Barrett, E. J. (1998). Extreme hyperinsulinemia unmasks insulin's effect to stimulate protein synthesis in the human forearm. *The American Journal of Physiology, 274*, E1067–E1074.

Hulmi, J.J., Kovanen, V., Selanne, H., Kraemer, W.J., Hakkinen, K., MERO, A.A. (2008). Acute and long-term effects of resistance exercise with or without protein ingestion on muscle hypertrophy and gene expression. *Amino Acids, 37*, 297–308.

Iglay, H. B., Thyfault, J. P., Apolzan, J. W., Campbell, W. W. (2007). Resistance training and dietary protein: effects on glucose tolerance and contents of skeletal muscle insulin signaling proteins in older persons. *The American Journal of Clinical Nutrition, 85*, 1005–1013.

Kadi, F. & Thornell, L. E. (2000). Concomitant increases in myonuclear and satellite cell content in female trapezius muscle following strength training. *Histochemistry and Cell Biology, 113*, 99–103.

Kadi, F., Charifi, N., Denis, C., Lexell, J. (2004a). Satellite cells and myonuclei in young and elderly women and men. *Muscle and Nerve, 29*, 120–127.

Kadi, F., Schjerling, P., Andersen, L. L., Charifi, N., Madsen, J. L., Christensen, L. R., Andersen, J. L. (2004b). The effects of heavy resistance training and detraining on satellite cells in human skeletal muscles. *Journal de Physiologie, 558*, 1005–1012.

Katsanos, C. S., Kobayashi, H., Sheffield-Moore, M., Aarsland, A., Wolfe, R. R. (2005). Aging is associated with diminished accretion of muscle proteins after the ingestion of a small bolus of essential amino acids. *The American Journal of Clinical Nutrition, 82*, 1065–1073.

Katsanos, C. S., Kobayashi, H., Sheffield-Moore, M., Aarsland, A., Wolfe, R. R. (2006). A high proportion of leucine is required for optimal stimulation of the rate of muscle protein synthesis by essential amino acids in the elderly. *American Journal of Physiology. Endocrinology and Metabolism, 291*, E381–E387.

Kim, J. S., Cross, J. M., Bamman, M. M. (2005a). Impact of resistance loading on myostatin expression and cell cycle regulation in young and older men and women. *American Journal of Physiology. Endocrinology and Metabolism, 288*, E1110–E1119.

Kim, J. S., Kosek, D. J., Petrella, J. K., Cross, J. M., Bamman, M. M. (2005b). Resting and load-induced levels of myogenic gene transcripts differ between older adults with demonstrable sarcopenia and young men and women. *Journal of Applied Physiology, 99*, 2149–2158.

Kimball, S. R. & Jefferson, L. S. (2004). Regulation of global and specific mRNA translation by oral administration of branched-chain amino acids. *Biochemical and Biophysical Research Communications, 313*, 423–427.

Kimball, S. R., Farrell, P. A., Jefferson, L. S. (2002). Invited review: role of insulin in translational control of protein synthesis in skeletal muscle by amino acids or exercise. *Journal of Applied Physiology, 93*, 1168–1180.

Koopman, R. & van Loon, L.J. (2009) Aging, exercise and muscle protein metabolism. *Journal of Applied Physiology, 106*, 2040–2048.

Koopman, R., Pannemans, D. L., Jeukendrup, A. E., Gijsen, A. P., Senden, J. M., Halliday, D., Saris, W. H., van Loon, L. J., Wagenmakers, A. J. (2004). Combined ingestion of protein and carbohydrate improves protein balance during ultra-endurance exercise. *American Journal of Physiology. Endocrinology and Metabolism, 287*, E712–E720.

Koopman, R., Wagenmakers, A. J., Manders, R. J., Zorenc, A. H., Senden, J. M., Gorselink, M., Keizer, H. A., van Loon, L. J. (2005). Combined ingestion of protein and free leucine with carbohydrate increases postexercise muscle protein synthesis in vivo in male subjects. *American Journal of Physiology. Endocrinology and Metabolism, 288*, E645–E653.

Koopman, R., Verdijk, L., Manders, R. J., Gijsen, A. P., Gorselink, M., Pijpers, E., Wagenmakers, A. J., van Loon, L. J. (2006). Co-ingestion of protein and leucine stimulates muscle protein synthesis rates to the same extent in young and elderly lean men. *The American Journal of Clinical Nutrition, 84*, 623–632.

Koopman, R., Beelen, M., Stellingwerff, T., Pennings, B., Saris, W. H., Kies, A. K., Kuipers, H., van Loon, L. J. (2007a). Co-ingestion of carbohydrate with protein does not further augment post-exercise muscle protein synthesis. *American Journal of Physiology. Endocrinology and Metabolism, 293*, E833–E842.

Koopman, R., Pennings, B., Zorenc, A. H., van Loon, L. J. (2007b). Protein ingestion further augments S6K1 phosphorylation in skeletal muscle following resistance type exercise in males. *The Journal of Nutrition, 137*, 1836–1842.

Koopman, R., Verdijk, L. B., Beelen, M., Gorselink, M., Kruseman, A. N., Wagenmakers, A. J., Kuipers, H., van Loon, L. J. (2008). Co-ingestion of leucine with protein does not further augment post-exercise muscle protein synthesis rates in elderly men. *The British Journal of Nutrition, 99*, 571–580.

Koopman, R., Crombach, N., Gijsen, A. P., Walrand, S., Fauquant, J., Kies, A. K., Lemosquet, S., Saris, W. H., Boirie, Y., van Loon, L. J. (2009a). Ingestion of a protein hydrolysate is accompanied by an accelerated in vivo digestion and absorption rate when compared with its intact protein. *The American Journal of Clinical Nutrition, 90*, 106–115.

Koopman, R., Walrand, S., Beelen, M., Gijsen, A.P., Kies, A.K., Boirie, Y., Saris, W.H., van Loon, L.J. (2009b) Dietary protein digestion and absorption rate and the subsequent muscle protein synthetic response are not different between young and elderly men. *Journal of Nutrition, 139*(9), 1707–1713.

Kosek, D. J., Kim, J. S., Petrella, J. K., Cross, J. M., Bamman, M. M. (2006). Efficacy of 3 days/wk resistance training on myofiber hypertrophy and myogenic mechanisms in young vs. older adults. *Journal of Applied Physiology, 101*, 531–544.

Kumar, V., Selby, A., Rankin, D., Patel, R., Atherton, P., Hildebrandt, W., Williams, J., Smith, K., Seynnes, O., Hiscock, N., Rennie, M. J. (2009). Age-related differences in dose response of muscle protein synthesis to resistance exercise in young and old men. *Journal de Physiologie, 587*, 211–217.

Landers, K. A., Hunter, G. R., Wetzstein, C. J., Bamman, M. M., Weinsier, R. L. (2001). The interrelationship among muscle mass, strength, and the ability to perform physical tasks of daily living in younger and older women. *The Journals of Gerontology. Series A: Biological Sciences and Medical Sciences, 56*, B443–B448.

Larsson, L. (1978). Morphological and functional characteristics of the ageing skeletal muscle in man. A cross-sectional study. *Acta Physiologica Scandinavica. Supplementum, 457*, 1–36.

Larsson, L. & Karlsson, J. (1978). Isometric and dynamic endurance as a function of age and skeletal muscle characteristics. *Acta Physiologica Scandinavica, 104*, 129–136.

Larsson, L., Sjodin, B., Karlsson, J. (1978). Histochemical and biochemical changes in human skeletal muscle with age in sedentary males, age 22–65 years. *Acta Physiologica Scandinavica, 103*, 31–39.

Lexell, J. (1995). Human aging, muscle mass, and fiber type composition. *The Journals of Gerontology. Series A: Biological Sciences and Medical Sciences, 50*(Spec No), 11–16.

Lexell, J., Taylor, C. C., Sjostrom, M. (1988). What is the cause of the ageing atrophy? Total number, size and proportion of different fiber types studied in whole vastus lateralis muscle from 15- to 83-year-old men. *Journal of the Neurological Sciences, 84*, 275–294.

Lexell, J., Downham, D. Y., Larsson, Y., Bruhn, E., Morsing, B. (1995). Heavy-resistance training in older Scandinavian men and women: short- and long-term effects on arm and leg muscles. *Scandinavian Journal of Medicine & Science in Sports, 5*, 329–341.

Lindle, R. S., Metter, E. J., Lynch, N. A., Fleg, J. L., Fozard, J. L., Tobin, J., Roy, T. A., Hurley, B. F. (1997). Age and gender comparisons of muscle strength in 654 women and men aged 20-93 yr. *Journal of Applied Physiology, 83*, 1581–1587.

Mackey, A. L., Esmarck, B., Kadi, F., Koskinen, S. O., Kongsgaard, M., Sylvestersen, A., Hansen, J. J., Larsen, G., Kjaer, M. (2007). Enhanced satellite cell proliferation with resistance training in elderly men and women. *Scandinavian Journal of Medicine and Science in Sports, 17*, 34–42.

Martel, G. F., Roth, S. M., Ivey, F. M., Lemmer, J. T., Tracy, B. L., Hurlbut, D. E., Metter, E. J., Hurley, B. F., Rogers, M. A. (2006). Age and sex affect human muscle fibre adaptations to heavy-resistance strength training. *Experimental Physiology, 91*, 457–464.

Mauro, A. (1961). Satellite cell of skeletal muscle fibers. *The Journal of Biophysical and Biochemical Cytology, 9*, 493–495.

Mckay, B. R., O'Reilly, C. E., Phillips, S. M., Tarnopolsky, M. A., Parise, G. (2008). Co-expression of IGF-1 family members with myogenic regulatory factors following acute damaging muscle-lengthening contractions in humans. *Journal de Physiologie, 586*, 5549–5560.

Melton, L. J., 3RD Khosla, S., Crowson, C. S., O'Connor, M. K., O'Fallon, W. M., Riggs, B. L. (2000). Epidemiology of sarcopenia. *Journal of the American Geriatrics Society, 48*, 625–630.

Meredith, C. N., Frontera, W. R., O'Reilly, K. P., Evans, W. J. (1992). Body composition in elderly men: effect of dietary modification during strength training. *Journal of the American Geriatrics Society, 40*, 155–162.

Miller, S. L., Tipton, K. D., Chinkes, D. L., Wolf, S. E., Wolfe, R. R. (2003). Independent and combined effects of amino acids and glucose after resistance exercise. *Medicine and Science in Sports and Exercise, 35*, 449–455.

Morais, J. A., Chevalier, S., Gougeon, R. (2006). Protein turnover and requirements in the healthy and frail elderly. *The Journal of Nutrition, Health and Aging, 10*, 272–283.

Moss, F. P. & Leblond, C. P. (1970). Nature of dividing nuclei in skeletal muscle of growing rats. *The Journal of Cell Biology, 44*, 459–462.

Moss, F. P. & Leblond, C. P. (1971). Satellite cells as the source of nuclei in muscles of growing rats. *The Anatomical Record, 170*, 421–435.

Musaro, A., Cusella de Angelis, M. G., Germani, A., Ciccarelli, C., Molinaro, M., Zani, B. M. (1995). Enhanced expression of myogenic regulatory genes in aging skeletal muscle. *Experimental Cell Research, 221*, 241–248.

Nair, K. S. (1995). Muscle protein turnover: methodological issues and the effect of aging. *The Journals of Gerontology. Series A: Biological Sciences and Medical Sciences, 50*(Spec No), 107–112.

Nair, K. S. (2005). Aging muscle. *The American Journal of Clinical Nutrition, 81*, 953–963.

Nofziger, D., Miyamoto, A., Lyons, K. M., Weinmaster, G. (1999). Notch signaling imposes two distinct blocks in the differentiation of C2C12 myoblasts. *Development, 126*, 1689–1702.

Norton, L. E. & Layman, D. K. (2006). Leucine regulates translation initiation of protein synthesis in skeletal muscle after exercise. *The Journal of Nutrition, 136*, 533S–537S.

Olsen, S., Aagaard, P., Kadi, F., Tufekovic, G., Verney, J., Olesen, J. L., Suetta, C., Kjaer, M. (2006). Creatine supplementation augments the increase in satellite cell and myonuclei number in human skeletal muscle induced by strength training. *Journal de Physiologie, 573*, 525–534.

Paddon-Jones, D., Sheffield-Moore, M., Zhang, X. J., Volpi, E., Wolf, S. E., Aarsland, A., Ferrando, A. A., Wolfe, R. R. (2004). Amino acid ingestion improves muscle protein synthesis in the young and elderly. *American Journal of Physiology. Endocrinology and Metabolism, 286*, E321–E328.

Paddon-Jones, D., Sheffield-Moore, M., Katsanos, C. S., Zhang, X. J., Wolfe, R. R. (2006). Differential stimulation of muscle protein synthesis in elderly humans following isocaloric ingestion of amino acids or whey protein. *Experimental Gerontology, 41*, 215–219.

Petrella, J. K., Kim, J. S., Tuggle, S. C., Hall, S. R., Bamman, M. M. (2005). Age differences in knee extension power, contractile velocity, and fatigability. *Journal of Applied Physiology, 98*, 211–220.

Petrella, J. K., Kim, J. S., Cross, J. M., Kosek, D. J., Bamman, M. M. (2006). Efficacy of myonuclear addition may explain differential myofiber growth among resistance-trained young and older men and women. *American Journal of Physiology. Endocrinology and Metabolism, 291*, E937–E946.

Phillips, S. M., Tipton, K. D., Aarsland, A., Wolf, S. E., Wolfe, R. R. (1997). Mixed muscle protein synthesis and breakdown after resistance exercise in humans. *The American Journal of Physiology, 273*, E99–E107.

Psilander, N., Damsgaard, R., Pilegaard, H. (2003). Resistance exercise alters MRF and IGF-I mRNA content in human skeletal muscle. *Journal of Applied Physiology, 95*, 1038–1044.

Rand, W. M., Pellett, P. L., Young, V. R. (2003). Meta-analysis of nitrogen balance studies for estimating protein requirements in healthy adults. *The American Journal of Clinical Nutrition, 77*, 109–127.

Rasmussen, B. B., Tipton, K. D., Miller, S. L., Wolf, S. E., Wolfe, R. R. (2000). An oral essential amino acid-carbohydrate supplement enhances muscle protein anabolism after resistance exercise. *Journal of Applied Physiology, 88*, 386–392.

Rasmussen, B. B., Fujita, S., Wolfe, R. R., Mittendorfer, B., Roy, M., Rowe, V. L., Volpi, E. (2006). Insulin resistance of muscle protein metabolism in aging. *The FASEB Journal, 20*, 768–769.

Raue, U., Slivka, D., Jemiolo, B., Hollon, C., Trappe, S. (2006). Myogenic gene expression at rest and after a bout of resistance exercise in young (18–30 yr) and old (80–89 yr) women. *Journal of Applied Physiology, 101*, 53–59.

Raue, U., Slivka, D., Jemiolo, B., Hollon, C., Trappe, S. (2007). Proteolytic gene expression differs at rest and after resistance exercise between young and old women. *The Journals of Gerontology. Series A: Biological Sciences and Medical Sciences, 62*, 1407–1412.

Renault, V., Thornell, L. E., Eriksson, P. O., Butler-Browne, G., Mouly, V. (2002). Regenerative potential of human skeletal muscle during aging. *Aging Cell, 1*, 132–139.

Rennie, M. J. (2009). Anabolic resistance: the effects of aging, sexual dimorphism, and immobilization on human muscle protein turnover. *Applied Physiology, Nutrition, and Metabolism, 34*, 377–381.

Rennie, M. J., Edwards, R. H., Halliday, D., Matthews, D. E., Wolman, S. L., Millward, D. J. (1982). Muscle protein synthesis measured by stable isotope techniques in man: the effects of feeding and fasting. *Clinical Science (London), 63*, 519–523.

Rieu, I., Balage, M., Sornet, C., Giraudet, C., Pujos, E., Grizard, J., Mosoni, L., Dardevet, D. (2006). Leucine supplementation improves muscle protein synthesis in elderly men independently of hyperaminoacidaemia. *Journal de Physiologie, 575*, 305–315.

Rommel, C., Bodine, S. C., Clarke, B. A., Rossman, R., Nunez, L., Stitt, T. N., Yancopoulos, G. D., Glass, D. J. (2001). Mediation of IGF-1-induced skeletal myotube hypertrophy by PI(3)K/Akt/mTOR and PI(3)K/Akt/GSK3 pathways. *Nature Cell Biology, 3*, 1009–1013.

Rooyackers, O. E., Adey, D. B., Ades, P. A., Nair, K. S. (1996). Effect of age on in vivo rates of mitochondrial protein synthesis in human skeletal muscle. *Proceedings of the National Academy of Sciences of the United States of America, 93*, 15364–15369.

Roth, S. M., Martel, G. F., Ivey, F. M., Lemmer, J. T., Metter, E. J., Hurley, B. F., Rogers, M. A. (2000). Skeletal muscle satellite cell populations in healthy young and older men and women. *The Anatomical Record, 260*, 351–358.

Roth, S. M., Martel, G. F., Ivey, F. M., Lemmer, J. T., Tracy, B. L., Metter, E. J., Hurley, B. F., Rogers, M. A. (2001). Skeletal muscle satellite cell characteristics in young and older men and women after heavy resistance strength training. *The Journals of Gerontology. Series A: Biological Sciences and Medical Sciences, 56*, B240–B247.

Roth, S. M., Martel, G. F., Ferrell, R. E., Metter, E. J., Hurley, B. F., Rogers, M. A. (2003). Myostatin gene expression is reduced in humans with heavy-resistance strength training: a brief communication. *Experimental Biology Medicine (Maywood), 228*, 706–709.

Roy, B. D., Tarnopolsky, M. A., Macdougall, J. D., Fowles, J., Yarasheski, K. E. (1997). Effect of glucose supplement timing on protein metabolism after resistance training. *Journal of Applied Physiology, 82*, 1882–1888.

Sajko, S., Kubinova, L., Cvetko, E., Kreft, M., Wernig, A., Erzen, I. (2004). Frequency of M-cadherin-stained satellite cells declines in human muscles during aging. *The Journal of Histochemistry and Cytochemistry, 52*, 179–185.

Schubert, C. (2004). Notch, the new muscle booster. *Natural Medicines, 10*, 24.

Shefer, G., van de Mark, D. P., Richardson, J. B., Yablonka-Reuveni, Z. (2006). Satellite-cell pool size does matter: defining the myogenic potency of aging skeletal muscle. *Developmental Biology, 294*, 50–66.

Sheffield-Moore, M., Yeckel, C. W., Volpi, E., Wolf, S. E., Morio, B., Chinkes, D. L., Paddon-Jones, D., Wolfe, R. R. (2004). Postexercise protein metabolism in older and younger men following moderate-intensity aerobic exercise. *American Journal of Physiology. Endocrinology and Metabolism, 287*, E513–E522.

Short, K. R. & Nair, K. S. (2000). The effect of age on protein metabolism. *Current Opinion in Clinical Nutrition and Metabolic Care, 3*, 39–44.

Short, K. R., Vittone, J. L., Bigelow, M. L., Proctor, D. N., Rizza, R. A., Coenen-Schimke, J. M., Nair, K. S. (2003). Impact of aerobic exercise training on age-related changes in insulin sensitivity and muscle oxidative capacity. *Diabetes, 52*, 1888–1896.

Short, K. R., Vittone, J. L., Bigelow, M. L., Proctor, D. N., Nair, K. S. (2004). Age and aerobic exercise training effects on whole body and muscle protein metabolism. *American Journal of Physiology. Endocrinology and Metabolism, 286*, E92–E101.

Smith, K., Barua, J. M., Watt, P. W., Scrimgeour, C. M., Rennie, M. J. (1992). Flooding with L-[1-13C]leucine stimulates human muscle protein incorporation of continuously infused L-[1-13C]valine. *The American Journal of Physiology, 262*, E372–E376.

Snijders, T., Verdijk, L. B., van Loon, L. J. (2009). The impact of sarcopenia and exercise training on skeletal muscle satellite cells. *Ageing Research Reviews, 8*(4), 328–338.

Tang, J. E., Perco, J. G., Moore, D. R., Wilkinson, S. B., Phillips, S. M. (2008). Resistance training alters the response of fed state mixed muscle protein synthesis in young men. *American Journal of Physiology: Regulatory, Integrative and Comparative Physiology, 294*, R172–R178.

Thornell, L. E., Lindstrom, M., Renault, V., Mouly, V., Butler-Browne, G. S. (2003). Satellite cells and training in the elderly. *Scandinavian Journal of Medicine and Science in Sports, 13*, 48–55.

Timmerman, K. L. & Volpi, E. (2008). Amino acid metabolism and regulatory effects in aging. *Current Opinion in Clinical Nutrition and Metabolic Care, 11*, 45–49.

Tipton, K. D., Ferrando, A. A., Phillips, S. M., Doyle, D., Jr., Wolfe, R. R. (1999a). Postexercise net protein synthesis in human muscle from orally administered amino acids. *The American Journal of Physiology, 276*, E628–E634.

Tipton, K. D., Gurkin, B. E., Matin, S., Wolfe, R. R. (1999b). Nonessential amino acids are not necessary to stimulate net muscle protein synthesis in healthy volunteers. *The Journal of Nutritional Biochemistry, 10*, 89–95.

Tipton, K. D., Rasmussen, B. B., Miller, S. L., Wolf, S. E., Owens-Stovall, S. K., Petrini, B. E., Wolfe, R. R. (2001). Timing of amino acid-carbohydrate ingestion alters anabolic response of muscle to resistance exercise. *American Journal of Physiology. Endocrinology and Metabolism, 281*, E197–E206.

Trumbo, P., Schlicker, S., Yates, A. A., Poos, M. (2002). Dietary reference intakes for energy, carbohydrate, fiber, fat, fatty acids, cholesterol, protein and amino acids. *Journal of the American Dietetic Association, 102*, 1621–1630.

Verdijk, L. B., Koopman, R., Schaart, G., Meijer, K., Savelberg, H. H., van Loon, L. J. (2007). Satellite cell content is specifically reduced in type II skeletal muscle fibers in the elderly. *American Journal of Physiology. Endocrinology and Metabolism, 292*, E151–E157.

Verdijk, L. B., Gleeson, B. G., Jonkers, R. A. M., Meijer, K., Savelberg, H. H. C. M., Dendale, P., van Loon, L. J. C. (2009a). Skeletal muscle hypertrophy following resistance training is accompanied by a fiber type-specific increase in satellite cell content in elderly men. *The Journals of Gerontology. Series A: Biological Sciences and Medical Sciences, 64*, 332–339.

Verdijk, L. B., Jonkers, R. A. M., Gleeson, B. G., Beelen, M., Meijer, K., Savelberg, H. H. C. M., Wodzig, K. W. H., Dendale, P., van Loon, L. J. C. (2009b). Protein supplementation before and after exercise does not further augment skeletal muscle hypertrophy following resistance training in elderly men. *The American Journal of Clinical Nutrition, 89*, 608–616.

Verney, J., Kadi, F., Charifi, N., Feasson, L., Saafi, M. A., Castells, J., Piehl-Aulin, K., Denis, C. (2008). Effects of combined lower body endurance and upper body resistance training on the satellite cell pool in elderly subjects. *Muscle and Nerve, 38*, 1147–1154.

Vincent, K. R., Braith, R. W., Feldman, R. A., Magyari, P. M., Cutler, R. B., Persin, S. A., Lennon, S. L., Gabr, A. H., Lowenthal, D. T. (2002). Resistance exercise and physical performance in adults aged 60 to 83. *Journal of the American Geriatrics Society, 50*, 1100–1107.

Volpi, E., Ferrando, A. A., Yeckel, C. W., Tipton, K. D., Wolfe, R. R. (1998). Exogenous amino acids stimulate net muscle protein synthesis in the elderly. *Journal of Clinical Investigation, 101*, 2000–2007.

Volpi, E., Mittendorfer, B., Wolf, S. E., Wolfe, R. R. (1999). Oral amino acids stimulate muscle protein anabolism in the elderly despite higher first-pass splanchnic extraction. *The American Journal of Physiology, 277*, E513–E520.

Volpi, E., Mittendorfer, B., Rasmussen, B. B., Wolfe, R. R. (2000). The response of muscle protein anabolism to combined hyperaminoacidemia and glucose-induced hyperinsulinemia is impaired in the elderly. *The Journal of Clinical Endocrinology and Metabolism, 85*, 4481–4490.

Volpi, E., Sheffield-Moore, M., Rasmussen, B. B., Wolfe, R. R. (2001). Basal muscle amino acid kinetics and protein synthesis in healthy young and older men. *Jama, 286*, 1206–1212.

Volpi, E., Kobayashi, H., Sheffield-Moore, M., Mittendorfer, B., Wolfe, R. R. (2003). Essential amino acids are primarily responsible for the amino acid stimulation of muscle protein anabolism in healthy elderly adults. *The American Journal of Clinical Nutrition, 78*, 250–258.

Walker, K. S., Kambadur, R., Sharma, M., Smith, H. K. (2004). Resistance training alters plasma myostatin but not IGF-1 in healthy men. *Medicine and Science in Sports and Exercise, 36*, 787–793.

Walrand, S., Short, K. R., Bigelow, M. L., Sweatt, A. J., Hutson, S. M., Nair, K. S. (2008). Functional impact of high protein intake on healthy elderly people. *American Journal of Physiology. Endocrinology and Metabolism, 295*, E921–E928.

Welle, S. & Thornton, C. A. (1998). High-protein meals do not enhance myofibrillar synthesis after resistance exercise in 62- to 75-yr-old men and women. *The American Journal of Physiology, 274*, E677–E683.

Welle, S., Thornton, C., Jozefowicz, R., Statt, M. (1993). Myofibrillar protein synthesis in young and old men. *The American Journal of Physiology, 264*, E693–E698.

Welle, S., Thornton, C., Statt, M. (1995). Myofibrillar protein synthesis in young and old human subjects after three months of resistance training. *The American Journal of Physiology, 268*, E422–E427.

WHO (2008) Ageing (online) http://www.who.int/topics/ageing/en

Wilkes, E., Selby, A., Patel, R., Rankin, D., Smith, K., Rennie, M. J. (2008). Blunting of insulin-mediated proteolysis in leg muscle of elderly subjects may contribute to age-related sarcopenia (Abstract). *Proceedings of the Nutrition Society, 67*, E153.

Wilkinson, S. B., Tarnopolsky, M. A., Macdonald, M. J., Macdonald, J. R., Armstrong, D., Phillips, S. M. (2007). Consumption of fluid skim milk promotes greater muscle protein accretion after resistance exercise than does consumption of an isonitrogenous and isoenergetic soy-protein beverage. *The American Journal of Clinical Nutrition, 85*, 1031–1040.

Wilkinson, S. B., Phillips, S. M., Atherton, P. J., Patel, R., Yarasheski, K. E., Tarnopolsky, M. A., Rennie, M. J. (2008). Differential effects of resistance and endurance exercise in the fed state on signalling molecule phosphorylation and protein synthesis in human muscle. *Journal de Physiologie, 586*, 3701–3717.

Willoughby, D. S. & Nelson, M. J. (2002). Myosin heavy-chain mRNA expression after a single session of heavy-resistance exercise. *Medicine and Science in Sports and Exercise, 34*, 1262–1269.

Wolfson, L., Judge, J., Whipple, R., King, M. (1995). Strength is a major factor in balance, gait, and the occurrence of falls. *The Journals of Gerontology. Series A: Biological Sciences and Medical Sciences, 50*(Spec No), 64–67.

Yang, Y., Creer, A., Jemiolo, B., Trappe, S. (2005). Time course of myogenic and metabolic gene expression in response to acute exercise in human skeletal muscle. *Journal of Applied Physiology, 98*, 1745–1752.

Yarasheski, K. E., Zachwieja, J. J., Bier, D. M. (1993). Acute effects of resistance exercise on muscle protein synthesis rate in young and elderly men and women. *The American Journal of Physiology, 265*, E210–E214.

Yarasheski, K. E., Welle, S., Nair, K. S. (2002). Muscle protein synthesis in younger and older men. *Jama, 287*, 317–318.

Reactive Oxygen Species Generation and Skeletal Muscle Wasting – Implications for Sarcopenia

Anne McArdle and Malcolm J. Jackson

Abstract Frailty in the elderly is largely caused by loss of muscle mass and strength, increased susceptibility to injury, and impaired recovery following damage, particularly contraction-induced damage. The mechanisms responsible for the age-related loss of muscle mass and function are unclear although modified generation of Reactive Oxygen and Nitrogen Species (RONS) have been implicated in age-related tissue dysfunction. Many studies have provided evidence for the pivotal role of ROS in signal transduction and recognized these molecules as second messengers. Aberrant generation of RONS in the mitochondria and cytosol of cells and tissues of old mammals leads to an altered activation of crucial redox-responsive transcription factors at rest, following acute stress or during the regenerative process. Data suggest that targeted interventions to suppress altered mitochondrial ROS generation in muscle of old individuals are necessary to restore the signal for adaptive responses to contractions. Interventions based on antioxidant supplementation will suppress ROS signals in both mitochondrial and cytosolic compartments and hence be ineffective at prevention of age-related loss of muscle mass and function.

Keywords Skeletal muscle • Ageing • ROS • RONS • HSPs • Adaptive responses • Mitochondria • Cytosol

A. McArdle (✉) and M.J. Jackson
School of Clinical Sciences, University of Liverpool, UK
e-mail: mdcr02@liverpool.ac.uk

G.S. Lynch (ed.), *Sarcopenia – Age-Related Muscle Wasting and Weakness*,
DOI 10.1007/978-90-481-9713-2_14, © Springer Science+Business Media B.V. 2011

1 Skeletal Muscle Atrophy and Weakness Contribute to Physical Frailty in the Elderly

Frailty in the elderly (Hadley et al. 1993) is largely caused by loss of muscle mass and strength, increased susceptibility to injury, and impaired recovery following damage, particularly contraction-induced damage (Faulkner et al. 2007; Marcell 2003). By the age of 70, the cross-sectional area (CSA) of muscle is reduced by 25–30% (Porter et al. 1995) associated with a loss in absolute force generation (Grimby and Saltin 1983) and a decrease in specific force (per unit CSA) generation (Morse et al. 2005). After 70, strength continues to fall and power in the lower leg declines at ~3.5% per year (Skelton et al. 1994). These deficits profoundly impact on the *quality of life* of even healthy older people, as many are at, or near thresholds that limit the ability to carry out everyday tasks (Young and Skelton 1994). This age-related muscle weakness significantly increases the risk for elderly falling. Approximately 20% of community-dwelling elderly fall each year (Prudham and Evans 1981). Many elderly who fall suffer loss of independence and some never re-enter the community. One half of the accidental deaths in those over 65 are related to falls. While regular exercise can modify the rate of muscle deficits, even active elderly people show significant age-related declines in muscle mass and function (Wiswell et al. 2001). The mechanisms responsible for the age-related loss of muscle mass and function are unclear although modified generation of Reactive Oxygen and Nitrogen Species (RONS) have been implicated in age-related tissue dysfunction (Harman 2003).

2 The Source and Nature of the RONS Generated by Skeletal Muscle

The source and nature of the RONS generated by muscle of young or adult mammals during contractions has been studied since the 1980s. Initial studies demonstrated increased generation of free radicals by contracting skeletal muscles (Davies et al. 1982; Jackson et al. 1983). The main reactive oxygen species (ROS) produced in the cell are free radical species, such as the superoxide anion and hydroxyl radical, and non-radical species, such as hydrogen peroxide (H2O2) (Palomero and Jackson 2010). The generation of specific RONS, including superoxide, nitric oxide and hydroxyl radicals by muscle were then described (Reid et al. 1992a,b; O'Neill et al. 1996; Balon and Nadler 1994; Kobzik et al 1994).

Cells are required to preserve a delicate balance between ROS generation and elimination to maintain the correct redox status necessary to carry out vital functions. In excess, ROS can attack cellular structures, such as lipids, proteins, and DNA, thereby inducing irreversible changes that can lead to the disruption of cellular functions and integrity. Under normal physiologic conditions, the reactive nature of ROS allows their incorporation into the structure of macromolecules in a reversible fashion. Such reversible oxidative modifications play a critical role in different signalling pathways that regulate different cellular functions and the fate of the cell (Sies and Jones 2007). Many studies have provided evidence for the pivotal role of ROS in signal transduction and recognized these molecules as second messengers (Powers and Jackson 2008).

Most authors have assumed that the ROS generated by contractions are predominantly generated by mitochondria due to the increased demand for energy, but recent data argue against this possibility (for discussion see Jackson 2008). In order to evaluate the relative magnitude of the increase in ROS activity that occurs in skeletal muscle fibres in response to contractions, Palomero et al (2008a) applied a protocol of electrically stimulated, isometric contractions to single isolated fibres from the mouse Flexor digitorum brevis (FDB) muscle. Fibres were loaded with 5- (and 6-) chloromethyl-2′,7′-dichlorodihydrofluorescein diacetate (CM-DCFH DA) and measurements of 5- (and 6-) chloromethyl-2′,7′-dichlorofluorescin (CM-DCF) fluorescence from individual fibers were obtained by microscopy to study ROS in skeletal muscle. This technique has advantages because of the maturity of the fibres compared with muscle cells in culture and the analysis of single cells prevents contributions from non-muscle cells. The contraction protocol used has been shown to (1) to induce release of superoxide and nitric oxide from muscle cells in culture and muscles of mice in vivo (McArdle et al. 2001; Pattwell et al. 2004), (2) to lead to a fall in muscle glutathione and protein thiol content (Vasilaki et al. 2006c) and (3) to stimulate redox-regulated adaptive responses (Vasilaki et al. 2006b) when applied to intact muscles in vivo. The increase in intracellular DCF fluorescence induced by the contraction protocol was less than that following exposure of the fibres to 1 uM hydrogen peroxide. We (Palomero et al. 2008a) calculated that the likely change in intracellular hydrogen peroxide following addition of 1 uM to the extracellular medium is ~0.1 uM. Thus it can be inferred that the absolute level of cytosolic ROS activity in muscle fibres that was achieved following contractile activity was potentially equivalent to ~0.1 uM hydrogen peroxide. Such levels of hydrogen peroxide have traditionally been associated with a signalling role for the oxidant and our recent data indicate that the ROS generated by contractions are reduced by inhibitors of NADPH oxidase enzymes. The increase in ROS activity with contractions is also observed where dihydroethidium (DHE) is used as a probe. This probe is predominantly located in the cytosol, but when DHE is modified to locate within mitochondria (as a probe called Mito-HE or MitoSox) no increase in mitochondrial fluorescence was seen during contractions. We conclude that the source of ROS that acts as a signal for adaptive responses to contractions is not mitochondria, but is associated with the cytosol. Inhibitor studies indicate that this is likely to be a membrane-located NADPH oxidase that is activated during contractions to generate superoxide (which is converted to hydrogen peroxide) and these ROS activate adaptive responses to contractions.

3 Modified ROS Generation Activates Redox-Sensitive Transcription Factors in Contracting Muscle

Skeletal muscles of adult mice and humans adapt rapidly to contractile activity. Numerous proteins show adaptive responses to contraction, including the antioxidant defence enzymes and Heat Shock Proteins (HSPs) that protect against subsequent cellular damage (Hollander et al. 2003; McArdle et al. 2004, 2005). ROS have become increasingly recognised to mediate some adaptive responses of skeletal muscle to contractile activity through activation of redox-sensitive transcription factors (Jackson et al. 2002; McArdle et al. 2004; Jackson 2005; Ji et al. 2006; Gomez-Cabrera et al. 2008; Ristow et al. 2009). Nuclear factor kappa B (NFκB) is one such factor, along with Activator Protein-1 (AP-1) and Heat Shock Factor 1 (Cotto and Morimoto 1999). These transcription factors are involved in remodelling, production of other cytoprotective proteins and production of inflammatory cytokines. ROS are principal regulators of NFκB activation in many situations (Moran et al. 2001). NFκB family members expressed in skeletal muscle play critical roles in modulating the specificity of NFκB (Bar-Shai et al. 2005; Hayden and Ghosh 2008). In skeletal muscle, NFκB modulates expression of genes associated with myogenesis (Bakkar et al. 2008; Dahlman et al. 2009), catabolism-related genes (Bar-Shai et al. 2005; Peterson and Guttridge 2008; Van Gammeren et al. 2009) and cytoprotective proteins during adaptation to contractile activity (Vasilaki et al. 2006b). Moreover, skeletal muscle has been identified as an endocrine organ producing cytokines via NFκB activation following stresses such as systemic inflammation or physical strain (Lee et al. 2007). The specificity of the responses of skeletal muscle cells to NFκB activation is likely to be largely due to subtle differences in NFκB activation such as B binding sequences and NFκB dimer formation that regulate expression of specific genes (Bakkar et al. 2008). Activation of NFκB by ROS involves oxidation of key cysteine residues in upstream activators of NFκB and the process can be inhibited by antioxidants or reducing agents (Hansen et al. 2006) and more recently by HSPs (Chen and Currie 2006).

Evidence from our laboratory and others have demonstrated that the HSP content of skeletal muscles increases rapidly following a demanding but non-damaging period of isometric contractions and this is termed the stress response (McArdle et al. 2001; Vasilaki et al. 2006b) and this increased HSP content is part of a more widespread adaptive response in transcription of cellular proteins (McArdle F et al. 2004). Data also demonstrated that this was associated with significant protection against subsequent damage (McArdle F et al. 2004). Definitive data demonstrating a functional role of HSPs in protection against damage and rapid recovery from damage was provided by a study using HSP70 overexpressor mice whereby muscles of these mice were protected against the secondary deficit characteristic of lengthening contraction – induced damage in mice and resulted in a more rapid recovery of maximum force generation (McArdle et al. 2004).

The signal for increased HSP production following exercise has been a topic of interest for some time and oxidative stress, hyperthermia and modified energy supplies have all been proposed to play a role. Data from our laboratory has provided evidence that the primary signal for activation of transcription of HSPs in skeletal muscle in both rodents and humans following isometric contractions is an increased production of reactive oxygen species (ROS). Studies in mice have demonstrated that increased HSP production is associated with increased detection of ROS in the muscle extracellular space (McArdle et al. 2001) and a transient fall in protein sulphydryl groups and this occurred in the absence of any significant change in muscle temperature. Supplementation of human subjects with nutritional antioxidants abolished the exercise-induced increase in muscle HSP content (Khassaf et al. 2001). Further studies in humans have demonstrated that although the production of HSPs is dependent upon the intensity of exercise, exercise conditions and the training status of the individuals (Morton et al. 2008; Palomero et al. 2008b), an equivalent rise in muscle temperature without exercise did not result in increased muscle content of HSPs (Morton et al. 2007) although the cumulative effect of heat and ROS production may result in a reduction in threshold for ROS-induced HSP production.

Changes in HSP content of muscle can play a direct role in modification of ROS production and thus feedback to modify the activation of the stress response. Neuronal nitric oxide synthase (nNOS) produces nitric oxide but also produces superoxide at low levels of L-arginine (Heinzel et al. 1992; Pou et al. 1992). nNOS is localised to the plasma membrane of muscle cells, associated with the dystrophin glycoprotein complex (Vranić et al. 2002) and HSP90 is also associated with nNOS. HSP90 is thought to modify the action of nNOS since the presence of HSP90 dose-dependently inhibits the superoxide anion radical generation from nNOS. At lower levels of L-arginine where marked superoxide anion radical generation occurred, HSP90 caused a more dramatic enhancement of NO synthesis from nNOS as compared to that under normal L-arginine (Song et al. 2002). The balance of production of NO and/or superoxide anion radical by nNOS may also be linked to the cellular localisation of nNOS since it has also been proposed that, in certain pathological conditions including Duchenne muscular dystrophy, delocalised nNOS produces altered patterns of NO/superoxide although the role of HSP90 in this production is unknown. The interaction between HSPs and other ROS generating systems is yet to be determined.

4 HSPs Interact with and Mediate Activation of Transcription Factors

The dependence of a stress response in muscles following non-damaging exercise on the initial level of HSPs in the quiescent muscle seems to be due to a feedback mechanism by which increased cellular HSPs deactivate Heat Shock Factor 1 (HSF1), the main transcription factor thought to be responsible for the acute stress response (Pirkkala et al. 2001). It is also possible that other adaptations to exercise may play a role in this lack of response, such as an increase in ROS defences (McArdle et al. 2001) which would also reduce the ROS signal. Thus, the threshold for activation of the stress response changes in muscles with altered HSP content or altered oxidant/antioxidant status (termed redox status).

Data have demonstrated interactions between cellular HSPs and the activation of other transcription factors, particularly NFκB and AP-1, which are involved in remodelling, production of other cytoprotective proteins and production of inflammatory cytokines. These studies have concentrated on the protective role of HSPs in ameliorating the activation of the pro-inflammatory pathways of NFκB whereby high levels of HSP70 and HSP27 have been shown to suppress the pro-inflammatory pathway of NFκB (Chen and Currie 2006). Heat shock treatment suppresses NFκB activation in mucosal cells of endotoxin treated mice by inhibiting the phosphorylation and degradation of the NFκB inhibitor, IkB-α and prior heat shock treatment also inhibits IκB kinase (IKK) activation and results in a decreased cytoplasmic level of IKK-α and IKK complex insolublisation (Pritts et al. 2000; Yoo et al. 2000; Chen et al. 2004). In non-muscle cells, HSP70 and HSP27 have been found to interact directly with NFκB, IkB-α, IKK-α, and IKK-β in suppress, resulting in the suppression of NFκB (Shimizu et al. 2002; Guzhova et al. 1997; Park et al. 2003).

It is entirely feasible that a similar interaction is present in skeletal muscle cells and that this interaction not only modulates cytokine production by skeletal muscle, but other pathways in which NFκB and AP-1 may be involved. The pattern and time course of HSP production in skeletal muscles to different forms of exercise and other stresses differs and our data have shown that different HSPs provide specific protection to various aspects of damage and regeneration. It is likely that there is some specificity in these interactions with specific HSPs modulating different aspects of transcription factor activation or inhibitor degradation and the induction of the stress response in skeletal muscle may act as a shut-down mechanism of NFκB - mediated cytokine production by muscle cells.

The interaction of HSPs with AP-1 and NFκB is further complicated since several HSPs are known to contain promoters which can be regulated by both NFκB and AP-1. For example, HSP90 contains a promoter which is regulated by NFκB and downregulation of the p65 component of NFκB resulted in reduced constitutive expression of HSP90 (Ammirante et al. 2008). HSPs can also contain an AP-1 promoter (e.g. Hosokawa et al. 1993).

5 Changes in and HSP Content and Redox Status of Muscles Facilitate Successful Myogenesis and Rapid Regeneration Following Damage

Controlled changes in transcription factor activation and deactivation are crucial to successful myogenesis and regeneration. During myoblast proliferation and fusion, the HSP content of cells is relatively high and this is primarily due to the expression and activation of the developmental Heat Shock Factor 2 (HSF2). HSP content then falls gradually with maturation of the cells, along with expression of HSF2 (McArdle et al. 2006). Expression of HSF1 increases at the later stages of myogenesis once myotubes have been formed (McArdle et al. 2006) such that these cells are now stress responsive. Data from our laboratory examining NFκB activation in vivo following muscle damage have demonstrated a secondary and relatively late phase of NFκB activation at 14 and 28 days post-damage, a time associated with a secondary phase of remodelling, maturation and reinnervation of skeletal muscle fibres.

Myogenesis and regeneration are dependent on changes in ROS generation since muscle cells with altered ROS production demonstrate a failure of successful myogenesis in culture. This may be associated with aberrant activation of redox-responsive transcription factors. For example, primary myoblasts from glutathione peroxidase 1 null mice do not fuse to form multinuclear myotubes in culture (Lee et al. 2006).

Thus, it is clear that RONS generation plays a major role in determining transcriptional activation in skeletal muscle during contraction-induced adaptive responses and alteration of such generation results in adaptive and functional deficits.

6 Modified Generation of Reactive Oxygen Species (ROS) Have Been Implicated in Age-Related Skeletal Muscle Dysfunction

The mechanisms responsible for the age-related loss of muscle mass and function are unclear. Initial studies implicated an increase in oxidative damage in all tissues, including skeletal muscle, in the functional decline of those tissues (Sastre et al. 2003; Drew et al. 2003; Vasilaki et al. 2006b,c).

Detrimental roles of ROS in tissues have been widely studied and a chronic increase in the production of ROS has been implicated in a number of pathological conditions such as cancer and ageing (Jackson et al. 2002). In contrast, it is now accepted that acute changes in ROS generation are essential for physiological signalling processes. These include ROS acting as short-lived messengers in signal transduction pathways such as those involved in cellular differentiation, proliferation, maturation and programmed cell death via activation of redox-responsive transcription factors (Jackson et al. 2002). However, these processes are still poorly defined and in particular there is a lack of information on the magnitude, time course and localisation of such redox changes in tissues. A chronic accumulation of oxidative damage has been postulated as a major component of the ageing process for over 50 years (Harman 1956). Mitochondria have been claimed to be the major site of reactive oxygen species (ROS) generation that contributes to increased oxidative damage during ageing (see Sanz et al. 2006 for a review) and isolated skeletal muscle mitochondria from old organisms release a greater amount of hydrogen peroxide that is attributable to increased superoxide generation by electron transport chain complexes (Lass et al. 1998; Mansouri et al. 2006; Vasilaki et al. 2006c). Studies of the mutations in mitochondrial DNA in a number of cell types have shown that these accumulate with age (Shah et al. 2009; Taylor et al. 2003). Mutations in mitochondrial DNA can theoretically disrupt the function of the respiratory chain thereby compromising the production of ATP from oxidative phosphorylation. Although much of the current data has concentrated on mitochondria as a predominant site for ROS generation during ageing, alternative cellular sites for ROS generation are receiving increasing attention. For example, copper, zinc superoxide dismutase (SOD1) is normally located in the cytosol and mitochondrial intermembrane space and mice lacking SOD1 show a shortened lifespan and an acceleration of the normal age-related changes in structure and function of several tissues (Muller et al. 2007). It must be noted however that, although the oxidative stress theory of ageing is by far the most popular theory on ageing, data in support of this theory in mammalian systems is sparce (Pérez et al. 2009).

We have undertaken a number of studies to define the site of the defect in adaptive responses following contractile activity in muscle from aged mice. We examined the effect of contractile activity on various indicators of ROS activity in muscle from old compared with adult mice. A protocol of contractile activity caused a significant fall in the total glutathione content of contracting muscles from adult mice, but less of a fall in muscles from old animals and this was associated with a diminished release of extracellular superoxide from the muscles of old mice (Vasilaki et al. 2006c). Vasilaki et al. (2007) also reported a contraction-induced increase in the 3-nitrotyrosine content of muscle from adult mice that was not seen in the muscle from old mice. These data all suggest that the contraction-induced increase in ROS activities is reduced in muscle from old mice compared with that from muscle of adult mice.

The chronic increase in the activities of regulatory enzymes for ROS (SOD1 and SOD2 and catalase) and HSP content seen in muscle from old mice (Kayani et al. 2008b) appears to reflect an attempt to adapt to a chronic increase in ROS activities. Despite this attempted adaptation, increased muscle oxidation remains evident in the muscle from old mice (Broome et al. 2006). The effects of these changes on the ROS signals that normally stimulate adaptations to contractions are unknown.

7 The Altered Generation of ROS in Muscles of Old Mice is Associated with an Inability of Muscles of Old Individuals to Adapt to Stress

Activation of redox-responsive transcription factors in response to an acute stress such as exercise is aberrant in muscles of old humans and mice. These muscles demonstrate both chronic constitutive activation of redox-sensitive transcription factors (Vasilaki et al. 2006b; Cuthbertson et al. 2005) and an inability to further activate these transcription factors following an acute non-damaging contraction protocol (Vasilaki et al. 2006b). The chronic activation of transcription factors such as NFκB in muscles of old mice is associated with chronic increases in the expression of a number of genes. For example, increased content and activities of antioxidant defence enzymes such as the superoxide dismutases and catalase (Broome et al. 2006), increased content of HSPs (Vasilaki et al. 2006b; Kayani et al. 2008b) and increased production of cytokines and chemokines by muscle cells (Febbraio and Pedersen 2005).

The inability to further activate NFκB in response to an acute contraction protocol is associated with severe attenuation of normal changes in expression of cytoprotective genes (Demirel et al. 2003; Heydari et al. 2000; Locke and Tanguay 1996; Muramatsu et al. 1996; Rao et al. 1999; Vasilaki et al. 2006b). We have shown that the increases in HSP content and antioxidant enzyme activities stimulated by isometric contractions in muscles of adult rodents were abolished in muscles of old rodents (Vasilaki et al. 2002, 2006b). These severely blunted adaptive responses to acute contractions in muscles from old rodents contribute to age-related muscle dysfunction (McArdle et al. 2004a; Broome et al. 2006) and can be overcome by activation of the transcription factor through alternative, pharmacological routes (Kayani et al. 2008a). Transgenic overexpression of HSP70 in skeletal muscle throughout life partially preserved muscle function in old mice and prevented the age-related chronic activation of transcription factors and changes in muscle content of cytoprotective proteins at rest (McArdle et al. 2004b; Broome et al. 2006). The mechanisms by which an increased muscle content of HSP70 exerts these effects on NFκB are unclear although overexpression of HSP70 throughout life also prevented the accumulation of markers of oxidative damage in muscle from old mice (Broome et al. 2006).

A diminished ability to respond to the stress of contractions plays an important role in other age-related defects in muscle function and adaptation. Ljubicic and Hood (2008) reported a severe attenuation of the signalling pathways involved in mitochondrial biogenesis in type II muscle fibres of old rats following contractions compared with that seen in fibres from young rats. ROS play an important role in the activation of these signalling cascades (Irrcher et al. 2009). These authors suggest that ROS affect mitochondrial biogenesis via the upregulation of transcriptional regulators as peroxisome proliferator-activated receptor-gamma coactivator-1 protein-alpha (PGC-1alpha), suggesting that an aberrant activation of ROS generation following contractions may be responsible for the diminished mitochondrial biogenesis in muscles of old rats. This blunted or absent adaptation to stress in muscle of old humans and mice is not limited to the exercise response. Skeletal muscle of healthy elderly humans demonstrates a reduction in anabolic sensitivity and responsiveness of muscle protein synthesis pathways. Cuthbertson et al. (2005) demonstrated a reduction in the phosphorylation of mTOR and downstream translational regulators in response to essential amino acid (EAA) ingestion when compared with the young despite a greater plasma EAA availability in elderly subjects. The authors concluded that the nutrient signal was not transduced as well by old as by young muscle, resulting in a lower protein synthesis response to the same stimulus.

8 Future Directions

In the light of these data we hypothesise that attenuation of the adaptive responses to contractions is a key factor leading to age-related loss of muscle mass and function; ROS generated during contractions are important stimulators of adaptive responses and they originate from a source associated with the cytosol; Mitochondria release increased amounts of hydrogen peroxide and the resulting chronic oxidation blocks the normal adaptations to contractile activity through either: (1) inducing upregulation of ROS defence systems (SODs, catalase and HSPs) that suppress the cytosolic ROS signal that normally stimulates adaptive responses to contractions or (2) preventing activation of the cytosol-associated ROS generating system that are activated by contractions.

Targeted interventions to suppress mitochondrial H2O2 generation are necessary to restore adaptive responses to contractions in old mice since interventions based on antioxidant supplementation will suppress ROS signals in both mitochondrial and cytosolic compartments and hence be ineffective at prevention of age-related changes.

Acknowledgements The authors would like to thank The Biotechnology and Biological Sciences Research Council, The Medical Research Council, The Wellcome Trust, Research into Ageing, The United States National Institutes on Aging (PO1, AG20591) and The Dowager Countess Eleanor Peel Trust for financial support and current and past collaborators.

References

Ammirante, M., Rosati, A., Gentilella, A., Festa, M., Petrella, A., Marzullo, L., Pascale, M., Belisario, M. A., Leone, A., Turco, M. C. (2008). The activity of hsp90 alpha promoter is regulated by NF-kappa B transcription factors. *Oncogene, 27*, 1175–1178.

Bakkar, N., Wang, J., Ladner, K. J., Wang, H., Dahlman, J. M., Carathers, M., Acharyya, S., Rudnicki, M. A., Hollenbach, A. D., Guttridge, D. C. (2008). IKK/NF-kappaB regulates skeletal myogenesis via a signaling switch to inhibit differentiation and promote mitochondrial biogenesis. *The Journal of Cell Biology, 180*, 787–802.

Balon, T. W. & Nadler, J. L. (1994). Nitric oxide release is present from incubated skeletal muscle preparations. *Journal of Applied Physiology, 77*(6), 2519–2521.

Bar-Shai, M., Carmeli, E., Reznick, A. Z. (2005). The role of NF-kappaB in protein breakdown in immobilization, aging, and exercise: from basic processes to promotion of health. *Annals of the New York Academy of Sciences, 1057*, 431–447.

Broome, C. S., Kayani, A. C., Palomero, J., Dillmann, W. H., Mestril, R., Jackson, M. J., McArdle, A. (2006). Effect of lifelong overexpression of HSP70 in skeletal muscle on age-related oxidative stress and adaptation after nondamaging contractile activity. *The FASEB Journal, 20*, 1549–1551.

Chen, Y. & Currie, R. W. (2006). Small interfering RNA knocks down heat shock factor-1 (HSF-1) and exacerbates pro-inflammatory activation of NF- B and AP-1 in vascular smooth muscle cells. *Cardiovascular Research, 69*, 66–75.

Chen, Y., Arrigo, A. P., Currie, R. W. (2004). Heat shock treatment suppresses Angiotensin II-induced activation of NF- B pathway and heart inflammation: a role for IKK depletion by heat shock? *The American Journal of Physiology, 287*, H1104–H1114.

Cotto, J. J. & Morimoto, R. I. (1999). Stress-induced activation of the heat-shock response: cell and molecular biology of heat-shock factors. *Biochemical Society Symposia, 64*, 105–118.

Cuthbertson, D., Smith, K., Babraj, J., Leese, G., Waddell, T., Atherton, P., Wackerhage, H., Taylor, P. M., Rennie, M. J. (2005). Anabolic signaling deficits underlie amino acid resistance of wasting, aging muscle. *The FASEB Journal, 19*, 422–424.

Dahlman, J. M., Wang, J., Bakkar, N., Guttridge, D. C. (2009). The RelA/p65 subunit of NF-kappaB specifically regulates cyclin D1 protein stability: implications for cell cycle withdrawal and skeletal myogenesis. *Journal of Cellular Biochemistry, 106*, 42–51.

Davies, K. J., Quintanilha, A. T., Brooks, G. A., Packer, L. (1982). Free radicals and tissue damage produced by exercise. *Biochemical and Biophysical Research Communications, 107*, 1198–1205.

Demirel, H. A., Hamilton, K. L., Shanely, R. A., Tümer, N., Koroly, M. J., Powers, S. K. (2003). Age and attenuation of exercise-induced myocardial HSP72 accumulation. *American Journal of Physiology. Heart and Circulatory Physiology, 285*, H1609–H1615.

Drew, B., Phaneuf, S., Dirks, A., Selman, C., Gredilla, R., Lezza, A., Barja, G., Leeuwenburgh, C. (2003). Effects of aging and caloric restriction on mitochondrial energy production in gastrocnemius muscle and heart. *The American Journal of Physiology, 284*, R474–R480.

Faulkner, J. A., Larkin, L. M., Claflin, D. R., Brooks, S. V. (2007). Age-related changes in the structure and function of skeletal muscles. *Clinical and Experimental Pharmacology and Physiology, 34*, 1091–1096.

Febbraio, M. A. & Pedersen, B. K. (2005). Contraction-induced myokine production and release: is skeletal muscle an endocrine organ? *Exercise and Sport Sciences Reviews, 33*, 114–119.

Gomez-Cabrera, M. C., Domenech, E., Romagnoli, M., Arduini, A., Borras, C., Pallardo, F. V., Sastre, J. Viña-Ribes, J. (2008). Oral administration of vitamin C decreases muscle mitochondrial biogenesis and hampers training-induced adaptations in endurance performance. *American Journal of Clinical and Nutrition, 87*, 142–149.

Grimby, G. & Saltin, B. (1983). The ageing muscle. *Clinical Physiology, 3*, 209–218.

Guzhova, I. V., Darieva, Z. A., Melo, A. R., Margulis, B. A. (1997). Major stress protein Hsp70 interacts with NF- B regulatory complex in human T-lymphoma cells. *Cell Stress and Chaperones, 2*, 132–139.

Hadley, E. C., Ory, M. G., Suzman, R., Weindruch, R., Fried, L. (1993). Physical frailty: A treatable cause of dependence in old age. *Journal of Gerontology, 48*, 1–88.

Hansen, J. M., Zhang, H., Jones, D. P. (2006). Mitochondrial thioredoxin-2 has a key role in determining tumor necrosis factor-alpha-induced reactive oxygen species generation, NF-kappaB activation, and apoptosis. *Toxicological Sciences, 91*, 643–650.

Harman, D. (1956). Aging: a theory based on free radical and radiation chemistry. *Journal of Gerontology, 11*, 298–300.

Harman, D. (2003). The free radical theory of aging. *Antioxidants Redox Signaling, 5*, 557–561.

Hayden, M. S. & Ghosh, S. (2008). Shared principles in NF-kappaB signalling. *Cell, 132*, 344–362.

Heinzel, B., John, M., Klatt, P., Bohme, E., Mayer, B. (1992). Ca2+/calmodulin-dependent formation of hydrogen peroxide by brain nitric oxide synthase. *The Biochemical Journal, 281*, 627–630.

Heydari, A. R., You, S., Takahashi, R., Gutsmann-Conrad, A., Sarge, K. D., Richardson, A. (2000). Age-related alterations in the activation of heat shock transcription factor 1 in rat hepatocytes. *Experimental Cell Research, 256*, 83–93.

Hollander, J. M., Lin, K. M., Scott, B. T., Dillmann, W. H. (2003). Overexpression of PHGPx and HSP60/10 protects against ischemia/reoxygenation injury. *Free Radical Biology and Medicine, 35*, 742–751.

Hosokawa, N., Takechi, H., Yokota, S., Hirayoshi, K., Nagata, K. (1993). Structure of the gene encoding the mouse 47-kDa heat-shock protein (HSP47). *Gene, 126*, 187–193.

Irrcher, I., Ljubicic, V., Hood, D. A. (2009). Interactions between ROS and AMP kinase activity in the regulation of PGC-1alpha transcription in skeletal muscle cells. *American Journal of Physiology. Cell Physiology, 296*, C116–C123.

Jackson, M. J. (2005). Reactive oxygen species and redox-regulation of skeletal muscle adaptations to exercise. *Philosophical Transactions of the Royal Society of London. Series B: Biological Sciences, 360*(1464), 2285–2291.

Jackson, M. J. (2008). Redox regulation of skeletal muscle. *IUBMB Life, 60*(8), 497–501.

Jackson, M. J., Jones, D. A., Edwards, R. H. (1983). Vitamin E and skeletal muscle. *Ciba Foundation Symposium, 101*, 224–239.

Jackson, M. J., Papa, S., Bolaños, J., Bruckdorfer, R., Carlsen, H., Elliott, R. M., Flier, J., Griffiths, H. R., Heales, S., Holst, B., Lorusso, M., Lund, E., Øivind Moskaug, J., Moser, U., Di Paola, M., Polidori, M. C., Signorile, A., Stahl, W., Viña-Ribes, J., Astley, S. B. (2002). Antioxidants, reactive oxygen and nitrogen species, gene induction and mitochondrial function. *Molecular Aspects of Medicine, 23*, 209–285.

Ji, L. L., Gomez-Cabrera, M. C., Vina, J. (2006). Exercise and hormesis: activation of cellular antioxidant signaling pathway. *Annals of the New York Academy of Sciences, 1067*, 425–435.

Kayani, A. C., Close, G. L., Broome, C. S., Jackson, M. J., McArdle, A. (2008). Enhanced recovery from contraction-induced damage in skeletal muscles of old mice following treatment with the heat shock protein inducer 17-(allylamino)-17-demethoxygeldanamycin. *Rejuvenation Research, 11*, 1021–1030.

Kayani, A. C., Close, G. L., Jackson, M. J., McArdle, A. (2008). Prolonged treadmill training increases HSP70 in skeletal muscle but does not affect age-related functional deficits. *American Journal of Physiology: Regulatory, Integrative and Comparative Physiology, 294*, R568–R576.

Khassaf, M., Child, R. B., McArdle, A., Brodie, D. A., Esanu, C., Jackson, M. J. (2001). Time course of responses of human skeletal muscle to oxidative stress induced by nondamaging exercise. *Journal of Applied Physiology, 90*, 1031–1035.

Kobzik, L., Reid, M. B., Bredt, D. S., Stamler, J. S. (1994). Nitric oxide in skeletal muscle. *Nature, 372*(6506), 546–548.

Lass, A., Sohal, B. H., Weindruch, R., Forster, M. J., Sohal, R. S. (1998). Caloric restriction prevents age-associated accrual of oxidative damage to mouse skeletal muscle mitochondria. *Free Radical Biology and Medicine, 25*, 1089–1097.

Lee, S., Shin, H. S., Shireman, P. K., Vasilaki, A., Van Remmen, H., Csete, M. E. (2006). Glutathione-peroxidase-1 null muscle progenitor cells are globally defective. *Free Radical Biology and Medicine, 41*(7), 1174–1184.

Lee, C. E., McArdle, A., Griffiths, R. D. (2007). The role of hormones, cytokines and heat shock proteins during age-related muscle loss. *Clinical Nutrition, 26*, 524–534.

Ljubicic, V. & Hood, D. A. (2008). Kinase-specific responsiveness to incremental contractile activity in skeletal muscle with low and high mitochondrial content. *American Journal of Physiology. Endocrinology and Metabolism, 295*, E195–E204.

Locke, M. & Tanguay, R. M. (1996). Diminished heat shock response in the aged myocardium. *Cell Stress and Chaperones, 1*(4), 251–260.

Mansouri, A., Muller, F. L., Liu, Y., Ng, R., Faulkner, J., Hamilton, M., Richardson, A., Huang, T. T., Epstein, C. J., Van Remmen, H. (2006). Alterations in mitochondrial function, hydrogen peroxide release and oxidative damage in mouse hind-limb skeletal muscle during aging. *Mechanisms of Ageing and Development, 127*, 298–306.

Marcell, T. J. (2003). Sarcopenia: causes, consequences, and preventions. *The Journals of Gerontology. Series A: Biological Sciences and Medical Sciences, 58*, M911–M916.

McArdle, A., Pattwell, D., Vasilaki, A., Griffiths, R. D., Jackson, M. J. (2001). Contractile activity-induced oxidative stress: cellular origin and adaptive responses. *American Journal of Physiology (Cell), 280*, C621–C627.

McArdle, F., Spiers, S., Aldemir, H., Vasilaki, A., Beaver, A., Iwanejko, L., McArdle, A., Jackson, M. J. (2004a). Preconditioning of skeletal muscle against contraction-induced damage: the role of adaptations to oxidants in mice. *Journal de Physiologie, 561*, 233–244.

McArdle, A., Dillmann, W. H., Mestril, R., Faulkner, J. A., Jackson, M. J. (2004b). Overexpression of HSP70 in mouse skeletal muscle protects against muscle damage and age-related muscle dysfunction. *The FASEB Journal, 18*, 355–357.

McArdle, A., Broome, C. S., Kayani, A. C., Tully, M. D., Close, G. L., Vasilaki, A., Jackson, M. J. (2006). HSF expression in skeletal muscle during myogenesis: implications for failed regeneration in old mice. *Experimental Gerontology, 41*, 497–500.

McArdle, F., Pattwell, D. M., Vasilaki, A., McArdle, A., Jackson, M. J. (2005).Intracellular generation of reactive oxygen species by contracting muscle cells. *Free Radial Biology and Medicine, 39*, 651–657.

Moran, L. K., Gutteridge, J. M., Quinlan, G. J. (2001). Thiols in cellular redox signalling and control. *Current Medicinal Chemistry, 8*(7), 763–772.

Morton, J. P., Maclaren, D. P., Cable, N. T., Campbell, I. T., Evans, L., Bongers, T., Griffiths, R. D., Kayani, A. C., McArdle, A., Drust, B. (2007). Elevated core and muscle temperature to levels comparable to exercise do not increase heat shock protein content of skeletal muscle of physically active men. *Acta Physiologica (Oxford), 190*(4), 319–327.

Morton, J. P., Maclaren, D. P., Cable, N. T., Campbell, I. T., Evans, L., Kayani, A. C., McArdle, A., Drust, B. (2008). Trained men display increased Basal heat shock protein content of skeletal muscle. *Medicine and Science in Sports and Exercise, 40*(7), 1255–1262.

Morse, C. I., Thom, J. M., Reeves, N. D., Birch, K. M., Narici, M. V. (2005). In vivo physiological cross-sectional area and specific force are reduced in the gastrocnemius of elderly men. *Journal of Applied Physiology, 99*, 1050–1055.

Muller, F. L., Song, W., Jang, Y. C., Liu, Y., Sabia, M., Richardson, A., Van Remmen, H. (2007). Denervation-induced skeletal muscle atrophy is associated with increased mitochondrial ROS production. *American Journal of Physiology: Regulatory, Integrative and Comparative Physiology, 293*, R1159–R1168.

Muramatsu, T., Hatoko, M., Tada, H., Shirai, T., Ohnishi, T. (1996). Age-related decrease in the inductability of heat shock protein 72 in normal human skin. *The British Journal of Dermatology, 134*, 1035–1038.

O'Neill, C. A., Stebbins, C. L., Bonigut, S., Halliwell, B., Longhurst, J. C. (1996). Production of hydroxyl radicals in contracting skeletal muscle of cats. *Journal of Applied Physiology, 81*(3), 1197–1206.

Palomero, J., Pye, D., Kabayo, T., Spiller, D. G., Jackson, M. J. (2008a). In situ detection and measurement of intracellular reactive oxygen species in single isolated mature skeletal muscle fibers by real time fluorescence microscopy. *Antioxidants Redox Signaling, 10*, 1463–1474.

Palomero, J., Broome, C. S., Rasmussen, P., Mohr, M., Nielsen, B., Nybo, L., McArdle, A., Drust, B. (2008b). Heat shock factor activation in human muscles following a demanding intermittent exercise protocol is attenuated with hyperthermia. *Acta Physiologica (Oxford), 193*(1), 79–88.

Palomero, J., Jackson, M.J. (2010). Redox regulation in skeletal muscle during contractile activity and aging. *Journal of Animal Science, 88*, 1307–1313.

Park, K. J., Gaynor, R. B., Kwak, Y. T. (2003). Heat shock protein 27 association with the I kappa B kinase complex regulates tumor necrosis factor alpha-induced NF-kappa B activation. *The Journal of Biological Chemistry, 278*, 35272–35278.

Pattwell, D. M., McArdle, A., Morgan, J. E., Patridge, T. A., Jackson, M. J. (2004). Release of reactive oxygen and nitrogen species from contracting skeletal muscle cells. *Free Radical Biology and Medicine, 37*, 1064–1072.

Pérez, V. I., Bokov, A., Van Remmen, H., Mele, J., Ran, Q., Ikeno, Y., Richardson, A. (2009). Is the oxidative stress theory of aging dead? *Biochimica et Biophysica Acta, 1790*(10), 1005–1014.

Peterson, J. M. & Guttridge, D. C. (2008). Skeletal muscle diseases, inflammation, and NF-kappaB signaling: insights and opportunities for therapeutic intervention. *International Reviews of Immunology, 27*, 375–387.

Pirkkala, L., Nykanen, P., Sistonen, L. (2001). Roles of the heat shock transcription factors in regulation of the heat shock response and beyond. *The FASEB Journal, 15*, 1118–1131.

Porter, M. M., Vandervoort, A. A., Lexell, J. (1995). Aging of human muscle: structure, function and adaptability. *Scandinavian Journal of Medicine and Science in Sports, 5*, 129–142.

Pou, S., Pou, W. S., Bredt, D. S., Snyder, S. H., Rosen, G. M. (1992). Generation of superoxide by purified brain nitric oxide synthase. *The Journal of Biological Chemistry, 267*, 24173–24176.

Powers, S. K. & Jackson, M. J. (2008). Exercise-induced oxidative stress: cellular mechanisms and impact on muscle force production. *Physiological Reviews, 88*, 1243–1276.

Pritts, T. A., Wang, Q., Sun, X., Moon, M. R., Fischer, D. R., Fischer, J. E., Wong, H. R., Hasselgren, P. O. (2000). Induction of the stress response in vivo decreases nuclear factor-kappa B activity in jejunal mucosa of endotoxemic mice. *Archives of Surgery, 135*, 860–866.

Prudham, D. & Evans, J. G. (1981). Factors associated with falls in the elderly: a community study. *Age and Ageing, 10*, 141–146.

Rao, D. V., Watson, K., Jones, G. L. (1999). Age-related attenuation in the expression of the major heat shock proteins in human peripheral lymphocytes. *Mechanisms of Ageing and Development, 107*, 105–118.

Reid, M. B., Haack, K. E., Franchek, K. M., Valberg, P. A., Kobzik, L., West, M. S. (1992). Reactive oxygen in skeletal muscle. I. Intracellular oxidant kinetics and fatigue in vitro. *Journal of Applied Physiology, 73*(5), 1797–1804.

Reid, M. B., Shoji, T., Moody, M. R., Entman, M. L. (1992). Reactive oxygen in skeletal muscle. II. Extracellular release of free radicals. *Journal of Applied Physiology, 73*(5), 1805–1809.

Ristow, M., Zarse, K., Oberbach, A., Klöting, N., Birringer, M., Kiehntopf, M., Stumvoll, M., Kahn, C. R., Blüher, M. (2009). Antioxidants prevent health-promoting effects of physical exercise in humans. *Proceedings of the National Academy of Sciences of the United States of America, 106*, 8665–8670.

Sanz, A., Pamplona, R., Barja, G. (2006). Is the mitochondrial free radical theory of aging intact? *Antioxidants and Redox Signaling, 8*, 582–599.

Sastre, J., Pallardo, F. V., Vina, J. (2003). The role of mitochondrial oxidative stress in aging. *Free Radical Biology and Medicine, 35*, 1–8.

Shah, V. O., Scariano, J., Waters, D., Qualls, C., Morgan, M., Pickett, G., Gasparovic, C., Dokladny, K., Moseley, P., Raj, D. S. (2009). Mitochondrial DNA deletion and sarcopenia. *Genetics in Medicine, 11*(3), 147–152.

Shimizu, M., Tamamori-Adachi, M., Arai, H., Tabuchi, N., Tanaka, H., Sunamori, M. (2002). Lipopolysaccharide pretreatment attenuates myocardial infarct size: A possible mechanism involving heat shock protein 70-inhibitory kappaBalpha complex and attenuation of nuclear factor kappaB. *The Journal of Thoracic and Cardiovascular Surgery, 124*, 933–941.

Sies, H. & Jones, D. P. (2007). In G. Fink (Ed.), *Oxidative stress in Encyclopedia of stress* (pp. 45–48). San Diego, CA: Elsevier.

Skelton, D. A., Greig, C. A., Davies, J. M., Young, A. (1994). Strength, power and related functional ability of healthy people aged 65-89 years. *Age and Ageing, 23*, 371–377.

Song, Y., Cardounel, A. J., Zweier, J. L., Xia, Y. (2002). Inhibition of superoxide generation from neuronal nitric oxide synthase by heat shock protein 90: implications in NOS regulation. *Biochemistry, 41*(34), 10616–10622.

Taylor, R. W., Barron, M. J., Borthwick, G. M., Gospel, A., Chinnery, P. F., Samuels, D. C., Taylor, G. A., Plusa, S. M., Needham, S. J., Greaves, L. C., Kirkwood, T. B., Turnbull, D. M. (2003). Mitochondrial DNA mutations in human colonic crypt stem cells. *Journal of Clinical Investigation, 112*, 1351–1360.

Van Gammeren, D., Damrauer, J. S., Jackman, R. W., Kandarian, S. C. (2009). The IkappaB kinases IKKalpha and IKKbeta are necessary and sufficient for skeletal muscle atrophy. *The FASEB Journal, 23*, 362–370.

Vasilaki, A., Jackson, M, J., McArdle, A. (2002). Attenuated HSP70 response in skeletal Muscle of aged rats following contractile activity. *Muscle Nerve, 25*, 902–905.

Vasilaki, A., Csete, M., Pye, D., Lee, S., Palomero, J., McArdle, F., Van Remmen, H., Richardson, A., McArdle, A., Faulkner, J. A., Jackson, M. J. (2006a). Genetic modification of the manganese superoxide dismutase/glutathione peroxidase 1 pathway influences intracellular ROS generation in quiescent, but not contracting, skeletal muscle cells. *Free Radical Biology and Medicine, 41*, 1719–1725.

Vasilaki, A., McArdle, F., Iwanejko, L. M., McArdle, A. (2006b). Adaptive responses of mouse skeletal muscle to contractile activity: the effect of age. *Mechanisms of Ageing and Development, 127*, 830–839.

Vasilaki, A., Mansouri, A., Remmen, H., van der Meulen, J. H., Larkin, L., Richardson, A. G., McArdle, A., Faulkner, J. A., Jackson, M. J. (2006c). Free radical generation by skeletal muscle of adult and old mice: effect of contractile activity. *Aging Cell, 5*, 109–117.

Vasilaki, A., McArdle, F., McLean, L., Simpson, D., Beynon, R. J., Van Remmen, H., Richardson, A. G., McArdle, A., Faulkner, J. A., Jackson, M. J. (2007). Formation of 3-nitrotyrosines in carbonic anhydrase III is a sensitive marker of oxidative stress in skeletal muscle. *Proteomics (Clinical Applications), 1*, 362–372.

Vranić, T. S., Bobinac, D., Jurisić-Erzen, D., Muhvić, D., Sandri, M., Jerković, R. (2002). Expression of neuronal nitric oxide synthase in fast rat skeletal muscle. *Collegium Antropologicum, 26*(Suppl), 183–188.

Wiswell, R. A., Hawkins, S. A., Jaque, S. V., Hyslop, D., Constantino, N., Tarpenning, K., Marcell, T., Schroeder, E. T. (2001). Relationship between physiological loss, performance decrement, and age in master athletes. *The Journals of Gerontology. Series A: Biological Sciences and Medical Sciences, 56*, M618–M626.

Yoo, C. G., Lee, S., Lee, C. T., Kim, Y. W., Han, S. K., Shim, Y. S. (2000). Anti-inflammatory effect of heat shock protein induction is related to stabilization of I kappa B alpha through preventing I kappa B kinase activation in respiratory epithelial cells. *Journal of Immunology, 164*, 5416–5423.

Young, A. & Skelton, D. A. (1994). Applied physiology of strength and power in old age. *International Journal of Sports Medicine, 15*, 149–151.

Exercise as a Countermeasure for Sarcopenia*

Donato A. Rivas and Roger A. Fielding

Abstract The aging process is characterized by the gradual decrease in muscle mass, strength and power leading to a decline in physical functioning, increased frailty and disability. This age related loss of muscle mass and function has been termed sarcopenia. The mechanisms that underlie sarcopenia are only beginning to be elucidated. However, specific modes and intensities of physical activity can both act to preserve and also increase skeletal muscle mass, strength, power in healthy and functionally limited older individuals. This effect appears to be pervasive throughout the lifespan and there is evidence for similar responses in men and women. The focus of this chapter is on the role of exercise as a therapeutic intervention for the prevention and treatment of sarcopenia. This will be accomplished by (1) reviewing the epidemiology on physical activity and sarcopenia (2) summarizing the molecular mechanisms associated with sarcopenia and exercise, (3) discussing the efficacy of resistance and endurance exercise or multi-modal exercise, such as the combination of aerobic and resistance exercise for the management of sarcopenia.

Keywords Sarcopenia • Anabolic stimuli • Molecular signaling • Exercise • Muscle mass

---

*This chapter is based upon work supported by the U.S. Department of Agriculture, under agreement No. 58-1950-7-707. Any opinions, findings, conclusion, or recommendations expressed in this publication are those of the author(s) and do not necessarily reflect the view of the U.S. Department of Agriculture.

D.A. Rivas and R.A. Fielding (✉)
Nutrition Exercise Physiology and Sarcopenia Laboratory, Jean Mayer USDA
Human Nutrition Research Center on Aging, Tufts University,
711 Washington Street, Boston, MA 02111, USA
e-mail: roger.fielding@tufts.edu

G.S. Lynch (ed.), *Sarcopenia – Age-Related Muscle Wasting and Weakness*,
DOI 10.1007/978-90-481-9713-2_15, © Springer Science+Business Media B.V. 2011

1 Introduction

The aging process is characterized by the gradual decrease in muscle mass and strength leading to a decline in physical functioning, increased frailty and disability. This age related loss of muscle mass and function has been termed sarcopenia (Rosenberg 1997). The prevalence of sarcopenia between the ages of 60–70 years is between 5% and 13% and increases to between 11% and 50% at 80 years of age (Morley 2008). The large variability in the data is the result of how sarcopenia is defined and measured. Additionally, it has been observed that a loss in muscle mass is associated with metabolic alterations such as, insulin resistance, type 2 diabetes, dyslipidaemia, and obesity that are coupled with an increase in mortality (Evans 1997). The total cost of sarcopenia to the American Health System has been reported to be approximately $18.4 billion (Morley 2008; Janssen et al. 2004). Individuals over the age of 69 years are the largest growing segment of the American population (Manton and Vaupel 1995). Therefore, therapeutic interventions that treat sarcopenia may have profound effects on the independence and physical functioning in the elderly.

There is compelling evidence that increased physical activity in older adults is associated with decreased risk of functional limitation, disability, frailty and metabolic disease states (DiPietro 2001; Fielding 1995; Tanaka and Seals 2008; Kohrt and Holloszy 1995; Sugawara et al. 2002; Chin et al. 2008). Therefore, exercise may be a highly effective treatment for preventing the loss of muscle mass associated with ageing (Chin et al. 2008; Fielding 1995). The focus of this chapter is on the role of exercise as a therapeutic intervention for the prevention and treatment of sarcopenia. This will be accomplished by (1) reviewing the epidemiology on physical activity and sarcopenia (2) summarizing the molecular mechanisms associated with sarcopenia and exercise, (3) discussing the efficacy of resistance and endurance exercise or multi-modal exercise, such as the combination of aerobic and resistance exercise for the management of sarcopenia.

2 Role of Lifelong Habitual Physical Activity with Changes in Muscle Mass

There are several parallels between the physiological effects of aging and the adaptation as a result of disuse and inactivity (Lynch et al. 2007; Bortz 1982; Corcoran 1991; Timiras 1994). For example, aging, disuse and inactivity all have adverse effects on the cardiovascular system such as, lowering maximal oxygen uptake and stroke volume and raising blood pressure. Body composition and metabolism are also similarly affected by aging, disuse and inactivity as seen by decreased lean body mass, increased fat mass and impaired glucose tolerance. The effects of aging, disuse and inactivity on the cardiovascular system, body composition and muscle composition are very difficult to differentiate. For example, it has been reported that there are changes in muscle fiber type composition as a result of aging, disuse and inactivity. However, while disuse is shown to mostly decrease the number of type 1 muscle fibers, studies on aging have revealed a reduction on the number of both Type 1 and Type 2 fibers and the specific size of type 2 fibers (Lexell et al. 1988; Larsson 1983; Larsson et al. 1978).

A sedentary lifestyle during aging is associated with decreased lean body mass and increased fat mass leading to increased mortality and functional limitations (Baumgartner et al. 1999; Dziura et al. 2004; DiPietro 2001; Fielding 1995; Evans 1997). This is demonstrated in studies showing a decrease in the relative risk of cardiovascular and all cause mortality in highly active compared to moderately active and sedentary individuals (Lakatta and Levy 2003; Singh 2004; Chodzko-Zajko et al. 2009). Declines in exercise capacity throughout an individual's life span can affect functional capacity and impinge on the ability to perform activities of daily living. Recently, Sugawara et al. (2002) observed that appendicular muscle mass relative to body mass declines with advancing age regardless of physical activity status, but is significantly higher in endurance-trained men at any age than their sedentary peers (Sugawara et al. 2002). Both aerobic and resistance exercise have been shown to increase protein synthesis, while also increasing the cross-sectional area of both myosin heavy chain (MHC) I and II, respectively (Harber et al. 2009a, b; Short et al. 2004). The decreased cardiorespiratory function and reduced muscle mass and strength observed with advancing age and a sedentary lifestyle resemble the change in these variables which occur with disuse, bedrest or reduced activity (Saltin and Rowell 1980; Bortz 1982; Chopard et al. 2009a, b).

Despite the evidence demonstrating the benefits of increased physical activity on healthy aging; the Centers for Disease Control (CDC) reported that three of four older adults do not meet the minimum recommendation of a brisk walk, or similar activity, of at least 5 days each week. Studies have reported that increased physical activity during aging is associated with decreased body fat, increased relative muscle mass, reduced coronary risk profile (i.e. better insulin sensitivity and glucose homeostasis etc.), slower development of disability in old age, and athletes that resistance trained (RET) are ~50–60% stronger than their peers (Going et al. 1995; Sugawara et al. 2002; Hagberg et al. 1985; Seals et al. 1984a, b; Hunter et al. 2000, 2002; Klitgaard et al. 1990). The Yale Health and Aging Study, an epidemiological study conducted over 12 years, showed that physical activity had the ability to attenuate age related weight-loss among the elderly with chronic disease (Dziura et al. 2004). Furthermore, Baumgartner and colleagues observed that physical activity was positively correlated with muscle mass and negatively correlated with body-fat in a cross-sectional study among older men and women (Baumgartner et al. 1999).

Currently it is projected that the number of elderly will double worldwide from 11% of the population to 22% by 2050 (UN 2007). Because of the rapidly expanding population of older adults and the accumulation of evidence showing the benefits of increased physical activity for healthy older adults and older adults with chronic disease, a number of guidelines and recommendations on physical activity have been introduced for this population in the last few years. For the first time, in

2007, the American College of Sports Medicine (ACSM)/American Heart Association (AHA) released a joint recommendation on physical activity and public health recommendations for older adults, the Department of Health and Human Services (DHHS)/Center for Disease Control (CDC) released the "2008 Physical Activity Guidelines for Americans" and in 2009 the ACSM updated and expanded their position stand on "Exercise and Physical Activity for Older Adults". These recommendations and guidelines affirm that regular physical activity reduces the risk of many adverse health outcomes and there are additional benefits as the amount of physical activity increases with higher intensity, greater frequency and/or longer duration (see Table 1).

3 Mechanisms of Muscle Atrophy Associated with Sarcopenia

3.1 Protein Synthesis and Degradation

The maintenance of muscle mass is regulated by a balance between protein synthesis and protein degradation and is associated with rates of anabolic and catabolic processes, respectively. In conditions of atrophy, there is evidence for a shift toward myofibrillar and non-myofibrillar protein degradation (Mitch and Goldberg 1996) and a corresponding reduction in protein synthesis (Munoz et al. 1993). When protein synthesis exceeds protein degradation there is increased muscle mass (hypertrophy). In contrast, if protein degradation exceeds protein synthesis there is muscle loss (atrophy). During muscle atrophy as a result of disease processes, disuse or aging there is a preferential degradation of intermittently used white muscle (Type 2) fibers rather than continually used red muscle (Type 1) fibers (Tomlinson et al. 1969; Larsson 1983; Aniansson et al. 1986). Lexell et al. (1988), when studying 15–83 year old previously healthy men, reported that after the age of 25 years there is both a loss in the number and size of muscle fibers (Lexell et al. 1988). These researchers concluded that the fiber size reduction can be explained mostly by the smaller Type 2 fibers. However, it has been recently reported that there is a disproportionate loss of muscle function relative to muscle loss (Goodpaster et al. 2006; Haus et al. 2007). Therefore, the loss of muscle mass during aging could be the result in a decline of protein synthesis, increase in protein degradation or a combination of both. There is some contention regarding whether the decrease in protein synthesis associated with aging occurs solely during anabolic stimulation (Volpi et al. 2001; Cuthbertson et al. 2005; Rennie 2009) or also in the basal state (Nair 1995; Welle et al. 1993; Rooyackers et al. 1996; Yarasheski et al. 1993). It was originally reported that old subjects had decreased rates of basal muscle protein synthesis (Rooyackers et al. 1996; Yarasheski et al. 1993; Welle et al. 1993). However, others have been unable to reproduce these results and have observed a decrease only during anabolic stimulation (Rennie 2009; Volpi et al. 2001; Cuthbertson et al. 2005).

Table 1 Summary of physical activity recommendations for older adults from the American College of Sports Medicine/American Heart Association and the U.S. Centers for Disease Control and Prevention/ Department of Health and Human Services (Adapted from Nelson et al. 2007; Chodzko-Zajko et al. 2009; DHHS 2008)

ACSM/AHA Physical activity recommendations for older adults:
Aerobic exercise:
Frequency: For moderate-intensity activities, accumulate at least 30 or up to 60 (for greater benefit) min/day in bouts of at least 10 min each to total 150–300 min/week, at least 20–30 min/day or more of vigorous-intensity activities to total 75–150 min/week, an equivalent combination of moderate and vigorous activity.
Intensity*:* On a scale of 0–10 for level of physical exertion, 5–6 for moderate-intensity and 7–8 for vigorous intensity.
Duration: For moderate-intensity activities, accumulate at least 30 min/day in bouts of at least 10 min each or at least 20 min/day of continuous activity for vigorous-intensity activities.
Type: Any modality that does not impose excessive orthopedic stress; walking is the most common type of activity. Aquatic exercise and stationary cycle exercise may be advantageous for those with limited tolerance for weight bearing activity.
Strength exercise:
Frequency: At least 2 days/week.
Intensity: Between moderate- (5–6) and vigorous- (7–8) intensity on a scale of 0–10.
Type: Progressive weight training program or weight bearing calisthenics (eight to ten exercises involving the major muscle groups of 8–12 repetitions each), stair climbing, and other strengthening activities that use the major muscle groups.
Flexibility exercise:
Frequency: At least 2 days/week.
Intensity: Moderate (5–6) intensity on a scale of 0–10.
Type: Any activities that maintain or increase flexibility using sustained stretches for each major muscle group and static rather than ballistic movements.
Balance exercise: recommended for frequent fallers or individuals with mobility problems.
CDC/DHHS Physical activity recommendations for older adults:
All adults should avoid inactivity. Some physical activity is better than none, and adults who participate in any amount of physical activity gain some health benefits.
Aerobic exercise:
Frequency: For moderate-intensity exercise, perform 30 min/day for 5 days/week or vigorous-intensity exercise, perform 20 min/day for 3 days/week. You can do moderate- or vigorous-intensity aerobic activity, or a mix of the two each week.
Intensity: On a scale of 0–10 for level of physical exertion, 5–6 for moderate-intensity and 7–8 for vigorous intensity.
Duration: For moderate-intensity activities, accumulate at least 30 min/day in bouts of at least 10 min each.
Strength exercise:
Frequency: Ten strength-training exercises, 10–15 repetitions of each exercise 2–3/week.
Balance exercises: perform if at risk of falling.

The concept of aging is also strongly associated with increased protein degradation leading to muscle atrophy. The effects of aging on protein degradation are difficult to quantify. This is because in adult humans and animals only 60–70% of skeletal muscle proteins are made up of myofibrillar protein and these turn over very slowly making their quantification very difficult [see review: (Attaix et al. 2005)].

In keeping with this idea, Volpi et al. (2001) were only able to observe a small increase in basal protein degradation in old versus young humans (Volpi et al. 2001).

There are three known major proteolytic pathways that are revealed to have a role in skeletal muscle: the lysosomal pathway, the Ca^{2+}-dependent pathway comprising the μ– and m-calpains, and the ubiquitin-proteasome dependent proteolytic pathway (Attaix et al. 2005). Of these, the pathway that has recently received the most interest is the ubiquitin-proteasome pathway. In skeletal muscle this pathway is involved in the breakdown of long-lived myofibrillar proteins. In a variety of conditions such as cancer, diabetes, denervation, disuse, and fasting, skeletal muscles atrophy through degradation of myofibrillar proteins via the ubiquitin–proteasome pathway (Edstrom et al. 2006; Attaix et al. 2005; Cao et al. 2005). The induction of the muscle-specific ubiquitin E3-ligases (atrophy gene-1/muscle atrophy F-box (Atrogin-1/MAFbx) and muscle ring-finger protein 1 (MuRF1)) are thought to be the common mechanism associated with these diseases (Cao et al. 2005). The roles of Atrogin-1 and MuRF-1 in aging related muscle loss are not as clear cut. For example, some studies reported a small increase (Pattison et al. 2003), no change (Welle et al. 2003) or even a downregulation of Atrogin-1 and MuRF-1 mRNA in aged muscle (Edstrom et al. 2006). Of interest, Raue et al. (2007) observed that older women who are experiencing a large degree of sarcopenia express the MuRF-1 gene at higher levels compared to young adults, but this is reversed with resistance exercise (Raue et al. 2007). Although there was no difference in Atrogin-1 expression between the old and young subjects, after resistance exercise there was a pronounced upregulation of this gene in older women (Raue et al. 2007).

3.2 *Anabolic Resistance*

Anabolic stimulators, such as insulin, insulin-like growth factors (IGF1), amino acids (AA) and muscle contraction, rapidly and significantly increase skeletal muscle protein synthesis in young healthy tissue. Increased rates of protein synthesis are a key feature of hypertrophy driving muscle growth. The effect of essential amino acids on the dose-dependent stimulation of muscle protein synthesis is even observed when circulating insulin concentrations were clamped (10 μIU/mL) (Cuthbertson et al. 2005) or when somatostatin was used to inhibit insulin and insulin-like growth factors in human subjects (Greenhaff et al. 2008). The aging-induced "resistance" to amino acids to the stimulation of muscle protein synthesis has previously been observed in humans and rodents (Guillet et al. 2004; Cuthbertson et al. 2005; Rasmussen et al. 2006; Prod'homme et al. 2005). Rennie and colleagues (Cuthbertson et al. 2005) termed the age-related inability of nutrients to induce an appropriate anabolic response as "anabolic resistance". Cuthbertson et al. (2005) observed in older humans, following introduction of essential amino acids (EAA), there was a reduced increase in skeletal muscle protein synthesis that was correlated with increased concentrations of circulating and intramuscular EAA (leucine) compared to their young counterparts (Cuthbertson et al. 2005). The authors hypothesized that this was related to "anabolic resistance" that is distinguishable in aging muscle (Cuthbertson et al. 2005).

Aging is associated with an inability of insulin to stimulate muscle protein synthesis and amino acid uptake in otherwise healthy, glucose-tolerant persons (Rasmussen et al. 2006; Guillet et al. 2004; Bell et al. 2006; Fujita et al. 2009). The decline in muscle protein anabolic response to insulin is likely to be responsible for the observed reduction in postprandial muscle protein anabolism in older people. Rasmussen et al. (2006) observed that protein synthesis does not increase in response to hyperinsulinemia in older adults, in contrast to young subjects (Rasmussen et al. 2006). Prod'homme (2005) reported that insulin and EAA had differential effects on muscle protein synthesis in aging animals (Prod'homme et al. 2005). These researchers observed that young and old animals had a similar response to insulin, while anabolic stimulation by EAA was completely abolished in the older animals. Insulin resistance of muscle protein metabolism with ageing may induce a slow but progressive decline in muscle protein content thereby contributing to the development of sarcopenia in older.

It is well established that within a few hours of muscle contraction there is an increase in protein synthesis even in the fasted state. The contraction-induced effects on muscle protein synthesis have been previously shown to be decreased in older compared to young humans (Kumar et al. 2009; Sheffield-Moore et al. 2004). Welle et al. (1995) even observed this effect after a 3 week strength exercise program in male and female human subjects (Welle et al. 1995). We (Funai et al. 2006; Parkington et al. 2004) and others (Thomson and Gordon 2005, 2006; Thomson et al. 2009) have also observed an inhibition of an anabolic signaling in response to muscle contraction and/or overload in aging skeletal muscle. Funai et al. (2006) reported that anabolic signaling was increased in skeletal muscle after a single bout of in situ muscle contractile activity induced by high-frequency electrical stimulation (HFES) in adult animals, but these responses were attenuated in aged animals (Funai et al. 2006). However, the anabolic resistance attributed to aging muscle has not been observed in all studies (Reynolds et al. 2004; Paddon-Jones et al. 2004; Volpi et al. 2003; Drummond et al. 2009a; Short et al. 2003, 2004). Therefore, more study is needed to elucidate the significance of anabolic resistance to sarcopenia.

3.3 Anabolic Signaling

The mammalian target of rapamycin (mTOR) signaling kinase, which can be activated by Akt/Protein Kinase B (PKB), has emerged as a necessary effector of skeletal muscle growth in response to contraction and anabolic agents (for review see: Wang and Proud 2006; Bodine et al. 2001; Rommel et al. 2001). Insulin, amino acids and acute contractile activity have all been observed to increase the phosphorylation of mTOR and its downstream targets, p70 ribosomal protein S6 kinase 1

(S6K1) and 4E binding protein 1 (4EBP1). mTOR is a highly conserved, serine/threonine kinase of the phosphatidylinositol kinase-related kinase family and is a key regulatory protein for a multiplicity of cell processes including, but not limited to, cell growth and differentiation, protein synthesis, and actin cytoskeletal organization. The primary phosphorylation targets of mTOR are the threonine (Thr)389 site of S6K1 and the Thr37/46 sites of 4EBP1 that mediate translational initiation.

The observation of decreased protein synthesis in response to anabolic stimulation with aging is believed to occur as a result of the inhibition of mTOR signaling (Wang and Proud 2006). Multiple studies that have utilized rapamycin, a highly potent inhibitor of mTOR activation, have observed decreased protein synthesis *in vivo* and *in vitro* (Drummond et al. 2009; Kubica et al. 2005, 2008; Fluckey et al. 2004; Kimball et al. 2000; Anthony et al. 2000; Grzelkowska et al. 1999). The inhibitory effect of rapamycin on mTOR activation and protein synthesis can even occur despite an effective anabolic stimulation (Vary et al. 2007; Rivas et al. 2009; Kubica et al. 2005; Anthony et al. 2000). Cuthbertson et al. (2005) hypothesized that the "anabolic resistance" that was observed in their older subjects was related to the reduced phosphorylation of mTOR and its downstream substrate S6K1 (Cuthbertson et al. 2005). We and others have observed an age induced attenuation of the Akt/mTOR signaling pathway in response to contractile stimulation and overload (Funai et al. 2006; Hwee and Bodine 2009; Thomson and Gordon 2006; Parkington et al. 2004). Recently, Drummond et al. (2009) demonstrated that the contraction-induced increase of mTOR signaling, protein synthesis and extracellular related kinase signaling (ERK1/2) are reduced with prior rapamycin treatment in humans (Drummond et al. 2009b). These results provide some understanding for the role of mTOR in the initiation of protein synthesis in response to anabolic stimuli, such as muscle contraction. However, there is some disagreement whether the phosphorylation of mTOR is responsible for the changes in the protein synthetic rates in response to an anabolic stimulation (Greenhaff et al. 2008). Greenhaff et al. (2008) recently demonstrated that changes in signaling protein phosphorylation can be almost completely be disconnected from protein synthesis with an anabolic stimulus such as insulin.

3.4 Skeletal Muscle Attenuation

It is well understood that with advancing age there is a change in the composition of skeletal muscle. Lean muscle mass normally contributes up to 50% of total body weight in young adults but declines with age to 25% at 75–80 years (Koopman and van Loon 2009; Short et al. 2004). The loss in lean muscle mass is usually offset by gains in fat mass. Longitudinal studies have shown that fat mass increases with age peaking at about 60–75 years (Rissanen et al. 1988; Droyvold et al. 2006). Aging is associated with the increased accumulation of intramuscular fat as well as with an increase in the incidence of metabolic disorders such as insulin resistance (Tucker and Turcotte 2003; Nakagawa et al. 2007). Impaired lipid metabolism and increased visceral adiposity associated with aging are thought to contribute to the muscle atrophy associated with sarcopenia (Nakagawa et al. 2007). Researchers have observed the defects in lipid metabolism, such as increased intramuscular and circulating lipids, even in lean and otherwise healthy elderly persons (Nakagawa et al. 2007). Furthermore, studies have found significant difference in protein metabolism between obese and non-obese humans (Guillet et al. 2009; Nair et al. 1983; Jensen and Haymond 1991; Luzi et al. 1996). Goodpaster et al. (2000) observed that increased mid-thigh muscle attenuation (a marker of intramuscular lipids with CT scan) was related to the loss of muscular specific strength in 2,627 older men and women (Goodpaster et al. 2000). The concomitant age-related changes in body composition, obesity, impaired metabolism and low muscle mass have lead to the hypothesis that there may be a causal link between obesity and low strength.

Growth factors (i.e. insulin and IGF1), AA and muscle contraction are known modulators of muscle protein synthesis and inhibitors of protein degradation and their capacity to stimulate muscle protein synthesis is impaired in both aging and obesity (Rasmussen et al. 2006; Guillet et al. 2009). Insulin resistance is also highly coupled with obesity and aging and results in decreased insulin-stimulated glucose uptake, protein synthesis and the inability to inhibit lipid uptake (Corcoran et al. 2007; Tucker and Turcotte 2003; Hawley and Lessard 2008; Rasmussen et al. 2006; Anderson et al. 2008; Guillet et al. 2009). Guillet et al. (2009) recently observed that obese humans had a decreased fractional synthetic rate during an amino acid infusion and insulin clamp in the basal and insulin-stimulated state compared to their age matched controls (Guillet et al. 2009). In addition to the evidence showing that high-fat feeding and obesity inhibit protein synthesis in response to an anabolic stimulus, there is also evidence of altered mTOR signaling in the basal and insulin-stimulated state (Guillet et al. 2004, 2009; Rivas et al. 2009; Anderson et al. 2008; Khamzina et al. 2005; Katta et al. 2009). For example, Katta et al. (2009) demonstrated in obese Zucker rats that mTOR signaling was inhibited in response to in situ HFES muscle contraction compared to their lean litter mates (Katta et al. 2009). However, studies report there is no relationship between acutely increased circulating free fatty-acids (artificially-induced with heparin treatment) and decreased protein synthesis (Katsanos et al. 2009) or impaired mTOR signaling (Lang 2006) in skeletal muscle. Although there is some contention regarding role of increased circulating free-fatty acids and reduced protein synthesis, the increased storage of fat in muscle during aging has been clearly demonstrated to have role in reduced muscle mass and functional impairment.

3.5 Skeletal Muscle Regeneration

Aging skeletal muscle displays a significant reduction in regenerative capacity this leads to the inability to adapt to an increased load and is therefore less responsive to injury. The regenerative capacity of muscle fibers depends on a pool of myogenically specified undifferentiated mononuclear precursor stem cells called ‘satellite’ cells that appear to function as “reserve” myoblasts (for review see: Wagers and Conboy (2005), Gopinath and Rando (2008). Satellite cells (SC) are the primary stem cells in adult skeletal muscle, and are responsible for postnatal muscle growth, hypertrophy, regeneration and repair. SC were identified ultrastructurally and were named for their peripheral location beneath the basal lamina of the myofiber (Mauro 1961). SC are primarily in a quiescent, non-differentiating state, dividing infrequently under normal conditions in the adult but activated (reenter the cell cycle) by regenerative cues such as injury or exercise. Once activated, the cells will proliferate, increase in number and the daughter cells (myoblasts) will repair damaged skeletal muscle by fusing to existing myofibers or generating new myofibers by fusing together (Hawke 2005).

It is believed that muscle hypertrophy requires the addition of nuclei to existing myofibers (Adams 2006). This follows the premise that increases in fiber size must be associated with a proportional increase in myonuclei for the control of mRNA and protein production per volume of cytoplasm (Hawke 2005). Growth factors such as, interleukin (IL) 6, testosterone, IGF1 and the IGF isoform, mechanogrowth factor, have been identified as having a role in post-exercise hypertrophy (Vierck et al. 2000; Adams 2002; Sinha-Hikim et al. 2003). Of interest, Machida and Booth (2004) recently demonstrated a key role for the PI3K/Akt pathway in IGF induced SC proliferation (Machida and Booth 2004).

The potential role of SC in age-induced muscle atrophy is not clear cut. Studies have either shown a similar (Dreyer et al. 2006a; Roth et al. 2000; Sinha-Hikim et al. 2006) or lower (Kadi et al. 2004; Renault et al. 2002) SC proportion in older adults when compared with young adults. It has been demonstrated that SC in aged muscle display a delayed response to activating stimuli and reduced proliferative expansion (Schultz and Lipton 1982; Conboy et al. 2003). Verdijk and colleagues have reported marked decreases in Type 2 versus Type 1 muscle fiber myonuclear domain size and a specific decrease in the Type 2 fiber satellite cell content in elderly humans (Verdijk et al. 2007). In a follow up study, these researchers observed that Type 2 muscle fiber atrophy and the associated lower satellite cell proportion in Type 2 versus Type 1 muscle fibers in older adults can be reversed by prolonged resistance type exercise training (Verdijk et al. 2009). Roth et al. (2001) have also reported that satellite cell proportion in young and older men and women was significantly increased as a result of 9 weeks of strength training (Roth et al. 2001b). Interestingly, older women demonstrated a significantly greater increase in SC content and the largest increase in the number of active satellite cells in response to strength training. Therefore, because of the significant role of SC in skeletal muscle regeneration, repair and hypertrophy unraveling their role in sarcopenia remains a high priority.

Some possible mechanisms that contribute to sarcopenia are outlined in Fig. 1. Sarcopenia is a multifactorial process and the mechanisms that underlie it are only beginning to be elucidated. More research is needed determine their roles in the onset of sarcopenia.

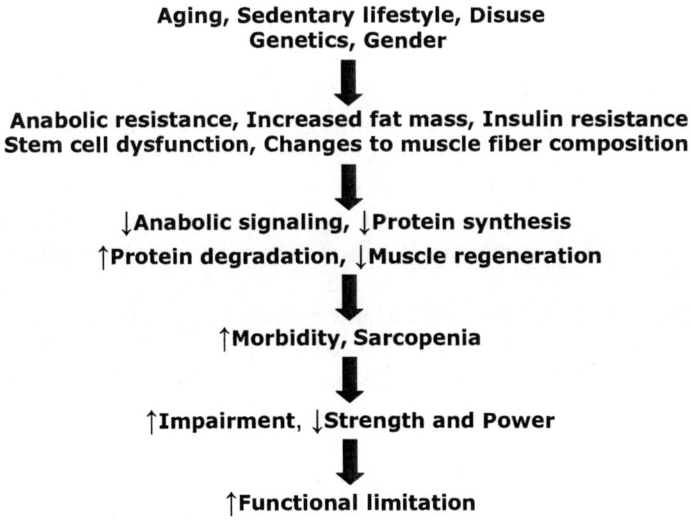

Fig. 1 A few possible mechanistic contributors to sarcopenia and its consequences

4 Exercise as an Intervention for the Modulation of Sarcopenia

As discussed earlier in this chapter, life-long habitual physical activity is the most effective preventative treatment for age-induced sarcopenia. Multiple groups have studied the effects of exercise training on energy metabolism and as a treatment for metabolic disorders such as, insulin resistance, obesity and type 2 diabetes (Hawley and Lessard 2008; Berger and Berchtold 1979; Wallberg-Henriksson and Holloszy 1984, 1985; Zierath 2002; Goodyear and Kahn 1998; Kelley and Goodpaster 2001; Musi and Goodyear 2006). Exercise training, with respect to substrate metabolism, is associated with enhanced oxidative capacity and insulin sensitivity, decreased intramuscular lipid storage and improved body composition (Hawley and Lessard 2008; Toledo et al. 2007; Richter and Ruderman 2009; Tanaka and Seals 2003; Lessard et al. 2007).

There is growing evidence demonstrating the benefits of exercise late in life as a countermeasure for sarcopenia and its related functional limitations (Keysor 2003; Henwood and Taaffe 2005; Galvao and Taaffe 2005; Galvao et al. 2005). Regular physical activity is associated with greater functional capacity, increased appendicular muscle mass and reduced incidence of metabolic diseases and this is particularly observed in middle-aged and older adults (Sugawara et al. 2002; Harber et al. 2009b). Since the 1980s, numerous intervention studies have reported the benefits of resistance, aerobic and a combination (aerobic and resistance) of these exercise modalities for the treatment muscle loss and disability as a result of aging (Frontera et al. 1988; Tanaka and Seals 2003). The purpose of this section is to review the molecular events and whole-body benefits of the different modes of exercise for the treatment of sarcopenia.

4.1 Aerobic Exercise

Aerobic exercise is a widely recommended therapeutic agent for older adults because of its beneficial effects on cardiovascular and metabolic health, body composition and improved function. Endurance exercise is based on movements performed with a high number of repetitions and low resistance. Maximal aerobic capacity (VO_2 max) is generally thought to be the best indicator of the capacity to perform aerobic exercise. Maximal oxygen consumption declines about 1% per year after the age of 25 in sedentary individuals. This is important since low aerobic capacity has been highly correlated with increased rates of all-cause mortality in numerous epidemiological studies (Paffenbarger et al. 1993, 1970; Leon et al. 1987; Morris et al. 1953a, b). However, in master athletes who participate in regular aerobic activity the decline in VO_2 max is only 0.5% per year (Tanaka and Seals 2003, 2008; Paffenbarger et al. 1993).

It is thought that the key contributors to a decline in maximal aerobic capacity in sedentary individuals are a decrease in maximal cardiac output (Ogawa et al. 1992; Proctor et al. 1998), a decrease in muscle oxidative capacity (Ljubicic et al. 2009; Short et al. 2003; Conley et al. 2000a, b; Harber et al. 2009) and a decrease in metabolically active muscle mass with a concomitant increase in metabolically inactive fat mass (Paffenbarger et al. 1970; Goodpaster et al. 2000, 2006; Short et al. 2003; Proctor and Joyner 1997; Fleg and Lakatta 1988). When measuring VO_2 max normalized to muscle mass (as indexed by 24 h urinary creatinine excretion) in old and young men and women, Fleg and Lakkata (1988) reported that the age-induced decrease in VO_2 max is explained by the selective loss of muscle mass that accompanies aging. Recently, Proctor and Joyner (1997), when examining the effect of reduced muscle mass (and increased fat mass) on VO_2 max in the elderly, expressed maximal oxygen consumption relative to appendicular muscle mass (Proctor and Joyner 1997). They observed that 50% of the decline in VO_2 max, as a result of aging, was attributed to the age-induced decreases in muscle mass and increases in fat mass. Therefore, understanding the possible benefits from aerobic exercise for increasing maximal oxidative capacity and/or muscle mass in older adults could have implications for healthy aging.

4.1.1 Improving Oxidative Capacity

Aerobic exercise of sufficient intensity and duration can significantly increase VO_2 max in middle aged and older adults (Huang et al. 2005; Malbut et al. 2002; Lanza et al. 2008). It has been hypothesized that increases in mitochondrial number, increases in the expression of mitochondrial proteins and/or an increase in the expression of transcription factors involved in mitochondrial biogenesis are mechanisms for the enhancement in post-exercise VO_2 max. Lanza et al. (2008) observed increases in mitochondrial ATP production rate (MAPR), citrate synthase (CS) activity, pparγ-coactivator 1 α (PGC1α), mtDNA abundance. Of interest, the researchers also reported an increase in sirtuin 3 (SIRT3), a protein deacetylase that has been associated with the life prolonging benefits of caloric restriction, in aerobically trained older individuals compared to their sedentary peers (Lanza et al. 2008).

There is some evidence of a reduction in the activation of the energy sensor, AMP-activated protein kinase (AMPK), in response to endurance exercise in aging muscle (Reznick et al. 2007). However, Ljubicic and Hood (2009) observed no difference with endurance-like contraction induced AMPK activation in high-oxidative red muscle between old and young animals (Ljubicic and Hood 2009). The researchers did observe an inhibition of AMPK activity in the less oxidative white muscle in the acute response to endurance-like contraction. This may be an important consequence because of AMPK has recently been observed to have a critical role in the regulation of muscle hypertrophy as a result of muscle overload (McGee et al. 2008; Thomson et al. 2009).

4.1.2 Increased Muscle Mass

There has been minimal study on aerobic exercise and its effects on improving muscle function, increasing muscle mass and protein synthesis in the elderly. Some researchers have provided evidence that aerobic exercise was as proficient as resistance training at improving functional limitations associated with aging (Wood et al. 2001; Davidson et al. 2009; Coggan et al. 1992; Verney et al. 2006). For example, Davidson et al. (2009) reported that 6 months of resistance and aerobic exercise was associated with similar improvements in functional limitation in 136 previously sedentary, obese older men and women (Davidson et al. 2009).

Researchers have previously reported that aerobic exercise does not alter muscle size in older individuals (Ferrara et al. 2006; Verney et al. 2006; Short et al. 2004; Weiss et al. 2007). However, Harber et al. (2009) have recently shown that a 12 week aerobic training intervention induced a 16.5% increase in single fiber cross sectional area (CSA) and a 20% increase in quadriceps muscle volume that was accompanied by improvements in whole muscle power and force production in healthy older women (Harber et al. 2009). The investigators hypothesized that their results differed from previous studies because their subjects were in good health and the body weights of their subjects were maintained throughout the intervention. Also, habitually endurance-trained elderly males have higher appendicular muscle mass, relative to body mass, compared to their sedentary controls (Sugawara et al. 2002). The increased muscle hypertrophy and appendicular muscle mass observed in these studies could be as a result of increases in protein synthesis observed after aerobic exercise (Harber et al. 2009a, b; Short et al. 2004; Fujita et al. 2007). Short et al. (2004) reported that men and women have a decline in whole-body protein metabolism as a result of aging. A 4 month aerobic exercise program had no effect on whole-body protein turnover but, significantly increased mixed muscle protein synthesis in the older subjects (Short et al. 2004). Fujita et al. (2007) have further shown an increase in insulin-stimulated muscle protein turnover as a result of an acute aerobic exercise bout in aging humans. This was related to increased endothelial function and an increase in the insulin-stimulated phosphorylation of mTOR, S6K1 and Akt (Fujita et al. 2007).

4.2 Resistance Exercise

In contrast to aerobic exercise, resistance exercise is based on movements performed with high resistance and a small number of repetitions over a short period of time. The purpose of resistance exercise is to provide an overload stimulus that strengthens muscles. This is usually a training stimulus in the range of 60–70% of a single repetition maximum (1RM), which can lift 8–12 times to failure and is performed repeatedly with progressive intensity that induces hypertrophy (Phillips 2007). How resistance exercise is performed is no different between young and older individuals. Furthermore, studies comparing aerobic exercise-trained (AE) and resistance exercise-trained (RE) older athletes to sedentary age-matched controls have reported many physiological advantages associated with the prevention of age-induced diseases (Klitgaard et al. 1990). In a seminal cross-sectional study of elderly men with different training backgrounds; Klitgaard et al. (1990) reported older men (69 years), who had been strength-training approximately 12–17 years before being studied, maximal isometric torque and muscle mass, as measured by computed tomography (CT) scan of the upper arm and mid-thigh, were significantly greater than those in age-matched swimmers or runners and were similar to young sedentary controls (Klitgaard et al. 1990). The subjects examined in this study exercised an average of three times per week at approximately 70–90% of their 1RM (Klitgaard et al. 1990). Klitgaard and colleagues provide some evidence that resistance exercise is possibly a superior intervention to aerobic training for the treatment of sarcopenia.

The effects of resistance exercise in young healthy men and women have been well described (for early review see Kraemer and Ratamess 2004). Briefly, high intensity progressive resistance training in young adults has resulted in significant increases in dynamic strength, explosive power, and muscle mass (McCall et al. 1996; Staron et al. 1991, 1994; Anderson and Kearney 1982). The effect of resistance exercise on the older humans is not a potent when compared to the young. More recent studies have confirmed these findings (Kraemer et al. 2004; Glowacki et al. 2004; Campos et al. 2002; McCaulley et al. 2009; Luden et al. 2008). Aged skeletal muscle does not respond as effectively to resistance exercise, particularly at the genetic and protein-signaling level, as young skeletal muscle (Kosek et al. 2006; Petrella et al. 2005, 2006; Mayhew et al. 2009; Bamman et al. 2004; Slivka et al. 2008; Raue et al. 2009). Although there is some attenuation of the effect of resistance exercise in the old, it is established that resistance exercise is a practical and effective intervention to increase muscle strength, power and mass in the elderly even into the ninth decade of life (McCartney et al. 1995, 1996; Kostek et al. 2005; Valkeinen et al. 2005; Fiatarone et al. 1990; Ferri et al. 2003;

Adams et al. 2001). These measures are important to the elderly population because they may reduce the relative stress imposed by activities of daily living.

4.2.1 Improving Strength

A decline in dynamic, isokinetic and static muscle strength has been noted with advancing age. Muscle strength is defined as the maximum force generation capacity of an individual, it reaches its peak at about the third decade of life and decreases about 12–15% per decade after the age of 50 years (Larsson et al. 1979). Direct comparisons of young and older sedentary individuals have shown that older persons of around 70 years have approximately 60% of the force-generating ability of their younger peers of 20–30 years (Klitgaard et al. 1990). This loss of strength with aging is observed in both men and women (Lindle et al. 1997). Several studies of the elderly have suggested that muscle strength is closely associated with functional activities of daily living (Bassey et al. 1988; Jette and Branch 1981; Rantanen et al. 1994). The declines in muscle strength with age are related to impairment in function even in otherwise healthy older individuals.

Investigators have documented gains in strength as a direct result of resistance training regimens throughout the lifespan (Korpelainen et al. 2006). In the young, a 2-week isokinetic resistance training program in men effectively increased isokinetic and isometric right quadricep muscle peak torque at both 60° and 240° (Akima et al. 1999). In another report in men, a 12-week high resistance strength training program resulted in an increase in isokinetic concentric (quadriceps) knee joint strength at a velocity of 30° and eccentric (hamstring) knee joint strength at velocities of 30°, 120° and 240° (Aagaard et al. 1996). The hamstring/quadriceps ratio also increased. A dynamic resistance training protocol of similar duration in men and women resulted in isometric torso rotation strength gains in men and women who exercised twice weekly (DeMichele et al. 1997). Significant gains in both upper- and lower-body strength have also been reported for studies of 6 months duration (Kirk et al. 2007).

Strength gains have been reported for shorter (8–12 weeks) duration studies in older adults. Studies that have utilized similar resistance training protocols (10–12 weeks, 3 days/week, 80% of 1RM), have shown a the mean improvement in muscle strength of ~80%, post-exercise training (Balagopal et al. 2001; Brown et al. 1990; Fiatarone et al. 1994; Frontera et al. 1988; Trappe et al. 2000, 2001; Campbell et al. 1994, Harridge et al. 1999). These studies provide evidence there is a substantial increase in muscle strength in older individuals who resistance exercise. Larsson et al. (1979) first reported that a selective loss of Type 2 (fast twitch) muscle fibers is associated with a decline in strength. Frontera et al. (1988) examined the effects of a high-intensity dynamic-resistance training program in healthy older men (mean age 64 years; (Frontera et al. 1988)). Their subjects performed knee flexion and extension exercises 3 days/week at 80% of the 1RM (eight to ten repetitions) for 12 weeks. They found a 107% increase in knee extensor strength and a 226% increase in knee flexor strength. In addition, they observed an 11% increase in mid-thigh cross-sectional area as assessed by CT. Muscle biopsy analysis revealed a 33% and 27% increase in Type 1 and 2 fiber area respectively. This was the first study to demonstrate that in healthy older men dynamic high-intensity strength-training can result in marked increases in muscle strength and muscle hypertrophy (Frontera et al. 1988).

Researchers using a progressive resistance training protocol in older adults observed a linear increase in dynamic strength at different time points of a 12-week study (Sousa and Sampaio 2005). In an 8-week comparison between a combined resistance/gymnasium based functional training regimen and high- and a moderate-velocity resistance training protocols, significant dynamic strength gains were reported for the combined resistance/gymnasium indicating a synergistic effect of exercise (Henwood and Taaffe 2006). However, others report a dose–response relationship between high-intensity progressive resistance training and functional capacity that may explain the preponderant use of this type of resistance training (Galvao and Taaffe 2005; Seynnes et al. 2004). Gains in strength also occur with low- (Tsutsumi et al. 1997) and variable-intensity resistance training (6 months) (Hunter et al. 2001).

4.2.2 Increasing Power

Although physical activity interventions that increase or maintain of muscle strength have important health implications, there is emerging evidence that muscle power generating capacity (the rate at which muscle force can be generated) may play a more important role in functional independence and fall prevention, particularly among older adults. Peak muscle power has only recently been examined in older individuals as a variable distinct from strength and has been shown to decline earlier and more precipitously throughout the life span (Metter et al. 1997). Lower extremity muscle power is a strong predictor of physical performance, functional mobility and risk of falling among older adults (Bean et al. 2002, 2003). Muscle power is also inversely associated with self-reported disability status in community-dwelling older adults with mobility limitations (Foldvari et al. 2000; Suzuki et al. 2001) and is a better discriminator of mobility limitations than muscle strength (Bean et al. 2003). In particular, in two separate studies of older individuals with self-reported functional limitations, peak lower extremity power has been shown to be more closely associated with gait speed than strength (Bean et al. 2002; Cuoco et al. 2004).

Exercise interventions targeted at improving lower extremity muscle power in the elderly have been well-tolerated and effective (Henwood and Taaffe 2005; Earles et al. 2001; Miszko et al. 2003). Indeed, we have previously reported that an exercise regimen of high-force, high-velocity progressive resistance training resulted in a twofold increase in muscle power in older women with self-reported functional limitations, compared to traditional high-force, slow-velocity progressive resistance training (Fielding et al. 2002). Despite the observed improvements in musculoskeletal strength, few studies have examined the specific velocity of training and its subsequent physiological and functional effects. Fiatarone et al. (1994) have noted in their nursing home study only a 28% increase in stair climbing power in response to progressive resistance training despite over a 100% increase in strength, suggesting a disproportionate and specific rise in strength versus power with traditional resistance training (Fiatarone et al. 1994).

Skelton et al. (1995) also examined changes in peak leg extensor power in response to 12 weeks of resistance training in older women (Skelton et al. 1995). They observed increases in strength of 22–27% with a non-significant increase in leg extensor power. Recently, Jozsi et al. (1999) noted a modest improvement (30%) in leg extensor power in response to 12 weeks of RE in healthy older men and women (Jozsi et al. 1999). These studies suggest that RE results in minimal improvements in peak power and that training interventions need to be designed to more closely maximize the capacity to improve peak power in older individuals. We have shown that a 16 week high velocity high force resistance training to maximize muscle power intervention is feasible, well tolerated, and can dramatically improve lower extremity muscle power in older women with self-reported disability (Fielding et al. 2002). These results have recently been confirmed in two recent randomized trials (Earles et al. 2001; Signorile et al. 2002). Recently, we have reported (Reid et al. 2008) that a short-term intervention of high-velocity high-power progressive resistance training was associated with similar improvements of lower extremity muscle power compared to traditional slow-velocity strength training in elderly adults with preexisting mobility impairments. Although both training modalities yielded similar increases of lower extremity strength in this population, high-velocity power training was associated with significant gains in specific muscle power. Future studies should directly quantify neural adaptations and physiological mechanisms to power training, and further randomized controlled trials are warranted to investigate the optimal training duration and volume required to elicit significant improvements of muscle power, strength and functional performance in elderly subjects who are at increased risk for subsequent disability.

4.2.3 Increasing Muscle Mass

The increase in size in response to resistance training is typically given as a change in the CSA of the muscle, as measured with magnetic resonance imaging, ultrasonography, or CT. Changes in muscle strength and size after resistance training are likely accompanied by alterations in the size of the muscle fibers that are determined by immunohistochemistry. Studies that have directly compared changes in muscle mass, CSA or protein synthesis in response to resistance exercise training have noted significant increases in both males and females (Burd et al. 2009; Staron et al. 1994; Pansarasa et al. 2009; Holm et al. 2008; Hubal et al. 2005). Increases in muscle CSA by CT scanning have also been shown to be similar between men (17.5%) and women (20.4%) in response to 16 week of upper and lower extremity high intensity resistance training (Cureton et al. 1988). However, one study employing elastic bands for resistance training noted significant increases muscle fiber cross sectional areas in men but not in women in response to 8 week of training two to three sessions per week (Hostler et al. 2001). More recently, assessment of fat free mass by dual energy x-ray absorptiometry and serial CT scans to measure muscle volume have confirmed similar increases in muscle mass and volume between young men and women in response to a 6 month whole body program of progressive resistance exercise training (Roth et al. 2001a). These results suggest that resistance exercise training can increase muscle strength and mass to similar extent in both men and women.

Several studies have assessed the optimal dose of resistance training required to maximize gains in muscle strength and mass in young adults. Campos et al. (2002) compared the responses to 8 weeks of progressive resistance training (Campos et al. 2002). Young healthy men were randomized to perform low repetition high intensity, intermediate repetition moderate intensity, or high repetition low intensity progressive resistance training of the lower extremities (leg press, squat, and knee extension). These authors found that there was greater muscle fiber hypertrophy and gains in muscle strength observed low repetition high intensity group and the intermediate repetition moderate intensity group compared to the high repetition low intensity group. In young women, Hisaeda et al. (1996) observed similar gains in peak torque and muscle cross sectional in response to 8 weeks of either high intensity/low repetition or high repetition/low intensity resistance training (Hisaeda et al. 1996). Studies have also examined the influence of the number of sets performed at each training session on changes in muscle strength and mass in response to resistance training. Ronnestad et al. (2007) demonstrated that three sets of lower body resistance exercise per session compared to one set per session was more effective in increasing muscle strength and CSA suggesting that the volume of training per session may drive the gains in muscle strength and mass (Ronnestad et al. 2007). In contrast, by varying the number of training days per week and the number of training sets performed while normalizing the total volume of work performed per week resulted in similar gains in muscle strength and CSA in young men and women (Candow and Burke 2007). The evidence from these randomized trials suggests that muscle hypertrophy from resistance training occurs in a dose-dependent manner that is primarily dependent on the intensity at which the training sessions are performed. In addition, the total volume of work performed during resistance training may also influence to magnitude of increase in muscle mass.

Early studies have demonstrated the positive effects of resistance training on muscle strength and size in healthy older men and women (For background review see: Fielding 1995). A number of randomized trials have now confirmed these initial findings (Sipila and Suominen 1996; Ferri et al. 2003; Suetta et al. 2004; Tsuzuku et al. 2007), and one study has demonstrated that muscle mass can continue to increase in older adults throughout 2 years of resistance training (McCartney et al. 1996). More recently, studies have examined the influence of resistance training on changes in muscle mass and the influence of age per se. Resistance exercise training interventions (RT) can increase both whole muscle and fiber CSA in older men and women. However, there is some evidence that this response may be attenuated with advancing age. Cross sectional studies of older bodybuilders who had been performing RE for 12–17 years were reported to have mid-thigh muscle CSA that were similar to young sedentary controls, suggesting that the ability to stimulate muscle growth is diminished with age (Klitgaard et al. 1990). In young men and women, the change in mid-thigh CSA after 4 months of high intensity resistance training is typically 16–23% (Cureton et al. 1988), compared to a 2.5–9.0% increase in institutionalized or frail older individuals in response to similar resistance interventions (Fiatarone et al. 1990, 1994; Binder et al. 2005).

Few studies have directly compared the effect of age on muscle hypertrophy utilizing a similar standardized training intervention. Welle et al. (1996) reported impaired responses of both knee and elbow flexors but not knee extensors after a whole body RE program in older compared to young men and women (Welle et al. 1996). Data from Hakkinen et al. (1998) suggest a decline in the adaptive response of the vastus lateralis from middle to old age of approximately 40% (Hakkinen et al. 1998). Lemmer et al. (2001) reported significant increase in thigh muscle CSA in both young and older adults following resistance training, however the magnitude of the increase was greater in the young (Lemmer et al. 2001). Similar results were also observed by Dionne et al. (2004) following 6 months of resistance training in young and older non-obese women (Dionne et al. 2004). In contrast, similar duration resistance training studies have examined changes in total thigh CSA and have reported similar responses in young and old (Ivey et al. 2000; Roth et al. 2001a). These findings suggest that progressive resistance training-induced increases in muscle mass can occur in older individuals but that the magnitude of this response may be attenuated, particularly in the oldest old.

Conflicting evidence has been presented on the effects of gender on the anabolic response to resistance training among older adults. Several studies that have enrolled both older men and women have reported similar increases in muscle mass with resistance training (Hakkinen et al. 1998; McCartney et al. 1996; Roth et al. 2001a; Wieser and Haber 2007). Nine weeks of high intensity resistance training resulted in lower muscle volume increases in women compared to men (Ivey et al. 2000) and similar findings were reported for whole body fat free mass in response to 12 week of high intensity resistance training in moderately overweight men and women (Joseph et al. 1999). Bamman et al. (2003) have also confirmed at the cellular level a greater degree of hypertrophy of both type I and II fibers in older men compared to older women in response to 26 weeks of high intensity resistance training (Bamman et al. 2003). However, in contrast to these reports Hakkinen et al. (1998) reported a smaller increase in muscle cross sectional area in older men compared to older women (Hakkinen et al. 1998).

4.3 Multi-modal Exercise Therapy

While the preferential mode for strength gains has been strength training (Keeler et al. 2001; Putman et al. 2004; Sarsan et al. 2006), with a bias towards eccentric exercises (Hilliard-Robertson et al. 2003), observations indicate that other modes or multi-modal training may also be highly effective in the aging population. Current guidelines stress the importance of multi-modal exercise for this cohort, including strengthening exercises, cardiovascular, flexibility, balance training and the combination of strength and endurance training (Cress et al. 2005; Baker et al. 2007; Chodzko-Zajko et al. 2009). These include but aren't limited to: Nordic training (Mjolsnes et al. 2004), circuit weight training (Harber et al. 2004), balance training (Heitkamp et al. 2001), a combination of strength and endurance as well as endurance only protocols (Binder et al. 2002; Putman et al. 2004; LaStayo et al. 2000; Englund et al. 2005; Izquierdo et al. 2004). Putman et al. (2004) reported that concurrent strength and endurance exercise training resulted in greater fast-to-slow fiber type transitions and attenuated hypertrophy of the type I fibers compared with strength training alone (Putman et al. 2004). Futhermore, multimodal exercise training was associated with a decreased lipid profile in older women compared to strength training alone (Marques et al. 2009).

In middle-aged men and women subjected to short duration physical activity interventions, strength gains were also improved with combinatory aerobics/weight (Tsourlou et al. 2003) training protocols. The gains in strength persist throughout longer duration studies (4–6 months) in this age group (Dornemann et al. 1997; Izquierdo et al. 2005) but demonstrate that greater gains in strength begin to occur after 8 weeks of a combined resistance and endurance exercise protocol (Izquierdo et al. 2005). In older adults, investigators have implemented longer duration (4–12 months) resistance training (Galvao and Taaffe 2005; Lord et al. 1996a, b) and combinatory resistance/endurance (King et al. 2000; Izquierdo et al. 2004; Tsourlou et al. 2006; Fahlman et al. 2007; Cress et al. 1999) type regimens to successfully increase strength in an effort to counteract the late-life decline in physical functioning. While resistance training induces muscle strength gains, functional-task exercises may be more effective at counteracting declines in function (de Vreede et al. 2005). Investigators have suggested that gains in isometric and dynamic muscle strength (Tsourlou et al. 2006) as well as in isokinetic muscle strength (Galvao and Taaffe 2005) are associated with improved physical functioning. However, the gains in strength may be muscle specific and translate into improvements in only select parameters of physical functioning as indicated in both long (Schlicht et al. 2001; Asikainen et al. 2006; Fahlman et al. 2007) and short duration exercise interventions (Topp et al. 1996). Although there have been some benefits associated with multimodal exercise regiments in young and older populations. Baker et al. (2007) recently reported, in a systematic review, that limited available data suggests that multi-modal exercise has a small effect on physical, functional and quality of life outcomes (Baker et al. 2007). However, more investigation is needed on the efficacy of simultaneous prescription of multi-modal training as a treatment for improving clinically relevant outcomes, and to establish whether multi-modal exercise at adequate volumes and intensities is feasible in older populations.

4.4 Anabolic Signaling/Protein Synthesis

Exercise of sufficient intensity and duration disrupts homeostasis initiating adaptive processes to generate new functional protein in skeletal muscle (Coffey and Hawley 2007). An essential process in the regulatory steps controlling protein synthesis is mRNA translation (Coffey and Hawley 2007). In this regard, mTOR has been implicated as an upstream mediator of protein synthesis via putative control of ribosomal biogenesis and Cap-dependent translation (Nader et al. 2005; Hannan et al. 2003; Besse and Ephrussi 2008). Moreover, the translation repressor eukaryotic translation initiation factor-4E binding protein 1 (4E-BP1) is a direct phosphorylation target of mTORC1 which de-represses 4E-BP1 inhibition of translation initiation (Besse and Ephrussi 2008). Intuitively, both endurance and resistance exercise would be expected to "switch on" translation following exercise and generate skeletal muscle adaptation, yet clearly identifying increased mTOR activation and subsequent 4E-BP1 phosphorylation during recovery from exercise has proved elusive. Indeed, there is limited evidence with regard to these exercise-induced phosphorylation events being associated with increased protein fractional synthetic rate (Fujita et al. 2007) likely due to the energy-consuming, catabolic state of skeletal muscle during and immediately following exercise in the fasted state. Nonetheless, as a nutrient sensor it is not surprising that amino-acid ingestion has been shown to augment exercise-induced mTOR activation and 4E-BP1 phosphorylation, and subsequent fractional synthetic rate in skeletal muscle (Koopman et al. 2007; Dreyer et al. 2008; Drummond et al. 2008; Rivas et al. 2009).

It is apparent that chronic endurance and resistance training generate specificity of adaptation and subsequent divergent phenotypes (Coffey and Hawley 2007). As such, the concomitant increase in mTOR-4E-BP1 mediated translation initiation with exercise likely contributes to the specificity of training adaptation. In support of this contention, novel findings by Wilkinson and co-workers (2008) showed increased translational signalling and fractional synthetic rate following both endurance and resistance training (Wilkinson et al. 2008). Notably, these workers observed specificity of adaptation with chronic endurance exercise only elevating the mitochondrial protein synthetic response, while resistance training increased myofibrillar but not mitochondrial fractional synthetic rate (Wilkinson et al. 2008). Therefore, exercise-induced mTOR translation initiation following endurance and resistance exercise may enhance skeletal muscle metabolism via alternate adaptation that promotes muscle quality (mitochondria) and quantity (cross-sectional area), respectively. Regardless, the apparent capacity of mTOR to promote global protein synthesis through translational processes in response to exercise is undoubtedly beneficial for the metabolic status of skeletal muscle.

Several reports have identified skeletal muscle cell signaling and protein synthesis inconsistencies between young and older subjects after an acute bout of resistance and aerobic exercise (Kim et al. 2005a, b; Raue et al. 2006; Fujita et al. 2007; Harber et al. 2009). For example, we have previously shown a decreased phosphorylation of mTOR and S6K1 in response to muscle contraction by in situ HFES in aged animals (Parkington et al. 2004; Funai et al. 2006). Rasmussen and colleagues have confirmed our findings in humans and have further reported that muscle protein synthesis was unchanged in older humans, after a single bout of resistance exercise (Dreyer et al. 2006b) and the ingestion of AA (Drummond et al. 2008), compared to young humans. Kumar et al. (2009) revealed that an acute bout of resistance exercise at different intensities stimulate myofibrillar protein synthesis and anabolic signalling in a dose-dependent manner in both young and old men during a fasting state (Kumar et al. 2009). The stimulatory effect of exercise peaked at 1–2 h post-exercise and was suppressed, but not delayed, in older men. Although the extent of S6K1 phosphorylation predicted the stimulation of myofibrillar protein synthesis in young men, older men did not appear to match the changes in anabolic signalling and myofibrillar protein synthesis, possibly explaining the deficiency in the muscle protein anabolic response.

Prolonged resistance or aerobic type exercise training represent an effective therapeutic strategy to augment skeletal muscle mass and improve functional performance in the elderly. Improvements associated with chronic exercise training are said to be a result of adaptation of skeletal muscle to the additive effect of an acute exercise bout over a period of time. Exercise training, in addition to the signaling events that occur with an acute bout of exercise, also leads to the increased expression of key proteins involved in the adaptation of the muscle. The compensatory overload model of synergist ablation is an attractive model because it quickly provides a large and fast hypertrophic response. This is a commonly used model to study the effects of resistance exercise training on skeletal muscle adaptations which overloads the plantaris muscle through the removal of the synergist muscles. This mechanical overload of the plantaris muscle results in significant inductions of muscle growth of ~30% after 7 days and ~100% after 35 days in young animals (Spangenburg and Booth 2006; Spangenburg et al. 2008). In an original study that demonstrated a role for mTOR phosphorylation in muscle hypertrophy; Reynolds et al. (2002) reported a ~100% increase in the phosphorylation of mTOR on Ser2448 after 14 days of muscle overload in young animals (Reynolds et al. 2002). Recently, it has been shown that anabolic signaling and muscle hypertrophy are impaired in aged skeletal muscle in response to functional overload (Thomson and Gordon 2006; Hwee and Bodine 2009; Blough and Linderman 2000; Degens and Alway 2003). Thomson and Gordon (2006) observed decreased translational signaling and muscle hypertrophy in aged skeletal muscle in response to 7 days of muscle overload (Thomson and Gordon 2006). Interestingly, the authors correlated the inhibition of translational signaling to the activation of AMPK in the aged muscle. We have recently reported, after 28 days of chronic overload, although there was an attenuation of hypertrophy in aged animals (30 months) this was not reflected in the phosphorylation of mTOR signaling components compared to adult animals (6 months) (Chale-Rush et al. in press).

5 Conclusion

Specific modes and intensities of physical activity can both act to preserve and also increase skeletal muscle mass, strength, power, protein synthesis and anabolic signalling. This effect appears to be pervasive throughout the lifespan and there is evidence for similar responses in men and women. In general, studies supported the concept that moderate to high intensity progressive resistance (strength) training exercise was most effective in improving muscle mass, strength, and power. There is extensive evidence that specific modes of physical activity can effectively increase fat free/lean body mass, strength, and power. In particular, there is extensive experimental evidence that performance of regular (two to four times per week) high intensity (60–80% of the one repetition maximum) progressive resistance (strength) training exercise can result in significant increases in muscle size, strength, and protein synthesis. Progressive resistance (strength) training has consistently been shown to results in improvements in skeletal muscle mass and muscle quality. However, resistance training-induced increases in muscle mass can occur in older individuals but the magnitude of this response may be attenuated, particularly in the oldest of the old. The directionality has been established and the observed physiological responses are improvements in muscle size, strength and power. Endurance/aerobic and other more non-traditional forms of physical activity have not been shown to consistently increase muscle mass or quality but may be associated with the prevention of loss.

Acknowledgements This chapter is based upon work supported by the U.S. Department of Agriculture, under agreement No. 58-1950-7-707. Any opinions, findings, conclusion, or recommendations expressed in this publication are those of the author(s) and do not necessarily reflect the view of the U.S. Department of Agriculture.

The authors would like to thank Dr. Sarah J. Lessard of the Joslin Diabetes Center / Harvard Medical School for the careful review and insightful comments of the manuscript.

References

Aagaard, P., Simonsen, E. B., Trolle, M., Bangsbo, J., Klausen, K. (1996). Specificity of training velocity and training load on gains in isokinetic knee joint strength. *Acta Physiologica Scandinavica, 156*, 123–9.

Adams, G. R. (2002). Invited review: Autocrine/paracrine IGF-I and skeletal muscle adaptation. *Journal of Applied Physiology, 93*, 1159–1167.

Adams, G. R. (2006). Satellite cell proliferation and skeletal muscle hypertrophy. *Applied Physiology, Nutrition, and Metabolism, 31*, 782–90.

Adams, K. J., Swank, A. M., Berning, J. M., Sevene-Adams, P. G., Barnard, K. L., Shimp-Bowerman, J. (2001). Progressive strength training in sedentary, older African American women. *Medicine and Science in Sports and Exercise, 33*, 1567–1576.

Akima, H., Takahashi, H., Kuno, S. Y., Masuda, K., Masuda, T., Shimojo, H., Anno, I., Itai, Y., Katsuta, S. (1999). Early phase adaptations of muscle use and strength to isokinetic training. *Medicine and Science in Sports and Exercise, 31*, 588–94.

Anderson, T. & Kearney, J. T. (1982). Effects of three resistance training programs on muscular strength and absolute and relative endurance. *Research Quarterly for Exercise Sport, 53*, 1–7.

Anderson, S. R., Gilge, D. A., Steiber, A. L., Previs, S. F. (2008). Diet-induced obesity alters protein synthesis: Tissue-specific effects in fasted versus fed mice. *Metabolism, 57*, 347–54.

Aniansson, A., Hedberg, M., Henning, G. B., Grimby, G. (1986). Muscle morphology, enzymatic activity, and muscle strength in elderly men: a follow-up study. *Muscle and Nerve, 9*, 585–591.

Anthony, J. C., Yoshizawa, F., Anthony, T. G., Vary, T. C., Jefferson, L. S., Kimball, S. R. (2000). Leucine stimulates translation initiation in skeletal muscle of postabsorptive rats via a rapamycin-sensitive pathway. *The Journal of Nutrition, 130*, 2413–9.

Asikainen, T. M., Suni, J. H., Pasanen, M. E., Oja, P., Rinne, M. B., Miilunpalo, S. I., Nygard, C. H., Vuori, I. M. (2006). Effect of brisk walking in 1 or 2 daily bouts and moderate resistance training on lower-extremity muscle strength, balance, and walking performance in women who recently went through menopause: A randomized, controlled trial. *Physical Therapy, 86*, 912–23.

Attaix, D., Mosoni, L., Dardevet, D., Combaret, L., Mirand, P. P., Grizard, J. (2005). Altered responses in skeletal muscle protein turnover during aging in anabolic and catabolic periods. *The International Journal of Biochemistry & Cell Biology, 37*, 1962–73.

Baker, M. K., Atlantis, E., Fiatarone Singh, M. A. (2007). Multi-modal exercise programs for older adults. *Age and Ageing, 36*, 375–81.

Balagopal, P., Schimke, J. C., Ades, P., Adey, D., Nair, K. S. (2001). Age effect on transcript levels and synthesis rate of muscle MHC and response to resistance exercise. *American Journal of Physiology. Endocrinology and Metabolism, 280*, E203–E208.

Bamman, M. M., Ragan, R. C., Kim, J. S., Cross, J. M., Hill, V. J., Tuggle, S. C. (2004). Myogenic protein expression before and after resistance loading in 26- and 64-yr-old men and women. *Journal of Applied Physiology, 97*, 1329–1337.

Bamman, M. M., Hill, V. J., Adams, G. R., Haddad, F., Wetzstein, C. J., Gower, B. A. (2003). Gender differences in resistance-training-induced myofiber hypertrophy among older adults. *The Journals of Gerontology. Series A, Biological Sciences and Medical Sciences, 58*, 108–116.

Bassey, E. J., Bendall, M. J., Pearson, M. (1988). Muscle strength in the triceps surae and objectively measured customary walking activity in men and women over 65 years of age. *Clin Sci (Lond), 74*, 85–9.

Baumgartner, R. N., Waters, D. L., Gallagher, D., Morley, J. E., Garry, P. J. (1999). Predictors of skeletal muscle mass in elderly men and women. *Mechanisms of Ageing and Development, 107*, 123–36.

Bean, J. F., Kiely, D. K., Herman, S., Leveille, S. G., Mizer, K., Frontera, W. R., Fielding, R. A. (2002). The relationship between leg power and physical performance in mobility-limited older people. *Journal of the American Geriatrics Society, 50*, 461–7.

Bean, J. F., Leveille, S. G., Kiely, D. K., Bandinelli, S., Guralnik, J. M., Ferrucci, L. (2003). A comparison of leg power and leg strength within the InCHIANTI study: Which influences mobility more? *The Journals of Gerontology. Series A: Biological Sciences and Medical Sciences, 58*, 728–33.

Bell, J. A., Volpi, E., Fujita, S., Cadenas, J. G., Sheffield-Moore, M., Rasmussen, B. B. (2006). Skeletal muscle protein anabolic response to increased energy and insulin is preserved in poorly controlled type 2 diabetes. *The Journal of Nutrition, 136*, 1249–55.

Berger, M. & Berchtold, P. (1979). The role of physical exercise and training in the management of diabetes mellitus. *Bibliotheca Nutritio et Dieta, 27*, 41–54.

Besse, F. & Ephrussi, A. (2008). Translational control of localized mRNAs: Restricting protein synthesis in space and time. *Nature Reviews. Molecular Cell Biology, 9*, 971–80.

Binder, E. F., Schechtman, K. B., Ehsani, A. A., Steger-May, K., Brown, M., Sinacore, D. R., Yarasheski, K. E., Holloszy, J. O. (2002). Effects of exercise training on frailty in community-dwelling older adults: Results of a randomized, controlled trial. *Journal of the American Geriatrics Society, 50*, 1921–8.

Binder, E. F., Yarasheski, K. E., Steger-May, K., Sinacore, D. R., Brown, M., Schechtman, K. B., et al. (2005). Effects of progressive resistance training on body composition in frail older adults: Results of a randomized, controlled trial. *The Journals of Gerontology. Series A, Biological Sciences and Medical Sciences, 60*, 1425–1431.

Blough, E. R. & Linderman, J. K. (2000). Lack of skeletal muscle hypertrophy in very aged male Fischer 344 x Brown Norway rats. *Journal of Applied Physiology, 88*, 1265–70.

Bodine, S. C., Stitt, T. N., Gonzalez, M., Kline, W. O., Stover, G. L., Bauerlein, R., et al. (2001). Akt/mTOR pathway is a crucial regulator of skeletal muscle hypertrophy and can prevent muscle atrophy in vivo. *Nature Cell Biology, 3*, 1014–1019.

Bortz, W. M., 2nd. (1982). Disuse and aging. *The Journal of the American Medical Association, 248*, 1203–1208.

Brown, A. B., McCartney, N., Sale, D. G. (1990). Positive adaptations to weight-lifting training in the elderly. *Journal of Applied Physiology, 69*, 1725–1733.

Burd, N. A., Tang, J. E., Moore, D. R., Phillips, S. M. (2009). Exercise training and protein metabolism: influences of contraction, protein intake, and sex-based differences. *Journal of Applied Physiology, 106*, 1692–1701.

Campbell, W. W., Crim, M. C., Young, V. R., Evans, W. J. (1994). Increased energy requirements and changes in body composition with resistance training in older adults. *The American Journal of Clinical Nutrition, 60*, 167–175.

Campos, G. E., Luecke, T. J., Wendeln, H. K., Toma, K., Hagerman, F. C., Murray, T. F., et al. (2002). Muscular adaptations in response to three different resistance-training regimens: Specificity of repetition maximum training zones. *European Journal of Applied Physiology, 88*, 50–60.

Candow, D. G. & Burke, D. G. (2007). Effect of short-term equal-volume resistance training with different workout frequency on muscle mass and strength in untrained men and women. *Journal of Strength and Conditioning Research, 21*, 204–7.

Cao, P. R., Kim, H. J., Lecker, S. H. (2005). Ubiquitin-protein ligases in muscle wasting. *The International Journal of Biochemistry & Cell Biology, 37*, 2088–97.

Chale-Rush, A., Morris, E. P., Kendall, T. L., Brooks, N. E., Fielding, R. A. (2009). Effects of chronic overload on muscle hypertrophy and mTOR signaling in young adult and age rats. *J Gerontol A Biol Sci Med Sci, 64*(12), 1232–1239.

Chin, A. V., Robinson, D. J., O'Connell, H., Hamilton, F., Bruce, I., Coen, R. (2008). Vascular biomarkers of cognitive performance in a community-based elderly population: The Dublin Healthy Ageing study. *Age and Ageing, 37*, 559–564.

Chodzko-Zajko, W. J., Proctor, D. N., Fiatarone Singh, M. A., Minson, C. T., Nigg, C. R., Salem, G. J., Skinner, J. S. (2009). American College of Sports Medicine position stand. Exercise and physical activity for older adults. *Medicine and Science in Sports and Exercise, 41*, 1510–30.

Chopard, A., Lecunff, M., Danger, R., Lamirault, G., Bihouee, A., Teusan, R., et al. (2009a). Large-scale mRNA analysis of female skeletal muscles during 60 days of bed rest with and without exercise or dietary protein supplementation as countermeasures. *Physiological Genomics, 38*, 291–302.

Chopard, A., Hillock, S., Jasmin, B. J. (2009b). Molecular events and signalling pathways involved in skeletal muscle disuse-induced atrophy and the impact of countermeasures. *Journal of Cellular and Molecular Medicine, 13*, 3032–3050.

Coffey, V. G. & Hawley, J. A. (2007). The molecular bases of training adaptation. *Sports Medicine, 37*, 737–63.

Conboy, I. M., Conboy, M. J., Smythe, G. M., Rando, T. A. (2003). Notch-mediated restoration of regenerative potential to aged muscle. *Science, 302*, 1575–1577.

Conley, K. E., Esselman, P. C., Jubrias, S. A., Cress, M. E., Inglin, B., Mogadam, C., Schoene, R. B. (2000). Ageing, muscle properties and maximal O(2) uptake rate in humans. *Journal de Physiologie, 526*(Pt 1), 211–7.

Conley, K. E., Jubrias, S. A., Esselman, P. C. (2000). Oxidative capacity and ageing in human muscle. *Journal de Physiologie, 526*(Pt 1), 203–10.

Corcoran, M. P., Lamon-Fava, S., Fielding, R. A. (2007). Skeletal muscle lipid deposition and insulin resistance: Effect of dietary fatty acids and exercise. *The American Journal of Clinical Nutrition, 85*, 662–77.

Corcoran, P. J. (1991). Use it or lose it the hazards of bed rest and inactivity. *The Western Journal of Medicine, 154*, 536–538.

Cress, M. E., Buchner, D. M., Questad, K. A., Esselman, P. C., Delateur, B. J., Schwartz, R. S. (1999). Exercise: Effects on physical functional performance in independent older adults. *The Journals of Gerontology. Series A: Biological Sciences and Medical Sciences, 54*, M242–8.

Cress, M. E., Buchner, D. M., Prohaska, T., Rimmer, J., Brown, M., Macera, C., (2005). Best practices for physical activity programs and behavior counseling in older adult populations. *Journal of Aging and Physical Activity, 13*, 61–74.

Cuoco, A., Callahan, D. M., Sayers, S., Frontera, W. R. Fielding, R. A. (2004) Impact of muscle power and force on gait speed in disabled older men and women. *Journal of Gerontology, 59*, 1200–1205.

Cureton, K. J., Collins, M. A., Hill, D. W., Mcelhannon, F. M., JR. (1988). Muscle hypertrophy in men and women. *Medicine and Science in Sports and Exercise, 20*, 338–44.

Cuthbertson, D., Smith, K., Babraj, J., Leese, G., Waddell, T., Atherton, P., Wackerhage, H., Taylor, P. M., Rennie, M. J. (2005). Anabolic signaling deficits underlie amino acid resistance of wasting, aging muscle. *The Faseb Journal, 19*, 422–4.

Davidson, L. E., Hudson, R., Kilpatrick, K., Kuk, J. L., Mcmillan, K., Janiszewski, P. M., Lee, S., Lam, M., Ross, R. (2009). Effects of exercise modality on insulin resistance and functional limitation in older adults: a randomized controlled trial. *Archives of Internal Medicine, 169*, 122–31.

Degens, H. & Alway, S. E. (2003). Skeletal muscle function and hypertrophy are diminished in old age. *Muscle & Nerve, 27*, 339–47.

Demichele, P. L., Pollock, M. L., Graves, J. E., Foster, D. N., Carpenter, D., Garzarella, L., Brechue, W., Fulton, M. (1997). Isometric torso rotation strength: Effect of training frequency on its development. *Archives of Physical Medicine and Rehabilitation, 78*, 64–9.

De Vreede, P. L., Samson, M. M., Van Meeteren, N. L., Duursma, S. A., Verhaar, H. J. (2005). Functional-task exercise versus resistance strength exercise to improve daily function in older women: A randomized, controlled trial. *Journal of the American Geriatrics Society, 53*, 2–10.

DHHS (2008) *Physical activity guidelines for Americans*. In Services, D. O. H. A. H. (Ed.). Rockville, MD: U.S. Department of Health and Human Services.

Dionne, I. J., Melancon, M. O., Brochu, M., Ades, P. A., Poelhman, E. T. (2004). Age-related differences in metabolic adaptations following resistance training in women. *Experimental Gerontology, 39*, 133–8.

DiPietro, L. (2001). Physical activity in aging: changes in patterns and their relationship to health and function. *The Journals of Gerontology Series A: Biological Sciences and Medical Sciences, 56*(Spec No 2), 13–22.

Dornemann, T. M., Mcmurray, R. G., Renner, J. B., Anderson, J. J. (1997). Effects of high-intensity resistance exercise on bone mineral density and muscle strength of 40-50-year-old women. *Journal of Sports Medicine and Physical Fitness, 37*, 246–51.

Dreyer, H. C., Blanco, C. E., Sattler, F. R., Schroeder, E. T., Wiswell, R. A. (2006). Satellite cell numbers in young and older men 24 hours after eccentric exercise. *Muscle & Nerve, 33*, 242–53.

Dreyer, H. C., Fujita, S., Cadenas, J. G., Chinkes, D. L., Volpi, E., Rasmussen, B. B. (2006). Resistance exercise increases AMPK activity and reduces 4E-BP1 phosphorylation and protein synthesis in human skeletal muscle. *Journal de Physiologie, 576*, 613–24.

Dreyer, H. C., Drummond, M. J., Pennings, B., Fujita, S., Glynn, E. L., Chinkes, D. L., Dhanani, S., Volpi, E., Rasmussen, B. B. (2008). Leucine-enriched essential amino acid and carbohydrate ingestion following resistance exercise enhances mTOR signaling and protein synthesis in human muscle. *American Journal of Physiology. Endocrinology and Metabolism, 294*, E392–400.

Drøyvold, W. B., Nilsen, T. I., Krüger, O., Holmen, T. L., Krokstad, S., Midthjell, K., et al. (2006). Change in height, weight and body mass index: Longitudinal data from the HUNT Study in Norway. *International Journal of Obesity (London), 30*, 935–939.

Drummond, M. J., Dreyer, H. C., Pennings, B., Fry, C. S., Dhanani, S., Dillon, E. L., Sheffield-Moore, M., Volpi, E., Rasmussen, B. B. (2008). Skeletal muscle protein anabolic response to resistance exercise and essential amino acids is delayed with aging. *Journal of Applied Physiology, 104*, 1452–61.

Drummond, M. J., Fry, C. S., Glynn, E. L., Dreyer, H. C., Dhanani, S., Timmerman, K. L., Volpi, E., Rasmussen, B. B. (2009). Rapamycin administration in humans blocks the contraction-induced increase in skeletal muscle protein synthesis. *Journal de Physiologie, 587*, 1535–46.

Dziura, J., Mendes DE Leon, C., Kasl, S., Dipietro, L. (2004). Can physical activity attenuate aging-related weight loss in older people? The Yale Health and Aging Study, 1982–1994. *American Journal of Epidemiology, 159*, 759–67.

Earles, D. R., Judge, J. O., Gunnarsson, O. T. (2001). Velocity training induces power-specific adaptations in highly functioning older adults. *Archives of Physical Medicine and Rehabilitation, 82*, 872–878.

Edstrom, E., Altun, M., Hagglund, M., Ulfhake, B. (2006). Atrogin-1/MAFbx and MuRF1 are downregulated in aging-related loss of skeletal muscle. *The Journals of Gerontology. Series A: Biological Sciences and Medical Sciences, 61*, 663–74.

Englund, U., Littbrand, H., Sondell, A., Pettersson, U., Bucht, G. (2005). A 1-year combined weight-bearing training program is beneficial for bone mineral density and neuromuscular function in older women. *Osteoporosis International, 16*, 1117–23.

Evans, W. (1997). Functional and metabolic consequences of sarcopenia. *The Journal of Nutrition, 127*, 998S–1003S.

Fahlman, M., Morgan, A., Mcnevin, N., Topp, R., Boardley, D. (2007). Combination training and resistance training as effective interventions to improve functioning in elders. *J Aging Phys Act, 15*, 195–205.

Ferri, A., Scaglioni, G., Pousson, M., Capodaglio, P., Van Hoecke, J., Narici, M. V. (2003). Strength and power changes of the human plantar flexors and knee extensors in response to resistance training in old age. *Acta Physiologica Scandinavica, 177*, 69–78.

Fiatarone, M. A., Marks, E. C., Ryan, N. D., Meredith, C. N., Lipsitz, L. A., Evans, W. J. (1990). High-intensity strength training in nonagenarians. Effects on skeletal muscle. *JAMA: The Journal of the American Medical Association, 263*, 3029–3034.

Fiatarone, M. A., O'neill, E. F., Ryan, N. D., Clements, K. M., Solares, G. R., Nelson, M. E., Roberts, S. B., Kehayias, J. J., Lipsitz, L. A., Evans, W. J. (1994). Exercise training and nutritional supplementation for physical frailty in very elderly people. *The New England Journal of Medicine, 330*, 1769–75.

Fielding, R. A. (1995). Effects of exercise training in the elderly: Impact of progressive- resistance training on skeletal muscle and whole-body protein metabolism. *The Proceedings of the Nutrition Society, 54*, 665–75.

Fielding, R. A., Lebrasseur, N. K., Cuoco, A., Bean, J., Mizer, K., Fiatarone Singh, M. A. (2002). High-velocity resistance training increases skeletal muscle peak power in older women. *Journal of the American Geriatrics Society, 50*, 655–62.

Fleg, J. L. & Lakatta, E. G. (1988). Role of muscle loss in the age-associated reduction in VO2 max. *Journal of Applied Physiology, 65*, 1147–51.

Fluckey, J. D., Dupont-Versteegden, E. E., Knox, M., Gaddy, D., Tesch, P. A., Peterson, C. A. (2004). Insulin facilitation of muscle protein synthesis following resistance exercise in hindlimb-suspended rats is independent of a rapamycin-sensitive pathway. *American Journal of Physiology. Endocrinology and Metabolism, 287*, E1070–5.

Foldvari, M., Clark, M., Laviolette, L. C., Bernstein, M. A., Kaliton, D., Castaneda, C., et al. (2000). Association of muscle power with functional status in community-dwelling elderly women. *The Journals of Gerontology. Series A, Biological Sciences and Medical Sciences, 55*, M192–M199.

Frontera, W. R., Meredith, C. N., O'reilly, K. P., Knuttgen, H. G., Evans, W. J. (1988). Strength conditioning in older men: Skeletal muscle hypertrophy and improved function. *Journal of Applied Physiology, 64*, 1038–44.

Fujita, S., Rasmussen, B. B., Cadenas, J. G., Drummond, M. J., Glynn, E. L., Sattler, F. R., Volpi, E. (2007). Aerobic exercise overcomes the age-related insulin resistance of muscle protein metabolism by improving endothelial function and Akt/mammalian target of rapamycin signaling. *Diabetes, 56*, 1615–22.

Fujita, S., Glynn, E. L., Timmerman, K. L., Rasmussen, B. B., Volpi, E. (2009). Supraphysiological hyperinsulinaemia is necessary to stimulate skeletal muscle protein anabolism in older adults: Evidence of a true age-related insulin resistance of muscle protein metabolism. *Diabetologia, 52*, 1889–98.

Funai, K., Parkington, J. D., Carambula, S., Fielding, R. A. (2006). Age-associated decrease in contraction-induced activation of downstream targets of Akt/mTor signaling in skeletal muscle. *American Journal of Physiology: Regulatory, Integrative and Comparative Physiology, 290*, R1080–6.

Galvão, D. A., Newton, R. U., Taaffe, D. R. (2005). Anabolic responses to resistance training in older men and women: A brief review. *Journal of Aging and Physical Activity, 13*, 343–358.

Galvao, D. A. & Taaffe, D. R. (2005). Resistance exercise dosage in older adults: Single- versus multiset effects on physical performance and body composition. *Journal of the American Geriatrics Society, 53*, 2090–7.

Glowacki, S. P., Martin, S. E., Maurer, A., Baek, W., Green, J. S., Crouse, S. F. (2004). Effects of resistance, endurance, and concurrent exercise on training outcomes in men. *Medicine and Science in Sports and Exercise, 36*, 2119–2127.

Going, S., Williams, D., Lohman, T. (1995). Aging and body composition: Biological changes and methodological issues. *Exercise and Sport Sciences Reviews, 23*, 411–58.

Goodpaster, B. H., Thaete, F. L., Kelley, D. E. (2000). Thigh adipose tissue distribution is associated with insulin resistance in obesity and in type 2 diabetes mellitus. *The American Journal of Clinical Nutrition, 71*, 885–92.

Goodpaster, B. H., Park, S. W., Harris, T. B., Kritchevsky, S. B., Nevitt, M., Schwartz, A. V., Simonsick, E. M., Tylavsky, F. A., Visser, M., Newman, A. B. (2006). The loss of skeletal muscle strength, mass, and quality in older adults: The health, aging and body composition study. *The Journals of Gerontology. Series A: Biological Sciences and Medical Sciences, 61*, 1059–64.

Goodyear, L. J., Kahn, B. B. (1998). Exercise, glucose transport, and insulin sensitivity. *Annual Review of Medicine, 49*, 235–261.

Gopinath, S. D. & Rando, T. A. (2008). Stem cell review series: Aging of the skeletal muscle stem cell niche. *Aging Cell, 7*, 590–598.

Greenhaff, P. L., Karagounis, L. G., Peirce, N., Simpson, E. J., Hazell, M., Layfield, R., Wackerhage, H., Smith, K., Atherton, P., Selby, A., Rennie, M. J. (2008). Disassociation between the effects of amino acids and insulin on signaling, ubiquitin ligases, and protein turnover in human muscle. *American Journal of Physiology. Endocrinology and Metabolism, 295*, E595–604.

Grzelkowska, K., Dardevet, D., Balage, M., Grizard, J. (1999). Involvement of the rapamycin-sensitive pathway in the insulin regulation of muscle protein synthesis in streptozotocin-diabetic rats. *The Journal of Endocrinology, 160*, 137–45.

Guillet, C., Prod'homme, M., Balage, M., Gachon, P., Giraudet, C., Morin, L., Grizard, J., Boirie, Y. (2004). Impaired anabolic response of muscle protein synthesis is associated with S6K1 dysregulation in elderly humans. *The Faseb Journal, 18*, 1586–7.

Guillet, C., Delcourt, I., Rance, M., Giraudet, C., Walrand, S., Bedu, M., Duche, P., Boirie, Y. (2009). Changes in basal and insulin and amino acid response of whole body and skeletal muscle proteins in obese men. *Journal of Clinical Endocrinology Metabolism, 94*(8), 3044–3050.

Hagberg, J. M., Allen, W. K., Seals, D. R., Hurley, B. F., Ehsani, A. A., Holloszy, J. O. (1985). A hemodynamic comparison of young and older endurance athletes during exercise. *Journal of Applied Physiology, 58*, 2041–6.

Hakkinen, K., Kallinen, M., Izquierdo, M., Jokelainen, K., Lassila, H., Malkia, E., Kraemer, W. J., Newton, R. U., Alen, M. (1998). Changes in agonist-antagonist EMG, muscle CSA, and force during strength training in middle-aged and older people. *Journal of Applied Physiology, 84*, 1341–9.

Hannan, K. M., Brandenburger, Y., Jenkins, A., Sharkey, K., Cavanaugh, A., Rothblum, L., Moss, T., Poortinga, G., Mcarthur, G. A., Pearson, R. B., Hannan, R. D. (2003). mTOR-dependent regulation of ribosomal gene transcription requires S6K1 and is mediated by phosphorylation of the carboxy-terminal activation domain of the nucleolar transcription factor UBF. *Molecular and Cellular Biology, 23*, 8862–77.

Harber, M. P., Fry, A. C., Rubin, M. R., Smith, J. C., Weiss, L. W. (2004). Skeletal muscle and hormonal adaptations to circuit weight training in untrained men. *Scandinavian Journal of Medicine & Science in Sports, 14*, 176–85.

Harber, M. P., Konopka, A. R., Douglass, M. D., Minchev, K., Kamisky, L. A., Trappe, T. A., Trappe, S. W. (2009). Aerobic exercise training improves whole muscle and single myofiber size and function in older women. *American Journal of Physiology – Regulatory, Integrative, and Comparative Physiology, 297*, R1452–R1459.

Harber, M. P., Crane, J. D., Dickinson, J. M., Jemiolo, B., Raue, U., Trappe, T. A. (2009a). Protein synthesis and the expression of growth-related genes are altered by running in human vastus lateralis and soleus muscles. *American Journal of Physiology. Regulatory, Integrative and Comparative Physiology*, *296*, R708–R714.

Harber, M. P., Konopka, A. R., Douglass, M. D., Minchev, K., Kaminsky, L. A., Trappe, T. A. (2009b). Aerobic exercise training improves whole muscle and single myofiber size and function in older women. *American Journal of Physiology. Regulatory, Integrative and Comparative Physiology*, *297*, R1452–R1459.

Harridge, S. D., Kryger, A., Stensgaard, A. (1999). Knee extensor strength, activation, and size in very elderly people following strength training. *Muscle and Nerve*, *22*, 831–839.

Haus, J. M., Carrithers, J. A., Trappe, S. W., Trappe, T. A. (2007). Collagen, cross-linking, and advanced glycation end products in aging human skeletal muscle. *Journal of Applied Physiology*, *103*, 2068–2076.

Hawke, T. J. (2005). Muscle stem cells and exercise training. *Exercise and Sport Sciences Reviews, 33*, 63–8.

Hawley, J. A. & Lessard, S. J. (2008). Exercise training-induced improvements in insulin action. *Acta Physiol (Oxf), 192*, 127–35.

Heitkamp, H. C., Horstmann, T., Mayer, F., Weller, J., Dickhuth, H. H. (2001). Gain in strength and muscular balance after balance training. *International Journal of Sports Medicine, 22*, 285–90.

Henwood, T. R. & Taaffe, D. R. (2005). Improved physical performance in older adults undertaking a short-term programme of high-velocity resistance training. *Gerontology*, *51*, 108–115.

Henwood, T. R. & Taaffe, D. R. (2006). Short-term resistance training and the older adult: The effect of varied programmes for the enhancement of muscle strength and functional performance. *Clinical Physiology and Functional Imaging, 26*, 305–13.

Hilliard-Robertson, P. C., Schneider, S. M., Bishop, S. L., Guilliams, M. E. (2003). Strength gains following different combined concentric and eccentric exercise regimens. *Aviation Space and Environmental Medicine, 74*, 342–7.

Hisaeda, H., Miyagawa, K., Kuno, S., Fukunaga, T., Muraoka, I. (1996). Influence of two different modes of resistance training in female subjects. *Ergonomics, 39*, 842–52.

Holm, L., Reitelseder, S., Pedersen, T. G., Doessing, S., Petersen, S. G., Flyvbjerg, A., et al. (2008). Changes in muscle size and MHC composition in response to resistance exercise with heavy and light loading intensity. *Journal of Applied Physiology*, *105*, 1454–1461.

Hostler, D., Schwirian, C. I., Campos, G., Toma, K., Crill, M. T., Hagerman, G. R., Hagerman, F. C., Staron, R. S. (2001). Skeletal muscle adaptations in elastic resistance-trained young men and women. *European Journal of Applied Physiology, 86*, 112–8.

Hubal, M. J., Gordish-Dressman, H., Thompson, P. D., Price, T. B., Hoffman, E. P., Angelopoulos, T. J. (2005). Variability in muscle size and strength gain after unilateral resistance training. *Medicine and Science in Sports and Exercise*, *37*, 964–972.

Hunter, G. R., Wetzstein, C. J., Fields, D. A., Brown, A., Bamman, M. M. (2000). Resistance training increases total energy expenditure and free-living physical activity in older adults. *Journal of Applied Physiology, 89*, 977–84.

Hunter, G. R., Wetzstein, C. J., Mclafferty, C. L., JR, Zuckerman, P. A., Landers, K. A., Bamman, M. M. (2001). High-resistance versus variable-resistance training in older adults. *Medicine and Science in Sports and Exercise, 33*, 1759–64.

Hunter, G. R., Bryan, D. R., Wetzstein, C. J., Zuckerman, P. A., Bamman, M. M. (2002). Resistance training and intra-abdominal adipose tissue in older men and women. *Medicine and Science in Sports and Exercise, 34*, 1023–8.

Hwee, D. T. & Bodine, S. C. (2009). Age-related deficit in load-induced skeletal muscle growth. *The Journals of Gerontology. Series A: Biological Sciences and Medical Sciences, 64*, 618–28.

Ivey, F. M., Roth, S. M., Ferrell, R. E., Tracy, B. L., Lemmer, J. T., Hurlbut, D. E., Martel, G. F., Siegel, E. L., Fozard, J. L., Jeffrey Metter, E., Fleg, J. L., Hurley, B. F. (2000). Effects of age, gender, and myostatin genotype on the hypertrophic response to heavy resistance strength training. *The Journals of Gerontology. Series A: Biological Sciences and Medical Sciences, 55*, M641–8.

Izquierdo, M., Ibanez, Jkha, Kraemer, W. J., Larrion, J. L., Gorostiaga, E. M. (2004). Once weekly combined resistance and cardiovascular training in healthy older men. *Medicine and Science in Sports and Exercise, 36*, 435–43.

Izquierdo, M., Hakkinen, K., Ibanez, J., Kraemer, W. J., Gorostiaga, E. M. (2005). Effects of combined resistance and cardiovascular training on strength, power, muscle cross-sectional area, and endurance markers in middle-aged men. *European Journal of Applied Physiology, 94*, 70–5.

Janssen, I., Shepard, D. S., Katzmarzyk, P. T., Roubenoff, R. (2004). The healthcare costs of sarcopenia in the United States. *Journal of the American Geriatrics Society, 52*, 80–85.

Jensen, M. D. & Haymond, M. W. (1991). Protein metabolism in obesity: effects of body fat distribution and hyperinsulinemia on leucine turnover. *The American Journal of Clinical Nutrition, 53*, 172–176.

Jette, A. M. & Branch, L. G. (1981). The Framingham Disability Study: II. Physical disability among the aging. *American Journal of Public Health, 71*, 1211–6.

Joseph, L. J., Davey, S. L., Evans, W. J., Campbell, W. W. (1999). Differential effect of resistance training on the body composition and lipoprotein-lipid profile in older men and women. *Metabolism, 48*, 1474–80.

Jozsi, A. C., Campbell, W. W., Joseph, L., Davey, S. L., Evans, W. J. (1999). Changes in power with resistance training in older and younger men and women. *The Journals of Gerontology. Series A: Biological Sciences and Medical Sciences, 54*, M591–6.

Kadi, F., Charifi, N., Denis, C., Lexell, J. (2004). Satellite cells and myonuclei in young and elderly women and men. *Muscle & Nerve, 29*, 120–7.

Katsanos, C. S., Aarsland, A., Cree, M. G., Wolfe, R. R. (2009). Muscle protein synthesis and balance responsiveness to essential amino acids ingestion in the presence of elevated plasma free fatty acid concentrations. *The Journal of Clinical Endocrinology and Metabolism, 94*, 2984–90.

Katta, A., Karkala, S. K., WU, M., Meduru, S., Desai, D. H., Rice, K. M., Blough, E. R. (2009). Lean and obese Zucker rats exhibit different patterns of p70s6 kinase regulation in the tibialis anterior muscle in response to high-force muscle contraction. *Muscle & Nerve, 39*, 503–11.

Keeler, L. K., Finkelstein, L. H., Miller, W., Fernhall, B. (2001). Early-phase adaptations of traditional-speed vs. superslow resistance training on strength and aerobic capacity in sedentary individuals. *Journal of Strength and Conditioning Research/National Strength & Conditioning Association, 15*, 309–314.

Kelley, D. E. & Goodpaster, B. H. (2001). Effects of exercise on glucose homeostasis in Type 2 diabetes mellitus. *Medicine and Science in Sports and Exercise, 33*(6 Suppl), S495–S501.

Keysor, J. J. (2003). Does late-life physical activity or exercise prevent or minimize disablement? A critical review of the scientific evidence. *American Journal of Preventive Medicine, 25*(3 Suppl 2), 129–136.

Khamzina, L., Veilleux, A., Bergeron, S., Marette, A. (2005). Increased activation of the mammalian target of rapamycin pathway in liver and skeletal muscle of obese rats: Possible involvement in obesity-linked insulin resistance. *Endocrinology, 146*, 1473–81.

Kim, J. S., Cross, J. M., Bamman, M. M. (2005). Impact of resistance loading on myostatin expression and cell cycle regulation in young and older men and women. *American Journal of Physiology. Endocrinology and Metabolism, 288*, E1110–9.

Kim, J. S., Kosek, D. J., Petrella, J. K., Cross, J. M., Bamman, M. M. (2005). Resting and load-induced levels of myogenic gene transcripts differ between older adults with demonstrable sarcopenia and young men and women. *Journal of Applied Physiology, 99*, 2149–58.

Kimball, S. R., Jefferson, L. S., Nguyen, H. V., Suryawan, A., Bush, J. A., Davis, T. A. (2000). Feeding stimulates protein synthesis in muscle and liver of neonatal pigs through an mTOR-dependent process. *American Journal of Physiology. Endocrinology and Metabolism, 279*, E1080–7.

King, A. C., Pruitt, L. A., Phillips, W., Oka, R., Rodenburg, A., Haskell, W. L. (2000). Comparative effects of two physical activity programs on measured and perceived physical functioning and other health-related quality of life outcomes in older adults. *The Journals of Gerontology. Series A: Biological Sciences and Medical Sciences, 55*, M74–83.

Kirk, E. P., Washburn, R. A., Bailey, B. W., Lecheminant, J. D., Donnelly, J. E. (2007). Six months of supervised high-intensity low-volume resistance training improves strength independent of changes in muscle mass in young overweight men. *Journal of Strength and Conditioning Research, 21*, 151–6.

Klitgaard, H., Mantoni, M., Schiaffino, S., Ausoni, S., Gorza, L., Laurent-Winter, C., Schnohr, P., Saltin, B. (1990). Function, morphology and protein expression of ageing skeletal muscle: A cross-sectional study of elderly men with different training backgrounds. *Acta Physiologica Scandinavica, 140*, 41–54.

Kohrt, W. M. & Holloszy, J. O. (1995). Loss of skeletal muscle mass with aging: Effect on glucose tolerance. *The Journals of Gerontology Series A: Biological Sciences and Medical Sciences, 50*(Spec No), 68–72.

Koopman, R., Pennings, B., Zorenc, A. H., Van Loon, L. J. (2007). Protein ingestion further augments S6K1 phosphorylation in skeletal muscle following resistance type exercise in males. *The Journal of Nutrition, 137*, 1880–6.

Koopman, R. & van Loon, L. J. (2009). Aging, exercise, and muscle protein metabolism. *Journal of Applied Physiology, 106*, 2040–2048.

Korpelainen, R., Keinänen-Kiukaanniemi, S., Heikkinen, J., Väänänen, K., Korpelainen, J. (2006). Effect of impact exercise on bone mineral density in elderly women with low BMD: a population-based randomized controlled 30-month intervention. *Osteoporosis International, 17*, 109–118.

Kostek, M. C., Delmonico, M. J., Reichel, J. B., Roth, S. M., Douglass, L., Ferrell, R. E., et al. (2005). Muscle strength response to strength training is influenced by insulin-like growth factor 1 genotype in older adults. *Journal of Applied Physiology, 98*, 2147–2154.

Kosek, D. J., Kim, J. S., Petrella, J. K., Cross, J. M., Bamman, M. M. (2006). Efficacy of 3 days/wk resistance training on myofiber hypertrophy and myogenic mechanisms in young vs. older adults. *Journal of Applied Physiology, 101*, 531–544.

Kraemer, W. J., Ratamess, N. A. (2004). Fundamentals of resistance training: Progression and exercise prescription. *Medicine and Science in Sports and Exercise, 36*, 674–88.

Kraemer, W. J., Nindl, B. C., Ratamess, N. A., Gotshalk, L. A., Volek, J. S., Fleck, S. J., et al. (2004). Changes in muscle hypertrophy in women with periodized resistance training. *Medicine and Science in Sports and Exercise, 36*, 697–708.

Kubica, N., Bolster, D. R., Farrell, P. A., Kimball, S. R., Jefferson, L. S. (2005). Resistance exercise increases muscle protein synthesis and translation of eukaryotic initiation factor 2Bepsilon mRNA in a mammalian target of rapamycin-dependent manner. *The Journal of Biological Chemistry, 280*, 7570–80.

Kubica, N., Crispino, J. L., Gallagher, J. W., Kimball, S. R., Jefferson, L. S. (2008). Activation of the mammalian target of rapamycin complex 1 is both necessary and sufficient to stimulate eukaryotic initiation factor 2Bvarepsilon mRNA translation and protein synthesis. *The International Journal of Biochemistry & Cell Biology, 40*, 2522–33.

Kumar, V., Selby, A., Rankin, D., Patel, R., Atherton, P., Hildebrandt, W., Williams, J., Smith, K., Seynnes, O., Hiscock, N., Rennie, M. J. (2009). Age-related differences in the dose-response relationship of muscle protein synthesis to resistance exercise in young and old men. *Journal de Physiologie, 587*, 211–7.

Lakatta, E. G. & Levy, D. (2003). Arterial and cardiac aging: major shareholders in cardiovascular disease enterprises: Part I: aging arteries: a "set up" for vascular disease. *Circulation, 107*, 139–46.

Lang, C. H. (2006). Elevated plasma free fatty acids decrease basal protein synthesis, but not the anabolic effect of leucine, in skeletal muscle. *American Journal of Physiology. Endocrinology and Metabolism, 291*, E666–E674.

Larsson, L. (1983). Histochemical characteristics of human skeletal muscle during aging. *Acta Physiologica Scandinavica, 117*, 469–71.

Larsson, L., Sjodin, B., Karlsson, J. (1978). Histochemical and biochemical changes in human skeletal muscle with age in sedentary males, age 22–65 years. *Acta Physiologica Scandinavica, 103*, 31–9.

Larsson, L., Grimby, G., Karlsson, J. (1979). Muscle strength and speed of movement in relation to age and muscle morphology. *Journal of Applied Physiology, 46*, 451–6.

Lastayo, P. C., Pierotti, D. J., Pifer, J., Hoppeler, H., Lindstedt, S. L. (2000). Eccentric ergometry: Increases in locomotor muscle size and strength at low training intensities. *American Journal of Physiology: Regulatory, Integrative and Comparative Physiology, 278*, R1282–8.

Lemmer, J. T., Ivey, F. M., Ryan, A. S., Martel, G. F., Hurlbut, D. E., Metter, J. E., Fozard, J. L., Fleg, J. L., Hurley, B. F. (2001). Effect of strength training on resting metabolic rate and physical activity: Age and gender comparisons. *Medicine and Science in Sports and Exercise, 33*, 532–41.

Leon, A. S., Connett, J., Jacobs, D. R., Rauramaa, R., JR. (1987). Leisure-time physical activity levels and risk of coronary heart disease and death. The Multiple Risk Factor Intervention Trial. *JAMA, 258*, 2388–95.

Lessard, S. J., Rivas, D. A., Chen, Z. P., Bonen, A., Febbraio, M. A., Reeder, D. W., et al. (2007). Tissue-specific effects of rosiglitazone and exercise in the treatment of lipid-induced insulin resistance. *Diabetes, 56*, 1856–1864.

Lexell, J., Taylor, C. C., Sjostrom, M. (1988). What is the cause of the ageing atrophy? Total number, size and proportion of different fiber types studied in whole vastus lateralis muscle from 15- to 83-year-old men. *Journal of the Neurological Sciences, 84*, 275–94.

Lindle, R. S., Metter, E. J., Lynch, N. A., Fleg, J. L., Fozard, J. L., Tobin, J., Roy, T. A., Hurley, B. F. (1997). Age and gender comparisons of muscle strength in 654 women and men aged 20–93 yr. *Journal of Applied Physiology, 83*, 1581–7.

Ljubicic, V. & Hood, D. A. (2009). Diminished contraction-induced intracellular signaling towards mitochondrial biogenesis in aged skeletal muscle. *Aging Cell, 8*, 394–404.

Ljubicic, V., Joseph, A. M., Saleem, A., Uguccioni, G., Collu-Marchese, M., Lai, R. Y., Nguyen, L. M., Hood, D. A. (2009). Transcriptional and post-transcriptional regulation of mitochondrial biogenesis in skeletal muscle: Effects of exercise and aging. *Biochimica Biophysica Acta, 1800*(3), 223–234.

Lord, S. R., Lloyd, D. G., Nirui, M., Raymond, J., Williams, P., Stewart, R. A. (1996). The effect of exercise on gait patterns in older women: A randomized controlled trial. *The Journals of Gerontology. Series A: Biological Sciences and Medical Sciences, 51*, M64–70.

Lord, S. R., Ward, J. A., Williams, P. (1996). Exercise effect on dynamic stability in older women: A randomized controlled trial. *Archives of Physical Medicine and Rehabilitation, 77*, 232–6.

Luden, N., Minchev, K., Hayes, E., Louis, E., Trappe, T., Trappe, S. (2008). Human vastus lateralis and soleus muscles display divergent cellular contractile properties. *American Journal of Physiology. Regulatory, Integrative and Comparative Physiology, 295*, R1593–R1598.

Luzi, L., Castellino, P., DeFronzo, R. A. (1996). Insulin and hyperaminoacidemia regulate by a different mechanism leucine turnover and oxidation in obesity. *The American Journal of Physiology*, *270*, E273–E281.

Lynch, G. S., Schertzer, J. D., Ryall, J. G. (2007). Therapeutic approaches for muscle wasting disorders. *Pharmacology & Therapeutics*, *113*, 461–487.

Machida, S. & Booth, F. W. (2004). Insulin-like growth factor 1 and muscle growth: Implication for satellite cell proliferation. *The Proceedings of the Nutrition Society, 63*, 337–40.

Manton, K. G. & Vaupel, J. W. (1995). Survival after the age of 80 in the United States, Sweden, France, England, and Japan. *The New England Journal of Medicine, 333*, 1232–5.

Marques, E., Carvalho, J., Soares, J. M., Marques, F., Mota, J. (2009). Effects of resistance and multicomponent exercise on lipid profiles of older women. *Maturitas, 63*, 84–8.

Mauro, A. (1961). Satellite cell of skeletal muscle fibers. *The Journal of Biophysical and Biochemical Cytology, 9*, 493–5.

Mayhew, D. L., Kim, J. S., Cross, J. M., Ferrando, A. A., Bamman, M. M. (2009). Translational signaling responses preceding resistance training-mediated myofiber hypertrophy in young and old humans. *Journal of Applied Physiology*, *107*, 1655–1662.

McCall, G. E., Byrnes, W. C., Dickinson, A., Pattany, P. M., Fleck, S. J. (1996). Muscle fiber hypertrophy, hyperplasia, and capillary density in college men after resistance training. *Journal of Applied Physiology*, *81*, 2004–2012.

McCartney, N., Hicks, A. L., Martin, J., Webber, C. E. (1995). Long-term resistance training in the elderly: effects on dynamic strength, exercise capacity, muscle, and bone. *The Journals of Gerontology. Series A, Biological Sciences and Medical Sciences*, *50*, B97–B104.

McCartney, N., Hicks, A. L., Martin, J., Webber, C. E. (1996). A longitudinal trial of weight training in the elderly: Continued improvements in year 2. *The Journals of Gerontology. Series A: Biological Sciences and Medical Sciences, 51*, B425–33.

McCaulley, G. O., McBride, J. M., Cormie, P., Hudson, M. B., Nuzzo, J. L., Quindry, J. C., et al. (2009). Acute hormonal and neuromuscular responses to hypertrophy, strength and power type resistance exercise. *European Journal of Applied Physiology*, *105*, 695–704.

Metter, E. J., Conwit, R., Tobin, J., Fozard, J. L. (1997). Age-associated loss of power and strength in the upper extremities in women and men. *Journal of Gerontology, 52A*, B267–B276.

Miszko, T. A., Cress, M. E., Slade, J. M., Covey, C. J., Agrawal, S. K., Doerr, C. E. (2003). Effect of strength and power training on physical function in community-dwelling older adults. *The Journals of Gerontology. Series A, Biological Sciences and Medical Sciences*, *58*, 171–175.

Mitch, W. E. & Goldberg, A. L. (1996). Mechanisms of muscle wasting. The role of the ubiquitin-proteasome pathway. *The New England Journal of Medicine, 335*, 1897–905.

Mjolsnes, R., Arnason, A., Osthagen, T., Raastad, T., Bahr, R. (2004). A 10-week randomized trial comparing eccentric vs. concentric hamstring strength training in well-trained soccer players. *Scandinavian Journal of Medicine & Science in Sports, 14*, 311–7.

Morley, J. E. (2008). Sarcopenia: Diagnosis and treatment. *The Journal of Nutrition, Health & Aging, 12*, 452–456.

Morris, J. N., Heady, J. A., Raffle, P. A., Roberts, C. G., Parks, J. W. (1953a). Coronary heart-disease and physical activity of work. *Lancet, 265*, 1111–20. concl.

Morris, J. N., Heady, J. A., Raffle, P. A., Roberts, C. G., Parks, J. W. (1953b). Coronary heart-disease and physical activity of work. *Lancet, 265*, 1053–7. contd.

Munoz, K. A., Satarug, S., Tischler, M. E. (1993). Time course of the response of myofibrillar and sarcoplasmic protein metabolism to unweighting of the soleus muscle. *Metabolism, 42*, 1006–12.

Musi, N. & Goodyear, L. J. (2006). Insulin resistance and improvements in signal transduction. *Endocrine*, *29*(1), 73–80.

Nader, G. A., Mcloughlin, T. J., Esser, K. A. (2005). mTOR function in skeletal muscle hypertrophy: Increased ribosomal RNA via cell cycle regulators. *American Journal of Physiology. Cell Physiology, 289*, C1457–65.

Nair, K. S., Garrow, J. S., Ford, C., Mahler, R. F., Halliday, D. (1983). Effect of poor diabetic control and obesity on whole body protein metabolism in man. *Diabetologia, 25*, 400–403.

Nair, K. S. (1995). Muscle protein turnover: Methodological issues and the effect of aging. *The Journals of Gerontology. Series A: Biological Sciences and Medical Sciences, 50*, 107–12.

Nakagawa, Y., Hattori, M., Harada, K., Shirase, R., Bando, M., Okano, G. (2007). Age-related changes in intramyocellular lipid in humans by in vivo H-MR spectroscopy. *Gerontology, 53*, 218–23.

Nelson, M. E., Rejeski, W. J., Blair, S. N., Duncan, P. W., Judge, J. O., King, A. C., Macera, C. A., Castaneda-Sceppa, C. (2007). Physical activity and public health in older adults: Recommendation from the American College of Sports Medicine and the American Heart Association. *Circulation, 116*, 1094–105.

Paddon-Jones, D., Børsheim, E., Wolfe, R. R. (2004). Potential ergogenic effects of arginine and creatine supplementation. *Journal of Nutrition, 134*(10 Suppl), S2888–2894S.

Paffenbarger, R. S., JR, Laughlin, M. E., Gima, A. S., Black, R. A. (1970). Work activity of longshoremen as related to death from coronary heart disease and stroke. *The New England Journal of Medicine, 282*, 1109–14.

Paffenbarger, R. S., JR, Hyde, R. T., Wing, A. L., Lee, I. M., Jung, D. L., Kampert, J. B. (1993). The association of changes in physical-activity level and other lifestyle characteristics with mortality among men. *The New England Journal of Medicine, 328*, 538–45.

Pansarasa, O., Rinaldi, C., Parente, V., Miotti, D., Capodaglio, P., Bottinelli, R. (2009). Resistance training of long duration modulates force and unloaded shortening velocity of single muscle fibres of young women. *Journal of Electromyography and Kinesiology, 19*, e290–e300.

Parkington, J. D., Lebrasseur, N. K., Siebert, A. P., Fielding, R. A. (2004). Contraction-mediated mTOR, p70S6k, and ERK1/2 phosphorylation in aged skeletal muscle. *Journal of Applied Physiology, 97*, 243–8.

Pattison, J. S., Folk, L. C., Madsen, R. W., Childs, T. E., Booth, F. W. (2003). Transcriptional profiling identifies extensive downregulation of extracellular matrix gene expression in sarcopenic rat soleus muscle. *Physiological Genomics, 15*, 34–43.

Petrella, J. K., Kim, J. S., Tuggle, S. C., Hall, S. R., Bamman, M. M. (2005). Age differences in knee extension power, contractile velocity, and fatigability. *Journal of Applied Physiology, 98*, 211–220.

Petrella, J. K., Kim, J. S., Cross, J. M., Kosek, D. J., Bamman, M. M. (2006). Efficacy of myonuclear addition may explain differential myofiber growth among resistance-trained young and older men and women. *American Journal of Physiology. Endocrinology and Metabolism, 291*, E937–E946.

Phillips, S. M. (2007). Resistance exercise: Good for more than just Grandma and Grandpa's muscles. *Applied Physiology, Nutrition, and Metabolism, 32*, 1198–205.

Proctor, D. N. & Joyner, M. J. (1997). Skeletal muscle mass and the reduction of VO2max in trained older subjects. *Journal of Applied Physiology, 82*, 1411–5.

Prod'homme, M., Balage, M., Debras, E., Farges, M. C., Kimball, S., Jefferson, L., Grizard, J. (2005). Differential effects of insulin and dietary amino acids on muscle protein synthesis in adult and old rats. *The Journal of Physiology, 563*, 235–248.

Putman, C. T., XU, X., Gillies, E., Maclean, I. M., Bell, G. J. (2004). Effects of strength, endurance and combined training on myosin heavy chain content and fibre-type distribution in humans. *European Journal of Applied Physiology, 92*, 376–84.

Rantanen, T., Era, P., Heikkinen, E. (1994). Maximal isometric strength and mobility among 75-year-old men and women. *Age and Ageing, 23*, 132–7.

Rasmussen, B. B., Fujita, S., Wolfe, R. R., Mittendorfer, B., Roy, M., Rowe, V. L., Volpi, E. (2006). Insulin resistance of muscle protein metabolism in aging. *The FASEB Journal, 20*, 768–9.

Raue, U., Slivka, D., Jemiolo, B., Hollon, C., Trappe, S. (2006). Myogenic gene expression at rest and after a bout of resistance exercise in young (18–30 yr) and old (80–89 yr) women. *Journal of Applied Physiology, 101*, 53–9.

Raue, U., Slivka, D., Jemiolo, B., Hollon, C., Trappe, S. (2007). Proteolytic gene expression differs at rest and after resistance exercise between young and old women. *The Journals of Gerontology. Series A: Biological Sciences and Medical Sciences, 62*, 1407–12.

Raue, U., Slivka, D., Minchev, K., Trappe, S. (2009). Improvements in whole muscle and myocellular function are limited with high-intensity resistance training in octogenarian women. *Journal of Applied Physiology, 106*, 1611–1617.

Reid, K. F., Callahan, D. M., Carabello, R. J., Phillips, E. M., Frontera, W. R., Fielding, R. A. (2008). Lower extremity power training in elderly subjects with mobility limitations: A randomized controlled trial. *Aging Clinical and Experimental Research, 20*, 337–43.

Renault, V., Thornell, L. E., Eriksson, P. O., Butler-Browne, G., Mouly, V. (2002). Regenerative potential of human skeletal muscle during aging. *Aging Cell, 1*, 132–9.

Rennie, M. J. (2009). Anabolic resistance: The effects of aging, sexual dimorphism, and immobilization on human muscle protein turnover. *Applied Physiology, Nutrition, and Metabolism, 34*, 377–81.

Reynolds, T. H. 4th, Bodine, S. C., Lawrence, J. C., Jr. (2002). Control of Ser2448 phosphorylation in the mammalian target of rapamycin by insulin and skeletal muscle load. *The Journal of Biological Chemistry, 277*, 17657–17662.

Reynolds, T. H., 4th, Reid, P., Larkin, L. M., Dengel, D. R. (2004). Effects of aerobic exercise training on the protein kinase B (PKB)/mammalian target of rapamycin (mTOR) signaling pathway in aged skeletal muscle. *Experimental Gerontology, 39*, 379–385.

Richter, E. A. & Ruderman, N. B. (2009). AMPK and the biochemistry of exercise: implications for human health and disease. *The Biochemical Journal, 418*, 261–275.

Rissanen, A., Heliövaara, M., Aromaa, A. (1988). Overweight and anthropometric changes in adulthood: A prospective study of 17,000 Finns. *International Journal of Obesity, 12*, 391–401.

Rivas, D. A., Yaspelkis, B. B., 3RD, Hawley, J. A., Lessard, S. J. (2009). Lipid-induced mTOR activation in rat skeletal muscle reversed by exercise and 5′-aminoimidazole-4-carboxamide-1-beta-D-ribofuranoside. *The Journal of Endocrinology, 202*, 441–51.

Rivas, D. A., Lessard, S. J. Coffey, V. G. (2009). mTOR function in skeletal muscle: a focal point for over-nutrition and exercise *Appl Physiol Nutr Metab, 34*, 807–816.

Rommel, C., Bodine, S. C., Clarke, B. A., Rossman, R., Nunez, L., Stitt, T. N., et al. (2001). Mediation of IGF-1-induced skeletal myotube hypertrophy by PI(3)K/Akt/mTOR and PI(3)K/Akt/GSK3 pathways. *Nature Cell Biology, 3*, 1009–1013.

Ronnestad, B. R., Egeland, W., Kvamme, N. H., Refsnes, P. E., Kadi, F., Raastad, T. (2007). Dissimilar effects of one- and three-set strength training on strength and muscle mass gains in upper and lower body in untrained subjects. *Journal of Strength and Conditioning Research, 21*, 157–63.

Rooyackers, O. E., Adey, D. B., Ades, P. A., Nair, K. S. (1996). Effect of age on in vivo rates of mitochondrial protein synthesis in human skeletal muscle. *Proceedings of the National Academy of Sciences of the United States of America, 93*, 15364–9.

Rosenberg, I. H. (1997). Sarcopenia: Origins and clinical relevance. *The Journal of Nutrition, 127*, 990S–991S.

Roth, S. M., Martel, G. F., Ivey, F. M., Lemmer, J. T., Metter, E. J., Hurley, B. F., Rogers, M. A. (2000). Skeletal muscle satellite cell populations in healthy young and older men and women. *The Anatomical Record, 260*, 351–8.

Roth, S. M., Ivey, F. M., Martel, G. F., Lemmer, J. T., Hurlbut, D. E., Siegel, E. L., Metter, E. J., Fleg, J. L., Fozard, J. L., Kostek, M. C., Wernick, D. M., Hurley, B. F. (2001). Muscle size responses to strength training in young and older men and women. *Journal of the American Geriatrics Society, 49*, 1428–33.

Roth, S. M., Martel, G. F., Ivey, F. M., Lemmer, J. T., Tracy, B. L., Metter, E. J., Hurley, B. F., Rogers, M. A. (2001). Skeletal muscle satellite cell characteristics in young and older men and women after heavy resistance strength training. *The Journals of Gerontology. Series A: Biological Sciences and Medical Sciences, 56*, B240–7.

Roth, S. M., Martel, G. F., Ivey, F. M., Lemmer, J. T., Tracy, B. L., Metter, E. J., et al. (2001). Skeletal muscle satellite cell characteristics in young and older men and women after heavy resistance strength training. *The Journals of Gerontology. Series A, Biological Sciences and Medical Sciences, 56*, B240–B247.

Saltin, B. & Rowell, L. B. (1980). Functional adaptations to physical activity and inactivity. *Federation Proceedings, 39*, 1506–1513.

Sarsan, A., Ardiç, F., Ozgen, M., Topuz, O., Sermez, Y. (2006). The effects of aerobic and resistance exercises in obese women. *Clinical Rehabilitation, 20*, 773–782.
Schlicht, J., Camaione, D. N., Owen, S. V. (2001). Effect of intense strength training on standing balance, walking speed, and sit-to-stand performance in older adults. *The Journals of Gerontology. Series A: Biological Sciences and Medical Sciences, 56*, M281–6.
Schultz, E. & Lipton, B. H. (1982). Skeletal muscle satellite cells: Changes in proliferation potential as a function of age. *Mechanisms of Ageing and Development, 20*, 377–383.
Seals, D. R., Hagberg, J. M., Hurley, B. F., Ehsani, A. A., Holloszy, J. O. (1984a). Effects of endurance training on glucose tolerance and plasma lipid levels in older men and women. *JAMA, 252*, 645–9.
Seals, D. R., Hagberg, J. M., Hurley, B. F., Ehsani, A. A., Holloszy, J. O. (1984b). Endurance training in older men and women. I. Cardiovascular responses to exercise. *Journal of Applied Physiology, 57*, 1024–9.
Seynnes, O., Fiatarone Singh, M. A., Hue, O., Pras, P., Legros, P., Bernard, P. L. (2004). Physiological and functional responses to low-moderate versus high-intensity progressive resistance training in frail elders. *The Journals of Gerontology. Series A, Biological Sciences and Medical Sciences, 59*, 503–509.
Sheffield-Moore, M., Yeckel, C. W., Volpi, E., Wolf, S. E., Morio, B., Chinkes, D. L., et al. (2004). Postexercise protein metabolism in older and younger men following moderate-intensity aerobic exercise. *American Journal of Physiology. Endocrinology and Metabolism, 287*, E513–E522.
Short, K. R., Vittone, J. L., Bigelow, M. L., Proctor, D. N., Rizza, R. A., Coenen-Schimke, J. M., Nair, K. S. (2003). Impact of aerobic exercise training on age-related changes in insulin sensitivity and muscle oxidative capacity. *Diabetes, 52*, 1888–96.
Short, K. R., Vittone, J. L., Bigelow, M. L., Proctor, D. N., Nair, K. S. (2004). Age and aerobic exercise training effects on whole body and muscle protein metabolism. *American Journal of Physiology. Endocrinology and Metabolism, 286*, E92–101.
Signorile, J. F., Carmel, M. P., Czaja, S. J., Asfour, S. S., Morgan, R. O., Khalil, T. M., et al. (2002). Differential increases in average isokinetic power by specific muscle groups of older women due to variations in training and testing. *The Journals of Gerontology. Series A, Biological Sciences and Medical Sciences, 57*, M683–M690.
Singh, M. A. (2004). Exercise and aging. *Clinics in Geriatric Medicine, 20*, 201–21.
Sinha-Hikim, I., Roth, S. M., Lee, M. I., Bhasin, S. (2003). Testosterone-induced muscle hypertrophy is associated with an increase in satellite cell number in healthy, young men. *American Journal of Physiology. Endocrinology and Metabolism, 285*, E197–E205.
Sinha-Hikim, I., Cornford, M., Gaytan, H., Lee, M. L., Bhasin, S. (2006). Effects of testosterone supplementation on skeletal muscle fiber hypertrophy and satellite cells in community-dwelling older men. *The Journal of Clinical Endocrinology and Metabolism, 91*, 3024–33.
Sipilä, S. & Suominen, H. (1996). Quantitative ultrasonography of muscle: Detection of adaptations to training in elderly women. *Archives of Physical Medicine and Rehabilitation, 77*, 1173–1178.
Skelton, D. A., Young, A., Greig, C. A., Malbut, K. E. (1995). Effects of resistance training on strength, power, and selected functional abilities of women aged 75 and older. *Journal of the American Geriatrics Society, 43*, 1081–7.
Slivka, D., Raue, U., Hollon, C., Minchev, K., Trappe, S. (2008). Single muscle fiber adaptations to resistance training in old (>80 yr) men: Evidence for limited skeletal muscle plasticity. *American Journal of Physiology. Regulatory, Integrative and Comparative Physiology, 295*, R273–R280.
Sousa, N. & Sampaio, J. (2005). Effects of progressive strength training on the performance of the Functional Reach Test and the Timed Get-Up-and-Go Test in an elderly population from the rural north of Portugal. *American Journal of Human Biology, 17*, 746–51.
Spangenburg, E. E. & Booth, F. W. (2006). Leukemia inhibitory factor restores the hypertrophic response to increased loading in the LIF(–/–) mouse. *Cytokine, 34*, 125–30.

Spangenburg, E. E., LE Roith, D., Ward, C. W., Bodine, S. C. (2008). A functional insulin-like growth factor receptor is not necessary for load-induced skeletal muscle hypertrophy. *Journal de Physiologie, 586*, 283–91.

Staron, R. S., Leonardi, M. J., Karapondo, D. L., Malicky, E. S., Falkel, J. E., Hagerman, F. C., et al. (1991). Strength and skeletal muscle adaptations in heavy-resistance-trained women after detraining and retraining. *Journal of Applied Physiology, 70*, 631–640.

Staron, R. S., Karapondo, D. L., Kraemer, W. J., Fry, A. C., Gordon, S. E., Falkel, J. E., et al. (1994). Skeletal muscle adaptations during early phase of heavy-resistance training in men and women. *Journal of Applied Physiology, 76*, 1247–1255.

Suetta, C., Magnusson, S. P., Rosted, A., Aagaard, P., Jakobsen, A. K., Larsen, L. H., et al. (2004). Resistance training in the early postoperative phase reduces hospitalization and leads to muscle hypertrophy in elderly hip surgery patients – a controlled, randomized study. *Journal of the American Geriatrics Society, 52*, 2016–2022.

Sugawara, J., Miyachi, M., Moreau, K. L., Dinenno, F. A., Desouza, C. A., Tanaka, H. (2002). Age-related reductions in appendicular skeletal muscle mass: Association with habitual aerobic exercise status. *Clinical Physiology and Functional Imaging, 22*, 169–72.

Suzuki, T., Bean, J. F., Fielding, R. A. (2001). Muscle power of the ankle flexors predicts functional performance in community-dwelling older women. *Journal of the American Geriatrics Society, 49*, 1161–1167.

Tanaka, H. & Seals, D. R. (2003). Invited review: Dynamic exercise performance in Masters athletes: Insight into the effects of primary human aging on physiological functional capacity. *Journal of Applied Physiology, 95*, 2152–2162.

Tanaka, H. & Seals, D. R. (2008). Endurance exercise performance in Masters athletes: Age-associated changes and underlying physiological mechanisms. *The Journal of Physiology, 586*, 55–63.

Thomson, D. M., Gordon, S. E. (2005). Diminished overload-induced hypertrophy in aged fast-twitch skeletal muscle is associated with AMPK hyperphosphorylation. *Journal of Applied Physiology, 98*, 557–564.

Thomson, D. M. & Gordon, S. E. (2006). Impaired overload-induced muscle growth is associated with diminished translational signalling in aged rat fast-twitch skeletal muscle. *Journal de Physiologie, 574*, 291–305.

Thomson, D. M. & Brown, J. D., Fillmore, N., Ellsworth, S. K., Jacobs, D. L., Winder, W. W., et al. (2009). AMP-activated protein kinase response to contractions and treatment with the AMPK activator AICAR in young adult and old skeletal muscle. *The Journal of Physiology, 587*, 2077–2086.

Timiras, P. S. (1994). Disuse and aging: Same problem, different outcomes. *Journal of Gravitational Physiology, 1*, P5–P7.

Toledo, F. G., Menshikova, E. V., Ritov, V. B., Azuma, K., Radikova, Z., DeLany, J., et al. (2007). Effects of physical activity and weight loss on skeletal muscle mitochondria and relationship with glucose control in type 2 diabetes. *Diabetes, 56*, 2142–2147.

Tomlinson, B. E., Walton, J. N., Rebeiz, J. J. (1969). The effects of ageing and of cachexia upon skeletal muscle. A histopathological study. *Journal of the Neurological Sciences, 9*, 321–346.

Topp, R., Mikesky, A., Dayhoff, N. E., Holt, W. (1996). Effect of resistance training on strength, postural control, and gait velocity among older adults. *Clinical Nursing Research, 5*, 407–27.

Trappe, S., Williamson, D., Godard, M., Porter, D., Rowden, G., Costill, D. (2000). Effect of resistance training on single muscle fiber contractile function in older men. *Journal of Applied Physiology, 89*, 143–152.

Trappe, S., Godard, M., Gallagher, P., Carroll, C., Rowden, G., Porter, D. (2001). Resistance training improves single muscle fiber contractile function in older women. *American Journal of Physiology. Cell Physiology, 281*, C398–C406.

Tsourlou, T., Gerodimos, V., Kellis, E., Stavropoulos, N., Kellis, S. (2003). The effects of a calisthenics and a light strength training program on lower limb muscle strength and body composition in mature women. *Journal of Strength and Conditioning Research, 17*, 590–8.

Tsourlou, T., Benik, A., Dipla, K., Zafeiridis, A., Kellis, S. (2006). The effects of a twenty-four-week aquatic training program on muscular strength performance in healthy elderly women. *Journal of Strength and Conditioning Research, 20*, 811–8.

Tsutsumi, T., Don, B. M., Zaichkowsky, L. D., Delizonna, L. L. (1997). Physical fitness and psychological benefits of strength training in community dwelling older adults. *Appl Human Sci, 16*, 257–66.

Tsuzuku, S., Kajioka, T., Endo, H., Abbott, R. D., Curb, J. D., Yano, K. (2007). Favorable effects of non-instrumental resistance training on fat distribution and metabolic profiles in healthy elderly people. *European Journal of Applied Physiology, 99*, 549–555.

Tucker, M. Z. & Turcotte, L. P. (2003). Aging is associated with elevated muscle triglyceride content and increased insulin-stimulated fatty acid uptake. *American Journal of Physiology. Endocrinology and Metabolism, 285*, E827–35.

UN (2007) World Population Ageing 2007. In Department of Economic and Social Affairs, P. D. (Ed.), New York: United Nations.

Valkeinen, H., Häkkinen, K., Pakarinen, A., Hannonen, P., Häkkinen, A., Airaksinen, O., et al. (2005). Muscle hypertrophy, strength development, and serum hormones during strength training in elderly women with fibromyalgia. *Scandinavian Journal of Rheumatology, 34*, 309–314.

Vary, T. C., Anthony, J. C., Jefferson, L. S., Kimball, S. R., Lynch, C. J. (2007). Rapamycin blunts nutrient stimulation of eIF4G, but not PKCepsilon phosphorylation, in skeletal muscle. *American Journal of Physiology. Endocrinology and Metabolism, 293*, E188–96.

Verdijk, L. B., Koopman, R., Schaart, G., Meijer, K., Savelberg, H. H., Van Loon, L. J. (2007). Satellite cell content is specifically reduced in type II skeletal muscle fibers in the elderly. *American Journal of Physiology. Endocrinology and Metabolism, 292*, E151–7.

Verdijk, L. B., Gleeson, B. G., Jonkers, R. A., Meijer, K., Savelberg, H. H., Dendale, P., Vanloon, L. J. (2009). Skeletal muscle hypertrophy following resistance training is accompanied by a fiber type-specific increase in satellite cell content in elderly men. *The Journals of Gerontology. Series A: Biological Sciences and Medical Sciences, 64*, 332–9.

Vierck, J., O'Reilly, B., Hossner, K., Antonio, J., Byrne, K., Bucci, L., et al. (2000). Satellite cell regulation following myotrauma caused by resistance exercise. *Cell Biology International, 24*, 263–272.

Volpi, E., Sheffield-Moore, M., Rasmussen, B. B., Wolfe, R. R. (2001). Basal muscle amino acid kinetics and protein synthesis in healthy young and older men. *JAMA, 286*, 1206–12.

Volpi, E., Kobayashi, H., Sheffield-Moore, M., Mittendorfer, B., Wolfe, R. R. (2003). Essential amino acids are primarily responsible for the amino acid stimulation of muscle protein anabolism in healthy elderly adults. *The American Journal of Clinical Nutrition, 78*, 250–258.

Wagers, A. J. & Conboy, I. M. (2005). Cellular and molecular signatures of muscle regeneration: current concepts and controversies in adult myogenesis. *Cell, 122*, 659–667.

Wallberg-Henriksson, H. & Holloszy, J. O. (1984). Contractile activity increases glucose uptake by muscle in severely diabetic rats. *Journal of Applied Physiology, 57*, 1045–1049.

Wallberg-Henriksson, H. & Holloszy, J. O. (1985). Activation of glucose transport in diabetic muscle: Responses to contraction and insulin. *The American Journal of Physiology, 249*, C233–C237.

Wang, X. & Proud, C. G. (2006). The mTOR pathway in the control of protein synthesis. *Physiology (Bethesda), 21*, 362–9.

Weiss, E. P., Racette, S. B., Villareal, D. T., Fontana, L., Steger-May, K., Schechtman, K. B., et al. (2007). Lower extremity muscle size and strength and aerobic capacity decrease with caloric restriction but not with exercise-induced weight loss. *Journal of Applied Physiology, 102*, 634–640.

Welle, S., Thornton, C., Jozefowicz, R., Statt, M. (1993). Myofibrillar protein synthesis in young and old men. *The American Journal of Physiology, 264*, E693–8.

Welle, S., Thornton, C., Statt, M. (1995). Myofibrillar protein synthesis in young and old human subjects after three months of resistance training. *The American Journal of Physiology, 268*, E422–7.

Welle, S., Totterman, S., Thornton, C. (1996). Effect of age on muscle hypertrophy induced by resistance training. *The Journals of Gerontology. Series A: Biological Sciences and Medical Sciences, 51*, M270–5.

Welle, S., Brooks, A. I., Delehanty, J. M., Needler, N., Thornton, C. A. (2003). Gene expression profile of aging in human muscle. *Physiological Genomics, 14*, 149–59.

Wieser, M. & Haber, P. (2007). The effects of systematic resistance training in the elderly. *International Journal of Sports Medicine*, *28*, 59–65.

Wilkinson, S. B., Phillips, S. M., Atherton, P. J., Patel, R., Yarasheski, K. E., Tarnopolsky, M. A., Rennie, M. J. (2008). Differential effects of resistance and endurance exercise in the fed state on signalling molecule phosphorylation and protein synthesis in human muscle. *Journal de Physiologie, 586*, 3701–17.

Yarasheski, K. E., Zachwieja, J. J., Bier, D. M. (1993). Acute effects of resistance exercise on muscle protein synthesis rate in young and elderly men and women. *The American Journal of Physiology, 265*, E210–4.

Zierath, J. R. (2002). Invited review: Exercise training-induced changes in insulin signaling in skeletal muscle. *Journal of Applied Physiology*, *93*, 773–781.

Role of Contraction-Induced Injury in Age-Related Muscle Wasting and Weakness

John A. Faulkner, Christopher L. Mendias, Carol S. Davis, and Susan V. Brooks

Abstract In mammalian taxonomy, skeletal muscle constitutes a remarkable tissue not only in its innate capacity to generate force while shortening, remaining isometric, or lengthening, but in its capacity to adapt through atrophy or hypertrophy in response to decreased or increased loads, respectively and regenerate when injured. The chapter begins with Section 1 on the *Structure of Skeletal Muscles and Skeletal Muscle Fibers*. Section 2 describes *Types of Contractions*, shortening, isometric, and lengthening and the differences in the force development by each. The interactive roles of decreased usage and aging are covered in Section 3: *Age-Related Muscle Wasting and Muscle Weakness* and the condition of physical frailty is discussed. Section 4 focuses on *Late-Onset Muscle Soreness* described by Hough in 1902 and gaining widespread attention in the 1980s. The development of the concepts: *Contraction-Induced Injury and Force Deficit* are discussed in Section 5. Section 6 clarifies *The Cause of the Contraction Induced Injury* as a function of interactions between homogeneity of sarcomere strengths within a muscle and the severity of lengthening contraction protocols. Section 7 elaborates on the significance of the stability of the sarcomeres within fibers and the *Contribution of Lateral Transmission of Force to Contraction-Induced Injury*. Section 8, the *Role of Contraction-Induced Injury in Wasting and Weakness* contrasts the impact of contraction-induced injury on young and healthy and on elderly and frail subjects.

J.A. Faulkner (✉) and S.V. Brooks
Departments of Biomedical Engineering and Molecular and Integrative Physiology, University of Michigan, Ann Arbor, MI 48109-2200, USA
e-mail: jafaulk@umich.edu

C.L. Mendias
Departments of Orthopaedic Surgery and the School of Kinesiology, University of Michigan, Ann Arbor, MI 48109-2200, USA

C.S. Davis
Department of Molecular and Integrative Physiology, University of Michigan, Ann Arbor, MI 48109-2200, USA

G.S. Lynch (ed.), *Sarcopenia – Age-Related Muscle Wasting and Weakness*,
DOI 10.1007/978-90-481-9713-2_16,

The final Section 9: *Measures to Prevent Contraction-Induced Injury* emphasizes the positive aspects of utilizing lengthening contractions in training programs for both young and old participants.

Keywords Contraction-induced injury • Delayed-onset muscle soreness • Muscle wasting • Muscle weakness • Lengthening contraction • Eccentric contraction • Muscle repair • Muscle regeneration • Force deficit • Muscle conditioning

1 Structure of Skeletal Muscles and Skeletal Muscle Fibers

Skeletal muscles are composed of muscle fibers organized into motor units innervated by a motor nerve. In humans, single muscles range from small finger flexor muscles in the hands with fewer than 100 motor units and around 100 muscle fibers per motor unit on average to the gastrocnemius muscles in the lower leg composed of almost 600 motor units and close to 2,000 fibers per motor unit (Feinstein et al. 1955). Each individual muscle fiber within a motor unit contains myofibrils that consist of myosin filaments surrounded by and overlapping with thin actin filaments that are anchored in the z-discs at either end of sarcomeres. The globular head of the myosin molecules are capable of binding to sites on the thin actin filaments when a muscle fiber receives an action potential and there is a release of calcium from intracellular calcium stores. The myosin cross-bridges then proceed through a driving stroke that, under circumstances when the muscle is unloaded, or loaded with a resistance that can be moved, draws the thin filaments past the thick filaments in a movement that brings the z-discs together and shortens the length of sarcomeres. If the muscle is held at a fixed length and activated, cross-bridges cycle generating force without filament sliding and sarcomere shortening. Finally, if while activated, the muscle is stretched by a load greater than that generated by the cycling cross-bridges, cross-bridges are strained prior to release and re-attachment.

2 Types of Contractions

When a muscle is activated by action potentials, the muscle fibers in the activated motor units attempt to shorten. Whether the fibers actually shorten, remain at the same length, or are lengthened depends on the interaction between the force generated by the muscle and the load on the muscle. Consequently, skeletal muscles make three types of contractions – a *shortening contraction,* wherein the load on the muscle is less than the force generated by the muscle and the activated muscle fibers shorten (Fig. 1, Panel a); an *isometric contraction,* wherein the load on the muscle is either immoveable, or equivalent to the force generated by the muscle

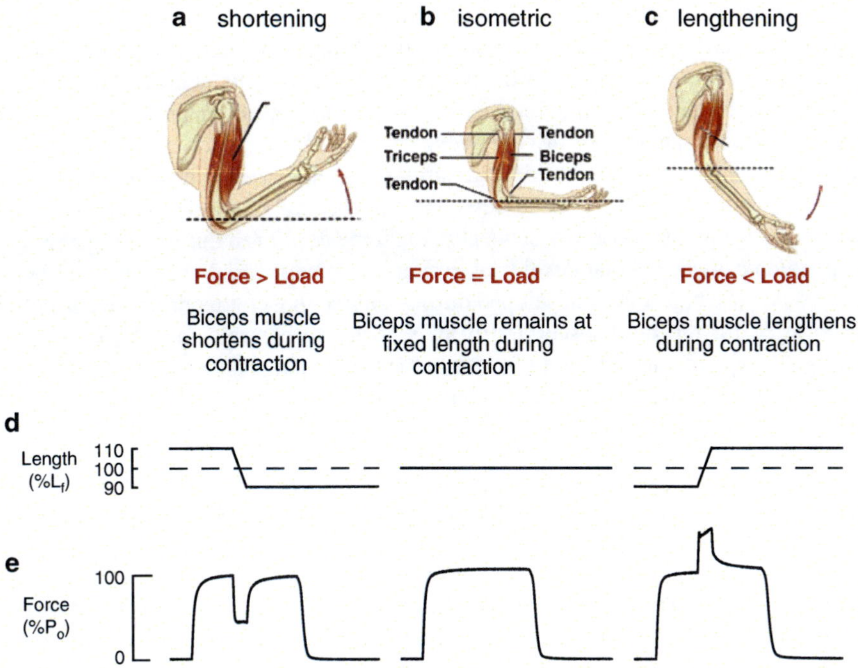

Fig. 1 The three types of contractions that single fibers, motor units and whole skeletal muscles are able to perform are dependent on the interaction of the force developed by the muscle and the load against which the muscle is attempting to shorten. A shortening contraction (**a**) occurs when the force is greater than the load. During a shortening contraction, the velocity of shortening is load dependent, with the greater the load the lower the velocity of shortening. During a shortening contraction, a muscle performs 'work'. An isometric contraction (**b**) occurs when the force developed by the muscle equals the load or under conditions when the load is immovable. A lengthening contraction (**c**) results when the load on the muscle is greater than the force developed by the muscle (Modified from Vander, Sherman 2001, Luciano *Human Physiology,* Figs 11–31, page 320, McGraw-Hill. Reproduced with permission of The McGraw-Hill Companies.) The changes in the lengths of the muscle are displayed during each of the three types of contractions. Tracings of the displacements initiated by a servo motor lever arm (**d**) and the forces developed (**e**) by a maximally activated muscle measured by a force transducer. L_f, fiber length that results in maximum force; P_o, maximum isometric tetanic force (Reprinted with permission from Faulkner et al. 2007, Wiley)

and the activated muscle remains activated at a fixed length (Fig. 1, Panel b); or a *lengthening contraction,* wherein the load on the muscle is greater than the force generated by the muscle and the muscle is lengthened (Fig. 1, Panel c). The terms concentric and eccentric contractions are now in wide usage for shortening and lengthening contractions, respectively. Although the terms concentric and eccentric contractions are useful clinically, these terms have no intrinsic meaning in terms of the characteristics of the contractions that limb muscles make. Thus, throughout this chapter the terms shortening, isometric, and lengthening will be used to describe the type of a specific contraction.

With maximum activation, the forces developed are greatest during lengthening contractions, intermediate during isometric contractions, and least during shortening contractions. The explanation for the greater force during lengthening contractions than during isometric contractions is that during isometric contraction only the cross-bridges that are in their driving stroke generate tension, but when a maximally activated muscle is stretched, additional strongly-bound cross-bridges that have not progressed into their 'driving stroke' resist the 'lengthening' of the skeletal muscle, are strained and generate force. Consequently, the force developed during a lengthening contraction can exceed that developed during an isometric contraction by as much as twofold. The high forces developed during lengthening contractions are partially responsible for the high susceptibility of muscles to contraction-induced injury during this type of contraction. In fact, only the lengthening contractions are capable of producing a contraction-induced injury.

3 Age-Related Muscle Wasting and Muscle Weakness

The *'wasting'* or *'atrophy'* of a skeletal muscle refers to a loss in the mass of the skeletal muscle, a condition that arises from a reduced usage of skeletal muscles at any age. The reduction in the daily usage may arise from: (a) sickness and imposed bed-rest, (b) disuse of a specific muscle due to immobilization by casting, or to the placement of an injured arm in a sling. In addition, by 70–80 years of age an outright loss of skeletal muscle fibers occurs that is estimated, based on data from vastus lateralis muscles, to be as high as 50% of the fibers (Lexell et al. 1988). The loss in the number of muscle fibers contributes significantly to the concurrent loss of muscle mass and myofibrillar protein. In contrast to atrophy, *'weakness'* of a muscle reflects an inability of a muscle to generate the normal or expected force when activated. As people age, particularly into advanced old age, the vast majority of humans, both men and women, become less physically active and invariably show signs of both muscle wasting and muscle weakness. Particularly in old age, the combined impact of decreased physical activity and muscle wasting and weakness lead to the debilitating condition of frailty (Hadley et al. 1993). The increase in physical frailty with old age has serious consequences in terms of the health and longevity of the elderly. Physical frailty invariably leads to a further decrease in physical activity as well contributing to respiratory and cardiovascular problems (Hadley et al. 1993). Despite the magnitude of the problem, even in the elderly, these conditions are at least partially reversible by re-establishing an increased level of physical activity, but such programs must be carefully designed with a slow progression and close supervision by highly trained exercise leaders. Although some amelioration of muscle atrophy is achievable through exercise, the component of muscle atrophy that is due to the loss of muscle fibers appears inevitable and irreversible. Consequently, the magnitude of the improvements attainable with physical training of the frail elderly must be realistic and kept in perspective with the limitations of the participants.

4 Late-Onset Muscle Soreness

The phenomenon of a contraction-induced injury to skeletal muscle fibers was first recognized inadvertently by Theodore Hough, during experiments on the fatigue of finger flexor muscles (Hough 1901, 1902). Hough's subjects performed a highly fatiguing muscle contraction protocol using a pulley-system that enabled lifting and lowering a weight with flexion and extension of the middle finger. Some of the participants complained of pain in the forearm between 8 and 12 h after the completion of the protocol, with the soreness increasing and reaching its highest level 48 or even 60 h afterward. In these experiments, it was not recognized that the soreness was initiated by the lowering of the weight. The phenomenon of muscle soreness encountered in the Hough studies was ignored for almost 80 years, and then re-surfaced as *'delayed onset muscle soreness'* in the early 1980s. Late onset muscle soreness has been observed after a number of different protocols that involved the lowering of a weight or the 'stretching' of the activated skeletal muscle fibers and a number of inventive protocols were developed to investigate the factors involved in the lengthening contractions that initiated the delayed soreness of the muscle. These early protocols involved repeatedly stepping up with one leg and down with the other leg on and off a fairly high stool (Newham et al. 1983a, b), raising and lowering a weight with forearm flexion and extension (Newham et al. 1987), and resisting the reverse-rotation of the pedals of a bicycle ergometer (Friden et al. 1983). Needle biopsy samples of both arms and legs indicated that these protocols of lengthening contractions invariably caused morphological evidence of injury to skeletal muscle fibers (Fig. 2a–c).

Lengthening contractions produce a decrease in maximum strength and assays of blood samples indicate a peak in plasma creatine kinase several days after the initial injury (Fig. 3a). Subjective assessments of pain indicate that the exact timing of the onset of muscle soreness varies somewhat with the individual and with the type of exercise, but typically peaks after ~2 days and is resolved within 5 days. The recovery of strength and reestablishment of pre-injury levels of circulating creatine kinase take anywhere from 1 to 2 weeks depending on the severity of the injury and repeated bouts of training with lengthening contractions reduce the occurrence of late onset muscle soreness (Newham et al. 1983a, b). The experiments on volitional lengthening contractions performed by human subjects were soon followed up with more definitive experiments on mice and rats (Armstrong et al. 1983; McCully and Faulkner 1985). The experiments on small mammals substantiated the time course of the injury to muscle fibers and that the magnitude of the injury was greatest approximately 3 days after the lengthening contraction protocol with complete recovery requiring 3–4 weeks (Fig. 3b). A number of factors have been cited as the likely causes of the late-onset muscle soreness. The most plausible of these factors are the actual damage to muscle fibers and connective tissue and inflammation (Cheung et al. 2003; Friden et al. 1983, 1986; Newham et al. 1983a, b; Jones et al. 1986; Schwane and Armstrong 1983). From the beginning, Hough (1902) cited the ruptures within the muscles as the cause of the

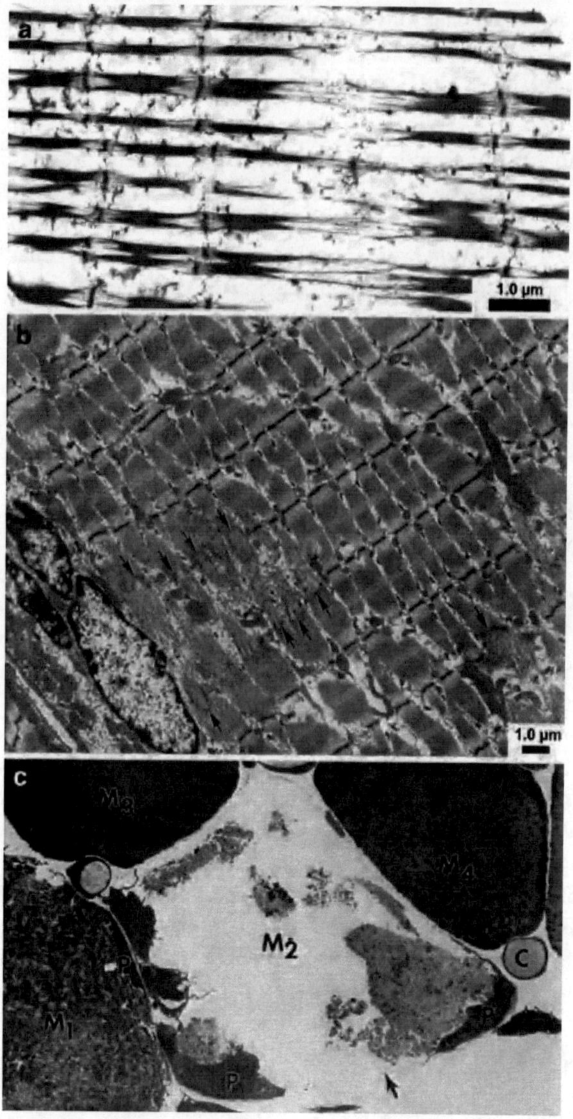

Fig. 2 Electron micrographs from an EDL muscle of a young mouse after a protocol of 75 lengthening contractions. (**a**) A longitudinal section of a single fiber at high magnification taken immediately after a severe lengthening contraction protocol. Note that some sarcomeres have actually shortened down to a 1.40 µm length, whereas the weaker sarcomeres have been damaged severely through a stretch out to a 3.80 µm that has displaced the thick filament to one end of the sarcomere or the other. This segment of this fiber will undergo the degenerative and regenerative stages shown in Fig. 6. This photomicrograph depicts a part of a single fiber in Stage 2. (**b**) A longitudinal section of a myofiber 10 min after a lengthening contraction protocol showing areas of focal damage (*) within single or small groups of sarcomeres. In some sarcomeres, the damage appears to be in the A-band region, with Z-lines remaining intact, whereas in other sarcomeres the damage involves the Z-lines. (**c**) Transverse sections of a muscle 3 days after the protocol. Muscle fibers range from those with intact myofibrils (M3 and M4) to those with degenerating myofibrils (M1) or devoid of cytosolic constituents (M2). Fiber M2 has phagocytes (P) within the basement membrane (arrows). C is a capillary (Figure 2b and c reproduced from Faulkner et al. 1995 with permission of Oxford University Press)

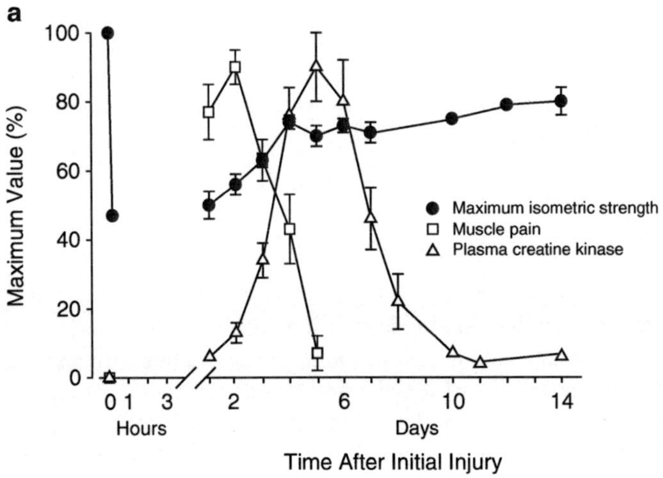

Fig. 3 Data are given for several indices of contraction-induced injury measured prior to and at selected time periods following a protocol of lengthening contractions administered to (**a**) the elbow flexor muscles of human beings and (**b**) the ankle dorsiflexor muscles of mice. The values indicated on the abscissa are the times in "hours" and "days" after the initiation of the contraction protocols. (**a**) Eight human subjects (age 24–43 years) performed maximal lengthening contractions of the elbow flexor muscles once every 15 s for 20 min. (**b**) The dorsiflexor muscle group of mice was exposed to a maximal lengthening contraction every 5 s for 30 min during plantar flexion of the ankle with the foot in a "shoe" apparatus. Data are shown for the maximum isometric forces developed by the tibialis anterior (TBA) and extensor digitorum longus (EDL) muscles measured in vitro following the injury protocol (n = 4–9 for each data point). All values are expressed as percentages of the maximum value for each variable. For isometric strength and maximum isometric force, the maximum values were achieved by all subjects prior to the exercise and are taken as 100%. For muscle pain and plasma creatine kinase, each subject did not reach his or her maximum values on the same day. Therefore, the peak values for these variables do not correspond to 100%. Values are given as means ± standard errors. When no error bars are shown, they are contained within the symbol (Modified from data in Newham et al. 1987; Faulkner et al. 1989; with permission. Reprinted from Faulkner et al. 1993; with permission of the American Physical Therapy Association. This material is copyrighted, and any further reproduction or distribution is prohibited)

soreness, although he had no direct evidence for this. Later needle biopsy studies of humans definitively demonstrated ultrastructual disruptions within muscle fibers associated with late-onset muscle soreness.

5 The Cause of the Contraction-Induced Injury

The concept of a contraction-induced injury that occurred only when skeletal muscle fibers were activated to produce high forces and then stretched was slow to evolve. Early investigations of lengthening contractions focused primarily on the absorption of the work done on the muscle and the 'heat of lengthening' (Abbott et al. 1951). A major advance occurred in the understanding of the physiological cost of positive and negative work with the Abbott et al. (1952) study utilizing the modified bicycle-ergometer that enabled both positive and negative work to be performed. Knuttgen and his colleagues (Knuttgen and Saltin 1972; Knuttgen et al. 1982) also modified a bicycle ergometer to enable subjects to pedal against the load and perform lengthening contractions with either the arms or the legs. The fourfold difference observed between the energy cost during the shortening compared with the lengthening contractions is rather amazing (Fig. 4) and the complex physiological implications of this difference in energy cost are still not understood. The focus of the research on lengthening contractions gradually shifted to the effects of the lengthening contraction protocols on muscle pain and damage. The prevailing view initially was that as long as a given protocol of contractions was sufficiently intense, select populations of fibers would be injured. Armstrong (1990) expressed this view at a Symposium on Muscle Injuries, when he wrote that "muscular

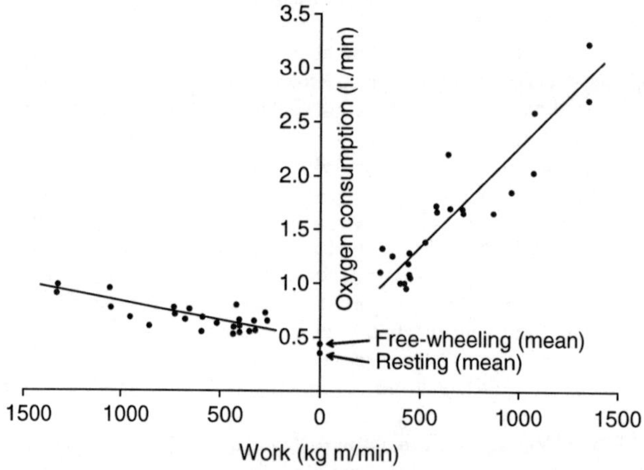

Fig. 4 Variation in the rate of oxygen consumption with the rate of work in pedaling for both positive and negative work (Reproduced from Abbott et al. 1952 with permission of Wiley)

exercise commonly results in injury to fibers in active muscles, particularly when the exercise is relatively intense, is of long duration, and/or includes lengthening contractions".

The hypothesis that "eccentric" exercise (exercise that involves lengthening contractions of muscles) preferentially damages fibers (Newham et al. 1987) was explored using comparable protocols of lengthening, shortening and isometric contractions of isolated muscles of mice (McCully and Faulkner 1985). With experiments on *in situ* single muscles of mice or rats (McCully and Faulkner 1985; Brooks and Faulkner 1990; Brooks et al. 1995) or single permeabilized fibers obtained from muscles of mice or rats (Macpherson et al. 1996; Brooks and Faulkner 1996; Lynch et al. 2008), precise protocols of lengthening contractions were designed to investigate the underlying mechanisms responsible for the injury associated with lengthening contractions. Such experiments demonstrated conclusively that injury was only observed following lengthening contractions regardless of the intensity of the shortening or isometric contraction protocol (McCully and Faulkner 1985). Furthermore, the magnitude of the injury induced by a given protocol of lengthening contractions was found to be a function of the force developed during the lengthening contraction, the magnitude of the stretches imposed, and the number of repetitions of the lengthening contractions in a given protocol (Brooks et al. 1995; Lynch et al. 2008; McCully and Faulkner 1986). Contraction-induced injury is thus most likely to occur during activities that involve a severe lengthening of a maximally activated muscle, such as lowering a very heavy object, or with multiple lengthening contractions of smaller groups of motor units as in distance running (Komi 2000). Running at relatively high speed, even on the level, involves stretching of the quadriceps muscles on the landing (Komi 2000), and running faster or longer distances than a runner is accustomed to may result in contraction-induced injury to fibers in the muscles involved. In any given activity, untrained participants are much more likely to experience a contraction-induced injury than trained subjects. Despite the protection provided by training, even trained athletes may sustain a contraction-induced injury during transition periods when training loads or work-outs are increased or modified.

After single lengthening contractions (Brooks and Faulkner 1990; Li et al. 2006) or a protocol of many lengthening contractions (McCully and Faulkner 1986), the severity of a contraction-induced injury is most accurately assessed by the deficit in force generation (Fig. 5). An immediate force deficit occurs when a maximally activated fast skeletal muscle fiber of a rat is stretched through a single 20% strain (Macpherson et al. 1996; Lynch and Faulkner 1998; Panchangam et al. 2008) or an *in situ* skeletal muscle is stimulated maximally and stretched through a 20% strain for three 5-min contraction periods separated by 5 min (McCully and Faulkner 1985). The single 20% lengthening contraction of the single fiber produced a 17% force deficit in fast fibers of rats (Macpherson et al. 1996; Panchangam et al. 2008), whereas the 450 lengthening contractions of extensor digitorum longus muscles of the mice produced a 60% force deficit immediately afterward (McCully and Faulkner 1985). Force deficits invariably cause a more severe initial injury in muscles of old compared with young or adult animals. When activated maximally and exposed to

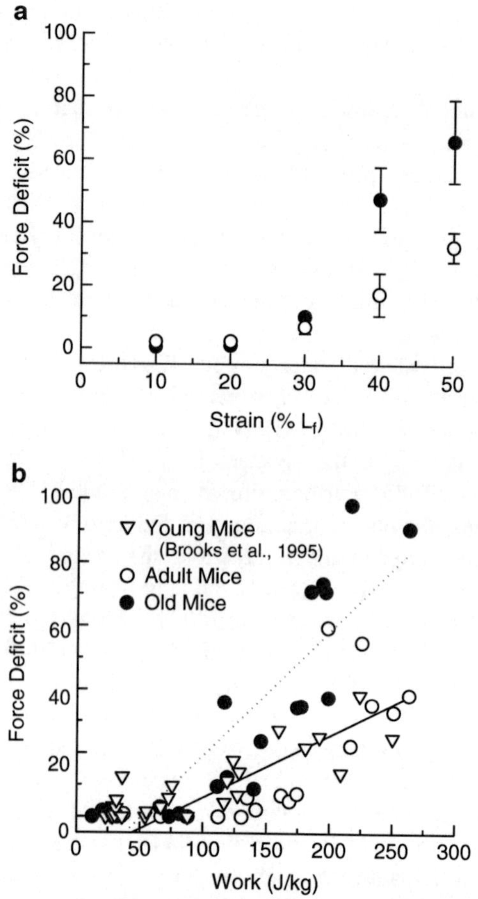

Fig. 5 The force deficits following single stretches of maximally activated muscles.Data are presented for single stretches varying in magnitude but not velocity($V = 2\ L_f\ s^{-1}$) for pooled young and adult mice (•) and old mice (•) in (**a**) and in situ EDL muscles of young (∇), adult (◦) and old (◦) mice in (**b**). The work input during the stretch is normalized by muscle wet mass (J kg $^{-1}$), strain is expressed as a percentage of optimum fiber length (L_f), and the force deficit observed 1 min after the stretch is expressed as a percentage of the isometric force developed just prior to the stretch. Each symbol in (**b**) indicates a data point from a single stretch. The coefficients of determination for the regression relationships for data from adult mice (*continuous line*) and old mice (*dashed line*) are 0.59 and 0.77, respectively. The slopes of the relationships, 0.20 for muscles in adult mice and 0.39 for muscles in old mice, are significantly different. Data for young mice ($r^2 = 0.73$; slope – 0.13) are reproduced from Brooks et al. 1995. Data in (**a**) are presented as means ± S.E.M. Sample size is from 3 to 12 for each point. *Significant difference (P <– .05) in the mean force deficits between the two groups (Reprinted from Brooks and Faulkner 1995)

a single stretch through 30% of fiber length, a small 8–10% force deficit was observed for *in situ* extensor digitorum longus (EDL) muscles of young and old mice, but 40% and 50% strains produced large force deficits with the muscles of the old experiencing twofold greater force deficits than those of the young and adult mice (Fig. 5a and b). For single permeabilized fibers from fast muscles of rats, force deficits immediately after single strains of 10% or greater were approximately twofold larger for single fibers from muscles of old compared with those from adult animals (Brooks and Faulkner 1996; Lynch et al. 2008). In combination, the whole muscle and single fiber experiments indicate a greater susceptibility of muscles in old animals to injury that is due at least in part to a mechanically compromised sarcomeric structure that is less able to withstand stretch.

6 Progression of the Injury

The severity of the contraction-induced injury is a direct function of how severely single fibers are injured and how many fibers are injured sufficiently to initiate the cascade of events associated with a secondary injury. This cascade of events involves phases of contraction-induced injury to skeletal muscles that can be broadly categorized as: (1) the initial lengthening contraction that triggers the injury; (2) an autogenic stage that includes degradation by proteolytic and lipolytic systems indigenous to the fibers, (3) a phagocytic stage from 4 to 6 h through 2–4 days including an inflammatory response, and (4) a regenerative stage beginning at 4–6 days and extending to 10–14 days depending on the severity of the injury (for review see Tidball 1995). These four phases match well with the seven phases depicted in Fig. 6, with Phases (c) and (d) the phagocytic stage and (e) and (f) depicting the regenerative phase. During lengthening contractions, the actual injury to sarcomeres in a myofibril appears to occur when thick filaments of single sarcomeres are displaced to one end of the sarcomere and some or all of the filaments fail to interdigitate properly within the myofibril when the sarcomere attempts to return to its resting length (Fig. 2a). Usually the injury occurs to a highly localized cluster of sarcomeres within a single fiber. Damage to the muscle fiber compromises the fiber's ability to maintain proper calcium homeostasis. The prolonged increase in intracellular calcium levels in damaged muscle fibers activates the m-calpain protease system. M-calpain and related proteases perform the initial disassembly of damaged myofibrils (Jackman and Kandarian 2004). Once the sarcomere has been disassembled, the damaged proteins are broken down into their constitutive amino acids by the ubiquitin-proteasome system. Within a few days following injury, protein synthesis pathways are activated and new sarcomeres are synthesized.

Following severe protocols of lengthening contractions, the large force deficits displayed by muscles from both young and old mice indicate that throughout the cross-sections of individual fibers a substantial number of sarcomeres have been injured and that portions of these fibers will undergo additional degeneration of the total cross-section of the injured fibers (Rader et al. 2006). The additional steps include: a sealing off of the damaged area accompanied by the infiltration of inflammatory cells, phagocytosis of the damaged tissues, and subsequent activation of satellite cells and regeneration of entirely new segment of fiber (Fig. 6). Satellite cells are muscle precursor cells that reside between the sarcolemma and the basal lamina in skeletal muscle fibers. Satellite cells normally exist in a quiescent state, but upon injury the satellite cells are activated, migrate to the site of injury, proliferate,

Initial injury
Hypercontraction by several hours
Vesicle clumping by 1 day
Secondary injury at 2 to 5 days
Regeneration begins by 5 days
Myotube by 1 to 2 weeks
Recovered by 1 month

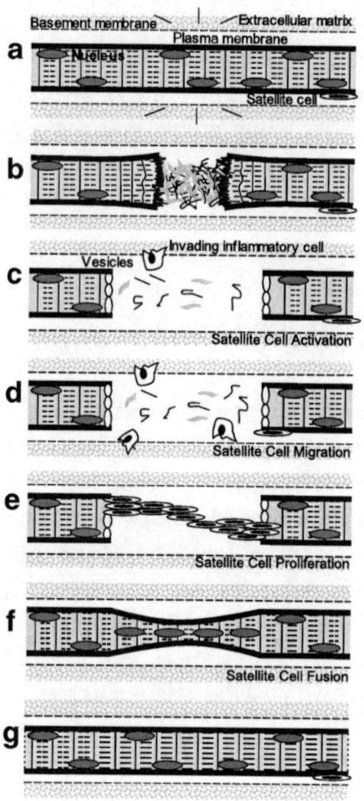

Fig. 6 Schematic diagram of the sequence of events for a typical muscle fiber following a severe LCP. Within several hours following focal injury, the plasma membrane is damaged, an influx of calcium activates proteases intrinsic to the muscle fiber, and myofibrils hypercontract, resulting in a zone of necrosis. The freely permeable basement membrane remains intact. By 1 day, the hypercontracted myofibrils degenerate while vesicles accumulate to seal off the viable portions from the necrotic segments of the fiber. Neutrophils infiltrate at this time. Between 2 and 5 days, macrophages infiltrate, releasing more cytotoxic substances such as ROS that break down damaged tissue further, as well as previously uninjured tissue, resulting in a secondary injury. Satellite cells migrate to the site of injury. At 5–30 days, satellite cells proliferate and fuse across the necrotic segment so that recovery takes place (Reproduced with modifications based on a previously published figure (Bischoff 1994) with permission of the McGraw-Hill Companies. Figure also published in Rader et al. 2006 with permission Wiley)

and fuse with the damaged fiber to replace the nuclei lost as a result of the injury. Mechanical disruption of the endomysium causes the release of inactive hepatocyte growth factor (HGF) (Tatsumi and Allen 2004). The HGF is activated within the injured tissue (Tatsumi et al. 2006) and binds to the c-met receptor on the plasma membrane of the resident satellite cells, which are thus activated from their quiescent state and migrate to the site of injury.

As satellite cells migrate to the site of injury, they also undergo several rounds of proliferation. The initial proliferation of satellite cells is brought about by an increase in the expression of the basic helix-loop-helix (bHLH) transcription factor MyoD. MyoD is one of four members of myogenic regulatory factor (MRF) family that also include Myf-5, myogenin and MRF-4. The MRFs induce the "myogenic program" in these proliferating satellite cells, causing the cells to begin to express skeletal muscle contractile proteins. Once in proximity of the damaged region of the muscle fiber, satellite cells fuse with each other to form multinucleated structures called myotubes. Myotubes fuse with the damaged muscle fiber and restore the nuclei lost after the initial injury. Some proportion of the satellite cells that underwent proliferation do not form myotubes, but instead resume a sub-basal lamina position, return to the quiescent state, and repopulate the satellite cell pool.

In addition to satellite cells, fibroblasts and inflammatory cells are attracted to the site of injury within the muscle. These cells assist in the removal of cellular debris and in the repair of the extracellular matrix (ECM). If there is a severe disruption of the ECM, fibroblasts respond with an overproduction of ECM resulting in the clinical condition of fibrosis, or scar tissue accumulation (Huard et al. 2002). The prevention of scar tissue accumulation is an important goal in the initial treatment of muscle injuries, as this scar tissue is disruptive to the normal function of muscle tissue and, once formed, is relatively permanent (Järvinen et al. 2005). Clear evidence shows that recovery from contraction-induced injury is impaired in muscles of old compared with adult animals (Brooks and Faulkner 1990; McArdle et al. 2004), but the basis for the regeneration defects remain an active area of investigation (Carlson et al. 2009; Conboy et al. 2003). Moreover, the impaired regenerative potential of skeletal muscle in old animals is associated with an increase in tissue fibrosis (Brack et al. 2007).

7 Contribution of Lateral Transmission of Force to Contraction-Induced Injury

A contraction-induced injury to a muscle fiber occurs when a segment, or segments, within the fiber contains groups of sarcomeres that are weaker than the sarcomeres in series with them (Fig. 6). The weaker sarcomeres normally receive lateral support from the adjacent sarcomeres in the myofibrils surrounding them through intermediate filament proteins, including desmin, located at the z-discs (Fig. 7a). The desmin anchors each of the z-discs of a myofibril to the z-lines of each of the surrounding myofibrils so that the force generated by each myofibril is transmitted laterally, providing stability for all of the myofibrils within a fiber. For the myofibrils that are immediately adjacent to the sarcolemma of a fiber, the z-discs are anchored into the sarcolemma by costameres (Fig. 7a). The costameres (Fig. 7) include the dystrophin-associated glycoprotein (DAG) complex, a portion of which extends into the ECM. The DAG appears to be situated in a position suitable for the transmission of the force laterally through the sarcolemma into the ECM. The lateral transmission of force continues without decrement through the intermediate filaments at each z-disc from myofibril to myofibril throughout the muscle fiber (Fig. 7b) and then through costameres from fiber to fiber throughout the muscle. This concept is supported by the successful demonstration of the lateral transmission of force from a maximally activated single fiber partially dissected free in a

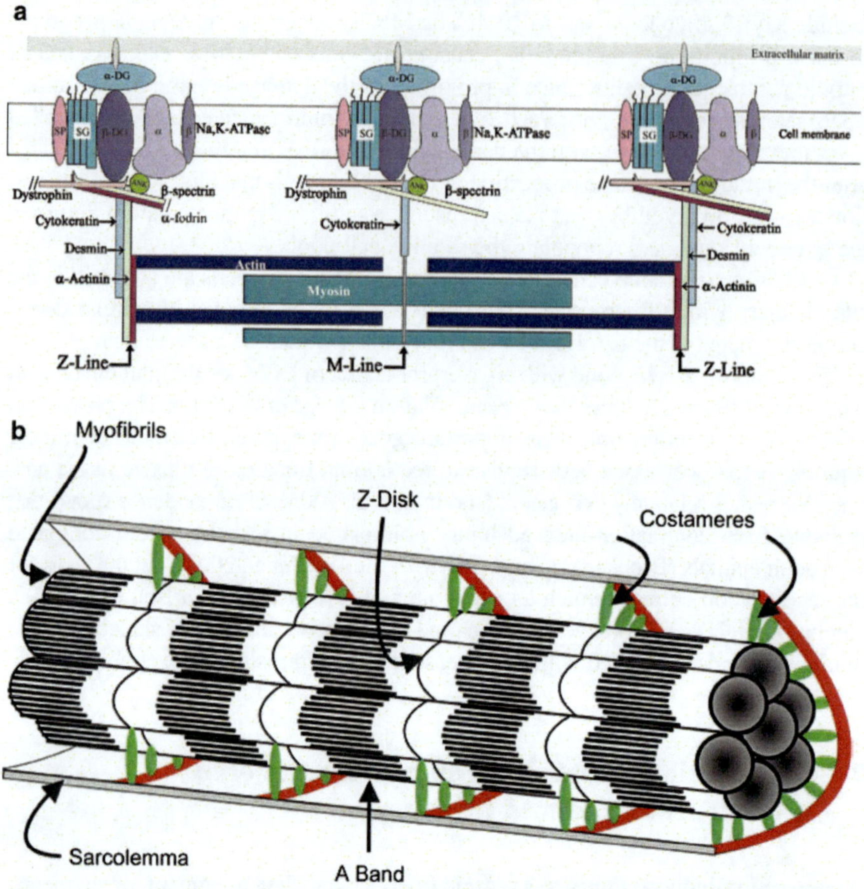

Fig. 7 **(a)** Model of the sarcolemmal membrane skeleton and its relationship to desmin and cytokeratin. This figure depicts a model of the organization of the muscle cell surface, from the extracellular space to the contractile apparatus. The membrane skeletal and intermediate filament proteins that we have studied at costameres are emphasized, whereas many proteins known to be at or near the sarcolemma or in the contractile structures have been omitted for clarity. Longitudinal domains, which are similar in composition to M line domains, and intercostameric regions are not illustrated. The only extracellular protein depicted is α-dystroglycan (α-DG). Integral proteins of the sarcolemma shown are the α and β chains of the Na,K-ATPase, β-dystropglycan (β-DG), sarcoglycans (SG), and sarcospan (SP). The membrane skeletal proteins illustrated are ankyrin 3 (Ank), dystrophin, αII-spectrin (α-fodrin), βIΣ2-spectrin (β-spectrin). Sarcomeric proteins shown are actin, myosin and α-actinin. Our results suggest that two sets of intermediate filaments connect the contractile apparatus to the costameres at the sarcolemma: desmin, which links the Z disks to the Z line domains of costameres, and cytokeratin, which links the contractile apparatus to all three costameric domains. Cytokeratin filaments were referred to as "connectors" in an earlier version of this cartoon (Williams et al. 2001). Not drawn to scale (Reprinted with permission) **(b)** Cellular location of costameres in striated muscle. Shown is a schematic diagram illustrating costameres as circumferential elements that physically couple peripheral myofibrils to the sarcolemma in periodic register with the Z-disk (Reprinted with permission Ervasti 2003. The American Society for Biochemistry and Molecular Biology)

frog semitendinous muscle to the epimysium of the muscle (Street 1983). The assumption is that the same process functions effectively in mammalian skeletal muscles (Patel and Lieber 1997; Monti et al. 1999).

The lateral transmission of force is absolutely vital to the stability of myofibers within a maximally-activated skeletal muscle, even during isometric contractions (Claflin and Brooks 2008), or of myofibrils within muscle fibers (Panchangam et al. 2008). The necessity for the lateral transmission of force within a maximally activated muscle fiber is that all the sarcomeres do not generate the same force while contracting (Panchangam et al. 2008). Stronger sarcomeres surrounding weaker sarcomeres laterally are able to provide some support by the balancing out of force through lateral transmission of force around the weaker sarcomeres during isometric or shortening contractions and even during short stretches of activated muscles. As with the myofibrils within a single fiber, when a whole skeletal muscle is activated maximally and fibers contract, all the fibers in the skeletal muscle do not generate exactly the same forces, because fibers vary in cross-sectional area and sarcomeres within the fibers vary in their intrinsic maximum strengths. Throughout a skeletal muscle, any given fiber has five to eight adjacent fibers around it, and each myofibril has about the same variability in lateral contacts with other myofibrils. This structure provides lateral stability for the sarcomeres throughout the myofibrils within a single muscle fiber, as well as for single fibers throughout the whole muscle. Consequently, for most people contraction-induced injuries to skeletal muscle fibers are not a frequent occurrence, but with maximum activation and a large strain, or even with smaller strains during repeated lengthening contractions, the lateral support system may break down. The result is that weaker sarcomeres are stretched excessively, and contraction induced injury occurs. The magnitudes of the force deficits attest to the severity of some contraction-induced injuries, but the magnitude and extent of the contraction-induced injury would be even greater were it not for the highly sophisticated system that has evolved for the lateral transmission of force in skeletal muscles. The extensive contraction-induced injury observed in the lumbrical muscles of dystrophin deficient *mdx* mice during isometric contractions compared with the lack of any sign of injury in the muscles of wild-type mice attests to the effectiveness of the system for the lateral transmission of force in control muscles (Claflin and Brooks 2008).

8 Role of Contraction-Induced Injury in Wasting and Weakness

For young, healthy men and women, even severe contraction-induced injuries are well-tolerated and recovery is fairly rapid and complete. Most athletes with well-defined competitive seasons expect to encounter some degree of discomfort as they transition into a period of more demanding training as their competitive season approaches. The already conditioned athlete is accustomed to regular, heavy training and they handle the transition into an increased training load with a minimum of discomfort. Under these circumstances, a severe contraction-induced injury is not likely to occur and moderate injuries are well-tolerated and rarely even disrupt the training schedule. For the elderly, the musculoskeletal system has been described as the *entry pathway* for the development of frailty (Bortz 2002). The timing of the onset and the rate of progression of frailty in the elderly is governed by both heredity and the degree of habitual physical activity in the life style (Bortz 2002). Immutable changes occur in skeletal muscles of humans that begin at about 50 years of age and initiate linear decreases in both the number of motor units (Campbell et al. 1973; Doherty and Brown 1993) and the number of fibers (Lexell et al. 1988) in skeletal muscles of humans. By age 80, these losses result in decreases of 75% in the number of motor units and 50% in the number of fibers. Due to these immutable changes, the skeletal muscles of the frail elderly are intrinsically weak and consequently highly susceptible to contraction-induced injury. Moreover, the frail elderly are neither accustomed to the rigors of training nor the inconvenience and discomfort that contraction-induced injuries may cause as a conditioning program is introduced into their daily schedule. Even more distressing is the inadvertent and often unexpected, slip, fall, or awkward movement that loads an unused muscle heavily and without preparation. The occurrence of severe injuries, from which the muscles of the elderly person may not recover, can further accelerate the rate of progression of worsening frailty.

9 Measures to Prevent Contraction-Induced Injury

Accepting that the musculoskeletal system constitutes a major/*entry pathway*/ for the development of frailty (Bortz 2002), it also qualifies as a potential/*exit pathway*/ to cure the elderly from the condition of frailty. An increase in daily physical activity that is carefully graded in intensity and highly selective as to the types of exercise can likely induce protective adaptations even in the frail elderly. Although protection from contraction-induced injury is achieved most effectively by training programs that include lengthening contractions through a full-range of motion and with at least a moderate load, contraction-induced injury and regeneration of a muscle are not required to increase resistance to subsequent injuries (Koh and Brooks 2001). Conditioning protocols that involved isometric contractions or even stretching of relaxed muscles provide some degree of protection for subsequent exposures to lengthening contractions protocols that have the potential to induce injuries to muscles in both young (Koh and Brooks 2001) and old (Koh et al. 2003) animals. Lengthening contraction exercises, although of considerable value for the elderly must be implemented with great care and with the involvement of a highly trained exercise leader well-versed in the physical training of frail elderly. Under these circumstances and with great attention to the details as to the intensity and types of physical activities involved, the benefits of exercise programs that involve lengthening contractions can be substantial.

References

Abbott, B. C., Aubert, X. M., Hill, A. V. (1951). The absorption of work by a muscle stretched during a single twitch or a short tetanus. *Proceedings of the Royal Society of London. Series B: Biological Sciences, 139*, 86–104.

Abbott, B. C., Bigland, B., Ritchie, J. M. (1952). The physiological cost of negative work. *Journal de Physiologie, 117*, 380–390.

Armstrong, R. B. (1990). Initial events in exercise-induced muscular injury. *Medicine and Science in Sports and Exercise, 22*, 429–435.

Armstrong, R. B., Ogilvie, R. W., Schwane, J. A. (1983). Eccentric exercise-induced injury to rat skeletal muscle. *Journal of Applied Physiology, 54*, 80–93.

Bischoff, R. (1994). The satellite cell and muscle regeneration. In Myology (eds) Engel, A.G., and Franzini-Armstrong, C. pp. 97-118. McGraw-Hill, New York.

Bortz, W. M. (2002). A conceptual framework of frailty: A review. *The Journals of Gerontology. Series A: Biological Sciences and Medical Sciences, 57*, M283–M288.

Brack, A. S., Conboy, M. J., Roy, S., Lee, M., Kuo, C. J., Keller, C., Rando, T. A. (2007). Increased Wnt signaling during aging alters muscle stem cell fate and increases fibrosis. *Science, 317*, 807–810.

Brooks, S. V. & Faulkner, J. A. (1990). Contraction-induced injury: Recovery of skeletal muscles in young and old mice. *American Journal of Physiology (Cell), 258*, C436–C442.

Brooks, S. V. & Faulkner, J. A. (1996). The magnitude of the initial injury induced by stretches of maximally activated muscle fibres of mice and rats increases in old age. *Journal of Physiology (London), 497*(Pt 2), 573–580.

Brooks, S. V., Zerba, E., Faulkner, J. A. (1995). Injury to muscle fibres after single stretches of passive and maximally stimulated muscles in mice. *Journal of Physiology (London), 488*(Pt 2), 459–469.

Campbell, M. J., McComas, A. J., Petito, F. (1973). Physiological changes in ageing muscles. *Journal of Neurology, Neurosurgery and Psychiatry, 36*, 174–182.

Carlson, M. E., Conboy, M. J., Hsu, M., Barchas, L., Jeong, J., Agrawal, A., Mikels, A. J., Agrawal, S., Schaffer, D. V., Conboy, I. M. (2009). Relative roles of TGF-beta1 and Wnt in the systemic regulation and aging of satellite cell responses. *Aging Cell, 8*, 676–689.

Cheung, K., Hume, P., Maxwell, L. (2003). Delayed onset muscle soreness: Treatment strategies and performance factors. *Sports Medicine, 33*, 145–164.

Claflin, D. R. & Brooks, S. V. (2008). Direct observation of failing fibers in muscles of dystrophic mice provides mechanistic insight into muscular dystrophy. *American Journal of Physiology (Cell), 294*, C651–C658.

Conboy, I. M., Conboy, M. J., Smythe, G. M., Rando, T. A. (2003). Notch-mediated restoration of regenerative potential to aged muscle. *Science, 302*, 1575–1577.

Ervasti, J. M. (2003). Costameres: the Achilles' heel of Herculean muscle. *J Biol Chem, 278*, 13591-13594.

Doherty, T. J. & Brown, W. F. (1993). The estimated numbers and relative sizes of thenar motor units as selected by multiple point stimulation in young and older adults. *Muscle & Nerve, 16*, 355–366.

Faulkner, J. A., Jones, D. A., Round, J. M. (1989). Injury to skeletal muscles of mice by forced lengthening during contractions. *Quarterly Journal of Experimental Physiology, 74*, 661–670.

Faulkner, J. A., Brooks, S. V., Opiteck, J. A. (1993). Injury to skeletal muscle fibers during contractions: conditions of occurrence and prevention. *Phys Ther, 73*, 911–921.

Faulkner, J. A., Brooks, S. V., Zerba, E. (1995). Muscle atrophy and weakness with aging: contraction-induced injury as an underlying mechanism. *J Gerontol A Biol Sci Med Sci*, 50 Spec No:124–129.

Faulkner, J. A., Larkin, L. M., Claflin, D. R., Brooks, S. V. (2007). Age-related changes in the structure and function of skeletal muscles. *Clin Exp Pharmacol Physiol, 34*:1091–1096.

Feinstein, B., Lindegard, B., Nyman, E., Wohlfart, G. (1955). Morphologic studies of motor units in normal human muscles. *Acta Anatomica Scandinavica (Basel), 23*, 127–142.

Friden, J., Sjostrom, M., Ekblom, B. (1983). Myofibrillar damage following intense eccentric exercise in man. *International Journal of Sports Medicine, 4*, 170–176.

Friden, J., Sfakianos, P. N., Hargens, A. R. (1986). Muscle soreness and intramuscular fluid pressure: Ccomparison between eccentric and concentric load. *Journal of Applied Physiology, 61*, 2175–2179.

Hadley, E. C., Ory, M. G., Suzman, R., Weindruch, R., Fried, L. (1993). Physical frailty: A treatable cause of dependence in old age. *Journal of Gerontology, 48*, 1–88.

Hough, T. (1901). Ergographic studies in neuro-muscular fatigue. *The American Journal of Physiology, 5*, 240–266.

Hough, T. (1902). Ergographic studies in muscular soreness. *The American Journal of Physiology, 7*, 76–92.

Huard, J., Li, Y., Fu, F. H. (2002). Muscle injuries and repair: Current trends in research. *The Journal of Bone and Joint Surgery, 84-A*, 822–832.

Jackman, R. W., Kandarian, S. C. (2004). The molecular basis of skeletal muscle atrophy. *American Journal of Physiology (Cell), 287*, C834–C843.

Järvinen, T. A., Järvinen, T. L., Kääriäinen, M., Kalimo, H., Jarvinen, M. (2005). Muscle injuries: Biology and treatment. *The American Journal of Sports Medicine, 33*, 745–764.

Jones, D. A., Newham, D. J., Round, J. M., Tolfree, S. E. (1986). Experimental human muscle damage: Morphological changes in relation to other indices of damage. *Journal de Physiologie, 375*, 435–448.

Knuttgen, H. G. & Saltin, B. (1972). Muscle metabolites and oxygen uptake in short-term submaximal exercise in man. *Journal of Applied Physiology, 32*, 690–694.

Knuttgen, H. G., Patton, J. F., Vogel, J. A. (1982). An ergometer for concentric and eccentric muscular exercise. *Journal of Applied Physiology, 53*, 784–788.

Koh, T. J. & Brooks, S. V. (2001). Lengthening contractions are not required to induce protection from contraction-induced muscle injury. *American Journal of Physiology: Regulatory, Integrative and Comparative Physiology, 281*, R155–R161.

Koh, T. J., Peterson, J. M., Pizza, F. X., Brooks, S. V. (2003). Passive stretches protect skeletal muscle of adult and old mice from lengthening contraction-induced injury. *The Journals of Gerontology. Series A: Biological Sciences and Medical Sciences, 58*, 592–597.

Komi, P. V. (2000). Stretch-shortening cycle: A powerful model to study normal and fatigued muscle. *Journal of Biomechanics, 33*, 1197–1206.

Lexell, J., Taylor, C. C., Sjostrom, M. (1988). What is the cause of the ageing atrophy? Total number, size and proportion of different fiber types studied in whole vastus lateralis muscle from 15- to 83-year-old men. *Journal of the Neurological Sciences, 84*, 275–294.

Li, S., Kimura, E., Ng, R., Fall, B. M., Meuse, L., Reyes, M., Faulkner, J. A., Chamberlain, J. S. (2006). A highly functional mini-dystrophin/GFP fusion gene for cell and gene therapy studies of Duchenne muscular dystrophy. *Human Molecular Genetics, 15*, 1610–1622.

Lynch, G. S., Faulkner, J. A. (1998). Contraction-induced injury to single muscle fibers: velocity of stretch does not influence the force deficit. *American Journal of Physiology, 275*, C1548-C1554.

Lynch, G. S., Faulkner, J. A., Brooks, S. V. (2008). Force deficits and breakage rates after single lengthening contractions of single fast fibers from unconditioned and conditioned muscles of young and old rats. *American Journal of Physiology (Cell), 295*, C249–C256.

Macpherson, P. C., Schork, M. A., Faulkner, J. A. (1996). Contraction-induced injury to single fiber segments from fast and slow muscles of rats by single stretches. *American Journal of Physiology (Cell), 271*, C1438–C1446.

McArdle, A., Dillmann, W. H., Mestril, R., Faulkner, J. A., Jackson, M. J. (2004). Overexpression of HSP70 in mouse skeletal muscle protects against muscle damage and age-related muscle dysfunction. *The FASEB Journal, 18*, 355–357.

McCully, K. K. & Faulkner, J. A. (1985). Injury to skeletal muscle fibers of mice following lengthening contractions. *Journal of Applied Physiology, 59*, 119–126.

McCully, K. K. & Faulkner, J. A. (1986). Characteristics of lengthening contractions associated with injury to skeletal muscle fibers. *Journal of Applied Physiology, 61*, 293–299.

Monti, R. J., Roy, R. R., Hodgson, J. A., Edgerton, V. R. (1999). Transmission of forces within mammalian skeletal muscles. *Journal of Biomechanics, 32*, 371–380.

Newham, D. J., Mills, K. R., Quigley, B. M., Edwards, R. H. (1983). Pain and fatigue after concentric and eccentric muscle contractions. *Clinical Science, 64*, 55–62.

Newham, D. J., McPhail, G., Mills, K. R., Edwards, R. H. (1983). Ultrastructural changes after concentric and eccentric contractions of human muscle. *Journal of the Neurological Sciences, 61*, 109–122.

Newham, D. J., Jones, D. A., Clarkson, P. M. (1987). Repeated high-force eccentric exercise: Effects on muscle pain and damage. *Journal of Applied Physiology, 63*, 1381–1386.

Panchangam, A., Claflin, D. R., Palmer, M. L., Faulkner, J. A. (2008). Magnitude of sarcomere extension correlates with initial sarcomere length during lengthening of activated single fibers from soleus muscle of rats. *Biophysical Journal, 95*, 1890–1901.

Patel, T. J. & Lieber, R. L. (1997). Force transmission in skeletal muscle: From actomyosin to external tendons. *Exercise and Sport Sciences Reviews, 25*, 321–363.

Rader, E. P., Song, W., Van Remmen, H., Richardson, A., Faulkner, J. A. (2006). Raising the antioxidant levels within mouse muscle fibres does not affect contraction-induced injury. *Experimental Physiology, 91*, 781–789.

Schwane, J. A. & Armstrong, R. B. (1983). Effect of training on skeletal muscle injury from downhill running in rats. *Journal of Applied Physiology, 55*, 969–975.

Street, S. F. (1983). Lateral transmission of tension in frog myofibers: A myofibrillar network and transverse cytoskeletal connections are possible transmitters. *Journal of Cellular Physiology, 114*, 346–364.

Tatsumi, R. & Allen, R. E. (2004). Active hepatocyte growth factor is present in skeletal muscle extracellular matrix. *Muscle & Nerve, 30*, 654–658.

Tatsumi, R., Liu, X., Pulido, A., Morales, M., Sakata, T., Dial, S., Hattori, A., Ikeuchi, Y., Allen, R. E. (2006). Satellite cell activation in stretched skeletal muscle and the role of nitric oxide and hepatocyte growth factor. *American Journal of Physiology (Cell), 290*, C1487–C1494.

Tidball, J. G. (1995). Inflammatory cell response to acute muscle injury. *Medicine and Science in Sports and Exercise, 27*, 1022–1032.

Williams, M. W., Resneck, W. G., Kaysser, T., Ursitti, J. A., Birkenmeier, C. S., Barker, J. E., Bloch, R. J. (2001). Na,K-ATPase in skeletal muscle: two populations of beta-spectrin control localization in the sarcolemma but not partitioning between the sarcolemma and the transverse tubules. *Journal of Cell Science, 114*, 751–762.

Role of IGF-1 in Age-Related Loss of Skeletal Muscle Mass and Function

Chris D. McMahon, Thea Shavlakadze, and Miranda D. Grounds

Abstract While insulin-like growth factor-1 (IGF-1) is closely involved in the growth, hypertrophy and maintenance of skeletal muscle mass, the role of IGF-1 in age-related muscle wasting (sarcopenia) is unclear: this is the focus of the present discussion. The complexity of the IGF-1 system that involves different IGF-1 isoforms, binding proteins and receptors, with modulation of systemic IGF-1 levels by growth hormone (GH) is first outlined. The classic IGF-1 signalling pathways in skeletal muscle with a focus on the central role of Akt in protein synthesis and degradation are presented and various conditions that can impair IGF-1 signalling are discussed with respect to inflammation (TNF), oxidative stress (ROS) and lipids. Complex interactions between other factors that influence the age-related decrease in IGF-1 activity are addressed, including GH, nutrition, caloric restriction, Klotho and Vitamin D. Finally, the potential for therapeutic interventions for sarcopenia related to IGF-1 signalling is considered. The big questions are '*to what extent does IGF-1 contribute to sarcopenia*' and '*can elevated IGF-1 prevent or reverse sarcopenia?*

Keywords Insulin like growth factor-1 (IGF-1) • Growth hormone • Skeletal muscle wasting • Muscle atrophy • Sarcopenia

---

M.D. Grounds (✉) and T. Shavlakadze
School of Anatomy & Human Biology, The University of Western Australia, Nedlands, WA, Australia 6009
e-mail: mgrounds@anhb.uwa.edu.au; tshavlakadze@anhb.uwa.edu.au

C.D. McMahon
AgResearch Limited, Ruakura Research Centre, Hamilton, New Zealand
e-mail: chris.mcmahon@agresearch.co.nz

G.S. Lynch (ed.), *Sarcopenia – Age-Related Muscle Wasting and Weakness*,
DOI 10.1007/978-90-481-9713-2_17,

1 Introduction

Insulin-like growth factor -1 (IGF-1) is, as the name implies, similar to insulin in its structure and some of its functions. For example, both IGF-1 and insulin can bind with different affinities to their respective receptors, and both similarly activate signalling pathways such as that mediated by Akt/mTOR. A key difference appears to be the distinct roles that insulin and IGF-1 play at different stages of life. IGF-1 is crucial for muscle formation and growth during embryogenesis and post-natal development, whereas insulin is more important for metabolism in the post-natal and adult states. In skeletal muscle, IGF-1 is closely involved in muscle growth, hypertrophy and maintenance of muscle mass (Fig. 1); however, the role of IGF-1 in age-related muscle wasting is unclear and is the focus of the present discussion. There are two isoforms of the insulin receptor A and B which vary by tissue and stage of development. Type A is more prevalent in developing tissue, and has a high affinity for IGF-2 as well as insulin. Activation of insulin receptor A by insulin leads primarily to metabolic effects, whereas its activation by IGF-2 leads primarily to mitogenic effects (Frasca et al. 1999). IGF-2 is expressed at high levels during fetal development in all species and is an important factor in overall growth regulation, acting through the type 1 IGF-1R and insulin receptor A. Indeed, the birth phenotype of IGF-2 knockout mice is more severe than for IGF-1R knockout mice (Accili et al. 1999; Dikkes et al. 2007). In rodents, IGF-2 is down-regulated at birth and has a small post-natal role; however, in humans IGF-2 expression is sustained throughout life and is believed to have important metabolic and anabolic functions. It is important to consider such species differences when extrapolating

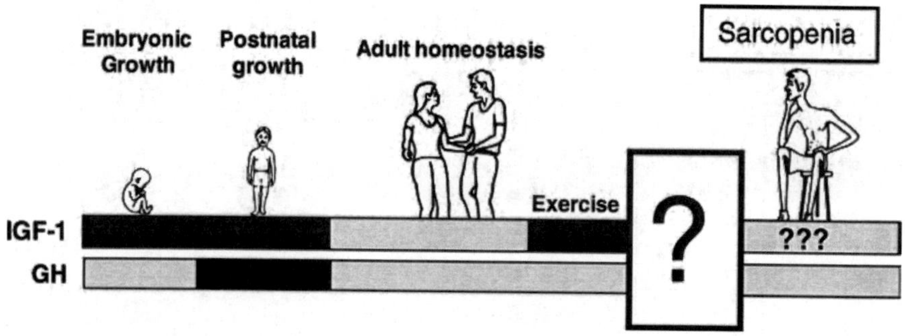

Fig. 1 Simplistic representation to indicate the relative importance of IGF-1 and growth hormone (GH) for maintenance of skeletal muscle mass throughout life. It is considered that IGF-1 is essential for normal skeletal muscle development and growth during embryogenesis and in post-natal life. When muscle mass reaches homeostasis in adults the role of IGF-1 decreases, although it is required for muscle maintenance and is important for increasing muscle mass and protein content during hypertrophy in response to loading/exercise. Growth hormone is especially important for postnatal growth and also regulates IGF-1 levels. The roles of IGF-1 and GH during muscle wasting with ageing (sarcopenia) remain to be fully defined. Dark bars indicate the relatively high importance for regulating muscle mass and the light bars indicate relatively low importance (Adapted from Shavlakadze and Grounds 2006)

rodent and other experimental data to the human condition. This review will focus on IGF-1.

The Chapter starts by introducing IGF-1, its isoforms, receptors and binding proteins, importance during growth and regulation by growth. The classic IGF-1 signalling is outlined and consequences of impaired IGF-1/insulin signalling related to diabetes, obesity and ageing are discussed. The focus then shifts to age-related muscle wasting (sarcopenia) and factors that may contribute to this. A wealth of information from animal studies related to modulation of levels of IGF-1 and related moleculesis presented. The impact of exercise and various therapies and molecular interventions (often involving IGF-1) that have been shown in animal models to slow sarcopenia are then critically discussed with respect to realistic applications to the human condition.

2 Complexity of the IGF-1 System and Importance in Skeletal Muscle

2.1 IGF-1 Isoforms and IGF-1 Availability in Muscle and Blood

IGF-1 plays a central role in skeletal muscle hypertrophy and atrophy (Grounds 2002) via promotion of protein synthesis and inhibition of protein degradation (Shavlakadze and Grounds 2006) and this protein balance is of critical importance for muscle wasting in ageing (sarcopenia), in inflammatory disorders (cachexia), denervation, disuse atrophy and also in the metabolic syndrome (Shavlakadze and Grounds 2006). The IGF-1 gene can be spliced in different ways to produce at least six mRNA isoforms although the specific biological function of these different isoforms of IGF-1 are still unknown (Winn et al. 2002; Shavlakadze et al. 2005b). The mechanisms by which these transcripts might exert different effects are unclear, since ultimately all are processed to produce the same 70 amino acid mature IGF-1 peptide (Fig. 2). While these various isoforms may exert distinct functions, another possibility is that transcription of these various isoforms may instead present the possibility for tissue specific regulation of IGF-1 expression. Available data from transgenic mice over-expressing the various isoforms only in skeletal muscle, indicate that the Ea isoforms (both Class 1 or Class 2) have hypertrophic effects in situations of growth (Shavlakadze et al., unpublished data), whereas the Eb isoform (in rodents and termed Ec in humans), also known as mechano-growth factor (MGF) may instead have early mitogenic and protective effects because mRNA is acutely increased and precedes an increase of IGF-IEa mRNA after injury to skeletal muscle (Yang and Goldspink 2002; Hill and Goldspink 2003). While transgenic studies are a powerful tool, it should be emphasised that this forced artificial over-expression may not accurately reflect the native in vivo situation, since different isoforms may instead normally be transcribed by tissues other than skeletal muscle. For example the Class 2 isoforms are expressed mainly by liver, whereas skeletal

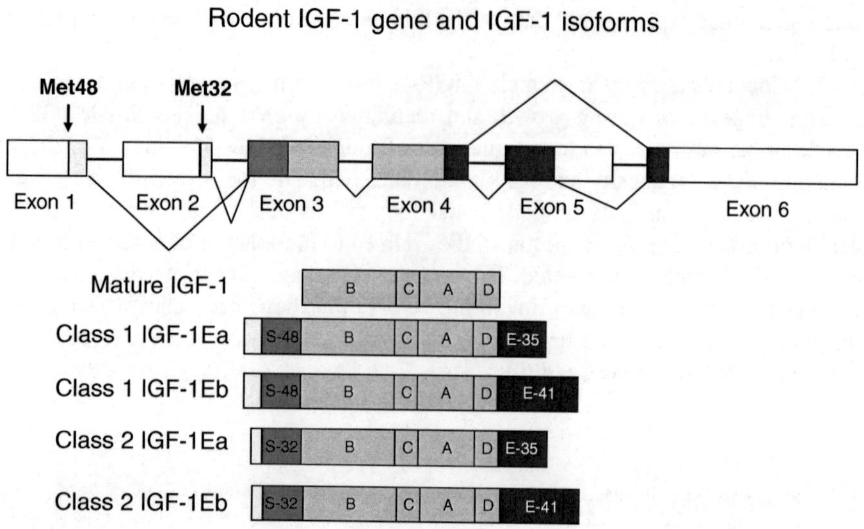

Fig. 2 Rodent IGF-1 gene and IGF-1 isoforms. Simplified diagram to indicate how different isoforms (only four are shown) result from alternative use of transcription start sites in Exon 1 (Class 1) or Exon 2 (Class 2) and from alternative splicing. Exons 3 and Exon 4 code for the 70 amino acids of the mature IGF-1 peptide. Exon 4 also contains code corresponding to the amino-terminal portion of the E-domain and Exon 5 and Exon 6 each encode distinct E extension peptides, termination codons and 3′-untranslated regions. Alternative splicing of the exons 4, 5 and 6 yields Ea and Eb variants. Subsequently these isoforms are cleaved to produce the mature functional 70 amino acid IGF-1 peptide. The significance of the initiation and signalling peptides of these IGF-1 isoforms remains to be fully defined (Based upon Shavlakadze et al. 2005a, b)

muscle expresses Class 1 isoforms (Shemer et al. 1992). Both sources of IGF-1 contribute to growth (see Section 2.2). A detailed critique of different genetically modified mouse models to investigate IGF-1 function lies beyond the scope of this Chapter but is reviewed elsewhere (Shavlakadze et al. 2005b; Le Roith et al. 2001b). This intriguing aspect of the potential roles of the different IGF-1 transcript isoforms awaits further clarification.

Meanwhile, the main focus of biological function is on the mature IGF-1 protein. Extracellular IGF-1 protein is sequestered and stabilised by binding to IGF-1 binding proteins (IGFBPs): the IGFBPs maintain control of IGF-1 binding to its receptor. There are six structurally related IGFBPs located in the vascular and interstitial spaces, they are modified by proteases and their tissue specific pattern of expression affects the bioavailability of IGF-1. In skeletal muscle, the most abundant are IGFBP-3 and -5, although -4 and -6 are also present: their availability is also influenced by gender and age (Oliver et al. 2005)

The action of IGF-1 is mediated by IGF-1 binding to specific receptors on the cell surface, especially the type 1 IGF-1 receptor (IGF-1R). Binding of IGF-1 to the receptor alters the configuration of the receptor subunits and brings the two intracellular motifs together to result in auto-phosphorylation (activation) of the receptor. This initiates a complexity of signalling pathways with effects on, not only protein synthesis and degradation resulting in atrophy/hypertrophy (Shavlakadze and Grounds 2006), but also on apoptosis, myoblast proliferation and muscle differentiation (Fig. 4). IGF-1 binds with greatest affinity to IGF-1R it also binds to the insulin receptor with decreased affinity. Thus there can be some redundancy and overlapping function between these molecules in certain situations.

2.2 *Growth Hormone and IGF-1 Activity*

Circulating IGF-1 (mainly produced by the liver) affects muscle, in addition to the locally produced IGF-1 that acts in an autocrine/paracrine way (Fig. 3). The circulating levels of IGF-1 are influenced by various factors including the IGFBPs (especially IGFBP-1 and -2) and Growth Hormone (GH) produced by the anterior

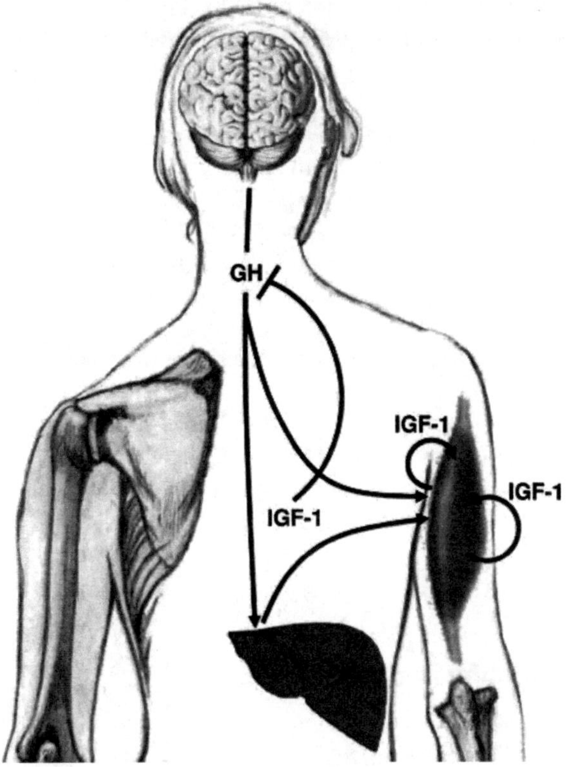

Fig. 3 Overview of systemic regulation of IGF-1 by growth hormone (GH). Growth and maintenance of skeletal muscle mass requires IGF-1 from both circulating (secreted from liver) and locally produced sources. The major regulation is attributed to locally produced IGF-1 acting in an autocrine/paracrine manner. Synthesis and secretion of IGF-1 is regulated by GH, which is secreted from the anterior pituitary gland. IGF-1 is able to regulate secretion of GH in a classical feedback mechanism (See text for details)

pituitary gland. The effects of GH are mediated in large part by IGF-1 in what has been termed the somatomedin hypothesis, although specific actions of GH to induce fusion of myoblasts independently of IGF-1 have been shown (Sotiropoulos et al. 2006). In the revised version of this hypothesis, GH stimulates synthesis and secretion of IGF-1 from the liver, which circulates in blood to downstream targets. In addition, GH stimulates autocrine and paracrine actions of IGF-1 in peripheral tissues, the most likely of which is skeletal muscle (Isgaard et al. 1989; Le Roith et al. 2001a; Kaplan and Cohen 2007). In fact autocrine/paracrine actions of IGF-1 predominate over endocrine originating from liver and this was convincingly demonstrated when liver specific deletion of IGF-1 failed to inhibit growth of mice despite a 75% reduction in concentrations of IGF-1 in blood (Sjogren et al. 1999; Yakar et al. 1999). A more recent study has demonstrated that endocrine derived IGF-1 is important and contributes about 30% to adult body size (Stratikopoulos et al. 2008). It is important to note that in mature organism, there is negative feedback of GH secretion by IGF-I (Giustina and Veldhuis 1998; McMahon et al. 2001) which in turn regulates liver production of IGF-I (Fig. 3).

GH activates the transcription factor Stat5b, but redundancy with Stat5a is also noted (Teglund et al. 1998; Herrington et al. 2000). Global deletion of Stat5b prevents sexually dimorphic growth in mice and a naturally occurring mutation retarded growth in a girl (Udy et al. 1997; Kofoed et al. 2003). The importance of GH acting via autocrine/paracrine stimulation of IGF-1 in skeletal muscle was demonstrated in two elegant studies. Targeted deletion of Stat5a and 5b from skeletal muscle resulted in reduced expression of IGF-1 in skeletal muscle and stunted growth of mice despite normal expression in, and availability of IGF-1 from liver (Klover and Hennighausen 2007). Furthermore, growth of skeletal muscle requires the presence of local IGF-1 and the IGF-1 receptor. When the IGF-1 receptor is absent in skeletal muscle, GH does not stimulate growth of skeletal muscle, despite an increase in circulating concentrations of IGF-1 (Kim et al. 2008).

Overall, it appears that IGF-1 plays a major role in growth of all tissues with both endocrine and paracrine sources of IGF-1 playing vital roles in hypertrophy of skeletal muscle (discussed below). In adult muscles, IGF-1 may play a lesser role in homeostasis of muscle mass. The big questions are '*to what extent does IGF-1 contribute to sarcopenia*' and '*can elevated IGF-1 prevent or reverse sarcopenia*'.

3 IGF-1 Signalling in Skeletal Muscle

3.1 Classic IGF-1 Signalling, with a Focus on Protein Synthesis and Degradation

IGF-1 acts via a transmembrane tyrosine kinase receptor to exert its anabolic effect: it is thought that IGF-1 stimulates muscle growth by promoting myoblast proliferation and their fusion into the myofibres as well as by increasing differentiation and protein accretion in the mature myofibres (Florini et al. 1991, 1996; Engert et al. 1996). Several intracellular signalling pathways mediate the pleiotropic effects of IGF-1. Studies using cultured muscle cells link the mitogenic effect of IGF-1 to the mitogen-activated protein kinase (MAPK) pathway (Coolican et al. 1997) and the anabolic effect of IGF-1 on protein accretion to the PI3K/Akt/mTOR pathway (Rommel et al. 2001). The effects mediated by these pathways in vivo are very complex and inter-connected.

Signalling through the PI3K/Akt pathway plays a fundamental role in controlling skeletal muscle mass and metabolism (Fig. 4). A particular emphasis has been placed on this pathway because it may increase protein synthesis as well as block protein degradation (reviewed in Glass 2005; Shavlakadze and Grounds 2006). Over-expression of constitutively active Akt increases myofibre cross sectional area

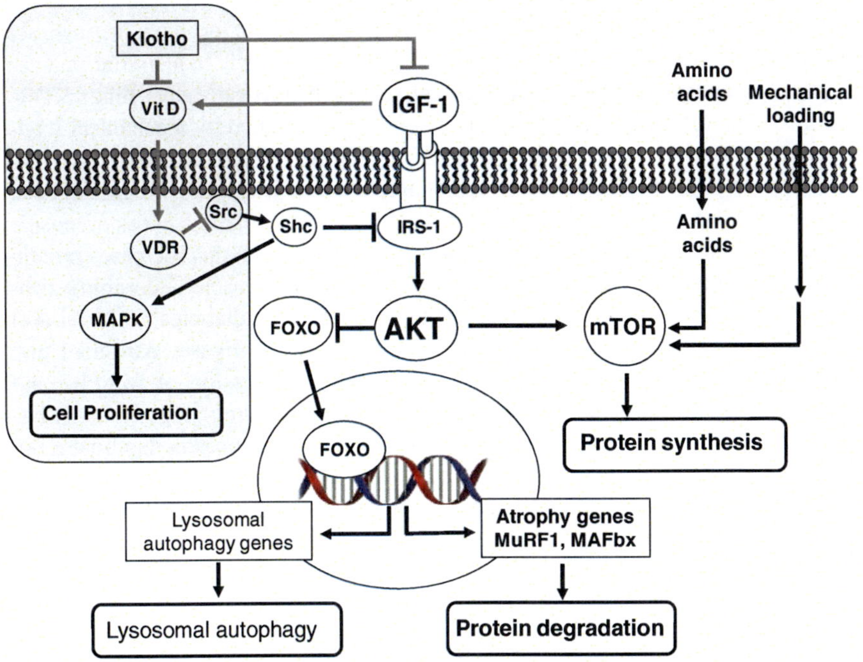

Fig. 4 Key molecules in IGF-1 signalling pathway in skeletal muscle. This highly simplified diagram indicates signalling downstream of the IGF-1 receptor. Akt plays a central role as activation (phosphorylation) results in increased protein synthesis and inhibition of protein degradation; this net signalling leads to muscle growth (hypertrophy). Exercise (loading and stretch) and amino acids (from ingested protein) also increase protein synthesis by direct activation of mTOR signalling. Muscle wasting (atrophy) is not a simple reversal of the Akt/mTOR signalling pathway. Instead, atrophy results from other pathways e.g. TNF-mediated (not shown) that directly activate the atrophy related genes in the nucleus (by mechanisms independent of FOXO) and also inhibit Akt phosphorylation, hence FOXO is not phosphorylated and remains in the nucleus to activate the atrophy related genes (MuRF1 and MAFbx). The insert tentatively indicates interactions of Klotho and vitamin D with IGF-1 signalling (Based in part on Shavlakadze and Grounds (2006) and Arthur et al. (2008))

caused by activation of the protein synthesis pathway (Bodine et al. 2001; Lai et al. 2004). In addition, Akt activation appears to antagonize signalling that leads to muscle atrophy: for example, over-expression of the constitutively active (genetic activation) Akt was sufficient to block muscle wasting following short term (7 days) denervation (Bodine et al. 2001). Not much is known about the regulation of protein synthesis and degradation pathways in old muscle. Some results suggest diminished responsiveness of old muscle to signalling stimuli controlling protein translation, which may determine a limited ability of old muscle to hypertrophy (Thomson and Gordon 2006; Hwee and Bodine 2009). In response to functional over-loading (caused by synergetic muscle ablation), old rat muscles upregulate Akt, however signal transmission to downstream targets involved in protein synthesis machinery is impaired (Hwee and Bodine 2009). Studies in humans demonstrate that protein synthesis rates decrease with age but can be dramatically stimulated by resistance exercise (Yarasheski 2003). The benefits of exercise for the elderly are widely recognised and such mechanical stress (Hornberger et al. 2004) can act downstream of IGF-1 via mTOR to increase protein synthesis in a similar manner to that of amino acids (Fig. 4). The extent to which IGF can boost this mTOR-mediated signalling stimulation of protein synthesis (initiated by mechanical loading or amino acids), especially in the elderly, remains to be determined.

Activation of the PI3K/Akt pathway not only increases protein synthesis, but can also counteract the protein degradation in catabolic states and reduce loss of muscle protein (myofibre atrophy). A common molecular mechanism that increases protein breakdown is revealed by microarray analysis of skeletal muscle undergoing atrophy induced by different factors (e.g. fasting, cancer, acute diabetes, renal failure) and involves induction of the muscle-specific ubiquitin E3-ligases Atrogin-1 and MuRF1, also referred to as atrophy related genes. Expression of MAFbx and MuRF1 is suppressed by activation of the PI3K/Akt signalling (Fig. 4) and it has been extensively shown that this pathway can counteract the protein degradation in catabolic states and reduce loss of muscle protein (myofibre atrophy) (Bodine et al. 2001). The extent to which age-related muscle mass loss is dependent on Atrogin-1 and MuRF1 gene expression is not known. While some studies report elevation of atrophy related genes in old muscle (Raue et al. 2007) others show suppression of their expression (Edstrom et al. 2006).

3.2 *Inflammation, TNF and ROS*

Various factors are known to inhibit IGF-1 signalling and these include factors associated with inflammation (e.g. TNF) and obesity (e.g. diglycerides). Many cytokines are altered during inflammation and the pro-inflammatory cytokines tumour necrosis factor (TNF) and interleukin-1 (IL-1) are strongly associated with catabolism and muscle atrophy. Such cytokines are responsible for muscle protein degradation in more severe cases of inflammation, such as cancer cachexia, sepsis and AIDS (Reviewed in (Tisdale 2005, 2009)). One of the main mechanisms by which TNF and IL-1 cause myofibre atrophy is by elevating protein degradation, due to increased atrogene expression mediated by nuclear factor–kappa beta (NFκB) (Tisdale 2005; Messina et al. 2006). Muscle wasting produced by TNF is associated with induction of oxidative stress (Tisdale 2005) that can modulate a complexity of interacting signalling pathways to result in muscle atrophy (Reviewed in (Arthur et al. 2008)). TNF can also directly interfere with IGF-1 signalling: TNF may inhibit IGF-1 dependent events by down-regulation of IGF-1 synthesis (Frost et al. 2003) and inhibition of signalling pathways downstream of the IGF-1 receptor leads to decreased protein synthesis and further up-regulation of atrophy related genes (Broussard et al. 2003, 2004; Strle et al. 2004). Activation of C-Jun N-Terminal Kinase (JNK) appears to play role in both of these processes (Frost et al. 2003; Grounds et al. 2008). Because of such cross-talk between TNF and IGF-1 signalling, changes in relative amounts of these cytokines during ageing are important to consider and inverse changes of TNF and IGF-1 are well documented with age. Ageing results in chronic low-grade increases in circulating inflammatory cytokines and high plasma levels of TNF and IL-6 are strongly associated with morbidity and mortality in elderly humans (Bruunsgaard and Pedersen 2003; Sandmand et al. 2003). However, in some situations IL-6 is clearly anti-inflammatory and can decrease systemic TNF levels and it is well documented that exercise increases muscle production of IL-6 and elevates systemic IL-6 (Pedersen 2006, 2007). The fine balance between these cytokines and others appears critical for modulating the precise inflammatory response. Human studies show that in the elderly, systemic low-grade inflammation with increased TNF and IL-6 can contribute to loss of muscle mass and strength (Visser et al. 2002; Schaap et al. 2006). In contrast, serum levels of GH and IGF-1 decrease in old humans and rats (Ullman et al. 1990; Grounds 2002) (discussed in more detail below). Thus, attempts to minimize muscle wasting in various clinical conditions have focused on both anti-inflammatory drugs to block TNF action and development of strategies to deliver IGF-1 to skeletal myofibres.

3.3 Lipids

Impaired IGF-1 signalling also results from high levels of lipids within muscles; this contributes to insulin resistance and type 2 diabetes that is of increasing prevelance in association with obesity and the ageing population (Reviewed (Shavlakadze and Grounds 2006)). It is suggested that a high fat diet activates S6K1 to inhibit signalling downstream of IRS1 (by phosphorylating IRS1 at Ser307 and Ser636/639) and thus suppresses insulin signalling and leads to insulin resistance (Um et al. 2004). In addition, increased lipid content within human myofibres correlates with skeletal muscle insulin resistance, and is independent of total body adiposity (Goodpaster and Brown 2005). This correlation is pronounced in patients with type 2 diabetes where myofibres display insulin resistance and significantly increased lipid content (Goodpaster et al. 2001). It is suggested that increased lipid deposition in myofibres per se does not affect insulin sensitivity, but rather represents a marker for the increase of other lipid molecules (such as ceramide, diglyceride, or long-chain acyl-CoA) that may induce defects in the insulin-signalling pathway and muscle insulin resistance [Reviewed (Goodpaster et al. 2001; Goodpaster and Brown 2005)]. Insulin sensitivity may also be influenced by the oxidative capacity of skeletal muscle (Reviewed (Goodpaster et al. 2001; Goodpaster and Brown 2005)). Ageing is associated with increased fat within myofibres, with healthy non-diabetic subjects showing increasing intramyocellular triacylgycerols with age and this correlates with insulin resistance (Cree et al. 2004).

4 Loss of IGF-1 in Ageing Animals

4.1 GH/IGF-1 Axis in Ageing

Concentrations of GH and IGF-1 in blood and GH receptor and IGF-1 mRNA in skeletal muscle decline steadily with age in humans, sheep and rodents (Oldham et al. 1996; Martin et al. 1997; Corpas et al. 1993; Dardevet et al. 1994; Florini et al. 1985; Maggio et al. 2006; O'Connor et al. 1998). In particular, secretion of GH becomes more irregular with age with a decrease in total secretion over a 24 h period and concentrations of IGF-1 decline at a rate of 2 ng per ml per year from 20 to 100 years (O'Connor et al. 1998; Maggio et al. 2006; Ho et al. 1987; Veldhuis et al. 1995). The decline in GH/IGF-1 axis is also correlated with a decline in cognitive function, suggesting a causal relationship and a role in neuroprotection (Ceda et al. 2005).

The impact of IGF-1 on the nervous (and other) systems must be considered with respect to maintenance of skeletal muscle mass and function, although detailed examination of this topic lies beyond the scope of this review. Loss of motorneurone function in the central nervous system will result in loss of axons and neuromuscular synapses, with subsequent denervation of myofibres (MacIntosh et al. 2006). To some extent this problem may be initially countered by sprouting of surviving motorneurones to form new neuromuscular synapses, but even this leads to some diminution of contractile capacity. Progressive motorneurone loss over time will result in permanent denervation with severe myofibre atrophy and loss of function (MacIntosh et al. 2006; Edstrom et al. 2007). This aspect of sarcopenia provides quite different potential targets for therapeutic interventions but will not be considered further in this Chapter.

While concentrations of IGF-1 decrease in blood, changes in expression of IGF-1 and IGF-1 receptor mRNA is less clear. IGF-1Ea and MGF mRNA were not changed in biopsy samples taken from vastus lateralis muscles of young (<30 years) and old (>60 years) humans before and shortly after leg extension exercises at 80% of one maximum repetition (RM) or before and after eccentric cycling exercise. Expression of MGF, but not IGF-1Ea mRNA increased in these muscles of young men after concentric, and in both young and old men after eccentric exercise (Hameed et al. 2003, 2008). In support, others confirm no change in expression of either IGF-1 transcripts with ageing, but show that expression of both is increased 24 h after concentric exercise at 1-RM and further increased after 16 weeks of exercise (three times per week) (Petrella et al. 2006). MGF is more sensitive to exercise and/or injury, which explains the observed increase measured within hours after an acute bout, while there is a delay in IGF-1Ea and the increase in both transcripts after 16 weeks is consistent with an adaptive response to resistance training (Goldspink 2005). In contrast, others do not distinguish between splice-variants of IGF-1 and have observed a decrease in IGF-1 mRNA in muscles of elderly compared with young men (Marcell et al. 2001; Welle et al. 2002; Dennis et al. 2008; Leger et al. 2008).

IGF-1 receptor numbers decrease in skeletal muscle over 12 months in rats. In addition, IGF-1 mRNA also decreased with age and the binding capacity of receptors was reduced, which is consistent with a decrease in the function of IGF-1 in muscle (Dardevet et al. 1994). Despite the progressive loss of skeletal muscle mass, ageing rat muscles retain the ability to recruit satellite cells and there is an increased density of cell nuclei with centrally located myonuclei in nascent myofibres. There was no downregulation of IGF-1 or of IGF-1 receptor mRNA and there was increased mRNA expression of myogenic regulatory factors (Edstrom and Ulfhake 2005): since neither the abundance of IGF-1 protein nor the receptor binding capacity were assessed in this study, it is possible that the bioavailability of IGF-1 was reduced in these aged rat muscles. A further factor affecting the ability to regenerate skeletal muscle during ageing is the decline in the number of motorneurons. Ageing motor neurons and the failure to innervate newly formed myofibres is consistent with the preferential loss of type II myofibres (Edstrom et al. 2007). IGF-1 also promotes angiogenesis and there is a 25% decline in the number of capillaries in elderly subjects. While it is unclear if the decline in capillary density with age is linked to IGF-1, the decline in IGF-1 bioavailability is associated with a reduction in multiple facets of skeletal muscle integrity which could, collectively, contribute to sarcopenia (Rogers and Evans 1993; Rabinovsky and Draghia-Akli 2004).

4.2 Nutrition

IGF-1 is secreted as it is synthesised and is directly regulated by nutrition and GH (Schwander et al. 1983). Fasting decreases the rate of transcription and the abundance of protein and IGF-1 mRNA (Isley et al. 1983; Hayden et al. 1994). In addition, undernutrition of sheep and cattle (30% of maintenance) is associated with reduced concentrations of IGF-1 in blood and reduced IGF-1 mRNA in skeletal muscle (Breier et al. 1986; Jeanplong et al. 2003). Paradoxically, there is increased secretion of GH in ruminants and humans during short-term fasting and undernutrition, yet secretion of IGF-1 is decreased and secretion in response to exogenous GH is blunted, which is consistent with refractoriness to GH (Breier et al. 1986, 1988; Thissen et al. 1994).

The influence of nutrition on the synthesis and secretion of IGF-1 may be a pivotal determinant of the loss of muscle mass during ageing. Appetite is progressively reduced at a linear rate of 0.5–1% per year from the age of 20–80 and, in conjunction, secretion of IGF-1 is reduced (Wurtman et al. 1988; Hallfrisch et al. 1990; Briefel et al. 1995; Morley 1997; Chapman et al. 2002; Chapman 2006, 2007). Despite the progressive decrease in quantity consumed, the proportions of fat, carbohydrate and protein in the diet remain similar (Wurtman et al. 1988). Both the energy and protein composition of a diet independently influence secretion of IGF-1. When diets are deficient in either component, secretion of IGF-1 is suppressed and secretion is further suppressed when both are inadequate (Isley et al. 1983). However, when the protein composition of the diet is restored to greater than 0.9 g per kg BWT per day concentrations of IGF-1 are increased in elderly (>50 years) subjects (Khalil et al. 2002; Dawson-Hughes et al. 2004). Increasing the protein intake beyond approximately 0.9 g per kg per day does not have any further effect on secretion of IGF-1, which is consistent with the RDA of 0.8 g per kg BWT per day (Roughead et al. 2003). However, it has been suggested that this figure is too low and should be increased to 1.6 g per kg BWT per day particularly for people who are active, athletes and the elderly (Evans 2004; Phillips 2006; Campbell and Leidy 2007; Wolfe et al. 2008; Paddon-Jones and Rasmussen 2009). In support, the incidence of protein-energy malnutrition in the elderly has been reported to be 15% in community-dwelling persons, up to 12% of homebound patients, up to 65% of hospitalised patients and up to 85% of institutionalised persons (Morley 1997). Therefore, it has been suggested that when overall food intake is marginal the elderly may benefit from acquiring a greater proportion of energy from the protein portion of the diet with each meal (Evans 2004; Wolfe et al. 2008; Paddon-Jones and Rasmussen 2009).

4.3 Calorie Restriction

Calorie restriction (CR) without malnutrition has been shown to prolong lifespan. While introduction of CR before skeletal maturity reduces body mass, there is a reduction in the rate of sarcopenia in rhesus monkeys (Colman et al. 2008). Moreover, CR extends lifespan and reduces (50% lower) the number of neoplasms and the incidence of cardiovascular disease in rhesus monkeys (Colman et al. 2009). CR of 8% prevented the reduction in CSA of the plantaris muscle in rats and the addition of exercise further protected the demise of this muscle and partially prevented the decline in secretion of IGF-1 (Kim et al. 2008). Typically, CR reduces concentrations of IGF-1 in rodents. In ad libitum fed mice, concentrations of IGF-1 in blood are increased at 24 weeks of age, while in CR mice, concentrations of IGF-1 in blood are unchanged from young controls (3 weeks). Lean mass was preserved and the mass of adipose tissue was reduced (Huffman et al. 2008). In contrast, CR (28%) for up to 6 years did not alter concentrations of IGF-1 in blood of healthy subjects (mean age 51 years). Only a decrease in protein content in the diet reduced concentrations of IGF-1 in blood (Fontana et al. 2008). A reduced intake of protein (0.45 g per kg BWT per day) was also shown to decrease concentrations of IGF-1 and reduce the CSA of type I myofibres over a 10 week period in elderly women (66–79 years) (Castaneda et al. 2000). This reinforces the view that the protein portion of the diet should be increased in the elderly given their reduced appetite (see Section 4.2).

Consistently, two interventions, repression of the GH/IGF-1/insulin axis and caloric restriction, have been shown to increase lifespan in both invertebrates and vertebrate animal model systems. Longevity is also associated with reduced levels of the active metabolite of thyroid hormone (T3), which affects metabolism and body temperature and this benefit is attributed to reduced oxidative stress (Buffenstein and Pinto 2009). Furthermore, since T3 influences myosin isoform composition and skeletal muscle function (the impact differs between various muscles and with gender), decreased concentrations of T3 probably contribute to age-related changes in myofibre types and loss of muscle function (Yu et al. 1999). The complex interactions between insulin/IGF-1, GH, T3, vitamin D and Klotho is the subject of an excellent review (Buffenstein and Pinto 2009).

4.4 Klotho

The ageing-suppressor gene Klotho was identified in 1997 and found to extend life-span by 20–30% when overexpressed, and to accelerate ageing when disrupted in mice (Kuro-o et al. 1997; Kurosu et al. 2005). The 1,014 amino acid protein is present in three isoforms. Firstly, as a single-pass transmembrane protein that serves as a receptor for multiple fibroblasts growth factors and a co-receptor for fibroblast growth factor-23 (FGF23), a bone-derived hormone that suppresses vitamin D synthesis. Secondly, the extracellular domain comprising the KL1 and KL2 domains can be enzymatically cleaved by the membrane-anchored proteases ADAM10 and ADAM17, which enables the KL1 and KL2 domains to be secreted to act in an endocrine manner to regulate insulin and IGF-1 signalling pathways along with Wnt (Chen et al. 2007). Thirdly, a splice-variant is translated that includes the KL1 domain, but lacks the KL2 domain, and is directly secreted into blood (Chen et al. 2007; Kuro-o 2009).

Klotho null mice grow normally to 3 weeks of age, then stop growing, age prematurely and die around 8–9 weeks of age. This pathology of accelerated ageing is attributed to elevated concentrations of vitamin D, which in turn, promote the absorption of phosphorus and calcium from food in the intestines. Consequently, concentrations of phosphate and calcium are also elevated in blood, which contribute to numerous histological changes. Notably, there is ectopic calcification in numerous soft tissues, pulmonary emphysema and decreased bone mineral density. The ageing process can be rescued by dietary restriction in vitamin D and phosphate, which implicates hypervitaminosis D and hyperphosphatemia as key regulators of ageing regulated by FGF23-Klotho signalling (Kuro-o et al. 1997): discussed in more detail below (see Section 4.5). While high levels of vitamin D can be deleterious, the opposite is also true since age-related lack of vitamin D may contribute to reduce signalling through the IGF-1 pathway (discussed in more detail below). Vitamin D deficiency is extremely prevalent in the elderly and can result in myopathy with loss of muscle strength and selective loss of fast myofibres similar to sarcopenia (Janssen et al. 2002).

A second and more direct influence of Klotho on ageing is via inhibition of insulin and IGF-1 signalling. Life-span is extended in *C. elegans* with loss-of-function mutations in the insulin receptor homologue *daf-2*, in the PI3-kinase homologue *age-1* and in the forkhead transcription factor (FOXO) homologue *daf-16*. Similarly, loss-of-function mutations in the insulin receptor homologue (*ins*) and insulin receptor substrate homologue (*chico*) in Drosophila result in an extension of life-span. Likewise, mutations that disrupt the GH/IGF-1 axis in mammals extend life-span despite causing dwarfism (Bartke et al. 2001). Klotho does not affect appetite and, therefore, the longevity of mice overexpressing Klotho occurs without calorie restriction, despite restriction of vitamin D and phosphates ameliorating the deleterious effect of absence of Klotho (Kurosu et al. 2005). Most likely, extension of life-span occurs via the inhibition of the insulin and GH/IGF-1 axes (Bartke 2006). Mutations in Klotho have been found in human populations that are associated with longevity. Specifically, a double mutation (V352 and S370 collectively termed KL-VS) confers an advantage in the heterozygous state in European, African-American, Ashkenazi Jews and Italians (Arking et al. 2002, 2005; Invidia et al. 2010). The mutation increases (1.6 fold) the secretion of Klotho in cultured cells and, therefore, longevity may be associated with increased concentrations of the secreted forms of Klotho in blood (Arking et al. 2002).

It remains unclear whether the effect of Klotho on vitamin D and hyperphosphatemia are independent of the actions on insulin/IGF-1 signalling (Fig. 4 insert) or if there is an interaction between these phenomena to influence ageing.

4.5 Vitamin D and Kidneys

Vitamin D is an essential hormone for maintaining muscle function and concentrations decline in blood with age in conjunction with frailty (Tuohimaa 2009). Restoring physiological concentrations of vitamin D help protect against frailty in old age (Janssen et al. 2002). Vitamin D is produced in the skin from cholesterol-derived precursors. Subsequent steps are hydroxylation to 25(OH) vitamin D3 in the liver and a final hydroxylation step occurs in the proximal tubules of the kidney to produce the active form $1,25(OH)_2$ vitamin D_3 (Ebert et al. 2006). The enzyme responsible for this final step in the kidney is 1α hydroxylase, which is downregulated in aged rats on low phosphate or calcium diets. Klotho decreases activity of 1α hydroxylase, which is consistent with increased concentrations of vitamin D in Klotho null mice (Imai et al. 2004). Activity and synthesis of vitamin D is restored by treatment with IGF-1, concentrations of which also decline with ageing, which suggests a causal link (Wong et al. 1997, 2000). In fact, there is a positive relationship between concentrations of vitamin D and IGF-1 in blood and higher concentrations of each hormone are associated with reduced prevalence of the metabolic syndrome in middle-aged subjects (Hypponen et al. 2008).

The role of vitamin D in ageing is not yet clear with conflicting points of view. Notably, elevated vitamin D is implicated in advanced senescence that occurs in Klotho- and FGF23-deficient mice (Kuro-o 2008). In contrast, absence of vitamin D signalling also induces enhanced ageing (Keisala et al. 2009; Tuohimaa 2009). More recent studies have clarified this discrepancy to some extent and show that hyperphosphatemia is the more likely causative factor underlying the etiology of premature senescence in Klotho-deficient ageing due to failure of FGF23 to activate the FGF/Klotho receptor complex. After binding to its cytosolic receptor, vitamin D directs IGF-1 to activate the MAPK pathway and this is mediated by PKC and Ca^{2+} (Morelli et al. 2000). The activated vitamin D receptor directly binds to and dephosphorylates Src, which allows Shc to be phosphorylated and outcompete the insulin receptor complex (IRS) for access to the IGF-1 receptor (Fig. 4) (Buitrago et al. 2000; Morelli et al. 2000; Sasaoka et al. 2001; Boland et al. 2002; Sekimoto and Boney 2003; Lieskovska et al. 2006). In support, pharmacological or transgenic inhibition of Src, prevents activation of MAPK (Boney et al. 2001; Sasaoka et al. 2001). Therefore, reduced or excessive concentrations of vitamin D could compromise MAPK signalling in skeletal muscle and tight regulation of this pathway is essential for normal health and ageing and both hypo- and hypervitaminosis D can accelerate ageing (Tuohimaa 2009).

Finally, it is worth noting that renal damage is prevalent in older subjects. Kidney function progressively declines with age and 11% of individuals older than 65 years without hypertension or diabetes had stage 3 or worse chronic kidney disease (Coresh et al. 2003, 2005). A similar increase in lesions occurs in aged rats, but these are reduced with a concomitant increase in life span in rats maintained on a calorie restricted diet and further improved (30%) when coupled with suppression of the GH/IGF-1 axis via hemizygote expression of an antisense transgene for GH (Zha et al. 2008). An emerging postulate is that renal function is crucial for health and, therefore, longevity and is consistent with the greatest expression of Klotho in the kidney and in close proximity to synthesis of vitamin D (Zha et al. 2008). Perhaps the etiology of sarcopenia can be explained, at least in part, by the progressive damage to the kidneys during ageing, which is accompanied by perturbed synthesis of vitamin D and Klotho. As a consequence, vitamin D and Klotho are not produced in sufficient quantities to regulate the insulin and IGF-1 signalling axes. In conjunction, the GH/IGF-1 axis is downregulated in the kidney in chronic kidney disease and the bioavailability of IGF-1 is further reduced due to resistance to GH and an increased abundance of IGFBP-1, -2, -4 and -6. This ageing-associated decline of the GH/IGF-1 axis may help reduce glomerular sclerosis and prolong glomerular function (Mak et al. 2008). Note that chronic kidney disease is associated with major disturbances in the GH/IGF axis (pre and post receptor with huge increases in IGFBPs (Mahesh and Kaskel 2008).

Clearly, regulation of the GH/IGF-1 axes is crucial for healthy ageing and there is a close and as yet unclear relationship between nutrition, Klotho and vitamin D (indicated in Fig. 4).

5 Therapies to Increase IGF-1 Signalling

Although it was a widely held notion that sarcopenia was directly related to an age-related decline in GH secretion, this view has been contested and numerous studies in humans do not support a benefit of GH administration on muscle protein synthesis (Lynch et al. 2005). Synthetic peptides that cause the release of GH (GH secretagogues, e.g. benzoazepines and their analogues) have been used clinically but their efficacy is unclear, as are the benefits of commercial hormone replacement therapies (Borst and Lowenthal 1997). Overall, simple hormone "top up" strategies to restore hormone levels in the elderly have not been successful, especially if they have not been performed in conjunction with a resistance exercise program (Lynch et al. 2005). There is some promise of treating elderly subjects (65–90 years) with GH together with testosterone, which increased concentrations of IGF-1 more than either treatment alone and increased strength (Huang et al. 2005; Sattler et al. 2009).

The benefits of over-expression of IGF-1 on age-related muscle wasting have started to be tested in transgenic animal models (Chakravarthy et al. 2001; Musaro et al. 2001). Transgenic over-expression of IGF-1, as well as its downstream target Akt, results in muscle hypertrophy (Bodine et al. 2001; Chakravarthy et al. 2001). In addition, genetic activation of Akt antagonizes signalling that leads to muscle atrophy: for example, over-expression of the constitutively active Akt was sufficient to block muscle wasting following short term (7 days) denervation (Bodine et al. 2001). While elevated IGF-1 is less efficient than Akt in blocking muscle atrophy, up-regulation of IGF-1 can slow down muscle atrophy in some but not all models of muscle wasting: e.g. transgenic muscle specific over-expression of IGF-1 reduced the rate of atrophy resulting from denervation at 1 month, but not at 2 months following nerve transaction (Shavlakadze et al. 2005a).

The challenge is to translate these experimental benefits of muscle restricted elevated IGF-1 to the clinical situation. It is noted that systemic administration of IGF-1 is not recommended as this results in hypertrophy of cardiac muscle and heart failure and also prostate cancer (Shavlakadze and Grounds 2003). In support, transgenic over-expression of human IGF-1 within skeletal muscle in Rskα-actin/hIGF-I mice that also elevated circulating IGF-1 throughout development had no hypertrophic effect on skeletal muscle (Shavlakadze et al. 2006) and resulted in enlarged seminal vesicles (Shavlakadze et al. 2005b). Since IGF-I and GH have overlapping as well as independent effects on somatic growth (Lupu et al. 2001) (and it is estimated that the overlapping GH/IGF-I effect makes 34% contribution to the total weight, IGF-I alone contributes 35% and GH alone 14%), it is possible that the absence of muscle hypertrophy in these Rskα-actin/hIGF-I mice is a result of decreased systemic GH. In addition, elevated systemic IGF-1 might down-regulate IGF-1 receptors and increase IGF-1 binding proteins to ablate the effects of elevated IGF-1 in these mice (Shavlakadze and Grounds 2003). Such experiments emphasise that systemic elevation of IGF-1 is not a promising approach. This accords with the view that the autocrine/paracrine mode of IGF-1 action is the principal mechanism for stimulating muscle growth/hypertrophy. The targeted elevation of IGF-1 selectively only within adult skeletal muscle remains an appealing option although this may only be effective as a hypertrophic agent to amplify a growth-related stimulus, that can be produced by regeneration or possibly resistance exercise in adults. The efficacy of this approach to prevent sarcopenia awaits validation.

6 Conclusions

Loss of skeletal muscle mass with ageing is associated with a decline in the GH/IGF-1 axis, however it is not know to what extent such decline contributes to sarcopenia. Evidence supports the role of IGF-1 to reduce the rate of muscle wasting in some atrophy models, but extensive data are not yet available for sarcopenia. Therapeutic strategies using GH or IGF-1 have mixed success unless administered in conjunction with androgens and exercise. While these strategies show some promise in reducing sarcopenia over a short term, there are contraindications suggesting that they may increase the risk of tumorigenesis and cardiovascular disease. We await with interest results from transgenic studies in which IGF-1 is increased in skeletal muscle alone, which may not incur the same pathologies as endocrine-mediated treatments. Currently, the protein content of a diet may be a safer strategy to increase concentrations of IGF-1 which, in conjunction with exercise may slow the rate of sarcopenia. In contrast, other studies suggest that the decline in the GH/IGF-1 axis may favour a slower decline in muscle mass and confer a longer, healthier life. Indeed, the evidence presented here from worms, insects and mammals suggests that there is an evolutionarily conserved pathway via which caloric restriction, Klotho and loss of function mutations in GH/INS/IGF axes act to regulate a common signal transduction pathway to confer a reduced rate of sarcopenia and longer life. Therefore, a pertinent question to ask in summary is 'should we intervene on the natural decline or should we manage the decline in the GH/IGF-1 axis to maintain healthy muscle in old age'? Part of this management might be to maintain expression of IGF-1 locally in muscle, but not systemically.

Acknowledgements Grateful acknowledgement is made by the authors for research funding from the National Health and Medical Research Council of Australia (MG, TS), and the Foundation for Research, Science and Technology (CM). We thank Marta Fiorotto (Baylor College of Medicine, Houston, USA) for reading the manuscript and her helpful and constructive comments on the manuscript.

References

Accili, D., Nakae, J., Kim, J. J., Park, B. C., Rother, K. I. (1999). Targeted gene mutations define the roles of insulin and IGF-I receptors in mouse embryonic development. *Journal of Pediatric Endocrinology & Metabolism, 12*, 475–485.

Arking, D. E., Krebsova, A., Macek, M., Sr., Macek, M., Jr., Arking, A., Mian, I. S., Fried L, Hamosh, A., Dey, S., Mcintosh, I., Dietz, H. C. (2002). Association of human aging with a functional variant of klotho. *Proceedings of the National Academy of Sciences of the United States of America, 99*, 856–861.

Arking, D. E., Atzmon, G., Arking, A., Barzilai, N., Dietz, H. C. (2005). Association between a functional variant of the KLOTHO gene and high-density lipoprotein cholesterol, blood pressure, stroke, and longevity. *Circulation Research, 96*, 412–418.

Arthur, P. G., Grounds, M. D., Shavlakadze, T. (2008). Oxidative stress as a therapeutic target during muscle wasting: considering the complex interactions. *Current Opinion in Clinical Nutrition and Metabolic Care, 11*, 408–416.

Bartke, A. (2006). Long-lived Klotho mice: new insights into the roles of IGF-1 and insulin in aging. *Trends in Endocrinology and Metabolism, 17*, 33–35.

Bartke, A., Brown-Borg, H., Mattison, J., Kinney, B., Hauck, S., Wright, C. (2001). Prolonged longevity of hypopituitary dwarf mice. *Experimental Gerontology, 36*, 21–28.

Bodine, S. C., Stitt, T. N., Gonzalez, M., Kline, W. O., Stover, G. L., Bauerlein, R., Zlotchenko, E., Scrimgeour, A., Lawrence, J. C., Glass, D. J., Yancopoulos, G. D. (2001). Akt/mTOR pathway is a crucial regulator of skeletal muscle hypertrophy and can prevent muscle atrophy in vivo. *Nature Cell Biology, 3*, 1014–1019.

Boland, R., DE Boland, A. R., Buitrago, C., Morelli, S., Santillan, G., Vazquez, G., Capiati, D., Baldi, C. (2002). Non-genomic stimulation of tyrosine phosphorylation cascades by 1,25(OH)(2)D(3) by VDR-dependent and -independent mechanisms in muscle cells. *Steroids, 67*, 477–482.

Boney, C. M., Sekimoto, H., Gruppuso, P. A., Frackelton, A. R., J. R. (2001). Src family tyrosine kinases participate in insulin-like growth factor I mitogenic signaling in 3T3-L1 cells. *Cell Growth & Differentiation, 12*, 379–386.

Borst, S. E. & Lowenthal, D. T. (1997). Role of IGF-I in muscular atrophy of aging. *Endocrine, 7*, 61–63.

Breier, B. H., Bass, J. J., Butler, J. H., Gluckman, P. D. (1986). The somatotrophic axis in young steers: influence of nutritional status on pulsatile release of growth hormone and circulating concentrations of insulin-like growth factor 1. *The Journal of Endocrinology, 111*, 209–215.

Breier, B. H., Gluckman, P. D., Bass, J. J. (1988). Influence of nutritional status and oestradiol-17 beta on plasma growth hormone, insulin-like growth factors-I and -II and the response to exogenous growth hormone in young steers. *The Journal of Endocrinology, 118*, 243–250.

Briefel, R. R., Mcdowell, M. A., Alaimo, K., Caughman, C. R., Bischof, A. L., Carroll, M. D., Johnson, C. L. (1995). Total energy intake of the US population: The third National Health and Nutrition Examination Survey, 1988–1991. *The American Journal of Clinical Nutrition, 62*, 1072S–1080S.

Broussard, S. R., Mccusker, R. H., Novakofski, J. E., Strle, K., Shen, W. H., Johnson, R. W., Freund, G. G., Dantzer, R., Kelley, K. W. (2003). Cytokine-hormone interactions: tumor necrosis factor alpha impairs biologic activity and downstream activation signals of the insulin-like growth factor I receptor in myoblasts. *Endocrinology, 144*, 2988–2996.

Broussard, S. R., Mccusker, R. H., Novakofski, J. E., Strle, K., Shen, W. H., Johnson, R. W., Dantzer, R., Kelley, K. W. (2004). IL-1beta impairs insulin-like growth factor I-induced differentiation and downstream activation signals of the insulin-like growth factor I receptor in myoblasts. *Journal of Immunology, 172*, 7713–7720.

Bruunsgaard, H. & Pedersen, B. K. (2003). Age-related inflammatory cytokines and disease. *Immunology and Allergy Clinics of North America, 23*, 15–39.

Buffenstein, R. & Pinto, M. (2009). Endocrine function in naturally long-living small mammals. *Molecular and Cellular Endocrinology, 299*, 101–111.

Buitrago, C., Vazquez, G., DE Boland, A. R., Boland, R. L. (2000). Activation of Src kinase in skeletal muscle cells by 1, 1,25-(OH(2))-vitamin D(3) correlates with tyrosine phosphorylation of the vitamin D receptor (VDR) and VDR-Src interaction. *Journal of Cellular Biochemistry, 79*, 274–281.

Campbell, W. W. & Leidy, H. J. (2007). Dietary protein and resistance training effects on muscle and body composition in older persons. *Journal of the American College of Nutrition, 26*, 696S–703S.

Castaneda, C., Gordon, P. L., Fielding, R. A., Evans, W. J., Crim, M. C. (2000). Marginal protein intake results in reduced plasma IGF-I levels and skeletal muscle fiber atrophy in elderly women. *The Journal of Nutrition, Health & Aging, 4*, 85–90.

Ceda, G. P., Dall'Aglio, E., Maggio, M., Lauretani, F., Bandinelli, S., Falzoi, C., Grimaldi, W., Ceresini, G., Corradi, F., Ferrucci, L., Valenti, G., Hoffman, A. R. (2005). Clinical implications of the reduced activity of the GH-IGF-I axis in older men. *Journal of Endocrinological Investigation, 28*, 96–100.

Chakravarthy, M. V., Fiorotto, M. L., Schwartz, R. J., Booth, F. W. (2001). Long-term insulin-like growth factor-I expression in skeletal muscles attenuates the enhanced in vitro proliferation ability of the resident satellite cells in transgenic mice. *Mechanisms of Ageing and Development, 122*, 1303–1320.

Chapman, I. M. (2006). Nutritional disorders in the elderly. *The Medical Clinics of North America, 90*, 887–907.

Chapman, I. M. (2007). The anorexia of aging. *Clinics in Geriatric Medicine, 23*, 735–756. v.

Chapman, I. M., Macintosh, C. G., Morley, J. E., Horowitz, M. (2002). The anorexia of ageing. *Biogerontology, 3*, 67–71.

Chen, C. D., Podvin, S., Gillespie, E., Leeman, S. E., Abraham, C. R. (2007). Insulin stimulates the cleavage and release of the extracellular domain of Klotho by ADAM10 and ADAM17. *Proceedings of the National Academy of Sciences of the United States of America, 104*, 19796–19801.

Colman, R. J., Beasley, T. M., Allison, D. B., Weindruch, R. (2008). Attenuation of sarcopenia by dietary restriction in rhesus monkeys. *The Journals of Gerontology. Series A: Biological Sciences and Medical Sciences, 63*, 556–559.

Colman, R. J., Anderson, R. M., Johnson, S. C., Kastman, E. K., Kosmatka, K. J., Beasley, T. M., Allison, D. B., Cruzen, C., Simmons, H. A., Kemnitz, J. W., Weindruch, R. (2009). Caloric restriction delays disease onset and mortality in rhesus monkeys. *Science, 325*, 201–204.

Coolican, S. A., Samuel, D. S., Ewton, D. Z., Mcwade, F. J., Florini, J. R. (1997). The mitogenic and myogenic actions of insulin-like growth factors utilize distinct signaling pathways. *The Journal of Biological Chemistry, 272*, 6653–6662.

Coresh, J., Astor, B. C., Greene, T., Eknoyan, G., Levey, A. S. (2003). Prevalence of chronic kidney disease and decreased kidney function in the adult US population: Third National Health and Nutrition Examination Survey. *American Journal of Kidney Diseases, 41*, 1–12.

Coresh, J., Byrd-Holt, D., Astor, B. C., Briggs, J. P., Eggers, P. W., Lacher, D. A., Hostetter, T. H. (2005). Chronic kidney disease awareness, prevalence, and trends among U.S. adults, 1999 to 2000. *Journal of the American Society of Nephrology, 16*, 180–188.

Corpas, E., Harman, S. M., Blackman, M. R. (1993). Human growth hormone and human aging. *Endocrine Reviews, 14*, 20–39.

Cree, M. G., Newcomer, B. R., Katsanos, C. S., Sheffield-Moore, M., Chinkes, D., Aarsland, A., Urban, R., Wolfe, R. R. (2004). Intramuscular and liver triglycerides are increased in the elderly. *The Journal of Clinical Endocrinology and Metabolism, 89*, 3864–3871.

Dardevet, D., Sornet, C., Attaix, D., Baracos, V. E., Grizard, J. (1994). Insulin-like growth factor-1 and insulin resistance in skeletal muscles of adult and old rats. *Endocrinology, 134*, 1475–1484.

Dawson-Hughes, B., Harris, S. S., Rasmussen, H., Song, L., Dallal, G. E. (2004). Effect of dietary protein supplements on calcium excretion in healthy older men and women. *The Journal of Clinical Endocrinology and Metabolism, 89*, 1169–1173.

Dennis, R. A., Przybyla, B., Gurley, C., Kortebein, P. M., Simpson, P., Sullivan, D. H., Peterson, C. A. (2008). Aging alters gene expression of growth and remodeling factors in human skeletal muscle both at rest and in response to acute resistance exercise. *Physiological Genomics, 32*, 393–400.

Dikkes, Pdbj, Guo, W. H., Chao, C., Hemond, P., Yoon, K., Zurakowski, D., Lopez, M. F. (2007). IGF2 knockout mice are resistant to kainic acid-induced seizures and neurodegeneration. *Brain Research, 1175*, 85–95.

Ebert, R., Schutze, N., Adamski, J., Jakob, F. (2006). Vitamin D signaling is modulated on multiple levels in health and disease. *Molecular and Cellular Endocrinology, 248*, 149–159.

Edstrom, E. & Ulfhake, B. (2005). Sarcopenia is not due to lack of regenerative drive in senescent skeletal muscle. *Aging Cell, 4*, 65–77.

Edstrom, E., Altun, M., Hagglund, M., Ulfhake, B. (2006). Atrogin-1/MAFbx and MuRF1 are downregulated in aging-related loss of skeletal muscle. *The Journals of Gerontology. Series A: Biological Sciences and Medical Sciences, 61*, 663–674.

Edstrom, E., Altun, M., Bergman, E., Johnson, H., Kullberg, S., Ramirez-Leon, V., Ulfhake, B. (2007). Factors contributing to neuromuscular impairment and sarcopenia during aging. *Physiology & Behavior, 92*, 129–135.

Engert, J. C., Berglund, E. B., Rosenthal, N. (1996). Proliferation precedes differentiation in IGF-I-stimulated myogenesis. *The Journal of Cell Biology, 135*, 431–440.

Evans, W. J. (2004). Protein nutrition, exercise and aging. *Journal of the American College of Nutrition, 23*, 601S–609S.

Florini, J. R., Prinz, P. N., Vitiello, M. V., Hintz, R. L. (1985). Somatomedin-C levels in healthy young and old men: relationship to peak and 24-hour integrated levels of growth hormone. *Journal of Gerontology, 40*, 2–7.

Florini, J. R., Ewton, D. Z., Roof, S. L. (1991). Insulin-like growth factor-1 stimulates terminal myogenic differentiation by induction of myogenin gene expression. *Molecular Endocrinology, 5*, 718–724.

Florini, J. R., Ewton, D. Z., Coolican, S. A. (1996). Growth hormone and the insulin-like growth factor system in myogenesis. *Endocrine Reviews, 17*, 481–517.

Fontana, L., Weiss, E. P., Villareal, D. T., Klein, S., Holloszy, J. O. (2008). Long-term effects of calorie or protein restriction on serum IGF-1 and IGFBP-3 concentration in humans. *Aging Cell, 7*, 681–687.

Frasca, F., Pandini, G., Scalia, P., Sciacca, L., Mineo, R., Costantino, A., Goldfine, I. D., Belfiore, A., Vigneri, R. (1999). Insulin receptor isoform A, a newly recognized, high-affinity insulin-like growth factor II receptor in fetal and cancer cells. *Molecular and Cellular Biology, 19*, 3278–3288.

Frost, R. A., Nystrom, G. J., Lang, C. H. (2003). Tumor necrosis factor-alpha decreases insulin-like growth factor-I messenger ribonucleic acid expression in C2C12 myoblasts via a Jun N-terminal kinase pathway. *Endocrinology, 144*, 1770–1779.

Giustina, A. & Veldhuis, J. D. (1998). Pathophysiology of the neuroregulation of growth hormone secretion in experimental animals and the human. *Endocrine Reviews, 19*, 717–797.

Glass, D. J. (2005). Skeletal muscle hypertrophy and atrophy signaling pathways. *The International Journal of Biochemistry & Cell Biology, 37*, 1974–1984.

Goldspink, G. (2005). Mechanical signals, IGF-I gene splicing, and muscle adaptation. *Physiology (Bethesda), 20*, 232–238.

Goodpaster, B. H. & Brown, N. F. (2005). Skeletal muscle lipid and its association with insulin resistance: What is the role for exercise? *Exercise and Sport Sciences Reviews, 33*, 150–154.

Goodpaster, B. H., HE, J., Watkins, S., Kelley, D. E. (2001). Skeletal muscle lipid content and insulin resistance: evidence for a paradox in endurance-trained athletes. *The Journal of Clinical Endocrinology and Metabolism, 86*, 5755–5761.

Grounds, M. D. (2002). Reasons for the degeneration of ageing skeletal muscle: a central role for IGF-1 signalling. *Biogerontology, 3*, 19–24.

Grounds, M. D., Radley, H. G., Gebski, B. G., Bogoyevitch, M. A., Shavlakadze, T. (2008). Implications of cross-talk between tumour necrosis factor and insulin-like growth factor-1 signalling in skeletal muscle. *Clinical and Experimental Pharmacology & Physiology, 35*, 846–851.

Hallfrisch, J., Muller, D., Drinkwater, D., Tobin, J., Andres, R. (1990). Continuing diet trends in men: the Baltimore longitudinal study of aging (1961–1987). *Journal of Gerontology, 45*, M186–M191.

Hameed, M., Orrell, R. W., Cobbold, M., Goldspink, G., Harridge, S. D. (2003). Expression of IGF-I splice variants in young and old human skeletal muscle after high resistance exercise. *Journal de Physiologie, 547*, 247–254.

Hameed, M., Toft, A. D., Pedersen, B. K., Harridge, S. D., Goldspink, G. (2008). Effects of eccentric cycling exercise on IGF-I splice variant expression in the muscles of young and elderly people. *Scandinavian Journal of Medicine & Science in Sports, 18*, 447–452.

Hayden, J. M., Marten, N. W., Burke, E. J., Straus, D. S. (1994). The effect of fasting on insulin-like growth factor-I nuclear transcript abundance in rat liver. *Endocrinology, 134*, 760–768.

Herrington, J., Smit, L. S., Schwartz, J., Carter-SU, C. (2000). The role of STAT proteins in growth hormone signaling. *Oncogene, 19*, 2585–2597.

Hill, M. & Goldspink, G. (2003). Expression and splicing of the insulin-like growth factor gene in rodent muscle is associated with muscle satellite (stem) cell activation following local tissue damage. *The Journal of Physiology, 549*, 409–418.

HO, K. Y., Evans, W. S., Blizzard, R. M., Veldhuis, J. D., Merriam, G. R., Samojlik, E., Furlanetto, R., Rogol, A. D., Kaiser, D. L., Thorner, M. O. (1987). Effects of sex and age on the 24-hour profile of growth hormone secretion in man: importance of endogenous estradiol concentrations. *The Journal of Clinical Endocrinology and Metabolism, 64*, 51–58.

Hornberger, T. A., Stuppard, R., Conley, K. E., Fedele, M. J., Fiorotto, M. L., Chin, E. R., Esser, K. A. (2004). Mechanical stimuli regulate rapamycin-sensitive signalling by a phosphoinositide 3-kinase-, protein kinase B- and growth factor-independent mechanism. *The Biochemical Journal, 380*, 795–804.

Huang, X., Blackman, M. R., Herreman, K., Pabst, K. M., Harman, S. M., Caballero, B. (2005). Effects of growth hormone and/or sex steroid administration on whole-body protein turnover in healthy aged women and men. *Metabolism, 54*, 1162–1167.

Huffman, D. M., Moellering, D. R., Grizzle, W. E., Stockard, C. R., Johnson, M. S., Nagy, T. R. (2008). Effect of exercise and calorie restriction on biomarkers of aging in mice. *American Journal of Physiology: Regulatory, Integrative and Comparative Physiology, 294*, R1618–R1627.

Hwee, D. T. & Bodine, S. C. (2009). Age-related deficit in load-induced skeletal muscle growth. *The Journals of Gerontology. Series A: Biological Sciences and Medical Sciences, 64*, 618–628.

Hypponen, E., Boucher, B. J., Berry, D. J., Power, C. (2008). 25-hydroxyvitamin D, IGF-1, and metabolic syndrome at 45 years of age: a cross-sectional study in the 1958 British Birth Cohort. *Diabetes, 57*, 298–305.

Imai, M., Ishikawa, K., Matsukawa, N., Kida, I., Ohta, J., Ikushima, M., Chihara, Y., Rui, X., Rakugi, H., Ogihara, T. (2004). Klotho protein activates the PKC pathway in the kidney and testis and suppresses 25-hydroxyvitamin D3 1alpha-hydroxylase gene expression. *Endocrine, 25*, 229–234.

Invidia, L., Salvioli, S., Altilia, S., Pierini, M., Panourgia, M. P., Monti, D., De Rango, F., Passarino, G., Franceschi, C. (2010). The frequency of Klotho KL-VS polymorphism in a large Italian population, from young subjects to centenarians, suggests the presence of specific time windows for its effect. *Biogerontology, 11*, 67–73.

Isgaard, J., Nilsson, A., Vikman, K., Isaksson, O. G. (1989). Growth hormone regulates the level of insulin-like growth factor-I mRNA in rat skeletal muscle. *The Journal of Endocrinology, 120*, 107–112.

Isley, W. L., Underwood, L. E., Clemmons, D. R. (1983). Dietary components that regulate serum somatomedin-C concentrations in humans. *Journal of Clinical Investigation, 71*, 175–182.
Janssen, H. C., Samson, M. M., Verhaar, H. J. (2002). Vitamin D deficiency, muscle function, and falls in elderly people. *The American Journal of Clinical Nutrition, 75*, 611–615.
Jeanplong, F., Bass, J. J., Smith, H. K., Kirk, S. P., Kambadur, R., Sharma, M., Oldham, J. M. (2003). Prolonged underfeeding of sheep increases myostatin and myogenic regulatory factor Myf-5 in skeletal muscle while IGF-I and myogenin are repressed. *The Journal of Endocrinology, 176*, 425–437.
Kaplan, S. A. & Cohen, P. (2007). The somatomedin hypothesis 2007: 50 years later. *The Journal of Clinical Endocrinology and Metabolism, 92*, 4529–4535.
Keisala, T., Minasyan, A., Lou, Y. R., Zou, J., Kalueff, A. V., Pyykko, I., Tuohimaa, P. (2009). Premature aging in vitamin D receptor mutant mice. *The Journal of Steroid Biochemistry and Molecular Biology, 115*, 91–97.
Khalil, D. A., Lucas, E. A., Juma, S., Smith, B. J., Payton, M. E., Arjmandi, B. H. (2002). Soy protein supplementation increases serum insulin-like growth factor-I in young and old men but does not affect markers of bone metabolism. *The Journal of Nutrition, 132*, 2605–2608.
Kim, J. H., Kwak, H. B., Leeuwenburgh, C., Lawler, J. M. (2008). Lifelong exercise and mild (8%) caloric restriction attenuate age-induced alterations in plantaris muscle morphology, oxidative stress and IGF-1 in the Fischer-344 rat. *Experimental Gerontology, 43*, 317–329.
Klover, P. & Hennighausen, L. (2007). Postnatal body growth is dependent on the transcription factors signal transducers and activators of transcription 5a/b in muscle: a role for autocrine/paracrine insulin-like growth factor I. *Endocrinology, 148*, 1489–1497.
Kofoed, E. M., Hwa, V., Little, B., Woods, K. A., Buckway, C. K., Tsubaki, J., Pratt, K. L., Bezrodnik, L., Jasper, H., Tepper, A., Heinrich, J. J., Rosenfeld, R. G. (2003). Growth hormone insensitivity associated with a STAT5B mutation. *The New England Journal of Medicine, 349*, 1139–1147.
Kuro-O, M. (2008). Klotho as a regulator of oxidative stress and senescence. *Biological Chemistry, 389*, 233–241.
Kuro-O, M. (2009). Klotho and aging. *Biochim Biophys Acta, 1790*(10), 1049–1058. [Epub ahead of print].
Kuro-O, M., Matsumura, Y., Aizawa, H., Kawaguchi, H., Suga, T., Utsugi, T., Ohyama, Y., Kurabayashi, M., Kaname, T., Kume, E., Iwasaki, H., Iida, A., Shiraki-Iida, T., Nishikawa, S., Nagai, R., Nabeshima, Y. I. (1997). Mutation of the mouse klotho gene leads to a syndrome resembling ageing. *Nature, 390*, 45–51.
Kurosu, H., Yamamoto, M., Clark, J. D., Pastor, J. V., Nandi, A., Gurnani, P., Mcguinness, O. P., Chikuda, H., Yamaguchi, M., Kawaguchi, H., Shimomura, I., Takayama, Y., Herz, J., Kahn, C. R., Rosenblatt, K. P., Kuro-O, M. (2005). Suppression of aging in mice by the hormone Klotho. *Science, 309*, 1829–1833.
Lai, K. M., Gonzalez, M., Poueymirou, W. T., Kline, W. O., NA, E., Zlotchenko, E., Stitt, T. N., Economides, A. N., Yancopoulo, G. D., Glass, D. J. (2004). Conditional activation of akt in adult skeletal muscle induces rapid hypertrophy. *Molecular and Cellular Biology, 24*, 9295–9304.
Le Roith, D., Bondy, C., Yakar, S., Liu, J. L., Butler, A. (2001a). The somatomedin hypothesis: 2001. *Endocrine Reviews, 22*, 53–74.
Le Roith, D., Scavo, L., Butler, A. (2001b). What is the role of circulating IGF-I? *Trends in Endocrinology and Metabolism, 12*, 48–52.
Leger, B., Derave, W., DE Bock, K., Hespel, P., Russell, A. P. (2008). Human sarcopenia reveals an increase in Socs-3 and myostatin and a reduced efficiency of Akt phosphorylation. *Rejuvenation Research, 11*, 163–175.
Lieskovska, J., Ling, Y., Badley-Clarke, J., Clemmons, D. R. (2006). The role of Src kinase in insulin-like growth factor-dependent mitogenic signaling in vascular smooth muscle cells. *The Journal of Biological Chemistry, 281*, 25041–25053.

Lupu, F., Terwilliger, J. D., Lee, K., Segre, G. V., Efstratiadis, A. (2001). Roles of growth hormone and insulin-like growth factor 1 in mouse postnatal growth. *Developmental Biology, 229*, 141–162.

Lynch, G. S., Shavlakadze, T., Grounds, M. D. (2005). Strategies to reduce age-related skeletal muscle wasting. In R. Sis (Ed.), *Ageing interventions and therapies*. Singapore: World Scientific Publishers.

Macintosh, B. R., Gardiner, P. F., Mccomas, A. J. (2006). *Skeletal muscle form and function (2nd edn)*. Champaign, IL: Human Kinetics.

Maggio, M., Ble, A., Ceda, G. P., Metter, E. J. (2006). Decline in insulin-like growth factor-I levels across adult life span in two large population studies. *The Journals of Gerontology. Series A: Biological Sciences and Medical Sciences, 61*, 182–183.

Mahesh, S. & Kaskel, F. (2008). Growth hormone axis in chronic kidney disease. *Pediatric Nephrology, 23*, 41–48.

Mak, R. H., Cheung, W. W., Roberts, C. T., Jr. (2008). The growth hormone-insulin-like growth factor-I axis in chronic kidney disease. *Growth Hormone & IGF Research, 18*, 17–25.

Marcell, T. J., Harman, S. M., Urban, R. J., Metz, D. D., Rodgers, B. D., Blackman, M. R. (2001). Comparison of GH, IGF-I, and testosterone with mRNA of receptors and myostatin in skeletal muscle in older men. *American Journal of Physiology. Endocrinology and Metabolism, 281*, E1159–E1164.

Martin, F. C., Yeo, A. L., Sonksen, P. H. (1997). Growth hormone secretion in the elderly: ageing and the somatopause. *Baillière's Clinical Endocrinology and Metabolism, 11*, 223–250.

Mcmahon, C. D., Radcliff, R. P., Lookingland, K. J., Tucker, H. A. (2001). Neuroregulation of growth hormone secretion in domestic animals. *Domestic Animal Endocrinology, 20*, 65–87.

Messina, S., Bitto, A., Aguennouz, M., Minutoli, L., Monici, M. C., Altavilla, D., Squadrito, F., Vita, G. (2006). Nuclear factor kappa-B blockade reduces skeletal muscle degeneration and enhances muscle function in Mdx mice. *Experimental Neurology, 198*, 234–241.

Morelli, S., Buitrago, C., Vazquez, G., De Boland, A. R., Boland, R. (2000). Involvement of tyrosine kinase activity in 1alpha, 25(OH)2-vitamin D3 signal transduction in skeletal muscle cells. *The Journal of Biological Chemistry, 275*, 36021–36028.

Morley, J. E. (1997). Anorexia of aging: physiologic and pathologic. *The American Journal of Clinical Nutrition, 66*, 760–773.

Musaro, A., Mccullagh, K., Paul, A., Houghton, L., Dobrowolny, G., Molinaro, M., Barton, E. R., Sweeney, H. L., Rosentha, N. (2001). Localized Igf-1 transgene expression sustains hypertrophy and regeneration in senescent skeletal muscle. *Nature Genetics, 27*, 195–200.

O'Connor, K. G., Tobin, J. D., Harman, S. M., Plato, C. C., Roy, T. A., Sherman, S. S., Blackman, M. R. (1998). Serum levels of insulin-like growth factor-I are related to age and not to body composition in healthy women and men. *The Journals of Gerontology. Series A: Biological Sciences and Medical Sciences, 53*, M176–M182.

Oldham, J. M., Martyn, J. A., Kirk, S. P., Napier, J. R., Bass, J. J. (1996). Regulation of type 1 insulin-like growth factor (IGF) receptors and IGF-I mRNA by age and nutrition in ovine skeletal muscles. *The Journal of Endocrinology, 148*, 337–346.

Oliver, W. T., Rosenberger, J., Lopez, R., Gomez, A., Cummings, K. K., Fiorotto, M. L. (2005). The local expression and abundance of insulin-like growth factor (IGF) binding proteins in skeletal muscle are regulated by age and gender but not local IGF-1 *in vivo*. *Endocrinology, 146*, 5455–5462.

Paddon-Jones, D. & Rasmussen, B. B. (2009). Dietary protein recommendations and the prevention of sarcopenia. *Current Opinion in Clinical Nutrition and Metabolic Care, 12*, 86–90.

Pedersen, B. K. (2006). The anti-inflammatory effect of exercise: its role in diabetes and cardiovascular disease control. *Essays in Biochemistry, 42*, 105–117.

Pedersen, B. K. (2007). IL-6 signalling in exercise and disease. *Biochemical Society Transactions, 35*, 1295–1297.

Petrella, J. K., Kim, J. S., Cross, J. M., Kosek, D. J., Bamman, M. M. (2006). Efficacy of myonuclear addition may explain differential myofiber growth among resistance-trained young and

older men and women. *American Journal of Physiology. Endocrinology and Metabolism, 291*, E937–E946.
Phillips, S. M. (2006). Dietary protein for athletes: from requirements to metabolic advantage. *Applied Physiology, Nutrition, and Metabolism, 31*, 647–654.
Rabinovsky, E. D. & Draghia-Akli, R. (2004). Insulin-like growth factor I plasmid therapy promotes in vivo angiogenesis. *Molecular Therapy, 9*, 46–55.
Raue, U., Slivka, D., Jemiolo, B., Hollon, C., Trappe, S. (2007). Proteolytic gene expression differs at rest and after resistance exercise between young and old women. *The Journals of Gerontology. Series A: Biological Sciences and Medical Sciences, 62*, 1407–1412.
Rogers, M. A. & Evans, W. J. (1993). Changes in skeletal muscle with aging: effects of exercise training. *Exercise and Sport Sciences Reviews, 21*, 65–102.
Rommel, C., Bodine, S. C., Clarke, B. A., Rossman, R., Nunez, L., Stitt, T. N., Yancopoulos, G. D., Glass, D. J. (2001). Mediation of IGF-1-induced skeletal myotube hypertrophy by PI(3)K/Akt/mTOR and PI(3)K/Akt/GSK3 pathways. *Nature Cell Biology, 3*, 1009–1013.
Roughead, Z. K., Johnson, L. K., Lykken, G. I., Hunt, J. R. (2003). Controlled high meat diets do not affect calcium retention or indices of bone status in healthy postmenopausal women. *The Journal of Nutrition, 133*, 1020–1026.
Sandmand, M., Bruunsgaard, H., Kemp, K., Andersen-Ranberg, K., Schroll, M., Jeune, B. (2003). High circulating levels of tumor necrosis factor-alpha in centenarians are not associated with increased production in T lymphocytes. *Gerontology, 49*, 155–160.
Sasaoka, T., Ishiki, M., Wada, T., Hori, H., Hirai, H., Haruta, T., Ishihara, H., Kobayashi, M. (2001). Tyrosine phosphorylation-dependent and -independent role of Shc in the regulation of IGF-1-induced mitogenesis and glycogen synthesis. *Endocrinology, 142*, 5226–5235.
Sattler, F. R., Castaneda-Sceppa, C., Binder, E. F., Schroeder, E. T., Wang, Y., Bhasin, S., Kawakubo, M., Stewart, Y., Yarasheski, K. E., Ulloor, J., Colletti, P., Roubenoff, R., Azen, S. P. (2009). Testosterone and growth hormone improve body composition and muscle performance in older men. *The Journal of Clinical Endocrinology and Metabolism, 94*, 1991–2001.
Schaap, L. A., Pluijm, S. M., Deeg, D. J., Visser, M. (2006). Inflammatory markers and loss of muscle mass (sarcopenia) and strength. *The American Journal of Medicine, 119*(526), e9–e17.
Schwander, J. C., Hauri, C., Zapf, J., Froesch, E. R. (1983). Synthesis and secretion of insulin-like growth factor and its binding protein by the perfused rat liver: dependence on growth hormone status. *Endocrinology, 113*, 297–305.
Sekimoto, H. & Boney, C. M. (2003). C-terminal Src kinase (CSK) modulates insulin-like growth factor-I signaling through Src in 3T3-L1 differentiation. *Endocrinology, 144*, 2546–2552.
Shavlakadze, T. & Grounds, M. D. (2003). Therapeutic interventions for age-related muscle wasting: importance of innervation and exercise for preventing sarcopenia. In S. Rattan (Ed.), *Modulating aging and longevity*. The Netherlands: Kluwer.
Shavlakadze, T. & Grounds, M. D. (2006). Of bears, frogs, meat, mice and men: insights into the complexity of factors affecting skeletal muscle atrophy/hypertrophy and myogenesis/adipogenesis. *BioEssays, 28*, 994–1009.
Shavlakadze, T., White, J. D., Davies, M., Hoh, J. F., Grounds, M. D. (2005a). Insulin-like growth factor I slows the rate of denervation induced skeletal muscle atrophy. *Neuromuscular Disorders, 15*, 139–146.
Shavlakadze, T., Winn, N., Rosenthal, N., Grounds, M. D. (2005b). Reconciling data from transgenic mice that overexpress IGF-I specifically in skeletal muscle. *Growth Hormone & IGF Research, 15*, 4–18.
Shavlakadze, T., Boswell, J. M., Burt, D. W., Asante, E. A., Tomas, F. M., Davies, M. J., White, J. D., Grounds, M. D., Goddard, C. (2006). Rskalpha-actin/hIGF-1 transgenic mice with increased IGF-I in skeletal muscle and blood: impact on regeneration, denervation and muscular dystrophy. *Growth Hormone & IGF Research, 16*, 157–173.
Shemer, J., Adamo, M. L., Roberts, C. T., J. R., Leroith, D. (1992). Tissue-specific transcription start site usage in the leader exons of the rat insulin-like growth factor-I gene: evidence for differential regulation in the developing kidney. *Endocrinology, 131*, 2793–2799.

Sjogren, K., Liu, J. L., Blad, K., Skrtic, S., Vidal, O., Wallenius, V., Leroith, D., Tornell, J., Isaksson, O. G., Jansson, J. O., Ohlsson, C. (1999). Liver-derived insulin-like growth factor I (IGF-I) is the principal source of IGF-I in blood but is not required for postnatal body growth in mice. *Proceedings of the National Academy of Sciences of the United States of America, 96*, 7088–7092.

Sotiropoulos, A., Ohanna, M., Kedzia, C., Menon, R. K., Kopchick, J. J., Kelly, P. A., Pende, M. (2006). Growth hormone promotes skeletal muscle cell fusion independent of insulin-like growth factor 1 up-regulation. *PNAS, 103*, 7315–7320.

Stratikopoulos, E., Szabolcs, M., Dragatsis, I., Klinakis, A., Efstratiadis, A. (2008). The hormonal action of Igf1 in postnatal mouse growth. *Proceedings of the National Academy of Sciences of the United States of America, 105*, 19378–19383.

Strle, K., Broussard, S. R., Mccusker, R. H., Shen, W. H., Johnson, R. W., Freund, G. G., Dantzer, R., Kelley, K. W. (2004). Proinflammatory cytokine impairment of insulin-like growth factor I-induced protein synthesis in skeletal muscle myoblasts requires ceramide. *Endocrinology, 145*, 4592–4602.

Teglund, S., Mckay, C., Schuetz, E., Van Deursen, J. M., Stravopodis, D., Wang, D., Brown, M., Bodner, S., Grosveld, G., Ihle, J. N. (1998). Stat5a and Stat5b proteins have essential and nonessential, or redundant, roles in cytokine responses. *Cell, 93*, 841–850.

Thissen, J. P., Ketelslegers, J. M., Underwood, L. E. (1994). Nutritional regulation of the insulin-like growth factors. *Endocrine Reviews, 15*, 80–101.

Thomson, D. M. & Gordon, S. E. (2006). Impaired overload-induced muscle growth is associated with diminished translational signalling in aged rat fast-twitch skeletal muscle. *Journal de Physiologie, 574*, 291–305.

Tisdale, M. J. (2005). The ubiquitin-proteasome pathway as a therapeutic target for muscle wasting. *The Journal of Supportive Oncology, 3*, 209–217.

Tisdale, M. J. (2009). Mechanisms of cancer cachexia. *Physiological Reviews, 89*, 381–410.

Tuohimaa, P. (2009). Vitamin D and aging. *The Journal of Steroid Biochemistry and Molecular Biology, 114*, 78–84.

Udy, G. B., Towers, R. P., Snell, R. G., Wilkins, R. J., Park, S. H., Ram, P. A., Waxman, D. J., Davey, H. W. (1997). Requirement of STAT5b for sexual dimorphism of body growth rates and liver gene expression. *Proceedings of the National Academy of Sciences of the United States of America, 94*, 7239–7244.

Ullman, M., Ullman, A., Sommerland, H., Skottner, A., Oldfors, A. (1990). Effects of growth hormone on muscle regeneration and IGF-1 concentration in old rats. *Acta Physiologica Scandinavica, 140*, 521–525.

UM, S. H., Frigerio, F., Watanabe, M., Picard, F., Joaquin, M., Sticker, M., Fumagalli, S., Allegrini, P. R., Kozma, S. C., Auwerx, J., Thomas, G. (2004). Absence of S6K1 protects against age- and diet-induced obesity while enhancing insulin sensitivity. *Nature, 431*, 200–205.

Veldhuis, J. D., Liem, A. Y., South, S., Weltman, A., Weltman, J., Clemmons, D. A., Abbott, R., Mulligan, T., Johnson, M. L., Pincus, S. (1995). Differential impact of age, sex steroid hormones, and obesity on basal versus pulsatile growth hormone secretion in men as assessed in an ultrasensitive chemiluminescence assay. *The Journal of Clinical Endocrinology and Metabolism, 80*, 3209–3222.

Visser, M., Pahor, M., Taaffe, D. R., Goodpaster, B. H., Simonsick, E. M., Newman, A. B., Nevitt, M., Harris, T. B. (2002). Relationship of interleukin-6 and tumor necrosis factor-alpha with muscle mass and muscle strength in elderly men and women: The Health ABC Study. *The Journals of Gerontology. Series A: Biological Sciences and Medical Sciences, 57*, M326–M332.

Welle, S., Bhatt, K., Shah, B., Thornton, C. (2002). Insulin-like growth factor-1 and myostatin mRNA expression in muscle: comparison between 62–77 and 21–31 yr old men. *Experimental Gerontology, 37*, 833–839.

Winn, N., Paul, A., Musaro, A., Rosenthal, N. (2002). Insulin-like growth factor isoforms in skeletal muscle aging, regeneration, and disease. *Cold Spring Harbor Symposia on Quantitative Biology, 67*, 507–518.

Wolfe, R. R., Miller, S. L., Miller, K. B. (2008). Optimal protein intake in the elderly. *Clinical Nutrition, 27*, 675–684.

Wong, M. S., Sriussadaporn, S., Tembe, V. A., Favus, M. J. (1997). Insulin-like growth factor I increases renal 1, 25(OH)2D3 biosynthesis during low-P diet in adult rats. *The American Journal of Physiology, 272*, F698–F703.

Wong, M. S., Tembe, V. A., Favus, M. J. (2000). Insulin-like growth factor-I stimulates renal 1, 25-dihydroxycholecalciferol synthesis in old rats fed a low calcium diet. *The Journal of Nutrition, 130*, 1147–1152.

Wurtman, J. J., Lieberman, H., Tsay, R., Nader, T., Chew, B. (1988). Calorie and nutrient intakes of elderly and young subjects measured under identical conditions. *Journal of Gerontology, 43*, B174–B180.

Yakar, S., Liu, J. L., Stannard, B., Butler, A., Accili, D., Sauer, B., Leroith, D. (1999). Normal growth and development in the absence of hepatic insulin-like growth factor I. *Proceedings of the National Academy of Sciences of the United States of America, 96*, 7324–7329.

Yang, S. Y. & Goldspink, G. (2002). Different roles of the IGF-I Ec peptide (MGF) and mature IGF-I in myoblast proliferation and differentiation. *FEBS Letters, 522*, 156–160.

Yarasheski, K. E. (2003). Exercise, aging, and muscle protein metabolism. *Journal of Gerontology: Medical Sciences, 58A*, 918–922.

Yu, F., Degens, H., Larsson, L. (1999). The influence of thyroid hormone on myosin isoform composition and shortening velocity of single skeletal muscle fibres with special reference to ageing and gender. *Acta Physiologica Scandinavica, 167*(4), 313–316.

Zha, Y., Taguchi, T., Nazneen, A., Shimokawa, I., Higami, Y., Razzaque, M. S. (2008). Genetic suppression of GH-IGF-1 activity, combined with lifelong caloric restriction, prevents age-related renal damage and prolongs the life span in rats. *American Journal of Nephrology, 28*, 755–764.

Role of Myostatin in Skeletal Muscle Growth and Development: Implications for Sarcopenia

Craig McFarlane, Mridula Sharma, and Ravi Kambadur

Abstract Myostatin is a secreted growth and differentiating factor that belongs to TGF-β super-family. Myostatin is expressed in skeletal muscle predominantly. Low levels of myostatin expression are seen in heart, adipose tissue and mammary gland. Naturally occurring mutations in bovine, ovine, canine and human *myostatin* gene or inactivation of the murine *myostatin* gene lead to an increase in muscle mass due to hyperplasia. Molecularly, myostatin has been shown to regulate muscle growth not only by controlling myoblast proliferation and differentiation during fetal myogenesis, but also by regulating satellite cell activation and self-renewal postnatally. Consistent with the molecular genetic studies, injection of several myostatin blockers including Follistatin, myostatin antibodies and the Prodomain of myostatin have all been independently shown to increase muscle regeneration and growth in muscular dystrophy mouse models of muscle wasting. Furthermore, prolonged absence of *myostatin* in mice has also been shown to reduce sarcopenic muscle loss, due to efficient satellite cell activation and regeneration of skeletal muscle in aged mice. Similarly, treatment of aged mice with Mstn-ant 1 also increased satellite cell activation and enhanced the efficiency of muscles to regenerate. Given that antagonism of myostatin leads to significant increase in postnatal muscle growth, we propose that myostatin antagonists have tremendous therapeutic value in alleviating sarcopenic muscle loss.

Keywords Myostatin • GDF-8 • Skeletal muscle • Smad • Wnt • Proliferation • Differentiation • Satellite cells • Muscle wasting • Atrophy • Cachexia • Sarcopenia

R. Kambadur (✉)
School of Biological Sciences, Nanyang Technological University, 60 Nanyang Drive, Singapore
e-mail: Kravi@ntu.edu.sg

C. McFarlane and R. Kambadur
Singapore Institute for Clinical Sciences, Singapore

M. Sharma
Department of Biochemistry, National University of Singapore, Singapore

G.S. Lynch (ed.), *Sarcopenia – Age-Related Muscle Wasting and Weakness*,
DOI 10.1007/978-90-481-9713-2_18,

1 Myostatin

1.1 The Myostatin Gene, Structure and Processing

Myostatin, or growth and differentiation factor-8 (GDF-8), is a TGF-β superfamily member that was initially characterised in 1997 as a specific regulator of skeletal muscle mass in mice (McPherron et al. 1997). Targeted disruption of the *myostatin* gene in mice (Fig. 1) resulted in a generalised increase in skeletal muscle mass (double-muscling); in particular a two to threefold increase in muscle weight was observed with no corresponding increase in adipose tissue (Fig. 1). The enhanced muscle phenotype in the *myostatin*-null mice was determined to result from a combination of both muscle hyperplasia and hypertrophy (McPherron et al. 1997).

Myostatin has a number of characteristics common to the TGF-β superfamily (Fig. 2). In particular, the precursor myostatin molecule contains an N-terminal (NH2) core of hydrophobic amino acids that functions as a signal sequence for secretion (McPherron et al. 1997). In addition, the C-terminal (COOH) region of myostatin contains nine conserved cysteine residues which are critical for homodimerisation and for the formation of the "cysteine knot" structure, a characteristic feature of the TGF-β superfamily (McPherron and Lee 1996; McPherron et al. 1997). Furthermore, myostatin is synthesised in myoblasts as a 376 amino acid precursor protein which, like other members of the TGF-β superfamily, is proteolytically cleaved at the RSRR site (Fig. 2), a process which occurs within the Golgi apparatus under the control of the serine protease furin or other members of the proprotein convertase family (Lee and McPherron 2001; McPherron et al. 1997; Sharma et al. 1999). Proteolytic processing of the myostatin 52 kDa precursor protein by furin results in the formation of a 36/40 kDa Latency-Associated Peptide (LAP) and a 12.5/26 kDa mature portion, which is suggested to correspond to a C-terminal monomer or dimer respectively (Lee and McPherron 2001; McFarlane et al. 2005; Thomas et al. 2000). The processed mature form of

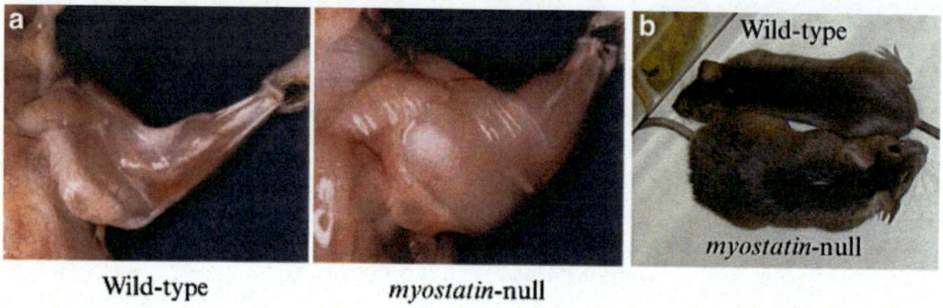

Fig. 1 Double-muscling in myostatin-null mice. **(a)** Photograph showing the difference between the forelimbs of wild-type and myostatin-null mice. A dramatic increase in skeletal muscle mass is observed in the myostatin-null mice compared to wild-type mice (Adapted from McPherron et al. 1997). **(b)** Photograph showing the size difference between wild-type and myostatin-null mice at the same age. Myostatin-null mice were generated by McPherron et al. (1997)

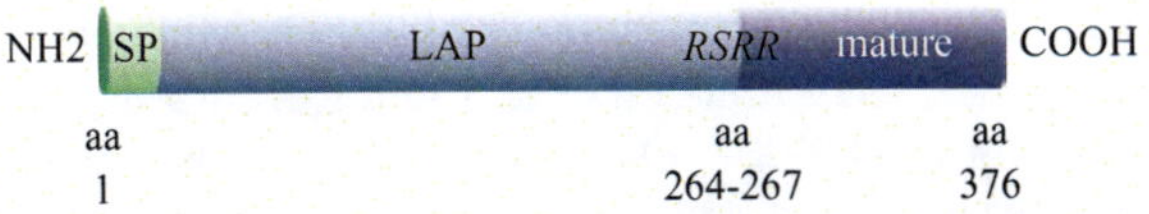

Fig. 2 The structure of myostatin. Schematic representation of the structure of myostatin. Myostatin shares characteristics common to the TGF-β superfamily, including a signal peptide (SP) for secretion and a RSRR proteolytic processing site. Proteolytic processing of myostatin gives rise to LAP and mature myostatin regions (Adapted from Joulia-Ekaza and Cabello [2006])

myostatin, together with LAP, is subsequently secreted from myoblasts and it is the C-terminal mature region that is able to bind to the receptor and elicit biological function. The importance of proteolytic processing is clear, as generation of a dominant-negative form of myostatin, through mutation of the RSRR site to the amino acids GLDG, results in widespread skeletal muscle hypertrophy (Zhu et al. 2000). Previously it has been demonstrated that processing of myostatin is developmentally regulated, whereby reduced myostatin processing is observed during fetal muscle development when comparted to post-natal stages of growth (McFarlane et al. 2005). Furthermore it was demonstrated that there is reduced proteolytic processing of myostatin during myogenic differentiation and more importantly myostatin has the ability to negatively regulate the expression of the serine protease furin. Myostatin inhibition of furin expression was proposed to be a mechanism through which myostatin negatively auto-regulates its processing during the critical periods of fetal growth, thereby facilitating the differentiation of myoblasts (McFarlane et al. 2005)

1.2 Expression of Myostatin

Myostatin is first detected in mice embryos at day 9.5 post-coitum, where it is specifically located within the most rostral somites (McPherron et al. 1997). By day 10.5 post-coitum, myostatin is expressed in the majority of the somites, specifically located in the myotome layer of developing somites (McPherron et al. 1997). In cattle, low levels of *myostatin* mRNA are detected in day 15 to day 29 embryos with increasing expression detected from day 31 onwards (Kambadur et al. 1997; Bass et al. 1999; Oldham et al. 2001). Furthermore, in the pig foetus *myostatin* mRNA expression is abundant at days 21 and 35 of gestation, with an increase in expression by day 49 (Ji et al. 1998). In the chicken *myostatin* expression is first detected as early as embryonic day 0 (the blastoderm stage) with relatively low levels detected through to embryonic day 6. From day 7, *myostatin* mRNA levels rapidly increase and level off through to day 16 (Kocamis et al. 1999). Post-natal skeletal muscle continues to express *myostatin*, although variation in *myostatin* expression is observed between individual muscles (Kambadur et al. 1997; McPherron et al. 1997). The expression of *myostatin* is primarily restricted to skeletal muscle (Kambadur et al. 1997; McPherron et al. 1997; Ji et al. 1998; Bass et al. 1999; Carlson et al. 1999; Kocamis et al. 1999; Sazanov et al. 1999; Jeanplong et al. 2001; Oldham et al. 2001), however, low levels of myostatin expression have been detected in various other tissues; in particular in the secretory lobules of lactating mammary glands (Ji et al. 1998), in adipose tissue (McPherron et al. 1997), and in cardiomyocytes and Purkinje fibres of the heart (Sharma et al. 1999). More recently it has been shown that both myostatin mRNA and protein are expressed in human placental tissue. The presence of myostatin in the placenta is suggested to be involved with uptake of glucose (Mitchell et al. 2006).

Myostatin expression may also be associated with specific fibre types in skeletal muscle. Carlson et al. have shown that higher amounts of myostatin mRNA and protein are detected in fast-twitch muscle (type-II fibres) as compared to slow-twitch muscle (type-I fibres) (Carlson et al. 1999). Furthermore, it has been shown that in *myostatin*-null mice there is an increase in fast fibres (type-II) in the typically slow fibre-dominated *M. soleus* muscle, and a switch from oxidative (type-IIA) to glycolytic fibres (type-IIB) in the predominantly fast-twitch EDL muscle (Girgenrath et al. 2005). Therefore, suggesting a fibre type-specific role for myostatin in regulation of muscle physiology.

1.3 Regulation of Myostatin

Myostatin is synthesised as a precursor protein, proteolytically processed and secreted to elicit its biological function. Studies have highlighted the importance of several proteins that interact with myostatin to regulate its action. Myostatin has been shown to interact with the sarcomeric protein Titin-cap (Nicholas et al. 2002); specifically, titin-cap interacts with the C-terminal mature portion of myostatin (Nicholas et al. 2002). Over-expression of titin-cap had no effect on myostatin synthesis and processing, however, increased titin-cap expression results in enhanced cell proliferation and accumulation of processed myostatin within myoblasts. Thus, titin-cap appears to function by regulating the secretion of mature myostatin (Nicholas et al. 2002). In addition, human small glutamine-rich tetratricopeptide repeat-containing protein (hSGT) has been shown to associate with intracellular myostatin (Wang et al. 2003). The C-terminal region of hSGT and the N-terminal signal peptide region of myostatin were shown to be critical for this interaction. It is suggested that hSGT likely plays a role in mediating myostatin secretion and activation (Wang et al. 2003). Latent TGF-β binding proteins (LTBPs) are extracellular matrix proteins which have been previously identified to interact with the TGF-β superfamily (Saharinen et al. 1999). LTBPs associate with TGF-β superfamily members to allow for secretion; once secreted, removal of LTBPs from the latent complex is essential for TGF-β activation (Saharinen et al. 1999). Although LTBPs play an essential role in the secretion and activation of TGF-β superfamily members, published results from this thesis suggest that LTBPs do not play a role in the regulation of myostatin (McFarlane et al. 2005). Following secretion, the majority of myostatin (>70%), like TGF-β, has been shown to exist in an inactive latent complex both in vitro and in vivo, whereby the mature processed portion of myostatin is bound non-covalently to the propeptide (LAP) region of myostatin (Lee and McPherron 2001; Thies et al. 2001; Yang et al. 2001). Recently it has been demonstrated that members of the bone morphogenetic protein-1/tolloid (BMP-1/TLD) family can cleave the myostatin LAP region from the latent myostatin complex, thus resulting in activation of mature myostatin (Wolfman et al. 2003). Furthermore, Wolfman et al. demonstrated that a mutation of LAP to confer resistance to cleavage by BMP/TLD resulted in enhanced muscle mass in vivo. Previous studies have demonstrated that follistatin is capable of binding and inhibiting various members of the TGF-β superfamily (Fainsod et al. 1997; Hemmati-Brivanlou et al. 1994; Michel et al. 1993). Follistatin has been shown to bind directly to the mature portion of myostatin blocking the ability of myostatin to bind with the ActRIIB receptor (Lee and McPherron 2001). Furthermore, interaction with follistatin interferes with the intrinsic ability of myostatin to inhibit muscle differentiation (Amthor et al. 2004). In support, mice over-expressing follistatin show a drastic increase in muscle mass, significantly greater than that of *myostatin*-null animals (Lee and McPherron 2001). Additionally, *follistatin*-null mice demonstrate reduced muscle mass at birth (Matzuk et al. 1995), consistent with increased myostatin activity. Follistatin-related gene (FLRG), like follistatin, is able to bind and inhibit members of the TGF-β superfamily (Tsuchida et al. 2000, 2001; Schneyer et al. 2001). In addition, FLRG has been shown to interact directly with the mature portion of myostatin, resulting in a dose-dependent reduction in the activity of myostatin, as assessed through reporter gene assay analysis (Hill et al. 2002). Growth and differentiation factor-associated serum protein-1 (GASP-1) has been shown to associate with myostatin in circulation; specifically associating with both mature and LAP regions of myostatin. Functionally GASP-1 has been shown to interfere with the activity of myostatin as determined by reporter gene analysis (Hill et al. 2003). More recently, decorin, a leucine-rich repeat extracellular proteoglycan, has been shown to interact with the mature region of myostatin, in a Zn^{2+}-dependent manner (Miura et al. 2006). This interaction was demonstrated to relieve the inhibitory effect of myostatin on myoblast proliferation in vitro. One of the intrinsic features of myostatin is its ability to negatively auto-regulate its expression. In particular, exogenous addition of recombinant myostatin protein results in both a decrease in *myostatin* mRNA and repression of *myostatin* promoter activity (Forbes et al. 2006). Furthermore, myostatin appears to signal through Smad7 to regulate its own activity (Forbes et al. 2006; Zhu et al. 2004). In support, addition of myostatin resulted in enhanced Smad7 expression, while over-expression of Smad7 resulted in repression of *myostatin* promoter activity and mRNA, an effect abolished through incubation with siRNA specific for Smad7 (Forbes et al. 2006; Zhu et al. 2004).

1.4 Mutations in Myostatin

In addition to the targeted disruption of *myostatin* in mice, several naturally occurring mutations have been identified in various double-muscled cattle breeds including Belgian Blue (Fig. 3a) and Piedmontese (Kambadur et al. 1997; McPherron and Lee 1997; Grobet et al. 1998). Specifically two separate mutations in the coding region of the *myostatin* gene have been reported to result in a non-functional myostatin product. The phenotype seen in Belgian Blue cattle (Fig. 3a) is caused by an

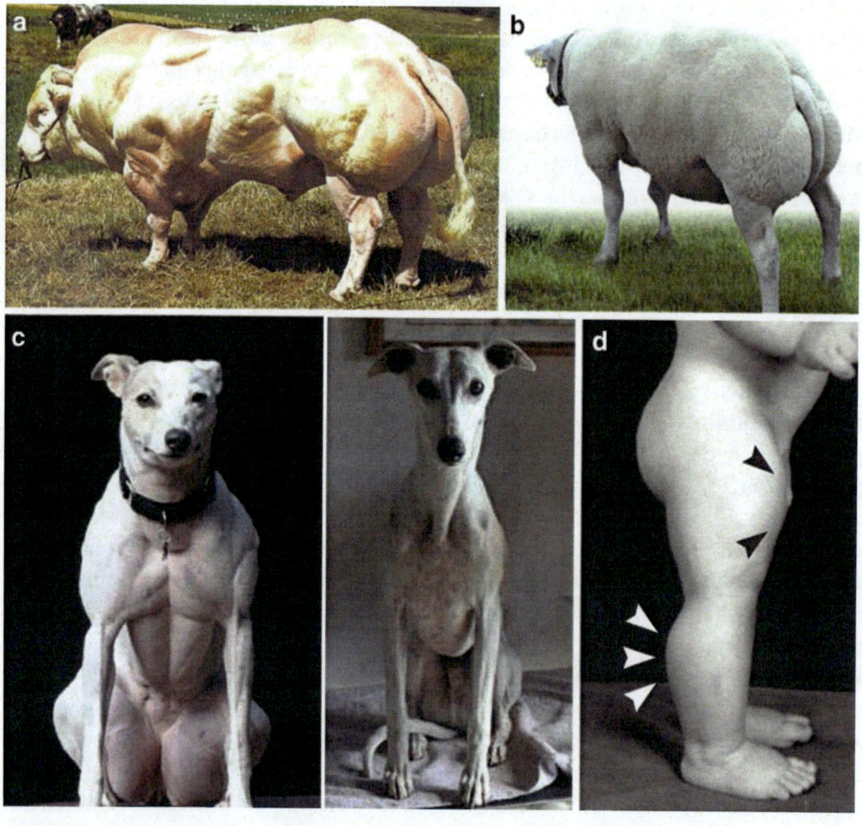

Fig. 3 Natural mutations in myostatin. (**a**) Photograph showing the heavy muscling observed in the Belgian Blue cattle breed (Reproduced from Haliba '96 Catalogue). (**b**) Photograph of a Texel sheep demonstrating the heavy muscle phenotype oberved in response to a G to A transition mutation in the 3′ UTR of the *myostatin* gene, which results in the formation of *mir1* and *mir206* miRNA sites (Reproduced from Skipper [2006]). (**c**) Photographs of a heavy muscled Whippet dog (*left*) and a Whippet dog demonstrating more typcial muscle mass (*right*) (Reproduced form Shelton and Engvall [2007]). (**d**) Photograph of a human child at 7 months of age possessing a G to A transition mutation in the *myostatin* gene, resulting in a non functional myostatin protein product. *Arrows* highlight protruding muscles from the boy's calf and thigh regions (Modified from Schuelke et al. [2004]).

11-nucleotide deletion, which ultimately results in expression of a non-functional truncated protein product (Kambadur et al. 1997). Conversely, the Piedmontese cattle express a non-functional myostatin protein through a missense mutation in the gene sequence, resulting in a G to A transition and substitution of cysteine for tyrosine (Kambadur et al. 1997; Berry et al. 2002). Furthermore, a mutation in the *myostatin* gene has been reported to result in the hyper-muscularity observed in compact (*Cmpt*) mice (Szabo et al. 1998). More recently, the heavy muscled phenotype of the Texel sheep breed has been traced to a mutation in the *myostatin* gene resulting in a G to A transition in the 3′ untranslated region (UTR) (Fig. 3b) (Clop et al. 2006). This mutation creates a target site for two microRNAs abundant in skeletal muscle, namely *mir1* and *mir206* (Clop et al. 2006). MicroRNAs are short non-coding RNAs which diminish gene activity post-transcriptionally by binding to target genes, resulting in destabilisation of mRNA and/or inhibition of protein translation (Tsuchiya et al. 2006). In addition to the Texel breed, a mutation in the *myostatin* gene has been demonstrated to result in the increased muscle mass phenotype observed in the Norwegian Spælsau sheep breed. Specifically a one base pair insertion mutation at nucleotide 120 from the translation start site (*c.120insA*) results in the formation of a premature stop codon at amino acid 49 resulting in the formation of a non-functional protein product (Boman and Vage 2009).

Recently a mutation in the *myostatin* gene has been shown to result in dramatic muscle hypertrophy in the Whippet racing dog breed (Fig. 3c) (Mosher et al. 2007). The pheotype results form a two base pair deletion in the third exon of the *myostatin* gene and leads to the formation of a premature stop codon at amino acid 313 resulting in a non-functional protein product. Interestingly, Whippet dogs heterozygote for the mutation are not only more muscular than wildtype but are significantly faster as well which, for the first time, demonstrates the utility of mutations in *myostatin* and enhanced atheletic performance (Mosher et al. 2007).

A mutation in the *myostatin* gene has also been shown to result in dramatic hypertrophy in a human child (Schuelke et al. 2004) (Fig. 3d). Cross-sectional measurements determined that the *M. quadriceps* muscle was more than twofold larger than age- and sex-matched controls, while the thickness of the sub-cutaneous fat pad was significantly lower than controls. The mutation was shown to result from a G to A transition within intron 1 of the *myostatin* gene. This transition resulted in mis-splicing of the precursor mRNA and insertion of the first 108 base pairs of intron 1 (Schuelke et al. 2004).

2 Physiological Actions of Myostatin

2.1 Myostatin Signaling

Members of the TGF-β superfamily elicit biological functions by binding to specific type-I and type-II serine/threonine kinase receptors. Studies have shown that myostatin specifically binds to the activin type-IIB (ActRIIB) receptor (Lee and McPherron 2001; Rebbapragada et al. 2003). Indeed, transgenic mice that over-express a dominant-negative form of the ActRIIB show a drastic increase in muscle weights, similar to that seen in *myostatin*-null mice (Lee and McPherron 2001). Myostatin-mediated type-II receptor activation results in the phosphorylation of the type-I receptor, either activin receptor-like kinase 4 (ALK4) or ALK5, which in turn initiates downstream signaling events (Rebbapragada et al. 2003).

TGF-β superfamily signalling is primarily mediated through substrates known as Smads (Piek et al. 1999). Smad proteins can be separated into three sub-groups: the receptor Smads (R-Smads; Smads 1, 2, 3, 5 and 8), the common Smad (Co-Smad; Smad 4) and the inhibitory Smads (I-Smads; Smads 6 and 7) (Piek et al. 1999). Phosphorylation of the R-Smads occurs at the type-I receptor, the now active R-Smad heterodimerises with the Co-Smad and translocates to the nucleus to regulate transcription (Nakao et al. 1997b; Souchelnytskyi et al. 1997; Zhang et al. 1997). Inhibitory Smads can compete with R-Smads for receptor binding and Co-Smad heterodimerisation, thus blocking Smad-mediated signaling (Hata et al. 1998; Hayashi et al. 1997; Nakao et al. 1997a). Consistent with other members of the TGF-β superfamily, myostatin has been shown to signal specifically through Smads 2/3 with the involvement of Smad 4 (Zhu et al. 2004). In addition, it appears that myostatin-mediated Smad signaling is negatively regulated by Smad 7 but not Smad 6 (Zhu et al. 2004). Furthermore, myostatin has also been shown to induce the expression of Smad 7. Interestingly, this induction of Smad 7 appears to provide an auto-regulatory mechanism through which myostatin negatively regulates its own activity (Forbes et al. 2006; Zhu et al. 2004).

In addition to canonical Smad signaling the Wnt pathway has been implicated in myostatin regulation of post-natal skeletal muscle growth. Microarray analysis of muscle isolated from wildtype and *myostatin*-null mice has identified differential expression of a number of genes involved in Wnt signaling (Steelman et al. 2006). In particular, it was identified that genes involved in the canonical β-catenin pathway were down regulated in muscle isolated from *myostatin*-null mice whereas genes involved in the Wnt/calcium pathway were up regulated. Furthermore, Steelman et al. identify that Wnt4 has a positive role in regulating satellite cell proliferation and further propose a mechanism whereby myostatin acts upstream of Wnt4 to block Wnt4-mediated satellite cell proliferation. In addition, myostatin is shown to enhance the expression of sFRP1 and -2, two known inhibitors of the Wnt signaling pathway (Steelman et al. 2006). Therefore myostatin may negatively regulate satellite cell proliferation through preceding regulation of the Wnt signaling pathway.

2.2 *Regulation of Proliferation and Differentiation*

It has been previously shown that myostatin is a negative regulator of skeletal muscle growth (Kambadur et al. 1997; McPherron et al. 1997). Several cell culture based studies have analysed the role of myostatin in the regulation of cell proliferation. Myostatin has been shown to negatively regulate skeletal muscle growth through inhibiting the proliferation of myoblast cell lines in a dose-dependent, reversible manner (Thomas et al. 2000). In support, primary myoblasts isolated from *myostatin*-null mice proliferate significantly faster than myoblast cultures from wild-type mice (McCroskery et al. 2003). More recently, myostatin has been demonstrated to reversibly inhibit the proliferation of Pax7-positive myogenic precursor cells in embryos injected with myostatin-coated beads (Amthor et al. 2006). Mechanistically, myostatin appears to interact with the cell cycle machinery, resulting in cell cycle exit during the gap phases (G_1 and G_2) (Thomas et al. 2000). Specifically, treatment with myostatin results in up-regulation of the cyclin-dependent kinase inhibitor (CKI), p21 (Thomas et al. 2000). p21 is a member of the Cip/Kip family of CKIs which, as their name suggests, block the action of cyclin-dependent kinases and their cyclin partners (Harper et al. 1993; Xiong et al. 1993). Consistent with this, treatment with recombinant myostatin protein has been shown to decrease the expression and activity of cyclin-dependent kinase 2 (cdk2) (Thomas et al. 2000). The myostatin-mediated loss in cdk2 activity resulted in accumulation of hypophosphorylated retinoblastoma (Rb), which in turn induces cell cycle arrest in the G_1 phase. A recent report has highlighted a role for the p38 mitogen-activated protein kinase (MAPK) signaling pathway in myostatin regulation of myogenesis (Philip et al. 2005). In particular, myostatin has been shown to activate p38 MAPK; moreover this activation was shown to augment myostatin-mediated transcription. Furthermore, p38 MAPK was shown to play an important role in myostatin-mediated up-regulation of p21 and subsequent inhibition of cell proliferation (Philip et al. 2005). In addition, myostatin has been shown to inhibit the proliferation of the rhabdomyosarcoma cell line, RD (Langley et al. 2004). However, unlike normal myoblasts, treatment with myostatin did not up-regulate the expression of p21 or alter the phosphorylation or activity of Rb. Langley et al. demonstrated that treatment with myostatin resulted in a reduction in expression and activity of cdk2 and cyclin E. NPAT is a substrate of cdk2/cyclinE and is critical for the continuation of the cell cycle at the G1/S checkpoint. Thus treatment of the RD cell line with myostatin also reduced the phosphorylation of NPAT, concomitant with a reduction in the expression of the NPAT target histone-H4 (Langley et al. 2004).

In addition to the intrinsic ability of myostatin to regulate myoblast proliferation, myostatin has been shown to negatively regulate myogenic differentiation. (Rios et al. 2002; Langley et al. 2002). In particular, treatment of myoblasts with recombinant myostatin protein resulted in a dose-dependent reversible inhibition of differentiation (Langley et al. 2002). Furthermore, treatment of differentiating myoblasts with myostatin inhibited the mRNA and protein expression of MyoD, Myf5, myogenin and MHC (Rios et al. 2002; Langley et al. 2002). Langley et al. further demonstrated that during differentiation, treatment with myostatin increased the phosphorylation of Smad 3 and enhanced Smad 3•MyoD interaction. MyoD is critical for the successful commitment to myogenic differentiation, and furthermore MyoD has been shown to induce cell cycle arrest and induce differentiation through up-regulation of p21. Thus, Langley et al. proposed that myostatin blocked myogenic differentiation by inhibiting the expression and activity of MyoD in a Smad 3-dependent manner. Recently a role for the extracellular signal-regulated kinase 1/2 (Erk1/2) MAPK signaling pathway has been identified in myostatin regulation of myogenesis (Yang et al. 2006). Indeed, inhibition of the Erk1/2 pathway suppressed myostatin-mediated inhibition of myoblast proliferation and differentiation and further interfered with the ability of myostatin to inhibit the expression of genes critical to myogenic differentiation, including MyoD, myogenin and Myosin Heavy Chain (MHC) (Yang et al. 2006).

2.3 *Post-Natal Muscle Growth and Repair*

Myostatin expression is detected during embryonic and foetal growth and is maintained through into adult muscle tissue, thus myostatin may be an important mediator of skeletal muscle mass throughout myogenesis. Indeed myostatin appears to play a critical role in the regulation of post-natal muscle growth and repair. Several studies have analysed the effect of post-natal modification of myostatin on skeletal muscle mass. Over-expression of a dominant-negative myostatin, whereby the RSRR processing site was mutated to GLDG, resulted in a 25–30% increase in skeletal muscle mass in mice; specifically resulting from increased hypertrophy rather than hyperplasia (Zhu et al. 2000). In contrast, recapitulation of the Piedmontese cattle C313Y mis-sense mutation in mice results in skeletal muscle hyperplasia without muscle hypertrophy (Nishi et al. 2002). Furthermore, injection of the JA16 monoclonal myostatin-neutralising antibody into mice resulted in an increase in skeletal muscle mass (Whittemore et al. 2003). It was determined that incubation with the JA16 antibody for 2–4 weeks was sufficient to induce an increase in muscle mass as compared to control mice. Concomitant to an effect on muscle mass, injection of the neutralising antibody increased the grip strength of treated mice, specifically a 10% increase in peak force was observed (Whittemore et al. 2003). Another study focused on the effect of conditionally targeting *myostatin* for inactivation using the cre-lox system. Subsequent inactivation of *myostatin* resulted in skeletal muscle hypertrophy phenotypically similar to that observed in *myostatin*-null mice (Grobet et al. 2003). More recently, an increase in muscle mass was observed following injection of a myostatin-specific short interfering RNA (siRNA) directly into the *M. tibialis anterior* (TA) muscle of rats (Magee et al. 2006). The siRNA-mediated knockdown resulted in a 27% decrease in *myostatin* mRNA and a 48% decrease in myostatin protein expression. Furthermore, myostatin inhibition resulted in an increase in TA muscle weight and myofibre area. Satellite cell number was also increased twofold, as quantified by the number of Pax7-positive cells (Magee et al. 2006). Thus inhibitors directed against myostatin may have therapeutic benefit in circumstances where skeletal muscle wasting enhances the morbidity or mortality of a disease.

Myostatin has been demonstrated to be involved in the regulation of skeletal muscle regeneration. A recent study has compared the regeneration process of skeletal muscle in *myostatin*-null mice versus wild-type controls following injection of the myotoxin, notexin (McCroskery et al. 2005). Following injury, satellite cell-derived myoblasts migrate to the site of injury to help repair the damage (Watt et al. 1987, 1994). Muscle damage is closely followed by a localised inflammatory response resulting in the influx of macrophages to the site of injury (Tidball 1995). Interestingly, McCroskery et al. found that lack of *myostatin* increased the rate of myogenic cell migration and the rate of macrophage infiltration to the site of injury, resulting in enhanced numbers of both. Furthermore, presence of recombinant myostatin protein in vitro significantly reduced the migration of both myoblasts and macrophages in chemotaxis chambers (McCroskery et al. 2005). McCroskery et al. subsequently proposed a mechanism for myostatin regulation of skeletal muscle regeneration, as shown in Fig. 4. The formation of scar tissue is a prominent feature of skeletal muscle injury. However, during the process of regeneration the presence of scar tissue was greatly reduced in regenerated muscle from *myostatin*-null as compared with muscle from wild-type mice. Thus, in addition to regulating the involvement of satellite cells and macrophages in muscle regeneration, myostatin may also contribute to skeletal muscle fibrosis (McCroskery et al. 2005).

Satellite cells are responsible for maintaining and repairing skeletal muscle mass following injury. Myostatin has been shown to play a role in regulating satellite cell activation, growth and self-renewal (McCroskery et al. 2003). Myostatin is expressed within muscle satellite cells and satellite cell-derived primary myoblasts. Specifically, satellite cells, characterised through positive Pax7 staining, were also positive for myostatin by immunocytochemistry. Furthermore, in situ hybridisation confirmed high expression of both *pax7* and *myostatin* mRNA in satellite cells (McCroskery et al. 2003). In addition, McCroskery et al. also demonstrated that abundant expression of myostatin could be detected by both RT-PCR and Western Blot analysis in isolated satellite cells and satellite cell-derived myoblasts. Functionally, myostatin appears to negatively regulate the activation and proliferation of satellite cells. In particular, increased satellite cell activation, quantified by percentage of BrdU positive cells, is observed in satellite cells isolated from *myostatin*-null mice as compared to wild-type controls (McCroskery et al. 2003; Siriett et al. 2006). In support, treatment of isolated single fibres with recombinant myostatin protein results in a dose-dependent decrease in BrdU-positive satellite cells, concomitant with a decrease in satellite cell migration (McCroskery et al. 2003, 2005). Furthermore, treatment of satellite cell-derived myoblasts with myostatin results in inhibition of proliferation (McCroskery et al. 2003; McFarland et al. 2006; Thomas et al. 2000). Conversely, primary myoblasts isolated from *myostatin*-null mice proliferate at a faster rate compared with cultures isolated from wild-type mice (McCroskery et al. 2003). A recent paper by Amthor et al. presents evidence to contradict the role of myostatin in regulating satellite cell biology. Specifically, Amthor et al. state that the hypertrophic phenotype observed in *myostatin*-null mice is mainly due to an increase in myonuclear domain rather than from a contribution of satellite cells (Amthor et al. 2009). In addition they observed fewer numbers of

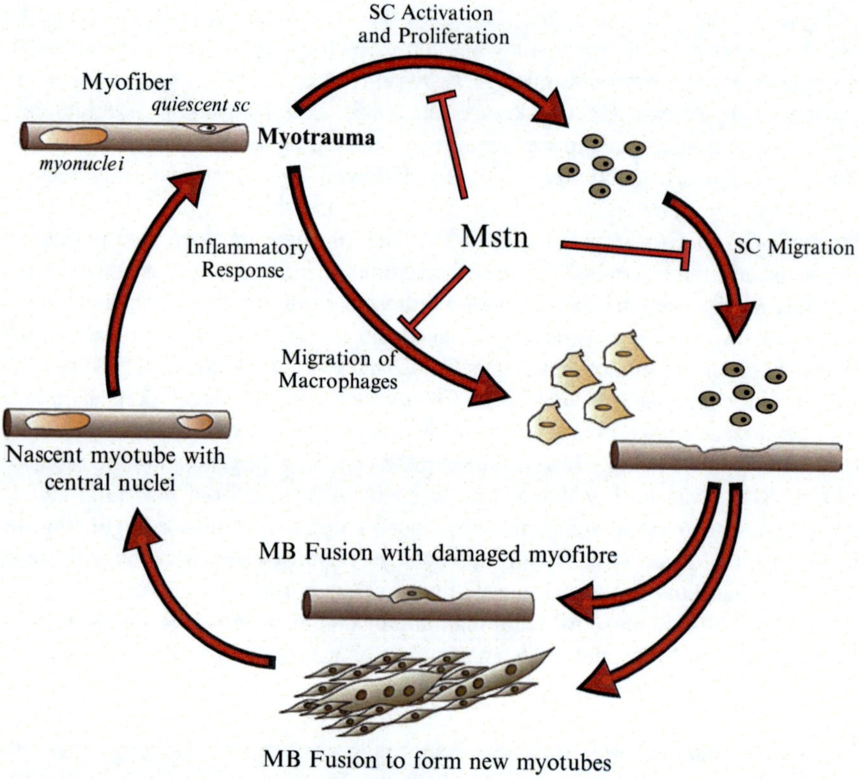

Fig. 4 A model for the role of myostatin in skeletal muscle regeneration. Muscle injury activates satellite cells (SC) and the inflammatory response. As a result, macrophages and satellite cells migrate to the site of injury. Myostatin (Mstn) negatively regulates satellite cell activation and inhibits migration of macrophages and satellite cells. Activated satellite cells proliferate at the site of injury and resulting myoblasts (MB) either fuse with the damaged myofiber or fuse to form new myotubes (Modified from McCroskery et al. [2005])

satellite cells in muscle isolated from *myostatin*-null as compared with wild type controls (Amthor et al. 2009), which is contradictory to what has been previously reported (McCroskery et al. 2003; Siriett et al. 2006). Furthermore they present evidence to suggest that treatment with myostatin has no significant effect on satellite cell proliferation in vitro (Amthor et al. 2009). However a recent paper from Gilson et al., studying the mechansim behind Follistatin induced muscle hypertrophy, demonstrates that Follistatin-induced hypertrophy is mediated by satellite cell proliferation, and inhibition of both myostatin and Activin (Gilson et al. 2009), a feature consistent with a role for myostatin in regulating satellite cell proliferation. Despite the conflicting reports the weight of evidence suggests that myostatin controls post-natal myogenesis through regulation of satellite cell activation and proliferation (McCroskery et al. 2003; McFarland et al. 2006; Siriett et al. 2006; Thomas et al. 2000).

Satellite cells, consistent with the term muscle stem cell, are able to self-renew their population. Myostatin has been implicated in regulation of satellite cell self-renewal; in fact, single fibres isolated from *myostatin*-null mice contain a greater proportion of satellite cells as compared with wild-type controls (McCroskery et al. 2003). In addition, a recent report has demonstrated that injection of myostatin-specific short hairpin interfering RNA (shRNA) into the TA muscle of rats results in an increase in satellite cell number, as assessed by Pax7 immunostaining (Magee et al. 2006). McCroskery et al. suggested that increased proliferation and increased satellite cell number per muscle fibre, in the *myostatin*-null mice, is indicative of increased self-renewal. The paired box transcription factor Pax7 is thought to play a role in the induction of satellite cell self-renewal. Indeed satellite cells, which maintain expression of Pax7 but lose MyoD exit the cell cycle, fail to differentiate, and adopt a quiescent phenotype (Olguin and Olwin 2004; Zammit et al. 2004). Recently published results highlight a possible Pax7-dependent mechanism behind myostatin regulation of satellite cell self-renewal (McFarlane et al. 2008). Treatment of primary myoblasts with recombinant myostatin protein resulted in a significant down-regulation of Pax7 via ERK1/2 signaling, while genetic inactivation or functional antagonism of myostatin results in enhanced expression of Pax7 (McFarlane et al. 2008). Furthermore, absence of *myostatin* increased the pool of quiescent reserve cells, a group of cells which share several characteristics with self-renewed satellite cells. It is therefore suggested that myostatin may regulate satellite cell self-renewal by negatively regulating Pax7 (McFarlane et al. 2008).

3 Myostatin and Muscle Wasting

3.1 Myostatin as a Cachexia-Inducing Growth Factor

Myostatin has been associated with the induction of cachexia, a severe form of muscle wasting that manifests as a result of disease. HIV-infected men undergoing muscle wasting have increased intramuscular and serum concentrations of myostatin protein as compared with healthy controls (Gonzalez-Cadavid et al. 1998). Thus myostatin may contribute to the muscle wasting pathology observed as a result of HIV-infection. Recent evidence highlights a role for myostatin in cancer-associated cachexia. Specifically, injection of the S-180 ascitic tumor into mice resulted in a 50% increase in *myostatin* mRNA expression concomitant with a reduction in muscle mass (Liu et al. 2008). Furthermore, Liu et al. demonstrated that antisense inactivation of myostatin in the S-180 tumor bearing mice resulted in increased muscle mass. Myostatin has also been associated with muscle wasting resulting from liver cirrhosis; Dasarathey et al. used the portacaval anastamosis rat, a model of human liver cirrhosis, to study the involvement of myostatin in the muscle wasting associated with this disease. Gene expression analysis demonstrated an increase in the mRNA and protein levels of myostatin and the myostatin receptor, activin type-IIb (Dasarathy et al. 2004). Patients suffering from Addison's disease (adrenal insufficiency) commonly experience skeletal muscle atrophy. Recently it was shown that active myostatin serum levels increased over time in adrenalectomized rats, a model of Addison's disease (Hosoyama et al. 2005). This increase in serum myostatin correlated with a decrease in muscle weights as compared with controls (Hosoyama et al. 2005). Cushing's syndrome is associated with an excessive increase in glucocorticoid production resulting in skeletal muscle wasting (Shibli-Rahhal et al. 2006). Ma et al. has demonstrated that injection of the glucocorticoid Dexamethasone into rats induces skeletal muscle atrophy, concomitant with a dose-dependent up-regulation of myostatin mRNA and protein. The Dexamethasone-induced up-regulation of myostatin was inhibited in the presence of glucocorticoid antagonist RU-486 (Ma et al. 2003). A separate study has demonstrated that, in addition to mRNA and protein, *myostatin* promoter activity is induced following Dexamethasone-induced muscle wasting (Salehian et al. 2006). The amino acid glutamine has been previously shown to antagonise glucocorticoid-induced skeletal muscle atrophy (Hickson et al. 1995, 1996). Consistent with this, injection of glutamine in conjunction with Dexamethasone into rats significantly reduced the muscle atrophy phenotype, concomitant with a down-regulation of myostatin expression (Salehian et al. 2006). In addition to an associative role in cachexia, myostatin has been shown to induce cachexia following administration to mice, specifically, injection of CHO-control cells and CHO cells over-expressing myostatin (CHO-Myostatin) resulted in the formation of tumors. However, in contrast to the gain in body weight observed in CHO-control mice, injection of CHO-Myostatin cells resulted in a 33% reduction in total body weight within 16 days (Zimmers et al. 2002). This severe body mass reduction was ameliorated by injection of CHO cells expressing the myostatin propeptide (LAP) region or follistatin, two identified antagonists of myostatin function. Furthermore, injection of CHO-Myostatin cells resulted in a significant reduction in fat pad mass, consistent with cachexia (Zimmers et al. 2002). Recently, Hoenig et al. has hypothesized that myostatin also contributes to cardiac cachexia. This hypothesis is based on the following findings. Firstly, increased myostatin expression was detected in the peri-infarct zone of the heart having undergone myocardial infarction (Sharma et al. 1999), and secondly, in a rat model of congestive heart failure, myostatin levels were up-regulated with a significant number of rats demonstrating signs of muscle wasting (Shyu et al. 2006).

3.2 Mechanism Behind Myostatin Regulation of Muscle Wasting

Myostatin-mediated induction of muscle wasting results in the down-regulation of myogenic gene expression. Over-expression of myostatin in post-natal skeletal muscle reduced the expression of several myogenic structural genes, including MHC and desmin (Durieux et al. 2007). Furthermore, myostatin-mediated muscle wasting results in a reduction in the expression of key myogenic regulatory factors, including MyoD and myogenin (Durieux et al. 2007; McFarlane et al. 2006). One could imagine that a reduction in these key myogenic genes would only serve to exacerbate the wasting phenotype through potentially impaired post-natal myogenesis and muscle regeneration. Concomitant with down-regulation of key genes involved with myogenesis, myostatin-mediated muscle wasting in vitro and in vivo results in the up-regulation of genes involved with the ubiquitin-proteasome proteolytic pathway including *atrogin-1*, *MuRF-1* and $E2_{14k}$ (McFarlane et al. 2006). In the same study it was demonstrated that treatment of C2C12 myotubes with recombinant myostatin protein antagonised the IGF-1/PI3-K/AKT pathway, resulting in enhanced activation of the transcription factor FoxO1 and subsequent activation of atrophy-related genes (McFarlane et al. 2006). It was further delineated that myostatin signals independently of NF-κB during the induction of muscle wasting. In support, myostatin and NF-κB have been previously shown to signal through separate pathways to regulate myogenesis (Bakkar et al. 2005). The proposed mechanism(s) through which myostatin promotes skeletal muscle wasting are summarised in Fig. 5. In contrast to this, a recent paper by Trendelenburg et al. presents data which indicates that myostatin induces atrophy through a mechanism involving inhibition of the Akt/TORC1/p70S6K signaling pathway (Trendelenburg et al. 2009). It was further demonstrated that myostatin-induced atrophy in myotube populations was dependent on Smad2 and Smad3 signaling and did not result in the up-regulation of components of the ubiquitin-proteasome pathway, and in fact, myostatin treatment was shown to inhibit the expression of Atrogin-1 and MuRF-1 (Trendelenburg et al. 2009). Another recent paper by Sartori et al., demonstrates that activation of the myostatin pathway, through transfection of constitutively active ALK5 into adult muscle fibres, results in muscle atrophy (Sartori et al. 2009). Interestingly, Sartori et al. further demonstrate that the myostatin-induced atrophy is dependent on Smad2 and Smad3 signaling and results in enhanced Atrogin-1, but not MuRF-1, promoter activity (Sartori et al. 2009). While there is conflicting evidence for myostatin-regulation of protein degradation and the ubiquitin-proteasome pathway it is clear that myostatin has a critical role in regulating post-natal skeletal muscle growth and the progression of skeletal muscle wasting. Recently it has been demonstrated that FoxO1 can regulate the expression of myostatin; in particular, over-expression of constitutively active FoxO1 increased the expression of *myostatin* mRNA and promoter reporter activity. Allen and Unterman suggest that FoxO1 up-regulation of myostatin may contribute to skeletal muscle atrophy (Allen and Unterman 2007). In addition, RNA oligonucleotide mediated down-regulation of FoxO1 has been shown to reduce the expression of myostatin (Liu et al. 2007). Moreover, the RNA-mediated reduction in FoxO1 expression promoted an increase in muscle mass in control mice and mice undergoing cancer-associated cachexia (Liu et al. 2007), a feature consistent with loss of myostatin function.

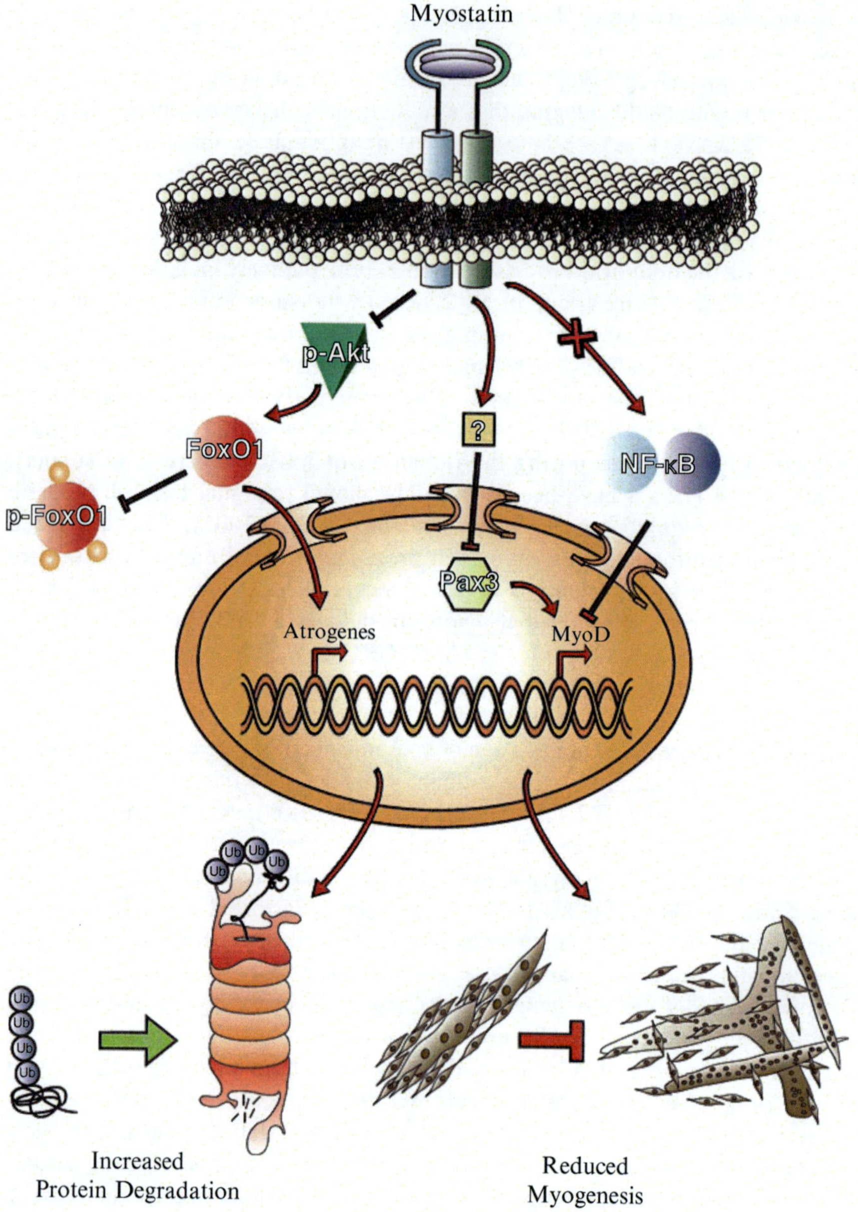

Fig. 5 Proposed mechanism behind myostatin induced cachexia. Unlike TNF-α, myostatin appears to induce cachexia independent of the NF-κB pathway. Myostatin blocks myogenesis by down-regulating the expression of *pax3* and *myoD*. In addition, myostatin appears to up-regulate components of the ubiquitin proteolysis system (Atrogenes) by hypo-phosphorylating FoxO1 through the inhibition of the PI3-K/AKT signalling pathway. *Arrows* represent activation while *blunt-ended lines* represent inhibition (Modified from McFarlane et al. [2006])

3.3 *Myostatin and Muscle Atrophy*

Muscle disuse or inactivity, such as that experienced during periods of prolonged bed rest, also contributes to skeletal muscle atrophy. Several studies have implicated myostatin in the muscle atrophy associated with disuse. The expression of myostatin was measured in a mouse model of hindlimb unloading. Carlson et al. showed that *myostatin* mRNA was significantly increased following 1 day of hindlimb unloading, however, no detectable difference in *myostatin* expression was observed at days 3 and 7 of unloading, as compared with controls (Carlson et al. 1999). In a separate study, hindlimb unloading in the rat resulted in a 16% decrease in *M. plantaris* muscle weight, concomitant with a 110% increase in *myostatin* mRNA and a 35% increase in myostatin protein (Wehling et al. 2000). A dramatic 30-fold increase in *myostatin* mRNA was observed in patients suffering from disuse atrophy as a result of chronic osteoarthritis of the hip (Reardon et al. 2001). In addition, a significant negative correlation was observed between expression of *myostatin* and type-IIA and type-IIB fibre area, suggesting that myostatin may target type-IIA and IIB fibres during disuse atrophy (Reardon et al. 2001). Furthermore, a 25 day period of bedrest increased the levels of serum myostatin-immunoreactive protein to 12% above that observed in baseline measurements (Zachwieja et al. 1999). In addition, myostatin has been associated with skeletal muscle loss during space flight (Lalani et al. 2000). In particular, exposing rats to the microgravity environment of space resulted in muscle weight loss, with an associated increase in both myostatin mRNA and protein (Lalani et al. 2000).

3.4 *Myostatin and Muscular Dystrophy*

The most common forms of muscular dystrophy are Duchenne muscular dystrophy (DMD) and Becker muscular dystrophy (BMD) (Zhou et al. 2006). Both DMD and BMD are X-linked recessive disorders that can be traced back to mutations in the *dystrophin* gene (*DMD*) (Flanigan et al. 2003; Sironi et al. 2003). BMD results from in-frame mutations in the *DMD* gene, resulting in a partially functional protein product (Hoffman et al. 1988; Koenig et al. 1989), however in DMD patients, frame-shift mutations result in very low levels or complete absence of the dystrophin (Hoffman et al. 1987; Koenig et al. 1987). DMD and BMD afflict about one in every 3,500 and one in 18,500 newborn males respectively (Darin and Tulinius 2000; Emery 1991; Peterlin et al. 1997; Siciliano et al. 1999; Zhou et al. 2006). Myostatin is a well-characterised negative regulator of skeletal muscle mass: as such, several studies have been performed looking at the role of myostatin in the severe muscular dystrophy phenotype. The expression of myostatin has been shown to decrease by fourfold in regenerated *mdx* muscle (Tseng et al. 2002). It is suggested that a reduction in myostatin may be an adaptive response to aid in the maintenance and rescue of *mdx* skeletal muscle. Antibody-mediated blockade of myostatin results in both enhanced body mass and skeletal muscle hypertrophy in the *mdx* mouse model of DMD (Bogdanovich et al. 2002). Furthermore, antagonising myostatin resulted in increased muscle strength, as measured through grip strength experiments. Bogdanovich et al. further demonstrated that blocking myostatin, through injection of an Fc-fusion stabilised myostatin propeptide region (LAP), resulted in improvement of the *mdx* DMD phenotype. Consistent with antibody-mediated myostatin blockade, propeptide injection resulted in enhanced growth, increased muscle mass and grip strength (Bogdanovich et al. 2005). They further showed that this blockade resulted in enhanced muscle specific force, over and above that shown by antibody-mediated inhibition of myostatin. Recently, transgenic *mdx* mice containing a dominant negative activin type-IIB receptor gene (ActRIIB) showed phenotypic improvement over wild-type *mdx* mice (Benabdallah et al. 2005). Indeed, increased skeletal muscle mass was observed in conjunction with increased resistance to exercise-induced muscle damage. More recently, Minetti et al. have examined the effect of deacetylase inhibitors on the *mdx* phenotype. Treatment of *mdx* mice with deacetylase inhibitors resulted in an improvement in muscle quality and function with an increase in myofibre size (Minetti et al. 2006). Interestingly, addition of the deacetylase inhibitors TSA or MS 27-275 resulted in enhanced expression of the myostatin antagonist follistatin (Minetti et al. 2006). In addition to disruption in dystrophin, muscular dystrophy can result from mutations in several genes involved in the formation of the dystrophin-associated protein complex, including laminin-II. Crossing of the *myostatin*-null mice with the *dy* mice, a model of laminin-II-associated dystrophy, resulted in increased muscle mass and enhanced regeneration (Li et al. 2005). However, elimination of *myostatin* in the *dy* mice was unable to correct the severe dystrophic pathology associated with loss of laminin-II, moreover, deletion of *myostatin* resulted in an increase in post-natal mortality (Li et al. 2005). Further work described by Ohsawa et al. demonstrates that inhibition of myostatin through either, introduction of the myostatin prodomain by genetic crossing, or intraperitoneal injection of the soluble Activin type IIB receptor, improves muscle atrophy associated with autosomal dominant limb-girdle muscular dystrophy 1C (LGMD1C), which results from mutations in the caveolin-3 gene (Ohsawa et al. 2006). Furthermore, inhibition of myostatin in the mouse model of LGMD1C also resulted in the suppression of p-Smad2 and p21, two known targets of myostatin signaling (Ohsawa et al. 2006). More recently, a study by Bartoli et al. demonstrated that antagonizing myostatin, through viral introduction of a mutated myostatin pro-peptide, improved muscle mass and force in the LGMD2A animal model of limb-girdle muscular dystrophy, a dystrophy resulting from mutations in calpain 3 (Bartoli et al. 2007). However, in the same study introduction of the pro-preptide into a mouse model of LGMD2D limb-girdle muscular dystrophy, resulting from mutations in the α-sarcoglycan gene, failed to improve muscle mass (Bartoli et al. 2007). In addition, Bogdanovich et al. demonstrated that antibody-mediated disruption of myostatin in the LGMD2C mouse model of limb-girdle muscular dystrophy, resulting from a deficiency in δ-sarcoglycan, enhanced muscle mass, muscle fiber area and muscle strength. However, the antibody-mediated disruption of myostatin failed to significantly improve the dystrophic pathology observed in the α-sarcoglycan deficient mice (Bogdanovich et al. 2007). Therefore, the validity and robustness of myostatin as a target for treatment of all forms of dystrophy remains a matter of contention.

In conclusion, recent research suggests that myostatin is a potent inducer of muscle wasting. Furthermore, additional cachectic agents, such as Dexamethasone, may also signal muscle wasting via mechanisms involving the up regulation of *myostatin* gene expression. Therefore, myostatin appears to be a key molecule during the induction of muscle wasting. In the future, myostatin antagonists could be a viable therapeutic option for alleviating the severe symptoms associated with numerous muscle wasting conditions.

4 Myostatin and Sarcopenia

Myostatin protein levels have been shown to change with aging in humans. Several studies have indicated that there is a significant increase in both myostatin mRNA and/or protein levels during aging in humans and rodents (Baumann et al. 2003; Leger et al. 2008; Raue et al. 2006; Yarasheski et al. 2002). However, some studies have also reported that *myostatin* mRNA levels were unchanged during aging (Welle et al. 2002). Using *myostatin*-null mice, it has been recently reported that myostatin inactivation enhances bone density, insulin sensitivity and heart function in old mice (Morissette et al. 2009).

In our laboratory we have investigated the role of myostatin during sarcopenia using *myostatin*-null mice and myostatin antagonists. Some of the important observations are described below.

4.1 Prolonged Absence of Myostatin Alleviates Sarcopenic Muscle Loss

One of the most striking effects of aging in muscle is the associated loss in muscle mass resulting in loss of strength and endurance. Furthermore, aging muscle has a marked reduction in its regenerative capabilities after muscle damage. It has been difficult to establish a primary cause and to formulate a unified theory explaining the molecular basis behind the aging muscle phenotype. Although the roles of several positive regulators have been extensively studied (Allen et al. 1995; Barton-Davis et al. 1998; Marsh et al. 1997; Mezzogiorno et al. 1993; Yablonka-Reuveni et al. 1999), the role of negative regulators during age-related muscle wasting is not known. In this chapter we explore the involvement of myostatin, a known negative regulator of muscle growth, during the aging process. Well-established effects of aging on muscle are: atrophy of the muscle and its individual fibres, a shift towards oxidative fibres, and impairment of satellite cell activation and subsequent muscle regeneration. In the *myostatin*-null mice, the prolonged absence of myostatin reduces fibre atrophy associated with aging (Siriett et al. 2006). Currently, satellite cells are believed to be largely responsible for muscle growth and maintenance throughout life (see Hawke and Garry (2001) for review). Previously it has been suggested that satellite cell numbers decline during aging (Gibson and Schultz 1983; Shefer et al. 2006) while others report no change (Conboy et al. 2003; Nnodim 2000). Myostatin has been shown to be involved in the maintenance of satellite cell quiescence (McCroskery et al. 2003) and that a lack of *myostatin* results in increased activation of satellite cells. Myostatin acts by inhibiting cell cycle progression from G0 to S phase. In its absence, cell cycle progression can proceed resulting in an increase in satellite cell activation and proliferation as observed in the young *myostatin*-null mice. This increased cell number and activation would provide a mechanism for greater myoblast recruitment and subsequent fibre formation and enlargement leading to the fibre hypertrophy observed in the young *myostatin*-null mice. The prolonged absence of myostatin maintains the increased satellite cell number and activation even in aged muscle (Siriett et al. 2006). The increased cell number and activation would provide an essential resource during aging, when a significant pressure on the maintenance of the fibres would be present in response to the aging process. Therefore we propose that lack or inactivation of myostatin would lead to increased self-renewal of satellite cells and efficient replacement of lost muscle fibres, leading to increased muscle growth and reduced muscle wasting. With aging, murine muscle undergoes specific fibre type switches, with functional and metabolic consequences. Specifically, numerous reports suggest a shift from glycolytic fibres to oxidative fibres with increasing age (Alnaqeeb and Goldspink 1987; Grimby et al. 1982; Larsson et al. 1993). In contrast, all *myostatin*-null muscles displayed minimal type IIA fibres in aged muscles. This indicates an alteration in the fibre type composition with the loss of *myostatin*, as well as a resistance to an increase of type IIA fibres, which was associated with aging in the wild-type mice (Siriett et al. 2006). The role played by myostatin in the determination of fibre types is still unclear. Regardless of the mechanism, increased type IIB fibres would cause the muscle to remain predominantly glycolytic during aging.

Aging is also thought to negatively influence satellite cell behavior. These cells are heavily involved in the regenerative process after muscle injury. Aging has a significant effect on the muscle regenerative capacity, since the proliferative potential of satellite cells in skeletal muscles of aged rodents is decreased as compared with young adults (Schultz and Lipton 1982). Furthermore, some reports also suggest that the poor regenerative capacity of skeletal muscle is also due to a decrease in the number of satellite cells (Snow 1977). Since inactivation of *myostatin* leads to increased satellite cell activation, it was no surpirse that even during aging *myostatin*-null muscles showed remarkable ability to regenerate. Nascent fibres formed faster, muscle and fibre hypertrophy and fibre type composition were preserved, and the formation of scar tissue was greatly reduced (Siriett et al. 2006). Interestingly, senescent *myostatin*-null mice were virtually able to recapitulate the enhanced regeneration seen in young adult *myostatin*-null mice. In common with the prevention of fibre atrophy during the aging process, the subsequent muscle regeneration following notexin damage would be heavily reliant on satellite cell availability and activation. Undoubtedly, an increased number of satellite cells and activation propensity, as observed in the *myostatin*-null mice, would be advantageous during this regenerative process.

4.2 *Antagonism of Myostatin Enhances Muscle Regeneration during Sarcopenia*

Since lack of *myostatin* increases the propensity of satellite cell activation and regeneration of skeletal muscle even during aging, our laboratory examined the effect of a short-term antagonism of myostatin. For this purpose we developed a peptide antagonist to myostatin (Mstn-ant1) and screened for its ability to neutralize myostatin function. Cultured myoblasts express and secrete myostatin, which regulates the proliferation rate of myoblasts (McFarlane et al. 2005; Thomas et al. 2000). Thus, antagonism of myostatin by Mstn-ant1 would result in an increase in the myoblast proliferation rate. Indeed, a C2C12 myoblast proliferation assay indicated that Mstn-ant1 effectively increased the proliferation of the myoblasts above that of the control (Siriett et al. 2007), thus confirming its biological activity. In addition, administration of Mstn-ant1 immediately after notexin injury was able to enhance muscle healing in aging mice (Siriett et al. 2007). In addition, Mstn-ant1 treated muscles also displayed reduced levels of collagen suggesting myostatin antagonist reduces scar tissue formation. Collectively, these results indicate that a short-term blockade of myostatin during sarcopenia is sufficient to enhance the regeneration during aging. During muscle regeneration, MyoD is expressed earlier and at higher levels in *myostatin*-null muscle as compared with wild-type muscle (McCroskery et al. 2005). Similarly, Western blot analysis performed on the regenerating muscle from mice treated with Mstn-ant1 showed increased levels of MyoD during regeneration, suggesting increased myogenesis directly resulting from a myostatin blockade by Mstn-ant1 (Siriett et al. 2007). In addition, Pax7, which is expressed in quiescent and proliferating cells (Seale et al. 2000), was higher with Mstn-ant1 treatment throughout the trial period suggesting an increase in satellite cell number, activation and/or self renewal compared to saline treated mice (Siriett et al. 2007). These higher Pax7 and MyoD levels could be due to increased numbers of satellite cells and the subsequent myogenesis, and increased satellite cell self renewal due to myostatin antagonist. Collectively, the results presented here suggest that short-term blockade of myostatin and its function through antagonist treatment can effectively enhance muscle regeneration in aged mice after injury and during age-related muscle wasting. The ramifications of antagonist treatment for human health are potentially extensive. The antagonism of myostatin is a viable option for treatment of deficient muscle regeneration and sarcopenia in humans, through a restoration of myogenic and inflammatory responses and decreased fibrosis.

References

Allen, D. L. & Unterman, T. G. (2007). Regulation of myostatin expression and myoblast differentiation by FoxO and SMAD transcription factors. *American Journal of Physiology. Cell Physiology, 292*, C188–C199.

Allen, R. E., Sheehan, S. M., Taylor, R. G., Kendall, T. L., Rice, G. M. (1995). Hepatocyte growth factor activates quiescent skeletal muscle satellite cells in vitro. *Journal of Cellular Physiology, 165*, 307–312.

Alnaqeeb, M. A. & Goldspink, G. (1987). Changes in fibre type, number and diameter in developing and ageing skeletal muscle. *Journal of Anatomy, 153*, 31–45.

Amthor, H., Nicholas, G., Mckinnell, I., Kemp, C. F., Sharma, M., Kambadur, R., Patel, K. (2004). Follistatin complexes myostatin and antagonises myostatin-mediated inhibition of myogenesis. *Developmental Biology, 270*, 19–30.

Amthor, H., Otto, A., Macharia, R., Mckinnell, I., Patel, K. (2006). Myostatin imposes reversible quiescence on embryonic muscle precursors. *Developmental Dynamics, 235*, 672–680.

Amthor, H., Otto, A., Vulin, A., Rochat, A., Dumonceaux, J., Garcia, L., Mouisel, E., Hourde, C., Macharia, R., Friedrichs, M., Relaix, F., Zammit, P. S., Matsakas, A., Patel, K., Partridge, T. (2009). Muscle hypertrophy driven by myostatin blockade does not require stem/precursor-cell activity. *Proceedings of the National Academy of Sciences of the United States of America, 106*, 7479–7484.

Bakkar, N., Wackerhage, H., Guttridge, D. C. (2005). Myostatin and NF-κB regulate skeletal myogenesis through distinct signaling pathways. *Signal Transduction, 5*, 202–210.

Bartoli, M., Poupiot, J., Vulin, A., Fougerousse, F., Arandel, L., Daniele, N., Roudaut, C., Noulet, F., Garcia, L., Danos, O., Richard, I. (2007). Aav-mediated delivery of a mutated myostatin propeptide ameliorates calpain 3 but not alpha-sarcoglycan deficiency. *Gene Therapy, 14*, 733–740.

Barton-Davis, E. R., Shoturma, D. I., Musaro, A., Rosenthal, N., Sweeney, H. L. (1998). Viral mediated expression of insulin-like growth factor I blocks the aging-related loss of skeletal muscle function. *Proceedings of the National Academy of Sciences of the United States of America, 95*, 15603–15607.

Bass, J., Oldham, J., Sharma, M., Kambadur, R. (1999). Growth factors controlling muscle development. *Domestic Animal Endocrinology, 17*, 191–197.

Baumann, A. P., Ibebunjo, C., Grasser, W. A., Paralkar, V. M. (2003). Myostatin expression in age and denervation-induced skeletal muscle atrophy. *Journal of Musculoskeletal & Neuronal Interactions, 3*, 8–16.

Benabdallah, B. F., Bouchentouf, M., Tremblay, J. P. (2005). Improved success of myoblast transplantation in mdx mice by blocking the myostatin signal. *Transplantation, 79*, 1696–1702.

Berry, C., Thomas, M., Langley, B., Sharma, M., Kambadur, R. (2002). Single cysteine to tyrosine transition inactivates the growth inhibitory function of Piedmontese myostatin. *American Journal of Physiology, 283*, C135–C141.

Bogdanovich, S., Krag, T. O., Barton, E. R., Morris, L. D., Whittemore, L. A., Ahima, R. S., Khurana, T. S. (2002). Functional improvement of dystrophic muscle by myostatin blockade. *Nature, 420*, 418–421.

Bogdanovich, S., Perkins, K. J., Krag, T. O., Whittemore, L. A., Khurana, T. S. (2005). Myostatin propeptide-mediated amelioration of dystrophic pathophysiology. *The FASEB Journal, 19*, 543–549.

Bogdanovich, S., Mcnally, E. M., Khurana, T. S. (2007). Myostatin blockade improves function but not histopathology in a murine model of limb-girdle muscular dystrophy 2C. *Muscle & Nerve, 37*, 308–316.

Boman, I. A. & Vage, D. I. (2009). An insertion in the coding region of the myostatin (MSTN) gene affects carcass conformation and fatness in the Norwegian Spaelsau (Ovis aries). *BMC Res Notes, 2*, 98.

Carlson, C. J., Booth, F. W., Gordon, S. E. (1999). Skeletal muscle myostatin mRNA expression is fiber-type specific and increases during hindlimb unloading. *The American Journal of Physiology, 277*, R601–R606.

Clop, A., Marcq, F., Takeda, H., Pirottin, D., Tordoir, X., Bibe, B., Bouix, J., Caiment, F., Elsen, J. M., Eychenne, F., Larzul, C., Laville, E., Meish, F., Milenkovic, D., Tobin, J., Charlier, C., Georges, M. (2006). A mutation creating a potential illegitimate microRNA target site in the myostatin gene affects muscularity in sheep. *Nature Genetics, 38*, 813–818.

Conboy, I. M., Conboy, M. J., Smythe, G. M., Rando, T. A. (2003). Notch-mediated restoration of regenerative potential to aged muscle. *Science, 302*, 1575–1577.

Darin, N. & Tulinius, M. (2000). Neuromuscular disorders in childhood: A descriptive epidemiological study from western Sweden. *Neuromuscular Disorders, 10*, 1–9.

Dasarathy, S., Dodig, M., Muc, S. M., Kalhan, S. C., Mccullough, A. J. (2004). Skeletal muscle atrophy is associated with an increased expression of myostatin and impaired satellite cell function in the portacaval anastamosis rat. *American Journal of Physiology Gastrointestinal and Liver Physiology, 287*(6), G1124–G1130.

Durieux, A. C., Amirouche, A., Banzet, S., Koulmann, N., Bonnefoy, R., Pasdeloup, M., Mouret, C., Bigard, X., Peinnequin, A., Freyssenet, D. (2007). Ectopic expression of myostatin induces atrophy of adult skeletal muscle by decreasing muscle gene expression. *Endocrinology, 148*, 3140–3147.

Emery, A. E. (1991). Population frequencies of inherited neuromuscular diseases – a world survey. *Neuromuscular Disorders, 1*, 19–29.

Fainsod, A., Deissler, K., Yelin, R., Marom, K., Epstein, M., Pillemer, G., Steinbeisser, H., Blum, M. (1997). The dorsalizing and neural inducing gene follistatin is an antagonist of BMP-4. *Mechanisms of Development, 63*, 39–50.

Flanigan, K. M., Von Niederhausern, A., Dunn, D. M., Alder, J., Mendell, J. R., Weiss, R. B. (2003). Rapid direct sequence analysis of the dystrophin gene. *American Journal of Human Genetics, 72*, 931–939.

Forbes, D., Jackman, M., Bishop, A., Thomas, M., Kambadur, R., Sharma, M. (2006). Myostatin auto-regulates its expression by feedback loop through Smad7 dependent mechanism. *Journal of Cellular Physiology, 206*, 264–272.

Gibson, M. C. & Schultz, E. (1983). Age-related differences in absolute numbers of skeletal muscle satellite cells. *Muscle & Nerve, 6*, 574–580.

Gilson, H., Schakman, O., Kalista, S., Lause, P., Tsuchida, K., Thissen, J. P. (2009). Follistatin induces muscle hypertrophy through satellite cell proliferation and inhibition of both myostatin and activin. *American Journal of Physiology. Endocrinology and Metabolism, 297*, E157–E164.

Girgenrath, S., Song, K., Whittemore, L. A. (2005). Loss of myostatin expression alters fiber-type distribution and expression of myosin heavy chain isoforms in slow- and fast-type skeletal muscle. *Muscle & Nerve, 31*, 34–40.

Gonzalez-Cadavid, N. F., Taylor, W. E., Yarasheski, K., Sinha-Hikim, I., Ma, K., Ezzat, S., Shen, R., Lalani, R., Asa, S., Mamita, M., Nair, G., Arver, S., Bhasin, S. (1998). Organization of the human myostatin gene and expression in healthy men and HIV-infected men with muscle wasting. *Proceedings of the National Academy of Sciences of the United States of America, 95*, 14938–14943.

Grimby, G., Danneskiold-Samsoe, B., Hvid, K., Saltin, B. (1982). Morphology and enzymatic capacity in arm and leg muscles in 78–81 year old men and women. *Acta Physiologica Scandinavica, 115*, 125–134.

Grobet, L., Poncelet, D., Royo, L. J., Brouwers, B., Pirottin, D., Michaux, C., Menissier, F., Zanotti, M., Dunner, S., Georges, M. (1998). Molecular definition of an allelic series of mutations disrupting the myostatin function and causing double-muscling in cattle. *Mammalian Genome, 9*, 210–213.

Grobet, L., Pirottin, D., Farnir, F., Poncelet, D., Royo, L. J., Brouwers, B., Christians, E., Desmecht, D., Coignoul, F., Kahn, R., Georges, M. (2003). Modulating skeletal muscle mass by postnatal, muscle-specific inactivation of the myostatin gene. *Genesis, 35*, 227–238.

Harper, J. W., Adami, G. R., Wei, N., Keyomarsi, K., Elledge, S. J. (1993). The p21 Cdk-interacting protein Cip1 is a potent inhibitor of G1 cyclin- dependent kinases. *Cell, 75*, 805–816.

Hata, A., Lagna, G., Massague, J., Hemmati-Brivanlou, A. (1998). Smad6 inhibits BMP/Smad1 signaling by specifically competing with the Smad4 tumor suppressor. *Genes & Development, 12*, 186–197.

Hawke, T. J. & Garry, D. J. (2001). Myogenic satellite cells: Physiology to molecular biology. *Journal of Applied Physiology, 91*, 534–551.

Hayashi, H., Abdollah, S., Qiu, Y., Cai, J., Xu, Y. Y., Grinnell, B. W., Richardson, M. A., Topper, J. N., Gimbrone, M. A., Jr, Wrana, J. L., Falb, D. (1997). The MAD-related protein Smad7 associates with the TGFbeta receptor and functions as an antagonist of TGFbeta signaling. *Cell, 89*, 1165–1173.

Hemmati-Brivanlou, A., Kelly, O. G., Melton, D. A. (1994). Follistatin, an antagonist of activin, is expressed in the Spemann organizer and displays direct neuralizing activity. *Cell, 77*, 283–295.

Hickson, R. C., Czerwinski, S. M., Wegrzyn, L. E. (1995). Glutamine prevents downregulation of myosin heavy chain synthesis and muscle atrophy from glucocorticoids. *The American Journal of Physiology, 268*, E730–E734.

Hickson, R. C., Wegrzyn, L. E., Osborne, D. F., Karl, I. E. (1996). Alanyl-glutamine prevents muscle atrophy and glutamine synthetase induction by glucocorticoids. *The American Journal of Physiology, 271*, R1165–R1172.

Hill, J. J., Davies, M. V., Pearson, A. A., Wang, J. H., Hewick, R. M., Wolfman, N. M., Qiu, Y. (2002). The myostatin propeptide and the follistatin-related gene are inhibitory binding proteins of myostatin in normal serum. *The Journal of Biological Chemistry, 277*, 40735–40741.

Hill, J. J., Qiu, Y., Hewick, R. M., Wolfman, N. M. (2003). Regulation of myostatin in vivo by growth and differentiation factor-associated serum protein-1: A novel protein with protease inhibitor and follistatin domains. *Molecular Endocrinology, 17*, 1144–1154.

Hoffman, E. P., Brown, R. H., JR, Kunkel, L. M. (1987). Dystrophin: The protein product of the Duchenne muscular dystrophy locus. *Cell, 51*, 919–928.

Hoffman, E. P., Fischbeck, K. H., Brown, R. H., Johnson, M., Medori, R., Loike, J. D., Harris, J. B., Waterston, R., Brooke, M., Specht, L. (1988). Characterization of dystrophin in muscle-biopsy specimens from patients with Duchenne's or Becker's muscular dystrophy. *The New England Journal of Medicine, 318*, 1363–1368.

Hosoyama, T., Tachi, C., Yamanouchi, K., Nishihara, M. (2005). Long term adrenal insufficiency induces skeletal muscle atrophy and increases the serum levels of active form myostatin in rat serum. *Zoology Science, 22*, 229–236.

Jeanplong, F., Sharma, M., Somers, W. G., Bass, J. J., Kambadur, R. (2001). Genomic organization and neonatal expression of the bovine myostatin gene. *Molecular and Cellular Biochemistry, 220*, 31–37.

JI, S., Losinski, R. L., Cornelius, S. G., Frank, G. R., Willis, G. M., Gerrard, D. E., Depreux, F. F., Spurlock, M. E. (1998). Myostatin expression in porcine tissues: Tissue specificity and developmental and postnatal regulation. *The American Journal of Physiology, 275*, R1265–R1273.

Joulia-Ekaza, D. & Cabello, G. (2006). Myostatin regulation of muscle development: Molecular basis, natural mutations, physiopathological aspects. *Experimental Cell Research, 312*, 2401–2414.

Kambadur, R., Sharma, M., Smith, T. P., Bass, J. J. (1997). Mutations in myostatin (GDF8) in double-muscled Belgian Blue and Piedmontese cattle. *Genome Research, 7*, 910–916.

Kocamis, H., Kirkpatrick-Keller, D. C., Richter, J., Killefer, J. (1999). The ontogeny of myostatin, follistatin and activin-B mRNA expression during chicken embryonic development. *Growth, Development, and Aging, 63*, 143–150.

Koenig, M., Hoffman, E. P., Bertelson, C. J., Monaco, A. P., Feener, C., Kunkel, L. M. (1987). Complete cloning of the Duchenne muscular dystrophy (DMD) cDNA and preliminary genomic organization of the DMD gene in normal and affected individuals. *Cell, 50*, 509–517.

Koenig, M., Beggs, A. H., Moyer, M., Scherpf, S., Heindrich, K., Bettecken, T., Meng, G., Muller, C. R., Lindlof, M., Kaariainen, H. (1989). The molecular basis for Duchenne versus Becker muscular dystrophy: Correlation of severity with type of deletion. *American Journal of Human Genetics, 45*, 498–506.

Lalani, R., Bhasin, S., Byhower, F., Tarnuzzer, R., Grant, M., Shen, R., Asa, S., Ezzat, S., Gonzalez-Cadavid, N. F. (2000). Myostatin and insulin-like growth factor-I and -II expression in the muscle of rats exposed to the microgravity environment of the NeuroLab space shuttle flight. *The Journal of Endocrinology, 167*, 417–428.

Langley, B., Thomas, M., Bishop, A., Sharma, M., Gilmour, S., Kambadur, R. (2002). Myostatin inhibits myoblast differentiation by down regulating MyoD expression. *The Journal of Biological Chemistry, 18*, 18.

Langley, B., Thomas, M., Mcfarlane, C., Gilmour, S., Sharma, M., Kambadur, R. (2004). Myostatin inhibits rhabdomyosarcoma cell proliferation through an Rb-independent pathway. *Oncogene, 23*, 524–534.

Larsson, L., Biral, D., Campione, M., Schiaffino, S. (1993). An age-related type IIB to IIX myosin heavy chain switching in rat skeletal muscle. *Acta Physiologica Scandinavica, 147*, 227–234.

Lee, S. J. & Mcpherron, A. C. (2001). Regulation of myostatin activity and muscle growth. *Proceedings of the National Academy of Sciences of the United States of America, 98*, 9306–9311.

Leger, B., Derave, W., de Bock, K., Hespel, P., Russell, A. P. (2008). Human sarcopenia reveals an increase in SOCS-3 and myostatin and a reduced efficiency of Akt phosphorylation. *Rejuvenation Research, 11*, 163–175.

Li, Z. F., Shelton, G. D., Engvall, E. (2005). Elimination of myostatin does not combat muscular dystrophy in dy mice but increases postnatal lethality. *The American Journal of Pathology, 166*, 491–497.

Liu, C. M., Yang, Z., Liu, C. W., Wang, R., Tien, P., Dale, R., Sun, L. Q. (2007). Effect of RNA oligonucleotide targeting Foxo-1 on muscle growth in normal and cancer cachexia mice. *Cancer Gene Therapy, 14*, 945–952.

Liu, C. M., Yang, Z., Liu, C. W., Wang, R., Tien, P., Dale, R., Sun, L. Q. (2008). Myostatin antisense RNA-mediated muscle growth in normal and cancer cachexia mice. *Gene Therapy, 15*, 155–160.

Ma, K., Mallidis, C., Bhasin, S., Mahabadi, V., Artaza, J., Gonzalez-Cadavid, N., Arias, J., Salehian, B. (2003). Glucocorticoid-induced skeletal muscle atrophy is associated with upregulation of myostatin gene expression. *American Journal of Physiology. Endocrinology and Metabolism, 285*, E363–E371.

Magee, T. R., Artaza, J. N., Ferrini, M. G., Vernet, D., Zuniga, F. I., Cantini, L., Reisz-Porszasz, S., Rajfer, J., Gonzalez-Cadavid, N. F. (2006). Myostatin short interfering hairpin RNA gene transfer increases skeletal muscle mass. *The Journal of Gene Medicine, 8*(9), 1171–1181.

Marsh, D. R., Criswell, D. S., Hamilton, M. T., Booth, F. W. (1997). Association of insulin-like growth factor mRNA expressions with muscle regeneration in young, adult, and old rats. *The American Journal of Physiology, 273*, R353–R358.

Matzuk, M. M., Lu, N., Vogel, H., Sellheyer, K., Roop, D. R., Bradley, A. (1995). Multiple defects and perinatal death in mice deficient in follistatin. *Nature, 374*, 360–363.

Mccroskery, S., Thomas, M., Maxwell, L., Sharma, M., Kambadur, R. (2003). Myostatin negatively regulates satellite cell activation and self-renewal. *The Journal of Cell Biology, 162*, 1135–1147.

McCroskery, S., Thomas, M., Platt, L., Hennebry, A., Nishimura, T., Mcleay, L., Sharma, M., Kambadur, R. (2005). Improved muscle healing through enhanced regeneration and reduced fibrosis in myostatin-null mice. *Journal of Cell Science, 118*, 3531–3541.

McFarland, D. C., Velleman, S. G., Pesall, J. E., Liu, C. (2006). Effect of myostatin on turkey myogenic satellite cells and embryonic myoblasts. *Comparative Biochemistry and Physiology. Part A: Molecular & Integrative Physiology, 144*, 501–508.

McFarlane, C., Langley, B., Thomas, M., Hennebry, A., Plummer, E., Nicholas, G., Mcmahon, C., Sharma, M., Kambadur, R. (2005). Proteolytic processing of myostatin is auto-regulated during myogenesis. *Developmental Biology, 283*, 58–69.

McFarlane, C., Plummer, E., Thomas, M., Hennebry, A., Ashby, M., Ling, N., Smith, H., Sharma, M., Kambadur, R. (2006). Myostatin induces cachexia by activating the ubiquitin proteolytic system through an NF-kappaB-independent, FoxO1-dependent mechanism. *Journal of Cellular Physiology, 209*, 501–514.

McFarlane, C., Hennebry, A., Thomas, M., Plummer, E., Ling, N., Sharma, M., Kambadur, R. (2008). Myostatin signals through Pax7 to regulate satellite cell self-renewal. *Experimental Cell Research, 314*, 317–329.

McPherron, A. C. & Lee, S. (1996). The transforming growth factor-b superfamily. *Growth Factors Cytokines Health Diseases, 1B*, 357–393.

McPherron, A. C. & Lee, S. J. (1997). Double muscling in cattle due to mutations in the myostatin gene. *Proceedings of the National Academy of Sciences of the United States of America, 94*, 12457–12461.

McPherron, A. C., Lawler, A. M., Lee, S. J. (1997). Regulation of skeletal muscle mass in mice by a new TGF-beta superfamily member. *Nature, 387*, 83–90.

Mezzogiorno, A., Coletta, M., Zani, B. M., Cossu, G., Molinaro, M. (1993). Paracrine stimulation of senescent satellite cell proliferation by factors released by muscle or myotubes from young mice. *Mechanisms of Ageing and Development, 70*, 35–44.

Michel, U., Farnworth, P., Findlay, J. K. (1993). Follistatins: More than follicle-stimulating hormone suppressing proteins. *Molecular and Cellular Endocrinology, 91*, 1–11.

Minetti, G. C., Colussi, C., Adami, R., Serra, C., Mozzetta, C., Parente, V., Fortuni, S., Straino, S., Sampaolesi, M., di Padova, M., Illi, B., Gallinari, P., Steinkuhler, C., Capogrossi, MC., Sartorelli, V., Bottinelli, R., Gaetano, C., Puri, P. L. (2006). Functional and morphological recovery of dystrophic muscles in mice treated with deacetylase inhibitors. *Natural Medicines, 12*, 1147–1150.

Mitchell, M. D., Osepchook, C. C., Leung, K. C., Mcmahon, C. D., Bass, J. J. (2006). Myostatin is a human placental product that regulates glucose uptake. *The Journal of Clinical Endocrinology and Metabolism, 91*, 1434–1437.

Miura, T., Kishioka, Y., Wakamatsu, J., Hattori, A., Hennebry, A., Berry, C. J., Sharma, M., Kambadur, R., Nishimura, T. (2006). Decorin binds myostatin and modulates its activity to muscle cells. *Biochemical and Biophysical Research Communications, 340*, 675–680.

Morissette, M. R., Stricker, J. C., Rosenberg, M. A., Buranasombati, C., Levitan, E. B., Mittleman, M. A., Rosenzweig, A. (2009). Effects of myostatin deletion in aging mice. *Aging Cell, 8*, 573–583.

Mosher, D. S., Quignon, P., Bustamante, C. D., Sutter, N. B., Mellersh, C. S., Parker, H. G., Ostrander, E. A. (2007). A mutation in the myostatin gene increases muscle mass and enhances racing performance in heterozygote dogs. *PLoS Genetics, 3*, e79.

Nakao, A., Afrakhte, M., Moren, A., Nakayama, T., Christian, J. L., Heuchel, R., Itoh, S., Kawabata, M., Heldin, N. E., Heldin, C. H., Ten Dijke, P. (1997a). Identification of Smad7, a TGFbeta-inducible antagonist of TGF-beta signalling. *Nature, 389*, 631–635.

Nakao, A., Imamura, T., Souchelnytskyi, S., Kawabata, M., Ishisaki, A., Oeda, E., Tamaki, K., Hanai, J., Heldin, C. H., Miyazono, K., Ten Dijke, P. (1997b). Tgf-beta receptor-mediated signalling through Smad2, Smad3 and Smad4. *The EMBO Journal, 16*, 5353–5362.

Nicholas, G., Thomas, M., Langley, B., Somers, W., Patel, K., Kemp, C. F., Sharma, M., Kambadur, R. (2002). Titin-cap associates with, and regulates secretion of, myostatin. *Journal of Cellular Physiology, 193*, 120–131.

Nishi, M., Yasue, A., Nishimatu, S., Nohno, T., Yamaoka, T., Itakura, M., Moriyama, K., Ohuchi, H., Noji, S. (2002). A missense mutant myostatin causes hyperplasia without hypertrophy in the mouse muscle. *Biochemical and Biophysical Research Communications, 293*, 247–251.

Nnodim, J. O. (2000). Satellite cell numbers in senile rat levator ani muscle. *Mechanisms of Ageing and Development, 112*, 99–111.

Ohsawa, Y., Hagiwara, H., Nakatani, M., Yasue, A., Moriyama, K., Murakami, T., Tsuchida, K., Noji, S., Sunada, Y. (2006). Muscular atrophy of caveolin-3-deficient mice is rescued by myostatin inhibition. *Journal of Clinical Investigation, 116*, 2924–2934.

Oldham, J. M., Martyn, J. A., Sharma, M., Jeanplong, F., Kambadur, R., Bass, J. J. (2001). Molecular expression of myostatin and MyoD is greater in double-muscled than normal-muscled cattle fetuses. *American Journal of Physiology: Regulatory, Integrative and Comparative Physiology, 280*, R1488–R1493.

Olguin, H. C. & Olwin, B. B. (2004). Pax-7 up-regulation inhibits myogenesis and cell cycle progression in satellite cells: A potential mechanism for self-renewal. *Developmental Biology, 275*, 375–388.

Peterlin, B., Zidar, J., Meznaric-Petrusa, M., Zupancic, N. (1997). Genetic epidemiology of Duchenne and Becker muscular dystrophy in Slovenia. *Clinical Genetics, 51*, 94–97.

Philip, B., Lu, Z., Gao, Y. (2005). Regulation of GDF-8 signaling by the p38 MAPK. *Cellular Signalling, 17*, 365–375.

Piek, E., Heldin, C. H., Ten Dijke, P. (1999). Specificity, diversity, and regulation in TGF-beta superfamily signaling. *The FASEB Journal, 13*, 2105–2124.

Raue, U., Slivka, D., Jemiolo, B., Hollon, C., Trappe, S. (2006). Myogenic gene expression at rest and after a bout of resistance exercise in young (18–30 yr) and old (80–89 yr) women. *Journal of Applied Physiology, 101*, 53–59.

Reardon, K. A., Davis, J., Kapsa, R. M., Choong, P., Byrne, E. (2001). Myostatin, insulin-like growth factor-1, and leukemia inhibitory factor mRNAs are upregulated in chronic human disuse muscle atrophy. *Muscle & Nerve, 24*, 893–899.

Rebbapragada, A., Benchabane, H., Wrana, J. L., Celeste, A. J., Attisano, L. (2003). Myostatin signals through a transforming growth factor beta-like signaling pathway to block adipogenesis. *Molecular and Cellular Biology, 23*, 7230–7242.

Rios, R., Carneiro, I., Arce, V. M., Devesa, J. (2002). Myostatin is an inhibitor of myogenic differentiation. *American Journal of Physiology. Cell Physiology, 282*, C993–C999.

Saharinen, J., Hyytiainen, M., Taipale, J., Keski-Oja, J. (1999). Latent transforming growth factor-beta binding proteins (LTBPS) – structural extracellular matrix proteins for targeting TGF-beta action. *Cytokine & Growth Factor Reviews, 10*, 99–117.

Salehian, B., Mahabadi, V., Bilas, J., Taylor, W. E., MA, K. (2006). The effect of glutamine on prevention of glucocorticoid-induced skeletal muscle atrophy is associated with myostatin suppression. *Metabolism, 55*, 1239–1247.

Sartori, R., Milan, G., Patron, M., Mammucari, C., Blaauw, B., Abraham, R., Sandri, M. (2009). Smad2 and 3 transcription factors control muscle mass in adulthood. *American Journal of Physiology. Cell Physiology, 296*, C1248–C1257.

Sazanov, A., Ewald, D., Buitkamp, J., Fries, R. (1999). A molecular marker for the chicken myostatin gene (GDF8) maps to 7p11. *Animal Genetics, 30*, 388–389.

Schneyer, A., Tortoriello, D., Sidis, Y., Keutmann, H., Matsuzaki, T., Holmes, W. (2001). Follistatin-related protein (FSRP): A new member of the follistatin gene family. *Molecular and Cellular Endocrinology, 180*, 33–38.

Schuelke, M., Wagner, K. R., Stolz, L. E., Hubner, C., Riebel, T., Komen, W., Braun, T., Tobin, J. F., Lee, S. J. (2004). Myostatin mutation associated with gross muscle hypertrophy in a child. *The New England Journal of Medicine, 350*, 2682–2688.

Schultz, E. & Lipton, B. H. (1982). Skeletal muscle satellite cells: Changes in proliferation potential as a function of age. *Mechanisms of Ageing and Development, 20*, 377–383.

Seale, P., Sabourin, L. A., Girgis-Gabardo, A., Mansouri, A., Gruss, P., Rudnicki, M. A. (2000). Pax7 is required for the specification of myogenic satellite cells. *Cell, 102*, 777–786.

Sharma, M., Kambadur, R., Matthews, K. G., Somers, W. G., Devlin, G. P., Conaglen, J. V., Fowke, P. J., Bass, J. J. (1999). Myostatin, a transforming growth factor-beta superfamily member, is expressed in heart muscle and is upregulated in cardiomyocytes after infarct. *Journal of Cellular Physiology, 180*, 1–9.

Shefer, G., Van de Mark, D. P., Richardson, J. B., Yablonka-Reuveni, Z. (2006). Satellite-cell pool size does matter: Defining the myogenic potency of aging skeletal muscle. *Developmental Biology, 294*, 50–66.

Shelton, G. D. & Engvall, E. (2007). Gross muscle hypertrophy in whippet dogs is caused by a mutation in the myostatin gene. *Neuromuscular Disorders, 17*, 721–722.

Shibli-Rahhal, A., Van Beek, M., Schlechte, J. A. (2006). Cushing's syndrome. *Clinics in Dermatology, 24*, 260–265.

Shyu, K. E., Lu, M. J., Wang, B. W., Sun, H. Y., Chang, H., (2006) Myostatin expression in ventricular myocardium in a rat model of volume-overload heart failure. *European Journal of Clinical Investigation, 36*, 713–719.

Siciliano, G., Tessa, A., Renna, M., Manca, M. L., Mancuso, M., Murri, L. (1999). Epidemiology of dystrophinopathies in North-West Tuscany: A molecular genetics-based revisitation. *Clinical Genetics, 56*, 51–58.

Siriett, V., Platt, L., Salerno, M. S., Ling, N., Kambadur, R., Sharma, M. (2006). Prolonged absence of myostatin reduces sarcopenia. *Journal of Cellular Physiology, 209*, 866–873.

Siriett, V., Salerno, M. S., Berry, C., Nicholas, G., Bower, R., Kambadur, R., Sharma, M. (2007). Antagonism of myostatin enhances muscle regeneration during sarcopenia. *Molecular Therapy, 15*, 1463–1470.

Sironi, M., Cagliani, R., Comi, G. P., Pozzoli, U., Bardoni, A., Giorda, R., Bresolin, N. (2003). Trans-acting factors may cause dystrophin splicing misregulation in BMD skeletal muscles. *FEBS Letters, 537*, 30–34.

Skipper, M. (2006). A lean feat – microRNAs and muscle mass. *Nature Reviews. Genetics, 7*, 491.

Snow, M. H. (1977). The effects of aging on satellite cells in skeletal muscles of mice and rats. *Cell and Tissue Research, 185*, 399–408.

Souchelnytskyi, S., Tamaki, K., Engstrom, U., Wernstedt, C., Ten Dijke, P., Heldin, C. H. (1997). Phosphorylation of Ser465 and Ser467 in the C terminus of Smad2 mediates interaction with Smad4 and is required for transforming growth factor-beta signaling. *The Journal of Biological Chemistry, 272*, 28107–28115.

Steelman, C. A., Recknor, J. C., Nettleton, D., Reecy, J. M. (2006). Transcriptional profiling of myostatin-knockout mice implicates Wnt signaling in postnatal skeletal muscle growth and hypertrophy. *The FASEB Journal, 20*, 580–582.

Szabo, G., Dallmann, G., Muller, G., Patthy, L., Soller, M., Varga, L. (1998). A deletion in the myostatin gene causes the compact (Cmpt) hypermuscular mutation in mice. *Mammalian Genome, 9*, 671–672.

Thies, R. S., Chen, T., Davies, M. V., Tomkinson, K. N., Pearson, A. A., Shakey, Q. A., Wolfman, N. M. (2001). GDF-8 propeptide binds to GDF-8 and antagonizes biological activity by inhibiting GDF-8 receptor binding. *Growth Factors, 18*, 251–259.

Thomas, M., Langley, B., Berry, C., Sharma, M., Kirk, S., Bass, J., Kambadur, R. (2000). Myostatin, a negative regulator of muscle growth, functions by inhibiting myoblast proliferation. *The Journal of Biological Chemistry, 275*, 40235–40243.

Tidball, J. G. (1995). Inflammatory cell response to acute muscle injury. *Medicine and Science in Sports and Exercise, 27*, 1022–1032.

Trendelenburg, A. U., Meyer, A., Rohner, D., Boyle, J., Hatakeyama, S., Glass, D. J. (2009). Myostatin reduces Akt/TORC1/p70S6K signaling, inhibiting myoblast differentiation and myotube size. *American Journal of Physiology. Cell Physiology, 296*, C1258–C1270.

Tseng, B. S., Zhao, P., Pattison, J. S., Gordon, S. E., Granchelli, J. A., Madsen, R. W., Folk, L. C., Hoffman, E. P., Booth, F. W. (2002). Regenerated mdx mouse skeletal muscle shows differential mRNA expression. *Journal of Applied Physiology, 93*, 537–545.

Tsuchida, K., Arai, K. Y., Kuramoto, Y., Yamakawa, N., Hasegawa, Y., Sugino, H. (2000). Identification and characterization of a novel follistatin-like protein as a binding protein for the TGF-beta family. *The Journal of Biological Chemistry, 275*, 40788–40796.

Tsuchida, K., Matsuzaki, T., Yamakawa, N., Liu, Z., Sugino, H. (2001). Intracellular and extracellular control of activin function by novel regulatory molecules. *Molecular and Cellular Endocrinology, 180*, 25–31.

Tsuchiya, S., Okuno, Y., Tsujimoto, G. (2006). MicroRNA: Biogenetic and functional mechanisms and involvements in cell differentiation and cancer. *Journal of Pharmacological Sciences, 101*, 267–270.

Wang, H., Zhang, Q., Zhu, D. (2003). hsgt interacts with the N-terminal region of myostatin. *Biochemical and Biophysical Research Communications, 311*, 877–883.

Watt, D. J., Morgan, J. E., Clifford, M. A., Partridge, T. A. (1987). The movement of muscle precursor cells between adjacent regenerating muscles in the mouse. *Anatomy and Embryology (Berlin), 175*, 527–536.

Watt, D. J., Karasinski, J., Moss, J., England, M. A. (1994). Migration of muscle cells. *Nature, 368*, 406–407.

Wehling, M., Cai, B., Tidball, J. G. (2000). Modulation of myostatin expression during modified muscle use. *The FASEB Journal, 14*, 103–110.

Welle, S., Bhatt, K., Shah, B., Thornton, C. (2002). Insulin-like growth factor-1 and myostatin mRNA expression in muscle: Comparison between 62–77 and 21–31 yr old men. *Experimental Gerontology, 37*, 833–839.

Whittemore, L. A., Song, K., Li, X., Aghajanian, J., Davies, M., Girgenrath, S., Hill, J. J., Jalenak, M., Kelley, P., Knight, A., Maylor, R., O'hara, D., Pearson, A., Quazi, A., Ryerson, S., Tan, X. Y., Tomkinson, K. N., Veldman, G. M., Widom, A., Wright, J. F., Wudyka, S., Zhao, L., Wolfman, N. M. (2003). Inhibition of myostatin in adult mice increases skeletal muscle mass and strength. *Biochemical and Biophysical Research Communications, 300*, 965–971.

Wolfman, N. M., Mcpherron, A. C., Pappano, W. N., Davies, M. V., Song, K., Tomkinson, K. N., Wright, J. F., Zhao, L., Sebald, S. M., Greenspan, D. S., Lee, S. J. (2003). Activation of latent myostatin by the BMP-1/tolloid family of metalloproteinases. *Proceedings of the National Academy of Sciences of the United States of America, 100*, 15842–15846.

Xiong, Y., Hannon, G. J., Zhang, H., Casso, D., Kobayashi, R., Beach, D. (1993). p21 is a universal inhibitor of cyclin kinases. *Nature, 366*, 701–704.

Yablonka-Reuveni, Z., Seger, R., Rivera, A. J. (1999). Fibroblast growth factor promotes recruitment of skeletal muscle satellite cells in young and old rats. *The Journal of Histochemistry and Cytochemistry, 47*, 23–42.

Yang, J., Ratovitski, T., Brady, J. P., Solomon, M. B., Wells, K. D., Wall, R. J. (2001). Expression of myostatin pro domain results in muscular transgenic mice. *Molecular Reproduction and Development, 60*, 351–361.

Yang, W., Chen, Y., Zhang, Y., Wang, X., Yang, N., Zhu, D. (2006). Extracellular signal-regulated kinase 1/2 mitogen-activated protein kinase pathway is involved in myostatin-regulated differentiation repression. *Cancer Research, 66*, 1320–1326.

Yarasheski, K. E., Bhasin, S., Sinha-Hikim, I., Pak-Loduca, J., Gonzalez-Cadavid, N. F. (2002). Serum myostatin-immunoreactive protein is increased in 60–92 year old women and men with muscle wasting. *The Journal of Nutrition, Health & Aging, 6*, 343–348.

Zachwieja, J. J., Smith, S. R., Sinha-Hikim, I., Gonzalez-Cadavid, N., Bhasin, S. (1999). Plasma myostatin-immunoreactive protein is increased after prolonged bed rest with low-dose T3 administration. *Journal of Gravitational Physiology, 6*, 11–15.

Zammit, P. S., Golding, J. P., Nagata, Y., Hudon, V., Partridge, T. A., Beauchamp, J. R. (2004). Muscle satellite cells adopt divergent fates: A mechanism for self-renewal? *The Journal of Cell Biology, 166*, 347–357.

Zhang, Y., Musci, T., Derynck, R. (1997). The tumor suppressor Smad4/DPC 4 as a central mediator of Smad function. *Current Biology, 7*, 270–276.

Zhou, G. Q., Xie, H. Q., Zhang, S. Z., Yang, Z. M. (2006). Current understanding of dystrophin-related muscular dystrophy and therapeutic challenges ahead. *Chinese Medical Journal (England), 119*, 1381–1391.

Zhu, X., Hadhazy, M., Wehling, M., Tidball, J. G., Mcnally, E. M. (2000). Dominant negative myostatin produces hypertrophy without hyperplasia in muscle. *FEBS Letters, 474*, 71–75.

Zhu, X., Topouzis, S., Liang, L. F., Stotish, R. L. (2004). Myostatin signaling through Smad2, Smad3 and Smad4 is regulated by the inhibitory Smad7 by a negative feedback mechanism. *Cytokine, 26*, 262–272.

Zimmers, T. A., Davies, M. V., Koniaris, L. G., Haynes, P., Esquela, A. F., Tomkinson, K. N., Mcpherron, A. C., Wolfman, N. M., Lee, S. J. (2002). Induction of cachexia in mice by systemically administered myostatin. *Science, 296*, 1486–1488.

Role of β-Adrenergic Signalling in Skeletal Muscle Wasting: Implications for Sarcopenia

James G. Ryall and Gordon S. Lynch

Abstract While the importance of β-adrenergic signalling in the heart has been well documented for more than half a century and continues to receive significant attention, it is only more recently that we have begun to understand the importance of this signalling pathway in skeletal muscle. There is considerable evidence regarding the stimulation of the β-adrenergic system with β-adrenoceptor agonists (β-agonists) in animals and humans. Although traditionally used for the treatment of bronchospasm, it became apparent that some β-agonists, such as clenbuterol, had the ability to increase skeletal muscle mass and decrease body fat (Ricks et al. 1984; Beerman et al. 1987). These so-called "repartitioning effects" proved desirable for those working in the livestock industry trying to improve feed efficiency and meat quality (Sillence 2004). Not surprisingly, β_2-agonists were soon being used by those engaged in competitive bodybuilding and by other athletes, especially those engaged in strength- and power-related sports (Lynch 2002; Lynch and Ryall 2008).

As a consequence of their muscle anabolic actions, the effects of β-agonist administration on skeletal muscle have been examined in a number of animal models (and in humans) with the hope of discovering therapeutic applications, particularly for muscle wasting conditions including sarcopenia (age-related muscle wasting and associated weakness), cancer cachexia, sepsis, and other forms of metabolic stress, denervation, disuse, inactivity, unloading or microgravity, burns, HIV-acquired immunodeficiency syndrome (AIDS), chronic kidney or heart failure, chronic obstructive pulmonary disease, muscular dystrophies, and other neuromuscular disorders. For many of these conditions, the anabolic properties of β-agonists have the potential to attenuate (or potentially reverse) the muscle

G.S. Lynch (✉)
Department of Physiology, Basic and Clinical Myology Laboratory,
The University of Melbourne, Victoria, Australia
e-mail: gsl@unimelb.edu.au

J.G. Ryall
The Laboratory of Muscle Stem Cells and Gene Regulation,
National Institute of Arthritis, Musculoskeletal and Skin Diseases,
National Institutes of Health (NIH), Bethesda, MD, USA
e-mail: ryallj@mail.nih.gov

G.S. Lynch (ed.), *Sarcopenia – Age-Related Muscle Wasting and Weakness*,
DOI 10.1007/978-90-481-9713-2_19,

wasting, muscle fibre atrophy, and associated muscle weakness. β-agonists also have clinical significance for enhancing muscle repair and restoring muscle function after injury or following reconstructive surgery.

In addition to having anabolic effects on skeletal muscle, β-agonists have also been associated with some undesirable side effects, including increased heart rate (tachycardia) and muscle tremor, which have so far limited their therapeutic potential. In this chapter we describe the physiological significance of β-adrenergic signalling in skeletal muscle and discuss the therapeutic potential of β-adrenergic stimulation for age-related muscle wasting and weakness. We describe the effects of current β-agonists on skeletal muscle and identify novel research strategies to minimize the unwanted side-effects associated with systemic β-adrenergic stimulation.

Keywords β-adrenoceptor agonist • β-adrenergic signalling • cardiac muscle • fibre type • G-protein couple receptor • heart • muscle hypertrophy • muscle wasting • skeletal muscle

1 Overview of β-Adrenergic Signalling

Before discussing the therapeutic potential of β-adrenergic stimulation for sarcopenia, it is important to characterize the role of this important signalling pathway in normal healthy skeletal muscle.

β-adrenoceptors belong to the guanine nucleotide-binding G-protein coupled receptor (GPCR) family (Fredriksson et al. 2003), and are activated endogenously via adrenaline (epinephrine) and/or noradrenaline (norepinephrine). One of the defining features of the GPCR superfamily is that all of the receptors couple to heterotrimeric guanine-nucleotide-binding regulatory proteins (G-proteins). These molecules received their name from the typical three subunit composition (designated 'αβγ').

All GPCRs (including β-adrenoceptors) have a conserved seven transmembrane α-helical structure forming three extracellular loops; including an amino-terminus and three intracellular loops, including a carboxy-terminus (Johnson 2006; Morris and Malbon 1999). The third-fifth intramembranous regions are believed to be important in ligand binding, while the third intracellular loop of the GPCR has a central role in G-protein coupling (Johnson 2006).

The G-proteins are located in the cytoplasmic space and act intracellularly, interacting with an intracellular loop of the GPCR (Fig. 1). The G-protein βγ subunits (Gβγ) form a tightly interacting dimer which is bound to the intracellular plasma membrane via an isoprenyl moiety located on the C-terminus of the γ subunit, whereas the G-protein α subunit (Gα), in its inactive state, remains attached to the Gβγ dimer (Bockaert and Pin 1999). Activation of the GPCR causes a profound change in the conformation of the intracellular loops and uncovers a previously masked G-protein binding site (Filipek et al. 2004; Klco et al. 2005; Meng and Bourne 2001). Specifically, the third intracellular loop of the GPCR is involved in G-protein binding (Kobilka et al. 1988). Upon binding of a ligand to

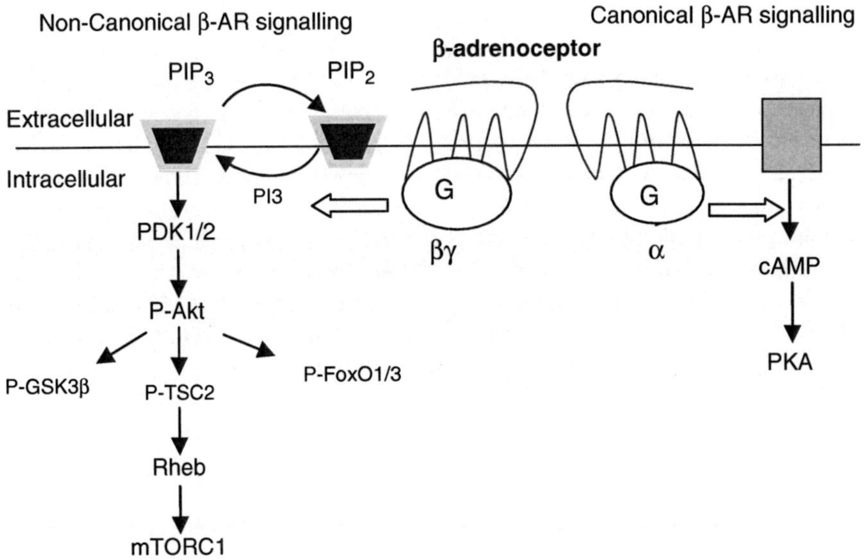

Fig. 1 β-adrenergic signalling in skeletal muscle. Traditionally, the stimulated β-adrenoceptor has been thought to couple with the stimulatory Gα subunit ($G\alpha_s$) of the heterotrimeric G-protein (Gαβγ) and adenylate cyclase (AC), resulting in conversion of ATP to cAMP and the activation of protein kinase A (PKA). Stimulation of this pathway has been linked to the inhibition of proteolytic pathways and possibly to protein synthesis. In the non-canonical signalling pathway β-adrenoceptors signal via the G-protein Gβγ subunits to promote phosphorylation of phosphatidylinositol-4,5-bisphosphate (i.e. PIP_2 becomes PIP_3) by phosphatidylinositol 3-kinase (PI3-K), leading to Akt activation. These events trigger the downstream activators, glycogen synthase kinase 3β (GSK3β), tuberous sclerosis complex 2 (TSC2, an activator of mammalian target of rapamycin complex-1, mTORC1) and the forkhead box O (FoxO) family of transcription factors. Thus, β-adrenoceptor stimulation can influence protein synthesis and degradation by several mechanisms

the GPCR, guanosine diphosphate (GDP) is released from the Gα subunit, and subsequent guanosine triphosphate (GTP) binding occurs, which activates the Gα subunit and exposes effector-interaction sites in the Gβγ dimer (Bockaert and Pin 1999; Gilman 1995; Hampoelz and Knoblich 2004; Rodbell et al. 1971).

The Gα-subunits can be divided into four main families, based on their primary sequence: $G\alpha_s$, $G\alpha_{i/o}$, $G\alpha_{q/11}$ and $G\alpha_{12}$, which regulate the activity of many different second messenger systems (Lohse 1999; Wilkie et al. 1992). β-adrenoceptors couple predominantly with $G\alpha_s$ and $G\alpha_i$ isoforms to initiate downstream effector pathways including adenylyl cyclase (AC), transmembrane protein kinases, and phospholipases (Dascal 2001; Wenzel-Seifert and Seifert 2000).

Three subtypes of β-adrenoceptors have been identified and cloned; β_1-, β_2- and β_3-adrenoceptors (Dixon et al. 1986; Emorine et al. 1989; Frielle et al. 1987), each with a 65–70% homology in their amino acid composition (Kobilka et al. 1987). Skeletal muscle contains a significant proportion of β-adrenoceptors, mostly of the β_2-subtype, but also include approximately 7–10% β_1-adrenoceptors (Kim et al. 1991; Williams et al. 1984) and a smaller population of α-adrenoceptors, usually in higher proportions in slow-twitch muscles (Rattigan et al. 1986). Slow-twitch muscles like the soleus have a greater density of β-adrenoceptors than fast-twitch muscles, such as the extensor digitorum longus (EDL) (Martin et al. 1989; Ryall et al. 2002, 2004). Although the functional significance of this difference in β-adrenoceptor density is not yet understood fully, the response to β-agonist administration appears to be greater in fast-, than in slow-twitch skeletal muscles (Ryall et al. 2002, 2006).

The $G\alpha_s$-AC-cyclic AMP (cAMP) is the most well characterized of the β_2-adrenoceptor signalling pathways and is generally thought to be, at least partially, responsible for the β_2-adrenoceptor mediated hypertrophy in skeletal muscle (Hinkle et al. 2002; Navegantes et al. 2000). The production of cAMP results in the activation of numerous downstream signalling pathways, including the well-described protein kinase A (PKA) signalling pathways.

Following cAMP activation, PKA is thought to phosphorylate and regulate the activity of numerous proteins. In addition, PKA is capable of diffusing passively into the nucleus, where it can regulate the expression of many target genes via direct phosphorylation of the cAMP response element (CRE) binding protein (CREB), or via a modulator that acts on second generation target genes (Carlezon et al. 2005; Mayr and Montminy 2001).

The CRE binding protein is a nuclear transcription factor that is expressed ubiquitously and has been implicated in many processes, including cell proliferation, differentiation, adaptation, and survival (Mayr and Montminy 2001). CREB forms a homodimer and binds to a conserved CRE-region on DNA. Nuclear entry of PKA, phosphorylates CREB at a single serine residue site (Ser^{133}) (Hagiwara et al. 1993). Phosphorylation of Ser^{133} promotes transcription at the CRE-region through recruitment of the transcriptional co-activators CREB-binding protein (CBP) and p300, which mediate transcriptional activity through their association with RNA Polymerase II (Goodman and Smolik 2000; Mayr and Montminy 2001). CREB-phosphorylation promotes activation of genes containing a CRE-region, of which there are >4,000 in the human genome (Pourquié 2005; Zhang et al. 2005). Finally, CRE-gene activation is terminated by dephosphorylation of CREB, a process regulated by the serine/threonine phosphatases PP-1 and PP-2A (Hagiwara et al. 1992; Wadzinski et al. 1993).

One target for β-adrenoceptor mediated CRE activation in skeletal muscle is the promoter region of the orphan nuclear receptor, NOR-1 (NR4A3) (Ohkura et al. 1998; Pearen et al. 2006). β_2-adrenoceptor activation is associated with an increased expression of NOR-1 and the related orphan nuclear receptor nur-77 (NR4A1) (Maxwell et al. 2005; Pearen et al. 2006). Interestingly, Pearen and colleagues (2006) found that siRNA mediated inhibition of NOR-1 expression was associated with a dramatic increase (>65 fold) in the levels of myostatin mRNA in C2C12 cells. Myostatin is a member of the transforming growth factor-β superfamily and a potent negative regulator of muscle mass (McPherron et al. 1997). Thus, β-adrenoceptor activation, through increased NOR-1 expression, may inhibit myostatin expression and hence promote skeletal muscle growth.

The transcriptional adapters, CBP and p300, promote skeletal muscle myogenesis via the coactivation of a number of myogenic basic helix-loop-helix (bHLH) proteins (Eckner et al. 1996; McKinsey et al. 2002; Sartorelli et al. 1997). The family of myogenic bHLH proteins, including MyoD, myogenin, myf5 and MRF4, activate muscle gene transcription via pairing with the ubiquitously expressed E-box consensus sequence in the control regions of muscle-specific genes (McKinsey et al. 2002; Molkentin and Olson 1996). Sartorelli and colleagues (1997) found that p300 and CBP may positively influence myogenesis by acting as a 'bridge' between the myogenic bHLH and the myocyte enhancer factor 2 (MEF2) family of proteins.

In addition to transcriptional coactivation, CBP and p300 have intrinsic histone acetyltransferase (HAT) activity (Goodman and Smolik 2000; Roth et al. 2003; Thompson et al. 2004). Histone acetyltransferases are believed to play an important role in transcription, since they catalyze the transfer of acetyl groups from acetyl-coenzyme A to the ε-amino group of lysine side chains of specific proteins, including several transcriptional regulatory proteins (Yang 2004). Therefore, the β-adrenoceptor mediated actions of CBP and p300 could increase the accessibility of docking sites for transcriptional proteins and regulators (Ogryzko et al. 1996; Thompson et al. 2004).

Chen and colleagues (2005) identified an unexpected role for PKA/CREB signalling during myogenesis, proposing that myogenic gene expression of Pax3, MyoD, and Myf5 is dependent on AC/cAMP mediated phosphorylation of PKA and subsequent activation of CREB. The authors demonstrated the importance of CREB in the developing myotome, since $CREB^{-/-}$ mice did not express Pax3, MyoD, or Myf5 and myotome formation was defective (Chen et al. 2005). It remains to be determined whether β-adrenoceptor mediated activation of PKA/CREB signalling has a similar response during myogenesis.

Berdeaux and colleagues (2007) demonstrated a novel role of CREB in mediating the activity of MEF2. They showed that β-adrenergic stimulated CREB modulated the phosphorylation status of the class II histone deacetylase HDAC5 in mouse skeletal muscle, by increasing the expression of salt inducible kinase 1 (SIK1). Activated SIK1 phosphorylated HDAC5, resulting in its nuclear exclusion and subsequent activation of the MEF2 myogenic program (Berdeaux et al. 2007). These exciting results demonstrated the complexity of the downstream activators of the β-adrenergic signalling pathway and highlighted the previously unappreciated role of this pathway in skeletal muscle.

In addition to the well-described $G\alpha_s$-cAMP signalling pathways, studies have implicated the Gβγ subunits in various cell signalling processes, which may also play important roles in β-adrenoceptor signalling in skeletal muscle (Crespo et al. 1994; Dascal 2001; Diversé-Pierluissi et al. 2000; Ford et al. 1998; Mirshahi et al. 2002). Specifically, in vitro cell culture experiments have revealed that the $G\alpha_i$ linked Gβγ subunits activate the phosphoinositol 3-kinase (PI3K)-AKT signalling pathway (Lopez-Ilasaca et al. 1997; Murga et al. 1998, 2000; Schmidt et al. 2001).

The PI3K-AKT signalling pathway has been implicated in protein synthesis, gene transcription, cell proliferation, and cell survival (Bodine et al. 2001b; Glass 2003, 2005; Kline et al. 2007; Pallafacchina et al. 2002; Rommel et al. 2001). Although there are three distinct isoforms of AKT, the predominant skeletal muscle isoform is AKT1 (Nader 2005). Activation of PI3K phosphorylates the membrane bound PIP_2, creating a lipid-binding site on the cell membrane for both AKT1 and 3′-phoshphoinositide-dependent protein kinase 1 (PDK). PDK then phosphorylates AKT1 at the membrane (Nicholson and Anderson 2002). Akt activation, in turn, results in the phosphorylation of numerous downstream activators, including glycogen synthase kinase 3β (GSK3β), tuberous sclerosis complex 2 (TSC2, leading to the subsequent activation of mammalian target of rapamycin complex1, mTORC1) (Garami et al. 2003; Latres et al. 2005) and members of the forkhead box O (FOXO) family of transcription factors (Sandri et al. 2004; Stitt et al. 2004).

Kline and colleagues (2007) found that stimulation of the β-adrenoceptor signalling pathway resulted in AKT phosphorylation and subsequent activation of mTORC1. Initiation of mTORC signalling phosphorylates and subsequently activates $p70^{s6}$ kinase ($p70^{S6K}$), while concomitantly inactivating 4EBP-1 (also termed PHAS-1). $p70^{S6K}$ mediates the phosphorylation of the 40S ribosomal S6 protein, resulting in the upregulation of mRNA translation encoding for ribosomal proteins and elongation factors (Jefferies et al. 1997). Inactivation of 4EBP-1 removes its inhibitory action on the protein initiation factor eukaryotic initiation factor 4E (eIF-4E) (Lai et al. 2004; Nave et al. 1999). These findings supported those of Sneddon and colleagues (2001) who reported an increased phosphorylation of 4E-BP1 and $p70^{S6K}$ in rat plantaris muscle after 3 days of clenbuterol treatment.

GSK-3β is reported to be a negative regulator of protein translation and gene expression in cardiac (Hardt and Sadoshima 2002) and skeletal muscle (Childs et al. 2003; Bossola et al. 2008). Following β-adrenoceptor stimulation, GSK3β is phosphorylated and subsequently inactivated by AKT1 (Yamamoto et al. 2007), resulting in the expression of a dominant negative form of GSK3β. Since GSK3β normally acts to inhibit the translation initiation factor eIF2B, blockade of GSK3β by AKT1 might promote protein synthesis (Bodine et al. 2001b; Rommel et al. 2001).

AKT1 signalling is not only involved in the signalling pathways responsible for muscle hypertrophy, but it has been implicated in the inhibition of signalling pathways responsible for "muscle atrophy". AKT1 inactivation of FOXO leads to nuclear exclusion and inhibition of the forkhead transcriptional program. The DNA displacement and subsequent nuclear exclusion of FOXO requires the involvement of 14-3-3 proteins, which bind to FOXO following AKT1-mediated phosphorylation (Tran et al. 2003). 14-3-3 proteins are among a family of chaperone proteins that interact with specific phosphorylated protein ligands (Tran et al. 2003).

Activation of the forkhead transcriptional program is necessary for induction of both muscle RING finger 1 (muRF1) and muscle atrophy F-box (MAFbx, also called atrogin-1) (Sandri et al. 2004; Stitt et al. 2004). Both muRF1 and MAFbx encode ubiquitin ligases which conjugate ubiquitin to protein substrates, and are upregulated in numerous models of muscle atrophy (Bodine et al. 2001a; Tintignac et al. 2005). Thus, by phosphorylating and inactivating FOXO, AKT1 blocks the induction of FOXO-mediated atrophy signalling via muRF1 and MAFbx. β-Adrenoceptor activation reduces the expression of muRF1 and MAFbx in skeletal muscle from denervated and hindlimb-suspended rats, an effect possibly mediated via AKT1-initiated inhibition of the forkhead transcriptional program (Kline et al. 2007).

It is interesting to note that while FOXO1 regulates the expression of both MAFbx and muRF1 (Stitt et al. 2004), FOXO3a appears only to activate the MAFbx promoter (Sandri et al. 2004). In addition, while measurable levels of FOXO4 have been identified in skeletal muscle (Furuyama et al. 2002), very little is known about its role in skeletal muscle atrophy. Furuyama and colleagues (2002) characterized the expression pattern of FOXO1, FOXO3a and FOXO4 with ageing and caloric restriction in rats. FOXO4 mRNA expression increased from 3 to 12 months and then decreased from 12 to 26 months. A similar pattern was observed for FOXO3a expression (Furuyama et al. 2002). Interestingly, FOXO1 mRNA expression remained unchanged. In contrast, caloric restriction resulted in an increase in the expression levels of both FOXO4 and FOXO1, but not FOXO3a (Furuyama et al. 2002). These results indicate the complexity of the forkhead transcriptional program in the regulation of skeletal muscle atrophy (Kandarian and Jackman 2006).

Several studies have identified a role for FOXO1 in binding to the promoter region of 4EBP-1 which resulted in increased mRNA and protein expression (Léger et al. 2006; Wu et al. 2008). Associated with the increase in 4EBP-1 was a reduction in mTORC activation and $p70^{S6K}$. Thus, in addition to previously reported roles in atrophic signalling pathways, FOXO1 also plays an active role in inhibiting protein synthesis (Yang et al. 2008).

A number of researchers have identified genes that are activated by β-adrenoceptor stimulation, but the mechanism for their activation remains unclear. For example McDaneld and colleagues (2004) examined differential gene expression in skeletal muscle after β-agonist administration to evaluate the role of genes thought responsible for muscle growth. Decreased mRNA abundance following β-adrenoceptor stimulation was confirmed for DD143 identified as ASB15, a bovine gene encoding an ankyrin repeat and a suppressor of cytokine signalling (SOCS) box protein, in both cattle and rats (McDaneld et al. 2004, 2006; Spangenburg 2005). The authors reported that ASB15 was a member of an emerging gene family involved in a variety of cellular processes including cellular proliferation and differentiation (McDaneld et al. 2004).

Similarly, Spurlock and colleagues (Spurlock et al. 2006) examined gene expression changes in mouse skeletal muscle 24 hours and 10 days after β-adrenoceptor stimulation and identified genes involved in processes important to skeletal muscle growth, including regulators of transcription and translation, mediators of cell-signalling pathways, and genes involved in polyamine metabolism. They reported changes in mRNA abundance of multiple genes associated with myogenic differentiation relevant to the effect of β-adrenoceptor stimulation on the proliferation, differentiation, and/or recruitment of satellite cells into muscle fibres to promote muscle hypertrophy. Similarly, they showed an upregulation of translational initiators responsible for increasing protein synthesis (Spurlock et al. 2006).

More recently, Pearen and colleagues (2009) profiled skeletal muscle gene expression in mouse tibialis anterior muscles at 1 and 4 h after systemic administration of formoterol and revealed significant expression changes in genes associated with skeletal muscle hypertrophy, myoblast differentiation, metabolism, circadian rhythm, transcription, histones, and oxidative stress. With respect to formoterol's anabolic effects, differentially expressed genes relevant to the regulation of muscle mass and metabolism were validated by quantitative RT-PCR to examine gene expression after acute (1–24 h) and chronic administration (1–28 days) of formoterol. Following acute and chronic formoterol administration there was an attenuation of myostatin signalling (differential expression of myostatin, activin receptor IIB, and phospho-Smad3) which was a previously unreported effect of β-adrenoceptor signalling in skeletal muscle. Acute (but not chronic) formoterol administration induced expression of genes involved in oxidative metabolism, including hexokinase 2, sorbin and SH3 domain containing 1, and uncoupling protein 3. Interestingly, formoterol administration also appeared to influence some genes associated with the peripheral regulation of circadian rhythm (including nuclear factor interleukin 3 regulated, D site albumin promoter binding protein, and cryptochrome 2) indicating crosstalk between β-adrenoceptor signalling and circadian cycling in skeletal muscle. This was the first study showing regulation of the peripheral circadian regulators in skeletal muscle by β-adrenoceptor signalling, possibly implicating β-adrenoceptor (sympathetic) signalling as a pathway coordinating communication between central and peripheral circadian clocks in skeletal muscle (Pearen et al. 2009).

2 Changes in Skeletal Muscle β-Adrenergic Signalling with Aging

While there has been much conjecture as to the exact changes in catecholamine levels as a consequence of ageing, it is now accepted that there is an increase in the plasma level of noradrenaline and a decrease in adrenaline, in rats and humans (Esler et al. 1995; Kaye and Esler 2005; Larkin et al. 1996). In addition, work from our laboratory has demonstrated an age-related change in β-adrenoceptor signalling in skeletal muscle (Ryall et al. 2007). Chronic administration of the β-adrenoceptor agonist, formoterol, for 4 weeks increased the mass of the slow-twitch soleus muscle in young (3 months), but not in adult (16 months) or old (27 months) rats. In contrast, formoterol increased the mass of the fast-twitch EDL muscle of rats in all three age groups tested (Ryall et al. 2007). These findings suggest that the β-adrenergic signalling pathway and especially that pathway leading to striated muscle hypertrophy, is altered by age in slow- but not in fast-twitch skeletal muscles, an effect independent of β-adrenoceptor density.

There is currently a dearth of knowledge regarding how ageing affects this important signalling pathway with most of our current knowledge based on studies conducted on the ageing myocardium. However, due to the differences in β-adrenergic signalling between these two tissues it is important that future studies focus on skeletal muscle.

3 Therapeutic Potential of β-Adrenoceptor Agonists for Sarcopenia

There have been numerous studies on animals and several studies on humans investigating the effects of β-agonists on skeletal muscle (for review see Lynch and Ryall 2008). In relation to attenuating the loss of muscle mass and protein content or hastening the restoration of these parameters in the elderly during periods of malnutrition or extended periods of inactivity, three early studies by Carter and Lynch (1994a, b, c) provided encouraging evidence that β-agonists could find therapeutic application for these conditions. To examine the anabolic effects of low-dose salbutamol or clenbuterol administration on aged rats, Carter and Lynch (1994b) showed that in old rats, *s.c.* delivery by osmotic minipumps (at daily doses of 1.03 mg/kg or 600 μg/kg) for 3 weeks, increased combined hindlimb muscle mass by 19% and 25%, respectively. Gastrocnemius muscle mass and protein content were increased by 19% and 23%, respectively, in old rats. Overall, this study found that salbutamol and clenbuterol increased skeletal muscle protein content and reduced carcass fat content, suggesting that both β-agonists could potentially stimulate muscle growth in frail elders (Carter and Lynch 1994b).

In a related experiment, Carter and Lynch (1994c) studied the effect of clenbuterol on recovery of muscle mass and carcass protein content after protein malnutrition in aged rats. The rats were subjected to 3-weeks of dietary protein restriction that reduced overall body mass by 21%. During the recovery period, the rats were fed a normal diet with clenbuterol (10 mg/kg) added to the feed. The addition of clenbuterol to the diet increased hindlimb muscle mass by 30% and protein content by 25%, in aged rats (Carter and Lynch 1994c). In another experiment (Carter and Lynch 1994a), aged rats were injected daily with thyroid hormone (4–6.5 mg of triiodothyronine per 100 g body mass) for 3 weeks to cause an ~20% reduction in body mass and hindlimb muscle mass. Feeding the rats a diet containing 10 mg clenbuterol per kg during a 3-week recovery period restored body mass and muscle mass to euthyroid control levels, whereas feeding the rats a control diet did not (Carter and Lynch 1994a). Taken together, these findings suggested that clenbuterol, or other β-agonists, could find application in hastening recovery of muscle mass as a consequence of malnutrition in frail, elderly humans (Carter and Lynch 1994a, c).

In aged rats, clenbuterol treatment (2 mg/kg) via daily injection for 4 weeks restored the age-associated decline in the mass and specific force (i.e. normalized force or force per muscle cross-sectional area) of diaphragm muscle strips (Smith et al. 2002). A much lower dose of clenbuterol (10 μg/kg per day), attenuated the loss of specific force in the soleus muscle only slightly (i.e. by 8%) and reduced fatigue (in response to repeated stimulation) by approximately 30% in aged rats, with considerable muscle atrophy having been subjected to 21 days of hindlimb suspension (Chen and Alway 2001). However, low-dose clenbuterol treatment did not attenuate the loss of specific force in the soleus of adult rats or in the plantaris muscles of old or adult rats. The study concluded that clenbuterol could reduce muscle fatigue in slow muscles during disuse with some clinical implications for reducing fatigue in muscles of the elderly. Findings from this and a related study (Chen and Alway 2000), indicated that low-dose clenbuterol treatment did not attenuate atrophy of fast muscles and only modestly attenuated the atrophy of slow muscles, making it largely ineffective for preventing muscle wasting from disuse atrophy in aged rats.

In a study from our laboratory (Ryall et al. 2004) old rats were treated daily with a relatively high dose of the β-agonist, fenoterol (1.4 mg/kg/day, *i.p.*), or saline for 4 weeks. At 28 months of age, untreated old F344 rats exhibited a loss of skeletal muscle mass and a decrease in force-producing capacity, in both fast and slow muscles. Interestingly, the muscle mass, fibre size, and force-producing capacity of EDL and soleus muscles from old rats treated with fenoterol was equivalent to, or greater than, untreated adult rats (Ryall et al. 2004). Fenoterol treatment caused a small increase in the fatigability of both EDL and soleus muscles due to a decrease in oxidative metabolism. The findings highlighted the clinical potential of β-agonists to increase muscle mass and function to levels that exceeded those in adult rats.

Schertzer and colleagues (2005) found that treating aged rats with fenoterol (1.4 mg/kg/day, *i.p.*) for 4 weeks, reversed the slowing of (twitch) relaxation in slow- and fast-twitch skeletal muscle due to increased SERCA activity and SERCA protein levels (Fig. 2). That study provided evidence for an age-related alteration in the environment of the nucleotide binding domain and/or a selective nitration of the SERCA2a isoform, which was associated with depressed SERCA activity.

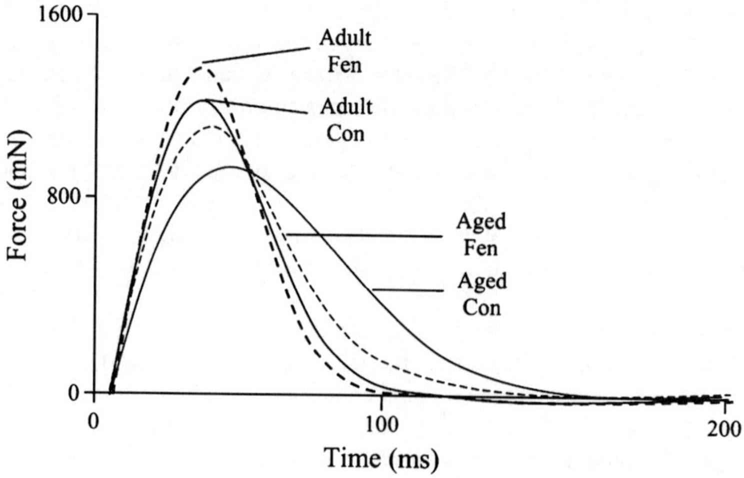

Fig. 2 Sample recordings of twitch characteristics in the predominately fast-twitch extensor digitorum longus muscles of adult (16 mo) and aged (28 mo) Fischer 344 rats that had been treated for 4 weeks with with fenoterol (Fen; *dashed line*) or saline vehicle only (control, Con; *solid line*) (see Ryall et al. 2004; Schertzer et al. 2005 for details). Reprinted with permission

Fenoterol treatment ameliorated the age-related decrease in nucleotide binding affinity and reversed the age-related accumulation of nitrotyrosine residues on the SERCA2a isoform. These changes, in combination with increases in SERCA1 protein levels, appeared to be the underlying mechanisms of fenoterol treatment reversing age-related decreases in the V_{max} of SERCA (Schertzer et al. 2005).

In a later study (Ryall et al. 2006), we demonstrated that 'newer' generation β-agonists, formoterol and salmeterol, could exert significant anabolic actions on skeletal muscle even at micromolar doses, compared with the millimolar doses required to elicit similar responses with older generation β-agonists such as fenoterol or clenbuterol. Using this information, we investigated the potential of formoterol, one of these newer generation β-agonists, to increase muscle mass and force producing capacity of EDL and soleus muscles in aged rats (Ryall et al. 2007). In addition, we studied the effects of formoterol withdrawal on parameters such as muscle mass and strength. Rats were similarly treated with either formoterol (25 μg/kg/day, *i.p.*), or saline vehicle for 4 weeks, and another group of rats were similarly treated with formoterol, followed by a period of withdrawal for 4 weeks. Formoterol treatment increased EDL muscle mass and the force producing capacity of both EDL and soleus muscles, without a concomitant increase in heart mass. The hypertrophy and increased force of EDL muscles persisted for 4 weeks after withdrawal of treatment. This study was important because it demonstrated significant improvements in muscle function in old rats after β-agonist administration, at a dose 1/50th that of other β-agonists that had been used previously (Ryall et al. 2004). These findings have important implications for clinical trials that might utilize β-agonists for muscle wasting conditions (Fowler et al. 2004; Kissel et al. 1998, 2001).

We and others have found that exogenous administration of clenbuterol, fenoterol and formoterol can result in a dramatic shift in the muscle fibre phenotype from slow-oxidative to fast oxidative-glycolytic fibres (Figs. 1 and 2; Ryall et al. 2002, 2007; Zeman et al. 1988). Although previous studies have identified the mechanisms underlying a shift from a fast to a slow muscle phenotype (Handschin et al. 2007; Kim et al. 2008; Oh et al. 2005), less is known about the pathways responsible for shifts from a slow to a fast muscle phenotype (Grifone et al. 2004; Ryall et al. 2008a, b). This is relevant if β-agonists are to be considered for therapeutic application for sarcopenia since age-related losses of fast motor units have important consequences for the preservation of fast muscles fibres during advancing age. Studies in rats and mice have shown that a significant shift in slow to fast fibre proportions within skeletal muscles as a consequence of chronic β-agonist administration can dramatically affect function, particularly shortening the duration of the isometric twitch response (Schertzer et al. 2005), increasing velocity of shortening (Dodd et al. 1996), and increasing muscle fatigability (Dupont-Versteegden 1996). In our hands, these effects are largely dependent on the type and dose of β-agonist employed (Harcourt et al. 2007).

Whether β-adrenergic signalling is implicated in the preservation of motor units has not been determined specifically but Zeman and colleagues (2004) reported that treating motor neuron degeneration (*mnd*) mice with clenbuterol enhanced regeneration of motor neuron axons and reduced the proportion of motor neurons with

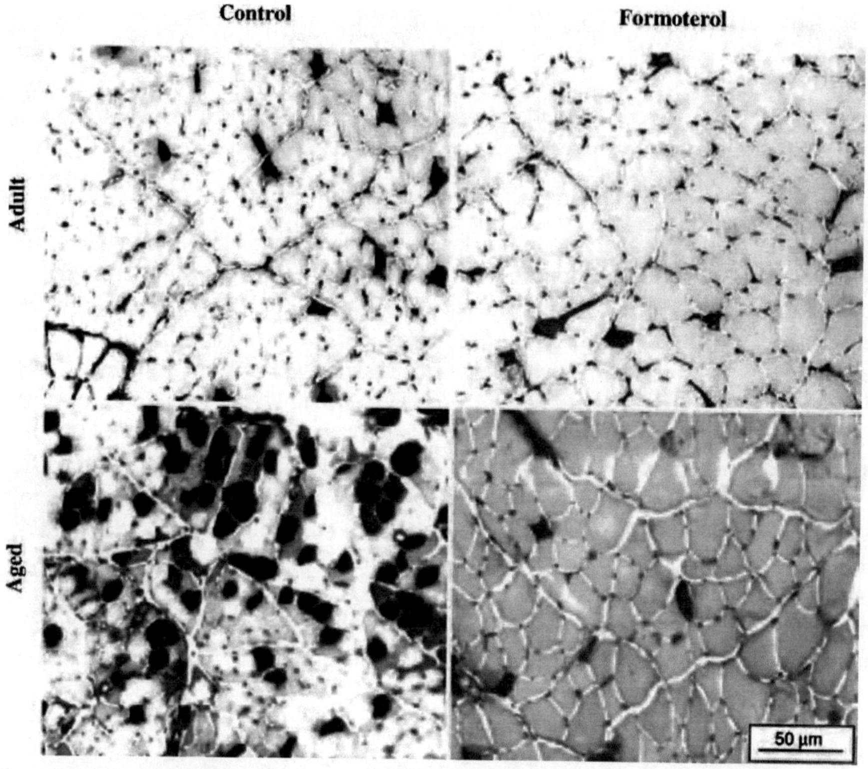

Fig. 3 Extensor digitorum longus (EDL) muscle sections from adult and old control rats and formoterol-treated rats reacted for mATPase at a preincubation pH of 4.3. Strongly reacting (dark) fibres are slow type I, and light gray fibres are fast-type II isoforms. EDL muscles from old control rats had a greater proportion of type I fibres, and formoterol treatment resulted in a decreased proportion of type I fibres. Note also the significant fibre hypertrophy in muscles from formoterol treated rats. Reprinted with permission (Ryall et al. 2007)

eccentric nuclei, a characteristic of axonal injury and subsequent compensatory axonal sprouting. These effects were consistent with improved synaptic function and an attenuated progression of motor deficits such as the decline in grip strength (Fig. 3) (Zeman et al. 2004).

4 Novel β-Adrenoceptor Therapeutic Strategies

Some of the most serious consequences of chronic β-agonist administration relate to the systemic responses to β-adrenoceptor activation (Gregorevic et al. 2005; Ryall et al. 2008b). Much research is currently focused on developing new methods of drug administration that limit unwanted systemic effects, with many having potential to improve the safe delivery of β-agonists to skeletal muscles.

4.1 Intramuscular Administration

We have examined whether direct intramuscular (*i.m.*) injection of the β-agonist formoterol can localize its effects to skeletal muscle directly and so minimize potential deleterious systemic effects (Ryall et al. 2008a). Two days after a single *i.m.* injection of formoterol, the force producing capacity of regenerating rat EDL muscles was two-fold higher than that of regenerating EDL muscles that received a single *i.m.* injection of saline. Importantly, *i.m.* administration of formoterol was not associated with cardiac hypertrophy. However, it should be noted that the increase in muscle mass and force-producing capacity after *i.m.* administration was lost within 5 days, and was still associated with a number of changes in cardiovascular function, including a transient increase in heart rate and a decrease in blood pressure. Furthermore, this mode of administration would prove problematic in a condition such as sarcopenia, where the loss of muscle mass and strength is not limited to a single muscle. More likely, this approach could find application in sports medicine and rehabilitation where functional impairments might be limited to a single muscle or muscle group.

4.2 Co-administration with a β_1-Adrenoceptor Antagonist

Blocking stimulation of the β_1-adenoceptors is possible with highly selective β_1-adrenoceptor antagonists such as CGP 20712A (Sillence and Matthews 1994) and the importance of blocking β_1-adrenoceptors in heart failure to abrogate cardiotoxic β_1-adrenoceptor-mediated effects is also well known (Ahmet et al. 2008; Molenaar and Parsonage 2005). Previous clinical trials of the older generation β-agonist, albuterol, for patients with neuromuscular disorders revealed some cardiovascular complications, including palpitations and tachycardia (Fowler et al. 2004). The fact that formoterol is highly selective for the β_2-adrenoceptor compared with older generation agonists such as albuterol and clenbuterol (Anderson 1993), and that it is efficacious in eliciting skeletal muscle anabolic effects even at micromolar doses (Ryall et al. 2006), offers the considerable advantage that simultaneous β_1-adrenoceptor blockade may prevent or attenuate many of these cardiovascular side effects. Molenaar and colleagues (2006) have suggested that the use of highly selective β_2-agonists, in conjunction with a selective β_1-blocker, could prevent unintended β_1-adrenoceptor activation and thus prevent unwanted cardiovascular effects while maintaining the desirable effects on skeletal muscle. This is particularly important for β_1-adrenoceptors in the cardiovascular system, where chronic activation of β_1-adrenoceptors is contraindicated for prevalent cardiac and vascular disorders including hypertension, ischemic heart disease, arrhythmias and heart failure where β-blockers are indicated. A pathological role of the β_1-adrenoceptor was confirmed in transgenic mice where 15-fold overexpression led to progressive deterioration of heart function, hypertrophy and heart failure (Engelhardt et al. 1999). The importance of blocking β_1-adrenoceptors in heart failure to abolish cardiotoxic β_1-mediated effects have been reported previously (Ahmet et al. 2008; Molenaar and Parsonage 2005).

4.3 Phosphodiesterase Inhibitors

Phosphodiesterase (PDE) is the enzyme responsible for the degradation of cAMP into 5′-AMP, and it therefore plays an important role in terminating the PKA-cAMP signaling cascade (for review see Omori and Kotera 2007). Skeletal muscle contains numerous isoforms of PDE, including: PDE4, PDE7, and PDE8, however, PDE4 is believed to be predominantly responsible for cAMP degradation in this tissue (Bloom 2002).

Selective inhibitors of PDE have been used to treat a diverse range of pathological conditions, including chronic obstructive pulmonary disorder, erectile dysfunction, and hypertension (Benedict et al. 2007; Burnett 2008; Kass et al. 2007). However, the potential of PDE inhibitors to treat skeletal muscle wasting and weakness has received only limited attention. Some of the earliest studies in skeletal muscle utilized the non-selective PDE inhibitor, pentoxifylline. Hudlická and Price (1990) found that 5 weeks of tri-daily administration of pentoxifylline (3mg/kg, *i.p.*) to rats increased the proportion of glycolytic fibres in EDL muscles. Breuillé and colleagues (1993) demonstrated that a single injection of pentoxifylline (100mg/kg, *i.p.*) to rats could attenuate the atrophy of the gastrocnemius muscle associated with 6 days of induced sepsis. More recently, Hinkle and colleagues (2005) administered either rolipram or Ariflo (both selective PDE4 inhibitors) or pentoxifylline via twice-daily *s.c.* injections to rats and mice after denervation or during disuse atrophy (limb-casting), respectively. PDE4 selective or PDE non-selective inhibition had little or no effect on muscle mass and strength in control muscles, while all three pharmacological inhibitors prevented the loss of muscle mass associated with denervation or disuse by ~20% to 40%. The results from these studies suggested a role for PDEs in proteolytic processes, and this was confirmed by Baviera et al. (2007) who found that pentoxifylline administration to diabetic rats reduced the activity of the Ca^{2+}-dependent and ATP proteasome-dependent proteolytic pathways.

An attractive hypothesis is that selective PDE inhibitors may be sufficient to prevent, attenuate, or reverse muscle wasting and weakness, without the complicating cardiac side-effects associated with β-agonist administration. However, it must be noted that chronic administration of the non-selective PDE pentoxifylline is associated with a rightward shift of the left ventricular end-diastolic pressure-volume relationship, thinning of the left ventricular wall, and infiltration of collagen in the myocardium (Anamourlis et al. 2006).

4.4 Engineered GPCRS, RASSLs, and DREADDs

An exciting avenue of research that may lead to ways that can obviate unwanted side-effects involves the use of designer GPCRs that allow for tight spatiotemporal control of GPCR signalling. This involves the development of both a synthetic receptor and an activator (neither of which activates or impairs endogenous GPCR signalling) and which therefore limits signalling to the tissue/region of interest – a result that current β-adrenoceptor agonists cannot achieve (Small et al. 2001). Roth and colleagues (in particular) are creating specific designer drug-designer receptor complexes to isolate the effects of GPCR activation (Dong et al. 2010; Conklin et al. 2008; Pei et al. 2008) recognising that exogenous ligands have off-target effects and endogenous ligands constantly modulate the activity of the native receptors (Dong et al. 2010). These include 'Receptors Activated Solely by Synthetic Ligands' (RASSLs) and Designer Receptors Exclusively Activated by Designer Drugs (DREADDs) (Nichols and Roth 2009) and represent tools for investigating biological function with a high degree of specificity. Although still in development, such approaches may yet lead to the successful separation of the effects of β-agonists on skeletal and cardiac muscle, thus promoting desirable effects that can improve the functional capacity of skeletal muscles without producing cardiovascular complications.

5 Conclusions

This chapter has provided evidence for the importance of β-adrenergic signalling in skeletal muscle and implicated this pathway as a potential target for the treatment of age-related muscle wasting and weakness. Although we are only beginning to understand the significance of the β-adrenergic signaling pathway in skeletal muscle, especially in relation to its role in sarcopenia, a wealth of information exists regarding the stimulation of the β-adrenergic system with β-agonists. Although there is great promise that β-agonists can be used for treating sarcopenia, and other conditions where muscle wasting is indicated, their clinical application has been limited by cardiovascular side effects, especially when β-agonists are administered chronically and at high doses. Newer generation β-agonists (such as formoterol) can elicit an anabolic response in skeletal muscle even when administered at very low doses and this has renewed enthusiasm for their clinical application, especially because they exhibit reduced effects on the heart and cardiovascular system compared with older generation β-agonists (such as fenoterol and clenbuterol). However, the potentially deleterious cardiovascular side effects associated with β-agonist administration have not been obviated completely and so it is important to refine their development and investigate novel strategies to limit β-adrenoceptor activation to skeletal muscle. If successful, these beneficial effects of β-adrenoceptor stimulation on skeletal muscle would find application for treating sarcopenia, where muscle wasting impacts not only upon the ability to perform the tasks of daily living, and quality of life, but ultimately on life itself, since the maintenance of functional muscle mass is critical for survival.

Acknowledgments Supported by research grants from the National Health & Medical Research Council (NHMRC, Australia; project grant 509313) and the Association Française contre les Myopathies (France). JGR is supported by a Biomedical Overseas Research Fellowship from the National Health and Medical Research Council of Australia (520034).

References

Ahmet, I., Krawczyk, M., Zhu, W., Woo, A. Y., Morrell, C., Poosala, S., Xiao, R. P., Lakatta, E. G., Talan, M. I. (2008). Cardioprotective and survival benefits of long-term combined therapy with β_2 AR agonist and β_1 AR blocker in dilated cardiomyopathy post-myocardial infarction. *The Journal of Pharmacology and Experimental Therapeutics, 325*, 491–499.

Anamourlis, C., Badenhorst, D., Gibbs, M., Correia, R., Veliotes, D., Osadchii, O., Norton, G. R., Woodiwiss, A. J. (2006). Phosphodiesterase inhibition promotes the transition from compensated hypertrophy to cardiac dilatation in rats. *Pflugers Archiv, 451*, 526–533.

Anderson, G. P. (1993). Formoterol: pharmacology, molecular basis of agonism, and mechanism of long duration of a highly potent and selective β_2-adrenoceptor agonist bronchodilator. *Life Sciences, 52*, 2145–2160.

Baviera, A. M., Zanon, N. M., Carvalho Navegantes, L. C., Migliorini, R. H., do Carmo Kettelhut, I. (2007). Pentoxifylline inhibits Ca^{2+}-dependent and ATP proteasome-dependent proteolysis in skeletal muscle from acutely diabetic rats. *American Journal of Physiology. Endocrinology and Metabolism, 292*, E702–E708.

Beerman, D. H., Butler, W. R., Hogue, D. E., Fishell, V. K., Dalrymple, R. H., Ricks, C. A., Scanes, C. G. (1987). Cimaterol-induced muscle hypertrophy and altered endocrine status in lambs. *Journal of Animal Science, 65*, 1514–1524.

Benedict, N., Seybert, A., Mathier, M. A. (2007). Evidence-based pharmacologic management of pulmonary arterial hypertension. *Clinical Therapeutics, 29*, 2134–2153.

Berdeaux, R., Goebel, N., Banaszynski, L., Takemori, H., Wandless, T., Shelton, G. D., Montminy, M. (2007). SIK1 is a class II HDAC kinase that promotes survival of skeletal myocytes. *Natural Medicines, 13*, 597–603.

Bloom, T. J. (2002). Cyclic nucleotide phosphodiesterase isozymes expressed in mouse skeletal muscle. *Canadian Journal of Physiology and Pharmacology, 80*, 1132–1135.

Bockaert, J. & Pin, J. P. (1999). Molecular tinkering of G protein-coupled receptors: an evolutionary success. *The EMBO Journal, 18*, 1723–1729.

Bodine, S. C., Latres, E., Baumhueter, S., Lai, V. K., Nunez, L., Clarke, B. A., Poueymirou, W. T., Panaro, F. J., Na, E., Dharmarajan, K., Pan, Z. Q., Valenzuela, D. M., DeChiara, T. M., Stitt, T. N., Yancopoulos, G. D., Glass, D. J. (2001a). Identification of ubiquitin ligases required for skeletal muscle atrophy. *Science, 294*, 1704–1708.

Bodine, S. C., Stitt, T. N., Gonzalez, M., Kline, W. O., Stover, G. L., Bauerlein, R., Zlotchenko, E., Scrimgeour, A., Lawrence, J. C., Glass, D. J., Yancopoulos, G. D. (2001b). Akt/mTOR pathway is a crucial regulator of skeletal muscle hypertrophy and can prevent muscle atrophy in vivo. *Nature Cell Biology, 3*, 1014–1019.

Bossola, M., Pacelli, F., Tortorelli, A., Rosa, F., Doglietto, G. B. (2008). Skeletal muscle in cancer cachexia: the ideal target of drug therapy. *Current Cancer Drug Targets, 8*, 285–298.

Breuillé, D., Farge, M. C., Rosé, F., Arnal, M., Attaix, D., Obled, C. (1993). Pentoxifylline decreases body weight loss and muscle protein wasting characteristics of sepsis. *The American Journal of Physiology, 265*, E660–E666.

Burnett, A. L. (2008). Molecular pharmacotherapeutic targeting of PDE5 for preservation of penile health. *Journal of Andrology, 29*, 3–14.

Carlezon, W. A., Jr., Duman, R. S., Nestler, E. J. (2005). The many faces of CREB. *Trends in Neurosciences, 28*, 436–445.

Carter, W. J. & Lynch, M. E. (1994a). Effect of clenbuterol on recovery of muscle mass and carcass protein content following experimental hyperthyroidism in old rats. *Comparative Biochemistry and Physiology. Comparative Physiology, 108*, 387–394.

Carter, W. J. & Lynch, M. E. (1994b). Comparison of the effects of salbutamol and clenbuterol on skeletal muscle mass and carcass composition in senescent rats. *Metabolism, 43*, 1119–1125.

Carter, W. J. & Lynch, M. E. (1994c). Effect of clenbuterol on recovery of muscle mass and carcass protein content following dietary protein depletion in young and old rats. *Journal of Gerontology, 49*, B162–B168.

Chen, K. D. & Alway, S. E. (2000). A physiological level of clenbuterol does not prevent atrophy or loss of force in skeletal muscle of old rats. *Journal of Applied Physiology, 89*, 606–612.

Chen, K. D. & Alway, S. E. (2001). Clenbuterol reduces soleus muscle fatigue during disuse in aged rats. *Muscle & Nerve, 24*, 211–222.

Chen, A. E., Ginty, D. D., Fan, C. M. (2005). Protein kinase A signalling via CREB controls myogenesis induced by Wnt proteins. *Nature, 433*, 317–322.

Childs, T. E., Spangenburg, E. E., Vyas, D. R., Booth, F. W. (2003). Temporal alterations in protein signaling cascades during recovery from muscle atrophy. *American Journal of Physiology. Cell Physiology, 285*, C391–C398.

Conklin, B. R., Hsiao, E. C., Claeysen, S., Dumuis, A., Srinivasan, S., Forsayeth, J. R., Guettier, J. M., Chang, W. C., Pei, Y., McCarthy, K. D., Nissenson, R. A., Wess, J., Bockaert, J., Roth, B. L. (2008). Engineering GPCR signaling pathways with RASSLs. *Nat Methods, 5*, 673–678.

Crespo, P., Xu, N., Simonds, W. F., Gutkind, J. S. (1994). Ras-dependent activation of MAP kinase pathway mediated by G-protein βγ subunits. *Nature, 369*, 418–420.

Dascal, N. (2001). Ion-channel regulation by G proteins. *Trends in Endocrinology and Metabolism, 12*, 391–398.

Diversé-Pierluissi, M., McIntire, W. E., Myung, C. S., Lindorfer, M. A., Garrison, J. C., Goy, M. F., Dunlap, K. (2000). Selective coupling of G protein βγ complexes to inhibition of Ca^{2+} channels. *The Journal of Biological Chemistry, 275*, 28380–28385.

Dixon, R. A. F., Kobilka, B. K., Strader, D. J., Benovic, J. L., Dohlman, H. G., Frielle, T., Bolanowski, M. A., Bennett, C. D., Rands, E., Diehl, R. E., Mumford, R. A., Slater, E. E., Sigal, I. S., Caron, M. G., Lefkowitz, R. J., Strader, C. D. (1986). Cloning of the gene and cDNA for mammalian β-adrenergic receptor and homology with rhodopsin. *Nature, 321*, 75–79.

Dodd, S. L., Powers, S. K., Vrabas, I. S., Criswell, D., Stetson, S., Hussain, R. (1996). Effects of clenbuterol on contractile and biochemical properties of skeletal muscle. *Medicine and Science in Sports and Exercise, 28*, 669–676.

Dong, S., Rogan, S. C., Roth, B. L. (2010). Directed molecular evolution of DREADDs: a generic approach to creating next-generation RASSLs. *Nature Protocols, 5*, 561–573.

Dupont-Versteegden, E. E. (1996). Exercise and clenbuterol as strategies to decrease the progression of muscular dystrophy in *mdx* mice. *Journal of Applied Physiology, 80*, 734–741.

Eckner, R., Yao, T. P., Oldread, E., Livingston, D. M. (1996). Interaction and functional collaboration of p300/CBP and bHLH proteins in muscle and B-cell differentiation. *Genes & Development, 10*, 2478–2490.

Emorine, L. J., Marullo, S., Briend-Sutren, M. M., Patey, G., Tate, K., Delavier-Klutchko, C., Strosberg, A. D. (1989). Molecular characterization of the human β_3-adrenergic receptor. *Science, 245*, 1118–1121.

Engelhardt, S., Hein, L., Wiesmann, F., Lohse, M. J. (1999). Progressive hypertrophy and heart failure in β_1-adrenergic receptor transgenic mice. *Proceedings of the National Academy of Sciences of the United States of America, 96*, 7059–7064.

Esler, M., Kaye, D., Thompson, J., Jennings, G., Cox, H., Turner, A., Lambert, G., Seals, D. (1995). Effects of aging on epinephrine secretion and regional release of epinephrine from the human heart. *The Journal of Clinical Endocrinology and Metabolism, 80*, 435–442.

Filipek, S., Krzysko, K. A., Fotiadis, D., Liang, Y., Saperstein, D. A., Engel, A., Palczewski, K. (2004). A concept for G protein activation by G protein-coupled receptor dimers: the transducin/rhodopsin interface. *Photochemical & Photobiological Sciences, 3*, 628–638.

Ford, C. E., Skiba, N. P., Bae, H., Daaka, Y., Reuveny, E., Shekter, L. R., Rosal, R., Weng, G., Yang, C. S., Iyengar, R., Miller, R. J., Jan, L. Y., Lefkowitz, R. J., Hamm, H. E. (1998). Molecular basis for interactions of G protein βγ subunits with effectors. *Science, 280*, 1271–1274.

Fowler, E. G., Graves, M. C., Wetzel, G. T., Spencer, M. J. (2004). Pilot trial of albuterol in Duchenne and Becker muscular dystrophy. *Neurology, 62*, 1006–1008.

Fredriksson, R., Lagerström, M. C., Lundin, L. G., Schiöth, H. B. (2003). The G-protein-coupled receptors in the human genome form five main families. Phylogenetic analysis, paralogon groups, and fingerprints. *Molecular Pharmacology, 63*, 1256–1272.

Frielle, T., Collins, S., Daniel, K. W., Caron, M. G., Lefkowitz, R. J., Kobilka, B. K. (1987). Cloning of the cDNA for the human β_1-adrenergic receptor. *Proceedings of the National Academy of Sciences of the United States of America, 84*, 7920–7924.

Furuyama, T., Yamashita, H., Kitayama, K., Higami, Y., Shimokawa, I., Mori, N. (2002). Effects of aging and caloric restriction on the gene expression of Foxo1, 3, and 4 (FKHR, FKHRL1, and AFX) in the rat skeletal muscles. *Microscopy Research and Technique, 59*, 331–334.

Garami, A., Zwartkruis, F. J., Nobukuni, T., Joaquin, M., Roccio, M., Stocker, H., Kozma, S. C., Hafen, E., Bos, J. L., Thomas, G. (2003). Insulin activation of Rheb, a mediator of mTOR/S6K/4E-BP signaling, is inhibited by TSC1 and 2. *Molecular Cell, 11*, 1457–1466.

Gilman, A. G. (1995). Nobel Lecture. G proteins and regulation of adenylyl cyclase. *Bioscience Reports, 15*, 65–97.

Glass, D. J. (2003). Signalling pathways that mediate skeletal muscle hypertrophy and atrophy. *Nature Cell Biology, 5*, 87–90.

Glass, D. J. (2005). Skeletal muscle hypertrophy and atrophy signaling pathways. *The International Journal of Biochemistry & Cell Biology, 37*, 1974–1984.

Goodman, R. H. & Smolik, S. (2000). CBP/p300 in cell growth, transformation, and development. *Genes & Development, 14*, 1553–1577.

Gregorevic, P., Ryall, J. G., Plant, D. R., Sillence, M. N., Lynch, G. S. (2005). Chronic β agonist administration affects cardiac function of adult but not old rats, independent of β-adrenoceptor density. *American Journal of Physiology. Heart and Circulatory Physiology, 289*, H344–H349.

Grifone. R., Laclef, C., Spitz, F., Lopez, S., Demignon, J., Guidotti, J. E., Kawakami, K., Xu, P. X., Kelly, R., Petrof, B. J., Daegelen, D., Concordet, J. P., Maire, P. (2004). Six1 and Eya1 expression can reprogram adult muscle from the slow-twitch phenotype into the fast-twitch phenotype. *Molecular Cell Biology, 24*, 6253–6267.

Hagiwara, M., Alberts, A., Brindle, P., Meinkoth, J., Feramisco, J., Deng, T., Karin, M., Shenolikar, S., Montminy, M. (1992). Transcriptional attenuation following cAMP induction requires PP-1-mediated dephosphorylation of CREB. *Cell, 70*, 105–113.

Hagiwara, M., Brindle, P., Harootunian, A., Armstrong, R., Rivier, J., Vale, W., Tsien, R., Montminy, M. R. (1993). Coupling of hormonal stimulation and transcription via the cyclic AMP-responsive factor CREB is rate limited by nuclear entry of protein kinase A. *Molecular and Cellular Biology, 13*, 4852–4859.

Hampoelz, B. & Knoblich, J. A. (2004). Heterotrimeric G proteins: new tricks for an old dog. *Cell, 119*, 453–456.

Handschin, C., Chin, S., Li, P., Liu, F., Maratos-Flier, E., Lebrasseur, N. K., Yan Z, Spiegelman BM. (2007). Skeletal muscle fiber-type switching, exercise intolerance, and myopathy in PGC-1α muscle-specific knock-out animals. *Journal of Biological Chemistry, 282*, 30014–30021.

Harcourt, L. J., Schertzer, J. D., Ryall, J. G., Lynch, G. S. (2007). Low dose formoterol administration improves muscle function in dystrophic *mdx* mice without increasing fatigue. *Neuromuscular Disorders, 17*, 47–55.

Hardt, S. E. & Sadoshima, J. (2002). Glycogen synthase kinase-3: a novel regulator of cardiac hypertrophy and development. *Circulation Research, 90*, 1055–1063.

Hinkle, R. T., Hodge, K. M., Cody, D. B., Sheldon, R. J., Kobilka, B. K., Isfort, R. J. (2002). Skeletal muscle hypertrophy and anti-atrophy effects of clenbuterol are mediated by the β_2-adrenergic receptor. *Muscle & Nerve, 25*, 729–734.

Hinkle, R. T., Dolan, E., Cody, D. B., Bauer, M. B., Isfort, R. J. (2005). Phosphodiesterase 4 inhibition reduces skeletal muscle atrophy. *Muscle & Nerve, 32*, 775–781.

Hudlická, O. & Price, S. (1990). Effects of torbafylline, pentoxifylline and buflomedil on vascularisation and fibre type of rat skeletal muscles subjected to limited blood supply. *British Journal of Pharmacology, 99*, 786–790.

Jefferies, H. B., Fumagalli, S., Dennis, P. B., Reinhard, C., Pearson, R. B., Thomas, G. (1997). Rapamycin suppresses 5′TOP mRNA translation through inhibition of $p70^{s6k}$. *The EMBO Journal, 16*, 3693–3704.

Johnson, M. (2006). Molecular mechanisms of β_2-adrenergic receptor function, response, and regulation. *The Journal of Allergy and Clinical Immunology, 117*, 18–24. quiz 25.

Kandarian, S. C. & Jackman, R. W. (2006). Intracellular signaling during skeletal muscle atrophy. *Muscle & Nerve, 33*, 155–165.

Kass, D. A., Champion, H. C., Beavo, J. A. (2007). Phosphodiesterase type 5: expanding roles in cardiovascular regulation. *Circulation Research, 101*, 1084–1095.

Kaye, D. & Esler, M. (2005). Sympathetic neuronal regulation of the heart in aging and heart failure. *Cardiovascular Research, 66*, 256–264.

Kim, M. S., Fielitz, J., McAnally, J., Shelton, J. M., Lemon, D. D., McKinsey, T. A., Richardson J. A., Bassel-Duby, R., Olson, E. N. (2008). Protein kinase D1 stimulates MEF2 activity in skeletal muscle and enhances muscle performance. *Molecular Cell Biology, 28*, 3600–3609.

Kim, Y. S., Sainz, R. D., Molenaar, P., Summers, R. J. (1991). Characterization of β_1- and β_2-adrenoceptors in rat skeletal muscles. *Biochemical Pharmacology, 42*, 1783–1789.

Kissel, J. T., McDermott, M. P., Natarajan, R., Mendell, J. R., Pandya, S., King, W. M., Griggs, R. C., Tawil, R. (1998). Pilot trial of albuterol in facioscapulohumeral muscular dystrophy. *Neurology, 50*, 1402–1406.

Kissel, J. T., McDermott, M. P., Mendell, J. R., King, W. M., Pandya, S., Griggs, R. C., Tawil, R. (2001). Randomized, double-blind, placebo-controlled trial of albuterol in facioscapulohumeral dystrophy. *Neurology, 57*, 1434–1440.

Klco, J. M., Wiegand, C. B., Narzinski, K., Baranski, T. J. (2005). Essential role for the second extracellular loop in C5a receptor activation. *Nature Structural & Molecular Biology, 12*, 320–326.

Kline, W. O., Panaro, F. J., Yang, H., Bodine, S. C. (2007). Rapamycin inhibits the growth and muscle-sparing effects of clenbuterol. *Journal of Applied Physiology, 102*, 740–747.

Kobilka, B. K., Dixon, R. A., Frielle, T., Dohlman, H. G., Bolanowski, M. A., Sigal, I. S., Yang-Feng, T. L., Francke, U., Caron, M. G., Lefkowitz, R. J. (1987). cDNA for the human β_2-adrenergic receptor: a protein with multiple membrane-spanning domains and encoded by a gene whose chromosomal location is shared with that of the receptor for platelet-derived growth factor. *Proceedings of the National Academy of Sciences of the United States of America, 84*, 46–50.

Kobilka, B. K., Kobilka, T. S., Daniel, K., Regan, J. W., Caron, M. G., Lefkowitz, R. J. (1988). Chimeric α_2-, β_2-adrenergic receptors: delineation of domains involved in effector coupling and ligand binding specificity. *Science, 240*, 1310–1316.

Lai, K. M., Gonzalez, M., Poueymirou, W. T., Kline, W. O., Na, E., Zlotchenko, E., Stitt, T. N., Economides, A. N., Yancopoulos, G. D., Glass, D. J. (2004). Conditional activation of akt in adult skeletal muscle induces rapid hypertrophy. *Molecular and Cellular Biology, 24*, 9295–9304.

Larkin, L. M., Halter, J. B., Supiano, M. A. (1996). Effect of aging on rat skeletal muscle β-AR function in male Fischer 344 × brown Norway rats. *The American Journal of Physiology, 270*, R462–R468.

Latres, E., Amini, A. R., Amini, A. A., Griffiths, J., Martin, F. J., Wei, Y., Lin, H. C., Yancopoulos, G. D., Glass, D. J. (2005). Insulin-like growth factor-1 (IGF-1) inversely regulates atrophy-induced genes via the phosphatidylinositol 3-kinase/Akt/mammalian target of rapamycin (PI3K/Akt/mTOR) pathway. *The Journal of Biological Chemistry, 280*, 2737–2744.

Léger, B., Cartoni, R., Praz, M., Lamon, S., Dériaz, O., Crettenand, A., Gobelet, C., Rohmer, P., Konzelmann, M., Luthi, F., Russell, A. P. (2006). Akt signalling through GSK-3beta, mTOR and Foxo1 is involved in human skeletal muscle hypertrophy and atrophy. *Journal de Physiologie, 576*, 923–933.

Lohse, M. J. (1999). G-Proteins and their regulators. *Naunyn Schmiedeberg's Archives of Pharmacology, 360*, 3–4.

Lopez-Ilasaca, M., Crespo, P., Pellici, P. G., Gutkind, J. S., Wetzker, R. (1997). Linkage of G protein-coupled receptors to the MAPK signaling pathway through PI 3-kinase γ. *Science, 275*, 394–397.

Lynch, G. S. & Ryall, J. G. (2008). Role of β-adrenoceptor signaling in skeletal muscle: implications for muscle wasting and disease. *Physiological Reviews, 88*, 729–767.

Martin, W. H., 3rd, Murphree, S. S., Saffitz, J. E. (1989). β-Adrenergic receptor distribution among muscle fiber types and resistance arterioles of white, red, and intermediate skeletal muscle. *Circulation Research, 64*, 1096–1105.

Maxwell, M. A., Cleasby, M. E., Harding, A., Stark, A., Cooney, G. J., Muscat, G. E. (2005). Nur77 regulates lipolysis in skeletal muscle cells. Evidence for cross-talk between the beta-adrenergic and an orphan nuclear hormone receptor pathway. *The Journal of Biological Chemistry, 280*, 12573–12584.

Mayr, B. & Montminy, M. (2001). Transcriptional regulation by the phosphorylation-dependent factor CREB. *Nature Reviews. Molecular Cell Biology, 2*, 599–609.

McDaneld, T. G., Hancock, D. L., Moody, D. E. (2004). Altered mRNA abundance of ASB15 and four other genes in skeletal muscle following administration of β-adrenergic receptor agonists. *Physiological Genomics, 16*, 275–283.

McDaneld, T. G., Hannon, K., Moody, D. E. (2006). Ankyrin repeat and SOCS box protein 15 regulates protein synthesis in skeletal muscle. *American Journal of Physiology: Regulatory, Integrative and Comparative Physiology, 290*, R1672–R1682.

McKinsey, T. A., Zhang, C. L., Olson, E. N. (2002). Signaling chromatin to make muscle. *Current Opinion in Cell Biology, 14*, 763–772.

McPherron, A. C., Lawler, A. M., Lee, S. J. (1997). Regulation of skeletal muscle mass in mice by a new TGF-β superfamily member. *Nature, 387*, 83–90.

Meng, E. C. & Bourne, H. R. (2001). Receptor activation: what does the rhodopsin structure tell us? *Trends in Pharmacological Sciences, 22*, 587–593.

Mirshahi, T., Mittal, V., Zhang, H., Linder, M. E., Logothetis, D. E. (2002). Distinct sites on G protein βγ subunits regulate different effector functions. *The Journal of Biological Chemistry, 277*, 36345–36350.

Molenaar, P. & Parsonage, W. A. (2005). Fundamental considerations of β-adrenoceptor subtypes in human heart failure. *Trends in Pharmacological Sciences, 26*, 368–375.

Molenaar, P., Chen, L., Parsonage, W. A. (2006). Cardiac implications for the use of β_2-adrenoceptor agonists for the management of muscle wasting. *British Journal of Pharmacology, 147*, 583–586.

Molkentin, J. D. & Olson, E. N. (1996). Combinatorial control of muscle development by basic helix-loop-helix and MADS-box transcription factors. *Proceedings of the National Academy of Sciences of the United States of America, 93*, 9366–9373.

Morris, A. J. & Malbon, C. C. (1999). Physiological regulation of G protein-linked signaling. *Physiological Reviews, 79*, 1373–1430.

Murga, C., Laguinge, L., Wetzker, R., Cuadrado, A., Gutkind, J. S. (1998). Activation of Akt/protein kinase B by G protein-coupled receptors. A role for α and βγ subunits of heterotrimeric G proteins acting through phosphatidylinositol-3-OH kinase γ. *The Journal of Biological Chemistry, 273*, 19080–19085.

Murga, C., Fukuhara, S., Gutkind, J. S. (2000). A novel role for phosphatidylinositol 3-kinase β in signaling from G protein-coupled receptors to Akt. *The Journal of Biological Chemistry, 275*, 12069–12073.

Nader, G. A. (2005). Molecular determinants of skeletal muscle mass: getting the "AKT" together. *The International Journal of Biochemistry & Cell Biology, 37*, 1985–1996.

Nave, B. T., Ouwens, M., Withers, D. J., Alessi, D. R., Shepherd, P. R. (1999). Mammalian target of rapamycin is a direct target for protein kinase B: identification of a convergence point for opposing effects of insulin and amino-acid deficiency on protein translation. *The Biochemical Journal, 344*(Pt 2), 427–431.

Navegantes, L. C., Resano, N. M., Migliorini, R. H., Kettelhut, I. C. (2000). Role of adrenoceptors and cAMP on the catecholamine-induced inhibition of proteolysis in rat skeletal muscle. *American Journal of Physiology. Endocrinology and Metabolism, 279*, E663–E668.

Nichols, C. D. & Roth, B. L. (2009). Engineered G-protein coupled receptors are powerful tools to investigate biological processes and behaviors. *Frontiers in Molecular Neuroscience, 2*, 16.

Nicholson, K. M. & Anderson, N. G. (2002). The protein kinase B/Akt signalling pathway in human malignancy. *Cellular Signalling, 14*, 381–395.

Ogryzko, V. V., Schiltz, R. L., Russanova, V., Howard, B. H., Nakatani, Y. (1996). The transcriptional coactivators p300 and CBP are histone acetyltransferases. *Cell, 87*, 953–959.

Ohkura, N., Ito, M., Tsukada, T., Sasaki, K., Yamaguchi, K., Miki, K. (1998). Alternative splicing generates isoforms of human neuron-derived orphan receptor-1 (NOR-1) mRNA. *Gene, 211*, 79–85.

Omori, K. & Kotera, J. (2007). Overview of PDEs and their regulation. *Circulation Research, 100*, 309–327.

Oh, M., Rybkin, I. I., Copeland, V., Czubryt, M. P., Shelton, J. M., van Rooij, E., Richardson, J. A., Hill, J. A., De Windt, L. J., Bassel-Duby, R., Olson, E. N., Rothermel, B. A. (2005). Calcineurin is necessary for the maintenance but not embryonic development of slow muscle fibers. *Molecular Cell Biology, 25*, 6629–6638.

Pallafacchina, G., Calabria, E., Serrano, A. L., Kalhovde, J. M., Schiaffino, S. (2002). A protein kinase B-dependent and rapamycin-sensitive pathway controls skeletal muscle growth but not fiber type specification. *Proceedings of the National Academy of Sciences of the United States of America, 99*, 9213–9218.

Pearen, M. A., Ryall, J. G., Maxwell, M. A., Ohkura, N., Lynch, G. S., Muscat, G. E. (2006). The orphan nuclear receptor, NOR-1, is a target of β-adrenergic signaling in skeletal muscle. *Endocrinology, 147*, 5217–5227.

Pearen, M. A., Ryall, J. G., Lynch, G. S., Muscat, G. E. (2009). Expression profiling of skeletal muscle following acute and chronic β2-adrenergic stimulation: implications for hypertrophy, metabolism and circadian rhythm. *BMC Genomics, 10*, 448.

Pei, Y., Rogan, S. C., Yan, F., Roth, B. L. (2008). Engineered GPCRs as tools to modulate signal transduction. *Physiology, 23*, 313–321.

Pourquié, O. (2005). Signal transduction: a new canon. *Nature, 433*, 208–209.

Rattigan, S., Appleby, G. J., Edwards, S. J., McKinstry, W. J., Colquhoun, E. Q., Clark, M. G., Richter, E. A. (1986). α-adrenergic receptors in rat skeletal muscle. *Biochemical and Biophysical Research Communications, 136*, 1071–1077.

Ricks, C. A., Dalrymple, R. H., Baker, P. K., Ingle, D. L. (1984). Use of a β-agonist to alter fat and muscle deposition in steers. *Journal of Animal Science, 59*, 1247–1255.

Rodbell, M., Birnbaumer, L., Pohl, S. L., Krans, H. M. (1971). The glucagon-sensitive adenyl cyclase system in plasma membranes of rat liver. V. An obligatory role of guanylnucleotides in glucagon action. *The Journal of Biological Chemistry, 246*, 1877–1882.

Rommel, C., Bodine, S. C., Clarke, B. A., Rossman, R., Nunez, L., Stitt, T. N., Yancopoulos, G. D., Glass, D. J. (2001). Mediation of IGF-1-induced skeletal myotube hypertrophy by PI3K/Akt/mTOR and PI(3)K/Akt/GSK3 pathways. *Nature Cell Biology, 3*, 1009–1013.

Roth, J. F., Shikama, N., Henzen, C., Desbaillets, I., Lutz, W., Marino, S., Wittwer, J., Schorle, H., Gassmann, M., Eckner, R. (2003). Differential role of p300 and CBP acetyltransferase during myogenesis: p300 acts upstream of MyoD and Myf5. *The EMBO Journal, 22*, 5186–5196.

Ryall, J. G., Gregorevic, P., Plant, D. R., Sillence, M. N., Lynch, G. S. (2002). β_2-Agonist fenoterol has greater effects on contractile function of rat skeletal muscles than clenbuterol. *American Journal of Physiology: Regulatory, Integrative and Comparative Physiology, 283*, R1386–R1394.

Ryall, J. G., Plant, D. R., Gregorevic, P., Sillence, M. N., Lynch, G. S. (2004). β_2-Agonist administration reverses muscle wasting and improves muscle function in aged rats. *Journal de Physiologie, 555*, 175–188.

Ryall, J. G., Sillence, M. N., Lynch, G. S. (2006). Systemic administration of β_2-adrenoceptor agonists, formoterol and salmeterol, elicit skeletal muscle hypertrophy in rats at micromolar doses. *British Journal of Pharmacology, 147*, 587–595.

Ryall, J. G., Schertzer, J. D., Lynch, G. S. (2007). Attenuation of age-related muscle wasting and weakness in rats after formoterol treatment: therapeutic implications for sarcopenia. *The Journals of Gerontology. Series A: Biological Sciences and Medical Sciences, 62*, 813–823.

Ryall, J. G., Schertzer, J. D., Alabakis, T. M., Gehrig, S. M., Plant, D. R., Lynch, G. S. (2008a). Intramuscular β_2-agonist administration enhances early regeneration and functional repair in rat skeletal muscle after myotoxic injury. *Journal of Applied Physiology, 105*, 165–172.

Ryall, J. G., Schertzer, J. D., Murphy, K. T., Allen, A. M., Lynch, G. S. (2008b). Chronic β_2-adrenoceptor stimulation impairs cardiac relaxation via reduced SR Ca^{2+}-ATPase protein and activity. *American Journal of Physiology. Heart and Circulatory Physiology, 294*, H2587–H2595.

Sandri, M., Sandri, C., Gilbert, A., Skurk, C., Calabria, E., Picard, A., Walsh, K., Schiaffino, S., Lecker, S. H., Goldberg, A. L. (2004). Foxo transcription factors induce the atrophy-related ubiquitin ligase atrogin-1 and cause skeletal muscle atrophy. *Cell, 117*, 399–412.

Sartorelli, V., Huang, J., Hamamori, Y., Kedes, L. (1997). Molecular mechanisms of myogenic coactivation by p300: direct interaction with the activation domain of MyoD and with the MADS box of MEF2C. *Molecular and Cellular Biology, 17*, 1010–1026.

Schertzer, J. D., Plant, D. R., Ryall, J. G., Beitzel, F., Stupka, N., Lynch, G. S. (2005). β_2-Agonist administration increases sarcoplasmic reticulum Ca^{2+}-ATPase activity in aged rat skeletal muscle. *American Journal of Physiology. Endocrinology and Metabolism, 288*, E526–E533.

Schmidt, P., Holsboer, F., Spengler, D. (2001). β_2-adrenergic receptors potentiate glucocorticoid receptor transactivation via G protein $\beta\gamma$-subunits and the phosphoinositide 3-kinase pathway. *Molecular Endocrinology, 15*, 553–564.

Sillence, M. N. (2004). Technologies for the control of fat and lean deposition in livestock. *The Veterinary Journal, 167*, 242–257.

Sillence, M. N. & Matthews, M. L. (1994). Classical and atypical binding sites for β-adrenoceptor ligands and activation of adenylyl cyclase in bovine skeletal muscle and adipose tissue membranes. *British Journal of Pharmacology, 111*, 866–872.

Small, K. M., Brown, K. M., Forbes, S. L., Liggett, S. B. (2001). Modification of the β_2-adrenergic receptor to engineer a receptor-effector complex for gene therapy. *The Journal of Biological Chemistry, 276*, 31596–31601.

Smith, W. N., Dirks, A., Sugiura, T., Muller, S., Scarpace, P., Powers, S. K. (2002). Alteration of contractile force and mass in the senescent diaphragm with β_2-agonist treatment. *Journal of Applied Physiology, 92*, 941–948.

Sneddon, A. A., Delday, M. I., Steven, J., Maltin, C. A. (2001). Elevated IGF-II mRNA and phosphorylation of 4E-BP1 and $p70^{S6k}$ in muscle showing clenbuterol-induced anabolism. *American Journal of Physiology. Endocrinology and Metabolism, 281*, E676–E682.

Spangenburg, E. E. (2005). SOCS-3 induces myoblast differentiation. *The Journal of Biological Chemistry, 280*, 10749–10758.

Spurlock, D. M., McDaneld, T. G., McIntyre, L. M. (2006). Changes in skeletal muscle gene expression following clenbuterol administration. *BMC Genomics, 7*, 320.

Stitt, T. N., Drujan, D., Clarke, B. A., Panaro, F., Timofeyva, Y., Kline, W. O., Gonzalez, M., Yancopoulos, G. D., Glass, D. J. (2004). The IGF-1/PI3K/Akt pathway prevents expression of muscle atrophy-induced ubiquitin ligases by inhibiting FOXO transcription factors. *Molecular Cell, 14*, 395–403.

Thompson, P. R., Wang, D., Wang, L., Fulco, M., Pediconi, N., Zhang, D., An, W., Ge, Q., Roeder, R. G., Wong, J., Levrero, M., Sartorelli, V., Cotter, R. J., Cole, P. A. (2004). Regulation of the p300 HAT domain via a novel activation loop. *Nature Structural & Molecular Biology, 11*, 308–315.

Tintignac, L. A., Lagirand, J., Batonnet, S., Sirri, V., Leibovitch, M. P., Leibovitch, S. A. (2005). Degradation of MyoD mediated by the SCF (MAFbx) ubiquitin ligase. *The Journal of Biological Chemistry, 280*, 2847–2856.

Tran, H., Brunet, A., Griffith, E. C., Greenberg, M. E. (2003). The Many Forks in FOXO's Road. Sci STKE 2003, RE5.

Wadzinski, B. E., Wheat, W. H., Jaspers, S., Peruski, L. F., Jr., Lickteig, R. L., Johnson, G. L., Klemm, D. J. (1993). Nuclear protein phosphatase 2A dephosphorylates protein kinase A-phosphorylated CREB and regulates CREB transcriptional stimulation. *Molecular and Cellular Biology, 13*, 2822–2834.

Wenzel-Seifert, K. & Seifert, R. (2000). Molecular analysis of β_2-adrenoceptor coupling to G_s-, G_i-, and G_q-proteins. *Molecular Pharmacology, 58*, 954–966.

Wilkie, T. M., Gilbert, D. J., Olsen, A. S., Chen, X. N., Amatruda, T. T., Korenberg, J. R., Trask, B. J., de Jong, P., Reed, R. R., Simon, M. I. (1992). Evolution of the mammalian G protein alpha subunit multigene family. *Nature Genetics, 1*, 85–91.

Williams, R. S., Caron, M. G., Daniel, K. (1984). Skeletal muscle β-adrenergic receptors: variations due to fiber type and training. *The American Journal of Physiology, 246*, E160–E167.

Wu, A. L., Kim, J. H., Zhang, C., Unterman, T. G., Chen, J. (2008). Forkhead box protein O1 negatively regulates skeletal myocyte differentiation through degradation of mammalian target of rapamycin pathway components. *Endocrinology, 149*, 1407–1414.

Yamamoto, D. L., Hutchinson, D. S., Bengtsson, T. (2007). β_2-Adrenergic activation increases glycogen synthesis in L6 skeletal muscle cells through a signalling pathway independent of cyclic AMP. *Diabetologia, 50*, 158–167.

Yang, X. J. (2004). Lysine acetylation and the bromodomain: a new partnership for signaling. *Bioessays, 26*, 1076–1087.

Yang, X., Yang, C., Farberman, A., Rideout, T. C., de Lange, C. F., France, J., Fan, M. Z. (2008). The mammalian target of rapamycin-signaling pathway in regulating metabolism and growth. *Journal of Animal Science, 86*(14 Suppl), E36–E50.

Zeman, R. J., Ludemann, R., Easton, T. G., Etlinger, J. D. (1988). Slow to fast alterations in skeletal muscle fibers caused by clenbuterol, a β_2-receptor agonist. *The American Journal of Physiology, 254*, E726–E732.

Zeman, R. J., Peng, H., Etlinger, J. D. (2004). Clenbuterol retards loss of motor function in motor neuron degeneration mice. *Experimental Neurology, 187*, 460–467.

Zhang, X., Odom, D. T., Koo, S. H., Conkright, M. D., Canettieri, G., Best, J., Chen, H., Jenner, R., Herbolsheimer, E., Jacobsen, E., Kadam, S., Ecker, J. R., Emerson, B., Hogenesch, J. B., Unterman, T., Young, R. A., Montminy, M. (2005). Genome-wide analysis of cAMP-response element binding protein occupancy, phosphorylation, and target gene activation in human tissues. *Proceedings of the National Academy of Sciences of the United States of America, 102*, 4459–4464.

Index

G.S. Lynch (ed.), *Sarcopenia – Age-Related Muscle Wasting and Weakness*,
DOI 10.1007/978-90-481-9713-2, © Springer Science+Business Media B.V. 2011

N